ALGORITHMS AND ARCHITECTURES FOR REAL-TIME CONTROL

*Preprints of the IFAC Workshop, Seoul, Korea,
31 August - 2 September 1992*

Edited by

P.J. FLEMING
Department of Automatic Control and Systems Engineering, University of Sheffield, UK

and

W.H. KWON
Engineering Research Center for Advanced Control and Instrumentation, Seoul National University, Korea

Published for the

INTERNATIONAL FEDERATION OF AUTOMATIC CONTROL

by

PERGAMON PRESS

OXFORD • NEW YORK • SEOUL • TOKYO

UK	Pergamon Press Ltd, Headington Hill Hall, Oxford OX3 0BW, England
USA	Pergamon Press, Inc., 660 White Plains Road, Tarrytown, New York 10591-5153, USA
KOREA	Pergamon Press Korea, KPO Box 315, Seoul 110-603, Korea
JAPAN	Pergamon Press Japan, Tsunashima Building Annex, 3-20-12 Yushima, Bunkyo-ku, Tokyo 113, Japan

First edition 1992

Library of Congress Cataloging in Publication Data

A catalogue record for this book is available from the Library of Congress

British Library Cataloguing in Publication Data

A catalogue record for this book is available from the British Library

ISBN 0-08-042050-8

These preprints were reproduced by means of the photo-offset process using the manuscripts supplied by the authors of the different papers. The manuscripts have been typed using different typewriters and typefaces. The lay-out, figures and tables of some papers did not agree completely with the standard requirements: consequently the reproduction does not display complete uniformity. To ensure rapid publication this discrepancy could not be changed: nor could the English be checked completely. Therefore, the readers are asked to excuse any deficiencies of this publication which may be due to the above mentioned reasons.

The Editor

Printed in Seoul, Korea

IFAC WORKSHOP ON ALGORITHMS AND ARCHITECTURES FOR REAL-TIME CONTROL

Sponsored by

International Federation of Automatic Control (IFAC) Technical Committee on Computers
Institution of Electric Engineers(IEE)
Korea Association of Automatic Control(KAAC)
Automation and Systems Research Institute of Seoul National University(ASRI)

Organized by

Engineering Research Center for Advanced Control and Instrumentation at Seoul National University on behalf of Korea Association of Automatic Control (KAAC)

International Program Committee

Peter Fleming (UK) (Chairman)
D. P. Atherton (UK)
R. Doraiswami (Canada)
R. J. Evans (Australia)
K. Furuta (Japan)
W. K. Giloi (Germany)
H. Hanselmann (Germany)
V. Hernandez (Spain)
D. J. Holding (UK)
C. C. Hang (Singapore)
A. Inamoto (Japan)
G. W. Irwin (UK)
D. I. Jones (UK)
K. C. Kim (Korea)
R. Kulhavy (Czechoslovakia)
W. H. Kwon (Korea)
R. Lauwereins (Belgium)
Yong Zai Lu (PRC)
L. Motus (Estonia)
A. P. Pashkevich (Russia)
L. Pun (France)
M. G. Rodd (UK)
E. T. A. Rogers(UK)
P. L. Shaffer (USA)
K. G. Shin (USA)
G. J. Suski (USA)
D. Tabak (USA)
K. B. Woo (Korea)

National Organizing Committee

Wook Hyun Kwon (Chairman)
Byung Joon Chang
Chong-Ho Choi
Byung Kook Kim
Kwang Bae Kim
Shin Jung Kim
Kye Young Lim
Se Young Oh
Chan Bin Park
Wonyong Sung
Hai-Won Yang

PREFACE

Modern controls are becoming more complex in order to meet better performance requirements in tracking, robustness, adaptation, and learning, etc. In many situations, the complexity of existing control algorithms may cause long delays in computation and cannot be implemented in real-time systems with timing constraints.

However, recent advances in computer technologies, especially in VLSI, have made it possible to realize several architectures for special purpose processors and multiprocessor systems which consist of such designs as systolic arrays, neural networks, hypercubes, and these have enabled massive parallel processing and computing. Therefore, either totally new control algorithms such as ones utilizing neural networks can now be devised or existing complex control algorithms can be modified to be used with above architectures.

This Workshop focuses on such issues as control algorithms which are suitable for real-time use, computer architectures which are suitable for real-time control algorithms, and applications for real-time control. Three keynote papers start on each of three days with special extended presentations. In the first keynote paper Professor Halang gives an inspired insight into real time control. By identifying severe deficiencies of contemporary computer systems with respect to predictability, data conversion, inherent delays, simultaneous operation, timing, and inappropriate performance evaluation, he points out that contemporary computer systems are inappropriate for real time control. Professor Halang finally stresses the need and some ideas for special real-time computer architectures to eliminate the above deficiencies.

In the second keynote paper, Professor Irwin describes the need for parallel algorithms in control. Taking real-time Kalman filtering as a unifying theme, he suggests two approaches for parallel algorithms; one is based on systolic arrays and the other on analogue flow diagrams. Implementation using transputers is also discussed.

In the third keynote paper, Professor Shin analyzes effects of failure in controller computers in real-time control systems. While most papers deal with algorithms and architectures for controller computers, this paper discusses analytically the hard deadlines of controlled processes which destroy system stability.

In addition to these, 56 papers from 22 countries are included in this volume. The papers cover real-time control issues in the areas of parallel algorithms, multiprocessor systems, neural networks, fault-tolerant systems, real time robot control, identification, real time filtering algorithms, control algorithms, fuzzy control, adaptive and self-tuning control, and real time control applications.

The Workshop is the second in a series, the first one being held in Bangor, UK, 11-13 September 1991. 187 papers were submitted from 35 different countries, from which 73 papers were selected through review procedures and finally 59 papers including 3 keynote papers are included in this volume.

The Workshop is held in Seoul, Korea, and organized by Engineering Research Center for Advanced Control and Instrumentation at Seoul National University, on behalf of Korean Association of Automatic Control(KAAC). This Workshop is the first one to which the new IFAC publication policy is applied, under which the Proceedings will not be published by IFAC. The Workshop secretariat are to be complimented on their fine efforts in producing these high-quality Proceedings.

Peter Fleming *Wook Hyun Kwon*

CONTENTS

Neural Networks

Identification

Real-Time Robot Control

Real-Time Filtering Algorithms

Fuzzy Control

CONTEMPORARY COMPUTERS CONSIDERED INAPPROPRIATE FOR REAL-TIME CONTROL

Wolfgang A. Halang

*University of Groningen, Department of Computing Science,
P.O. BOX 800, 9700 AV Groningen, The Netherlands*

Abstract: The thesis of this contribution — which may appear provocative for many readers — is substantiated by comparing the requirements for real time control with the state of the art in real time computing. By identifying severe deficiencies with respect to predictable behaviour, data sampling, delays between sensing and actuation, simultaneous operation, timing, dependability, and way of system design and application, it is shown that contemporary computer systems employed for real time control are inappropriate to a surprisingly high extent. The paper stresses the need for special real time computer architectures, points out how the present deficiencies can be eliminated with available technology, and concludes with a more suitable conceptual foundation of real time computing.

Keywords: Real time control, real time computing, fundamental concepts.

INTRODUCTION

Having a background in real time computing, I was rather surprised about the term "real time control" when I heard it for the first time — as a matter of fact it was in the title of this conference series. I was also confused, because I used to think that control necessarily takes place in real time. If there was a real time control, then there had to be a non-real-time control as well. I was, however, unable to imagine what this could be and which practical significance it could have.

So I turned to colleagues with expertise in control and asked for an explanation of the term real time control. I learnt that real time control was considered to be the activity of actually controlling, i.e., regulating and manipulating, technical systems and processes, inherently being a real time task in contrast to, for instance, control systems design, which takes place off-line and does not know hard timing limits. The discussions revealed that the definition of real time computing, with which I am familiar, also applies to real time control. Since the term real time is, unfortunately, often misused, it is useful to mention here the precise definition of the real time operating mode as given in the German industry standard document DIN 44 300 (1985):

> **The operating mode of a computer system in which the programs for the processing of data arriving from the outside are permanently ready, so that their results will be available within predetermined periods of time; the arrival times of the data can be randomly distributed or be already a priori determined depending on the different applications.**

It is, therefore, the task of digital computers working in this operating mode to execute programs, which are associated with external technical processes. The program processing must be temporally synchronised with events occurring in the external processes and must keep pace with them. Hence, real time systems are always to be considered as embedded in a larger environment and, correspondingly, also called "embedded systems".

In the sequel, we shall elaborate on real time operation to work out a sound conceptual basis for real time control in form of a set of fundamental requirements. Then, we shall consider a number of features of contemporary computing systems employed for real time control or, respectively, of the prevailing way computers are used in control to determine whether and how well the fundamental requirements are met. It will turn out that computing systems applied for control purposes do not deserve the adjective "real time", yet, establishing the thesis stated in the title of this contribution. This state of affairs is due to both inherent incompatibilities between the requirements of control and the operation of digital computers as well as to deficiencies which could be removed. For each of these features, after having discussed its inadequacies and shortcomings, we shall outline a possible way for improvement in order to show constructively either how one can adjust to the intrinsic deficiencies of computers — if one wishes to continue using computers for control purposes — or that new architectures apt for computers used in real time control can be devised avoiding the non-intrinsic deficiencies. The paper closes with a personal perspective on the future development.

REAL TIME REQUIREMENTS

Real time operation distinguishes itself from other forms of data processing by the explicit involvement of the dimension *time*. This is expressed by the following two fundamental user requirements, which real time systems must fulfill even under extreme load conditions:

- **timeliness** and
- **simultaneity**.

As we shall see a little bit later, these requirements are supplemented by two further ones of equal importance:

- **predictability** and
- **dependability**.

Upon request from the external process, data acquisition, evaluation, and appropriate reactions must be performed on time. Not the mere processing speed is decisive for this, but the timeliness of the reactions within predefined and *predictable* time bounds. Hence, it is characteristical for real time systems that their functional correctness does not only depend upon the processing results, but also upon the instants, when these results become available. Correct instants are determined by the environment, which cannot be forced to yield to the computer's processing speed, as is the case with batch and time-sharing systems. According to the external processes, into which real time systems may be embedded, there are environments with *hard* and with *soft* time constraints. They are distinguished by the consequences of violating the timeliness requirement: whereas soft real time environments are characterised by costs rising with increasing lateness of results, such lateness is not permitted under any circumstances in hard real time environments, because late computer reactions are either useless or dangerous. In other words, the costs for missing deadlines in hard real time environments are — from the application's point of view — (almost) infinitely high. Hard time constraints can be determined precisely and typically result from the physical laws governing the technical processes being controlled (Herrtwich, 1989).

External technical processes to be controlled usually consist of a number of more or less independent subprocesses which operate in parallel. This implies the requirement for simultaneous processing of external process requests. To meet it, real time systems must be essentially distributed and must also provide parallel processing capabilities.

The definition of the real time operating mode has important consequences for the ***dependability*** of real time systems, since the demanded permanent readiness of computers working in this mode can only be provided by fault tolerant and — especially with respect to inadequate handling — robust systems. These dependability requirements hold both for hardware and software. They are particularly important for those applications, for which computer malfunctions do not only cause loss of data, but also endanger people and major investments. Naturally, expecting high dependability does not mean the unrealistic demand for a system which will never fail, since no technical system is absolutely reliable. However, using appropriate measures, one must strive to make the violation of deadlines and corresponding damages quantifiable and as unlikely as possible. In doing so, the individual limitations of a real time system must be recognised and the risk of its utilisation in process control must be carefully pondered (Herrtwich, 1989).

Data to be processed may arrive randomly distributed in time, according to the definition of the real time operating mode. This fact has lead to the widespread conclusion, that the behaviour of real time systems cannot and may not be deterministic. This conclusion, which found an expression in the non-deterministic selection of a rendezvous partner as semantics of the selective-wait statement in the Ada language, ***is based on a thinking error!*** Indeed, the external technical process may be so complex, that its behaviour appears to be at random. The reactions to be carried out by the computer, however, must be precisely planned and fully predictable. This holds in particular for the case of the simultaneous occurrence of several events, leading to a situation of competition for service, and also includes transient overload and other error situations. Then, the user expects that the system degrades its performance gracefully in a way transparent and foreseeable to him. Only a fully deterministic system behaviour will ultimately enable the safety licensing of programmable devices for safety critical applications. The notion of predictability and determinism appropriate in this context is to be illustrated by an example: we do not know when it will burn, but we expect the arrival of the fire brigade after being called within some short duration. Predictability of system behaviour is, as we have seen, of central importance for the real time operating mode. It supplements the timeliness demand, for the latter can only be guaranteed, if the system behaviour is precisely predictable, both in time and with respect to external event reactions.

UNPREDICTABLE BEHAVIOUR

Today, conventional computers are employed to run real time control applications. They do not really deserve to be called real time, since they do not use (and support) the notion of time. Rather, with a minimum of features, viz., process peripherals, user accessible interrupt lines, and slightly modified multitasking operating systems that allocate tasks on the basis of (often fixed) priorities, and handle real time exceptions in a sudden, unpredictable manner, they are adapted to work as embedded systems. This approach is primitive and leads to a number of problems (Halang and Stoyenko, 1991). Specifically, almost all hardware and software features of conventional computers, while improving traditional, *average* performance, do little for — or even worsen — the prospects for predictable real time performance. These features include virtual storage management, direct memory access cycle stealing, pipelining, caching, communication protocols and bus arbitration methods based on probabilistic concepts, dynamic data structures, recursion, and exception handling. They introduce unpredictable delays, unbounded contention and other plagues to predictable program execution and must, therefore, like all other "dynamic" or "virtual" features be *considered harmful.* Furthermore, the recognition of interrupts may be inhibited for arbitrarily long periods, and reactions may starve, i.e., may be delayed indefinitely, due to the use of the inadequate fixed priority scheduling concept. Program execution and system response times are, therefore, completely unpredictable and generally unknown. As the consequence of this, the state of the art in performance verification of real time computing systems is the "HAP" (Hope and Pray) method. We can conclude that the state of affairs in using conventional von Neumann architectures in real time control is unacceptable.

Only quite recently, the lack of predictability of the behaviour of real time computer systems has been identified as a serious problem. Halang and Stoyenko (1991) and Colnarič (1992) have shown in a constructive way that it is possible to design computer architectures apt for real time control. Such architectures avoid the harmful features identified above and, instead, employ other means for performance enhancement, which allow worst case timing analyses.

DATA CONVERSIONS

The introduction of a digital system into the path of a closed loop system necessitates two data conversions. Computers have two inherent problems with the handling of continuous, analogue signals frequently delivered by process sensors: the signals need to be sampled to enable discrete processing, and the sampled values need to be converted from analogue to digital representations. Sampling always involves a number of errors, viz., information loss by discretisation in time and quantisation in the range of values, limited representation accuracy, and aliasing. If a sampler is included into a control system, the actual state of the system is unknown between successive samples. Naturally, the sampling time should be chosen in a way that the actual state can be estimated from the samples with small error. However, this cannot always be achieved, because the computer needs the time between two samples to generate its actuating signal. In other words, there is a minimum sampling time, which equals the execution time of the control algorithm. In case the required sampling time is smaller than the execution time, the latter must be reduced in one way or the other (cp. next section).

The practically available sampling methods do not meet the requirements of the corresponding idealised theory based upon Shannon's sampling theorem. For, there is an inherent reconstruction error since, in contrast to the preconditions of Shannon's theorem, signals can only be observed during finite time intervals. Real signals are also not band-limited, because time-limited signals can never be band-limited. Hence, there is always aliasing regardless of the sampling frequency. Finally, the ideal samplers considered in theory do not exist, because any real measuring device integrates the signal over a (small) time interval. If we want to continue using digital computers in control and, hence, need to sample analogue signals, we should turn to more realistic local integral sampling and should develop a new theory as its foundation.

A second data conversion process is necessary on the computers' output side to generate actuation signals. The disadvantages of digital to analogue conversion are accuracy errors in the analogue part and the form of the output signal, i.e., a step functions.

The problems described in this section are intrinsic and can only be avoided by renouncing the use of digital computers, since strictly speaking, ***the use of computers inhibits real time control.***

INHERENT DELAYS

It is difficult to guarantee punctuality of a real time computing system, because the response times depend on the actual workload of the computer, and may differ from one request of an external source to the next one. Thus, the duration between the arrival of signals or data from the process, and the output of corresponding control action by the computer, is generally unpredictable, since it is a function of operating system overhead and workload, caused by the application tasks both being not foreseeable at the time a certain control loop program is written.

In control theory it is often assumed that the pure time delay between the measurement of a controlled variable and the resulting actuation, which is caused by the computer processing of a control algorithm, is negligible. *This is erroneous.* Even a small delay can have a marked destabilising influence on the control loop (Marshall, 1979) and cause the loop to lack robustness with respect to parameter variations or disturbance inputs (El-Sakkary and Gough, 1986). It is common practice in digital control to close one or more feedback loops through a computer, which generates the control laws. In order to share its services between several control loops and, generally, various other tasks, the computer intermittently turns unavailable to immediately handle the requests of a single control loop. Such irregularities in computer service result in the deterioration of control quality and may even render control systems unstable. It is rather surprising, that no attempt has been made, yet, to solve this problem by designing a suitable computer architecture, which would be quite natural. Instead, the deficiencies of contemporary computers are obviously regarded as unalterable facts and, for instance, by Rekasius (1985) system theoretical methods are applied to determine only, under which conditions the deterioration of control quality can still be tolerated.

In (Halang, 1990) it was shown that the problems of unpredictable response times of non-negligible length can easily be solved by endowing real time computers with input/output facilities making use of exact time specifications. This is achieved by stating, for every data exchange with the outside process, a time when it is to take place and also by furnishing the peripheral units with some additional hardware, to carry through the time related functions. The instants of data exchanges become part of the control information, such as the device address, to be sent to the input/output controllers by the driving software. The advantage of this feature is the predictability it provides. The application programmer can select an appropriate delay, between the input of measurements and the output of corresponding control information, and can take this delay into consideration, or even compensate for it, by using a suitable algorithm. For the control engineer, a predictable and precisely realised time delay is often much more important than high speed of the applied computer. Many methods for measurement and control are based on equidistantly spaced data points. Employing accurately timed peripherals, biased results, caused by random errors in the timing of data transfers, can be prevented. A second fundamental requirement, viz., timeliness, can be met if the computer is able to service the external requests within time limits feasible for the given technical process. If these two conditions are fulfilled, a process control computer is equivalent to any other physical system, because there is no technical device that does not show some kind of a delay.

For externally triggered input operations (and any other interrupt occurrences as well), in addition to the entered data item itself, it is necessary to know the time instant, when it was received by the computer's input controller, to be made available to the application program. This is because in contemporary real time computers, due to the operating system overhead and varying workload, there is a noticeable delay of unpredictable length between an (data available) interrupt from a peripheral controller (or another source), and the execution of a corresponding reaction, e.g., an input statement, in an application task waiting for these data. The interrupt arrival times generally remain unknown, since reading the system clock immediately as the first activity in the reacting software is not sufficiently accurate.

In the preceding section we have seen, that the sampling time of a signal is bounded from below by the processing time of the control algorithm chosen. If the sampling time needs to be shorter than the execution time, either the control algorithm must be simplified so that it takes less time to process, or a faster computer must be used. Both alternatives have consequences. Reducing the complexity of the control algorithm will, in general, affect the system performance with negative effects up to instability. If, for instance, integer arithmetic is used instead of floating point arithmetic, the quantisation error will increase and the system response will change. Use of a faster system will certainly affect the costs, and is restricted by the available technology.

The problem can also be addressed in a different way, viz., by explicitly taking computational and other delays into account. Suppose, for instance, that a signal is sampled with sampling time δt, the execution time of a control algorithm (including all input/output and operating system overhead) is τ, and that the last sample was taken at time t_k. In a system without computational delay, an actuating signal is generated at t_k as well, whereas a computer will generate its response not before the time $t_k + \tau$. The actuating signal is based on the last sample and a number of previous ones, the number depending on the order of the controller. If this output signal was present in time, the controller could be considered as without computational delay, which simplified analysis and design. In order to achieve this, the computer should start executing its algorithm at $t_k - \tau$. The problem remains that the computer must know in advance the value to be sampled at time t_k, a demand that cannot be met in real time. Hence, the required value has to be extrapolated from the past samples in some way ("predictive control"). In general, we have to leave the unrealistic concept of digital control without computational delays and need to develop new control algorithms taking care of these and other delays.

MULTI-VARIABLE CONTROL

Control systems usually take care of more than one variable, because most processes consist of a number of simultaneously operating subsystems. In multi-variable control applications it is necessary to acquire the state variables simultaneously and also to send out the control signals together at one time. Therefore, classical, i.e., non-computerised, automation systems have parallel architectures. Owing to their sequential way of operation, however, computer based control systems do not allow simultaneous signal sampling and actuation. In general, and especially when employing single processor computers of the von Neumann type, the simultaneous manipulation of parallel processes can only be realised approximately. In a macroscopic view, simultaneous execution is only provided if the time constants of the processes and the corresponding computing times differ sufficiently. As was shown in (Halang, 1990), the simultaneity problem can easily be solved with the timed peripherals as outlined in the last section, provided that they work fully in parallel and independently from each other, which is a common property of input/output controllers. Thus, even a computer with only a single processor can appear, to the outside world, as a device that provides a processor of its own for each control loop to be serviced.

TIME

Since time is the independent variable of all physical processes and the behaviour of technical processes is planned as a function of time, it is indispensable to have accurate time readings available in a real time computer. In fact, an architecture should support the problem and user oriented concept of time throughout. In contrast to this, computer science has no well developed concept of time. As a matter of fact, time even appears to be systematically suppressed. Therefore, it is necessary to reflect on the rôle time is playing. A clear understanding of the reasons why and the manner in which time is involved in the design of real time systems is needed as a prerequisite for a sound methodology.

Many hardware and software features heavily rely on the availability of precise timing information. The application of these features in distributed systems therefore necessitates the synchronous operation of all its clocks. Clock synchronisation in distributed environments is assumed to be a serious problem by many researchers, and is usually approached with rather complicated methods. The opposite is true, because in each node of a distributed system a common reliable time base of highest precision can be provided in form of a radio receiver for standard time signals, which are derived from the official atomic time and which are being broadcast by official agencies through the satellites of the General Positioning System (GPS) (Lange; Robinson) or through terrestric stations in various countries. The long wave station DCF 77 of Physikalisch-Technische Bundesanstalt in Braunschweig, Germany, is an example of such a continuous radio service (Becker, 1975; PTB, 1989), which is already used for many years by industry, government, railway etc. GPS disseminates the official atomic time of the (U.S.) National Bureau of Standards over the entire planet via 14 satellites (18 in the final stage). The system has already proven its usefulness under difficult conditions (Gulf war 1991). Based on it, local clocks can be synchronised with the international time standard within an accuracy of 100 nsec. Recently, even wrist watches became available using the DCF 77 signal as time reference. Thus it has been demonstrated, that the internationally coordinated and, therefore, only global time base can be provided cheaply in electronic systems by the integration of receiver, local quartz controlled clock, and an appropriate synchronisation logic in a single chip. Consequently the question arises why, when computers are concerned in contrast to wrist watches, people still content themselves with very inaccurate internal clocks, which solely rely on interval timers and perform all other functions in software. We, at least, see no reason not to furnish any real time computer, which is to deserve this name, either stand-alone or being a node in a distributed system, with a receiver for official time signals, as they are now entering the market. Thus, precise timing of specified activities, such as the input/output operations as mentioned in the penultimate section, become possible and feasible. The same holds for the application oriented task synchronisation constructs based on the concept of time, which were introduced in (Hoogeboom and Halang, 1992).

INAPPROPRIATE PRACTICE

The definition of real time operation cited above implies that some prevailing misconceptions about real time systems need to be overcome: neither time-sharing nor just fast systems are necessarily real time systems. Commencing the processing of tasks in a timely fashion is much more significant than (average) speed. The thinking in probabilistic or statistical terms, which is common in computer science with respect to questions of performance evaluation, is as inappropriate in the real time domain as is the notion of fairness for the handling of competing requests or the minimisation of average reaction times as an optimality criterion of system design. Instead, worst cases, maximum run-times, and maximum delays need to be considered, deadlocks must be prevented, and deadlines must be observed under any circumstances. A system may have no features that take arbitrarily long to operate. For the realisation of predictable and dependable real time systems, reasoning in static terms and the acceptance of physical constraints is a must.

Despite the best planning of a system, there is always the possibility of a transient overload in a node resulting from an emergency situation. To handle such cases, load sharing schemes have been devised, which migrate tasks between the nodes of distributed systems. In industrial embedded real time systems, however, such ideas are generally not applicable, because they only hold for computing tasks. In contrast to this, control tasks are highly input/output bound and the permanent wiring of the peripherals to certain nodes makes load sharing impossible.

Maximum processor utilisation is a major issue by the thinking criteria of classical computer science, and still the subject of many articles. For embedded real time systems, however, it is totally irrelevant whether the processor utilisation is only suboptimal, as costs have to be seen in a larger context, viz., in the framework of the controlled external process and with regard to the latter's safety requirements. Taking the costs of a technical process and the possible damage into account, which a processor overload may cause, the cost of a processor is usually negligible and, in the light of the still constantly declining hardware costs, also in absolute figures its price is of decreasing significance. (E.g., one hour production stoppage of a medium size chemical facility due to a computer malfunction and the required clean-up cost some $ 50,000. This is about the price of the computer controlling the process itself. A processor board costs only a minor fraction of that amount.) Hence, processor utilisation is not a feasible design criterion for embedded real time systems. Lower processor utilisation is a — cheap — price to be paid for system and software simplicity (an experienced system engineer costs some $ 1,600 per day) as prerequisite to achieve dependability and predictability.

In line with the (inappropriate) high processor utilisation concept, in most real time computers, field stations of distributed control systems, and programmable logic controllers several control loops are usually allocated to a single CPU. This is detrimental for system dependability, because a failure in the computing system affects the entire process being controlled or, at least, a larger part thereof. If, instead, failures remained restricted to single loops, the inherent stability of well-designed processes would usually cope with such situations in an effective way. Consequently, the architecture of an automation system has to mirror the parallelism inherent to the external real time environment itself. Then, in contrast to other parallel computing systems, the parallelism does not lead to an increase in software complexity. By allocating intrinsically independent functions to different hardware devices working in parallel, not only the speed is increased, but also and much more importantly, predictability and dependability are enhanced.

SAFETY CRITICAL CONTROL

Depending on national legislation and practice, the licensing authorities are very reluctant or even refuse to approve safety related control systems, whose behaviour is exclusively program-controlled, because state of the art software verification measures cannot guarantee the correctness of larger programs with mathematical rigour. In general, safety licensing is denied for highly safety critical systems relying on software with non-trivial complexity. Therefore, safety related control systems are mostly based on hardwired processing elements designed according to well-established engineering procedures and employing proven discrete or relay logic.

Naturally, owing to their greater flexibility and higher information processing capability, it is desirable to let programmable electronic systems also take care of safety relevant functions. This requires that the corresponding control software and its time behaviour can be safety licensed. The greatest obstacle which needs to be overcome for reaching this goal is to cope with complexity.

To provide a remedy for this unsatisfactory situation, in (Halang and Jung, 1992) a low-complexity architecture for a customised real time computer control system was developed, which can carry out safety related functions within the framework of distributed process control systems or programmable logic controllers. It explicitly supports sequence controls, since many automation programs including safety relevant tasks are of that kind. A specification level programming paradigm is architecturally supported, which is based on the interconnection of application oriented standard software function modules. By restricting the flexibility in program formulation, this paradigm inherently reduces the potential to commit errors and to be unsafe. Thus, by design, there is no more semantic gap between the programming and machine execution levels enabling the safety licensing of application software by an extremely simple and understandable, but as well rigorous method, viz., diverse inverse documentation (Krebs and Haspel, 1984).

In available programmable logic controllers, the execution time for a step generally varies from one cycle to the next depending upon the program logic performed and the external conditions evaluated each time. Therefore, the measurement of external signals and the output of values to the process is usually not carried out at equidistantly spaced points in time, although this may be intended in the control software. In contrast to this, the customised architecture features full temporal predictability and determinism. All inputs occurring in a step are performed en bloc at the beginning of the cycle and the thus obtained data are buffered until they will be processed. Likewise, all output data are first buffered and finally sent out together at the end of the cycle.

A PERSPECTIVE FOR THE FUTURE DEVELOPMENT

If future real time computers are not to fail as the very early, fully centralised ones, the VLSI technology driven hardware development needs to be accompanied by a consolidation process with respect to architectural and software issues:

> "Military missions and commercial applications involving real time systems at their base are becoming rapidly prevalent while the science and technology to support the credible design, construction, and enhancement of such systems is woefully deficient." (van Tilburg, 1989).

The consolidation effort must be centred around the quest of fulfilling the timeliness, predictability, and dependability requirements, because they have not been met, yet. These objectives can only be achieved by choosing *simplicity* as the fundamental design principle, i.e., by following Dijkstra's (1989) advice stressing the need for simplification:

> "Computing's core challenge is how not to make a mess of it."
> "... so we better learn how not to introduce complexity in the first place."
> "It is only too easy to design resource sharing systems with such intertwined allocation strategies that no amount of applied queueing theory will prevent most unpleasant performance surprises from emerging. The designer that counts performance predictability among his responsibilities tends to come up with designs that need no queueing theory at all."
> "The moral is clear: prevention is better than cure, in particular if the illness

> is unmastered complexity, for which no cure exists."
>
> "... both the final product and the design process (must) reflect a theory that suffices to prevent a combinatorial explosion of complexity from creeping in."
>
> "It is time to unmask the computing community as a Secret Society for the Creation and Preservation of Artificial Complexity."

Since there is practically no real time system without safety relevance in one way or the other, the non-temporal requirement of dependability and high integrity is of paramount importance. We must come to a stage that real time systems can be engineered with a sufficient degree of confidence in their dependability, which allows the licensing authorities to formally approve their utilisation for safety critical control purposes. At the present state of affairs, simplicity is a precondition for this. The fundamental importance of simplicity is established by its position in the following causal chain:

$$\textit{Simplicity} \Rightarrow \textit{(easy) Predictability} \Rightarrow \textit{Dependability}$$

At first, it is surprising to encounter the notion of predictability in connection with computers since, in principle, all digital computers work fully deterministically and are, therefore, predictable in their behaviour. To precisely express the special meaning of predictability, which is appropriate as a fundamental concept of real time computing, the adjective "easy" was used in the above implication. It qualifies the notion of predictability, defined by Stankovic and Ramamritham (1990) as

> "Predictability means that it should be possible to show, demonstrate, or prove that requirements are met subject to any assumptions made, for example, concerning failures and workloads",

by paying tribute to the economic and intellectual effort which needs to be invested in order to establish the property for a given real time system. If the system is simple, it can be easily understood, which is the main step towards the verification of its correct behaviour in the sense of Descartes: *"verum est quod valde clare et distincte percipio"*.

Future hardware functions will be even more readily available than today, working at higher speeds, and costing less. Hence, following the new thinking criteria (in contrast to obsolete ones such as, e.g., high processor utilisation) as outlined above, problem oriented architectures for real time systems achieving predictability based on simplicity must be designed. Measures for performance enhancements are only valuable, if their effects can be a priori analysed and quantified, and if they behave fully deterministically. As advantageous utilisations of cheap hardware we envisage the separation of functions, their encapsulation into standardised and safety licensed modules, and the provision of physical parallelism reflecting the parallelism inherent to the embedding processes. Thus, real time kernels may be implemented in firmware or hardware and interrupt processing may be delegated to separate units in order to solve a serious problem, which was described by Dijkstra (1989) as "... the real time interrupt ... its effect was the introduction of non-determinism and endless headaches ...". In addition new functions could be realised, such as the already mentioned radio sets in every processor for the reception of official time signals.

Following the simplicity argument, specialised small real time operating system kernels will be developed for the domain of safety sensitive embedded systems with tight deadlines. In this demanding application area the first real time systems fulfilling all our four fundamental requirements will be realised. Some dedicated kernels have already come rather close towards guaranteeing timely and predictable operation. In order to catch up with respect to dependability, the endeavour of formally proving a kernel's correctness has been initiated (Tol, 1992).

There will be a new generation of real time languages, whose characteristics will not only be user and application oriented concepts but also, and more importantly, simplicity and inherent safety (Dijkstra (1989) calls Ada a monstrum!). These languages will have to be the vehicles of a new programming paradigm: program design and the corresponding proof design, between which there is an extremely close connection, will always be carried out together and, due to the dependability requirement, there will be no more real time programs without proof. Evidently, this paradigm necessitates the development of correct compilers. The high costs involved will foster software re-usability, that did not make much progress yet, because of lacking (self) discipline and "artistic" programmer attitudes. To a large extent, future critical real time software will be assembled of standardised and safety licensed modules, which may even be marketed in "ROM-canned" form.

Once hardware and real time operating systems with predictable behaviour and well-defined new languages will have made possible the design of predictably behaving real time systems, it will be the task of comprehensive development suites to facilitate the design process. These tools are to accompany all realisation phases, viz., requirements engineering, design and implementation of combined hardware and software systems, system integration out of preferably standardised components, and automatical generation of accompanying documents. They are to allow for the implementation independent structuring and description of systems, for the precise formulation of timing constraints and, at the same time, restrict the freedom of choice in order to prevent errors. The major novel feature of these environments, however, will be that they will incorporate for each development phase appropriate analysers and formal verification tools. Being a consequence of the predictability and dependability requirements, this will bring about another paradigm change from empirical a posteriori validation (testing), which cannot prove the absence of errors, to design integrated verification having the quality of mathematical rigour.

The innovations in VLSI technology will continuously have the major impact on the entire field of computer design by providing higher levels of integration. Within less than ten years, with the advent of Gigascale integration, some 100 millions to 1 billion transistors should fit on a single die. Thus, it should be possible to accommodate up to four processors on one chip. These processors do not necessarily have to be equal. Besides a general purpose task processor, there could be dedicated processors for fast interrupt recognition and response, for the operating system kernel, and for input/output handling. Such an architecture would reflect the parallelism inherent to real time operation. A thorough analysis of the application domain has to reveal whether this approach is feasible, or if it would be more advantageous to put the processors on separate chips and to utilise the enormous transistor count for integrated memories and peripheral devices in order to prevent communication bottle-necks.

In general, real time systems research must, much more than other areas of computer science, be based on and oriented at a solid, realistic model of the application domain. The employed methods are generally process specific, because the process is part of the control loop closed by and in the computer. This holds, e.g., for overload handling and error recovery procedures, which can be designed by exploiting the inertia of the process and its corresponding typical time constant. A real time system is subjected to variable time conditions in dependence on the process speed. There may be the possibility for their relaxation in the overload case, e.g., by reducing the speed of a robot arm grabbing components. Adaptive, self correcting systems with carefully designed graceful performance degradation behaviour in response to error occurrences can only be constructed by full utilisation of the process characteristics. The comparison of two different designs or systems is, analogously, only possible on the basis of application specific benchmarks — mere MIPS figures do not say anything.

REFERENCES

Becker, G. (1975). Die Sekunde. *PTB-Mitteilungen*, 85, 14 - 28.

Colnarič, M. (1992). *Predictability of the Temporal Behaviour of Hard Real Time Systems.* Ph.D. thesis. University of Maribor.

Dijkstra, E. W. (1989). The next forty years. Personal note *EWD 1051.*

DIN 44 300 (1985). *Informationsverarbeitung.* No. 9.2.11. Beuth-Verlag, Berlin Cologne.

Halang, W.A. (1990). Simultaneous and Predictable Real Time Control Achieved by Accurately Timed Computer Peripherals. *Proc. 11th IFAC World Congress on Automatic Control*, Vol. 7, 279 - 284.

Halang, W.A., and A.D. Stoyenko (1991). *Constructing Predictable Real Time Systems.* Kluwer Academic Publishers, Boston Dordrecht London.

Halang, W.A., and S.-K. Jung (1992). An Architecture Enabling the Safety Licensing of Real Time Controller. *These Proceedings.*

Herrtwich, R.G. (1989). Echtzeit. *Informatik-Spektrum* 12, 93 - 96.

Hoogeboom, B., and W.A. Halang (1992). The Concept of Time in the Specification of Real-Time Systems. In: M. Schiebe and S. Pferrer (Eds.), *Real-Time Systems Engineering and Applications*, pp. 11 - 40. Kluwer Academic Publishers, Boston Dordrecht London.

Kershaw, J. (1987). *The VIPER Microprocessor.* Royal Signal and Radar Establishment, Malvern, England, Report No. 87014.

Krebs, H., and U. Haspel (1984). Ein Verfahren zur Software-Verifikation. *Regelungstechnische Praxis rtp*, 26, 73 — 78.

Lange, W.R. Synchronisation verteilter Rechner mit Hilfe von standardisierten Zeitcodes. LANGE-ELECTRONIC GmbH, Olching.

Marshall, J.E. (1979). *Control of Time Delay Systems.* Peter Peregrinus Ltd., Stevenage New York.

PTB (1989). *Zur Zeit.* Physikalisch-Technische Bundesanstalt, Braunschweig Berlin.

Rekasius, Z.V. (1985). Digital Control With Computer Interruptions. *Proc. 1985 American Control Conference*, Vol. 3, 1618 - 1621.

Robinson, D.C. Using Bus Level IRIG Time Code Translators to Time Tag Data and Synchronize Multiple Processing Stations. Bancomm, Division of Datum Inc., San Jose, CA.

El-Sakkary, A.K., and N.E. Gough (1986). *Proc. 25th IEEE CDC Conference.*

Stankovic, J.A., and K. Ramamritham (1990). Editorial: What is Predictability for Real-Time Systems? *Real-Time Systems*, 2, 247 - 254.

van Tilburg, A. (1989). Preface to the ONR Kickoff Workshop on Foundations of Real-Time Computing Research Initiative. IEEE Real-Time Systems Newsletter, 5, 6 - 7.

Tol, R. (1992). A Small Real-Time Kernel Proven Correct. Submitted to 13th IEEE Real-Time Systems Symposium, Phoenix, AZ.

HARD DEADLINES IN REAL-TIME CONTROL SYSTEMS

Kang G. Shin and Hagbae Kim

Real-Time Computing Laboratory Department of Electrical Engineering and Computer Science, The University of Michigan Ann Arbor, MI 48109-2122

Abstract: In a real-time control system where the control input is computed by a controller computer, the transient computer failures caused by an electromagnetic interference may seriously affect system stability. The faulty controller computer causes either a long delay in the feedback loop thus failing to update the control input for one or more sampling intervals, or control input disturbances by updating the control input incorrectly until the fault is handled properly. If the period of this abnormal behavior exceeds a certain limit called a *hard deadline*, either the necessary conditions for system stability will be violated or the system will leave the *allowed* state space. In such a case a *dynamic failure* is said to occur in the system.

We present a method for deriving hard deadlines for linear time-invariant control systems by examining the stability of the state difference equations resulting from the modification of the original state equations with an assumed maximum delay and several random sequences that represent the effects of stationary occurrences of the disturbances to, as well as the random delays of, the control input. Moreover, a one-shot event model, in which a single long-lasting fault causes a dynamic failure, is presented based on the state trajectory and the allowed state space.

1 Introduction

Most real-time control systems consist of two synergistic parts: the *controlled process* or *environment*, and the *controller computer*. Digital computers are commonly used in real-time control systems due mainly to their improved performance and reliability in dealing with increasingly complex controlled processes. The control programs, which are executed by a controller computer residing in the feedback loop, perform a set of functions using sensor readings from the controlled process and/or the environment at regular time intervals.

Since the controller computer is highly susceptible to transient electromagnetic interferences inducing functional errors (perhaps without damaging any components), it is usually equipped with some fault-tolerance mechanisms especially for life-, or safety-critical systems like aircraft or nuclear reactors. When the abnormality (component failure or environmental interference) of the controller computer occurs, the computation-time delay increases significantly, thus either failing to update the control input during the time taken for error detection, fault location, and recovery; or updating the control input incorrectly until the failure is handled successfully (i.e., detected and recovered). The stationary occurrences of these abnormalities may lead to the loss of system stability if their active duration exceeds a certain limit called the *hard deadline* [4]. Even one occurrence of the abnormality for a long period — called a *one-shot event model* in [3] — may drive the controlled process out of its allowed state space, or a *dynamic failure* occurs.

Most conventional analyses of computation-time delay effects have been based on the assumption that the feedback delay is fixed or constant [1, 2]. In [3], we derived hard deadlines for linear time-invariant control systems based on the fact that computation-time delays are stochastic in both their occurrence times and magnitudes reflecting the nature of computer failures. However, we did not consider control input disturbances under the assumption of perfect fault detection.

In this paper, we derive hard deadlines by examining the stability of the state difference equations modified with random sequences that represent the stationary occurrences of computer failures and the imperfect error coverage (with binomial distributions), the duration of failures/interferences (with multinomial distributions), and the magnitude of disturbances to the control input (with a normal distribution). The system dynamics are modified first according to the assumed maximum delay, NT_s, and the probability distribution of delays whose occurrence periods $\leq NT_s$, where N is changed from 1 to the actual maximum delay (or

The work reported was supported in part by the Office of Naval Research under Grant N00014-91-J-1115 and by the NASA under Grant NAG-1-1120.

hard deadline), denoted by DT_s. In addition, the state and input constraints are used to derive the allowed state space from which the hard deadline is derived as a function of time and the system state. This analysis is useful for the one-shot event model, where a single event — a long-lasting failure — may cause a dynamic failure.

Section 2 addresses the generic problem for analyzing the effects of computation-time delays and input disturbances. Section 3 presents the basic assumptions and the random sequences that characterize controller failures and input disturbances. Then, hard deadlines are derived with the modified state difference equations for linear time-invariant control systems for both stationary and one-shot event models. In Section 4, simple linear systems are examined to demonstrate our approach. Section 5 deals with the application of the hard-deadline information to the design of a reliable controller computer. The paper concludes with Section 6.

2 The Effects of Controller Computer Failures

Linear time-invariant controlled processes are generally represented by state-space models as shown in Eq. (2.1) and are equipped with well-designed controllers that stabilize the overall control system and optimize the control objectives:

$$\mathbf{x}(k+1) = \mathbf{A}\mathbf{x}(k) + \mathbf{B}\mathbf{u}(k) \qquad (2.1)$$

where k is the time index, one unit of time represents the sampling interval T_s, and $\mathbf{x} \in \mathcal{R}^n$ and $\mathbf{u} \in \mathcal{R}^\ell$ are the state and input vectors, respectively. The (digital) controller computer reads sensors, and calculates the control input once every T_s seconds according to a programmed control strategy. The control input, which is held constant within each sampling interval by a latch circuit, is applied to the controlled process.

When a fault[1] occurs in the controller computer, it will trigger functional error modes — which are computer failures — perhaps, without component damages. Suppose a computer failure is detected upon its occurrence at time k_0, and its recovery takes n sampling intervals. The control inputs during these intervals, $\mathbf{u}(k_0+1), \cdots, \mathbf{u}(k_0+n)$, will be held constant at $\mathbf{u}(k_0)$ by the D/A converter and latch circuits. Suppose a computer failure is detected n_1 sampling intervals after its occurrence at time k_0 and the subsequent recovery takes n_2 sampling intervals. The control inputs during this period are $\mathbf{u}(k_0+1)\mathbf{I}_\Delta, \mathbf{u}(k_0+2)\mathbf{I}_\Delta, \cdots, \mathbf{u}(k_0+n_1)\mathbf{I}_\Delta, \cdots, \mathbf{u}(k_0+n_1)\mathbf{I}_\Delta, \mathbf{u}(k_0+n_2), \mathbf{u}(k_0+n_2+1), \cdots$, where $\mathbf{I}_\Delta$ is a diagonal matrix with $Diag[\mathbf{I}_\Delta]_i = 1+\Delta u_i$ and Δu_i is a random sequence modeled as the output of a dynamic system with a white-noise input. Since faults occur randomly during the mission lifetime, they are considered to be random disturbances to the controlled process, which can be modeled based on the fault characteristics.

The hard deadline of a stationary model is defined as the maximum duration of the controller computer's failure without losing system stability. Thus, in linear time-invariant systems, the hard deadline is defined as:

$$D(N) = \inf_{C_{env}} \sup\{N \ : \ \|\lambda(N)\| < 1\}, \qquad (2.2)$$

where $\lambda(N)$ is an eigenvalue of the controlled process in the presence of computer failures of the maximum duration NT_s and C_{env} represents all the environmental characteristics that cause computer failures. Let $\mathbf{X}_A(k)$ and $\mathbf{U}_A$ be the allowed state space at time kT_s and the admissible input space, respectively. Suppose the state is evolved from time k_0 in the presence of a computer failure (disturbance/delay) which occurred at k_1T_s, was detected N_1 sample intervals later and is recovered within N_2 sample intervals of its detection, where $N = N_1 + N_2, \ \ 0 \le N_1, N_2 \le N$. The control input during this period ($k_0 \le k \le k_0 + N$) is:

$$\mathbf{u}_a^N(k) = \mathbf{u}(k_0)\Pi_{k_0}(N_1) + \mathbf{u}(k)\mathbf{I}_\Delta\Pi_{k_0+N_1}(N_2),$$

where $\Pi_m(n) = \xi(k-m) - \xi(k-m-n)$ is a rectangular function from m to $m+n$, and ξ is a unit step function. Then, the hard deadline of a task during the time interval $[k_0T_s, \ k_fT_s]$ is defined as:

$$D(N, \mathbf{x}(k_0)) = \inf_{\mathbf{u}_a^N(k)\in\mathbf{U}_A} \sup\{N : \phi(k, k_0, \mathbf{x}(k_0), \mathbf{u}_a^N(k)) \in \mathbf{X}_A(k), \ k_0 \le k \le k_f\}, \qquad (2.3)$$

where the state trajectory is governed by:

$$\mathbf{x}(k) = \phi(k, k_0, \mathbf{x}(k_0), \mathbf{u}_a^N(k)). \qquad (2.4)$$

3 Derivation of Hard Deadlines

Let NT_s and DT_s be the <u>assumed</u> maximum and <u>actual</u> maximum delays, respectively. Then, the hard deadline can be obtained by iteratively testing the necessary conditions for system stability and state residence in the allowed state space while changing N from 1 to D.

[1] Transient electromagnetic interferences (EMI) such as lighting, high intensity radio frequency fields (HIRF) and nuclear electromagnetic pulses (NEMP) are considered as the possible sources of the fault.

3.1 The Hard Deadline of the Stationary Model

The controlled processes described by Eq. (2.1) are unstable without any state feedback control. Thus, the state feedback control input necessary to stabilize such unstable systems can be calculated in the usual form of $\mathbf{u}(k) = -\mathbf{F}\mathbf{x}(k)$, depending on the control objectives, e.g., time-optimal control with energy constraint, optimal state tracking, and optimal linear regulator.

To derive hard deadlines, the given state equation is modified to include all the stochastic behaviors of computer failures based on the following random sequences and the basic assumptions for tractability.

Definition of Random Sequences:

1. p: the probability of a computer failure at each sampling instant.
2. d: the conditional probability of successful failure detection given that a computer failure had occurred.
3. q_i^d: the conditional probability of a delay (recovery duration) for i sampling intervals ($\sum_{i=1}^{N} q_i^d = 1$) if a computer failure occurred and is detected before generating any incorrect control input.
4. q_i^w: the conditional probability of a control input disturbance for i sampling intervals ($\sum_{i=1}^{N} q_i^w = 1$) if a computer failure occurred and is not detected till its disappearance.
5. $q_{\Delta u}$: the probability density function (pdf) of the magnitude, Δu, of the control input disturbance at time kT_s, i.e., $\mathbf{u}_{actual}(k) = \mathbf{u}_{desired}(k)\mathbf{I}_{\Delta}$. The mean and variance of $q_{\Delta u}$ are given *a priori* as $\mu_{\Delta u}$ and $\sigma^2_{\Delta u}$.

Basic Assumptions:

1. The control inputs calculated after recovering computer failures are always correct.
2. The probability that two transient failures occur sequentially within a small number, $N-i$, of sample intervals, where the delay (recovery duration) or duration of incorrect control inputs (active duration of a transient failure) is i sample intervals and N is the assumed maximum value of such intervals, i.e., $1 \le i \le N$ — is small enough to be ignored. That is, we consider only one computer failure possible during N sample intervals.
3. Every random sequence considered here is independent identically distributed ($i.i.d$) with respect to the time index k.

Suppose the control input have been updated correctly at $t = mNT_s$. In case an abnormality (delay/disturbance) is active for i sampling intervals since time $t = mNT_s$ as a result of a controller computer failure, where $1 \le i \le N$, let the control input at $(mN+i)T_s$ be denoted as $\mathbf{u}_a(mN+i)$ which is equal to either $\mathbf{u}(mN+i)\mathbf{I}_{\Delta}$ for disturbance or $\mathbf{u}(mN)$ for delay. The corresponding state equations for the group of intervals during which the system failed to update the control input correctly become:

$$
\begin{aligned}
\mathbf{x}(mN+1) &= \mathbf{A}\mathbf{x}(mN) + \mathbf{B}\mathbf{u}_a(mN)\\
\mathbf{x}(mN+2) &= \mathbf{A}\mathbf{x}(mN+1) + \mathbf{B}\mathbf{u}_a(mN+1)\\
&= \mathbf{A}^2\mathbf{x}(mN) + (\mathbf{A}+\mathbf{I})\mathbf{B}\mathbf{u}_a(mN+1)\\
&\ \ \vdots\\
\mathbf{x}(mN+i) &= \mathbf{A}^i\mathbf{x}(mN) + \sum_{j=0}^{i-1}\mathbf{A}^j\mathbf{B}\mathbf{u}_a(mN+j)\\
\mathbf{x}(mN+i+1) &= \mathbf{A}^{i+1}\mathbf{x}(mN) +\\
&\quad \sum_{j=1}^{i}\mathbf{A}^j\mathbf{B}\mathbf{u}_a(mN+j) + \mathbf{B}\mathbf{u}(mN+i)\\
&\ \ \vdots\\
\mathbf{x}((m+1)N) &= \mathbf{A}^N\mathbf{x}(mN) + \sum_{j=N-i}^{N-1}\mathbf{A}^j\mathbf{B}\mathbf{u}_a(mN+j)\\
&\quad + \sum_{j=0}^{N-i-1}\mathbf{A}^j\mathbf{B}\mathbf{u}(mN+N-j-1),
\end{aligned}
$$

where m is the time index for the groups of N sampling intervals each. Let $\mathbf{X}(m) = [\mathbf{x}_1, \mathbf{x}_2, \ldots, \mathbf{x}_N]^T \equiv [\mathbf{x}(mN+1), \mathbf{x}(mN+2), \ldots, \mathbf{x}((m+1)N)]^T$ and $\mathbf{U}(m) = [\mathbf{u}_1, \mathbf{u}_2, \ldots, \mathbf{u}_N]^T \equiv [\mathbf{u}(mN+1), \mathbf{u}(mN+2), \ldots, \mathbf{u}((m+1)N)]^T$; that is, $\mathbf{X}(m)$ and $\mathbf{U}(m)$ are respectively the augmented state and control vectors for the group of sampling intervals of NT_s. Then, we get the following augmented state equations:

$$
\begin{aligned}
\mathbf{X}(m+1) &= \mathbf{A}_D\mathbf{X}(m) + \mathbf{B}^1_{D_i}\mathbf{U}(m)\\
&\quad + \mathbf{B}^2_{D_i}\mathbf{U}(m+1), && (3.1)\\
\mathbf{U}(m) &= -\mathbf{F}_D\mathbf{X}(m), && (3.2)
\end{aligned}
$$

where $[\mathbf{B}^1_{D_i}, \mathbf{B}^2_{D_i}]$ becomes $[\mathbf{B}^{n1}_{D_0}, \mathbf{B}^{n2}_{D_0}]$ for the normal behavior, $[\mathbf{B}^{d1}_{D}, \mathbf{B}^{d2}_{D_i}]$ for delay, and $[\mathbf{B}^{w1}_{D_i}, \mathbf{B}^{w2}_{D_i}]$ for disturbance, respectively. In the above equations, $\mathbf{A}_D, \mathbf{B}^{nk}_{D_0}, \mathbf{B}^{wk}_{D_0}, \mathbf{B}^{d1}_{D}$, and $\mathbf{B}^{d2}_{D_i}$ are the augmented state and input transition matrices, where $k \in \{1, 2\}$, and

$\mathbf{F}_D$ is an augmented feedback gain matrix. Then, the state difference equation is modified to:

$$\begin{aligned}\mathbf{X}(m+1) &= \mathbf{A}_D\mathbf{X}(m) + \Big((1-\psi)\mathbf{B}^{n1}_{D_0} + \\ &\quad \psi(1-\varphi)\mathbf{B}^{w1}_{D} + \psi\varphi\sum_{i=1}^{N}\xi_i\mathbf{B}^{d1}_{D_i}\Big)\mathbf{U}(m) \\ &+ \Big((1-\psi)\mathbf{B}^{n2}_{D_0} + \psi(1-\varphi)\sum_{i=1}^{N}\zeta_i\mathbf{B}^{w2}_{D_i} \\ &+ \psi\varphi\sum_{i=1}^{N}\xi_i\mathbf{B}^{d2}_{D_i}\Big)\mathbf{U}(m+1) \qquad (3.3)\end{aligned}$$

where $\psi, \varphi \in \{0,1\}$ are binomially–distributed random sequences with probabilities p, d, and $\xi_i, \zeta_i \in \{0,1\}$ are multinomially–distributed random sequences with probabilities q_i^w, q_i^d, i.e., $\Pr[\xi_i = 1] = q_i^w$.

Similarly to the method used in [3], the deterministic value of the hard deadline is determined by examining the pole positions of the first moment (ensemble average) of Eq. (3.3). Although the resulting hard deadline has little practical meaning, it indicates the trend of the ensemble system behavior with the uncertainty (in the state and output) that can be measured by the second moment of Eq. (3.3). In addition, one can derive the probability mass function (*pmf*) of the hard deadline with respect to $q_{\Delta u}$ rather than the deterministic value of the hard deadline based on the mean of $q_{\Delta u}$. The mapping between the hard deadlines and the magnitudes of disturbances (Δu's) is not one–to–one and hard deadlines can be derived iteratively using a numerical method with respect to each sample value of Δu's. The sample values of Δu's are obtained by uniformly quantizing the $q_{\Delta u}$ continuum of the interval $[a,b]$, where $\int_a^b q_{\Delta u} d\Delta u = \gamma$. Let this quantization result in M equal–length subintervals (cells), where a, b, and M depend on the required accuracy of analysis. Then, points are allocated to the quantized intervals (cells). Let the point of the ith cell ($[a+(i-1)\frac{M}{b-a}, a+i\frac{M}{b-a}]$) be Δu_i which corresponds to the value of the midpoint of the cell, i.e., $\Delta u_i = a + \frac{(2i-1)M}{2(b-a)}$, then the probability of being at this point is calculated as $q^i_{\Delta u} = \int_{a+(i-1)\frac{M}{b-a}}^{a+i\frac{M}{b-a}} q_{\Delta u}(s)ds$. A hard deadline is derived for each Δu_i, and let it be D_i whose probablity is equal to that of the ith cell (i.e., $q^i_{\Delta u}$). Finally, the *pmf* of the hard deadline is derived numerically by multiplying D_i and $q^i_{\Delta u}$, $1 \le i \le M$.

3.2 The Hard Deadline of the One–Shot Event Model

The pole locations do not change in case of only one single failure with a relatively long ($> T_s$) active period. The (asymptotic or global) stability condition discussed thus far is therefore no longer applicable. Instead, the terminal state constraints can be used to test whether or not the system leaves its allowed state space. Note that every critical process must operate within the state space circumscribed by given constraints, i.e., the allowed state space. When the control input is not updated for a period exceeding the hard deadline, the system may leave the allowed state space, thus causing a dynamic failure. The allowed state space consists of two sets of states $\mathbf{X}^1_A$ and $\mathbf{X}^2_A$ defined as follows:

- $\mathbf{X}^1_A$: the set of states in which the system must stay to avoid an *immediate* dynamic failure, e.g., a civilian aircraft flying upside down is viewed as an immediate dynamic failure. This set can usually be derived *a priori* from the physical constraints.
- $\mathbf{X}^2_A$: the set of states that can lead to meeting the terminal constraints with appropriate control inputs. This set is determined by the terminal constraints, the dynamic equation, and the control algorithm used.

The system must not leave $\mathbf{X}^1_A$ nor $\mathbf{X}^2_A$ in order to prevent catastrophic failure.

Let Let k_0, k_f, N_1, and N_2 denote the indices for the failure occurrence time, the mission completion time, and the period of disturbance, the period of delay measured in sampling intervals, respectively, where $N = N_1 + N_2$, $0 \le N_1, N_2 \le N$. The dynamic equation of a one-shot event model is:

$$\begin{aligned}\mathbf{x}(k+1) &= \mathbf{A}\mathbf{x}(k) + \mathbf{B}\left[\mathbf{u}(k) + (\mathbf{u}(k_0) - \mathbf{u}(k))\Pi_{k_0}(N_1)\right. \\ &\quad \left. + \mathbf{u}(k)(\mathbf{I}_\Delta - \mathbf{I})\Pi_{k_0+N_1}(N_2)\right], \qquad (3.4)\end{aligned}$$

where $\Pi_{k_0}(N)$ is a rectangular function defined in Section 2, and N_1 and N_2 are random variables whose probabilities are determined by q_i^d and q_i^w. Then, the determinstic value or the *pmf* of hard deadline can be derived, similarly to the stationary model, by using the first moment or the samples of Eq. (3.4).

We can obtain the state trajectory, which is tested for the given constraints, by using the first moment or the samples of Eq. (3.4). $\mathbf{x}_f \in \mathbf{X}^f_A$ can be tested indirectly by the following relation:

$$\mathbf{x}(k_f) \in \mathbf{X}^f_A \Longleftrightarrow \mathbf{x}(k_0 + N) \in \mathbf{X}^2_A, \qquad (3.5)$$

where

$$\mathbf{X}^2_A = \{\, \mathbf{x} \mid [\mathbf{A}^{k_f - k_0 - N}\mathbf{x} + \sum_{i=k_0+N}^{k_f-1} \mathbf{A}^{k_f-i-1}\mathbf{B}\mathbf{u}(i)] \in \mathbf{X}^f_A \,\}.$$

N	3	4	5	6	7
$d=1$	0.7537	0.8945	0.9659	1.0021	1.0204
$d=0.9$	0.7579	0.9985	1.1698	1.2922	1.3798

Table 1: Maximum magnitude of eigenvalues, $|\lambda_{max}|$.

In practice, it is difficult to obtain $\mathbf{X}_A^2$. Although there may be a one–to–one mapping between $\mathbf{x}(k_0+N)$ and $\mathbf{x}(k_f)$, $\mathbf{X}_A^f$ is usually a continuum, which requires an excessive amount of computation due to the curse of dimensionality. The size of $\mathbf{X}_A^2$ will decrease as either N increases or k approaches k_f, but the size of $\mathbf{X}_A^2$ is usually larger than that of $\mathbf{X}_A^f$ due to the (asymptotic) stability of a controlled process.

4 Examples

To demonstrate the concept of hard deadline, we derive the deadlines for several simple example control systems, two of which are described for the stationary and one-shot event models.

Example 1: Consider a simple controlled process:

$$\begin{aligned} x_1(k+1) &= 11.02x_1(k) + 1.08x_2(k) + 10u_1(k) \\ x_2(k+1) &= 0.95x_2(k) + 10u_2(k), \end{aligned}$$

where the coefficient matrices of a quadratic performance index and the corresponding optimal (feedback) control gain matrix stabilizing the controlled process by discrete Riccati equation are given by:

$$\mathbf{R}_{xx} = 2\mathbf{R}_{uu} = \begin{bmatrix} 10 & 0 \\ 0 & 10 \end{bmatrix}; \ \mathbf{F} = \begin{bmatrix} 3.1251 & 0.3090 \\ -1.0791 & 0.5512 \end{bmatrix}$$

This feedback control changes the eigenvalues from $\{0.95, 11.02\}$ to $\{0.0777, 0.2101\}$. Then, the change of poles as a result of incrementing N is derived deterministically for the occurrence of the largest delay possible ($p = q_N = 0.045$) and is given in Table 1, where the first case is for the perfect coverage ($d = 1$) and the second case represents the existence of input disturbances ($d = 0.9$ and $\mu_{\Delta u} = -5$). The deterministic value of the hard deadline is $D = 6T_s$ in the absence of input disturbances with an instant failure detection, whereas it decreases to $D = 5T_s$ with some (infrequent) input disturbances. The *pmf* of hard deadline is given in Table 2 with the *pmf* of the magnitude of disturbances to the control input.

Example 2: The hard deadline of a one-shot event model is derived for the system of a double integrator which was also used for a one-shot delay model in [3].

D	3	4	5	7	9
Pr[D]	0.12	0.28	0.48	0.08	0.04
Δu	-5 ± 4	-5 ± 3	-5 ± 2	-5 ± 1	-5
$q_{\Delta u}$	0.0667	0.1334	0.2	0.2667	0.3334

Table 2: Probability mass functions of hard deadline and $q_{\Delta u}$.

The state difference equation of the discretized process with the sampling rate, $T_s = 0.01s$, is:

$$\mathbf{x}(k+1) = \begin{bmatrix} 1 & 1 \\ 0 & 1 \end{bmatrix} \mathbf{x}(k) + \begin{bmatrix} 0.5 \\ 1 \end{bmatrix} u(k).$$

With the same (state/terminal) constaints and the same feedback control input as those of [3], the *pmf* of hard deadline at time $T = 15T_s$ is derived for a Gaussian probability density function of Δu, $q_{\Delta u} = \frac{1}{\sqrt{2\pi 10}} e^{-\frac{(\Delta u - 10)^2}{200}}$, and is given in Table 3.

D	2	3	4	5	6
Pr[D]	0.0006	0.0905	0.1326	0.0152	0.0678
D	8	10	11	20	21
Pr[D]	0.0248	0.0159	0.0093	0.0096	0.3048

Table 3: *Pmf* of hard deadline.

5 Application of Hard Deadline Information

The information on hard deadlines is very useful for modeling system reliability and designing both the hardware and software of a controller computer. When designing a controller computer, one has to make many design decisions in the context of controlled processes that are characterized by their hard deadlines and cost functions [4], including:

- hardware design issues dealing with the number of processors and the type of interconnection network to be used, and the synchronization of processors,
- software design issues related to the implementation of control algorithms, task assignment and scheduling, redundancy management, error detection and recovery.

From the hard-deadline information, one can deduce the knowledge of system inertia, which can, in turn, be used to specify the fault-tolerance requirement of a real-time control system. This knowledge is required to estimate the system's ability of meeting timing constraints in the presence of controller-computer failures, which was characterized as the ***probability of dynamic failure***, P_{dyn} [4].

To illustrate the application of the knowledge of system inertia, let us consider a simple example of a triple modular redundant (TMR) controller computer in which three identical processors execute the same set of cyclic tasks. The TMR controller computer updates, once every T_s seconds or every sampling period, the control input to the controlled process (plant). That is, the period of each cyclic task is equal to T_s. The input of the cyclic task is a discretized output of the plant and the output of the cyclic task will be used to control the plant during the next sampling interval. The output of the TMR controller is correct for each task only if at least two of the three processors in the TMR controller produce correct outputs. A *TMR failure* is said to occur if more than one processor in the TMR controller fail during T_s. Thus, the output of the TMR controller would not be changed in case of a TMR failure. The condition for a system (dynamic) failure resulting from controller-computer failures[2] is derived from the hard deadline, which is the allowable maximum computation-time delay. In other words, this condition gives us the knowledge about the controlled system's inertia against controller-computer failures.

More than 90% of computer failures have been reported to be transient, especially with short active durations. Thus, the controller computer may recover from most failures in a few sampling intervals, and it can correctly update the control input without causing any dynamic failure, if the active duration of controller-computer failure is smaller than the hard deadline.

Suppose the hard deadline derived from the controlled system is three sampling periods and a TMR controller computer is used. That is, no dynamic failure occurs if the faults inducing computer failures disappear (or are recovered by a fault-tolerance mechanism) within three sampling intervals. Then, the reliability model for this controller computer is built by extending a Markov chain model with two additional states before the state of a dynamic failure, where the parameters of the Markov chain model are to be estimated at a given level of confidence from empirical data. The additional states account for the system inertia, i.e., a dynamic failure results from only three consecutive incorrect (missing the update of) outputs of the controller computer or for a period of $3T_s$, not immediately from one or two incorrect (missing the update of) outputs. Without the information of hard deadline, one can over-estimate the probability of a system failure under the assumption that the system has no delay-tolerance, i.e., one incorrect output can lead to a dynamic failure.

[2]The other sources of system failure(s), such as failures in actuators or sensors or mechanical parts and failures of A/D and D/A converters, are not considered in this paper, because our main intent is to analyze the coupling between a controlled process and a (fault-tolerant) controller computer.

6 Conclusion

The hard deadline for a critical control task is usually assumed to be given *a priori*. This presupposes the existence of a precise definition of the hard deadline and a method to derive it, which, however, have not been addressed in detail. We derived hard deadlines for linear time-invariant control systems in the presence of input disturbances due to imperfect detection coverage based on consideration of the intrinsic nature of computer failures.

The knowledge of hard deadlines, which must be derived from real control applications, is very important for task assignment and scheduling, specification and evaluation of fault-tolerant controller computers.

Acknowledgement: The authors would like to thank Allan White, Chuck Meissner, Felix Pitts of NASA Langley Research Center and James Smith of the Office of Naval Research for their support of the work reported in this paper.

References

[1] A. Gosiewski and A. Olbrot, "The effect of feedback delays on the performance of multivariable linear control systems," *IEEE Trans. on Automat. Contr.*, vol. AC-25, no. 4, pp. 729–734, August 1980.

[2] K. Hirai and Y. Satoh, "Stability of a system with variable time delay," *IEEE Trans. on Automat. Contr.*, vol. AC-25, no. 3, pp. 552–554, June 1980.

[3] K. G. Shin and H. Kim, "Derivation and application of hard deadlines for real–time control systems," *IEEE Trans. on Systems, Man, and Cybernetics*, September 1992 (in press).

[4] K. G. Shin, C. M. Krishna, and Y.-H. Lee, "A unified method for evaliating real–time computer controller and its application," *IEEE Trans. on Automat. Contr.*, vol. AC–30, no. 4, pp. 357–366, April 1985.

PARALLEL ALGORITHMS FOR CONTROL

George W. Irwin

Control Research Group, Department of Electrical Electronic Engineering, The Queen's University of Belfast BT9 5AH, N. Ireland

Abstract Two classes of concurrent algorithm for real-time Kalman filtering are presented in order to illustrate some algorithm engineering concepts. One is based on systolic computation, and fine-grained algorithms are derived for both regular and square-root covariance Kalman filtering. The second relies on an analogue flow diagram description of the differential equations for continuous-time, Kalman-Bucy filtering. Implementations on Inmos transputers are described in both cases.

Keywords Control engineering, Computational methods, Kalman filters, Systolic arrays, Analogue flow diagrams, Transputers.

INTRODUCTION

Advances in automatic control, both theoretical and practical, have been stimulated by corresponding developments in computer technology. The increasing use of digital computers in the 1960's space programme was accompanied by advances in state space control theory while, more recently, the introduction of low cost microprocessors facilitated widespread application of digital control loops and a corresponding development of adaptive control techniques to exploit the new technology. In each case the technology helped the transfer of advanced control theory into practical applications and, at the same time, stimulated further theoretical advances.

Parallel processing is likely to have the same impact as earlier developments in computer technology. This was recognised by the Institute of Electrical and Electronic Engineers (IEEE, 1987) in a strategy document which attempted to predict significant future trends and identify areas where effort was needed. Increased computational speed is of course the primary benefit of parallel processing, particularly for real-time control. This allows faster systems to be controlled and gives the control engineer the choice of added complexity in the control algorithm. Easy expansion, within a uniform hardware and software base, is another feature of concurrent control systems, since it is possible to add more processors as required, with significant implications for reduced development and maintenance costs. Parallel processing also offers a natural relationship between the control system description at the design stage, in the form of a block diagram, and the final hardware implementation. Finally fault tolerance can be realised in a parallel control system by organising the computation in a distributed sense, allowing performance degradation rather than a complete breakdown in the event of processor failure. These advantages have been recognised and an increasing number of successful implementations of concurrent controllers are being reported (Irwin and Fleming, 1992; Rogers and Li, 1992).

The remit of this paper was to introduce some recent research themes on parallel algorithms for control. Examination of the literature shows that effort in this area falls into the two broad catagories of CACSD and real-time control. The former is motivated by the need to produce efficient, reliable and portable tools for many of the common control design procedures. Thus Laub and Gardiner (1988) describe parallel algorithm implementations for a 32 processor Intel iPSC Hypercube, suitable for the analysis and control design of large scale systems with state dimensions of several hundreds or thousands. Other work is motivated by a desire to accelerate workstation based design tools, like Matlab, by producing parallel algorithms for computationally intensive bottlenecks. Thus Megson (1990) identified key mathematical techniques in H^∞ design, like Singular Value Decomposition and implemented parallel solutions on transputers. Fleming, Crummey and Chipperfield (1991) used transputers to calculate the objective function in multiobjective optimisation.

While CACSD involves off-line computation, on-line control is characterised by the time critical nature of the control signal calculations. It was the need for systematic procedures for mapping control algorithms onto concurrent architectures suitable for the implementation of real-time controllers which motivated the work to be presented below. Taking real-time Kalman filtering as a unifying theme, the paper presents two approaches to parallel algorithms. One is based on systolic arrays for fine-grained matrix calculations, the other is suitable for continuous time algorithms and follows from analogue flow diagram representations of a system. Reference is made to real-time transputer implementations in each case and an attempt is made to unify the results by discussion of algorithm engineering concepts.

SYSTOLIC ALGORITHMS FOR MATRIX CALCULATIONS

A systolic array (Kung and Leiserson, 1978) is an array of individual processing cells, each of which has some local memory, and is connected to its nearest neighbours in the form of a regular lattice. The term systolic refers to the rhythmic movement of data across the array which is analogous to the regular pumping action of the heart. Figure 1 shows a *linear array* of inner step processors for the matrix-vector product $\underline{y}=A\underline{x}$. The y_i, initialised at zero, are clocked from left to right and accumulate the necessary partial products formed by the interaction of the x_i and a_{ij} data streams. The extension to an *orthogonal array* for a matrix-matrix product, as shown in Fig. 2, is relatively straightforward.

The original concept was proposed in order to exploit the high speed switching potential of VLSI technology in fast, real-time, matrix-based, signal processing applications. Systolic structures thus combine the necessary concurrent processing with simplicity of circuit design through cell replication across the silicon. Systolic arrays can also be regarded as descriptions of fine-grained, concurrent algorithms. Thus, Fig. 3 illustrates a *triangular array* for the QR decomposition of a matrix A, using Given's rotations. The boundary cells calculate the rotation parameters which are passed to the internal cells which operate on the a_{ij}.

SYSTOLIC ALGORITHMS FOR DISCRETE KALMAN FILTERING

The basic systolic arrays above have been used to construct architectures for more complex computations. It is useful to omit the details of the cell descriptions, and to associate the function of the array with its geometrical shape, in order to gain an understanding of the algorithms involved.

This section will describe algorithms for covariance Kalman filtering. This is a key technique in both signal processing and modern control and is generic, in the sense that the algorithm is closely related to LQ optimal control, to least squares parameter estimation, to parameter tracking and to adaptive control. Thus the VLSI algorithms developed here are more generally applicable applicable.

The discrete-time system is described by:

$$\underline{x}(k+1) = A(k)\underline{x}(k) + B(k)\underline{u}(k) + \underline{w}(k) \quad (1)$$

$$\underline{z}(k) = C(k)\underline{x}(k) + \underline{v}(k) \quad (2)$$

where $\underline{w}(k)$, $\underline{v}(k)$ are zero mean, independent, Gaussian white noise sequences with covariance matrices of W(k) and V(k) respectively.

The optimal state estimation problem is to estimate the nx1 state vector $\underline{x}(k)$ from a sequence of noisy mx1 measurements ($m \leq n$), in such a way that the mean square estimation error is minimised.

Regular Covariance Filter (RCF)

The RCF algorithm updates the estimated state vector in two stages, prediction (or time update) followed by filtering (or measurement update) as follows:

Prediction or Time update

$$\hat{\underline{x}}(k+1/k) = A(k)\hat{\underline{x}}(k/k) + B(k)\underline{u}(k) \quad (3)$$

$$P(k+1/k) = A(k)P(k/k)A^T(k) + W(k) \quad (4)$$

Filtering or Measurement update

$$V_e(k+1) = C(k)P(k+1/k)C^T(k) + V(k) \quad (5)$$

$$K(k+1) = P(k+1/k)C^T(k+1)V_e^{-1}(k+1) \quad (6)$$

$$\hat{\underline{x}}(k+1/k+1)=\hat{\underline{x}}(k+1/k) + K(k+1)[\underline{z}(k+1)-C(k+1)\hat{\underline{x}}(k+1/k)] \quad (7)$$

$$P(k+1/k+1) = P(k+1/k) - K(k+1)C(k+1)P(k+1/k) \quad (8)$$

Although the matrix based nature of this algorithm means that it is computationally intensive, with $O(n^3)$ calculations to be performed in each iteration, it is therefore ideal for the application of systolic arrays.

The approach taken employs the Fadeev algorithm for calculating the Schur complement of a matrix A (Fadeev and Fadeeva, 1963). Thus suppose A,B,C and D are known matrices and the matrix quantity $D+CA^{-1}B$ is required. The compound matrix

$$\begin{bmatrix} A & B \\ -C & D \end{bmatrix} \quad (9)$$

is first formed and a linear combination of row 1 added to row 2 as follows.

$$\begin{bmatrix} A & B \\ -C+WA & D+WB \end{bmatrix} \quad (10)$$

Now, if W is chosen so that the (2,1) element is null

$$\begin{bmatrix} A & B \\ 0 & D+CA^{-1}B \end{bmatrix} \quad (11)$$

ie. the (2,1) element is reduced to zero by simple row manipulation, then the required result appears in the (2,2) position. For systolic computation it is convenient if the

$$\begin{bmatrix} A & B \\ -C & D \end{bmatrix} \xrightarrow{T} \begin{bmatrix} TA & TB \\ -C & D \end{bmatrix} \xrightarrow{W} \begin{bmatrix} TA & TB \\ -C+WTA & D+WTB \end{bmatrix}$$

triangularisation nullification

matrix A is triangularised first, prior to nullification of C.

The Fadeev algorithm can be performed on a *trapezoidal* array as shown in Fig. 4, which consists of the basic triangular array for operating on A, together with a rectangular section for performing the operations required on B. Note also the dual mode of operation of the cells.

Now the compound matrix above is 'programmable' in the sense that, by selecting appropriate values for B,C and D, a range of algebraic matrix expressions can be generated by the Fadeev algorithm. Yeh (1988) realised that the trapezoidal array could be used for generating the matrix terms required for a systolic Kalman filter. Table 1 defines the data required for the RCF (Gaston and Irwin, 1990a).

TABLE 1 RCF using the Fadeev Algorithm

STEP	A	B	C	D	RESULT
1	I	$\hat{\underline{x}}(k/k)$	-A(k)	$B(k)\underline{u}(k)$	$\hat{\underline{x}}(k+1/k)$
2	I	P(k/k)	-A(k)	0	A(k)P(k/k)
3	I	$A^T(k)$	-A(k)P(k/k)	W(k)	P(k+1/k)
4	I	$C^T(k)$	-P(k+1/k)	0	$P(k+1/k)C^T(k)$
5	I	$P(k+1/k)C^T(k)$	-C(k)	V(k)	$V_e(k+1)$
6	$V_e(k+1)$	I	$-P(k+1/k)C^T(k+1)$	0	K(k+1)
7	I	$[P(k+1/k)C^T(k+1)]^T$	-K(k+1)	P(k+1/k)	P(k+1/k+1)
8	I	$\hat{\underline{x}}(k+1/k)$	C(k+1)	$\underline{z}(k+1)$	$\Delta\underline{z}(k+1)$
9	I	$\Delta\underline{z}(k+1)$	-K(k+1)	$\hat{\underline{x}}(k+1/k)$	$\hat{\underline{x}}(k+1/k+1)$

Analysis of the computation involved with this systolic filter shows that $O(3/2n^2)$ systolic cells are needed to reduce the timesteps between measurements to O(9n). However, the systolic filter has some undesireable features. Each iteration involves 9 steps, with feedback of intermediate results. Also, closer examination of Table 1 shows that the A matrix is an identity for eight of those steps, implying that implicit inversion of an I matrix is occurring. Further, steps 2,4,6 involve addition of a null matrix. Essentially redundancy has been added to facilitate parallel processing.

Square Root Covariance Filter (SRCF)

Numerical errors with regular Kalman filters cause the error covariance matrix P to lose its symmetry and the filter can 'blow up' causing the state estimes to diverge. Square root Kalman filters force the P matrix to remain symmetric by propogating the triangular factors or square roots. The equations below define the SRCP due to Morf and Kailath (1975).

$$Q(k)\begin{bmatrix} P^{T/2}(k/k-1)C^{T}(k) & P^{T/2}(k/k-1)A^{T}(k) \\ V^{1/2}(k) & 0 \\ 0 & W^{T/2}(k) \end{bmatrix} = \begin{bmatrix} V_e^{T/2}(k) & V_e^{-1/2}(k)C(k)P(k/k-1)A^{T}(k) \\ 0 & P^{T/2}(k+1/k) \\ 0 & 0 \end{bmatrix} \quad (12)$$

$$\hat{x}(k+1/k)=A(k)\hat{x}(k/k-1)+B(k)u(k)+A(k)P(k/k-1)C^{T}(k)V_e^{-1}[z(k)-C(k)\hat{x}(k/k-1)] \quad (13)$$

where,

$$V_e^{1/2}V_e^{T/2} = C(k)P(k/k-1)C(k)^T+V(k)$$

The systolic algorithm is derived from the basic arrays in Figs. 1-4. The orthogonal decomposition required in equation (12) can be carried out as shown in Fig. 5. This illustrates the data to be entered and the results stored in memory. Since the order in which the data is entered does not affect the final result, the $W(k)^{T/2}$ and $V(k)^{T/2}$ terms can be preloaded.

The results from the decomposition, residing in memory, are now used to provide the state update. Taking the transpose of equation (13), and dropping the time indices for clarity, produces

$$\hat{\underline{x}}^T = \underbrace{(A\hat{\underline{x}} + B\underline{u})^T}_{D} + \underbrace{[\underline{z} - C\hat{\underline{x}}]^T}_{C}\ \underbrace{V_e^{-T/2}}_{A^{-1}}\ \underbrace{V_e^{-1/2}CP^TA^T}_{B}$$

This Schur complement can be formed by using the first m rows of the architecture in Fig. 5 as a trapezoidal array for the Fadeev algorithm. Further, since the 'A' term is already triangular, only the second or nullification mode needs to be performed. This produces the split architecture given in Fig. 6.

The matrix products required for the next iteration are formed by inserting a linear array between the trapezoidal and lower triangular sections. This captures the updated state vector, once available, and performs matrix-vector multiplications. The matrix-matrix products are produced by using the bottom triangular section as a multiplier array to give the complete SRCF array in Fig. 7.

This robust and compact systolic filter compares favourable with others in the literature (Gaston, Irwin and McWhirter 1990) requiring O(2n) timesteps per iteration.

ALGORITHM ENGINEERING

The term algorithm engineering describes the hybrid discipline involving the derivation of stable numerical algorithms suitable for parallel computation, the mapping of these algorithms onto dedicated parallel architectures and the interaction between algorithm and architecture.

As mentioned earlier, systolic arrays do not necessarily imply a specific hardware realisation. However, they are fundamental in the sense of providing representations of fine-grained parallel algorithms. They define the calculations to be performed, the order in which these are to be done and the input data required. This viewpoint allowed the detail to be ommitted from the basic arrays in order to concentrate on the formation of the more complex systolic SRCF algorithm from these building blocks.

Gaston and Irwin (1990a) have identified 4 ways in which the basic arrays can be used to form more complex algorithms:

(i) ***Off the peg*** Direct mapping of an algorithm onto a

basic array, with no changes to the array. The data must be manipulated so that the algorithm is executed correctly.
(ii) ***Cut-to-fit*** Algorithms are mapped onto a basic array which is customised to take account of the special structure of the data. Here the array is modified, as well as the data flow being specified.
(iii) ***Ensemble*** Several well known arrays are combined to implement one algorithm. These are linked together, with data flowing from one to the other, to complete the calculations.
(iv) ***Layer*** Two or more basic arrays are mapped into one, which switches mode from one operation to the next.
The RCF algorithm provides a good example of the first approach while the SRCF algorithm was formed by the 'layer' approach. These definitions were motivated by the notion of manipulating geomerical shapes. McWhirter (1989) takes a more formal appproach by associating the arrays with fixed matrix operators which do not change with the data being processed. The more complex systolic algorithms are then expressed in terms of these operators.

The next section reports on transputer implementations of the systolic RCF and SRCF algorithms in order to illustrate further the process of algorithm engineering.

TRANSPUTER IMPLEMENTATIONS

Transputer implementation of the systolic filters was studied by Maguire and Irwin (1991a) . The simple idea of assigning each systolic cell to one transputer is rejected because of the mis-match in granularity between the algorithm and the target hardware. Practical issues like load balancing, communications requirements and the efficient use of each processor dictate that cells are grouped into coarser processes.

The strategy proposed uses the inherent linear schedule of the systolic architecture, which is based on a set of equi-temporal hyperplanes, as shown in Fig. 8. The schedule vector $\underline{S}$, which defines the sequence of operations of the array, points in a direction normal to these hyperplanes since the computations along a hyperplane can be performed simultaneously. If the array is treated as a wavefront one, the data driven structure allows the schedule vector to be resolved into orthogonal components, $\underline{S}_h$ and $\underline{S}_v$ respectively. The architecture can then be mapped efficiently into a linear array of row- or column-processes by projecting parallel to one of these schedule vectors and using the concurrency in the resulting pipeline.

For these algorithms, projection was done in the horizontal direction due to the geometry of the arrays and the larger computational requirements of the boundary cells. Fig. 8 illustrates this for a 4th order RCF array. The resulting structure consists of a pipeline of row-processes in which the original systolic cells are coalesced, thereby removing their systolic operation. One advantage of this approach is that the communication requirements are reduced to passing vectors of data between row-processes, which can be implemented efficiently on transputers.

The second, and final, stage in the mapping strategy assigns these row-processes to transputers. Fig. 9 shows a one-to-one assignment for a RCF example, while Fig. 10 illustrates assigning multiple row-processes for a SRCF example. This method proved to be more efficient than a technique suggested by Stewart (1988) which involved folding of the systolic array, much like paper, down to the size of the target hardware architecture.

Transputer implementation results, on an example problem of varying increasing state and measurement dimensions, confirm that the conventional filters run faster than the systolic algorithm on a single transputer. The difference was more pronounced for the RCF because of the redundant calculations introduced into the systolic RCF.

Heuristic partitioning of the sequential algorithms is restricted to two processors because of the difficulty in extracting parallelism, and marginal speedups of about 1.1 can be obtained. However, the systolic array approach facilitates the introduction of more processors, allowing significant performance improvements over the single transputer implementations for the more computationally intensive SRCF filter. For both filters mapping efficiencies improve with higher problem orders because of increasing task granularity.

PARALLEL KALMAN-BUCY FILTERING

The continous-time Kalman-Bucy filter (Kalman and Bucy, 1961) is described by a set of ordinary differential equations as follows:

$$d/dt\ \hat{\underline{x}}(t) = A(t)\underline{x}(t) + B(t)\underline{u}(t) + K(t)[\underline{z}(t) - C(t)\hat{\underline{x}}(t)] \quad (14)$$

$$K(t) = P(t)C^T(t)V^{-1}(t) \quad (15)$$

$$d/dt\ P(t) = A(t)P(t) + P(t)A^T(t) - P(t)C^T(t)V^{-1}(t)C(t)P(t) + W(t) \quad (16)$$

Application of numerical integration procedures requires the solution of n^2+n differential equations, which constitutes a considerable burden for online implementation. Solution of the Ricatti differential equation for the error covariance matrix P(t) can be made more robust by forcing symmetry through propogation of the triangular factor. This also reduces the computation to the solution of $(n^2+3n)/2$ differential equations.

The analogue flow diagram provides a graphical description of a system for implementation on an electronic analogue computer. Further, it shows the parallelism inherent in a continuous time system by illustrating the relationship between the concurrent operation of the various blocks. Figs. 11 and 12 contain the analogue flow diagrams describing the differential equations for a Kalman-Bucy filter with n=4 and m=1.

In Fig. 11 the filter is operating in open loop and the state update involves the solution of n differential equations which can be implemented in parallel. Similarly, Fig. 12 shows the analogue flow diagram for the solution of the Ricatti equation. This is simplified since it neglects to show the output of the covariance elements, which provide the link to the state update equation. Also, the dimension of the measurement matrix is used to reduce the third term in equation (16) to the multiplication of two elements in P(t). The solution of the n^2 differential equations involved can be reduced by propogating the upper triangular section of the matrix to produce a triangular array of processing cells similar in structure to the systolic arrays discussed above. Note that here the communication is not restricted to nearest-neighbour because each processing element, associated with the calculation of an element of P(t), is connected to every other processing element in its row or column. This implies that the processing elements in each row or column can be executed simultaneously.

The concurrent description of the Kalman-Bucy filter contained in the analogue flow diagrams of Figs 11 and 12 can be mapped onto a transputer array by orthogonal projection to produce either interconnected row- or column- processes, similar to stage 1 for the systolic algorithms discussed previously. Stage 2, assigning row-processes to transputers, requires care to achieve load balancing. Thus, for a fourth order problem, row-processes 1 and 2 were placed on separate transputers, while row-processes 3 and 4 were combined and assigned to a third processor. The broadcasting of data facilitated the placing of non-neighbouring row-processes onto transputers in order to achieve balanced computation.

DISCUSSION AND CONCLUSIONS

In this paper two classes of parallel algorithms suitable for real-time Kalman filtering have been described.

The first relies on a systolic array representation of the fine-grained computations involved in discrete-time covariance filtering. The RCF and SRCF filters were used to illustrate how complex systolic algorithms could be formed from simpler arrays for basic matrix calculations. It was argued that the systolic arrays constitute a fundamental description of the concurrent computation and do not necessarily imply a VLSI implementation. In order to illustrate this point the systolic filters were realised on transputer hardware. The mapping strategy involved projecting into either row- or column-processes followed by placing these of processes onto processors.

The transputer implementations highlighted some interesting points about algorithm engineering. The regular Kalman filter involves less computation than the square root form where added robustness is achieved at the expense of extra computation. This makes it less attractive for real-time realisation on conventional Von-Neumann processors. The systolic array algorithms facilitated the introduction of parallel processing to achieve the speedups necessary for real-time performance. The speedup obtained for the systolic RCF was not large because of the redundant calculations added in order that the data should fit the trapezoidal array, as required for the Fadeev algorithm. The systolic SRCF algorithm on the other hand produced a more successful hardware implementation since the othogonal decomposition fitted more neatly and efficiently onto the triangular array.

The second class of algorithm involves an analogue flow diagram description of the differential equations arising from a continuous time Kalman-Bucy type filter and again a projection strategy was used to produce transputer implementations, although this case involved communication of data since each cell in the algorithm was connected to every other cell in the same row or column of the processing array.

Systolic algorithms have also been proposed for LQ optimal control (Gaston and Irwin, 1990b) for parameter estimation (Gaston and Irwin, 1991c) and applied to turbogenerator control (Maguire and Irwin, 1991b). Some results have also appeared on massively parallel algorithms where the nearest neighbour communication restriction of systolic computation is removed. The idea here is to combine the fast processing properties of current VLSI technology with the communication advantages of optical technology in the form of broadcast arrays (Miller and Irwin, 1991). Both the classes of algorithm discussed in this paper have also been derived for concurrent self-tuning control (Maguire, 1991).

REFERENCES

Fadeev, D.K. and Fadeeva, V.N. (1963). Computational Methods of Linear Algebra, W.H. Freeman and Co.

Fleming, P.J., Crummey, T.P. and Chipperfield, A.J. (1992). Computer assisted control system design and multiobjective optimisation, Proc. ISA Conf. on Industrial Automation, pp.7.23-7.26.

Gaston, F.M.F., Irwin, G.W. and McWhirter, J.G. (1990). The systolic approach to square root covariance Kalman filtering, J. VLSI Sig. Proc., Vol.2, No.1, pp.37-49.

Gaston, F.M.F. and Irwin, G.W. (1990a). Systolic Kalman Filtering: an overview, IEE Proc.-D, Vol. 137, No 4, pp.235-244.

Gaston, F.M.F. and Irwin, G.W. (1990b) A systolic linear quadratic optimal controller, Electronics Letters, Vol.26, No 14, pp.1000-1002.

Gaston, F.M.F. and Irwin, G.W. (1990c). A systolic parameter estimator for real-time feedback control, Proc. IEE Conf. IT '90, pp111-116.

IEEE (1987). Challenges to control: a collective view. Trans., Vol AC-32, No.4.

Irwin, G.W. and Fleming, P.J. (1992). Transputers in Real-Time Control, Research Studies Press (John Wiley and Sons). In Press.

Kalman, R.E. and Bucy, R.S. (1961). New results in linear filtering and prediction theory, J. Basic Eng., Vol. 83D, pp. 95-108.

Kung, H.T. and Leisesson, C.E., (1978). Systolic arrays for VLSI, in Sparse Matrix Proceedings, Soc. Indust. and Appl. Math., Philadelphia, PA, pp.245-282.

Laub, A.J. and Gardiner, J. D., Hypercube implementation of some parallel algorithms in control, in Advanced Computing Concepts in Control Engineering, M.J. Denham and A.J. Laub (eds.), NATO ISI Series, Vol. F47, pp. 362-390.

Maguire, L (1991), Parallel architectures for Kalman filtering and self-tuning control, PhD Dissertation, The Queen's University of Belfast.

Maguire, L. and Irwin, G.W. (1991a), Transputer

implementation of Kalman filters, IEE Proc., Vol.138, No4, pp.355-362.

Maguire, L and Irwin, G.W. (1991b). Parallel adaptive control, Proc. 1st European Control Conf., Grenoble, Vol. 1, pp. 590-595.

McWhirter, J.G., (1989) Algorithm engineering - an emerging discipline, SPIE Proc. on Advanced Algorithms and Architectures for Signal Processing IV, Vol.1152, pp. 2-15.

Megson, G.M., (1990), Transputer arrays and computer aided control system design, IEE Proc., Vol. 137, No. 4, pp. 197-210.

Miller, P and Irwin, G.W. (1991), Mapping algorithms onto processor arrays with data broadcasts, in VLSI Systems for DSP and Control, Woods, McCanny and Irwin (eds.), Woodhead Publishing, U.K., pp21-27.

Morf, M and Kailath, T (1975), Square root algorithms for least squares estimation, IEEE Trans. on Automatic Control, Vol.20, No.4, pp 487-497.

Rogers, E and Li, Y. (1992) Parallel Processing, Prentice Hall Int. In Press.

Stewart, R.W. (1988). Mapping signal processing algorithms onto fixed architectures, Proc. Int. Conf. on Acoustics, Speech and Signal Processing, pp.2037-2040.

Yeh, H.G. (1988), Systolic implementation on Kalman filters, IEEE Trans. ASSP, Vol.36, No.9, pp.1514-1517.

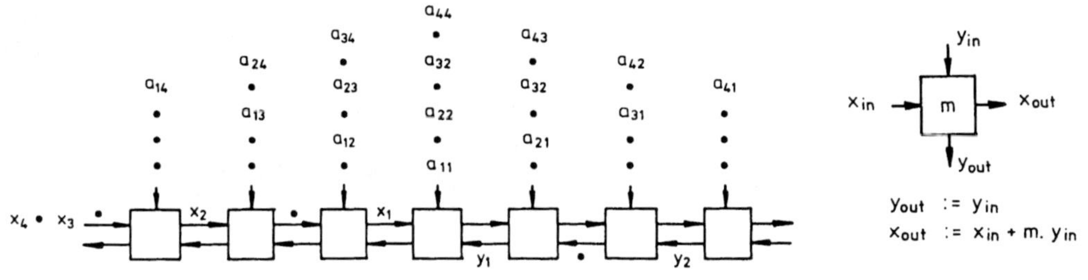

Fig. 1. Linear array for systolic matrix-vector product.

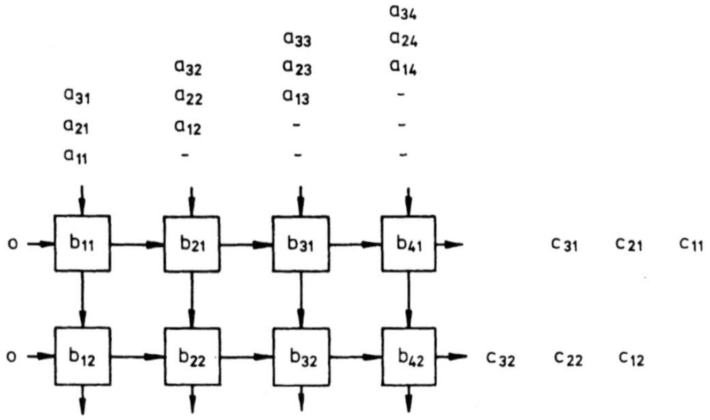

Fig. 2. Orthogonal array for systolic matrix-matrix product.

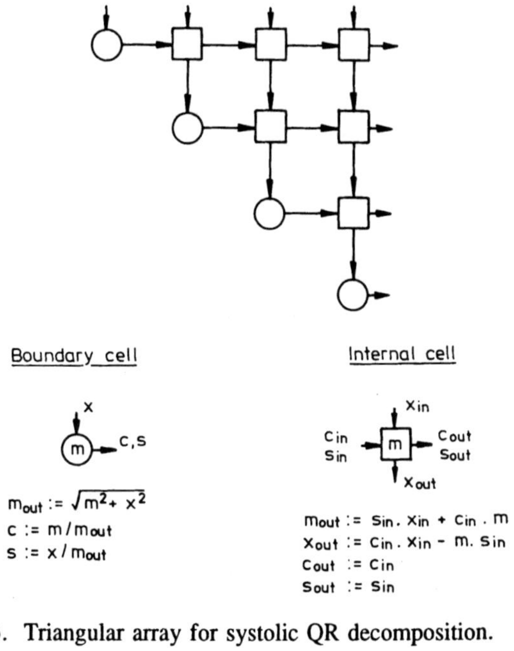

Fig. 3. Triangular array for systolic QR decomposition.

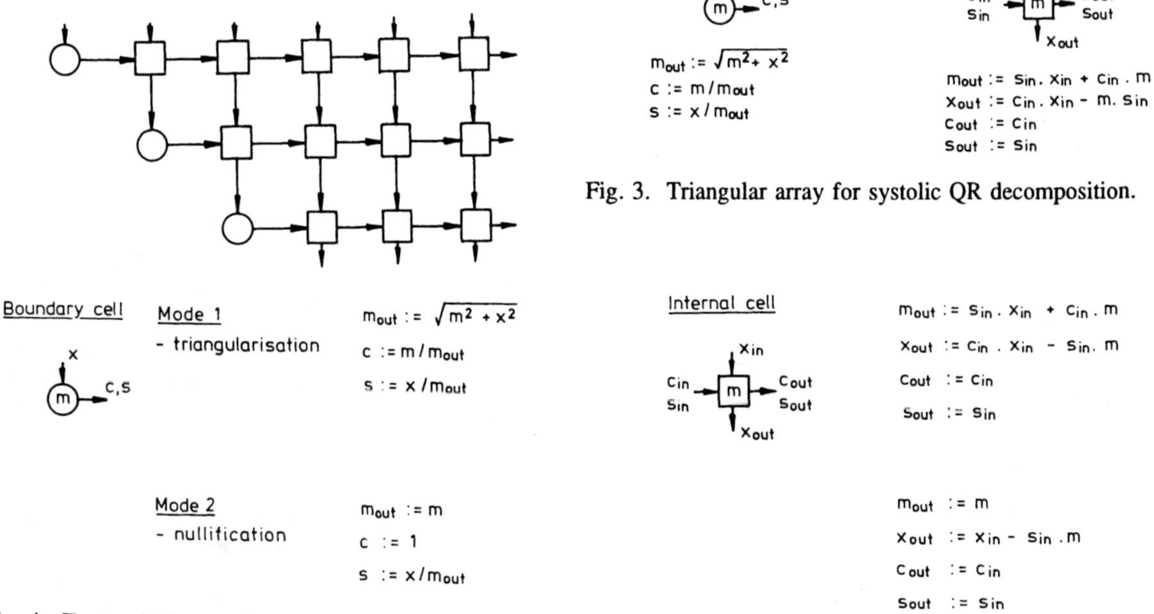

Fig. 4. Trapezoidal array for systolic Schur complement.

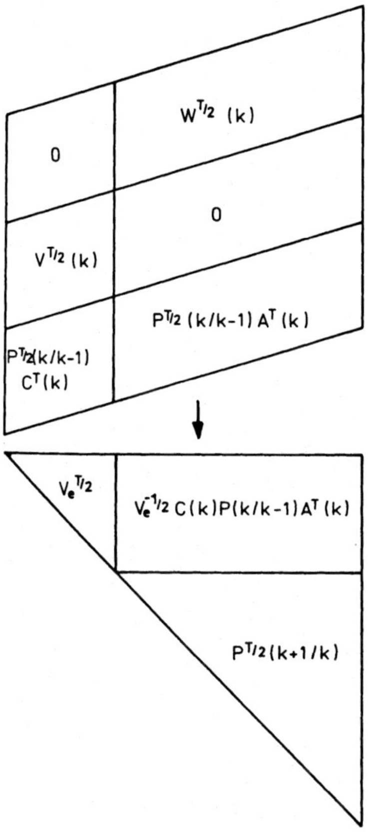

Fig. 5. Orthogonal decomposition for SRCF.

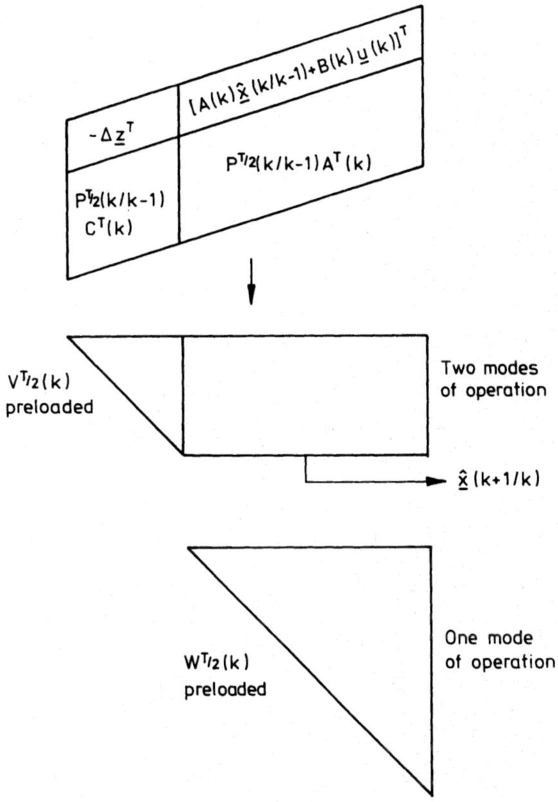

Fig. 6. Split architecture for SRCF.

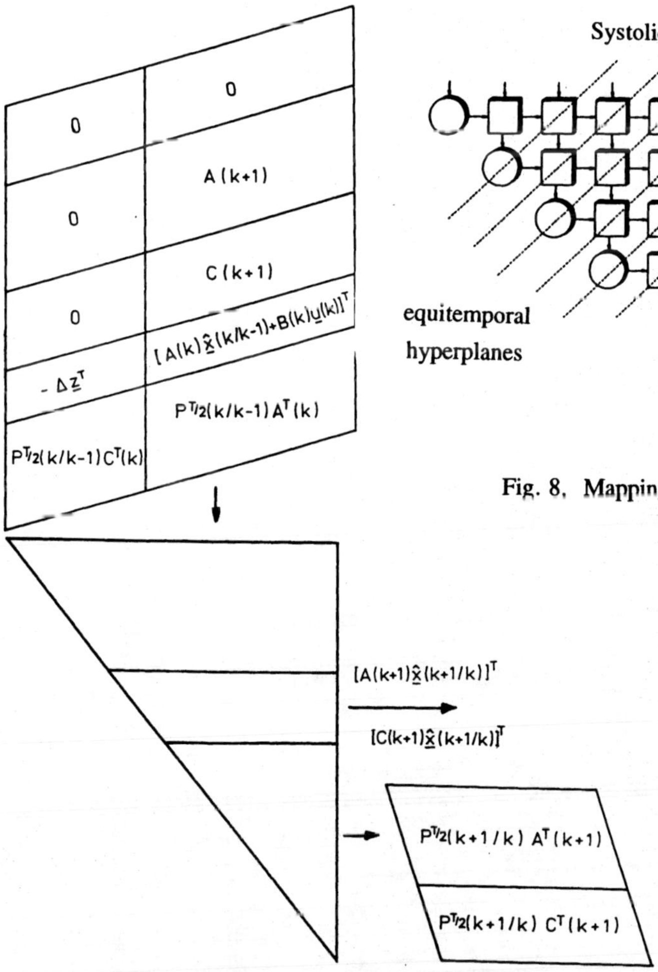

Fig. 7. Complete systolic SRCF.

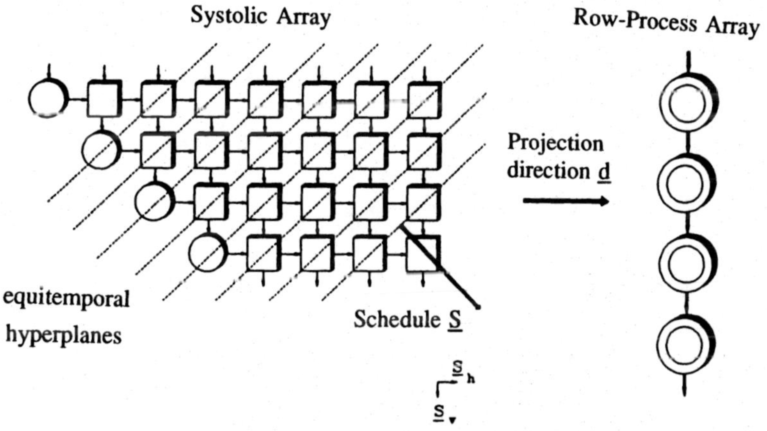

Fig. 8. Mapping systolic RCF into row-processes.

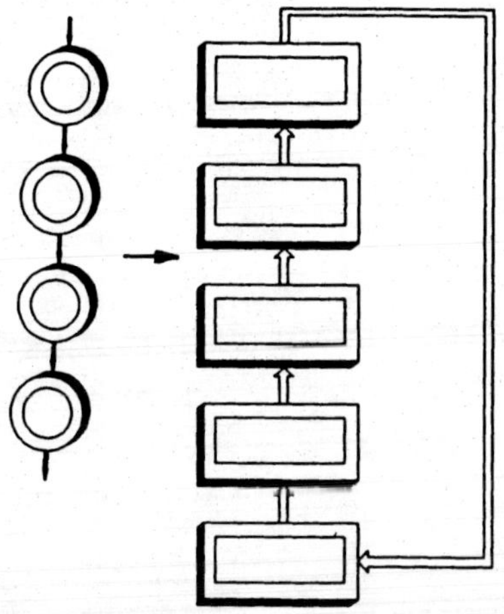

Fig. 9. Assignment of RCF row-processes to transputers.

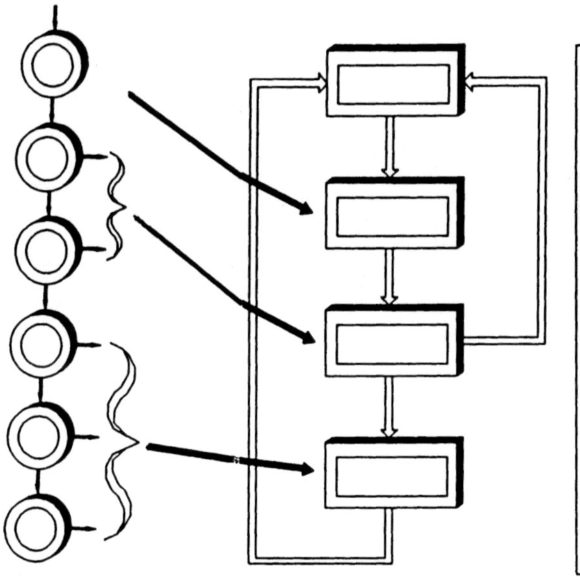

Fig. 10. Assignment of SRCF row-processes to transputers.

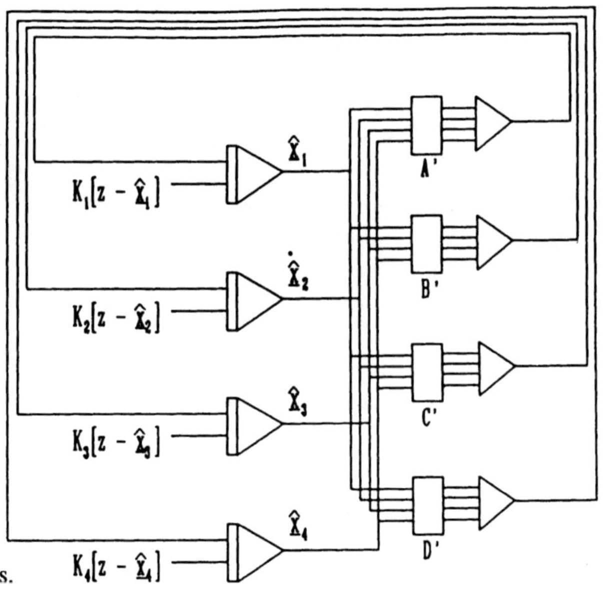

Fig. 11. Analogue flow diagram for state update (eqn.(14))

Fig. 12. Analogue flow diagram for covariance update (eqn.(16)).

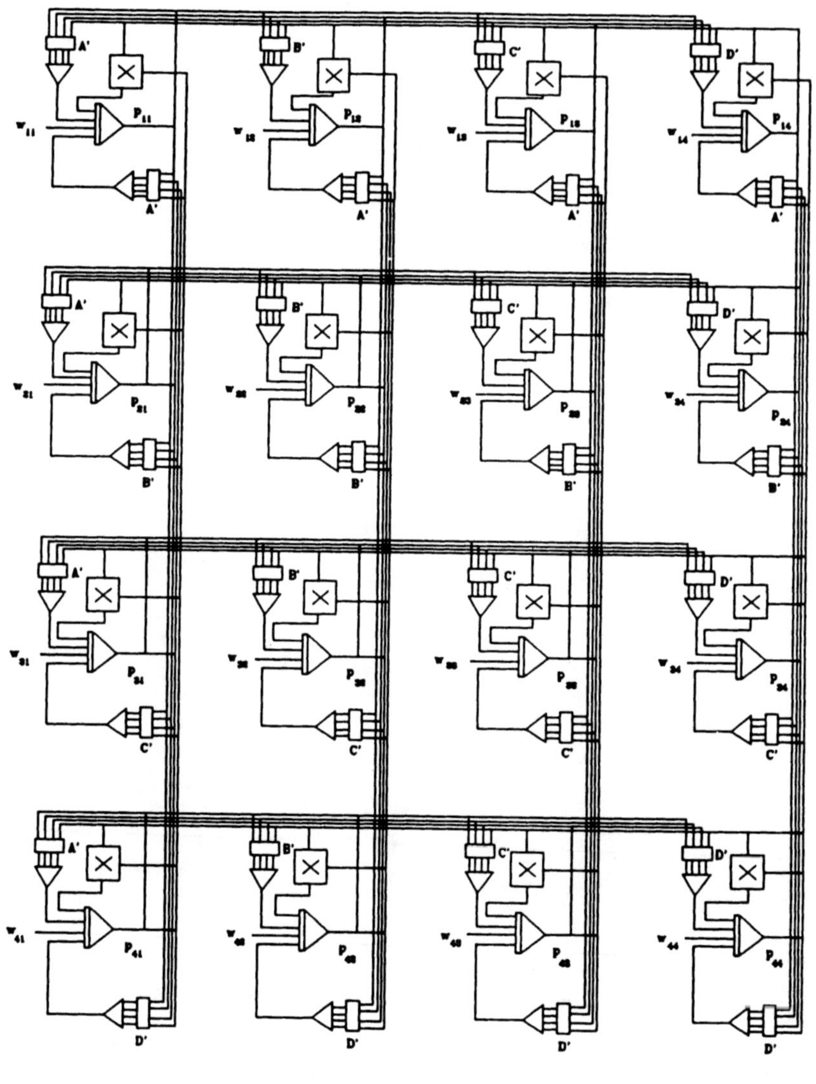

Copyright © IFAC Algorithms and Architectures
for Real-Time Control, Seoul, Korea, 1992

PARALLEL ALGORITHMS FOR THE QUASI-TRIANGULAR GENERALIZED SYLVESTER MATRIX EQUATION ON A SHARED MEMORY MULTIPROCESSOR

Mercedes Marqués, Vicente Hernández

Departamento de Sistemas Informáticos y Computación, Universidad Politécnica, Valencia, Spain

Abstract: In this paper we discuss the resolution of the generalized Sylvester matrix equation $AXB + CXD = E$, where $A, C \in R^{m\times m}, B, D \in R^{n\times n}, E \in R^{m\times n}$, and the unknown X is $m \times n$. This equation is related to different topics in Control Theory and Linear Algebra: perturbation analysis of the generalized eigenvalue problem (Stewart-Sun [19]), solution of implicit linear differential equations (Epton [4]), analysis and design of control problems for descriptor linear systems (Lewis [17]). Particular cases of this equation are the Sylvester equation $AX + XD = E$, and the Stein equation $AXB + X = E$. There are several sequential algorithms for solving both Sylvester and Stein equations. The most effective methods for the Sylvester matrix equation are the Bartels-Stewart algorithm [2] and the algorithm proposed by Golub, Nash and Van Loan [11]. Recently, some authors (Chu [3], Kågström-Westin [16], Gardiner and colleagues [9, 10]) have improved the Bartels-Stewart algorithm for solving the generalized Sylvester matrix equation. Our objective in this paper is the study of the parallel solution of the quasi-triangular matrix equation $AXB + CXD = E$, using coarse-grain algorithms on a shared memory multiprocessor with available BLAS routines [8]. Our approach extends the results obtained by Kågström, Nyström and Poromaa [14] for solving the triangular Sylvester matrix equation and by Marqués and Hernández [18] for the quasi-triangular Stein matrix equation.

Keywords: Control theory; linear algebra; linear matrix equations; parallel processing.

1. INTRODUCTION

Consider the generalized Sylvester matrix equation

$$\hat{A}\hat{X}\hat{B} + \hat{C}\hat{X}\hat{D} = \hat{E} \tag{1}$$

where $\hat{A}, \hat{C} \in R^{m\times m}, \hat{B}, \hat{D} \in R^{n\times n}, \hat{E} \in R^{m\times n}$. It is konwn that if matrix pencils $\hat{A} - \lambda\hat{C}$ y $\hat{D} - \lambda\hat{B}$ are regular pencils and have disjoint spectrums, then the solution of equation (1) exists and is unique (Hernández-Gassó [10]). Particular cases of equation (1) are the Sylvester equation

$$\hat{A}\hat{X} + \hat{X}\hat{D} = \hat{E}$$

and the Stein equation

$$\hat{A}\hat{X}\hat{B} + \hat{X} = \hat{E}.$$

There are several sequential algorithms for solving both Sylvester and Stein equations. One of the most effective methods for the Sylvester matrix equation is the Bartels-Stewart algorithm [2]. Key to this technique is the orthogonal reduction of A and D to quasi-triangular form (real Schur form), using the iterative QR algorithm. A quasi-triangular matrix is a block-triangular matrix, with 1-by-1 or 2-by-2 blocks along its diagonal. After the orthogonal reduction, the solution is obtained by solving several upper quasi-triangular linear systems. A faster algorithm is proposed by Golub, Nash and Van Loan [8], which differs from the Bartels-Stewart algorithm in that A is only reduced to Hessenberg form. Recently, some authors (Chu [3], Kågström-Westin [13], Gardiner and colleagues [6, 7]) presented a generalized Schur algorithm for the sequential solving of equation (1). The algorithm has four basic steps. In the first step, the two matrix pencils associated to (1) are reduced to the generalized Schur form by means of orthogonal equivalence transformations $P_1, P_2 \in R^{m\times m}$, $Q_1, Q_2 \in R^{n\times n}$

$$\begin{aligned} P_1(\hat{A} - \lambda\hat{C})P_2 &= A - \lambda C \\ Q_1(\hat{D} - \lambda\hat{B})Q_2 &= D - \lambda B. \end{aligned} \tag{2}$$

In (2), matrices $A = P_1\hat{A}P_2$ and $D = Q_1\hat{D}Q_2$ present the real Schur form

$$A = \begin{bmatrix} A_{11} & A_{12} & \cdots & A_{1p} \\ & A_{22} & \cdots & A_{2p} \\ & & \ddots & \vdots \\ & & & A_{pp} \end{bmatrix}, D = \begin{bmatrix} D_{11} & D_{12} & \cdots & D_{1q} \\ & D_{22} & \cdots & D_{2q} \\ & & \ddots & \vdots \\ & & & D_{qq} \end{bmatrix} \tag{3}$$

where A_{ii}, D_{jj} are 1×1 or 2×2, $i = 1,\ldots,p$, $j = 1,\ldots,q$, while $C = P_1\hat{C}P_2$, $B = Q_1\hat{B}Q_2$ are both upper triangular. The iterative QZ algorithm is used for the generalized Schur reduction [9].

In the second step, equation (1) is transformed in an equivalent equation, using expressions in (2),

$$AXB + CXD = E \tag{4}$$

where matrices $X = P_2^{-1}\hat{X}Q_1^{-1}$, $E = P_1\hat{E}Q_2$ are

$$X = \begin{bmatrix} X_{11} & \cdots & \cdots & X_{1q} \\ \vdots & & & \vdots \\ \vdots & & & \vdots \\ X_{p1} & \cdots & \cdots & X_{pq} \end{bmatrix}, E = \begin{bmatrix} E_{11} & \cdots & \cdots & E_{1q} \\ \vdots & & & \vdots \\ \vdots & & & \vdots \\ E_{p1} & \cdots & \cdots & E_{pq} \end{bmatrix},$$

and the upper triangular matrices C and B are

$$C = \begin{bmatrix} C_{11} & C_{12} & \cdots & C_{1p} \\ & C_{22} & \cdots & C_{2p} \\ & & \ddots & \vdots \\ & & & C_{pp} \end{bmatrix}, B = \begin{bmatrix} B_{11} & B_{12} & \cdots & B_{1q} \\ & B_{22} & \cdots & B_{2q} \\ & & \ddots & \vdots \\ & & & B_{qq} \end{bmatrix}.$$

From (4) we get

$$\begin{aligned} A_{ii}X_{ij}B_{jj} + C_{ii}X_{ij}D_{jj} = E_{ij} - \sum_{k=i+1}^{m}(A_{ik}\sum_{l=1}^{j}X_{kl}B_{lj}) - \\ -A_{ii}\sum_{l=1}^{j-1}X_{il}B_{lj} - \sum_{k=i+1}^{m}(C_{ik}\sum_{l=1}^{j}X_{kl}D_{lj}) - C_{ii}\sum_{l=1}^{j-1}X_{il}D_{lj}. \\ i = 1,\ldots,p\ ;\ \ j = 1,\ldots,q. \end{aligned} \tag{5}$$

In the third step, $X_{p1}, X_{p2},\ldots, X_{pq}; X_{p-1,1}, X_{p-1,2},\ldots$, are successively obtained, i.e., blocks of X are obtained from left to right, and from bottom to top, by solving a linear matrix equation like (5) for each 1×1 or 2×2 block.

In the last step, the solution of (1) is obtained by computing $\hat{X} = P_2XQ_1$. The arithmetic cost of the algorithm in flops is given by the expression

$$66(m^3+n^3)+2(m^2n+mn^2)+(m^2n^2+3m^2n+2mn^2)+2(m^2n+mn^2)$$

where each term represents the cost of each step, and one flop is the cost of one floating point operation. Assuming that input data matrices are overwritten, we need storage for $(4m^2 + 4n^2 + mn)$ real numbers.

[†]Supported by a research grant of the Department of Education of the Valencian Community Government.
[‡]This work has been supported by a Spanish CICYT Grant TIC91-1157-C03-02

Our objective in this paper is to present a parallel solution of (4) with coarse-grain algorithms for shared memory multiprocessors using the available BLAS routines [5]. This paper is structured as follows. In the next section, the case in which matrices A, B, C and D are upper triangular matrices is considered. This is the case in which the eigenvalues of pencils $A-\lambda C$ and $D-\lambda B$ are real. In section 3 we study the case in which A and D are in the upper quasi-triangular form, while B and C are upper triangular. In both cases, we present a row-oriented algorithm, a column-oriented algorithm and a block-oriented algorithm. Our work is an extension of the results by Kågström, Nyström and Poromaa [11] on the triangular Sylvester equation. In the fourth section we present results from a series of numerical experiments performed on an Alliant FX/80. Finally, conclusions of our work are commented.

2. SOLUTION OF THE TRIANGULAR GENERALIZED SYLVESTER EQUATION

Let us consider equation (4) where A, B, C and D are upper triangular matrices. We assume that they verify the existence and uniqueness condition

$$a_{ii}b_{jj}+c_{ii}d_{jj}\neq 0 \qquad i=1,\ldots,m\ ;\ \ j=1,\ldots,n. \tag{6}$$

Considering that all the blocks in (5) are 1×1 in the triangular case, the unknown X can be solved by back substitution by means of the expression

$$x_{ij}=\frac{e_{ij}-\sum_{k=i+1}^{m}(a_{ik}\sum_{l=1}^{j}x_{kl}b_{lj})-a_{ii}\sum_{l=1}^{j-1}x_{il}b_{lj}}{a_{ii}b_{jj}+c_{ii}d_{jj}}+$$
$$+\frac{-\sum_{k=i+1}^{m}(c_{ik}\sum_{l=1}^{j}x_{kl}d_{lj})-c_{ii}\sum_{l=1}^{j-1}x_{il}d_{lj}}{a_{ii}b_{jj}+c_{ii}d_{jj}} \tag{7}$$
$$i=m,\ldots,1\ ;\ \ j=1,\ldots,n$$

Data Dependency

We must know the information needed to compute an element $X(i,j)$ in order to find the parallelism of equation (4). We use the notation introduced by Golub-Van Loan in [9] for representing blocks of matrices.

In (7) we can see that to compute $X(i,j)$, we need:

- $A(i,:)$ and $C(i,:)$, i-th row of A and C
- $B(:,j)$ and $D(:,j)$, j-th column of B and D
- $E(i,j)$, element (i,j) of E
- $X(i:m,1:j)$, block of X formed by the elements $X(k,l)$ with $k=i,i+1,\ldots,m; l=1,2,\ldots j$.

This data dependency is shown in Fig. 1.

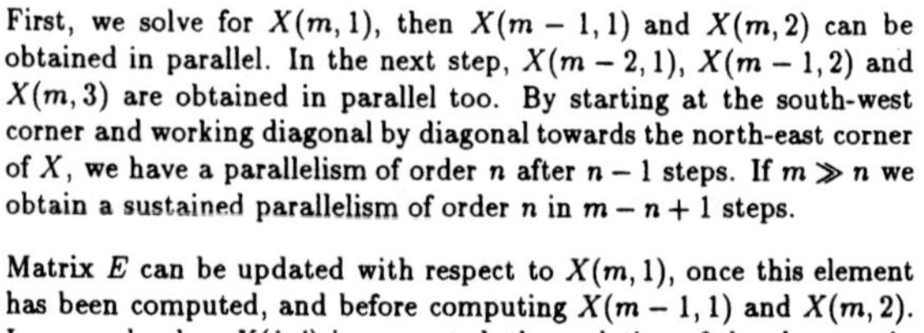

Figure 1: Data Dependency.

First, we solve for $X(m,1)$, then $X(m-1,1)$ and $X(m,2)$ can be obtained in parallel. In the next step, $X(m-2,1)$, $X(m-1,2)$ and $X(m,3)$ are obtained in parallel too. By starting at the south-west corner and working diagonal by diagonal towards the north-east corner of X, we have a parallelism of order n after $n-1$ steps. If $m\gg n$ we obtain a sustained parallelism of order n in $m-n+1$ steps.

Matrix E can be updated with respect to $X(m,1)$, once this element has been computed, and before computing $X(m-1,1)$ and $X(m,2)$. In general, when $X(i,j)$ is computed, the updating of the elements in E, $E(i,j+1:n)$ and $E(1:i-1,j:n)$, can be done in parallel.

In summary, equation (4) has two levels of parallelism:

1. The elements of X in the same diagonal can be computed in parallel.
2. The updatings of the elements of E can be computed in parallel.

Both levels of parallelism are fine-grained. When working on a system with several processors sharing a common memory, it is desirable to avoid excessive memory accesses. This can be done by working with coarse-grain algorithms. We are going to see how to obtain the solution X by working on rows, columns and blocks.

Row-Oriented Algorithm

First, we study a row-oriented algorithm. In Fig. 2 we can see that the last row of X satisfies the following equation

$$A(m,m)X(m,:)B+C(m,m)X(m,:)D=E(m,:)$$

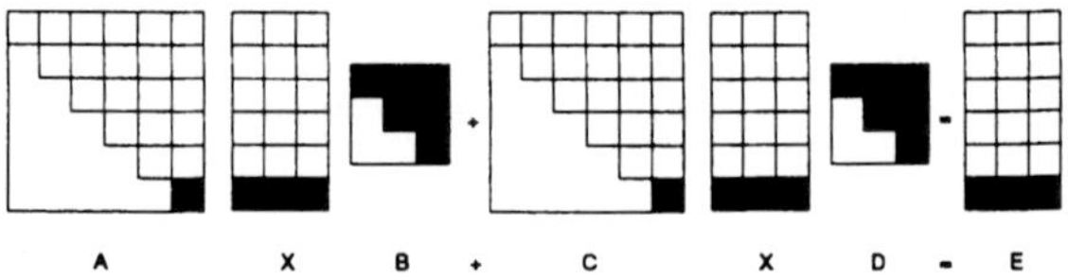

Figure 2: Solving for m-th row of X.

Then, $X(m,:)$ can be obtained by solving the upper triangular system

$$X(m,:)[A(m,m)B+C(m,m)D]=E(m,:).$$

Now we are going to compute the i-th row of X. Rows $m,m-1,\ldots,i+1$ of X have already been computed. In Fig. 3, we can see that these rows satisfy

$$\sum_{k=i}^{m}[A(i,k)X(k,:)B+C(i,k)X(k,:)D]=E(i,:)$$

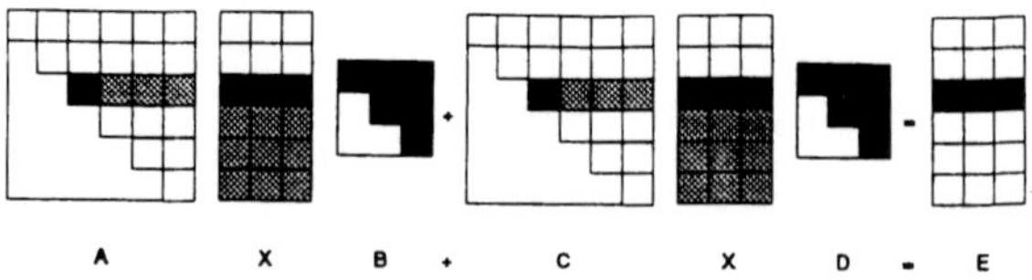

Figure 3: Solving for i-th row of X.

Then, $X(i,:)$ can be obtained by solving the upper triangular system

$$X(i,:)[A(i,i)B+C(i,i)D]=E(i,:)-\sum_{k=i+1}^{m}X(k,:)[A(i,k)B+C(i,k)D].$$

The right-hand side of the system suggests that we must update $E(i,:)$ with respect to rows $m,m-1,\ldots,i+1$ of X

$$E(i,:):=E(i,:)-\sum_{k=i+1}^{m}X(k,:)[A(i,k)B+C(i,k)D]$$

and then, solve the upper triangular system

$$X(i,:)[A(i,i)B+C(i,i)D]=E(i,:).$$

This can be done successively, until solving for the first row of X. We can see that once $X(i,:)$ has been computed, it is possible to update rows of E between the $(i-1)$-th and the first row

$$E(k,:):=E(k,:)-X(i,:)[A(k,i)B+C(k,i)D], \quad k=i-1:1.$$

Here we show the row-oriented algorithm.

Algorithm 1 (TGSYLV_R)

```
do i = m : 1
   solve for X(i,:) in
      X(i,:)[A(i,i)B + C(i,i)D] = E(i,:)
   {update rows of E}
   do k = i - 1 : 1
      E(k,:) := E(k,:) - X(i,:)[A(k,i)B + C(k,i)D]
   end do
end do.
```

Column-Oriented Algorithm

In the same way, we can obtain a column-oriented algorithm. As we can see in Fig. 4, solving for the first column of X can be done by solving the following upper triangular system

$$[A\ B(1,1) + C\ D(1,1)]X(:,1) = E(:,1).$$

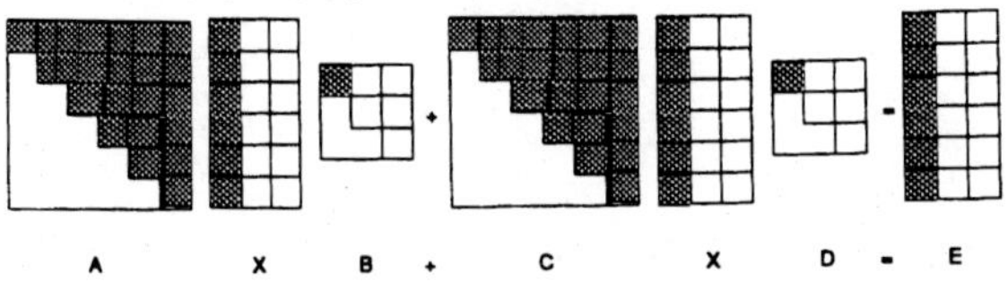

Figure 4: Solving for the first column of X.

Once columns $1, 2, \ldots, , j-1$ have been computed, it is possible to compute column j by forward elimination. For this, we must update column j of E

$$E(:,j) := E(:,j) - \sum_{k=1}^{j-1}[A\ B(k,j) + C\ D(k,j)]X(:,k)$$

and solve the following upper triangular system

$$[A\ B(j,j) + C\ D(j,j)]X(:,j) = E(:,j).$$

This can be done successively, until solving for the last column of X. As in the row-oriented algorithm, we can see that once $X(:,j)$ has been computed, it is possible to update the columns of E between the $(j+1)$-th and the last one

$$E(:,k) := E(:,k) - [A\ B(j,k) + C\ D(j,k)]X(:,j), \quad k = j+1:n.$$

We obtain the following column-oriented algorithm.

Algorithm 2 (TGSYLV_C)

do $j = 1 : n$
  solve for $X(:,j)$ in
    $[A\ B(j,j) + C\ D(j,j)]X(:,j) = E(:,j)$
  {update columns of E}
  do $k = j+1 : n$
    $E(:,k) := E(:,k) - [A\ B(j,k) + C\ D(j,k)]X(:,j)$
  end do
end do.

Block-Oriented Algorithm

Now we present a block-oriented algorithm, TGSYLV_B. If we want to compute blocks of X of size $M \times N$, we must partition matrices A, B, C and D in this way

$$A = [A[k,i]]_{r\times r},\ A[k,i] \in \Re^{M\times M}\ ;\ C = [C[k,i]]_{r\times r},\ C[k,i] \in \Re^{M\times M}$$

$$k,i = 1,\ldots,r\ ;\ rM = m$$

$$B = [B[j,l]]_{s\times s},\ B[j,l] \in \Re^{N\times N}\ ;\ D = [D[j,l]]_{s\times s},\ D[j,l] \in \Re^{N\times N}$$

$$j,l = 1,\ldots,s\ ;\ sN = n.$$

Note that diagonal blocks $A[k,k]$, $C[k,k]$ and $B[j,j]$, $D[j,j]$ are upper triangular matrices. The partitions in E and X are given by

$$X = [X[i,j]]_{r\times s},\ X[i,j] \in \Re^{M\times N}\ ;\ E = [E[i,j]]_{r\times s},\ E[i,j] \in \Re^{M\times N}$$

$$i = 1,\ldots,r\ ;\ j = 1,\ldots,s.$$

We are going to see how we can get a back substitution algorithm for computing the blocks of X from down to up, and from left to right. As we can see in Fig. 5, $X[r,1]$ is the first block to be computed. This can be done by solving the following equation with one of the former algorithms (TGSYLV_R or TGSYLV_C)

$$A[r,r]X[r,1]B[1,1] + C[r,r]X[r,1]D[1,1] = E[r,1].$$

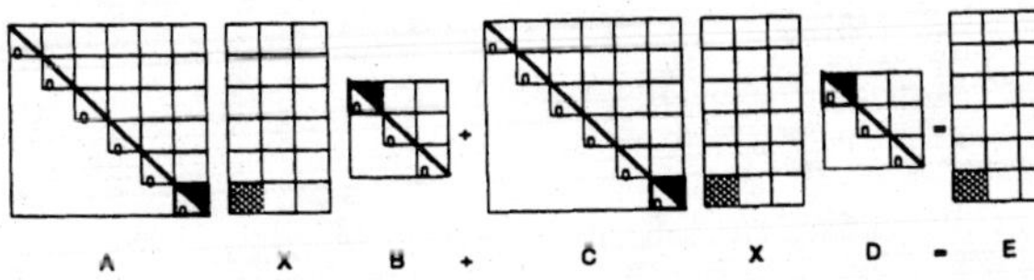

Figure 5: Solving for the $(r,1)$-th block of X.

Assume we have computed block columns $X[:,1], \ldots, X[:,j-1]$ and blocks $X[r,j], X[r-1,j], \ldots, X[i+1,j]$ of block column j. In Fig. 6 we can see that

$$\sum_{k=i}^{r}\sum_{l=1}^{j}\{A[i,k]X[k,l]B[l,j] + C[i,k]X[k,l]D[l,j]\} = E[i,j]$$

then $X[i,j]$ satisfies the equation

$$A[i,i]X[i,j]B[j,j] + C[i,i]X[i,j]D[j,j] = E[i,j] -$$
$$-\sum_{l=1}^{j-1}\{A[i,i]X[i,l]B[l,j] + C[i,i]X[i,l]D[l,j]\} -$$
$$-\sum_{k=i+1}^{r}\sum_{l=1}^{j}\{A[i,k]X[k,l]B[l,j] + C[i,k]X[k,l]D[l,j]\}.$$

The right-hand side of this equation suggests updating $E[i,j]$ with respect to the computed blocks of X

$$E[i,j] := E[i,j] - \sum_{l=1}^{j-1}\{A[i,i]X[i,l]B[l,j] + C[i,i]X[i,l]D[l,j]\} -$$
$$-\sum_{k=i+1}^{r}\sum_{l=1}^{j}\{A[i,k]X[k,l]B[l,j] + C[i,k]X[k,l]D[l,j]\}$$

and then, solve the triangular generalized Sylvester equation

$$A[i,i]X[i,j]B[j,j] + C[i,i]X[i,j]D[j,j] = E[i,j]$$

by means of TGSYLV_R or TGSYLV_C.

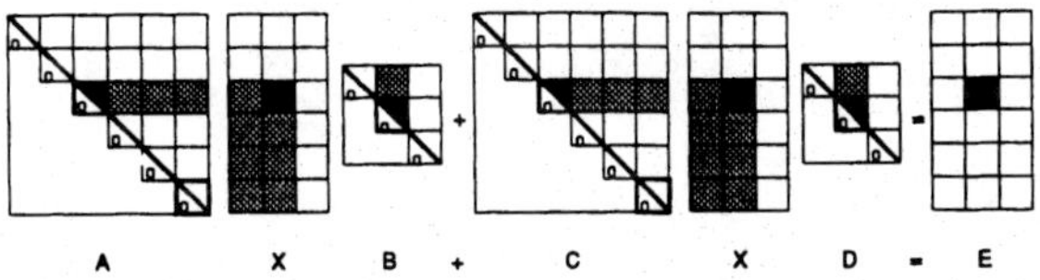

Figure 6: Solving for the (i,j)-th block of X.

Matrix X has been computed after solving $rs = (m/M)(n/N)$ triangular generalized Sylvester equations, where the unknown matrix is $M \times N$. Blocks are computed from down to up, and from left to right. Each computed block, except the last one, is used to update E with respect to A, B, C and D.

We obtain the following block-oriented algorithm.

Algorithm 3 (TGSYLV_B)

nb_AC:= number of diagonal blocks in A and C
nb_BD:= number of diagonal blocks in B and D
do $j = 1 : nb\_BD$
  do $i = nb\_AC : 1$
    solve for $X[i,j]$ using TGSYLV_R or TGSYLV_C in
      $A[i,i]X[i,j]B[j,j] + C[i,i]X[i,j]D[j,j] = E[i,j]$
    {update block column j of E}
    do $k = i-1 : 1$
      $E[k,j] := E[k,j] - A[k,i]X[i,j]B[j,j] - C[k,i]X[i,j]D[j,j]$
    end do
    {update block columns $j+1 : nb\_BD$ of E}
    do $k = 1 : i$
      do $l = j+1 : nb\_BD$
        $E[k,l] := E[k,l] - A[k,i]X[i,j]B[j,l] - C[k,i]X[i,j]D[j,l]$
      end do
    end do
  end do
end do.

3. SOLUTION OF THE QUASI-TRIANGULAR GENERALIZED SYLVESTER EQUATION

In this section we present coarse-grain algorithms for solving (4) when A and D are in the real Schur form (1×1 or 2×2 diagonal blocks). B and C are upper triangular.

Data Dependency

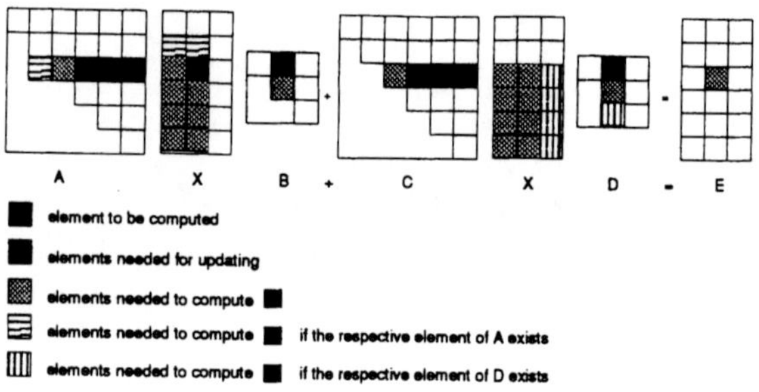

Figure 7: Data dependency.

To obtain the data dependency in this case, we need to know where the nonzero sub-diagonal elements of A and D are placed. To obtain element $X(i,j)$ in (4) we need:

- $A(i,:)$ y $C(i,:)$, i-th row of A and C
- $B(:,j)$ y $D(:,j)$, j-th column of B and D
- $E(i,j)$, element (i,j) of E
- $X(i:m, 1:j)$, block of X formed by elements $X(k,l)$ with $k = i, i+1, \ldots, m$; $l = 1, 2, \ldots, j$
  - if $A(i,i-1) \neq 0$, elements belonging to $X(i-1, 1:j)$ are needed too
  - if $D(j+1,j) \neq 0$, elements belonging to $X(i:m, j+1)$ are also needed.

This data dependency is shown with an example in Fig. 7.

Row-Oriented Algorithm

In the quasi-triangular case, we compute one or two rows simultaneously, depending on where sub-diagonal nonzero elements appear in matrix A. As in TGSYLV_R, we start at the bottom, so that

- if $A(m,m-1) = 0$, then $X(m,:)$ is computed
- if $A(m,m-1) \neq 0$, then $X(m-1,:)$ and $X(m,:)$ are computed simultaneously.

When computing row i, if $A(i,i-1) = 0$, we must

- solve the system $X(i,:)[A(i,i)B + C(i,i)D] = E(i,:)$ by Gaussian elimination with partial pivoting
- update rows $1 : i-1$ of E with respect to the computed row of X and A,B,C,D, as in TGSYLV_R.

Otherwise, $A(i,i-1) \neq 0$, there is a data dependency that forces us to compute rows i and $i-1$ simultaneously. Then we must

- compute

$$Q := \begin{bmatrix} A(i-1,i-1) & A(i-1,i) \\ A(i,i-1) & A(i,i) \end{bmatrix}^T \otimes B + \\ + \begin{bmatrix} C(i-1,i-1) & C(i-1,i) \\ 0 & C(i,i) \end{bmatrix}^T \otimes D$$

  where $\otimes$ is the Kronecker product
- solve the system

$$row[X(i-1,:), X(i,:)]Q = row[E(i-1,:), E(i,:)]$$

  by Gaussian elimination with partial pivoting
- update rows $1 : i-2$ of E with respect to the computed rows of X and A,B,C,D, as in TGSYLV_R.

This is the row-oriented algorithm for the quasi-triangular case:

Algorithm 4 (QTGSYLV_R)

```
i := m
while i ≥ 1
  if A(i, i−1) = 0
    solve for X(i,:) in
      X(i,:)[A(i,i)B + C(i,i)D] = E(i,:)
    {update rows of E}
    do k = i−1 : 1
      E(k,:) := E(k,:) − X(i,:)[A(k,i)B + C(k,i)D]
    end do
    i := i − 1
  else
    Q(1 : n, 1 : n) := A(i−1, i−1)B + C(i−1, i−1)D
    Q(1 : n, n+1 : 2n) := A(i, i−1)B
    Q(n+1 : 2n, 1 : n) := A(i−1, i)B + C(i−1, i)D
    Q(n+1 : 2n, n+1 : 2n) := A(i, i)B + C(i, i)D
    solve for X(i−1,:) y X(i,:) in
      row[X(i−1,:), X(i,:)]Q = row[E(i−1,:), E(i,:)]
    {update rows of E}
    do k = i−2 : 1
      E(k,:) := E(k,:) − X(i−1,:)[A(k,i−1)B + C(k,i−1)D]
      E(k,:) := E(k,:) − X(i,:)[A(k,i)B + C(k,i)D]
    end do
    i := i − 2
  end if
end while.
```

Column-Oriented Algorithm

Now, we compute one or two columns simultaneously, depending on where sub-diagonal nonzero elements appear in matrix D. As in TGSYLV_C, we start at the top, so that

- if $D(2,1) = 0$, then $X(:,1)$ is computed
- if $D(2,1) \neq 0$, then $X(:,1)$ and $X(:,2)$ are computed simultaneously.

When computing column j, if $D(j+1,j) = 0$, we must

- solve the system $[A\ B(j,j) + C\ D(j,j)]X(:,j) = E(:,j)$ by Gaussian elimination with partial pivoting
- update rows $j+1 : n$ of E with respect to the computed column of X and A,B,C,D, as in TGSYLV_C.

Otherwise, $D(j+1,j) \neq 0$, there is a data dependence that forces us to compute columns j and $j+1$ simultaneously. Then we must

- compute

$$Q := \begin{bmatrix} B(j,j) & B(j,j+1) \\ 0 & B(j+1,j+1) \end{bmatrix}^T \otimes A + \\ + \begin{bmatrix} D(j,j) & D(j,j+1) \\ D(j+1,j) & D(j+1,j+1) \end{bmatrix}^T \otimes C$$

- solve the system

$$Qcol[X(:,j), X(:,j+1)] = col[E(:,j), E(:,j+1)]$$

  by Gaussian elimination with partial pivoting
- update columns $j+2 : n$ of E with respect the computed columns of X and A,B,C,D, as in TGSYLV_C.

This is the column-oriented algorithm:

Algorithm 5 (QTGSYLV_C)

```
j := 1
while j ≤ n
  if D(j+1, j) = 0
    solve for X(:,j) in
      [A B(j,j) + C D(j,j)]X(:,j) = E(:,j)
    {update columns of E}
    do k = j+1 : n
      E(:,k) := E(:,k) − [A B(j,k) + C D(j,k)]X(:,j)
    end do
    j := j + 1
  else
    Q(1 : m, 1 : m) := A B(j,j) + C D(j,j)
    Q(1 : m, m+1 : 2m) := C D(j+1, j)
```

```
    Q(m + 1 : 2m, 1 : m) := A B(j, j + 1)
        +C D(j, j + 1)
    Q(m + 1 : 2m, m + 1 : 2m) := A B(j + 1, j + 1) + C D(j + 1, j + 1)
    solve for X(:, j) y X(:, j + 1) in
      Q col[X(:, j), X(:, j + 1)] = col[E(:, j), E(:, j + 1)]
    {update columns of E}
    do k = j + 1 : n
      E(:, k) := E(:, k) - [A B(j, k) + C D(j, k)]X(:, j)
      E(:, k) := E(:, k) - [A B(j + 1, k) + C D(j + 1, k)]X(:, j + 1)
    end do
    j := j + 2
  end if
end while.
```

Block-Oriented Algorithm

Now we describe the block-oriented method. First we consider the matrix partition. In the quasi-triangular case, the block size is variable. This is due to the existence of nonzero sub-diagonal elements in the real Schur form of A and D.

If $M \times N$ is the desired size of block $X[i, j]$, then matrix A is partitioned according to the following rules:

- the size of the first diagonal block is
  - $M \times M$ if $A(M + 1, M) = 0$
  - $(M + 1) \times (M + 1)$ if $A(M + 1, M) \neq 0$ (because rows M and $M + 1$ of X must be calculated simultaneously)
- repeat the previous step to obtain the size of the remaining diagonal blocks.

Matrix D is partitioned in the same way:

- the size of the first diagonal block is
  - $N \times N$ if $D(N + 1, N) = 0$
  - $(N + 1) \times (N + 1)$ if $D(N + 1, N) \neq 0$ (because columns N and $N + 1$ of X must be calculated simultaneously)
- repeat the previous step to obtain the size of the remaining diagonal blocks.

Once A and D have been partitioned according to its diagonal blocks, we must partition B,C,E and X in accordance with partitions in A and D.

This is the block-oriented algorithm:

Algorithm 6 (QTGSYLV_B)

```
partition A into blocks according to the previous rules, and store the
  number of diagonal blocks in nb_AC
  and the sizes of the diagonal blocks in b_AC(1 : nb_AC)
partition D into blocks according to the previous rules, and store the
  number of diagonal blocks in nb_BD
  and the sizes of the diagonal blocks in b_BD(1 : nb_BD)
do j = 1 : nb_BD
  do i = nb_AC : 1
    M := b_AC(i)
    N := b_BD(j)
    solve for X[i, j] (dim(X[i, j]) = M × N) with QTGSYLV_R or
        QTGSYLV_C, in
      A[i, i]X[i, j]B[j, j] + C[i, i]X[i, j]D[j, j] = E[i, j]
    {update block-column j of E}
    do k = i - 1 : 1
      {dim(E[k, j]) = b_AC(k) × N }
      E[k, j] := E[k, j] - A[k, i]X[i, j]B[j, j] - C[k, i]X[i, j]D[j, j]
    end do
    {update block-column j + 1 : nb_BD of E}
    do k = 1 : i
      do l = j + 1 : nb_BD
        {dim(E[k, l]) = b_AC(k) × b_BD(l) }
        E[k, l] := E[k, l] - A[k, i]X[i, j]B[j, l] - C[k, i]X[i, j]D[j, l]
      end do
    end do
  end do
end do.
```

4. ALLIANT FX/80 IMPLEMENTATION

In this section we explain how we have exploited the parallelism in the implementation of the block-oriented algorithm QTGSYLV_B. We also present performance results corresponding to this implementation in FX/Fortran on an Alliant FX/80.

Our objectives were:

- to use BLAS3 (matrix-matrix operations) at block level,
- to use BLAS2 (matrix-vector operations) in QTGSYLV_C and
- to parallelize E updatings at block level

These objectives can be achieved using BLAS routines in the following way:

Routine	BLAS	Operation	Used for
sgemm	3	$C \leftarrow \alpha AB + \beta C$	block updating
sgemv	2	$\alpha Ax + \beta y$	upt. w.r.t. A
strmv	2	$x \leftarrow Ax$	upt. w.r.t. C
saxpy	1	$y \leftarrow \alpha x + y$	upt. w.r.t $B(j, k)$, $D(j, k)$

We use the LINPACK routines *pdgefa* and *pdgesl* for the resolution of each system by Gaussian elimination with partial pivoting. The first one factorizes the matrix and the second one solves the system using the factors computed by *pdgefa*.

Additionally, compiler directives were used to allow concurrent calls to these routines in the block updating and column updating loops.

We show the performance results of the algorithm in fig. 8.

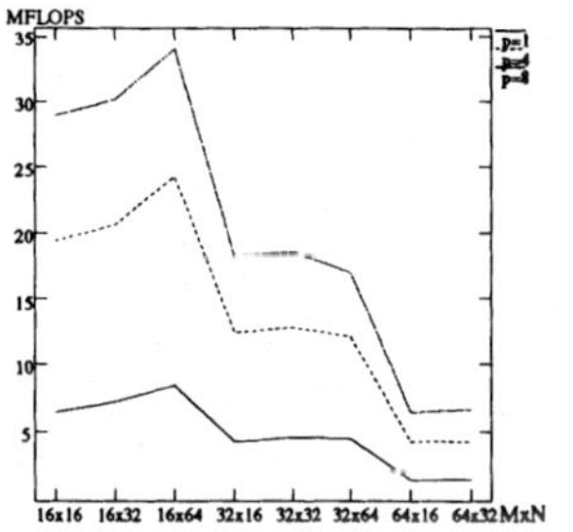

Figure 8: Chart of Performance in Mflops for Different Block Sizes and p=1,4,8 (m=n=512).

In this figure, we can see the performance in Mflops as a function of block size and number of processors, for $m = n = 512$. The best results are obtained when block sizes are 16×32 and 16×64. In both cases, the vector registers of the machine are optimally used. There are differences in performance for different block sizes on 1, 4 and 8 processors. They are due to the concurrent execution of the updating loops. We can see how this affects to the overall performance of the algorithm in the following table.

	$p = 4$		$p = 8$	
	speedup	efficiency	speedup	efficiency
no cncall	2.69	67%	3.86	48%
cncall	2.90	72%	4.17	52%

This table shows speedup and efficiency results when executing the algorithm with and without parallelizing the updating loops (*cncall* and *no cncall* in the table).

5. CONCLUSIONS

A new block-oriented method for solving the quasi-triangular generalized Sylvester matrix equation has been proposed in this paper. The method has been implemented on an Alliant FX/80 , a shared memory multiprocessor with 8 processors. For this computer, a 16×64 block size was shown to be optimal for matrix partitioning. For this size, efficiencies of 72 % and 52 % were obtained for 4 and 8 processors, respectively.

References

[1] Barnett S., Matrices in Control Theory, Van Nostrand Reinhold, London, 1971.

[2] Bartels R. and G.W. Stewart, Algorithm 432: The Solution of the Matrix Equation AX–XB=C, Communications of the ACM, Vol. 15, pp 820-826, 1972.

[3] Chu K.E., The Solution of the Matrix Equations AXB–CXD=E and (YA–DZ,YC–BZ)=(E,F), Linear Algebra and its Applications, No. 93, 1987, pp 93-105.

[4] Epton M.A., Methods for the Solution of AXD–BXC=E and its Application in the Numerical Solution of Implicit Ordinary Differential Equations, BIT 20, 1980, pp 341-345.

[5] Gallivan K.A., R.J. Plemmons and A.H. Sameh, Parallel Algorithms for Dense Linear Algebra Computations, SIAM REVIEW, Vol. 32, No. 1, pp 54-135, March 1990.

[6] Gardiner J.D., A.J. Laub, J.J. Amato and C.B. Moler, A Solution of the Sylvester Matrix Equation $AXB^T+CXD^T=E$, ACM Trans. Math. Software, 1991 (to appear).

[7] Gardiner J.D., R.W. Matthew, A.J. Laub, J.J. Amato and C.B. Moler, A FORTRAN-77 Software Package for Solving the Sylvester Matrix Equation $AXB^T+CXD^T=E$, ACM Trans. Math. Software, 1991 (to appear).

[8] Golub G.H., S. Nash and C. Van Loan, A Hessenberg - Schur Method for the Problem AX+XB=C, IEEE Transactions on Automatic Control, Vol. AC-24, No. 6, pp 909-913, 1979.

[9] Golub G.H. and C. Van Loan, Matrix Computations, Second Edition, The Johns Hopkins University Press, 1990.

[10] Hernández V. and M. Gassó, Explicit Solution of the Matrix Equation AXB–CXD=E, Linear Algebra and its Applications, No. 121, 1989, pp 333-343.

[11] Kågström B., L. Nyström and P. Poromaa, Parallel Shared Memory Algorithms for Solving the Triangular Sylvester Equation, Report UMINF 136-87, Inst. of Information Processing, Univ. of Umeå, 1987.

[12] Kågström B. and P. Poromaa, Distributed Block Algorithms for the Triangular Sylvester Equation with Condition Estimator, Hypercube and Distributed Computers, Elsevier Science Publishers B.V., pp 233-248, North-Holland, 1989.

[13] Kågström B. and L. Westin, Generalized Schur Methods with Condition Estimators for Solving the Generalized Sylvester Equation, IEEE Transactions on Automatic Control, Vol. 34, No. 7, 1988, pp 745-751.

[14] Lewis F.L., A Survey of Linear Singular Systems, Circuit Systems Signal Processing, Vol. 5, pp 3-36, 1986.

[15] Marqués M., V. Hernández, Parallel Algorithms for the Quasi-Trianguler Stein Matrix Equation on a Shared Memory Multiprocessor, Report DSIC-II/36/1991, Departamento de Sistemas Informáticos y Computación, Universidad Politécnica de Valencia, 1991.

[16] Stewart G.W. and J. Sun, Matrix Perturbation Theory, Academic Press, New York, 1990.

A SYSTOLIC ALGORITHM FOR THE TRIANGULAR SYLVESTER EQUATION IN A LINEAR ARRAY

José L. Hueso*, Gloria Martínez**, Vicente Hernández***

*Departamento de Matemática Aplicade Universidad Politécnica, Valencia,
Apartado 22012
46071 VALENCIA, SPAIN
**Departamento de Matemáticas e Informática
Universitat Jaume I
Campus Penyeta Roja, Edifici D
12071 CASTELLON, SPAIN
***Departamento de Sistemas Informáticos y Computación
Universidad Politécnica, Valencia
Apartado 22012
46071 VALENCIA, SPAIN

Abstract. The solution of the Sylvester equation, AX+XB=C, can be obtained by reducing it to triangular Schur form. Here we design systolic algorithms for solving this reduced equation in bidimensional arrays and in linear arrays. By using data partition techniques we obtain size independent algorithms in both cases.
Keywords. Systolic algorithms, Sylvester matrix equations, continuous-time linear systems.

INTRODUCTION

Let us consider the Sylvester equation

$$AX + XB = C \tag{1}$$

with A and B square matrices of order m and n, respectively, and C an mxn rectangular matrix. This equation is an extension of the well-known continuous-time algebraic Lyapunov equation

$$A^T X + XA = C.$$

The Sylvester and Lyapunov equations are related with Stability and Optimal Control Problems of Continuous-Time Dynamic Linear Systems (Laub, 1985). For instance, in the design of full-order or reduced-order state observers we need the solution of Sylvester-like equations (Kwakernaak and Sivan, 1972). Furthermore, we can solve the Continuous-Time Linear-Quadratic Optimal Control Problem by using a Newton-type algorithm (Kleinman, 1968 and Sandell, 1974). In each step of this algorithm we need to solve a continuous-time algebraic Lyapunov equation.

TRIANGULAR FORM OF SYLVESTER EQUATION

If A and B are real matrices with real eigenvalues, we can reduce them to triangular Schur form by orthogonal transformations (Golub, Van Loan, 1983). There exist orthogonal matrices U and V, an upper triangular matrix A' and a lower triangular matrix B', such that

$$A = UA'U^T, \qquad B = VB'V^T.$$

Then (1) becomes

$$(UA'U^T)X + X(VB'V^T) = C,$$

which can be written as

$$A'X' + X'B' = C'$$

where $X'=U^TXV$ and $C'=U^TCV$.

We also suppose that the conditions for existence and uniqueness of the solution of (1) are fulfilled

$$\lambda_i + \mu_j \neq 0, \quad \lambda_i \in \sigma(A), \quad \mu_j \in \sigma(B).$$

Some systolic algorithms have been developed in order to achieve the triangular Schur form of a real matrix (Torralba, Navarro, 1988a, 1988b). In this paper we will suppose that the matrices A and B in (1) are upper triangular and lower triangular, respectively, and will design a systolic algorithm to solve the triangular Sylvester equation. This equation has been studied by Kågström, Nyström and Poromaa (1988) and Kågström and Poromaa (1989). In these papers only parallel algorithms on multiprocessor systems with shared and distributed memory are considered.

SYSTOLIC ALGORITHM FOR THE TRIANGULAR SYLVESTER EQUATION

In the following we will consider the Sylvester equation (1) with triangular coefficient matrices A and B. The special structure of matrices A and B allows us to write this equation as a linear system

$$\left.\begin{array}{l} (a_{ii}+b_{jj})x_{ij} + \sum_{k=i+1}^{m} a_{ik}x_{kj} + \sum_{h=j+1}^{n} x_{ih}b_{hj} = c_{ij} \\ \qquad i=1,\dots,m \; ; \; j=1,\dots,n. \end{array}\right\} \tag{2}$$

The following algorithm solves it by applying back substitution. The indices in the innermost loops are taken in descending order, which will reflect better the order of computations in the systolic array.

(†) Supported by CICYT Grant TIC91-1157-C03-02

Triangular-Sylvester Algorithm

for j:=n **down to** 1

  for i:=m **down to** 1

    {compute x_{ij}}

      $x_{ij} := c_{ij}/(a_{ii}+b_{jj})$;

    {update column j of C}

      for k:=i-1 **down to** 1

        $c_{kj} := c_{kj} - a_{ki}x_{ij}$

      end *{for k}*

    {update row i of C}

      for h:=j-1 **down to** 1

        $c_{ih} := c_{ih} - x_{ij}b_{jh}$

      end *{for h}*

  end *{for i}*

end*{for j}*.

The arithmetic cost of the algorithm is

$$\sum_{i=1}^{m}\sum_{j=1}^{n}(i+j-1) = nm(\frac{m+n}{2}).$$

The variable x_{ij} depends on variables x_{ih}, with h>j, and on variables x_{kj}, with k>i, so that the elements in the same row or column must be computed sequentially. On the other hand, the elements with i+j=constant are independent and can be computed in parallel.

The systolic arrays we propose make use of this parallelism. In the bidimensional array the elements of X are computed by diagonals, starting with the lower right corner. Both the bidimensional array and the linear one exploit the parallelism of the updates of the elements of C.

Size-Dependent Bidimensional Array

Figure 1 shows the dataflow in the array for a triangular Sylvester problem with m=4 and n=3. Matrix B is stored in the lower part of the rhomboidal array before the beginning of the computations, as is shown in the figure. This storage suggests considering the case of m≥n. Otherwise the equation is easily reduced to this case by transposition.
As you see in Fig. 2, the cells in the lower edge of the rhomboid perform the computation of the elements of matrix solution X and send them to the left and upper neighbor cells. The remaining cells can update the elements of C simultaneously with respect to two computed values of X, using an element of A that enters the cell and the element of B stored in it as coefficients. As only the cells in the lower triangle have to update with respect to the b's, we introduce zeroes through the line E1-W1 of the cells in the upper part of the array.

Let us compute the cost of the systolic algorithm in an m×n array. First we have to consider the storage of matrix B, which is introduced into the array by rows, from right to left. This storage can be overlapped with the introduction of matrix C to minimize the idle time. A control indicates the arrival of the rows of C to their places and normalizes the function of the affected processors, as is shown in Fig. 3 in the case of the example.
After m-1 steps the lower right element of X is about to be computed in the corresponding cell. The process of C takes 2m-1 steps for the first row to arrive at the bottom of the rhomboid and 2n-2 for this row to end its exit.
Then, the cost of the algorithm is

$$C = (m-1)+(2m-1)+(2n-2) = 3m+2n-4.$$

Our algorithm provides a speed-up of

$$\frac{nm(m+n)}{6m+4n-8}$$

and the efficiency is not less than 1/5 when m=n.

Size-Independent Bidimensional Array

The aim of this section is to explain how to deal with problems of arbitrary size in a given bidimensional array.
The idea is to decompose the unknown matrix X into blocks of the same size as the array, say m×n. In this way, the problem splits into Sylvester problems and block updates that can be performed in the array, as we will show next.

Let us consider that matrices C and X are of size mp×nq and decompose them into p×q blocks of size m×n. Matrix A is decomposed into blocks of size m×m and matrix B into blocks of size n×n.

$$C = \underbrace{\begin{bmatrix} C_{11} & \cdots & C_{1q} \\ \vdots & & \vdots \\ C_{p1} & \cdots & C_{pq} \end{bmatrix}}_{n\ \cdots\ n}\begin{matrix} m \\ \vdots \\ m \end{matrix}, \quad X = \underbrace{\begin{bmatrix} X_{11} & \cdots & X_{1q} \\ \vdots & & \vdots \\ X_{p1} & \cdots & X_{pq} \end{bmatrix}}_{n\ \cdots\ n}\begin{matrix} m \\ \vdots \\ m \end{matrix},$$

$$A = \underbrace{\begin{bmatrix} A_{11} & \cdots & A_{1p} \\ & \ddots & \vdots \\ 0 & & A_{pp} \end{bmatrix}}_{m\ \cdots\ m}\begin{matrix} m \\ \vdots \\ m \end{matrix}, \quad B = \underbrace{\begin{bmatrix} B_{11} & & 0 \\ \vdots & \ddots & \\ B_{q1} & \cdots & B_{qq} \end{bmatrix}}_{n\ \cdots\ n}\begin{matrix} n \\ \vdots \\ n \end{matrix}$$

This results in a block matrix system of equations analogous to the scalar system (2):

$$\left.\begin{array}{c} \sum_{k=i}^{p} A_{ik}X_{kj} + \sum_{h=j}^{q} X_{ih}B_{hj} = C_{ij} \\ i=1,\ldots,p \ ; \ j=1,\ldots,q. \end{array}\right\} \qquad (3)$$

The sequence of operations is the same as in the scalar case. Solve first for X_{pq} and then update the block row p and the block column q. In the following step either $X_{p,q-1}$ or $X_{p-1,q}$ can be solved, and so on. It is not difficult to obtain the block version of the algorithm that solves the triangular Sylvester equation:

Block Triangular Sylvester Algorithm

for j:=q **down to** 1

  for i:=p **down to** 1

    Solve the triangular Sylvester equation for X_{ij}

      $A_{ii}X_{ij} + X_{ij}B_{jj} = C_{ij}$

    {Update block column j of C}

      for k:=i-1 **down to** 1

        $C_{kj} := C_{kj} - A_{ki}X_{ij}$

      end *{for k}*

    {update block row i of C}

      for h:=j-1 **down to** 1

        $C_{ih} := C_{ih} - X_{ij}B_{jh}$

      end *{for h}*

  end *{for i}*

end*{for j}*.

Our systolic array directly solves the triangular Sylvester equations that appear in this process, and can also be adapted to update with respect to the non-triangular blocks of A and B, as we will see in the next example.

Let us consider a triangular Sylvester problem of size 12x9. To treat it in a 4x3 array, the coefficient matrices are partitioned into the following blocks:

$$C = \begin{bmatrix} C_{11} & C_{12} & C_{13} \\ C_{21} & C_{22} & C_{23} \\ C_{31} & C_{32} & C_{33} \end{bmatrix} \begin{matrix} 4 \\ 4 \\ 4 \end{matrix}, \quad X = \begin{bmatrix} X_{11} & X_{12} & X_{13} \\ X_{21} & X_{22} & X_{23} \\ X_{31} & X_{32} & X_{33} \end{bmatrix} \begin{matrix} 4 \\ 4, \\ 4 \end{matrix}$$
$$\begin{matrix} 3 & 3 & 3 \end{matrix} \qquad\qquad \begin{matrix} 3 & 3 & 3 \end{matrix}$$

$$A = \begin{bmatrix} A_{11} & A_{12} & A_{13} \\ & A_{22} & A_{23} \\ & & A_{33} \end{bmatrix} \begin{matrix} 4 \\ 4 \\ 4 \end{matrix}, \quad B = \begin{bmatrix} B_{11} & & \\ B_{21} & B_{22} & \\ B_{31} & B_{32} & B_{33} \end{bmatrix} \begin{matrix} 3 \\ 3. \\ 3 \end{matrix}$$
$$\begin{matrix} 4 & 4 & 4 \end{matrix} \qquad\qquad \begin{matrix} 3 & 3 & 3 \end{matrix}$$

After solving the Sylvester equation

$$A_{33}X_{33} + X_{33}B_{33} = C_{33}$$

we have to update C_{23} and C_{13} with respect to blocks of A according to

$$C_{23} := C_{23} - A_{23}X_{33} \quad , \quad C_{13} := C_{13} - A_{13}X_{33}$$

and with respect to blocks of B

$$C_{32} := C_{32} - X_{33}B_{32} \quad , \quad C_{31} := C_{31} - X_{33}B_{31}.$$

The update with respect to a block of A is achieved by passing the block of X from the bottom to the top of the array and the blocks of C and A as in the solution of (1). During the updates, the cells in the lower edge of the rhomboid have to work as the other cells do: updating an element of matrix C instead of computing an element of matrix X.

As blocks of A are no longer triangular and are entered into the array by diagonals, we have to apply a DBT transformation to them (Navarro, Llabería and Valero, 1987), so that their bandwidth equals the number of rows in the array.

The dataflow corresponding to these updates is similar to that in the solution of the equation. Besides the blocks of C and B, we have to pass the computed block of X from the bottom to the top of the array.

As we see in Fig. 4 for block row 3 of the example, the updates with respect to blocks of A efficiently chain with the computation of the corresponding block of X.

In the general case, the cost of the computation of a block of X is the same as that of the basic algorithm: 3m+2n-4. For each computed block X_{ij} we have to update the i-1 blocks $C_{1j}, \dots C_{i-1,j}$ with respect to $A_{1i}, \dots, A_{i-1,i}$, respectively. These updates chain perfectly, costing only 2m cycles each. Then, the cost is of order 2mi. Adding this expression for each block of X we get

$$\sum_{i=1}^{p}\sum_{j=1}^{q} 2mi = mp(p+1)q.$$

In order to update with respect to a block of B, we store this block in the cells of the array, pass the block of X from right to left through line E1-W1 and C from the top to the bottom. Note that the algorithm is 1-slow because there is no contraflow.

Fig. 5 displays the computation of $C_{32} - X_{33}B_{32}$.

This update costs m+3n, but the last 2n steps can be overlapped with the processing of block C_{31} with respect to B_{31} and X_{33}. In the general case we have to do j-1 updates with respect to the blocks of B and block X_{ij}. The cost of all these updates is

$$\sum_{i=1}^{p}\sum_{j=1}^{q}(j-1)(m+n) = \frac{1}{2}pq(q-1)(m+n).$$

Adding this result to the cost of the updates with respect to the blocks of A, we obtain that the total cost of the block systolic algorithm is of order

$$pq(mp+\frac{m+n}{2}q).$$

If m is not very much greater than n, the cost approaches pq(mp+nq), which gives an efficiency of 1/2.

In the case of m=n we can treat the updates with respect to the blocks of B in the same way as that with respect to those of A, getting a more regular scheme in which all the data flow at the same speed and we do not have to store the rectangular blocks of B. The idea is to transpose the expressions

$$C_{ih} := C_{ih} - X_{ij}B_{jh}, \quad h = 1,2,\dots,j-1,$$

to obtain

$$C_{ih}^{T} := C_{ih}^{T} - B_{jh}^{T}X_{ij}^{T},$$

which can be computed and chained as are the updates

$$C_{kj} := C_{kj} - A_{ki}X_{ij}, \quad k = 1,2,\dots,i-1.$$

The cost will be of the same order for both types of update and the efficiency will again be 1/2.

Size-Dependent Linear Array

The linear systolic array we propose now computes the elements of a column of matrix X and updates the preceding columns with respect to the calculated values.

Figure 6 displays the funcionality of the cells and the dataflow for the processing of the last column of the 4x3 example problem.

The central cell computes an element of the solution and sends it to the right cells, that update elements of C with respect to A, and to the left cells, that perform the updates with respect to B.

With m+n+1 processors, the problem is solved after passing matrix C n times through the array. As the columns of C are of size m and the algorithm is 2-slow, the cost is 2mn and the efficiency 1/4.

Size-Independent Linear Array

By applying the same techniques as in the bidimensional case we are able to solve a problem of arbitrary size in a given linear array.

Let us consider, for example, a triangular Sylvester problem of size 12x9. If we decompose it into blocks of size 4x3, as in the size-independent bidimensional case, it can be solved in a 3+4 linear array. Fig. 8 shows the dataflow corresponding to the computation of block C_{33} and the update of blocks C_{23} and C_{13} with respect to A, and blocks C_{32} and C_{31} with respect to B.

There is an extra processor that updates elements of X with respect to B. During the updates, the central processor changes its function: it performs the same arithmetical operations as the processors on the right and sends an element of X to the processors on the left (see Fig. 7).

For a problem of size M×N = mp×nq we choose m and n so that p≈q, which minimizes the idleness of the processors. In that case the cost of the block systolic algorithm is $4mn(m+n)p^3/3$ and the efficiency rises to 3/8, closer than in the size-dependent case to the optimal efficiency for an algorithm with contraflow.

The fact that the efficiency of the size-independent linear array is lower than that of the bidimensional array may seem paradoxical. But if we analyze the cells, the reason will become apparent. Some of the processors in the bidimensional array perform simultaneously two updates, whereas in the linear array each cell performs only one computation. Thus, these cells are simpler, so that the clock of the linear array can be faster than that of the bidimensional array.

CONCLUSION

We have presented systolic algorithms for solving the triangular Sylvester equation in bidimensional and linear arrays.
In the size dependent case the efficiency of both arrays approaches 1/4, but the cells of the linear array are simpler and more regular, which can help the design of the array and lower the clock, getting an acceptable throughput. These low eficiencies are due to the triangular shape of the coefficient matrices and the startup time.
By applying data decomposition techniques, we are able to solve Sylvester problems of arbitrary size in a given array.
In the bidimensional case the efficiency is nearly doubled by partitioning, getting in theory the optimal efficiency for an array with contraflow, 1/2.
In the linear array only an increase of 50% is obtained, getting an efficiency of 3/8, closer to the optimal one.

REFERENCES

Barnett, S. (1971). *Matrices in Control Theory*. Van Nostrand Reinhold, London.

Bartels, R.H. and G.W. Stewart. (1972). Algorithm 432. Solution of the Matrix Equation AX+XB=C [F4], *Comm. of the ACM*, 15,9, 820-826.

Berger, C.S. (1971). A Numerical Solution of the Matrix Equation $P=\Phi P\Phi^T+S$. *IEEE Trans. Aut. Control*, 16, 381-382.

Golub, G.H., S. Nash and C. van Loan. (1979). A Hessemberg-Schur Method for the Problem AX+XB=C. *IEEE Trans. Aut. Control*. AC-24, 6 909-913.

Golub, G.H. and C. van Loan. (1983). *Matrix Computations*. John Hopkins University Press, Baltimore.

Hueso, J.L., G. Martínez and V. Hernández. (1991a). A Systolic Algorithm for the Triangular Stein Equation. *Proceedings 1991 Application-Specific Array Processors*. M. Valero and S.-Y. Kung (Eds.). IEEE Comp. Soc. Press, Los Alamitos, California. pp. 473-484.

Hueso, J.L., G. Martínez and V. Hernández. (1991b) A Linear Systolic Algorithm for the Triangular Stein Equation. *Proceedings of the IFAC Workshop Algorithms and Architectures for Real-Time Control*. Fleming, P.J. and D.I. Jones (Eds.). Oxford. pp. 43-48.

Kågström, B. and P. Poromaa. (1989). Distributed Block Algorithms for the Triangular Sylvester Equation with Condition Estimators. In F. André and J.P. Verjas (Eds.), *Hypercube and Distributed Computers*, Elsevier Science Publishers B.V. (North-Holland), pp. 233-249.

Kågström, B., L. Nyströn and P. Poromaa. (1988) Parallel Shared Memory Algorithms for the Triangular Sylvester Equation. *Report UMINF* 155-88, Inst of Information Processing, Univ. of Umeå, S-901 87 Umeå.

Kleinman, D.L. (1968). On an iterative technique for Ricati equation computations. *IEEE Trans. Autom. Control*, AC-13, 114-115.

Kung, H.T., and C.E. Leieserson. (1982). Systolic Arrays (for VLSI). In *Introduction to VLSI Systems*, Sec.8.3. Mead & Conway eds.

Kwakernaak, H and R Sivan. (1972). *Linerar Optimal Control Systems*. John Willey. New-York.

Laub, A.J. (1985). Numerical Linear Algebra Aspects of Control Design Computations. *IEEE Trans. Aut. Control*, AC-30, n.2, p.97.

Navarro, J.J., J.M. Llabería and M. Valero. (1987). Partitioning: an Essential Step in Mapping Algorithms into Systolic Array Processors. *IEEE Computer*, July 1987, 77-89.

Quinton, P. and Y. Robert. (1989). *Algorithmes et Architectures Systoliques*. Masson.

Sandell, N.R. (1974). On Newton's method for Ricati equation solution. *IEEE Trans. Autom. Control*, 254-255.

Torralba, N and J.J. Navarro. (1988a). A One Dimensional Systolic Array to solve the QR Iteration for Matrices of any size. *Proceedings of the 36th ISMM Int. Conf. on Mini and Microcomputers and their Applications*.

Torralba, N and J.J. Navarro. (1988b). A One Dimensional Systolic Array for Solving Arbitrarily Large Least Mean Square Problems. *Proceedings of the Int. Conf. on Systolic Arrays*, 103-112.

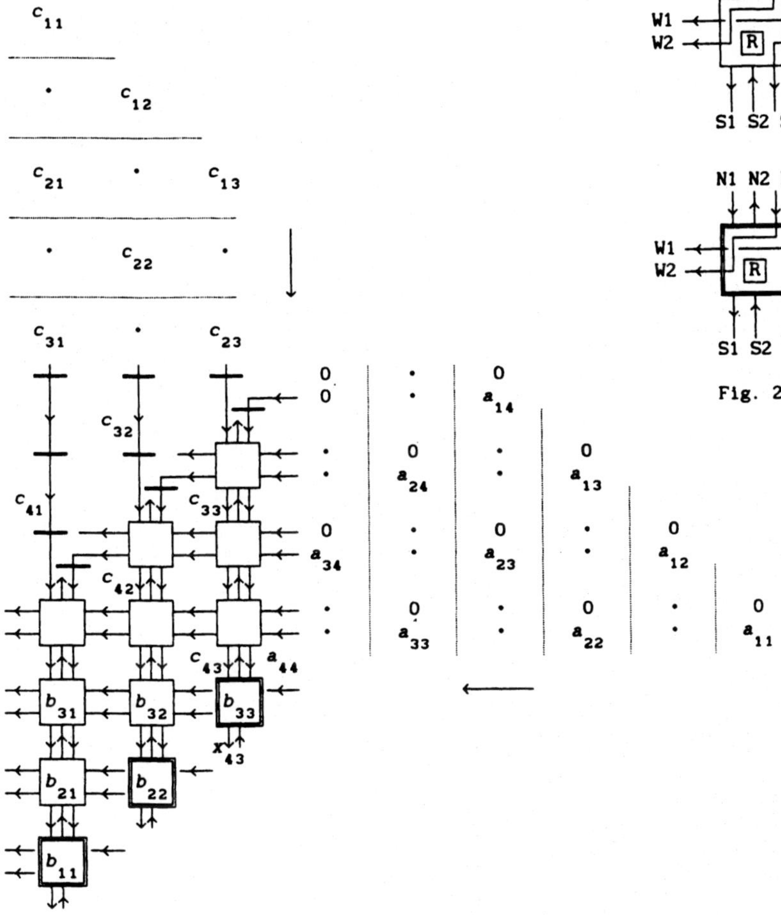

Fig. 1: Size-Dependent Bidimensional Array.

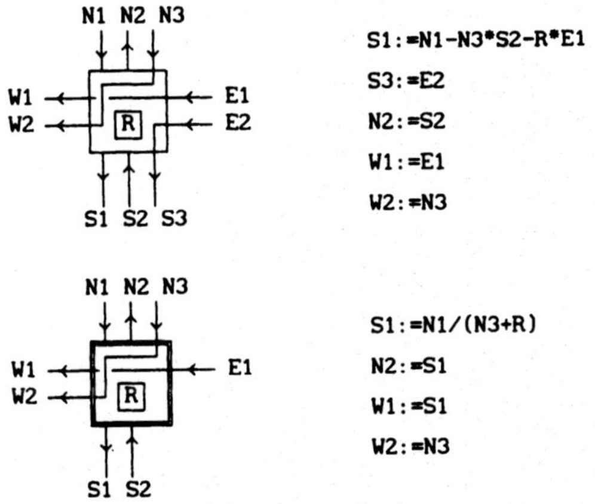

Fig. 2: Functionality of the cells.

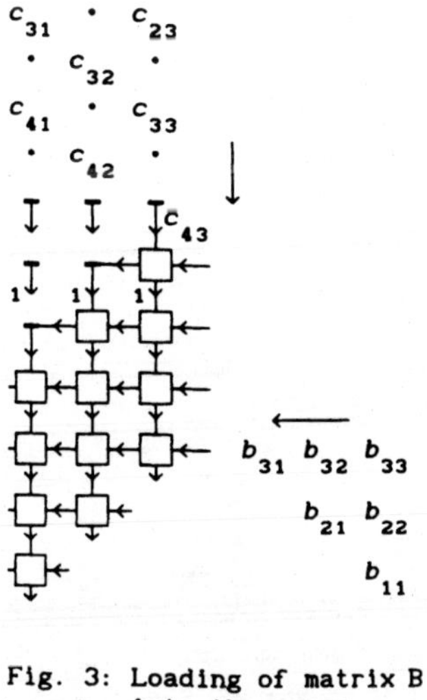

Fig. 3: Loading of matrix B into the array.

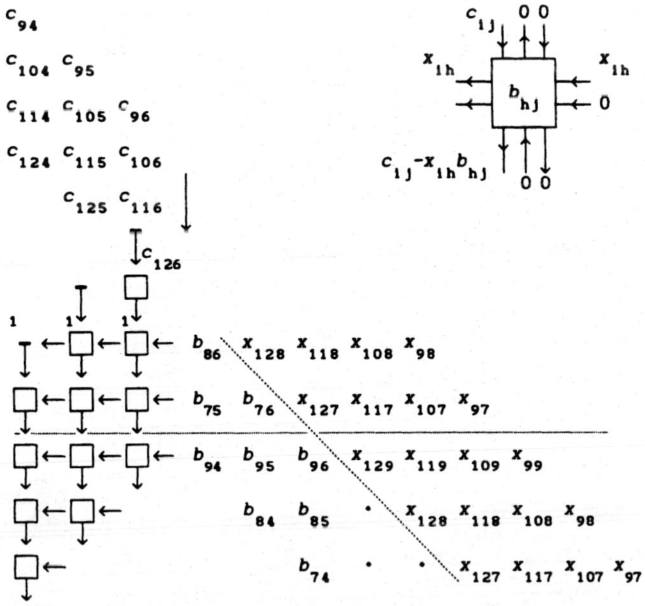

Fig. 5. Storage and update with respect to a block of B in the Size-Independent Bidimensional Array.

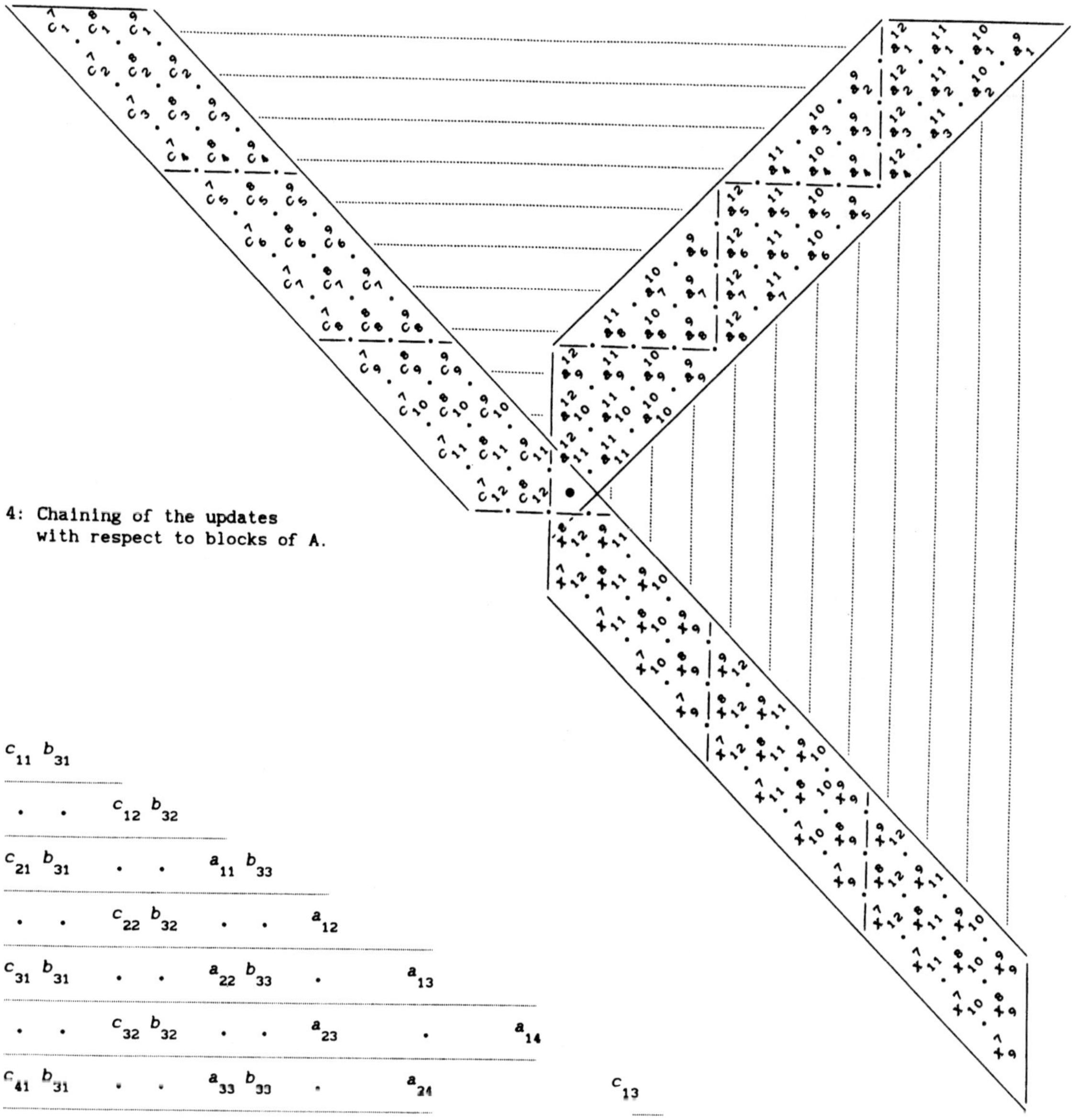

Fig. 4: Chaining of the updates with respect to blocks of A.

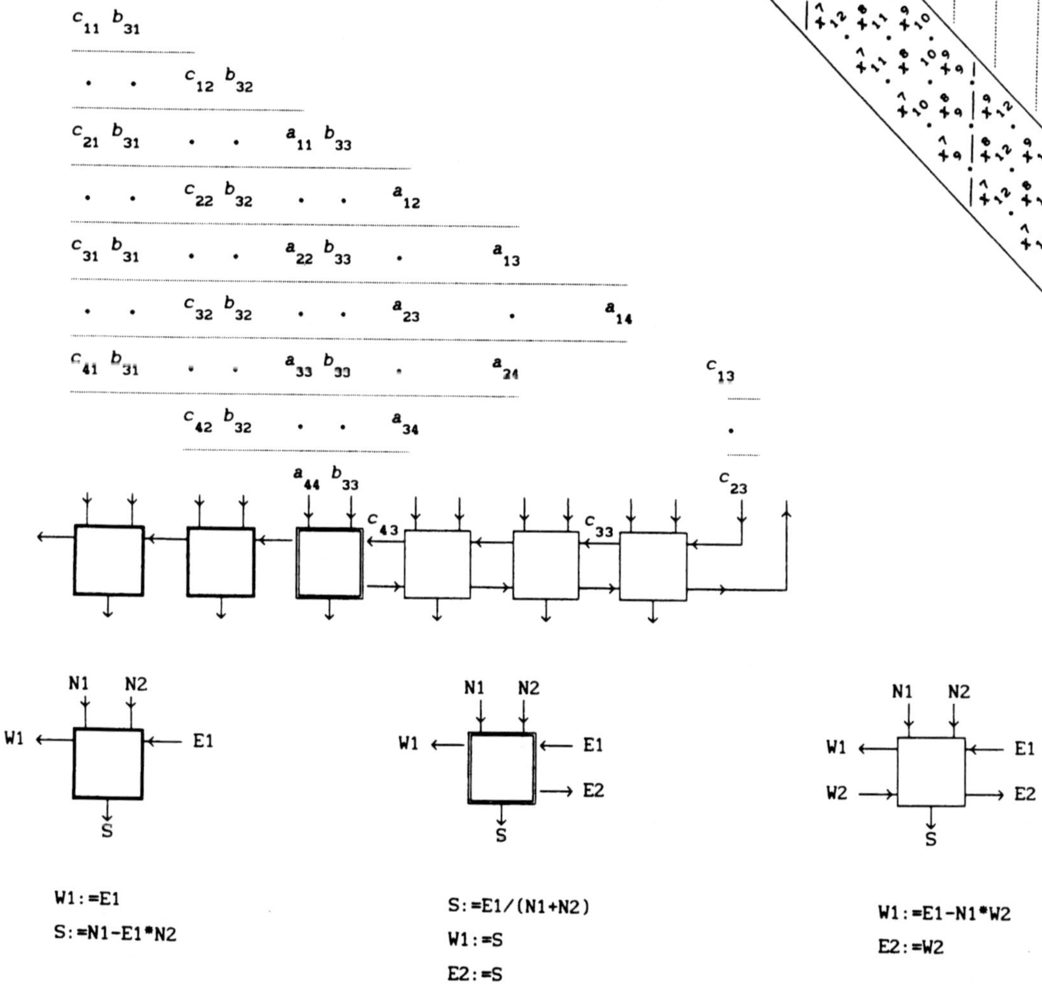

Fig. 5: Size-Independent Linear Array

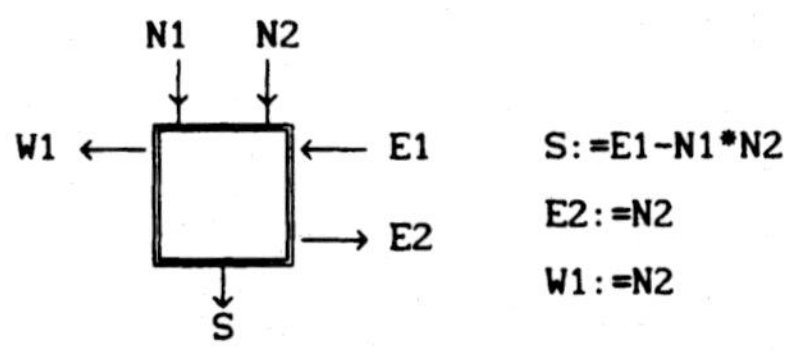

Fig. 6: Change of function of the central cell when it updates elements of C.

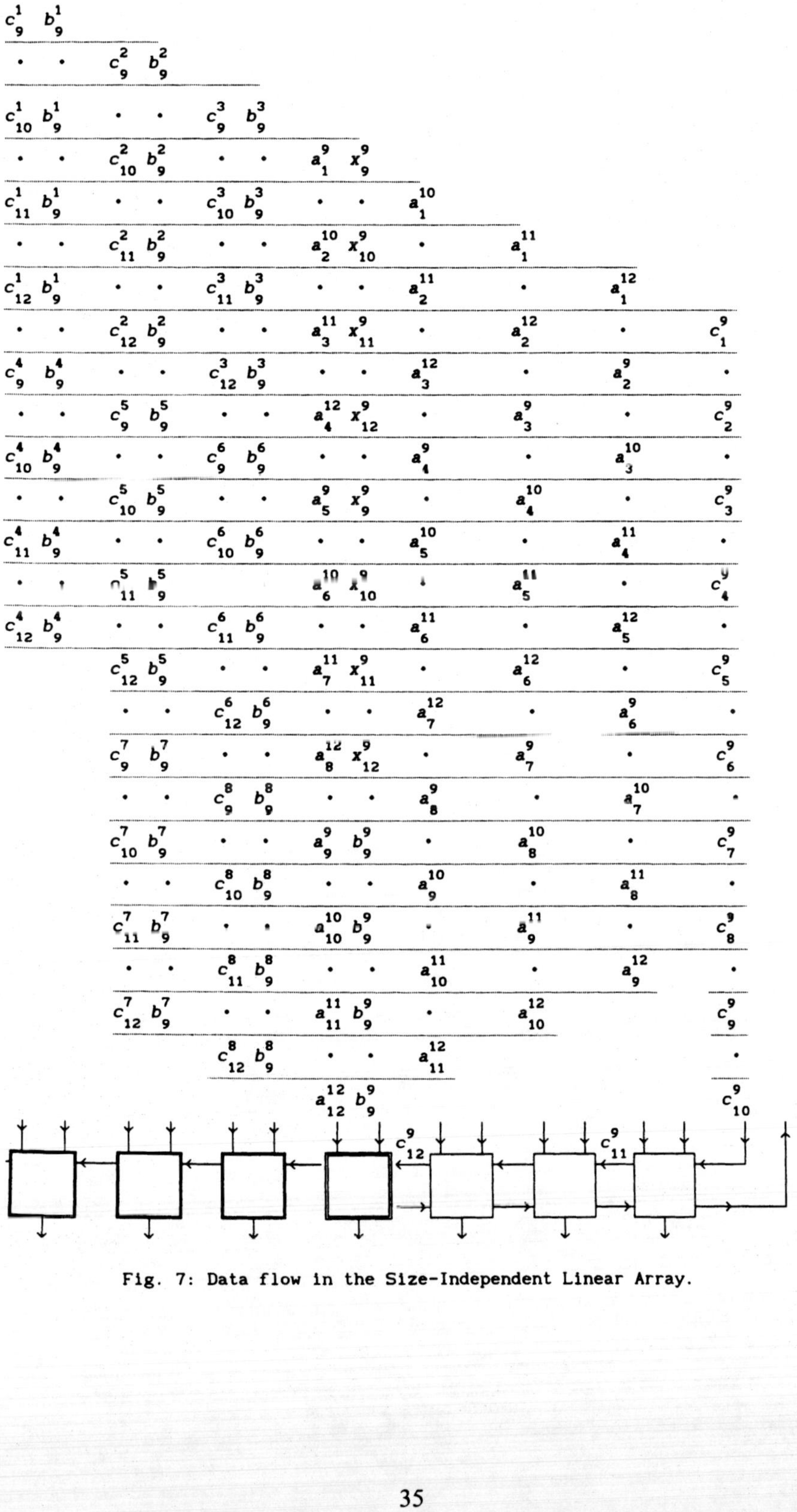

Fig. 7: Data flow in the Size-Independent Linear Array.

Real Time Control of a Multiple Arm System by Large Scale Multiprocessing

Tasuku Hoshino and Katsuhisa Furuta

Department of Control Engineering, Tokyo Institute of Technology, 2-12-1, Oh-okayama, Meguro-ku, Tokyo 152, Japan

Abstract This article describes a design of a high performance real time control system based on multiple processors. The hardware system is based on INMOS tranputers, and the software system is developed using occam and Transputer Development System. A main topic in this article is how to construct modular hardware or software system for developing a flexible, powerful, and convenient real time controller. A proposed method is applied for a robotic control system and good performance is examined.
Keywords Multiprocessing systems, Real time computer systems, Software development, Robots

INTRODUCTION

There are many concurrent phenomena which progress as an aggregation of their components' behavior in the world. Control system can be one of such phenomena, however, little attention has been paid to the concurrency so far in the actual implementation of control systems, and it has been realized by constructing sequential processes in a single processor. We offer here a few reasons for that: (1) Hardware design for oncurrent systems required higher level knowledge and techniques about computer systems. (2) There was few suitable computer languages for concurrent programming.

In robotics, especially in controlling intelligent robots, huge computation is unavoidable. It is likely that theoretical results obtained in robotics and its application are difficult to be realized in industries without any prospective solutions of the treatment of computation. A powerful computer developed for a special purpose may provide us a solution and we have used a bit slice computer to control multiple arms robot for three years as an example. There are several problems which should be solved to use such a special purpose computer, i.e., (1) Since the micro-code programming is required, software development takes much time and its maintenance is troublesome. (2) We can hardly extend the equipment. Though the system is fit for the current control plant very well, it is impossible to use the control system to the more complex plant to be controlled in the future. This is a fatal problem for a further research to use the equipment. (3) Since the system was developed for our special application, there are few interface devices for other applications.

These are problems we have to face whenever we realize a high performance controller constructed using only a single computer. Thus, another approach is necessary.

One solution is to construct the control system with multiple processors by taking account of the inherent concurrency of the control plant and control system. The multiple-processing-oriented processor, *transputer* and concurrent programming language *occam* enable us to overcome the difficulty.

There are four fundamental problems to construct the real time control system: (1) How to realize a control system in concurrent and distributed fashion. (2) How to divide a computational algorithm into parallel processes. (3) How to communicate with I/O peripheral devices. (4) How to establish a programming environment with which the system can be easily extended.

For the first problem we have proposed a virtual model following controller (Furuta, 1987; Kosuge, 1990) which composes of three concurrent components; Virtual Internal Model(VIM), Model following controller, and Non-linear feedback compensator. They are independent each other in a sense that they can be executed concurrently. In the robotic controller large amount of computation is required for the non-linear compensation in a servo execution layer rather than for the internal model. The computation of that part dominates the control period which should be shortened to improve the system performance.

The answer to this problem is dividing a computational algorithm itself into concurrent processes, which is the above problem two. We have solved the problem in the inverse dynamic calculation which requires large amount of computation by parallel computation of subtasks using predictions. (Yamakita, 1991)

As the two problems mentioned above are discussed in the references, we give solutions mainly to the third and fourth problems here. The solution is to develop an powerful computer environment for the concurrent and distributed computations, especially for real time applications.

TRANSPUTER FAMILY

Our target system is constructed on the transputer family. Some peripheral devices provides the system construction with flexibility and with its usefulness for practical applications. More details are described in (INMOS, 1989).

Brief description of transputer

The transputer is a high speed processor with its own local memories and four communication channels named ***serial links***. Using the communication channels, a network is realized by consisting of several processors, on which parallel multiprocessing, concurrent processing, is achieved.

The serial link makes the transputer as an unique processor. In the multiprocessing, the data should be shared by several processors. This is realized either by a shared memory or a *Point To Point* (PTP) communication. In the shared memory, simultaneous access by two processors is not possible. If the number of the processor increases, the access to the data will be a major factor of the performance.

On the other hand, several PTP channels can be used simultaneously, and the communication between the

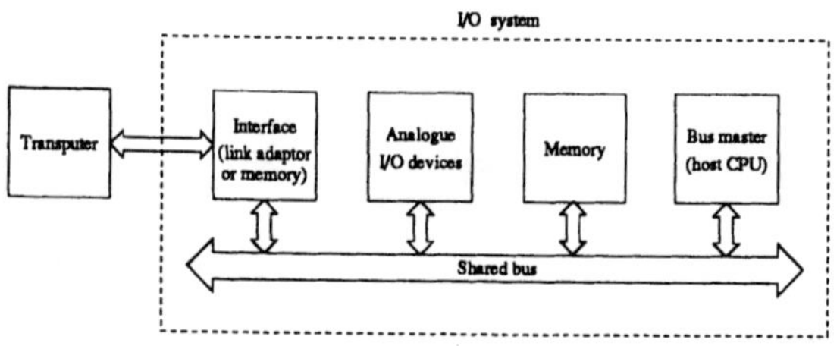

Fig. 1 Bus-based I/O system

processors does not affect the total performance. The networks realized by PTP connection can be easily modified for purposes and have flexibility.
The other unique characteristics of the transputer is to have a hardware scheduler to manage parallel processing. This emulates parallel processing for multiple processors on a single transputer chip. This makes a concurrent software executable either on multiple processors networks or a single chip.

Link adaptor and cross-bar switch

In order to connect transputer links to other processing system, the serial link interface device named *Link Adaptor* is provided by the INMOS corporation which developed transputers. It receives one byte data through a serial link from a transputer and generates parallel output, and conversely transfers one byte data from external devices through a serial link to a transputer.
Another member of the transputer family is C004, which is a *Transparent Programmable Link Switch* designed to provide a full crossbar switch between 32 link inputs and 32 link outputs. If a transputer network is constructed employing this device, its topology is easily changed to fit any application by software. It makes network configuration flexible.

PERIPHERAL INTERFACE

Because a transputer does not have any data acquisition mechanism in itself, it must be provided a large bandwidth, high-performance peripheral interface system. In this section, we consider how to design the interface system using available devices described in the previous section.

Interface architecture

Several peripheral handling strategies have been proposed and reported (Barlow, 1989; Milway, 1990; INMOS, 1989). These strategies are classified into three categories: one is a bus-based system, and the others are transputer-based systems.
In the bus-based system, transputers are connected by some combination of bus-mapped link adaptors and/or dual-ported memory to a bus on which peripheral devices reside. This architecture is shown in Fig. 1. Because standard commercial products for the bus can be used, they offer less cost systems. But the access speed is, in general, the slowest.
The second and the third are transputer-based systems. Systems in these classes make use of transputer links to interface peripheral devices. The resulted systems with multiple communication links has less bandwidth limitations than of the bus-based system because more than two communications are performed simultaneously. So they are inherently more suitable for time-critical input and output.
The second strategy is to use transputer based systems which do not contain a transputer. It is not normally programmable and typically consists of the peripheral devices, a link adaptor, and random discrete or one-time programmable logic elements which implement a fixed protocol for the communications over the link.
And the third one is a system which contains a transputer and the peripheral devices residing on the memory map of the transputer. Communication with the rest of the system may be done via any or all of the four serial links of the transputer and can be done in the form of more compact, high-level information, since the transputer allows a respectable degree of pre- and post-processing. These two architectures are shown in Fig. 2.

Design key features

We should consider what is a more important feature of a peripheral interfacing system when it is used in a multiple-processor system. Several properties may be required: (1) A whole system is easily constructed and extended. (2) The system is designed based on some standard specification in order to be used as long as the specification is being valid. (3) Large bandwidth, that is, less time to be required to access the peripheral device.
It is noted that less flexibility is required in the usage of the peripheral interface to be developed, while commercial products are designed intending to provide a variety of usage. The reason is that we use the peripheral almost in the same way every time, and on the other hand, commercial products should adapt to a variety of situation where they are used.
Taking these factors into account, we adopt the second strategy presented before. Using this architecture we can achieve following properties:

- The construction of the whole system is still easy because of its PTP link connection, even if peripheral interfaces are included in the network.
- The peripheral interface may be used as long as the system network is realized by using the transputer links. link compatibility,
- High speed access is achieved because the sequencer of the peripheral devices is composed of random discrete logic circuits.

Note that the disadvantage of the strategy is incapability of data processing inside the peripheral interface. This disadvantage might be avoided by adopting the third strategy, but the development would cost very high. So we choose the second strategy in this paper while the third one is reported in (Barlow, 1989).

Basic strategies of the design

To construct the interface devices with link adaptors, we propose the following structure which can be easily extended for future use. This structure is shown in Fig. 3.

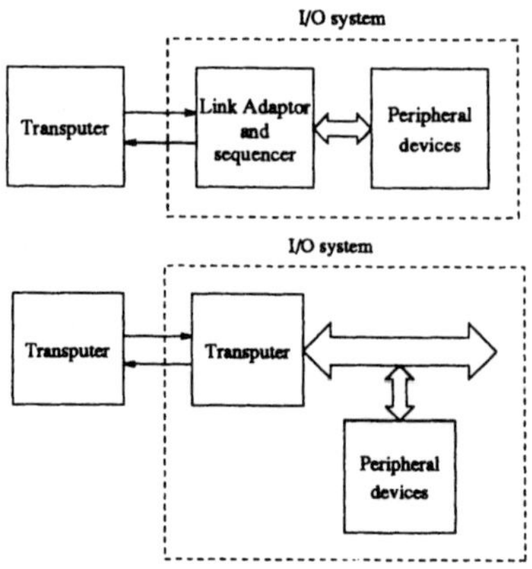

Fig. 2 Transputer-based I/O systems; a link adaptor driven (above), and memory mapped (below).

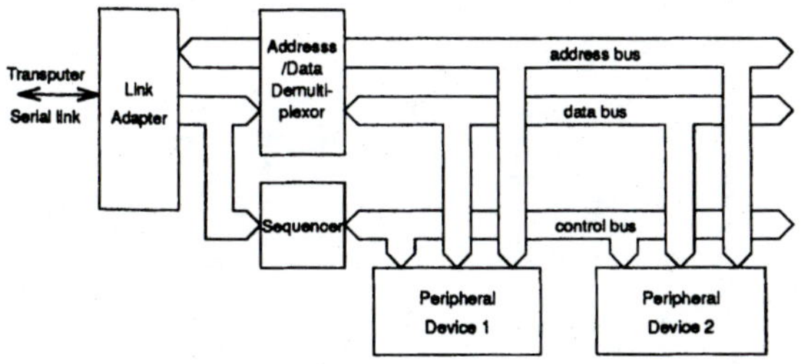

Fig. 3 Block diagram of peripheral interface system

R/W	C2	C1	C0	A3	A2	A1	A0

· R/W bit indicates read/write.
· C2, C1, and C0 indicate the number of accesses.
· A3, A2, A1, and A0 indicate the address of device to be accessed first.

Fig. 4 Format of extended address code

What we should consider next is the *protocol* to be used in this architecture. One may suggest a following one:

- First, specify on the I/O map the address of the device which is connected to the serial link and indicate also the type of operation (read or write) to be done.
- Then receive data for reading or send data for writing.

But this protocol is too primitive and has some disadvantages.

- Because not so much devices are available on one board in a practical use, so the address code is redundant.
- Furthermore, in the case that the same device is accessed successively, it is wasteful to send the same address repeatedly. It would be better to send firstly the information that how many times the address will be accessed so that we will not have to send the address information again.

These problems can be neglected if the communication speed is fast enough. But the communication speed is not so fast practically, so it is significant to consider these problems.
To cope with these problems, we extend the protocol by introducing the format of the extended address code which will be sent first. It is shown in Fig. 4. By sending this kind of extended address code, we need not send the redundant information and the communication time can be reduced.
The access time using extended protocols and basic protocols is shown in TABLE 1. These are performance of the A/D board described later and are measured using transputer's internal clock.

Developed peripheral interfaces

We developed three following peripheral interfaces under the idea discussed above; (1) Three-channel D/A converter and counter (TRIO-01-A), (2) six-channel A/D converter (TRIO-02-A), and (3) two-channel 16-bit parallel I/O port(TRIO-03-A). These peripheral interfaces can be connected in the PTP way to transputers through serial links to make a transputers access peripheral devices directly. The overall picture of interface is shown in Fig. 5.

Hardware construction

We constructed a real time controller using multiple transputers and three types of peripheral interface modules. These are connected via transputer serial links with each other. The processor cards and the interface cards are the same size as MultiBus-II standard, and are put in a system rack.

OCCAM/TDS

In this section, we briefly overview the programming language occam and *Transputer Development System* (TDS). As we construct our software system on the TDS, it is important to describe the basic architecture of the TDS. For more detail see (INMOS, 1988a; INMOS, 1988b).

occam

The programming style for transputers is defined by occam. occam enables a program to be described as a collection of concurrent processes, which communicate with each other and with peripheral devices through *channels*. occam programs are built from three *primitive processes*:

var := expr
Assigns expression **expr** to variable **var**.

chan ! expr
Outputs expression **expr** to channel **chan**.

chan ? var
Inputs variable **var** from channel **chan**.

The primitive processes are combined to form *constructs*: **SEQ**uential, **PAR**allel, **ALT**ernative, **IF**,**WHILE**, and **CASE**. Because a construct is itself a process, it may be used as a component of other construct.
Concurrent programs can be expressed with channels, inputs and outputs, which are combined in parallel and alternative constructs. The communication is synchronized and happens when both the input process and the output process are ready.
The following is a simple example of occam process:

```
ALT
  increment ? signal
    counter := counter + 1
  decrement ? signal
    counter := counter - 1
  value ? signal
    SEQ
      out ! counter
      counter := 0
```

This process has three input channels: **increment**, **decrement**, and **value**. If a signal is received via the channel **increment**, then the variable **counter** is increased by 1, and if a signal is received via the channel **decrement**, then the variable **counter** is decreased by 1. Otherwise, if a signal is received via the channel **value**, the current value of the counter is output, then the value is reset to zero. Thus this process works as a up/down counting process.

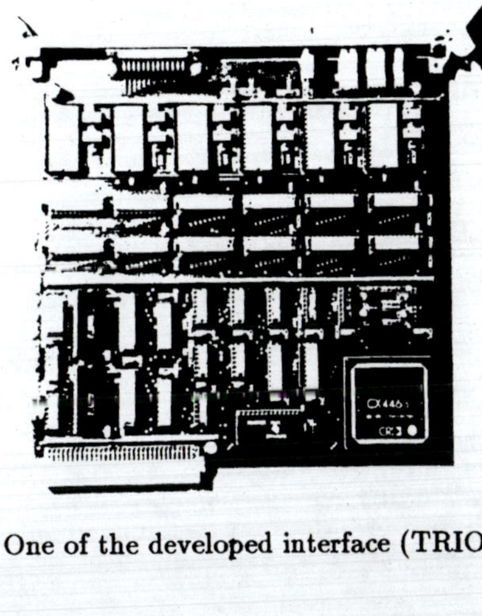

Fig. 5 One of the developed interface (TRIO-02-A)

TABLE 1 Access time required to get 6 analogue voltages in real numbers

Link speed [Mbits/sec]	Extended protocol		Basic protocol	
	T800, 20MHz	T414, 20MHz	T800, 20MHz	T414, 20MHz
10	69.0 μsec	171.0 μsec	86.8 μsec	189.7 μsec
20	64.7 μsec	170.4 μsec	78.9 μsec	184.5 μsec

Transputer Development System (TDS)

The TDS consists of a plug-in transputer board and development software which runs on the transputer board. This combination provides a complete, self-contained development environment in which programs can be developed, compiled and run.
There are variety of libraries provided by the TDS, for example, many mathematical functions and the input/output facilities that a programmer might require.

System interfaces

A several *system interfaces* are defined on the TDS. These are:

1. The terminal interfaces provided by the channels `keyboard` and `screen`.
2. The user file interface provided by the channel pairs of `to.user.filer` and `from.user.filer`.
3. The TDS file server interface and the host file server interface through the link to the host. These include the filer protocol for the channels `to.filer` and `from.filer`, and the kernel protocol available over the channels `to.kernel` and `from.kernel`.

All of these interfaces are available to an executable program running within the TDS. The provided input/output libraries are constructed based on these system interfaces. Users normally use the libraries to carry out sequences of system operations.
These interfaces have own interface *protocols*. One of the protocols used by the screen output channel `screen` is described below.
The channel `screen` communicate with the terminal of the host. The protocols used by the `screen` output channel are explicitly *tagged protocols*. A communication using such a protocol always starts with one of a limited set of constant values. The rest of the communication depends on the pre-defined value of the tag and may consist of zero or more further values of particular types. Arrays of arbitrary size may be transmitted, in this case their size precede.
For example, in order to display a string 'Hello' on the screen, a command out.string is used as follows:

```
SEQ
  screen ! tt.out.string
  screen ! SIZE "Hello"
  screen ! "Hello"
```

In this case, a tag `tt.out.string` is sent at first. This tag means that the following data are the length of the string and the string itself, and that the string should be displayed on the screen. In fact, the example above is a body of one of the provided I/O procedures.
Other system interfaces are also realized by such communications and the provided libraries are constructed based on these interface protocols.

WELL STRUCTURED SYSTEM SOFTWARE

To make effective use of the transputer systems, it is necessary to develop several concurrent processes which are run on the processor network. Such processes are a host file system handler, a terminal handler, peripheral interface process, and so on. We call them system software. As described in the previous section, occam is not only a programming language for concurrent process with simple syntax, but also a facilities to make use of all the features of a transputer efficiently. Therefore, using occam to develop the system software would be an excellent choice.

Routing problem over a transputer network

Concurrent processes on transputers each of which is a member of a processor network proceed communication each other. What should be verified in this situation is communications between any two processes are available or not. This issue is called *routing problem*, and always arise in a multiple processing system. To give a solution to this problem is the subject in this section.
In general, communication in a multiple transputer system can be classified into three cases; (1) internal communication between processes within a single transputer, (2) external communication between transputers which are directly connected, and (3) external communication between transputers which are not connected directly.
It can be observed that no communication problem occurs in the first case, since to exchange data within a single transputer is achievable via internal channels. In the second case, external communication among transputers which are directly connected is also achievable via the built-in serial links of each transputer. However, each transputer generally has only four built-in serial links, so that direct interprocessor connections are limited up to four per one transputer. This hardware limitation clearly leads to the third case whenever there is necessity of exchanging data between processes running in non-adjacent transputers. Consequently, development of software system is required to support such remote communication.
For a system of which the number of transputers is small, it is possible to develop a program such that we add some intermediate process between processes on the transputers, which constitute the path between the communication source and the destination. But this solution becomes more complicated as the number of communications increase. Moreover, each process on the transputers on the path must be revised even if one communication is added.
To overcome these difficulties, a routing process must be run on every processors in the network. The routing process and user processes are performed concurrently in each processor. Figure 6 illustrates this architecture. But unfortunately, developing a deadlock-free channel server program becomes difficult when the routing process is used to support two non-adjacent groups of transputers having heavy remote communication requirement. Moreover, the performance of the server becomes worse, so that the above technique is not appropriate for this case.
Some techniques to construct the routing process have been suggested (Knowles, 1989) though they can not be implemented in a real time control system. Here we propose a technique for construct a simple routing process which works well in a real time system.

Design of the communication protocols

In the network based on the routing process, communicated data on the network is attached a tag which designates the data type, the source processor, the destination processor, and the data size. The rout-

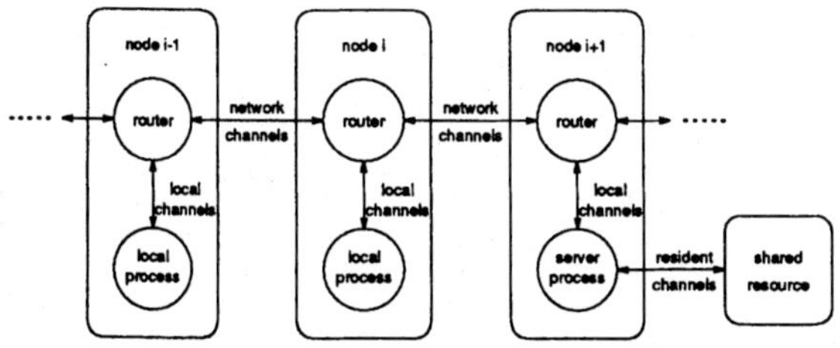

Fig. 6 Interprocess communication using routing processes

ing process sends the data to an appropriate adjacent processor in the network according to the information on the destination processor. If the destination is the router itself, the data is sent to the internal channel within the processor. The actual implementation of the tag and the format of a data packet is shown in Fig. 7.

The data type is one of array of real number, byte, null (null data for a synchronization), integer, string and error. If the data type is byte and the compact tag is used, an overhead for the communication will be suppressed since a packet for the communication contains only one word (4 bytes).

The routing process can receive a packet from any channel which is a internal channel or a physical channel(serial link), and it sends the packet to a proper channel recognizing the destination. Therefore, the routing process consists of an **ALT** construct. An example of the router is shown in the following.

```
ALT
  from.right ? packet
    ... output according to its destination
  from.left ? packet
    ... output according to its destination
  from.inside ? packet
    ... output according to its destination
```

User does not have to construct a transmitted data with the tag because the routing process automatically convert the user data to the transmitted data format shown in the figure. Moreover, the router extracts an original data from the received packet for the user to use the received data directly, i.e. the router has the coding and decoding functions. As the router and user processes are executed concurrently, the construction of software system on each processors is as follows:

```
PRI PAR
  WHILE TRUE
    ... routing process
  PAR
    ... user process 1
    ... user process 2
```

Since the router is run on every processors in a network, user can communicate with any process on any processor, and efficiency of the programming is enhanced and an extension of the system becomes easily. It should be noted that the developed network system

node	chan	type	data ···

node	chan	type	snode	schan	data ···

· node indicates a destination processor.
· chan is a destination channel.
· snode indicates the source processor.
· schan is the source channel.

Fig. 7 Two proposed formats of a transferred data packet; reduced format (above) and full format(below).

can be used for real time applications because of the simplicity.

Access to common resources in the network

Since a console and a file system are common resources of the system, it is desirable that they can be accessed from any processor in the network.

As explained in the previous section, we can perform a message passing between any two nodes using routing mechanism. So it is enough that we provide a server which manages the access to the host console and the file system, and we make it communicate with the router. When a request of the resource access occurs in the network, the request message is sent to the corresponding server on the root transputer via routers. Then the server receive the request message and converts it into the interface protocol which are described before. Thus the access is achieved by routing mechanism and the servers on the root processor.

In order to utilize the mechanism conveniently, we provided remote access procedures for the system. They are used in the situation that the routing mechanism and the servers are both available in the network.

One of such procedures, the remote screen access procedure rmt.out.string, is shown below as an example:

```
SEQ
  to.network ! tag.byte;
               root; screen; tt.out.string
  to.network ! tag.integer;
               root; screen; SIZE "Hello"
  to.network ! tag.string;
               root; screen; "Hello"
```

In this example, all these messages are sent to the **screen** server process running on the root transputer where the **screen** output channel is available. These data are received by a screen server process, and the server process convert the sequence of these messages into a out.string procedure.

Other server processes for common resource are also located on the root transputer, and corresponding procedures are developed. Figure 6 describes an system configuration including a server process.

Time management process

In order to realize a real time control, a periodic operation is required in general. In a transputer based system, it is achieved by using its own *timer*, which is available in a occam program by declaring a variable of **TIMER** type. Once the variable is declared and the process is executed, the variable increases automatically.

By the use of **TIMER** variable, we can program a process which generates successive periodic signals. An example is shown below.

```
TIMER clock :
... declaration of other variables
... definition some variables
SEQ
  operating := TRUE
  clock ? now                  -- set current time
  WHILE operating
    SEQ
      limit := now PLUS interval
                               -- set time limit
      to.slave ! op.code       -- job starts
      ALT
        clock ? AFTER limit    -- watch the time
          SEQ
            from.slave ? signal  -- job does not
                                 --    end in time
            operating := FALSE
                               -- terminate process
        from.slave ? signal    -- job ends in time
          ALT
            clock ? AFTER limit  -- idling
              now := limit  -- update current time
```

System libraries

We provides system libraries for our transputer-network based system. These are listed below.

usersys.tsr This is a *header library* in which several system constants are defined: (1) Number of transputers constitute the network. (2) Protocol definition of communications in the network and in a transputer. (3) Number of links and internal channels. (4) Identification code of transputers and processes in the network.

route.tsr This is also a header library where the topological data of the network for routing is defined.

rmtio.tsr This is a *code library* consists from remote I/O procedures, which is the systematic modification of userio.tsr, provided in the TDS.

rmtmsdos.tsr This is also a code library which contains procedures for remote file access. These are obtained by modify the original procedures in msdos.tsr in the TDS.

userdev.tsr This is a collection of device access procedures for developed peripheral interface boards.

APPLICATION TO ROBOT SYSTEM CONTROL

In this section, we describes an application of the developed system to real time control of a multiple-manipulator robotic system with high degree of freedom.

Hierarchical construction of controller

A *coordinated multiple-manipulator system* is one of advanced robotic systems and is used for dexterous manipulations. A *master-slave manipulator system* is also such a system. In a master-slave system, an operator indicates the position of the object by the master arm, and the force exerted to the object is fed back to him through the master arm. This enables the operator to manipulate what he cannot touch directly such as a radioactive object.
If we combine these two systems into one, a single-master multi-slave robot arm system can be constructed. This system may increase its operational performance and may have more dexterity. It should be noted that the target of our approach is to control such a large scale robotic system. If the components in the total system are not complexly interacted and compose a certain hierarchical structure, it can be implemented as a concurrent processing system.
We have proposed such a hierarchical control architecture, which has three control layers:

Organizing layer It specifies and plans the objective action of the whole system.

Coordination layer It generates sub-goals for each component of the system from the objective action, and coordinates these sub-goals so that an aggregate of these sub-goals meets the objective.

Execution layer It controls the local behavior of the components to achieve the sub-goal.

Evaluation of the system

We realized a bilateral-controlled single-master multi-slave manipulator system using the proposed controller. The slave arms are 2 industrial manipulators with a force sensor of six degree of freedom, and the master arm is a pair of 2 joy sticks each of which has three degree of freedom. There are 18 actuators in the system. The overview of the robotic system is described in Fig. 8.
The transputer network consists of a host computer, 29 processors, and the peripheral interface devices.
The control frequency proved to be 400 Hertz for the lower level controller, and 200 Hertz for the higher level controller.

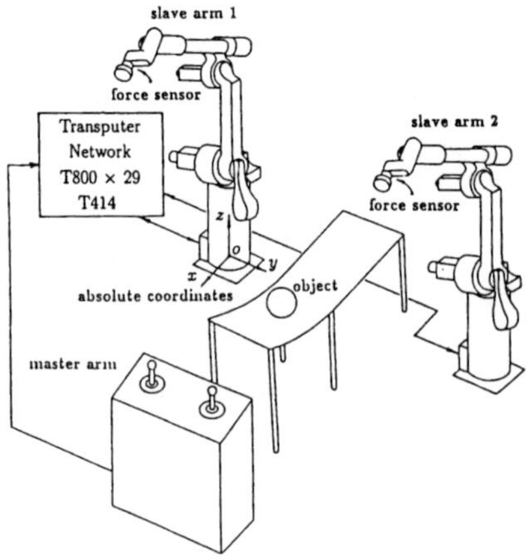

Fig. 8 Overview of the robotic system

CONCLUDING REMARKS

We described the construction of a real time controller based on a concurrent processing system. Transputers, the programming language occam and the TDS provide the flexible control system. In our approach, the modularity is enhanced to the peripheral interface system and also to the software system, which results in a complete modularity of a whole system. As a result, a high-performance concurrent real-time controller is constructed.

ACKNOWLEDGMENT

This article is written based on the research of a transputer based control system for a robotic system, which is partly supported by Nippon Steel Corporation and Nachi-Fujikoshi Corporation.

REFERENCES

Furuta, K. , K. Kosuge, Y. Shiote, and H. Hatano (1987). Master Slave Manipulator based on Virtual Internal Model Following Control Concept. IEEE Int. Conf. on Robot. and Auto. , 567-572.

Kosuge, K. , J. Ishikawa, K. Furuta, and M. Sakai (1990). Control of Single-Master Multi-Slave Manipulator System Using VIM. IEEE Int. Conf. on Robot. and Auto. , 1172-1177.

Yamakita, M. , T. Hoshino, K. Morimoto, and K. Furuta (1991). Parallel Implementation of Newton-Euler Algorithm with One Step Ahead Prediction. IEEE Int. Conf. on Robot. and Auto.

Knowles, A. and T. Kantchev (1989). Message passing in a transputer system. Microproc. and Microsys. , 13, 2, 113-123.

Barlow, M.I. , P. Konnanov, and S. E. Burge (1989). Analogue I/O strategies for transputers. Microproc. and Microsys. , 13, 6, 387-395.

Milway, M. , I. Gorton, J. Fulcher, and A. Zelinsky (1990). Interfacing transputer links to external devices. Microproc. and Microsys. , 14, 10, 644-652.

INMOS Ltd. (1988). occam2 Reference Manual. PRENTICE HALL INT. SER. IN COMPUTER SCIENCE, Prentice Hall.

INMOS Ltd. (1988). TRANSPUTER DEVELOPMENT SYSTEM. Pentice Hall.

INMOS Ltd. (1989). THE TRANSPUTER DATABOOK, first ed. INMOS document number: 72 TRN 203 00.

A MODULAR, MASSIVELY PARALLEL COMPUTER ARCHITECTURE FOR TRAINABLE REAL-TIME CONTROL SYSTEMS

Kenneth Nilsson*, Bertil Svensson*,** and Per-Arne Wiberg*

**Centre for Computer Science, Halmstad University, Box 823, S-30118 Halmstad, Sweden, **Department of Computer Engineering, Chalmers University of Technology, S-41296 Göteborg, Sweden*

E-mail:Kenneth. Nilsson@ite. hh. se,svensson@ce. chalmers. se, and Pre-Arne. Wiberg@cdv. hh. se

Abstract. A new system-architectural concept for trainable real-time control systems is presented. The concept is based on resource adequacy both in processing and communication. Cyclically executing programs in distributed nodes are communicating via a shared high-speed medium. The programs and the communication are statically scheduled by the designer, implying that the maximum possible work-load can always be handled in a time-deterministic manner. The use of Artificial Neural Network (ANN) algorithms and trainability implies a new system development strategy based on a Continuous Development paradigm. An implementation of the architectural concept is presented, in which SIMD (Single Instruction stream, Multiple Data streams)-nodes access their data from local real-time databases, updated via a shared optical link. The communication speed is measured in Gbps and the access method is TDMA. An implementation of the new system development strategy with a graphical developer's interface is also presented.

Keywords. Parallel processing; learning systems; neural nets; real time computer systems; control system design;

INTRODUCTION

The requirements for flexibility and adaptability to different circumstances and environments motivate research and development towards trainable systems rather than programmed ones. This is true particularly for "action-oriented systems", which interact with their environments by means of sophisticated sensors and actuators, often with a high degree of parallelism (Arbib, 1989). The ability to learn and adapt to different circumstances and environments are among the key characteristics of such systems. Development of applications based on action-oriented systems relies heavily on training, rather than progamming of the detailed behaviour.

Response-time requirements and demands to accomplish the training task point to massively parallel computer architectures. A system of homogeneous, highly parallel modules is foreseen. The modules perform perceptual tasks close to the sensors, advanced motoric control tasks close to the actuators, or complex calculations at "higher cognitive levels".

In this paper we introduce a new system-architectural concept for highly parallel real-time systems. It is based on the principle of resource adequacy (Lawson, 1992) in order to achieve predictability, meaning that enough processing and communication resources are designed into the system and statically allocated to guarantee that the maximum possible work-load can always be handled.

Not only do these trainable control systems require new architectural paradigms, they also require the acceptance of new system development philosophies. The traditional application-development model, characterized by a sequence of development phases, must be replaced by an interactive model based on training.

After introducing the architectural paradigm and system development model on the concept level we examplify by presenting implementations meeting the demands of typical advanced real-time control tasks. Specifically, this paper points on the possibilities based on multiple SIMD (Single Instruction stream, Multiple Data streams)arrays on which static allocation of processing tasks is made and on the power and appeal of graphical application development tools.

The work is part of REMAP³, the Real-Time, Embedded, Modular, Action-oriented, Parallel Processor Project, partly funded by STU/NUTEK, the Swedish National Board for Technical and Industrial Development, under contracts No. 9001583 and 9001585.

SYSTEM ARCHITECTURE: THE CONCEPT

We have shown, by own implementations and detailed studies, as well as by reviewing the implementation of others, that typical neural network algorithms used today map efficiently onto SIMD architectures (Nordström and Svensson, 1992). Based on this, and the discussion above, a hypothetical architecture for Artificial Neural Systems (ANSs) would look like the one shown in Fig. 1. Different modules (SIMD arrays) typically execute different Artificial Neural Network (ANN) models, or different instances of the same model. Full connectivity may be used within the modules, while the communication between modules is expected to be less intensive (although we will also devise solutions that satisfy the potential demand for tighter connections between pairs of modules).

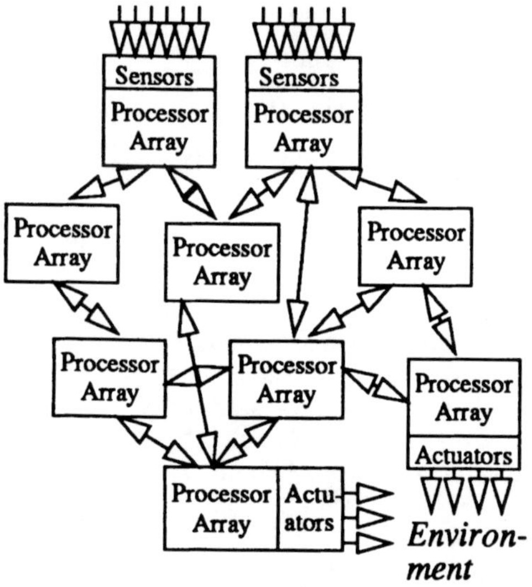

Fig. 1. A multi-module architecture for an action-oriented system

The architecture for action-oriented and other advanced, real-time control systems, which was outlined above, can be seen as an implementation of a more general architectural concept which we now present. It is based on the notions of nodes, channels, and local real-time databases.

Nodes, which differ in functionality are communicating via a *shared medium*. Input nodes deliver sensor data to the rest of the system and may perform perceptual tasks. Output nodes control actuators and may perform motoric control tasks. Processing nodes perform various kinds of calculations. I/O nodes and processing nodes may have great similarities but, because of their closeness to the environment, I/O nodes have additional circuits for interfacing to sensor and actuator signals.

Communication between nodes takes place via *channels*. A communication channel is a logical connection on the shared medium between a sending node and one or more listening nodes. The channels are statically scheduled so that the communication pattern required for the application is achieved. This is done by the designer. Two types of data are transported over the medium: *Program data* is distributed to the nodes to allow changes "on the fly" of the cyclically executed programs in the nodes. *Process data* informs the nodes about the status of the environment (including the states of other nodes). If the application requires intensive communication within a set of related nodes a hierarchical communication can be set up. The related nodes forms a cluster with more available bandwidth on the internal channels.

Rather than being individual signals, the process data exchanged between the nodes is more like patterns, often multi-dimensional. Therefore, the shared medium must be able to carry large amounts of information (Gigabits per second in a typical system).

Every node in the system executes its program in a cyclic manner. The cyclically executed program accesses its data from a *local real-time database* (LRTDB). This LRTDB is updated, likewise cyclically, via channels from the other nodes of the system.

The principle of resource adequacy, the cyclic paradigm and the statically scheduled communication via the LRTDBs imply the time-deterministic behaviour of the system which is so important in real-time applications.

SYSTEM DEVELOPMENT: THE CONCEPT

ANNs have been shown to have promising properties in making systems adaptable to new situations. Increased flexibility, adaptability, and the potential to solve some hard problems are the main reasons for introducing ANN in real-time control systems. Most implementations so far are specialized and can not stand up to industrial demands. Introducing ANN in industrial real-time control systems requires a new development philosophy, one that allows conventional control engineering and ANN principles to be mixed.

Trainability in real-time control systems

The most common development philosophy today in the domain of computer-based systems is the "sequence of phases" strategy. The development project is divided into a finite number of phases, e.g.: Analysis, conceptual design, detailed design, implementation, and testing. This is often referred to as the waterfall model (Fig. 2).

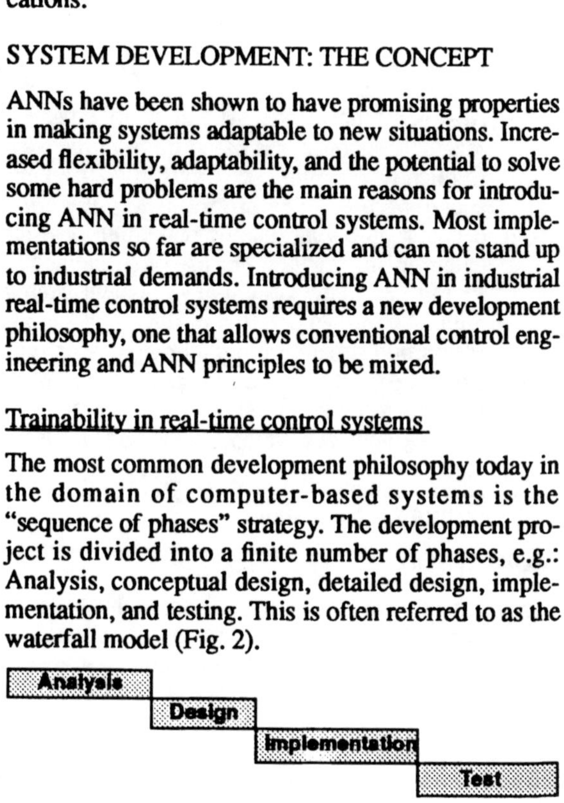

Fig. 2.The waterfall model.

The sequence of phases is no longer relevant when trainable systems are to be developed. A trainable ANN system can be considered as having two parts: structure and data. The structure is the ANN algorithms and the hardware architecture. The data is the information that the system gets from the environment and the stored information that yields the behaviour of the system (e.g., the weights). The structure is in most cases static in the sense that it is not changed by the system itself. The stored information can be static after a training session or dynamic meaning that the environment constantly influences the system's behaviour.

The dynamic system-behaviour calls for a new development model. The analysis activity has similarities to the waterfall model in sorting out the demands on the system, but turning these demands into, e.g., functions or objects is not relevant here. In contrast to programmed systems the main design task is to determine an adequate set of ANN-algorithms and a computer architecture. This does not give the system its function, which is an important difference to conventional systems. The function of the system is given by training, either in a special training session or by running the system in its proper environment.

To describe development of trainable systems we introduce a *circular development model* (see Fig. 3).

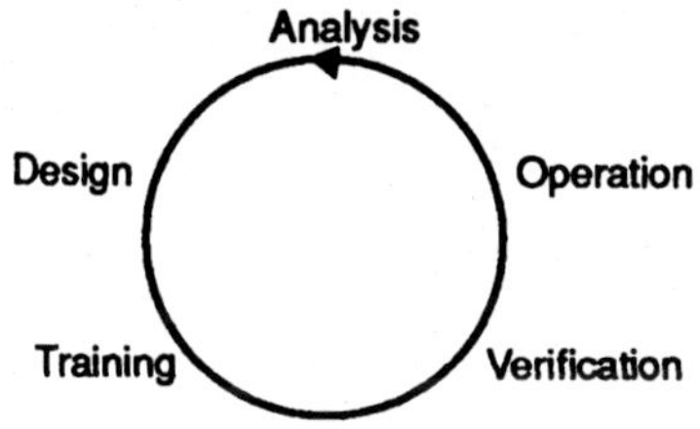

Fig. 3. The circular development model.

In contrast to the waterfall model, where system development is considered as a project and maintenance as a process, the circular development model incorporates development and maintenance as two activities in the same process. The parts of this process are:.

Analysis. Each instance of this activity handles a portion of the demands that the system is to fulfil.

Design. To meet the demands, existing algorithms are tested/modified or new ones developed. This design style can be compared to rapid prototyping to encourage the creativity of the developer. The activity leads to a structure which includes ANN-algorithms and conventional control algorithms.

Training. When the structure of the system is updated, the system is given its new properties by exposing it to environment data or a set of training data. Training may be a part of the operation activity but can also be a separate activity succeeded by verification.

Verification. In most cases the updated trained structure of the system has to be verified before letting it influence the environment. In this activity the developer can use own data or data from the environment and structures dedicated to verification.

Operation. There is no sharp distinction between operation and other activities. The behaviour of the system might change constantly during operation due to adaptation. The system might have only a fraction of its functionality implemented but still be a good test-bench for Analysis, Design and Training.

The Continuous Development Paradigm.

To support the development model we introduce the Continuous Development Paradigm (CD-paradigm). This paradigm can be expressed as "Development by changing and adding". This is a well-known approach in modern Software Engineering but here the aims are extended to include both hardware and software. A development environment to support the use of the CD-paradigm should share the following characteristics:

- Easy to change the system structure (hardware and software) and data "on the fly".
- Incremental Development using the running system as development platform.
- Operation of the running system is guaranteed.
- System data and structures can be viewed with emphasis on understandability.
- Developer gets immediate response to a change of the system.
- Developer can use concepts and symbols of the application domain.

In a later section we give an implementation based on the CD-paradigm.

After having introduced both system architecture and development strategy on the conceptual level we now present concrete implementations.

SYSTEM ARCHITECTURE: IMPLEMENTATION

We present a multiple SIMD-module architecture in which intermodule communication takes place over a high-speed medium shared by all nodes.

SIMD Node

A basic SIMD processor array is outlined in Fig. 4. We have performed detailed studies of the execution of the predominant ANN models on this kind of computing structures (Svensson and Nordström, 1990; Nordström, 1991a; Nordström, 1991b). A major conclusion is that broadcast or ring communication among the Processing Elements (PEs) of the array can be very efficiently utilized and actually provides the necessary means for communication within the array. Multiplication is the single most important operation in ANN computations. In bit-serial architectures, which have been our primary target, there is therefore much to gain if support for fast multiplication is added. In some of the ANN models, for example Sparse Distributed Memory, tailored hardware to support specific PE operations pays off well.

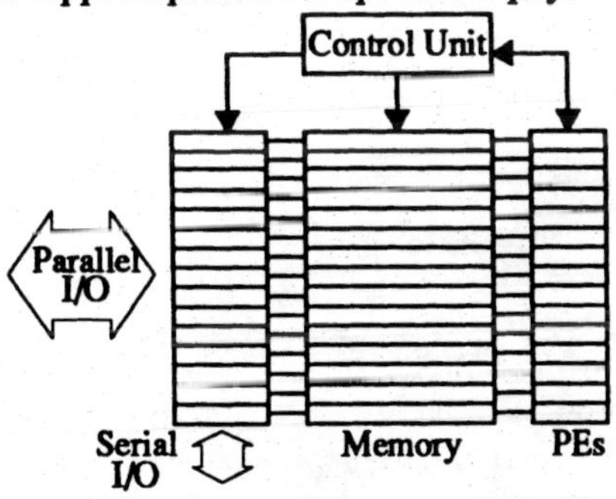

Fig. 4. SIMD Node

In the REMAP[3] project PE arrays along these lines are being developed (Bengtsson et al., 1991). A typical module (implemented as a few chips using common state-of-the-art technology) is a 256 PE array of bit-serial processors incorporating bit-serial multipliers. Such an array is capable of training at 90 MCUPS (Million Connection Updates Per Second) or recall at 160 MCPS (Million Connections Per Second) using 8-bit data at 25 MHz. A four-layered feedforward network with 256 neurons per layer would run at the speed of 340 training examples or 800 recall examples per second.

The architecture permits two or more modules to be linked together to form a larger module, if necessary.

This linking may be done either over the communication medium, in which case the intermodule communication shares time with all modules of the system, or over a separate medium. In the latter case the cooperating modules form a *cluster* with more available bandwidth for internal communication. Special "dual-port" nodes form the interface between the cluster and the main medium.

Communication

An all-optical network (the entire path between end-nodes is passive and optical) is used as the shared medium. Such a network has high bandwidth (25,000 GHz) and low bit-error rate (10^{-15}) (Green, 1991).

Communication channels between SIMD Nodes are established by time-multiplexing (TDMA) in a statical manner. In every scheduled time slot there is one sender and one or more listener (broadcast). In a time slot, which is one clock-cycle (and also the smallest time-unit) in the SIMD Module, a communication package is sent. A communication package is a bit-slice of a SIMD Module, e.g. 256 bits.

Broadcast implies that it is important to synchronize the communication very hard. The synchronization is done via a global distributed optical clock which gives a high clock-rate and an accurate synchronization of the nodes (12ps) (Electronic Design, 1991). Alternatively, a communication slot can be several time slots (Fig. 5), which gives a slower communication speed. For example, if one communication slot is three time slots (one ordinary and two security), a synchronization accuracy of 256 ns is required (assuming a communication package of 256 bits and a communication speed of 1 Gbps) which means about 50 meters difference in clock-path.

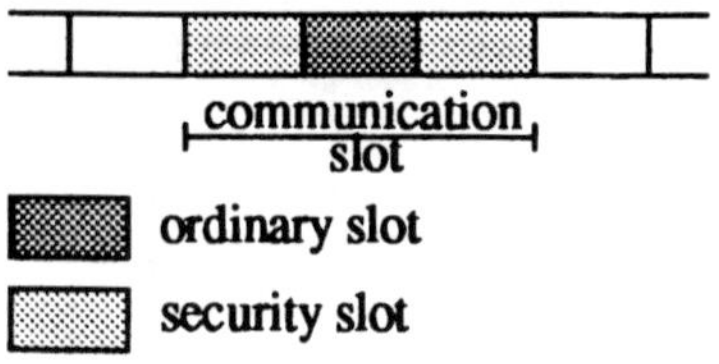

Fig. 5. A communication slot

In the communication interface of each SIMD Node (Fig. 6) a clock frequency reduction is done by a factor k by means of shiftregisters (k is the size of the PE array, e.g. k=256). It is important to synchronize the dataflow (Gbps) with the shift clock. This is done by sending the clock and the data in the same medium. Clock and data use different wavelengths (f_1 and f_2), implying that the communication interface must include two laserdiodes and two optical filters (F) for the flow of process data.

In addition to the exchange of process data there is also a distribution of program data caused by program changes made "on the fly". The program data flow has its own all-optical network in the same cable system. Access to this is managed by the Development Node in the system.

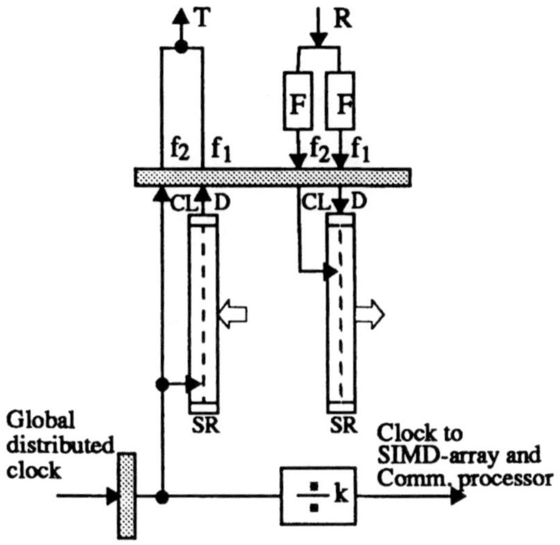

Fig. 6. Communication interface

Due to the high speed the communication interface must be integrated into one IC to work properly. Today there are shiftregisters available implemented in GaAs-technology for very high speed (some Gbps). The GaAs-technology also gives the possibility to integrate optical devices with logic. The topology of the all-optical network is a star,which has a decibel loss proportional to logN, while a bus topology has one proportional to N (N is the number of nodes in the system) (Green, 1991).

The SIMD Module accesses data from its own local real-time database (LRTDB), which reflects the status of the environment. The LRTDB is implemented as a dual-port memory. At one side the SIMD Module accesses data; at the other side the control unit in the communication module is updating the LRTDB via the communication interface. The control unit cyclically executes the statically scheduled send and receive commands necessary for carrying out the communication pattern of the node (Fig. 7).

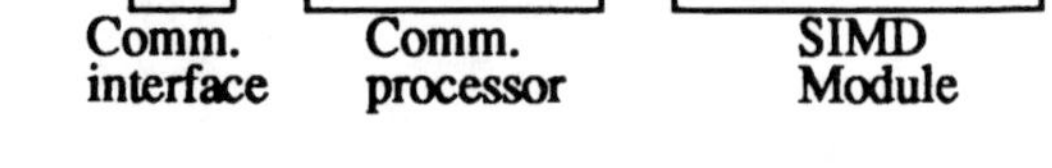

Fig. 7. A Node

SYSTEM DEVELOPMENT: IMPLEMENTATION

The implementation of the system development tool is based on the CD-paradigm described earlier. First a list of the most important features:

- Graphical developer's interface.
- Cyclic execution with temporal deterministic behaviour.

- Dynamic change of the running software.
- Dynamic inspection/change of data "on the fly".
- Change of the distributed hardware "on the fly".
- Use of symbols and concepts from the domain of control engineering and ANN.

As stated earlier the distributed nodes have cyclically executing programs. These cycles have two parts: the Monitor and the Work Process. The Monitor has the following functions:

- Start on a given time (a new *dt* has passed).
- Copy output data from the LRTDB to the communication buffer.
- Copy input data from the communication buffer to the LRTDB.
- Handle program changes.

A temporal view of the execution of one cycle is shown in Fig. 8, where the different paths of the Work Process are indicated. Continuous lines indicate processing that consumes time, dotted lines show idle processing, and lines splitting up means a selection in the control flow. The development tool guarantees that the worst case branch is within the cycle time, *dt*.

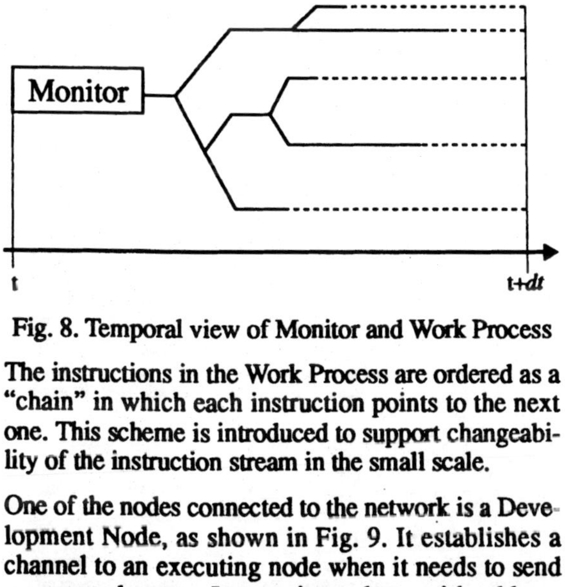

Fig. 8. Temporal view of Monitor and Work Process

The instructions in the Work Process are ordered as a "chain" in which each instruction points to the next one. This scheme is introduced to support changeability of the instruction stream in the small scale.

One of the nodes connected to the network is a Development Node, as shown in Fig. 9. It establishes a channel to an executing node when it needs to send program changes. Instructions along with address information are sent to the executing node where the monitor makes the change between two executions of the Work Process.

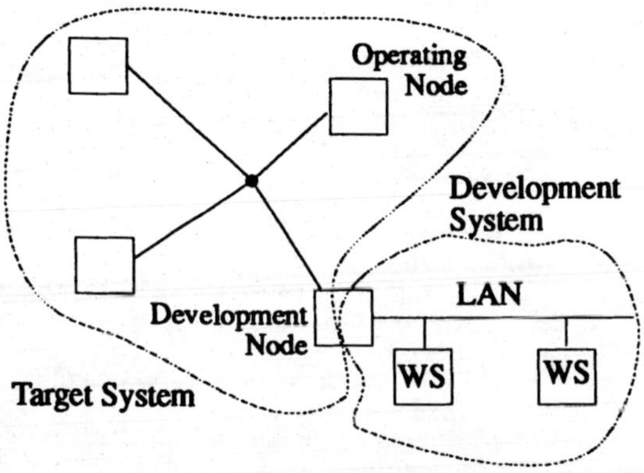

Fig. 9. Multi-node target system and multiple-workstation development system

The Development Node is connected to a Local Area Network (LAN) of workstations (WS) running the development system. The LAN connection can be removed without affecting the running system.

For inspection of the LRTDB and other local data the Development Node opens channels in the same way as when other process data is moved between nodes.

<u>A graphical developer's interface</u>

To support the CD-paradigm and demands of understandability the developer's interface to the system is an interactive graphical tool. The most basic properties of the tool are outlined below.

- *All development is done on a system in operation.* That is, a system operating in real-time but not necessarily affecting the system environment.
- *Hierarchical way of describing the application.* The levels of abstraction span from the instruction set of the node control unit to the abstract concepts of the application.
- *Support for reuse of system components.* Part of the tool is a browser where system components (processes, data, and connections) are stored.
- *Tools for viewing data.* Data can be viewed in various ways. The possibilities explored in this implementation are directed towards the domains of control engineering and ANN. In control engineering an often used datatype is the signal, i.e. a sequence of values. To support this the tool has bargraphs, diagrams, numeric presentation, and conditional recording. ANNs often handle patterns rather than single signals. To support this the designer can work with maps. A map is a two-dimensional array of activity values, which is viewed as a gray-scale image. Sequences of maps can be recorded on specified conditions.

<u>Symbols</u>

The symbols that the designer works with can vary, but a set of basic symbols, shown in Fig. 10, are always used. The *process* symbol is composed of an input data buffer, a processing box, and an output data buffer. The data buffers are shown with *data* symbols. The three parts are different icons and can be opened to windows for manipulation. The distance between the two data buffer icons indicates the time consumption of the process relative to a time scale in the Work Area on the screen.

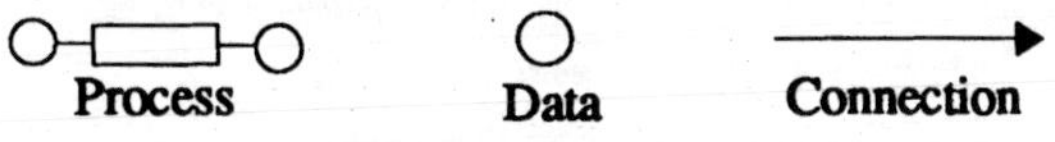

Fig. 10. Basic symbols.

The data in the system is statically typed and is built up from basic types (BOOLEAN, INTEGER, etc.) by the designer. This means that the tool can be equipped with the data types that fit the application; for the ANN-developer this can mean MATRICES and LAYERS. The *connection* symbol indicates the dataflow in a process. A connection has a data type that conforms to the connected data buffers. The control flow is given by the placement of the symbol in the Work Area.

System level.

On system level the developer sees a screen as outlined in Fig. 11. The arcs show the data exchange (i.e. the channels). The process symbol indicates a cyclic process running on a node. Here at the system level, the placement of the process symbol has no temporal meaning as it has at the node level and lower. Input and output data buffers are the input and output parts, respectively, of the LRTDB. By using the mouse, the developer can open up a process symbol and by that move to the node level.

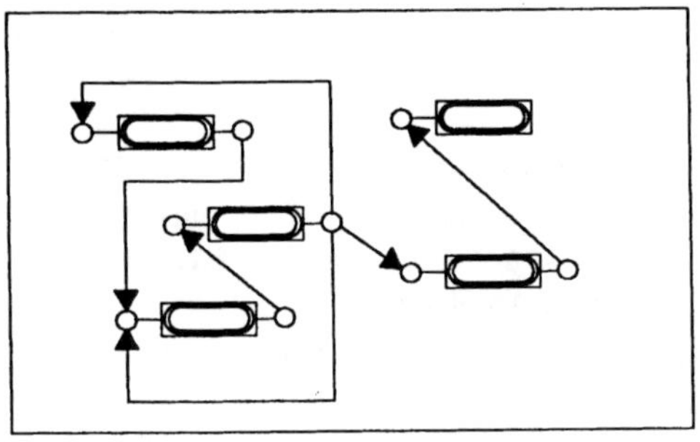

Fig. 11. System level display.

Node level and lower levels.

An example of a screen display at the node level is shown in Fig. 12. "I" is an icon that represents the input part of the LRTDB of the node. "O" is the output part, and WA is the Work Area where the designer can place symbols that specify the operation of the node. The placement of symbols in WA has temporal meaning relative to a time scale T that indicates the total time of the process. Every symbol in WA can be opened to move the designer one level of abstraction lower in the system hierarchy. When the designer places a symbol in WA, using the browser, the corresponding process will be added to the execution thread. The designer can then immediately use the inspection tools to verify the function of the added process. This is indicated in Fig. 12

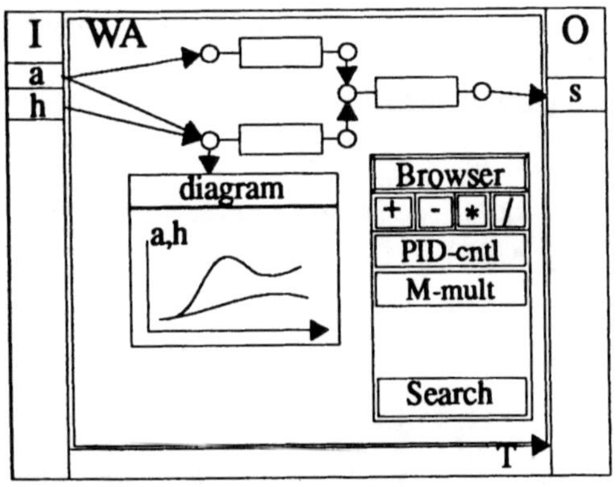

Fig. 12. Node level (or lower levels) display

DISCUSSION AND CONCLUSION

In control applications the security aspect is often emphasized. Letting ANN-based systems act on the environment without special precautions could lead to severe problems. A possible solution is to have a "security shell" which gives limits for the outputs from the ANN algorithms.

In the paper, TDMA is used as a simple implementation of the communication concept. If higher capacity is needed, WDMA (wavelength division multiple access) may be used instead. Then, scheduling of communication is not required. The nodes scan the wavelength spectrum to fill their LRTDBs. The scanning can be statically determined, trained, or a function of the internal state of the node.

This paper points to the strength of combining massively parallel architectures, trainability, and incremental development environments. The SIMD paradigm combines single-threaded programming with multiprocessing power and easy miniaturizing for embedded systems. We have presented a massively parallel system architecture based on multiple SIMD processor arrays to allow the implementation of real-time, ANN-based training using interaction-based system development tools.

The presented system architecture and development model are intended to be used in biologically inspired design of control systems (Kuperstein, 1991; Singer, 1990), where sensory, motoric, and higher cognitive functions are mapped onto nodes or clusters of nodes.

REFERENCES

Arbib, M.A. (1989). Schemas and neural networks for sixth generation computing. *J. Parallel and Distributed Computing*, Vol. 6, No. 2, pp. 185-216.

Bengtsson, L., A. Linde, T. Nordström, B. Svensson, M. Taveniku, and A. Åhlander (1991). Design and implementation of the REMAP³ software reconfigurable SIMD parallel computer", *Fourth Swedish Workshop on Computer Systems Architecture*, Linköping, Sweden, January, 1992. Available as Research Report CDv-9105 from Centre for Computer Science, Halmstad University, Halmstad, Sweden.

Electronic Design (1991). Technology Newsletter, March 28, p. 26.

Green, P.E. (1991). The future of fiber-optic computer networks. *Computer*, Vol. 24, No. 9, pp. 78-87

Kuperstein, M. (1991). INFANT neural controller for adaptive sensory-motor coordination. *Neural Networks*, Vol. 4, pp. 131-145.

Lawson, H.W., with contributions by B. Svensson and L. Wanhammar (1992). *Parallel Processing in Industrial Real-Time Applications*. Prentice-Hall, Englewood Cliffs.

Nordström T. (1991a). Sparse distributed memory simulation on REMAP³. Research Report No. TULEA 1991:16, Luleå University of Technology, Luleå, Sweden.

Nordström T. (1991b). Designing parallel computers for self organizing maps. Research Report No. TULEA 1991:17, Luleå University of Technology, Luleå, Sweden.

Nordström T. and B. Svensson (1992). Using and designing massively parallel computers for artificial neural networks. *J. Parallel and Distributed Computing*, Vol. 14, No. 3, pp. 260-285.

Singer, W. (1990). Search for coherence: a basic principle of cortical self-organization. *Concepts in Neuroscience*, Vol. 1, No. 1, pp. 1-26.

Svensson B. and T. Nordström (1990). Execution of neural network algorithms on an array of bit-serial processors. *Proceedings of 10th Internatinal Conference on Pattern Recognition – Computer Architectures for Vision and Pattern Recognition*, Atlantic City, NJ, USA, June 1990, Vol. II, pp. 501-505.

PARALLELIZED INVERSE DYNAMICS ALGORITHM FOR DYNAMIC CONTROL OF A ROBOT MANIPULATOR

Seungdo Kim, Hyung Kyi Yi, C. L. Roh and M. J. Chung

Dept. of Electrical Engineering,
Korea Advanced Institute of Science and Technology
373-1, Kusong-Dong, Yusong-Gu, Taejon, Korea 305-701

Abstract Real-time processing in robot dynamic control is known to be difficult because of heavy computational burden. To overcome this difficulty, In this paper, we propose a parallelized robot inverse dynamics algorithm by reconstruction of He's algorithm and implement a robot control system using transputers to show the validity of proposed algorithm for real-time control. Three transputers are used to build a high performance concurrent system and Occam language is used for programming the whole system. For the three axes manipulator, the actual computation time for executing the complete inverse dynamics is within 900 microseconds.
Keywords parallel processing, inverse dynamics, dynamic control

INTRODUCTION

Conventional industrial robots have been controlled with modified PD or PID control algorithm for following the desired trajectories. But, with these control algorithms we can not get a desired performance when the robot moves in high speed or when payload varies widely. Therefore, we need a dynamic control algorithm considering the robot dynamics such as nonlinearities, configurations, and torque characteristics (Yoo, 1991). Since the computational time to solve robot dynamics is very large, fast processor and/or parallel processing technique are needed. Recent advances in VLSI technology and parallel processing technique make it possible to use high performance chips with parallel processing architecture for real-time robot dynamic control.

Many researchers have proposed robot dynamics algorithms with reduced computation and parallel processing technique for real-time control. These may be summed up as the following points. Firstly, the number of subtasks is reduced as much as possible and subtasks are distributed properly to each processors. Without interaction with different processors (Luh, 1984; Kasahara, 1985), these methods will have an optimum performance. But, as robot dynamics algorithm based on Newton-Euler method has many communications due to sequential data dependency, it decreases the performance of parallel processing system. Secondly, new hardware architectures for the parallel algorithms are proposed (Nigam, 1985; Lathrop, 1985). Generally, these methods are difficult to implement and expensive. Thirdly, a few reseachers implemented a parallel algorithm using shared bus for interprocessor communication. But these methods have a defect of communication bottleneck (Wang, 1987). In this papaer, we use 4 transputers to implement parallel processing algorithm. We assign one axis to one processor for calculation of dynamics. And a network of transputers and peripheral controllers is constructed using point-to-point communication. To gain most benefit from the transputer architecture, the whole system is programmed in Occam which is a high level language for concurrent applications.

ROBOT INVERSE DYNAMICS

For dynamic control of a robot, we should first calculate the robot inverse dynamics to obtain desired torques at each joints. Various forms of motion equations describing a robot arm dynamics have been obtained from the Lagrange-Euler (L-E) and the Newton-Euler (N-E) formulations (Fu, 1987).

These algorithms are to reduce the number of arithmetic operations to calculate robot inverse dynamics. Table1 shows arithmetic operations of each algorithm of robot dynamics.

Table1 Arithmetic Operations of Robot Dynamics.

Method	Multiplication	n = 6	Addition	n = 6
Hollerbach	412n - 227	2195	320n - 201	1719
Luh,Walker,Paul	150n - 48	852	313n - 48	738
Wang,Ravani	165n - 33	903	113n - 24	654
Walker,Orin	137n - 22	800	101n - 11	595
McInnis,Liu	134n - 47	757	113n - 38	640
Kazerounian,Gupta	136n	816	118n	708
Khalil,Kleinfinger,Gautier	105n - 92	816	94n - 86	478
Cyril,Angeles,Misra	100n - 107	493	88n - 90	438
Balafoutis,Misra,Patel	93(n-1)+ 24	489	90(n-1) + 12	462
He,Goldenberg	91(n-1) - 6	449	86(n-1) - 10	420

He and Goldenberg (He, 1990) have proposed an efficient algorithm for robot dynamics using the concepts of "augmented body" and "barycenter". The utilization of "augmented body" makes it possible to increase the number of off-line calculations and decrease the number of on-line calculations. Also, barycenter concept makes it possible to compute the joint torques of revolute joints without calculating the joint forces. This algorithm has assumtions that the inertia matrix of each link are diagonal and all of robot joints are revolute type. He's algorithm of robot dynamics is depicted as following.

Off-line computation

$$M = \sum_{j=1}^{n} m_j$$

$$m^*_i = \sum_{j=i}^{n} m_j = m^*_{i+1} + m_i$$

$${}^i\bar{b}_{io} = Mb_{io} = m_i d_{ii} + (m_{i+1} + m_{i+2} + \cdots + m_n) d_{ii+1}$$

$${}^iK_i = {}^iJ_i - m_i {}^i\tilde{d}_{ii} {}^i\tilde{d}_{ii} - m_i {}^i\tilde{d}_{ii+1} {}^i\tilde{d}_{ii+1}$$

$${}^i\bar{K}_i = {}^iK_i - \frac{1}{2}(K_{xx} + K_{yy} + K_{zz})\, I$$

On-line computation

Forward Iteration

$${}^i\omega_i = A_i^T\, {}^{i-1}\omega_{i-1} + \dot{q}_i {}^iz_i$$

$${}^i\dot{\omega}_i = A_i^T\, {}^{i-1}\dot{\omega}_{i-1} + \ddot{q}_i {}^iz_i + {}^i\tilde{\omega}_{i-1}\dot{q}_i\, {}^iz_i$$

$${}^i\Omega_i = {}^i\tilde{\dot{\omega}}_i + {}^i\tilde{\omega}_i {}^i\tilde{\omega}_i$$

$${}^id^*_i = A_i^T({}^{i-1}d^*_{i-1} + {}^{i-1}\Omega_{i-1} {}^{i-1}d_{i-1i})$$

$${}^iu_i = \begin{bmatrix} {}^i\Omega_{yz} {}^i\bar{K}_{zz} - {}^i\Omega_{zy} {}^i\bar{K}_{yy} \\ {}^i\Omega_{zx} {}^i\bar{K}_{xx} - {}^i\Omega_{xy} {}^i\bar{K}_{zz} \\ {}^i\Omega_{xy} {}^i\bar{K}_{yy} - {}^i\Omega_{yx} {}^i\bar{K}_{xx} \end{bmatrix}$$

Backward Iteration

$${}^ib^*_i = A_{i+1} {}^{i+1}b^*_{i+1} + {}^i\Omega_i {}^i\bar{b}_{io}$$

$${}^in_i = A_{i+1} {}^{i+1}n_{i+1} + {}^iu_i + {}^i\tilde{d}_{il}({}^ib^*_{i+1} + {}^iP^*_{i+1}) + {}^i\tilde{\bar{b}}_{io} {}^id^*_i$$

$$\tau_i = {}^iz_i {}^in_i$$

PARALLELIZATION OF ROBOT INVERSE DYNAMICS

A mehtod of parallization should be different from the characteristics and architectures of parallel processing system. Previous works for parallel algorithm of robot dynamics used heuristic approaches to load distribution to each processors. In general, there should be three considerations for parallelization, such as

1. a sequence of subtask for data dpendency
2. interprocessor communication
3. external I/O bottleneck.

In parallel processing system, a task has data dependency with relation to the computational sequence of subtask. Data dependency means that a priory data is needed for computing current subtask. Thus, if we neglect the sequence, it will lead to dead-lock which is state of waiting for data transfer from each other processor. Due to the characteristics of microprocessor, external operations are more time-consuming than internal operations. Accessing the outer devices such as external memory or I/O devices frequently, the performance of parallel processing system decreases. If there is no communication between each processors, then previous works will be valid for real applications. But, until now, the ideal parallel processing system has not been developed, and the main factors for performance limitation of parallel processing system are the communication time between each processors.

Therefore, the more interporcessor communication burden is reduced, the better system performance is acquired. In this paper, we put the main emphasis on the communication and use the following strategies.

- *One processor computes for one axis.*

N-E method has complex data dependency. If one processor is assigned to one axis, then communication burden will be reduced as each processor uses its own data. Fig.1. shows the parallelization of He's dynamics algorithm by above philosophy.

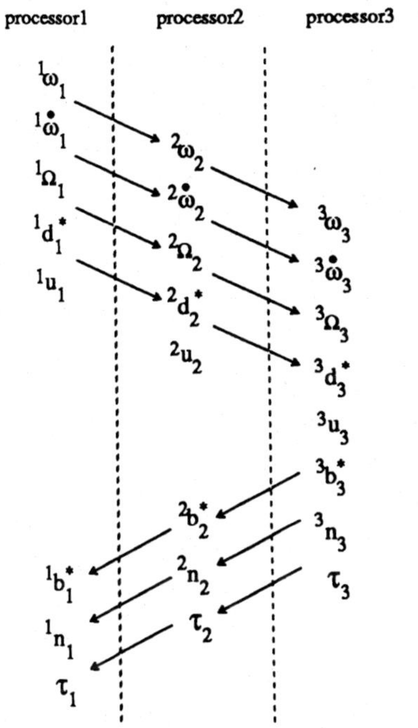

Fig.1. Data dependency of parallel dynamics.

- *Data communication happens aperiodically.*

In a periodic communication, all the processors transfer data at the same time. Therefore, the communication time must be decided according to a subtask with the largest computation time. As a result, the waiting time increases and the performance gets worse. So, we need a kind of processor that has communication port with handshake capability for checking the other's current state.

- *Computation and communication process is executed in parallel.*

This method is related to the chracteristics of communication port. Parallelization of computation and communication reduces communication time. Therefore, processors compute robot dynamics independently from the communication port.

- *Make the size of subtask as large as possible.*

In general, commmunication is more time consuming than computation. So, we changed from communication to computation and reduced total computational time. In the above case, ${}^{i-1}\Omega_{i-1}$ and ${}^{i-1}d^*_{i-1}$ requires data transfer of 3×3×4 bytes but product of the two requires 3×1×4 bytes. Byte communication takes triple time to floating-point multiplication.

From these strategies, we propose 4 methods of robot dynamics parallelization as follows.

1) implementing parallel algorithm of He's robot dynamics by mapping one axis to one processor with periodic communication.

2) same as 1) except with aperiodic communication.

3) same as 2) and also with parallelism of communication and computation.

4) same as 3) but with reduced communication and enlarged subtask.

Taking the fourth method, we propose a parallel algorithm based on the reconstruction of He's robot dynamics. It is depicted in Fig.2.

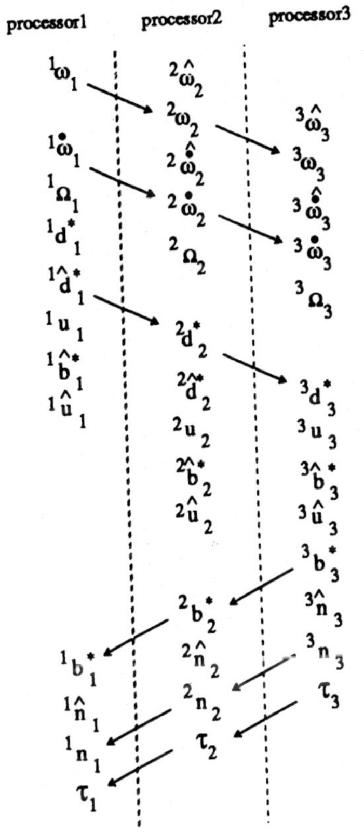

On-line Computation

Forward Iteration

$${}^i\hat{\omega}_i = \dot{q}_i\,{}^i z_i$$

$${}^i\omega_i = A_i^T\,{}^{i-1}\omega_{i-1} + {}^i\hat{\omega}_i$$

$${}^i\hat{\dot{\omega}}_i = \ddot{q}_i\,{}^i z_i + {}^i\bar{\omega}_{i-1}\dot{q}_i\,{}^i z_i$$

$${}^i\dot{\omega}_i = A_i^T\,{}^{i-1}\hat{\dot{\omega}}_{i-1}$$

$${}^i\Omega_i = {}^i\tilde{\dot{\omega}}_i + {}^i\tilde{\omega}_i\,{}^i\tilde{\omega}_i$$

$${}^i d^*_i = A_i^{T\,i-1}\hat{d}^*_{i-1}$$

$${}^i\hat{d}^*_i = {}^{i-1}d^*_{i-1} + {}^i\Omega_i\,{}^i d_{ii+1}$$

$${}^i u_i = \begin{bmatrix} {}^i\Omega_{yz}\,{}^i\bar{K}_{zz} - {}^i\Omega_{zy}\,{}^i\bar{K}_{yy} \\ {}^i\Omega_{zx}\,{}^i\bar{K}_{xx} + {}^i\Omega_{xy}\,{}^i\bar{K}_{zz} \\ {}^i\Omega_{xy}\,{}^i\bar{K}_{yy} + {}^i\Omega_{yx}\,{}^i\bar{K}_{xx} \end{bmatrix}$$

$${}^i\hat{b}^*_i = {}^i\Omega_i\,{}^i\bar{b}_{io}$$

$${}^i\hat{u}_i = {}^i u_i + {}^i\tilde{\bar{b}}_{io}\,{}^i d^*_i$$

Backward Iteration

$$ {}^{i}b^{*}_{i} = A_{i+1}{}^{i+1}b^{*}_{i+1} + {}^{\hat{b^{*}}}{}_{i} $$

$$ {}^{i}\hat{n}_{i} = {}^{i}\hat{u}_{i} + {}^{i}\tilde{d}_{il}{}^{i}b^{*}_{i+1} + {}^{i}\bar{\tilde{b}}_{io}{}^{i}d^{*}_{i} $$

$$ {}^{i}n_{i} = A_{i+1}{}^{i+1}n_{i+1} + {}^{i}\hat{n}_{i} $$

$$ \tau_{i} = {}^{i}z_{i}{}^{i}n_{i} $$

Fig.2. Proposed algorithm for parallel processing.

TRANSPUTER AND PARALLEL PROCESSING SYSTEM

Transputer T800 developed by INMOS is 32bit microprocessor. With the powerful communication capability, it can be aptly used in parallel processing systems. Transputer has on chip 4kbyte internal memory, four bidirectional communication links, floating point units, and a timer. It performs 10 MIPS (millions of instructions per second) and 1.5 MFLOPS (millions of floating point operations per second). The communication link is serial with 20 Mbit per second. It is possible to commnunicate directly between processors. Not only a parallel processing system is made easily with connection of serial links, but also internal devices of transputer operate in parallel (Homewood, 1987).
Transputer systems can be designed and programmed using the Occam language. The Occam is high level language developed for programming concurrent, distributed systems. It is simple, block structured, and also supports sequential (SEQ), parallel (PAR), and alternative (ALT) primitive processors on one or more transputers.
The parallel processing system for calculation of robot inverse dynamics is shown in Fig.3.

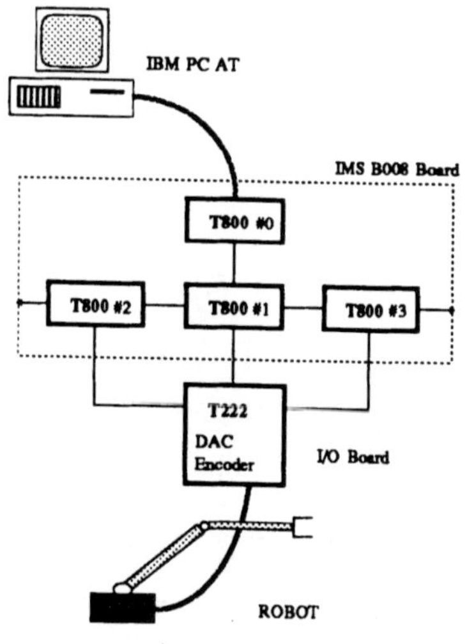

Fig.3. Parallel processing system.

IBM-PC/AT is used as a host computer that has some developement tools for transputer system for programming and debugging. Transputer #0 compiles Occam source, interfaces the host and other transputers, and supervises parallel processing system. Transputers #1,#2,#3 compute the robot dynamics and control the robot. T222 is a transputer with 16 bit and it is used only for interface with robot. This computes current robot position from the encoder readings then distributes it to each processor and relays outputs of robot to transputers. We used the first three axes of a robot arm, Movemaster RV-M1 that has 5 revolute joints.

EXPERIMENTAL RESULT

Table2 shows the computational time required for the proposed algorithm.

Table2 Computation Time for each Algorithms.

Method	Computation time
One processor	2439 usec
Case 1	1147 usec
Case 2	1036 usec
Case 3	930 usec
Case 4	883 usec

From this result, we see that the computation time of one processor is about 2.5 msec for dynamics of 3 axes revolute manipulator. If all of them are computed in parallel, the robot dynamics can be computed in 1msec. The modified robot dynamics algorithm can reduce this to 0.9 msec. Thus, reducing communication is proved to be effective for parallel processing system. The parallelism of communication and computation gives also performance enhancement. Table3 shows the performance of these algorithms.

Table3 Performance of each Algorithms.

Method	relative error	speed up	efficiency
1	0.43	2.12	0.70
2	0.29	2.35	0.78
3	0.16	2.62	0.87
4	0.10	2.76	0.92

With proposed algorithm and parallel processing system, we executed dynamic control to the robot. Desired trajectory was a circle movement with 7 cm diameter to impose some effect of robot dynamics. Fig.4. shows the result of dynamic control by feedforward method to three joints of robot and Fig.5. shows the result of proportional-derivative(PD) control after fine gain tuning. From these figures, we can conclude that the dynamic control gives better performance than PD control.

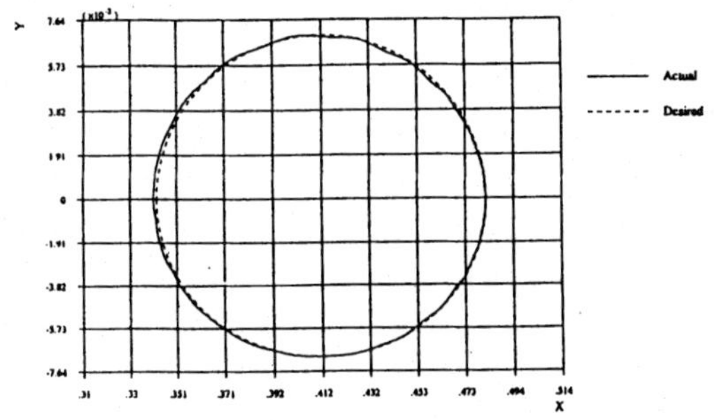

Fig.4. Feedforward control

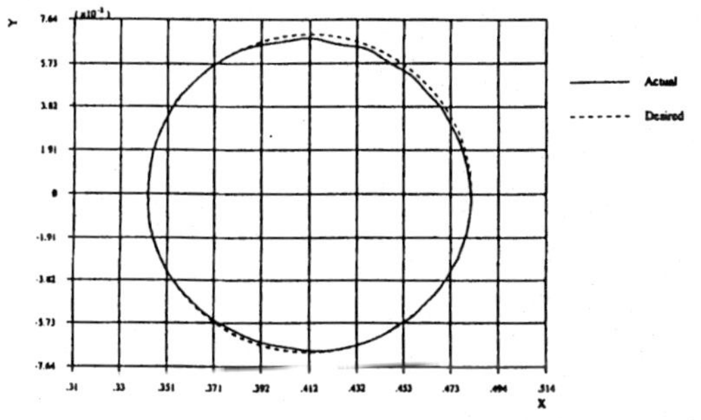

Fig.5. PD control

CONCLUSION AND FURTHER STUDIES

A parallel processing dynamic controller was implemented and applied to real robot with a desired control period of less than 1msec. The result shows it took 0.9 msec in computing the dynamics of 3 DOF robot moving in a planar circle path. In parallelizing the robot dynamics, it was found, the reduction in communication between the processors was more effective than reducing the computation burden on each of them. As a result, the performance of this controller can be improved approximately in proportion to the communication capacity. Even in heavy computation applications, this communication reduction approach can be applied successfully. As the internal gears of the robot used in the experiment absorbed quite sum of the dynamics effects, the proposed dynamic controller showed not so significant improvement with respect to the conventional PID controller. The direct drive manipulators are considered to be more appropriate objects for this controller, and the further study will be carried out on them.

REFERENCE

Dong Sang Yoo, Myung Jin Chung and Zeungnam Bien (1991). "Real-Time Implementation and Evaluation of Dynamic Control Algorithms for Industrial Manipulators", IEEE Trans. on Industrial Electronics Vol. 38, No. 1, pp. 26-31.

Xiaogene He and A. A. Goldenberg (1990). "An Algorithm for Efficient Computation of Dynamics of Robotic Manipulators", Jour. of Robotic Systems Vol.7, No.5, pp. 698-702.

J. Y. S. Luh and C. S. Lin (1984). "Scheduling of parallel computation for a computer-controlled mechanical manipulator", IEEE Trans. on SMC Vol. SMC-12, No 2, pp. 214-234.

Hironori Kasahara and Seinosuke Narita (1985). "Parallel Processing of Robot-Arm Control Computation on a Multimircoprocessor System", IEEE Jour. of Robotics and Automation Vol. RA-1, No. 2, pp. 104-113.

R. Nigam, C. S George Lee (1985). "A Multiprocessor-Based Controller for the Control of Mechanical Manipulators", IEEE Jour. of Robotics and Automation Vol. RA-1, No. 4, pp. 173-182.

R. H. Lathrop (1985). "Parallelism in Manipulator Dynamics", Inter Jour. of Robotics Research Vol.4, No.2, pp. 80-102.

Y. Wang and S. E. Butner (1987). "A New Architecture for Robot Control", Inter Conf. of Robotics and Automation pp. 664-670.

M. Homewood, D. May, and D. Shepherd, and R. shepherd (1987). "The IMS T800 Transputer", IEEE MICRO October.

K.S.Fu, R.C. Gonzalez and C.S.G.Lee (1987). "Robotics - Control, Sensing, Vision and Intelligence", Mcgraw-Hill

Zomaya A.Y. and A. Morris (1992). "Parallel Computation of robotic inverse dynamics for high speed motions", IEE Proceeding-D Vol.139, No. 2, March.

ROBOT POSITION CONTROL BASED ON A TRANSPUTER NETWORK WITH A CONTACTLESS SENSOR

K. H. Kim, P. Lauer and U. Kunz

Institute of Power Electronics and Electrical Drives
University of Siegen, Hölderlinstr. 3, D-5900 Siegen, Germany

Abstract : At the University of Siegen, a fast robot control is developed, which is completely based on a transputer network. This approach was made by interfacing each transputer directly to its associated drive. Every axis is controlled in a closed loop by its own transputer. The one-dimensional sensor is mounted on the end-effector and is also equipped with its own transputer, which is connected with the transputer-links to the robot control. In this contribution will be described the transputer network and a sensor system. The robot used is of the type Manutec r3. The control system is programmed in the programming language OCCAM.

Keywords : industrial robot, transputer, transputer network, axis controller, sensory information, laser ranging sensor, cartesian coordinate, real-time, transputer-links

Introduction

Nowadays the sensors in the industrial robots play a key role. The use of such sensors can reduce uncertainties of the robot and its environment. Therefore such sensors increase the precision of the position control to a high level. In general it is not easy in real-time to process the informations of the sensor. For this purpose the high speed INMOS transputer microprocessor family is qualified. It is possible to have a closed loop sensor control in real-time.

We have been engaged in path control studies for machine tools and robots over several years. We worked on high speed as well as high precision controls using the transputer network, because transputers provide the following benefits compared with conventional processors:

- high processing power per chip (10 MIPS + 2 MFLOPS)

- easy and inexpensive connection of the transputers through serial links

- low cost of processing nodes

- existing software for programming, testing and debugging of multi-processor systems

- inexpensive software and hardware development environment on the base of personal computers.

Transputer add-in boards for personal computers were available as development platform. This allowed us to build a homogeneous system, based exclusively on transputers. Today using transputers in control systems is a common practice, a detailed description of the basic transputer technology is therefore not included in this contribution.

The robot used is the Manutec r3, which is a six-axis robot from Manutec Siemens. This robot is equipped with brushloss DC drives, weighs 250 kg and has a maximum load of 15 kg. It is designed as an elbow manipulator with kinematic structure similar to the PUMA 560 robot. The dynamics data of the robot links have been measured and documented [7].

Control Architecture

Figure 1 shows the architecture of the robot control with a sensor. The path with its start and end points is specified via the user interface to the path generation unit.

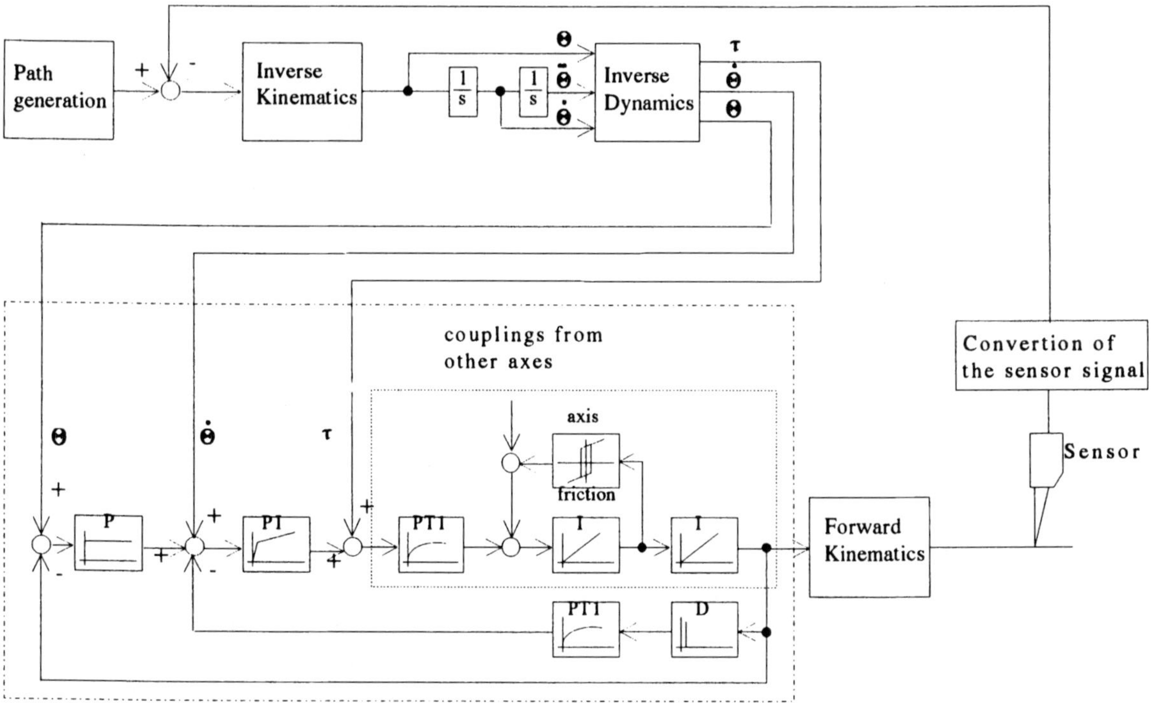

axis control system

Figure 1. Control Architecture

The path generation unit generates the path data considering the limitation of velocity and acceleration.

The constant acceleration characteristic of the suboptimal control rules in a trapezoidal velocity contour.

To limit jerking, an extended set of control rules is implemented. The rise of the acceleration follows a sine-square function. This results in continuous derivation and reduces oscillation of the robot arm. The maximum acceleration does not differ from that of the first case, ensuring high usage of the drive torque.

The path generation unit generates a cartesian point for every sample period, including the x, y, z coordinates and the roll, pitch, and yaw angles of the robot hand. This position (x, y, z coordinates) and orientation (roll, pitch, yaw angles) are added with the actual error from the sensors in order to generate the desired position and orientation. But in this contribution just one sensor is used, so that the normal distance offset between the end-effector of robot and the object surface can be measured. The joint angles will be transformed from the desired cartesian point through the inverse kinematic transformation. From the joint angle values supplied at discrete time intervals, the joint velocities and accelerations have to be re-determined. They are required to calculate the torque of the drives. The drive current controls the torque, which is the set value for the inner control loop of the power electronics actuator. The necessary torques are determined by using the inverse dynamic model.

The axis control loops control the position and the velocity. The joint velocity is controlled by a PI-controller. As mentioned above the output quantity of this controller works on the inner current or torque controller. Therefore the precomputed torque is added to the output of the velocity controller in order to reduce the influence of the mechanical system in motion. The velocity controller compensates only errors caused by friction torques. The actual velocity is computed from the actual position, because the robot is not equipped with velocity sensors. A P-controller is used for the position control loop.

Transputer Network

The architecture of the transputer network was designed according to the previously discussed control structure. There are different types of transputers. We used the T800 32-bit transputer with on-chip 64-bit floating point unit, which handles complex computing-intensive and memory-consuming tasks, for computation of complex algorithms, such as the inverse kinematic transformation and the inverse dynamic model. The T222 low cost 16-bit transputer performs I/O intensive tasks on the axis controllers and sensor. Figure 2 shows the architecture of the transputer network. The TA2 interface card, which is added in PC board, makes use of the PC keyboard, monitor and mass storage devices. The desired trajectory data are feeded into the system and the actual trajectory values are displayed in a graphic and also saved on the hard disk by using user-interface.

Transputer No. 11 runs the transputer development system and boots the transputer network. The user-interface transfers the commands to transputer No. 10. Transputer No. 9 executes the interpolation process and the inverse kinematic transformation. Every 2 ms (f = 500 Hz) a new cartesian point is evaluated and transformed into the joint angles. With some optimization the interpolation clock can be increased from 500 Hz to 1.5 kHz. This transputer No. 9 re-determines the joint velocity and acceleration values from the difference between two subsequent samples, respectively.

$$\dot{\Theta}_k = \frac{\Theta_k - \Theta_{k-1}}{\Delta t}, \quad \ddot{\Theta}_k = \frac{\dot{\Theta}_k - \dot{\Theta}_{k-1}}{\Delta t}$$

Δt: sampling rate (2ms)

After the position, velocity and acceleration values have been obtained, the torque values are computed using the methods described in the control structure section.

The transputers No. 1 to No. 6 are the joint processors. The number of serial links per node is limited to four. Networking the transputers No. 1 to No. 6 with transputer No. 9 requires two further nodes, the transputers No. 7 and No. 8. These transputers distribute and collect the data to be transferred from and to transputer No. 9. This involves no particular effort since only the two additional transputer chips had to be installed. The transputers No. 7, 8 and 9 are located on a single Euroboard. Every joint processor receives the joint angle, velocity and torque and executes the axis control loops. The sampling rate of the position control loop is 2ms, i.e. the same as that of the interpolation clock. All joint processors are synchronized via the link communication process. The velocity control loop of the digital controller is clocked at 4 kHz (t = 250μs). The control algorithm is implemented by using 32-bit integer values.

Figure 3 illustrates more details of the joint processor. It is designed as a single size Euroboard and was originally developed to control two drives. In the present network, each axis works with its own controller. The extra processing power is used for experiments with new control algorithms.

A 12-bit D/A converter supplies the signal to the joint inverter driving the joint motor. A 24-bit up/down counter, the incremental decoder interface, updates the current joint position from the signal generated by the incremental encoder. The actual velocity must be derived from the joint position. The T222 transputer communicates with its I/O ports via the local memory bus. 64 kbytes of external memory can be added on-board. The current version of the joint software requires less than 3.5 kbytes and is located in the 4 kbytes internal memory of the T222. A digital I/O port reads and sets additional signals. In every cycle, each axis receives three 32-bit set values and returns three 32-bit actual values to the superimposed transputer.

Sensor System

To obtain the sensor signal, the measuring sensor is mechanically mounted in such a way that it precedes the end-effector of the robot and transmits the signals to the sensor processor. The sensor processor (transputer No. 12) takes over the evaluation of the signals. In the transputer No. 9 the retrieved information for the robot is processed. It is easily coupled through the transputer links to the robot control. The information about the distance is obtained on the basis of the triangulation principle.

The sensor interface controls the robot in cartesian coordinates. Transputer No. 12 sends the information about the commanded path and the robot position vector of the robot end-effector to the sensor interface. With the evaluation of the sensor data and the trajectory of the robot, the sensor interface generates an offset vector in cartesian coordinates. Tansputer No. 12 uses the corrected position vector in the previously described control architecture. New values are computed every 2 ms, at the time base of the path generation clock. The stand-off distance of the sensor is 100 mm and the measurement range is $\pm$40 mm. The resolution of the sensor is 10 μm. The sensor outputs -4V to +4V (0.1V/mm).

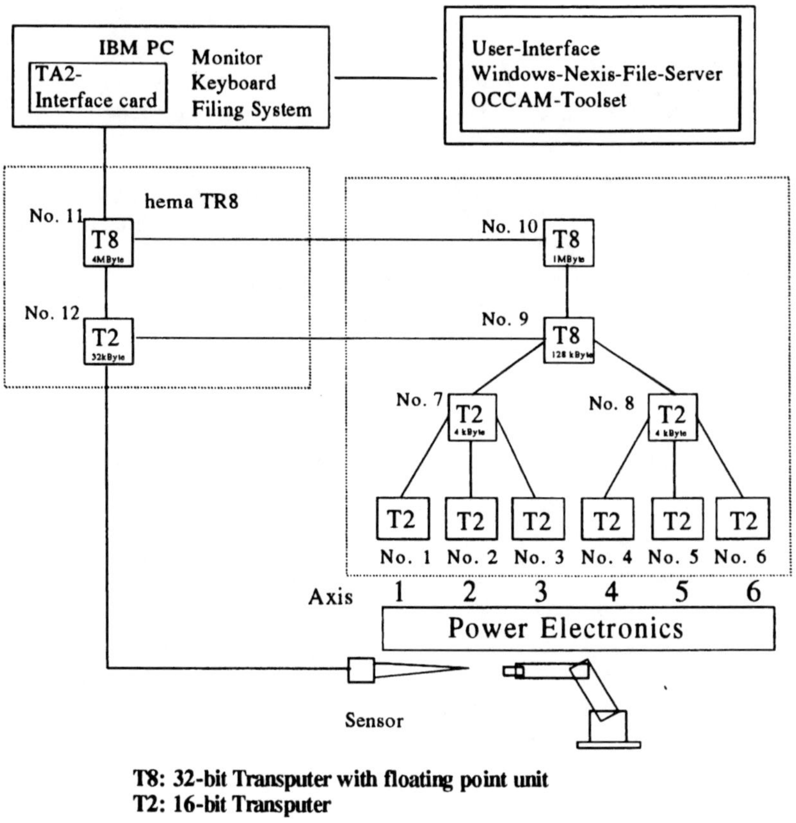

Figure 2. Structure of the transputer network

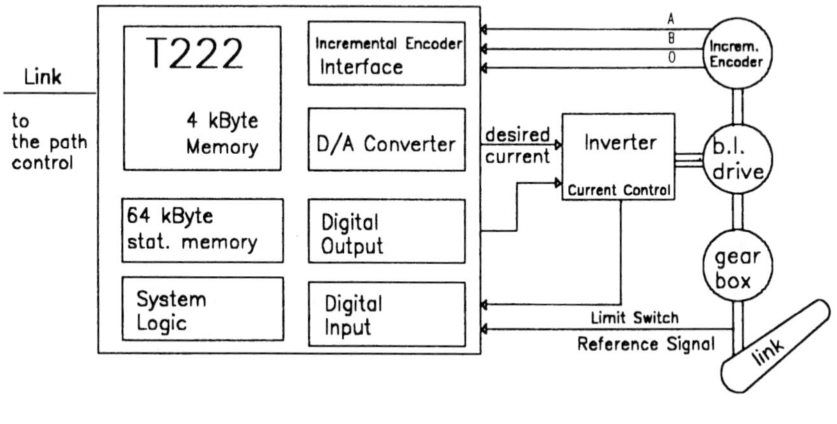

Figure 3. Axis Control Board

Results

The results of the experiments with cartesian motion of the Manutec r3 are shown by figure 4 to figure 8. The trajectory, a square in the x-y plane (P1-P2-P3-P4-P1), is illustrated in figure 4. First, the robot arm has to move from P1 to P2 and then changes the orientation of the end-effector in direction to the P3. This process is carried out to the final position (P1). The z position is kept constant (82.37 mm). The orientation of the end-effector is changed to the moving direction. All six axes are involved in the motion for this experiment. The maximum velocity was set to 1 m/s and the maximum acceleration to 3 m/s^2. The resulting error of the z-axis in cartesian coordinate system was calculated from the actual joint angles, which were measured by the incremental encoders. Figure 5 shows the tracking error in cartesian space for the z-axis in mm. We can see in the figure 5 that the robot arm has a steady state error of approximately 0.01 mm. In these points just the orientation of the end-effector of the robot arm is changed. The tracking error while moving the position remains below the 0.02 mm.

For testing the superimposed distance control with the laser sensor, the continuous path control generates a motion on a straight line parallel to the plane z = const. with a velocity of 0.3 m/s. The referece position of the sensor is 100 mm above this plane. On the plane a fest contour is shown in figure 6. The heights of the contour 1, 2 and 3 in figure 6 have 2.21 cm, 4.34 cm and 5.33 cm, respectively. At first the sensor based distance control is not in operation. When moving over the contour the sensor output signal is shown in figure 7. Since the heights of contour 2 and 3 are out of the measurement range of the sensor, the sensor gives at these contours the maximum signal of +4 V. In the next step the distance control is in operation and corrects the motion in order to hold constant distance to the contour. Figure 8 shows how the end-effector follows the contour when moving with a velocity of 0.3 m/s. The settling time of the sensor based distance control is below 100 ms.

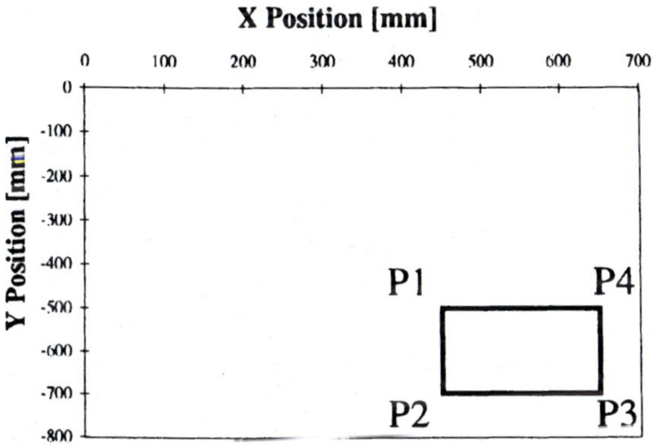

Figure 4. Test Contour in x-y plane

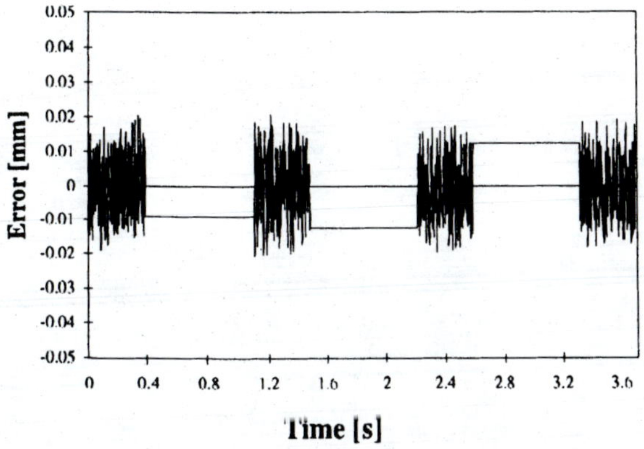

Figure 5. Position error of the z-axis

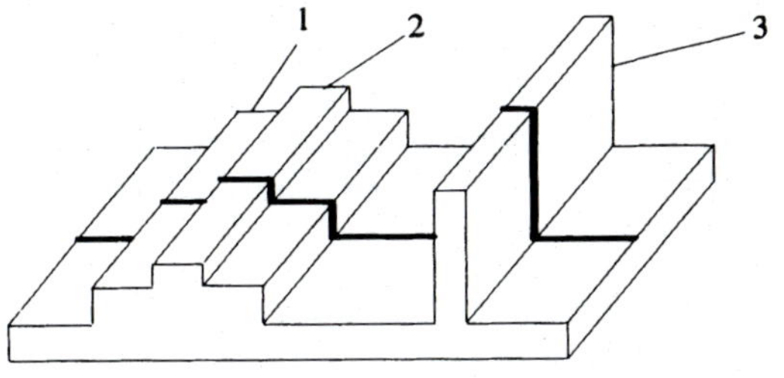

Figure 6. Test contour

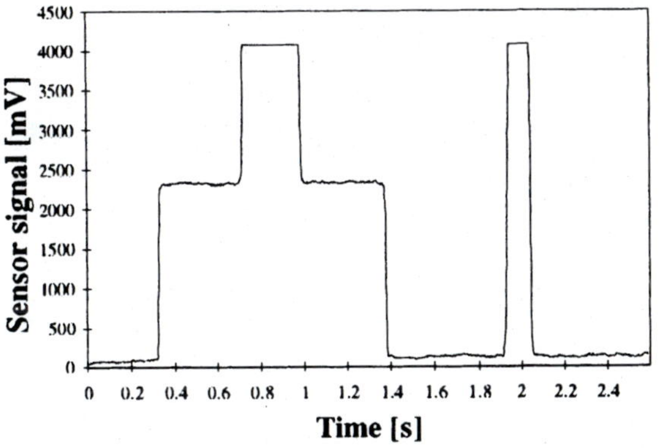

Figure 7. Sensor data in mV

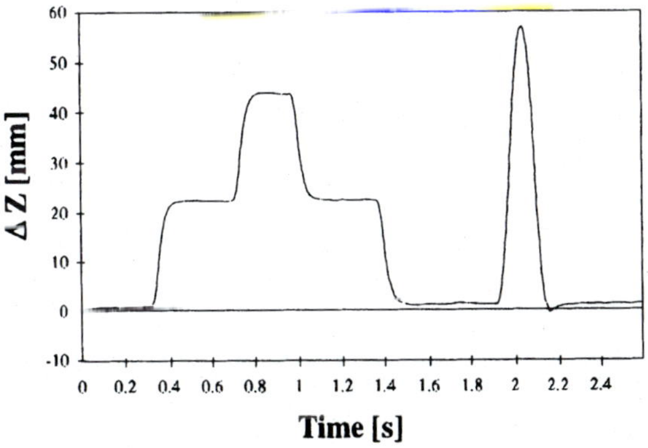

Figure 8. Position of the z-axis

Conclusion

This paper has discussed a robot control based on a transputer network with a contactless sensor. With the advantage of a transputer network system, we can control the robot arm in cartesian coordinate system in real time. The whole software is written in programming language OCCAM and its environment is very flexible, permitting to test quickly new algorithms. The sensory informations and processed data are stored in the personal computer in order to analyze the different control algorithms.

References

[1] Hirzinger, G. (1987). Sensory Feedback in Robotics - State-of-the-Art in Research and Industry. *IFAC 10th Triennial World Congress.* Munich FGR

[2] *INMOS Transputer reference manual.* (1988) Prentice Hall.

[3] Kunz, U., Oberschelp, W. (1983), High Accurate Continuous Contour Control by On-line-Simulation using Multiprocessors. *Proceedings of the PCI/MOTORCON Conference.* Geneva. pp. 665-675

[4] Lauer, P., Kunz, U. (1991), Transputer Network Controls Robot Axes. *Symposium on Robot Control.* Vienna Austria. pp. 589-594

[5] Olomski, J. (1989). *Bahnplanung undBahnführung von Industrierobotern*

[6] Paul, R. (1981). *Robot Manipulators: Mathematics, Programming and Control.* The MIT Press

[7] Türk, S. (1988). *Dynamische Roboter-modelle am Beispiel des Manutec r3.* DFVLR-Mitteilung 88-16. Oberpfaffen-hofen

A TRANSPUTER-BASED CONTROLLER FOR THE RTX ROBOT

P. Bauters, R. De Keyser, Y. Song, and P. Van Renterghem

Automatic Control Laboratory, University of Ghent,
Grotesteenweg Noord 6, B-9052 Zwijinaarde, Belgium

Abstract

This paper describes the design of a control interface for the RTX robot, motivates the decisions that were made and describes some of the applications. The interface was based on the transputer because of its high performance and its inherent support for multitasking.

Keywords : Robotics, Parallel Processing, Transputers, occam.

1 Introduction

The overall goal of our project is to build a demonstration system that illustrates in an integrated way several aspects of modern computer architecture, advanced control engineering and applied artificial intelligence. The evaluation of the potential of real-time trajectory generation, adaptive/predictive/neural and fuzzy kinematic robot control and computer vision for industrial applications are all incentives for and possible outcomes of this project.

The robot system, which is built around a standard PC, accelerated with transputers, should be able to track a randomly moving mechanical mouse. The intelligent robot system observes the environment via a camera, recognizes the mouse, predicts the future mouse path and controls its own links in order to catch the mouse when it enters a specified domain. Although at a first glance, this application may be blindly classified in the field of games and toys, there are several reasons to consider it as a challenging research project:

- many real-life human tasks consist of exactly the same actions: observation, target recognition, path forecasting in the case of moving objects, control of muscular movement. Operating and controlling production machines in factories can be broken down in above elementary steps, but even everyday common actions such as walking in a crowded street, driving a car, ... can be classified as such.
- many different disciplines from the field of modern automatic control are used and combined in an integrated way during the realisation of the project, such as parallel processing hardware, multitasking software, real-time image processing, artificial intelligence, neural control and identification, robot path planning and trajectory control, advanced model based control algorithms, rule based and heuristic control.
- several people specializing in these different disciplines are cooperating in one task-force instead of being isolated as individual researchers. This may lead to supplementary outcomes thanks to "cross-fertilization" (the whole is larger than the sum of the parts) and may also be a challenging experience from the point of view of project management.
- the final system will be appealing as a demonstration tool and will have a strong didactical value.

Mainly the low-level part of the control system will be described in this paper. Our RTX robot from Universal Machine Intelligence was completely stripped of all its original control hardware. It is now controlled at the low level via an in-house made interface between the robot and a transputer board. The transputer is a relatively fast microprocessor, which can be seen as a building block for multiprocessing systems. The interface receives motor control information from the transputer and sends joint position information back when requested.

To enable the use of this hardware interface by the developer of control algorithms, a set of basic software routines was developed on the transputer, in its low-level language occam. These routines turn motors on or off, reset the robot in a predefined state, transform the axis coordinates to cartesian coordinates (direct transform) and back (inverse transform), move the axes using a trapezoidal velocity profile, etc.

The transputer is an excellent and natural choice for the robot control, because it can be waiting for information on any of its links, without consuming CPU time, while it is calculating away at roughly 10 MIPS and 1.5 MFlops. Multiple parallel tasks are executed without any overhead. The transputer can do this thanks to its microcoded scheduler and on-chip real-time kernel.

2 The RTX Robot

The RTX robot is a 6-axis robot of the SCARA type, manufactured by the British company Universal Machine Intelligence (fig. 1). Most of the joints of the robot are named like a human arm (shoulder, elbow, wrist) and the wrist consists of 3 axis: yaw, roll and pitch. Each axis is driven by a DC servomotor, of which the speed is controlled by a continuously changing voltage between 0 and 24 V. The relative position of each axis is measured by two optical encoders which produce two square waves [7]. These are 90 degrees out of phase to each other to be able to detect the rotation direction. To enable the user to compute an absolute measurement, the robot has to be initialized in a predefined state using a simple control program (mechanical end stops) [8]. The robot was supplied with a dual microcontroller system (Intel 8031) to control the motors and to communicate with the PC through a serial line.

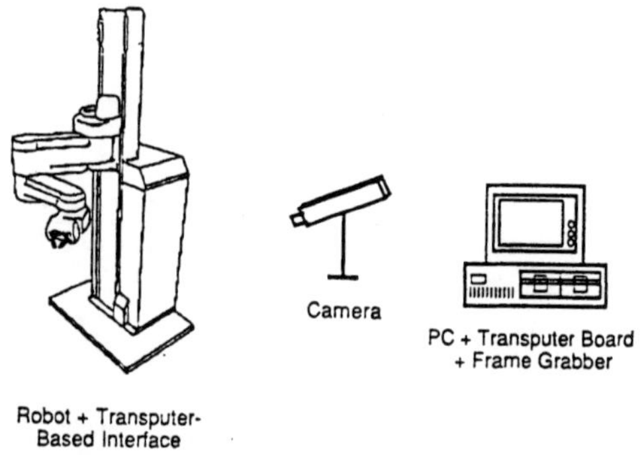

Figure 1: Robot control with vision feedback

3 The Transputer

The transputer family is a series of high-performance 16-bit and 32-bit microprocessors, which were specifically developed for the design of loosely coupled multiprocessor systems [1, 2, 5]. A transputer contains a fast central processing unit, an optional floating-point unit, some local memory and 4 high-speed communication links for the interconnection with other transputers.

The key innovation of the transputer is its inherent concurrency. With the transputer, a small multitasking kernel and scheduler are embedded in the transputer's microcode, resulting in very fast context switching times. Furthermore, the communication links work as DMA channels, enabling the processor to do computations while data is received or transmitted.

If the transputer is programmed in occam, one can get a stunning performance. E.g., on the floating-point intensive inverse kinematic transformation program for the robot, the T800-20 showed to be 6 times faster than a 25 MHz 386/387 PC. As these complex calculations are essential in path tracking algorithms, the use of a transputer will show very advantageous. The choice of a transputer as the heart of the robot controller was inspired by the inherent support of concurrency and communication by the transputer, its ease of use, expandable communication system, its computational power, its low power consumption and last but not least its simplicity in hardware design [9]. Currently, more high-level software is being developed in Turbo Pascal on a PC, which talks through a PC interface to the transputer which controls the robot and does all the dirty (i.e. time-consuming) work. Furthermore, research is done in the parallelization of the most time-costly algorithms.

4 The Interface – Hardware Aspects

The interface hardware consists of 2 parts: a driver part and a counter part (see fig. 2).

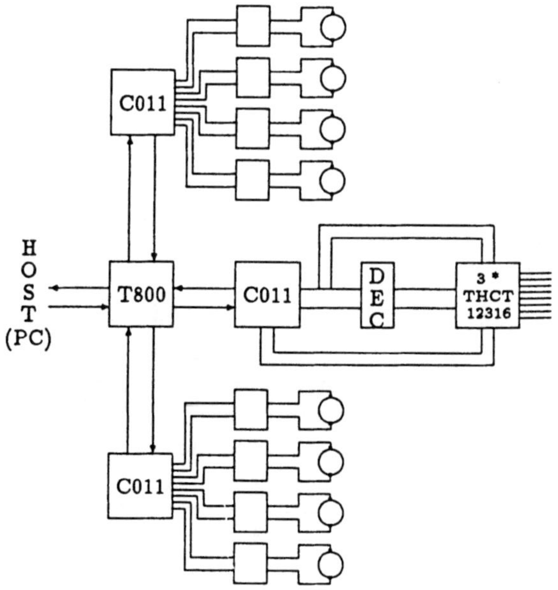

Figure 2: The transputer-based interface

The driver interface gets information from the transputer and translates it to an on/off and direction signal for each of the 8 DC servomotors. Although this interface could be realized using a D/A convertor, preferably mapped in the transputer's memory, we chose a chopper-oriented approach, where the speed is realized using pulse width modulation of the motor drive voltage. This was mainly done to simplify the hardware design, after thorough consideration that this interface would be fast enough. We also decided to use a link adaptor (IMS C011) to translate the serial link information (at 20 Mbit/s) to an 8-bit parallel port. As 8 motors need 16 signals (on/off and direction for each), we need 2 link adaptors, thus 2 serial transputer links.

The second part of the interface is responsible for passing the counter information of each of the optical encoders back to the transputer. Again, a transputer link (each transputer has 4 links) is used to transmit a counter identification number to the interface and to pass the counter information back to the transputer. An excellent chip for doing this sort of work is the THCT 12316 from Texas Instruments, which contains 3 16-bit counters and a direction discriminator, which implies that we don't need a control circuit to determine the rotation direction of the motor. We need three of these for the 8 motors. The transmission of 16 bits via the 8-bit C011 is a bit tricky, but it works.

5 The Interface – Software Aspects

The configuration of the software depends on the designed hardware structure. According to our hardware, the programming scheme adopts a master/server approach as shown in fig. 3. As master we have the PC and the transputer acts as server.

The master program on the PC talks through an interface to the transputer which actually controls the robot. The transputer runs an occam program with the basic routines, as well as the most computationally intensive parts of the master program, whilst the Pascal program provides the high-level features, such as the user interface. The intelligent use of the two processors has the extra advantage that the PC can be drawing pictures, waiting for user input, ..., while the transputer is working.

It can be noticed that the software interface, from the view-point of the user's application, consists of two main parts: the communication interface between both processors (master/server) and the basic functional programs (manual mode, point-to-point,..) which directly control the joints.

The communication software includes two programs. The first, running on the PC, written in Turbo PASCAL and compiled as a user's library unit, consists of an "interface" part and an "implementation" part (as standard in Turbo Pascal). The other runs on the transputer and is written in occam.

The interface part of the TP-library unit is open to the user and provides a variety of procedures such as ReadAngle(angle,motor). Calling this procedure from a user program invokes a functional call to the master module on the transputer which sends in this case the angle of the given motor. In this way, one can send data to the transputer, or receive data from the transputer, so the software interface can complete the bi-directional communication task.

The implementation part of the TP-library unit is hidden from the user which makes programming easier. Its main function is exchanging data between two different processors. This communication employs memory mapping on the PC, which is decided by the hardware, and software hand-shaking. When the interface program sends or receives the data, first it has to test the status of some special memory addresses of the PC. It then decides whether or not to send or receive the data according to the status flag. On the other hand, the interface program on the transputer receives the data (or commands) from the PC and translates these into other data. These are sent to the robot motors or encoders via one of the transputer links. We can state that the main function of the transputer program is to interpret the data received from the PC into the instructions or parameters according to the software protocol, and to send data back to the PC if the master program on the PC requires it.

The basic functional programs which are all written in occam have been developed in four kinds of modules: initialization, manual mode, point-to-point control and continuous path control (see section 6). Initialization means resetting the robot and moving the arms to a predefined state. In manual mode we can control the robot axes (motors) directly using some keys of the PC-keyboard or using some special function calls as explained earlier in this section. Transforming axis coordinates to cartesian coordinates (direct transformation) and back (inverse transformation) is a part of the point-to-point control. The continuous path control is explained in the next section.

6 Robot trajectory planning

The basic software we have been developing is based on the requirement of controlling the RTX robot to move from point to point (PTP control) or to follow a specified path with a given speed (CP control). PTP control ensures that the hand moves from one point in the cartesian coordinate

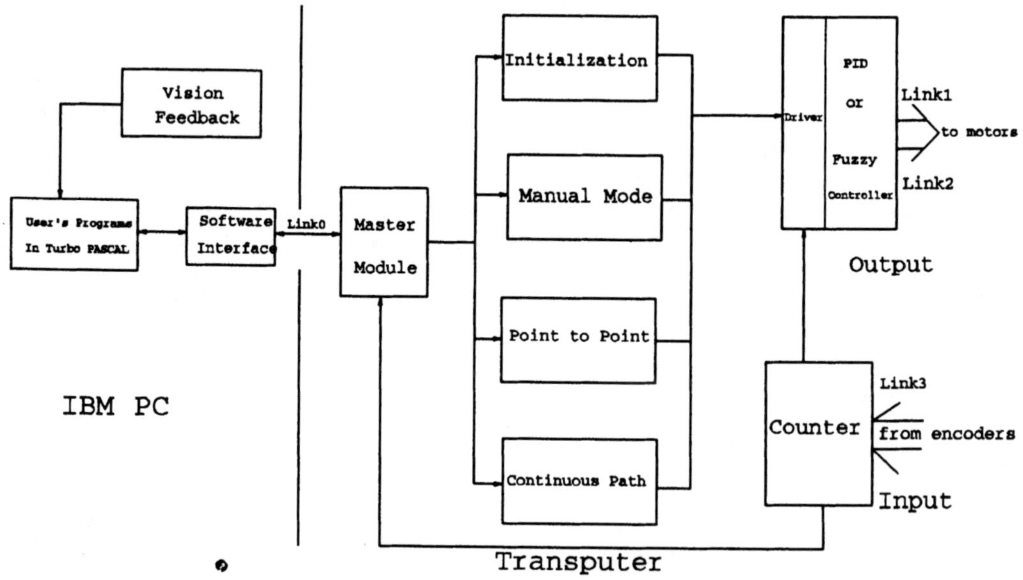

Figure 3: Overview of the software architecture

system to another at a maximum velocity while the locus of the hand between the two points is uncertain. CP control enables the robot to move continuously along a specified path in the cartesian space with or without controlling the speed. In our application, both ways of controlling the movement are relevant: PTP for initial large and fast motion, CP for final accurate tracking of the moving object path. Obviously, the continuous path robot control task is more important and more complicated. The task can briefly be divided in two parts: trajectory planning and trajectory tracking. Normally, the task is specified in a world cartesian coordinate system, but the robot only understands joint coordinates and the servo control is also at the joint level. Thus the direct and inverse kinematic transformations that link the two spaces of the control algorithm are necessary.

The trajectory planning can be done in joint space or in the world coordinate system.

6.1 Trajectory planning in joint space

If the trajectory planning is at the joint level, first a sequence of key points is selected in Cartesian coordinates, then these key points are inverse transformed into functionally equivalent joint coordinates. Finally n-cubic or quadric spline functions (n=DOF), one for each joint of the robot are solved and optimized to fit all the joint key points and result in a smooth fast Cartesian trajectory with limited geometric path deviation [4, 6].

When the robot is actuated, it samples these spline functions on-line at every sampling time and sends the sampled data to the n joint servo controllers as motion command signals.

Figure 4 illustrates an example of hand paths in a plane as a result of time optimized cubic and quadric spline function joint trajectories. Six key points are specified. After time optimization, the total motion time decreased from 6.01 to 5.18 second for the cubic spline joint trajectory and from 9.92 tot 7.82 second for the quadric spline joint trajectory. The maximum deviation decreased from 50.92 to 16.22 mm and from 24.85 to 3.63 mm for the cubic and quadric spline trajectory respectively. Further increase of the path accuracy needs more key points. In this approach the spline function joint trajectory has to be built off-line since too much computation is included in the procedure. It can't be used in our real-time tracking of the randomly moving object although it is valuable in some applications, such as obstacle avoidance in material handling.

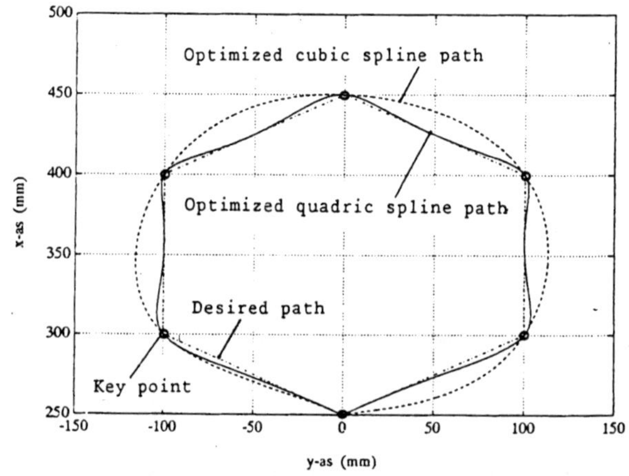

Figure 4: Simulation of a hexagonal path

6.2 Trajectory planning in the world coordinate system

For continuous path and velocity control, the trajectory is normally directly planned in the world coordinate system. In this case, the incremental interpolation strategy is employed. There are two approaches in this strategy [3].

The first approach uses the well known resolved motion rate control, which needs to build the Jacobian transformation matrix and solve the derivative motion equation:

$$\frac{d\mathbf{q}}{dt} = \mathbf{J}^{-1}(\mathbf{q}) \cdot \frac{d\mathbf{P}}{dt}$$

where **P** is the robot end effector position and orientation in the base frame X0, Y0, Z0 and **q** are the generalized joint coordinates:

$$\mathbf{P} = [X_0\ Y_0\ Z_0\ \theta_{xo}\ \theta_{yo}\ \theta_{zo}]^T$$

$$\mathbf{q} = [q_1\ q_2\ \ldots\ q_n]^T$$

We now solve the equation in an incremental way:

$$\Delta\mathbf{q} = \mathbf{J}^{-1}(\mathbf{q})\Delta\mathbf{p}$$

When the vision system sends the position of the moving object at every sampling time T_v, the trajectory planning routine plans the cartesian path and velocity for the robot to move. Then the interpolation routine samples the trajectory at the robot sampling time T_r ($T_v > T_r$) to obtain $\Delta\mathbf{p}$, and solves the inverse Jacobian equation to achieve $\Delta\mathbf{q}$.

The second approach obtains the incremental motion signal $\Delta\mathbf{q}$ indirectly. It applies absolute interpolation implemented by solving the inverse transformation equation:

$$\mathbf{q} = f^{-1}(\mathbf{P})$$

At every sampling time i, a new position $\mathbf{P}(i)$ is inverse transformed into joint coordinates $\mathbf{q}(i)$. The incremental motion command is obtained from the following equation

$$\Delta\mathbf{q}(i) = \mathbf{q}(i) - \mathbf{q}(i-1)$$

As an improvement, a correction loop is incorporated in our control strategy. The principle is shown in fig. 5.

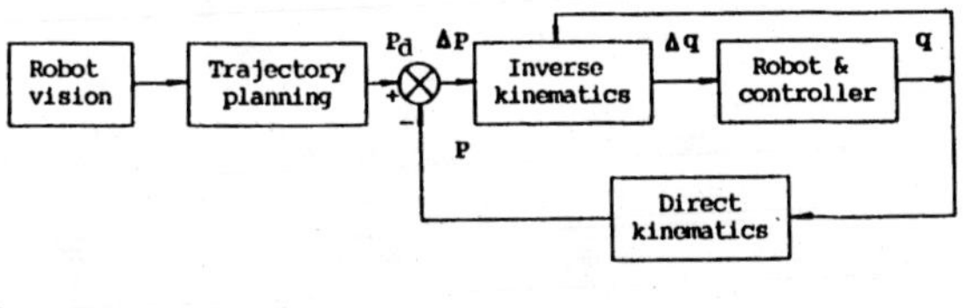

Figure 5: The principle of trajectory interpolation

This approach uses real position feedback, hence it can effectively compensate errors during trajectory tracking.

These approaches need intensive computation within every sampling period. Since transputers can compute at a much higher speed than normal microprocessors, they carry the heavy computational burden in the new system. The above mentioned calculations are written in standard elementary occam procedures. They can be called either from an occam program, or from a main program in Turbo Pascal through the software interface.

7 Conclusions

A challenging research and demonstration project concerning intelligent robot control has been formulated. Its long-term goal is to combine in an integrated way several aspects of advanced automatic control, i.e. modern computer architectures, new control algorithms, vision and applied artificial intelligence.

In this paper, mainly the low-level part of the control system has been described. It is built around a "PC-transputer-robot interface" hardware combination. More information on the latest status of the global project can be obtained from the authors.

References

[1] Inmos Ltd., *The Transputer Databook*, Second Edition, 1989

[2] Inmos Ltd., *The Transputer Applications Notebook*, Architecture and Software, Systems and Performance, 1989

[3] Koren Y., *Robotics for Engineers*, Mc Graw-Hill Book Company, 1985

[4] Lin C.-S. et al., *Formulation and optimization of cubic polynomial joint trajectories for industrial robots*, IEEE Trans. on Automatic Control, Vol. AC-28, No. 12, Dec. 1983

[5] Van Renterghem P. (1989), *Transputers for Industrial Applications*, Concurrency: Practice and Experience, John Wiley & Sons, Ltd.

[6] Song Y., De Keyser R., *Design of manipulator trajectories with minimum motion time and specified accuracy*. In: S. Tzafestas (Ed.), "Robotics and Intelligent Control Systems", Kluwer Academic Publishers, Dordrecht, 1991

[7] *Introducing RTX*, Universal Machine Intelligence Ltd., 1986

[8] *Inside RTX*, Universal Machine Intelligence Ltd., 1986

[9] Zhang Y., *Transputer-based behavioural module for multisensory robot control*, Parallel Processing and Artificial Intelligence, Wiley, 1989

Implementation of Parallel Logic Solving Algorithm for the PLC Based On Dataflow Architecture

Jaehyun Park, Naehyuck Chang, Gab Seon Rho, Wook Hyun Kwon,

Engineering Research Center for Advanced Control and Instrument
Dept. of Control and Instrumentation Engineering
Seoul National University, Seoul 151-742, Korea
FAX: +82-2-871-7010, Tel: +82-2-887-0040, E-mail: whkwon@krsnucc1. bitnet

Abstract. In this paper, an event driven parallel logic solving algorithm for PLC and a hardware structure to implement the proposed parallel algorithm are proposed. The proposed algorithm includes a parallel algorithm and a run-time optimization algorithm. The logic solving unit of the suggested PLC is designed based on the dataflow concept to implement event driven operations and fine grain parallelism.
With the suggested optimal algorithm, the proposed system solves only 1.33 instructions for each event.

The performance of the proposed hardware system is about 5 MOPS with a single processing element(LSU), and about 12.5 MOPS with four processing elements when operated at 20 MHz in the worst case logic program.

Keywords. Computer Architecture, Computer Control, Control equipment, Industrial Control

INTRODUCTION

A programmable logic controller(PLC) is widely used in factories to replace relays. Its applications can be found in many areas, such as chemical industries, manufacturing, and mining. In recent years, to decrease the processing time of a programmable logic controller, several new architectures for large scale programmable logic controllers have been proposed, and some of them are used in commercial programmable logic controllers.

The overall processing time of the programmable controller consists of logic program solving time and I/O update time. Between these, I/O update time can be reduced by using a high speed I/O processor which may have a multi-processor structure. Many outstanding large scale PLCs are based on the multi-processor architecture or array processor architecture. We developed a high performance Logic Solving Processor(LSP) based on the array processor architecture(Kim 1988, 1991). This LSP can solve a maximum of eight logic instructions in every other clock. Hence, the LSP can solve 1000 logic instructions in 22 μsec using eight cells with a 16 MHz clock. However, it can solve 1000 instructions in 180 μsec if block instructions are mixed. This means that the LSP based on the array architecture has serious performance degeneration when block instructions are mixed in. This occurs because the LSP uses the array processor structure for logic instructions and the multi-processor structure for block instructions. Thus, logic instructions and block instructions, can not be solved simultaneously in the fine grain level. This is a common characteristic of the multi-processor and array processor LSP.

Most of the existing commercial programmable controllers are control driven systems. In other words, a central computer controls the PLC. However, in this kind of system, there can be several problems. First, all of the logic program should be solved within each I/O scan time regardless of the input changes. However, in reality, most of the input points are unchanged in a small time interval. Hence, the existing PLC will solve many unnecessary logic solving steps in every scan interval. The second problem is an event ordering problem. In real plants, all of the events occur in a sequence. However, in the existing PLC's, input points are sampled periodically by a central processor. As a result, all of the events occurring in that I/O sampling interval are treated as if they occurred at the same time. This difference can make sequence ordering problems and a large time delay between events and response.

To eliminate these problems, the PLC itself should be operated by events rather than by a central computer. This event driven programmable controller has been studied in recent years. A Petri-net based PC is an example of data driven PC(Murakoshi, 1990, 1991). The suggested Petri-net PC is an event driven sequence control system, but it is hard to implement for a large scale general purpose PC because it uses an associating memory. Another example of a data driven PC is a dataflow logic solving processor based on a dataflow architecture(Park, 1991). This dataflow logic solving processor(DFLSP) is designed based on the dataflow concepts, while the logic solving algorithm is operated in data driven style. But in that system, only the logic solving unit is designed based on the dataflow architecture, so the overall system is not adequate for an event driven system.

In this paper, an event driven PLC is proposed and its operational strategy, parallel algorithm, and run-time optimization algorithm are detailed. Also an implementable hardware structure is proposed for implementing the suggested algorithm, and its performance is estimated.

EVENT DRIVEN SEQUENCE CONTROL

A large scale PLC can be divided into several blocks as in Fig.1. In the most of the existing PLC's, which are controlled by a central processor, the central processor scans all of the I/O points and solve logic programs with the scanned data. In this case, a time delay from the input change to the corresponding control output, t_p can be divided into several time delays, $t_1, t_2, \ldots t_7$ as shown in Fig.2. Among these time delays, only t_4 is the real program solving time. In the existing PLC's, to increase the performance of the PLC, i.e., to reduce t_p, all of the time delay, t_1, t_2, ...t_7, should be decreased uniformly. However, when only one input changes, only a small part of the programs related to the changed input needs to be solved. In this case, the unnecessary program solving time, t_3 and t_4, can be eliminated; and t_2, t_5, and t_6 can be reduced. Besides this performance profit, the event driven PLC can implement the real concept of sequence control in the plant. In the sample ladder program in Fig.3, if X3 and X1 change sequentially, but in one I/O scan period in the control driven system, X1 may be treated as if it occurs prior to X3. Alternatively, they may be treated as if they occurred at the same time. This sequence-order-inversion problem can cause a big problem in some plants and in some programs. In the event driven PLC, this sequence-order-inversion problem cannot happen because an event produces a data token, and the program solving sequence is determined by the produced data token.

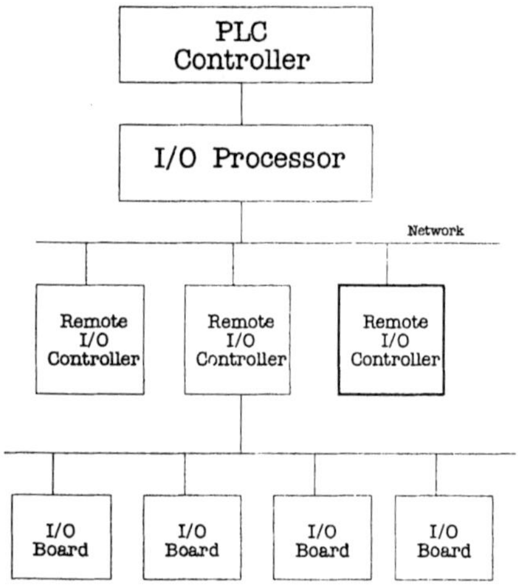

Fig. 1. Large sacle PLC

This event driven sequence control is a very useful concept for the PLC because the main purpose of the PLC is to control a real plant sequentially.

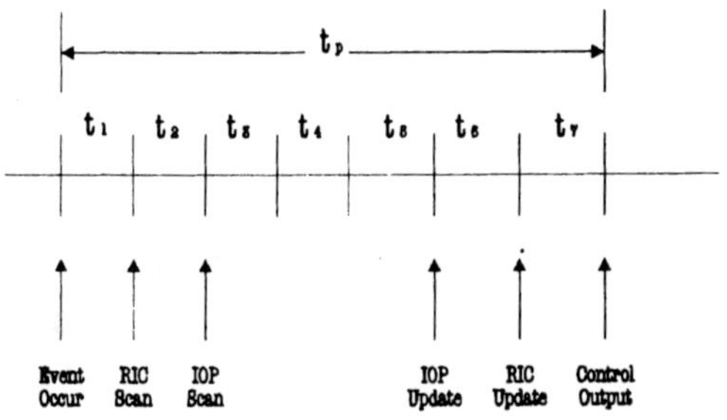

Fig. 2. Time delay of PLC

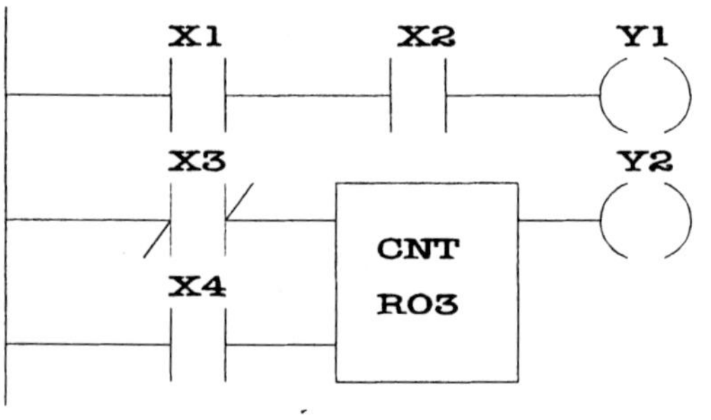

Fig. 3. Sample ladder program

Since a programmable logic controller(PLC) was originally developed as a sequential control device, the language for the PC should be well suitable for a sequential control problem. Currently, several kinds of programming language for PC's are used. Among them, a ladder language is the most widely used one. The typical ladder language is shown in Fig. 3. The logic instructions include AND, NOT-AND, OR, and STORE instructions; and the block instructions include the more complex instructions such as COUNTER, TIMER, SHIFT REGISTER, and DRUM. In real application programs, most of the instructions are logic instructions and the portion of the block instructions, is less than 25 percent. The ladder program is a graphical language and each instruction is solved sequentially from left to right. These ladder programs can be easily used to write small size programs, but this language is not adequate to represent the exact sequence of large programs. The reason is that in the ladder language, the precedence of two sub sequences can not be represented correctly in a large program. To write large sequential logic programs, a directed graph is more useful than a ladder program. The directed graph consists of two nodes connected by a directed branch, where each node has two inputs. With this representation, the ladder program in Fig. 4-a can be transformed into the directed graph in Fig. 4-b. This directed graph is often used for a dataflow computer and is called a dataflow graph. With this dataflow graph, we can represent the correct precedence of two instructions. Hence, for the event driven PLC, the dataflow graph is more adequate. To describe the operation of an event driven PLC systematically, it is necessary to define some notation and concepts using the dataflow graph.

LOGIC PROGRAMMING AND FORMALIZATION

To describe the architecture and operational strategy of the proposed PLC in detail, a formalized representation for sequential logic programming is needed. In this section, a formalized representation of sequential logic program is defined, and the basic operations of sequential control are described.

A dataflow graph is a directed graph composed of nodes(actors and links) and edges. This dataflow graph for the event driven PLC can be formalized using a similar method to that of Kavi(1986). The dataflow graph is defined as

$G = < A \cup L, E >$

where

$A = \{a_1, a_2, \cdots, a_n\}$ *is the set of instructions*

$L = \{l_1, l_2, \cdots, l_m\}$ *is the set of links*

$E \subseteq (A \times L) \cup (L \times A)$ *is the set of branches.*

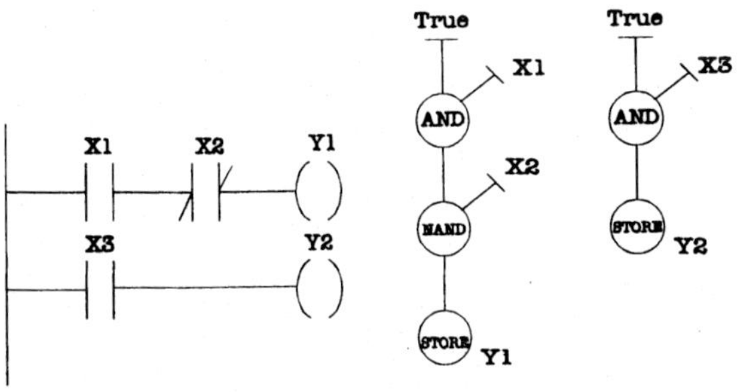

Fig. 4. Ladder program and directed graph

The instructions, a in this definition, include logic instructions and special function blocks like *fork* and *merge*. The links, l, are treated as place holders of data values(tokens) which flow from one instruction to the next instruction. For a new logic sequences, the fork instruction generates a new data token which is the same as input token. The merge instruction generates a single data token from two input tokens. This merge instruction is used for data synchronization of the STORE instruction. These fork and merge instructions are useful in writing large logic programs solving a data synchronization problem. When an event occurs in a PLC, a starting link set S is composed;

$$S = \{l \in L \mid (a,l) \notin E, \forall a \in A\}.$$

When a logic program are solved completely, the tokens are rested in terminating links, and the terminating link set T can be defined as

$$T = \{l \in L \mid (l,a) \notin E, \forall a \in A\}.$$

The logic program solving sequence can be thought as a sequence of partial mapping from the starting set S to the terminating link set T. We can define the input link set and output link set of each instruction a as

$$I(a) = \{l \in L \mid (l,a) \in E\}$$
$$O(a) = \{l \in L \mid (a,l) \in E\}.$$

To represent the existence of a token in l, a marking M is defined as a mapping

$$M : L \rightarrow \{0, 1\}.$$

A link l is said to contain a token in a marking M if $M(l) =$ 1.

OPERATION OF EVENT DRIVEN PLC

In this section, a model of an event driven PLC is proposed and the operational strategy of the proposed system is described using the notation defined in the previous section.

Assumption

The basic assumption of event driven PLC is that all of the events occur sequentially and none of them occur simultaneously. In real plants, this assumption is reasonable.

Operational Rule

The core part of our event driven PLC is operated by a dataflow concept. The dataflow machine is operated by data. It is very different from an ordinary computer, which is operated by a control processor. First, because it is operated by data, the data synchronization problem and processing order are managed by data. In our system, the processing order is managed by the events from the real plant. Second, it is possible to make a fine grain parallel processing system with our event driven PLC.

The basic operational strategy of an event driven PLC is that all of the program solving sequences are controlled by events. This control rule is that an instruction a can be solved at a marking M if the following conditions hold;

$$\begin{cases} M(l) = 1 & \text{for all } l \in I(a) \\ M(l) = 0 & \text{for all } l \in O(a). \end{cases} \quad (1)$$

A new marking M' resulting from solving one instruction a at marking M can be derived as follows.

$$M'(l) = \begin{cases} 0 & \text{if } l \in I(a) \wedge l \notin O(a) \\ 1 & \text{if } l \in O(a) \\ M(l) & \text{otherwise.} \end{cases} \quad (2)$$

In the suggested PLC, when an event occurs, the input/output processor generates a data token and forms a starting set S. From this starting set S, the event driven PLC solves the instructions which satisfy the condition (1). The event driven PLC continues solving process until no instruction satisfies the condition (1). After solving all of the possible instructions, the links that contain tokens form the terminating link set T.

As the event driven PLC continues solving instructions, the sequence of solved instructions forms a solving sequence, σ.

$$\sigma = \{ a_0, a_1, a_2, \cdots, a_n \}$$

For this solving sequence, the starting link set S is the input link set of the first instruction a_0, and the terminating link set T is the output link set of the last instruction a_n. The fork instruction splits the original solving sequence into two independent sub-sequences, and the merge instruction is combines two sub-sequences into one sequence. As a result, in our event driven PLC, there can be many sub-sequences starting from an event. These sub-sequences can be solved concurrently in the grain level in the independent solving units.

Parallel Algorithm

The solving sequence of the logic program σ can be split into several sub sequences $\sigma_1, \sigma_2, \ldots, \sigma_n$. Some of these sub-sequences can be solved concurrently while some of them cannot. We can say σ_i and σ_j are mutually exclusive when

$$\{O(\sigma_i) \cap R(\sigma_j) = \emptyset\} \text{ and } \{O(\sigma_j) \cap R(\sigma_i) = \emptyset\}$$

where

$$R(\sigma) = I(\sigma) - O(\sigma).$$

We can say σ_i and σ_j can be solved concurrently if they are mutually exclusive.

When a merge instruction is encountered, the two input sub-sequences should be solved prior to solving the resulting sub-sequence. This data synchronization problem should be manipulated by the hardware for fine grain parallelism. The operation algorithm for parallel processing of our event driven PLC is summarized below.

```
Algorithm 1)
    get an event from I/O and push it to the event queue
                /* generating starting link set S */
    do while (there are events in the event queue)
      begin
        get an instruction from the event queue
        if (merge instruction) then
            check all data are ready /* matching */
            if (ready) then
                solve it
            else
                save it for merge-waiting
            end if
        else if (fork instruction) then
            generate another event
            push the new event to the event queue
            generate a token for merge-waiting
        else if (STORE instruction)
            write output update memory
            generate a token for merge-waiting
        end if

        if (a token for merge-waiting occurs) then
            get the merge-waiting instruction
            push it to the event queue
        end if

    end while;
```

In reality, when an event occurs, it is very hard to know which sub-sequence should be solved first at run-time. Hence, in our suggested algorithm, the event driven PLC gets an arbitrary sub-sequence which starts from that event and solves it until a new sub-sequence is merged in. At the merge point, the processor checks whether the sub-sequences merged in have been solved or not. This is the matching process in the algorithm. For this matching process, the suggested parallel algorithm uses the merge-waiting concept to check whether the sub-sequences for merging has been solved or not. When a merge instruction is met, the PLC checks the sub-sequences, and if the sub-sequences has been solved already, it solves the merged sequence. And if the sub-sequences has not solved yet, it saves the instruction for later. The fork and STORE instructions generate a token the sub-sequences which is waiting for merging.

RUN TIME OPTIMIZATION ALGORITHM

In the suggested event driven PLC, the logic program can be optimized by run time optimization rules below.

(Rule 1) The possible logic solving sequence σ is divided into several sub-sequences which start from an input point.

(Rule 2) When an event occurs, only a sub-sequence σ_e stating from the event occurred, may be solved. This solving sequence σ_e is defined by

σ_e is a sub-sequence of possible solving sequence σ which satisfies

$$I(\sigma_e) - O(\sigma_e) = S$$

(Rule 3) When an event occurs, the instruction and data of the starting point are AND and logic FALSE, respectively; the occurred event can be ignored.

(Rule 4) The result of each instruction is compared to the previous data value. If they are same, stop running, otherwise, continue to next instruction. In the first case, the terminating link set T is adjusted to $O(a)$.

(Rule 5) At the merge point, if the merge-waiting situation can be avoided when then data value of the merging sub-sequence is logic TRUE.

When applying Rules 3 - 5 to the solving sequence σ_e, we can get a new minimal sub-sequence, σ_m. In our event driven PLC, the minimal sub-sequence, σ_m, is used rather than σ and σ_e.

ARCHITECTURE OF EVENT DRIVEN PLC

In this section, a hardware architecture is proposed to implement the event driven PLC suggested in previous sections.

The event driven PLC consists of four main parts: event driven IOP, event queue, logic solving unit(LSU), and merge unit. The block diagram of the overall system is shown in Fig. 5.

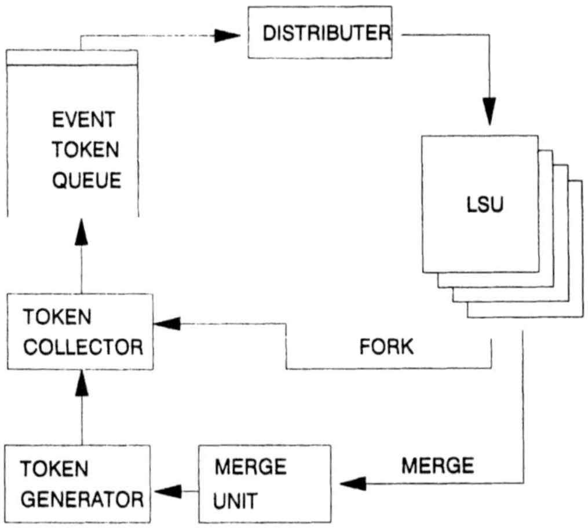

Fig. 5. Block diagram of our suggested PLC

Event driven IOP

To make an event driven PLC, it is necessary to detect an event with a very high resolution of sampling timebase. The event driven IOP scans all the input points of the PLC and checks whether any event occurs during every sampling interval. If an event occurs, it generates a data token and stores it in the token queue. To implement a high resolution event driven IOP, it is necessary to make a flash comparator with a very large input width. However, for a large scale PLC, this is very hard to make. Hence, in our event driven PLC, we design a small size, distributed form event detector.

Token Queue

In an event driven PLC, all of the solving operations are controlled by event token . As there can be many sub-sequences from an event, one event can generate multiple tokens through fork instructions. During the solving process, the tokens as many as LSUs can be enabled at the same time. The rest of the tokens are stored in the token queue and wait for an idle LSU.

Logic Solving Unit(LSU)

The LSU consists of six sub-parts: instruction MUX, operand MUX, instruction memory, operand memory, processing unit, and update control as shown in Fig. 6. An instruction a from the distributer is examined in the processing unit whether it can be solved immediately or not. If all links of the input link set have tokens, i.e. if

$$M(l) = 1 \quad \text{for all } l \in I(a),$$

the instruction is solved in the processing unit, and if one of the link doesn't have tokens, the token is sent to the merge unit. For run-time optimization, additional hardware which contains previous results of all actors and comparator is required.

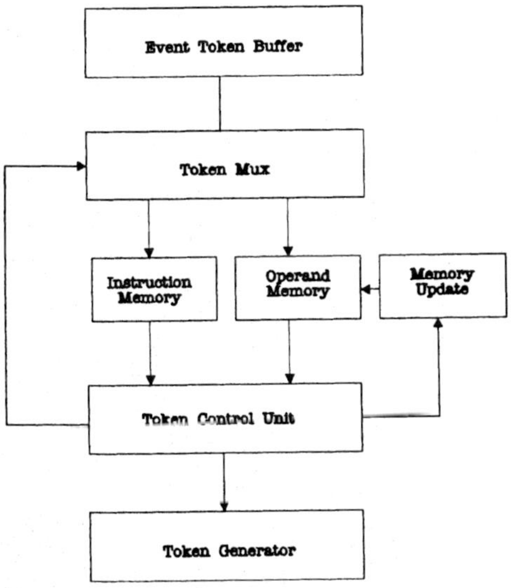

Fig. 6. Structure of the LSU

Merge Unit

At a merge point of the logic solving sequence, the sub-sequences are merged and form sub-sequences. In this case, to solve a new sub-sequence, the merged sub-sequences should be solved. Hence, the merge unit checks the sub-sequences and makes a token for the merged sub-sequence. All of the tokens which fail to match in the LSU are sent to the merge unit. If the LSU meets a STORE instruction, the LSU also sends that token to the merge unit. When re-matching occurs, the merge unit sends the matched tokens to the token queue in order to be solved by the LSU.

PERFORMANCE EVALUATION

As mentioned in previous sections, in the event driven PLC, there are three kind of solving sequence when an event occurs; σ, σ_e, and σ_m. In our event driven PLC, a minimal solving sequence σ_m is used. In this section, the performance of the event driven PLC and suggested machine are analyzed. The performance of the suggested hardware is measured by computer simulation.

Normal Case

In the normal case, all of the instructions of the logic program should be solved in every scan time. And in this case, the solving time elapsed for the program is constant regardless the number of events. The number of instructions solved per event, N_n, is

$$N_n = (\textit{length of } \sigma) \;/\; \mu$$
$$\textit{where } \mu \textit{ is the rate of event}$$

Event Driven PLC

In a logic program, we can say that the input point are used uniformly. This means that the average length of σ_e is the half of the whole logic program. In this case, for one event, the PLC should solve half of the logic program.

$$N_e = \textit{length of } \sigma_e = (\textit{length of } \sigma) \;/\; 2.$$

If there is only one event in every IO scan time of the existing PLC's, the suggested machine is twice as fast as the existing PLC's. But for a large number of input points, there can be more than two events in a IO scan time. In that case, the suggested machine is slower than existing PLC's. We can say that our suggested machine can control to the sequence event; however, for a system with a large number of input points, the performance of our system decrease.

Run-time Optimization

In the event driven case, for each event, all of the instruction in σ_e should be solved. But with the run-time optimization algorithm, only a sub-sequence, σ_m should be solved. From the run-time optimization rule, the expected number of instructions to be solved for an event is given by

$$N_m = E(\textit{length of } \sigma_m) = (1-p) + p(1-p)\cdot 2 + p^2(1-p)\cdot 3 + \cdots = \lim_{n\to\infty} \sum_{k=1}^{n} p^{k-1}(1-p)k$$

From the run-time optimization rule, the probability of the result change of logic instruction is 0.25. With this probability, the expected number of instructions are 1.33. This means that for an event, only 1.33 instruction may be solved.

Hardware Performance

To estimate the performance of the proposed hardware, a computer simulation is used with following parameters.

N_L	:	number of LSU's
T_L	:	elapsed time for solving one logic instruction by a LSU
D_M	:	delay time in merge unit
D_Q	:	delay time in event token queue
S	:	the portion of merge instruction in total instructions

In real systems, waiting time exists to seize common resources like the event token queue and the merge unit. That waiting time is D_M and D_Q. So the sustained performance can approximately be calculated by $(T_L + D_M + D_Q)/N_L$. In this equation, D_M and D_Q will vary according to N_L and the portion of the merge instruction.

The performance of the proposed hardware was evaluated by computer simulation with the SIMAN package[8] by varying three parameters, the number of processing elements(N_L), and the portion of merge instructions(S). Fig. 7 shows simulated processing time of DFLSP to the number of LSU. From the simulation results, the implementable DFLSP solves 1000 instructions in 195 μsec with one LSU, and in 75 μsec with four LSU's when they are operated at 20 MHz. From computer simulation results, we can say that the percentage of the store instruction hardly influences program solving time.

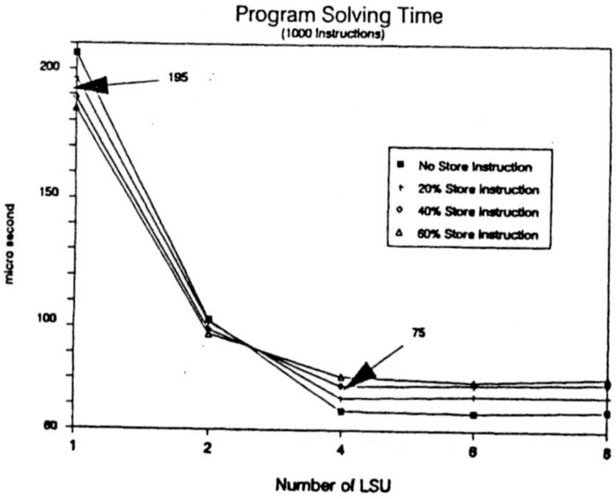

Fig. 7. Simulation results

CONCLUSION

The architecture of an event driven PLC is proposed in this paper. The proposed PLC is designed based on the event driven architecture in order to implement the real sense of sequence control. The logic solving unit of the proposed system is designed by dataflow machine which have the concurrent processing and the data synchronization capabilities. As proposed system used parallel algorithm and run-time optimal algorithms, the performance of the proposed system is very high. With our optimization rule, the average number of instructions to be solved for an event is only 1.33. This result shows the performance of our system can be better than other existing PLC. Furthermore, our system is operated by event driven, since event-order-inversion problems which can be happen in other systems do not occur. The SIMAN computer simulations show that the logic solving time of the proposed hardware is 195 μsec with one LSU and 75 μsec with four LSU's.

REFERENCES

Agerwala, T., and Arvind(1982). Data Flow Systems. IEEE Computer, 1982, Feb, 10-41.

Gurd, J.R., C. Kirkham and I. Watson (1985). The manchester prototype dataflow computer. *Communications of the ACM*, 28, 34-52.

Hwang, K. and F. Briggs(1988). Data flow computers and VLSI computations. In E. M. Munson(Ed.). *Computer architecture and parallel processing*, 4th ed. McGraw-Hill, New York. pp.732-768.

Kavi, K.M., Buckles, B.P., and Bhat, U.N.(1986). A Formal Definition of Data Flow Graph Models. *IEEE Trans. on Computers*, 35.

Kim, J.I., W.H. Kwon and J. Park(1988). Architecture of a logic solving processor for programmable controller. *Proceedings of '88 SICE.*

Kim, J.I., J. Park and W.H. Kwon(1991). Architecture of a ladder solving processor(LSP) for programmable controllers. *Proceedings of '91 IECON.*

Murakoshi, H. and Y. Dohi(1990). Petri net based high speed programmable controller by ASIC memory. Proceedings of '90 SICE

Watson I. and J. Gurd(1982). A practical data flow computer. *IEEE Computer*, 1982, Feb, 51-57.

AN ARCHITECTURE ENABLING THE SAFETY LICENSING OF REAL TIME CONTROLLERS

Wolfgang A. Halang* and Soon-Key Jung**

*University of Groningen, Department of Computing Science
P. O. Box 800, 9700 AV Groningen, The Netherlands
**Chungbuk National University, Department of Computer Engineering
360-763 Cheongju, Chungbuk, Korea

Abstract: A low-complexity, fault-detecting computer architecture for utilisation in programmable logic controllers is designed. The cyclic operating mode of PLCs and a specification level programming paradigm based on the interconnection of application oriented standard software function modules are architecturally supported. Thus, by design, there is no semantic gap between the programming and machine execution levels enabling the safety licensing of application software by an extremely simple, but rigorous method.

Keywords: Safety critical automation and process control, safety related computer architecture, safety licensing, real time computer control systems, programmable logic controllers (PLCs), multiprocessors, environment for high integrity software, fault tolerant systems.

INTRODUCTION

In society, there is an increasing awareness of and demand for dependable technical systems in order not to endanger human lifes and to prevent environmental desasters. Computer based technical systems have the special property that they consist of hardware and software. Hardware is subject to wear and faults occurring at random, which may be of a transient nature. These sources of non-dependability can (to a very large extend) successfully be coped with by applying a wide spectrum of redundancy and fault tolerance methods. The advent of the VIPER microprocessor (Kershaw, 1987), the correctness of whose design has been formally proven, signifies a leap forward in the direction of eliminating design errors in programmable electronic systems as well. In software, on the other hand, there are no faults caused by wear, environmental events etc. Instead, all errors are design errors, i.e., of systematic nature, and their causes are always (latently) present. Hence, dependability of software cannot be achieved by reducing the number of errors contained by testing, checks, or other heuristic methods to a low level, which is generally greater than zero, but only by rigorously proving that it is error-free. Taking the high complexity of software into account, only in exceptional cases this objective can be reached with the present state of the art. Whereas other researchers have the general situation in mind and, therefore, have to yield to the complexity problem, it is our intention to make an important step into the direction of realising a workable and useful programmable electronic system, which can be safety licensed in its entirety, by exploiting the intrinsic properties of a special, but not untypical case, that has been identified in industrial control problems. Here the complexity turns out to be manageable, because we restrict our attention to rather simple computing systems in the form of programmable logic controllers, and since application domains exist only demanding software of limited variability, which may be implemented in a well-structured way by interconnecting carefully designed and rigorously verified "software ICs".

There are already a number of established methods and guidelines, which have proven their usefulness for the development of high integrity software employed for the control of safety critical technical processes. Prior to its application, such software is further subjected to appropriate measures for its verification and validation. However, according to the present state of the art, these measures cannot guarantee the correctness of larger programs with mathematical rigour. For object code must be considered in the correctness proofs, since compilers — or even assemblers — are themselves far too complex software systems, as that their correct operation could be verified. Therefore, depending on national legislation and practice, the licensing authorities are still very reluctant or even refuse to approve safety related systems, whose behaviour is exclusively program-controlled. In general, safety licensing is denied for highly safety critical systems relying on software with non-trivial complexity.

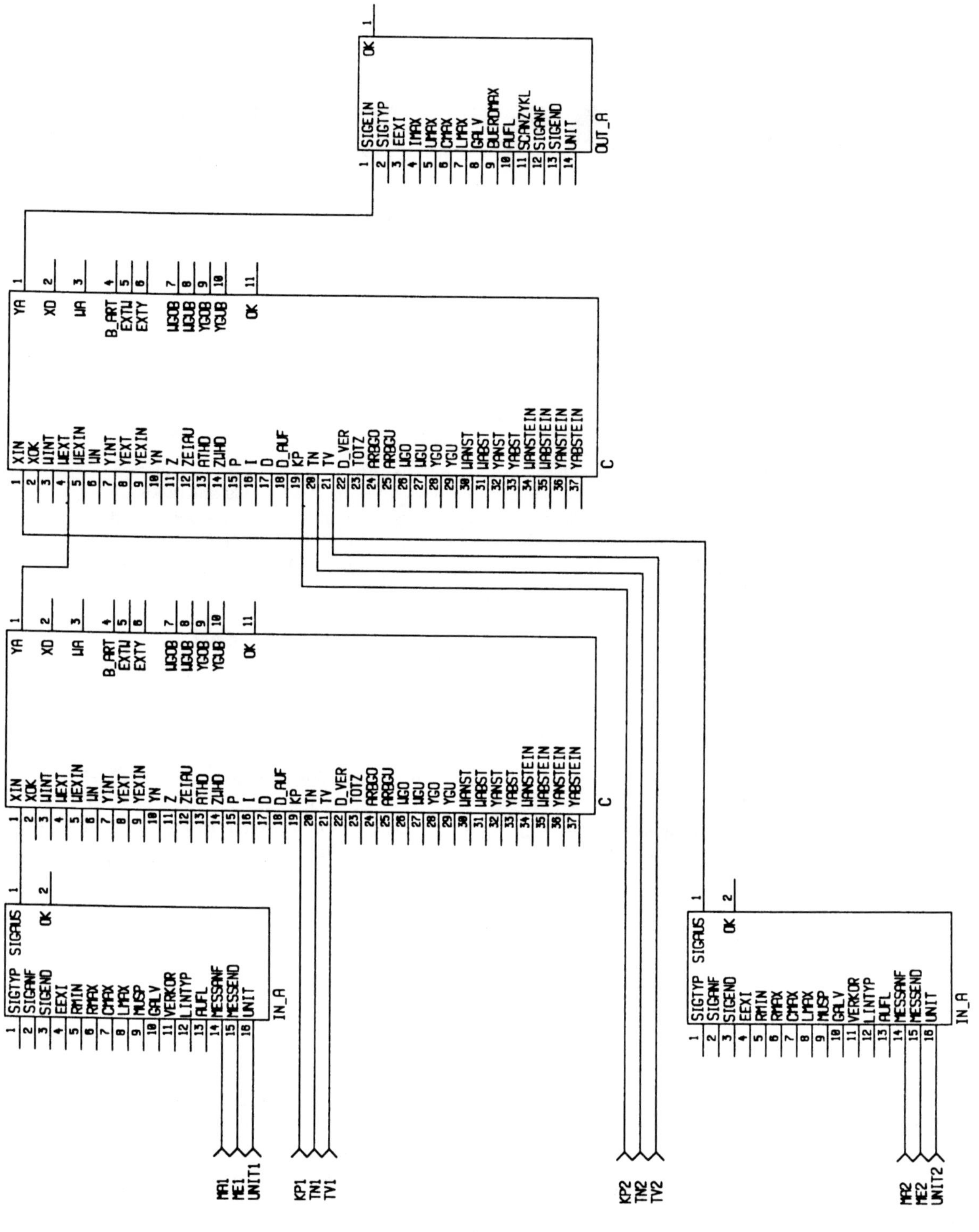

Fig. 1. A graphically formulated program

To provide a remedy for this unsatisfactory situation, the architecture of a customised real time computer control system was developed, which can carry out safety related functions within the framework of distributed process control systems or programmable logic controllers. It explicitly supports sequence controls, since many automation programs including safety relevant tasks are of that kind. The architecture features full temporal predictability, determinism, and supervision of program execution and of all other activities of the computer system, and supports the software verification method of diverse inverse documentation. The presented approach is unique as the first effort to provide support for software verification already in the architecture.

The leading idea followed throughout this design was to combine already existing software engineering and verification methods with novel architectural support. Thus, the semantic gap between software requirements and hardware capabilities is closed, relinquishing the need for not safety licensable compilers and operating systems. By keeping the complexity of each component in the system as low as possible, the safety licensing of the hardware in combination with application software is enabled on the basis of well-established and proven techniques.

A SOFTWARE ENGINEERING PARADIGM

A standardisation body of the German chemical industries has identified and defined a set of 64 application specific function modules suitable to formulate, on a very high level, the large majority of the occurring automation problems (Kempny and Maier, 1991). These software modules are usually quite short: not exceeding two pages of source code when written in a high level language. Therefore, their correctness can be formally proven, e.g., using predicate calculus, but also symbolic execution or, in some cases, complete test. Note that the rather costly safety licensing of a function module set needs to be carried out only once, after a certain set has been identified and standardised for a given application area. The licensing costs can be spread over many implementations leading to relatively low costs for each automation project. In order to give another typical example, the programming of emergency shutdown systems, which is usually performed graphically in form of functional logic diagrams, even requires as few as only four function modules, viz., three Boolean operators and a timer. For an in-depth treatment of the function module concept we refer to (Zöller, 1991).

The above analysis of process automation suggests to introduce a new programming paradigm, viz., to compose software out of high level user oriented building blocks instead out of low level machine oriented ones. Whereas a single machine instruction taken out of a program context does not reveal its purpose, the occurrence of a certain function module instance usually gives already a clue about the problem, its solution, and the module's rôle in it. Therefore, we introduce as the elementary units of application programming sets of basic function modules. Essentially, the sets of basic functions are specific to their respective application areas, although certain functions like analogue and digital input/output have general relevance.

For the formulation of automation applications with safety properties, the basic function modules are only interconnected with each other, i.e., single basic functions are invoked one after the other and, in the course of this, they pass parameters. Besides the provision of constants as external input parameters, the basic functions' instances and the parameter flows between them are the only language elements used on this programming level. The software development is carried out in graphical form: the instances of the basic functions are represented by rectangular symbols and the data flows are depicted as connecting lines (Fig. 1). Then, a compiler transforms the graphically represented program logic into object code. Owing to the simple structure, this logic is only able to assume, the generated programs contain no other features than sequences of procedure calls and some internal moves of data.

Many automation programs including safety relevant applications have the form of sequence controls composed of steps and transitions. With our architecture we support linear sequences of steps and alternative branches of such sequences (Fig. 2). While in a step, an associated program developed according to the above paradigm is being executed.

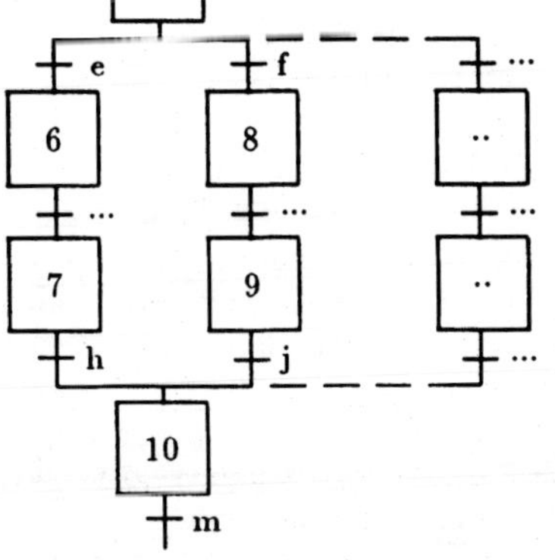

Fig. 2. Sequential function chart

THE ARCHITECTURAL CONCEPT

Since it is not our objective to save hardware costs, but to facilitate the conceivability of the implemented software and of its execution process, we design an architecture with, conceptually, two processors:

- a control flow processor (master) and
- a basic function block processor (slave).

Thus, we achieve a clear and physical separation of concerns: execution of the basic function modules in the slave processor and all other tasks, i.e., execution control, sequential function chart processing, and function module invocation, assigned to the master. This concept implies that the application code is restricted to the control flow processor, on which the project specific safety licensing can concentrate. There is special architectural support for the cyclic operating mode of programmable logic controllers implemented in the master processor. To enable the detection of faults in the hardware, a dual-channel configuration has been chosen, which supports diversity in form of different master processors and different slave processors. At least one of the master processors should have the extremely simple organisation described below.

As slave processors general purpose CPUs are required, preferably with a design that has been formally proven correct, such as for the VIPER. The master processor requests the slave processor to execute procedures by sending their identifications and parameters to the slave processor, which returns the results and, if applicable, internal states of the procedures. The basic function processor performs all data manipulations and takes care of the communication with the environment. It has no program RAM, but only executes the basic function modules whose code is contained in firmware ROMs. To recognise hardware faults, all processing is simultaneously performed on two processors each and all data communicated are subjected to comparison.

The master and slave processors communicate with each other through two FIFO-queues. They execute programs in co-ordination with each other as follows. The master processor lets the slave execute a function block by sending the latter's identification and the corresponding parameters and, if need be, also the block's internal state values via one of the FIFO-queues to the slave processor. Here the object program implementing the function block is performed and the generated results and new internal states are sent to the master processor through the other FIFO-queue. The elaboration of the function block ends with fetching these data from the output FIFO-queue and storing them in the master's memory. To avoid faults during operation, the function modules' object code is put in the slave's read-only program memory, after the correctness of the code has been established. The master/slave configuration has been chosen to physically separate two system parts from one another: one whose software only needs to be verified once, and the other one performing the application specific part of the software. Needless to say, that the latter requires indvidual safety licensing. All programs simultaneously run on the two master/slave pairs. A number of comparators checking the outputs from the master processors before they reach the slaves and vice versa completes a fault detecting two-channel configuration. Fig. 3 gives a conceptual diagram of the master/slave PLC architecture.

In available PLCs, the execution time for a step generally varies from one cycle to the next depending upon the program logic performed and the external conditions evaluated each time. Therefore, the measurement of external signals and the output of values to the process is usually not carried out at equidistantly spaced points in time, although this may be intended in the control software. To achieve full determinism of the time behaviour of programmable logic controllers, a basic cycle is introduced. The length of the cycle is selected in a way as to accommodate during its duration the execution of the most time-consuming step occurring in an application (class). It is supervised that the execution time of a step does not exceed this cycle period by awaiting, at the end of the step's program processing and after the evaluation of the corresponding transition condition(s), the occurrence of a clock signal, which marks the begin of the next cycle. An overload situation or a run time error, respectively, is encountered when the clock signal interrupts an active application program. In this case a suitable error handling has to be carried through. Although the introduction of the basic cycle exactly determines a priori the cyclic execution of the single steps, the processing instants of the various operations within a cycle, however, may still vary and, thus, remain undetermined. Since a precisely predictable timing behaviour is only important for input and output operations, temporal predicctability is achieved as follows. All inputs occurring in a step are performed en bloc at the beginning of the cycle and the thus obtained data are buffered until they will be processed. Likewise, all output data are first buffered and finally sent out together at the end of the cycle.

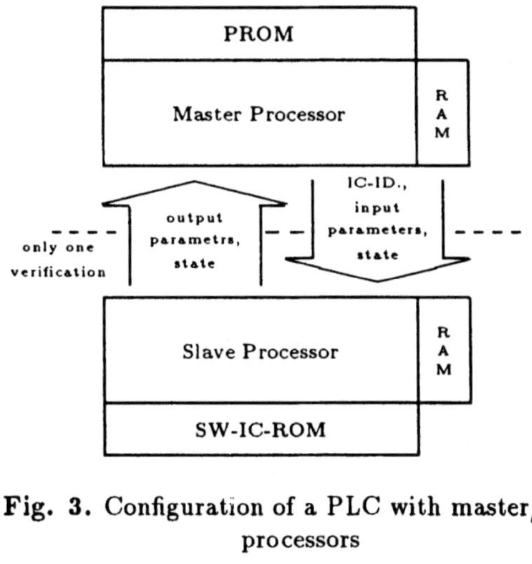

Fig. 3. Configuration of a PLC with master/slave processors

In the sequel, we discuss the hardware components

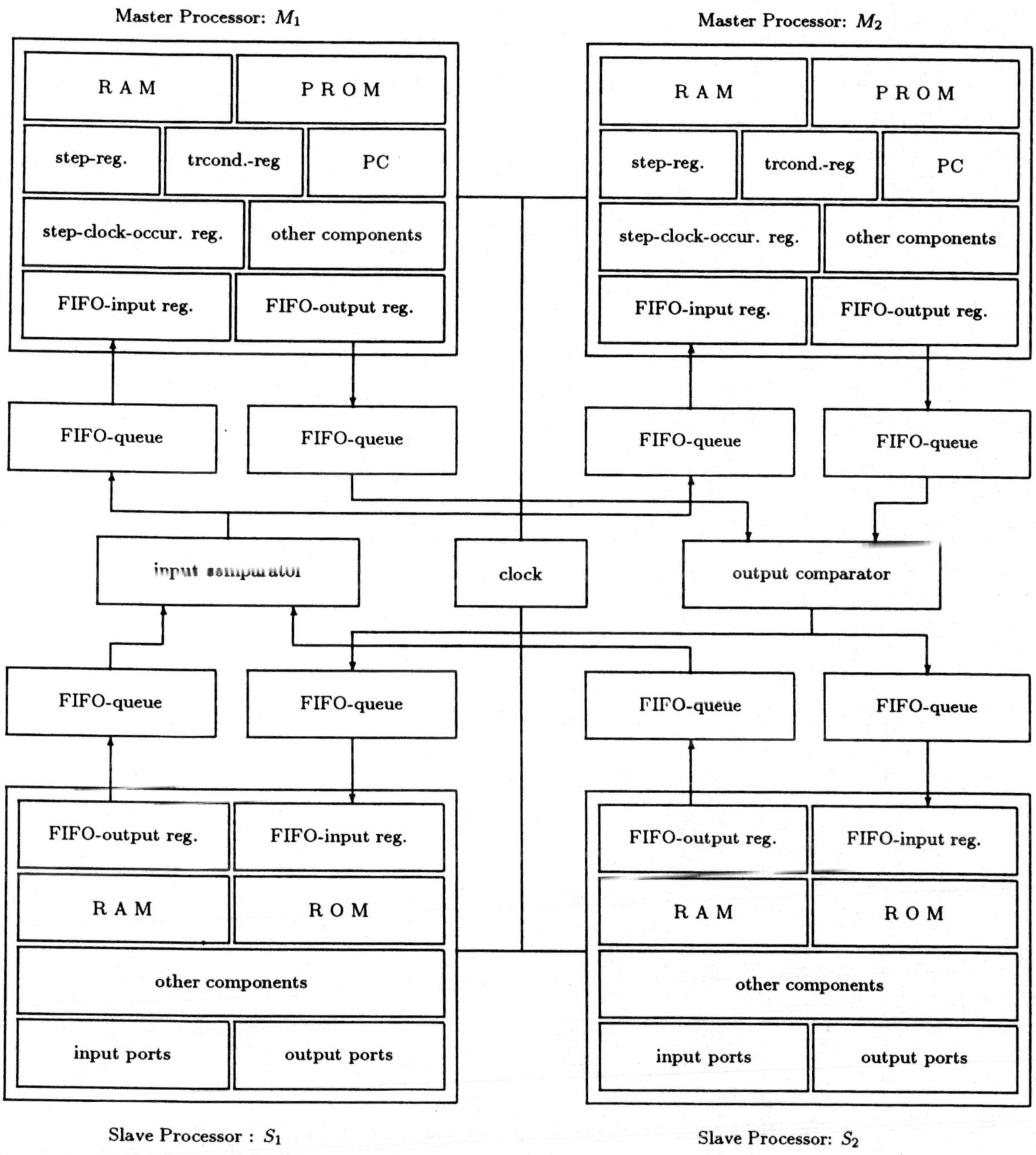

Fig. 4. Fault tolerant master/slave PLC architecture

- master processor,
- FIFO-queue,
- comparator, and
- system clock

of this special purpose computer in detail. The complete fault tolerant system architecture is depicted in Fig. 4.

The master processor executes the programs stored in its PROM memory. Besides this program memory, the master's address space also comprises a RAM memory and various registers. The latter are:

1. the FIFO-input register,
2. the FIFO-output register,
3. two step registers, viz., step identifier and step initial address, and
4. the transition condition register.

Furthermore, it has a program counter (PC) and a single bit step-clock-occurred register, which are not accessible to the programmer.

The master processor only performs two machine instructions, viz.,

- MOVE and
- STEP.

The instruction format is shown in Fig. 5. The PROM address of an instruction to be executed is given by the PC. The instruction type is identified by the instruction word's first bit.

S T E P	M O V E	Operand A	Operand B

Fig. 5. Instruction format

The MOVE instruction has two operands, which directly address locations in the address space of the processor. Thus, the memories and the above mentioned registers can be read and written. A read from the FIFO-input register implies that the processor has to wait when the input FIFO-queue register is empty. In case of writing into the output FIFO-queue register, the processor also has to wait when the register is full. Execution of a MOVE implies incrementation of the program counter.

The STEP instruction does not have operands. Since designed for the implementation of PLCs, the programs executed by the master processor consist of sequences of steps. Behind the program segment of each step a STEP instruction is inserted, which checks whether the segment was executed within a step cycle frame or not. The step cycle is a periodic signal generated by the system clock and establishing the basic time reference for the PLC operation. If the execution of a segment does not terminate within a step cycle, an error signal is generated and the program is stopped. Normally, the segment execution terminates before the instant of the next step cycle signal. Then the processor waits until the end of the present cycle period. When the clock signal finally occurs, according to the contents of the transition condition register it is decided whether the step segment is executed once more or whether the execution of the logically subsequent step is commenced, i.e., whether the program counter is re-loaded from the step-initial-address register or if another segment's initial program address is read from a memory location called next-step-address. Since program branching is only possible in this restricted form, erroneous access to code of inactive steps is prevented.

The design objective for providing the FIFOs is to implement easily synchronisable and understandable communication links, which decouple the master and slave processors with respect to their execution speeds. The FIFO-queues consist of a fall-through memory and two single bit status registers each, viz., FULL and EMPTY, which indicate the filling states of the FIFOs. The status registers are not user accessible. They are set and reset by the FIFO control hardware and, if set, they cause a MOVE to a FIFO's input port or from an output port, respectively, to wait until space in the FIFO becomes available or data arrives.

The comparison for equality of the outputs from the two master processors and of the inputs from the two slave processors, respectively, is carried out by the two comparators placed into the FIFO-queues. A comparator is connected with four FIFOs, two at its input and two at its output. The first data elements from each input queue are latched and subsequently compared with each other. If both latches do not hold the same value, then an error signal is generated, which stops the operation of the entire system. Otherwise, the value is transferred into both output FIFOs. The comparation of FIFO data is shown in Fig. 6.

The system timer generates, besides other periodic signals required for the operation of the hardware, a step cycle clock. This periodic signal marks the time frames dedicated for the execution of all steps (in the sense of PLCs). The step cycle is also applied to set the step-clock-occurred register. Fig. 7 shows the basic timing diagram for the complete system.

The PLC communicates with external technical processes through fault detecting input/output ports (cp. Fig. 8) attached to the slave processors. Output data words generated by the two slaves are first checked for equality in a comparator and, subsequently, they are latched in an output port. If output data are not identical, then an error signal is generated leading to a system stop. Output is not enabled, i.e., does not become effective to the environment, before the step-clock-occurred register is set by the step cycle signal. Also input data words arriving from the ex-

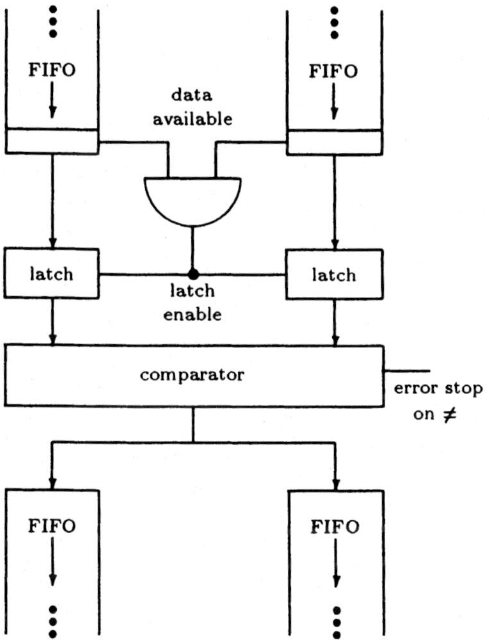

Fig. 6. Comparison of FIFO data

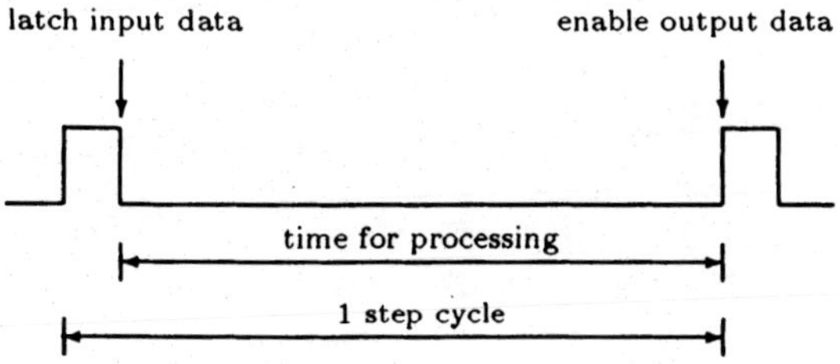

Fig. 7. Basic timing diagram

ternal technical process are first latched in an input port. With the next step cycle signal they are made available for input to both slave processors.

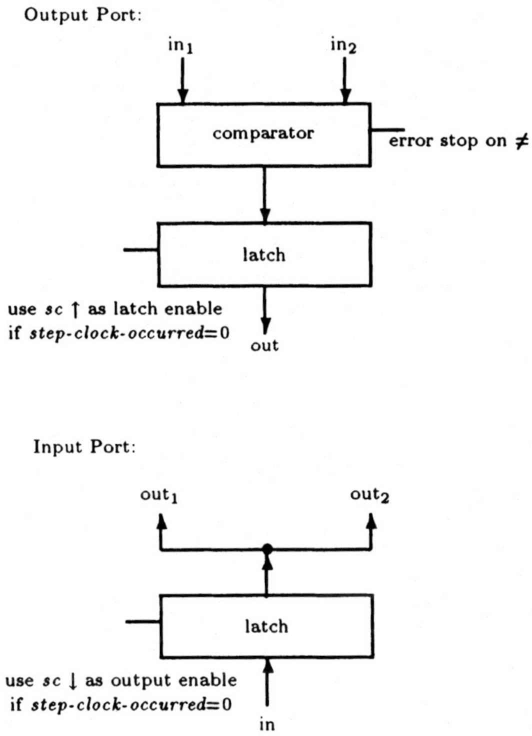

Fig 8. Output port and input port

SOFTWARE LICENSING

It is assumed that all elements of a function block set employed have been verified with appropriate formal methods. Hence, for any new application program, only the proper implementation of a particular interconnection pattern of invoked function block instances needs to be verified. For this purpose we subject the object code loaded into the master processor to diverse inverse documentation, a safety licensing method which was developed in the course of the Halden nuclear power plant project (Krebs and Haspel, 1984). This technique consists of reading machine programs out of computer memory and giving them to a number of teams working without any mutual contact. All by hand, these teams disassemble and decompile the code, from which they finally try to regain the specification. The software is granted a safety license if the original specification agrees with the inversely obtained re-specifications. Of course, in most circumstances the method is extremely cumbersome, time-consuming, and expensive. This is due to the semantic gap between a specification formulated in terms of user functions and the usual machine instructions carrying them out. Applying the programming paradigm of basic function modules, however, the specification is directly mapped onto sequences of subroutine invocations. The object code consists of just these calls and parameter passing. The implementation details of the function modules are part of our architecture. Thus, they are invisible from the application programming point of view and do not require safety licensing in this context.

Diverse inverse documentation is essentially informal, easily conceivable, and immediately applicable without any training. Thus, it is extremely well suited to be used on the application programming level by people with the most heterogeneous educational backgrounds. The ease of understanding and use inherently fosters error-free application of the method. Since graphical programming based on application oriented function blocks has the quality of specification level problem description, and because by design there is no semantic gap in our architecture between the levels interfacing to humans and to the machine, diverse inverse documentation leads back in one easy step from machine code to problem specification. For our architecture, the effort required to utilise diverse inverse documentation is by several orders of magnitude lower than for the von Neumann architecture.

As already its name implies, diverse inverse documentation is a verification method to the carried out with diverse redundancy. Originally, this called for different teams of human inspectors. Since in the case considered here there is only one rather simple inverse analysis step, we are optimistic that the licensing authorities will eventually accept the following procedure. Verification by inverse analysis is carried out by a number of different programs, which should be proven in practice but do not need to be formally verified. Such programs are to yield graphical outputs. An official licensor performs the inverse documentation as well, compares his results with the ones of the verification programs and with the original graphical application program under inspection and, upon coincidence, issues a safety license. Such a procedure is in line with the dependability requirements for diversely redundant programs demanded by the licensing authorities and necessitates only the minimum of highly expensive human involvement, viz., one licensor, who is always indispensable to take the legal responsibility for issuing a safety license.

CONCLUSION

Economical considerations impose stringent boundary conditions on the development and utilisation of technical systems. This holds for safety related systems as well. Since manpower is becoming increasingly expensive, also safety related systems need to be highly flexible, in order to be able to adjust them to changing requirements at low costs. In other words, safety related systems must be program controlled. Thus, we expect that the use of hard-wired safety systems will diminish in favour of computer based ones.

In our society there is a growing concern for safety (which comes hand in hand with the increasing awareness for the environment). This has important consequences for the assessment of computer controlled systems. One has begun to realise the inherent safety problems associated with software. Since it appears unrealistic to abandon the use of computers for safety relevant control purposes — on the contrary, for the reasons mentioned above, there is no doubt that their utilisation in such applications is going to increase considerably — the problem of software dependability will exacerbate severly.

In the situation as outlined above, this paper was timely to address a real problem. It did not present a solution to all open questions in safety related computing, but a beginning was made which is practically feasible and applicable to a wide class of common control problems. Hence, we hope that the concept presented here leads to the breakthrough of allowing to relinquish discrete or relay logic from taking care of safety critical functions in industrial processes by programmable electronic systems executing safety licensed high integrity software. Meeting the need of society for more dependable computing systems under the prevailing economic restrictions, we expect that the concept will give rise to workable industrial realisations.

In a constructive way, and only using presently available methods and hardware technology, for the first time a computer architecture was defined, which enables the safety licensing of complete programmable electronic systems including the software. Our solution deviates from the classical approach (as still followed by the mainstream in computer science) by using hardware as much as possible, but not necessarily in the most (hardware-) cost effective way, and by enforcing the (re-) use of pre-engineered off-the-shelf software modules. The former deviation is in line with the technological development: there is cheap hardware in abundance and it ought to be used to achieve our objective, viz., to realise inherently safe systems. The other deviation represents leaving the tradition of the von Neumann architecture allowing maximum flexibility — also to commit errors and to be unsafe. Later, when experiences will have been gained and a sound engineering methodology for safety related systems will have been developed, restrictions may be relaxed and more flexible approaches may be devised addressing wider application areas.

REFERENCES

Kempny, H.-P., and U. Maier (1991). *Vendor independent configuration of distributed process control systems.* Draft guideline, VDI/VDE-GMA VDI/VDE-GMA Working Group 5.3 and NAMUR Working Group 1.2. Personal communication.

Kershaw, J. (1987). *The VIPER Microprocessor.* Royal Signal and Radar Establishment, Malvern, England, Report No. 87014.

Krebs, H., and U. Haspel (1984). Ein Verfahren zur Software-Verifikation. *Regelungstechnische Praxis rtp*, 26, 73 — 78.

Zöller, H. (1991). *Wiederverwendbare Software-Bausteine in der Automatisierung.* VDI-Verlag, Düsseldorf.

Safe, Fault-Tolerant and Deterministic Algorithms for Real Time Control

Peter R. Croll

Parallel Processing Research Group, Department of Computer Science, The University of Sheffield, Tel: +44-(0)742-76855, E-mail: P. Croll@dcs. shef. ac. uk

Abstract: The use of parallel processing for real-time control applications is explored. How the programmer can be presented with a virtual environment which includes timing attributes is outlined. The inclusion of deterministic timings permits both a simple and dependable approach to processor failure and simplifies the formal analysis necessary for highly reliable applications. The use of CASE tools to automate this analysis and provide an integrated approach is discussed.

Keywords: Real-Time Control, Fault tolerance, MIMD, Message Passing, OCCAM, Formal Methods, CASE, Safety-Critical

Introduction

Stankovic [1] maintains that it is a common misconception to believe that many of the problems in real-time system design have already been solved. Solutions directed at guaranteeing *hard* real-time constraints based on MIMD message passing systems have been explored [2,3]. The results from this can benefit the many real-time control applications which exploit the versatility of the Inmos transputer, e.g. [4,5]. Many such applications can, if they fail, be hazardous to life, environment or machinery and should, therefore, be regarded as safety-critical. This necessitates a formal approach in order to provide a degree of rigour into the design process [6,7,8]. This paper will, therefore, outline the results of research which has produced *hard* real-time algorithms suitable for safety-critical control applications based on multi-transputer architectures.

Parallel Processing in Control

The use of parallel processing for control applications can offer the system designer many advantages. The transputer was designed to make high performance embedded systems which are scaleable, flexible and readily modelled. That is, it is possible to develop models which directly relate to the application in hand, as well as provide varying degrees of fault-tolerance by the addition of more processing elements. The real-time designer of such systems is now faced with a new set of problems. How can they be sure that external events will be serviced and the resultant messages will be passed around the parallel networks yet still meet their time critical constraints. That is, will the network ever **deadlock**, can all of the application processes avoid being **livelocked** out and will these conditions hold in the event of a processor failure?

Deterministic Timing Environments

Existing environments

The control engineer requires a development environment which permits deterministic timings to be calculated. Although many contemporary operating systems would claim to be real-time they would not facilitate *hard* real-time analysis. That is, the overheads introduced by scheduling and input/output operations would make calculation, in any practical system, not only very difficult but application dependent. Some specialised operating systems address this problem by offering 'light-weight' real-time kernels. These can allow the use of deterministic scheduling algorithms, for example, deadline or rate monotonic. Typically, an application will be divided up into a number of event-driven autonomous tasks which are executed until completion. This allows each task to be assigned a scheduling priority according to either its frequency of use or its time critical deadline. These mechanisms can make efficient utilisation of a single processor.

The parallel environment

With the distributed environment of a parallel processing system the above approaches do not necessarily apply. Efficiency may not be the key to developing a successful system since processing power can be enhanced by the addition of more processing nodes. For example, each task can be assigned to a separate processor or, alternatively, tasks can share processors up until the critical timing of any individual task on that processor is exceeded. With the single processor, the passing of data between tasks is facilitated by shared memory locations. This means that the time calculations would include data transfer in the total execution time of the instructions. Since this is not the case for a parallel processing system, a mechanism for passing the messages between distributed tasks within hard timing constraints will be needed. A control engineer would, therefore, need to include these overheads in the timing calculations.

The virtual programming environment.

The method advocated in this paper is to provide the control engineer with a virtual environment for distributing tasks which includes the timing aspects of message passing between tasks. Each task will have a timing associated with the instructions it executes plus the time it takes to pass data to other tasks, see figure 1. Communications are assumed to be synchronous, therefore, time delays for other tasks to respond to these communications will need to included in the calculations. The maximum response time it takes for a task to receive and process a message will be referred to as the *message latency*.

Since this is a virtual environment the programmer should not be concerned with the physical system. The next section shows how a message passing mechanism can be derived which will achieve this.

The Communication Harness

Many physical topologies are possible in parallel processing systems each with their own attributes, e.g. average communication distance, routing complexity. If the above virtual environment was to be implemented using transputers then point-to-point communications would be quickly limited by the four available links. This necessitates the use of a communication harness to route messages between non adjacent processing nodes.

The simple ring will be used here as an example of a scaleable topology. Figure 2 shows the data flows of the router process that will pass messages to and from the application task and around the ring. It is possible to calculate the worst case time delay for a message to be received by the router, i.e. the message latency. This was achieved by making the router and the event handler the only tasks to execute at high priority, which on the transputer is non-interruptable. Now, the message latency from the view point of the application task will be the sum of the latencies of all the routers that the message will pass through.

This assumes that the routers can never be delayed. In practice, due to the use of synchronous communications, it is possible for the communication harness to deadlock and prevent any further communication on the ring.

Deadlock can be prevented by using a mechanism which ensures at least one router process is able to continue passing messages round the ring. This can be achieved either by

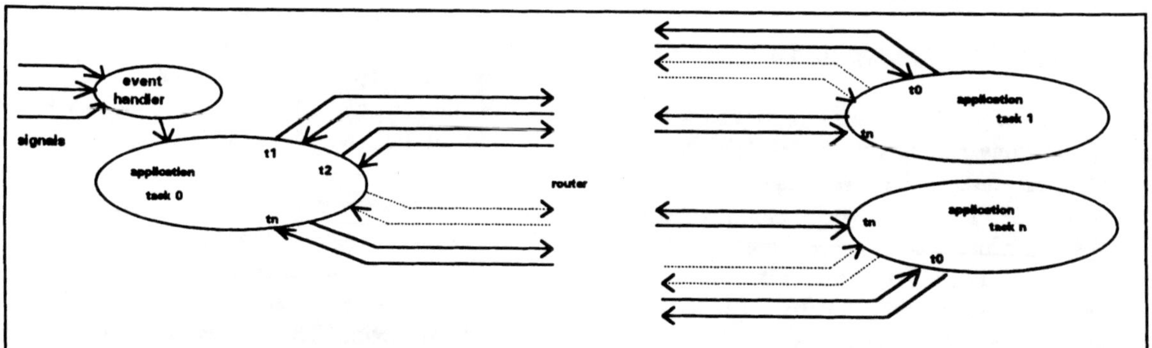

Figure 1 The Timing Model for the Application Tasks

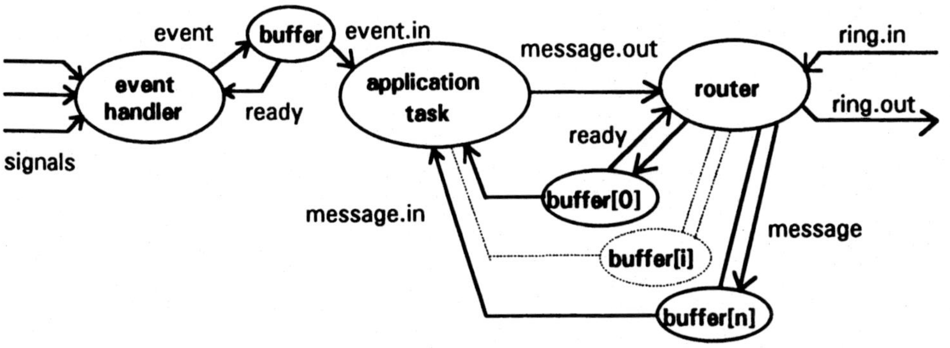

Figure 2 Data Flow of a Single Processing Node

adding extra buffering (which can only be used to free ring messages) or by keeping one router process in the ring free by assigning it a special flag. This flag, see figure 3, is referred to as a *blocking token* since it will block any attempt by the application task to add additional messages to the ring. Every time a router receives a message on the ring then it will send back an acknowledgement. If the router has the blocking token then this will be passed back as the acknowledgement. The blocking token will cycle round the ring to keep one router free and allows all the application tasks access to the ring.

This design has made the assumption that messages pass freely round the ring at all times. In practice, the generation of messages is nondeterministic, therefore, an application task can be livelocked by the blocking token for an indeterminable period of time. Obviously, this will not suit our hard real-time environment and measures will be need to be taken to combat this. The solution is to allow an application task to pre-empt the passing of the blocking token and wait for a message. If the previous node is currently processing a message it will behave exactly as before even though the acknowledge has already arrived and if it is idle then the blocking token can prompt for a dummy message to be generated. This would appear to satisfy the requirements of deadlock prevention and livelock avoidance. But, as will be shown below, when the design was put to the more rigorous analysis using formal methods a subtle, yet unlikely, livelock condition was revealed.

Fault Tolerance

This section will describe the mechanism by which faults can be detected in individual processing nodes and by-passed without deterioration to message passing latencies.

The transputer is supported by library procedures which can allow for faults on link communications, e.g. OutputOrFail(). These procedures allow the programmer to specify a time limit on a communication after which an alternative process can be activated. Often the time limit is a stochastic measurement after which an attempt to reset the link is made. The transputer links were never designed to be fault-tolerant and, therefore, the reset operation is not guaranteed under all conditions. It is safer to assume both the link and the processing node has failed completely and bypass this node until after a system reset and appropriate checking operations have been performed.

Figure 4 shows how a processing node can be bypassed as soon as the link communication times out. In the ring design above, the time out can be set precisely, since the worst case message latencies are deterministic. Whenever

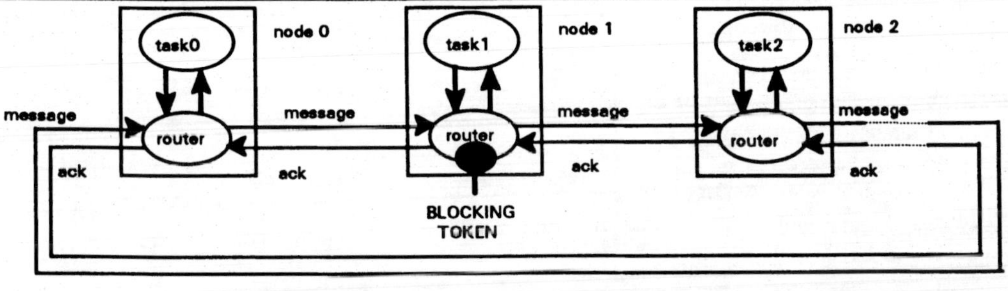

Figure 3 The Blocking Token

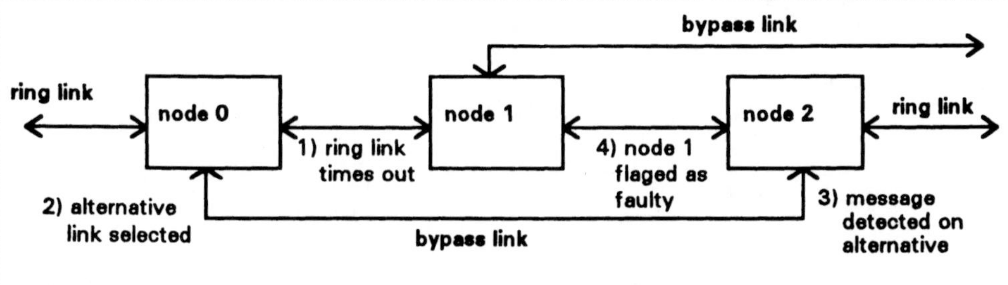

Figure 4 Bypassing a Failed Node

a message times out then an alternative route is selected and the adjacent processing node will be flagged as faulty to prevent any further communication. The next processor further down the ring will then receive a message on the alternative channel and will also flag the suspect processor as faulty. This bypass operation will not affect the message latency since it requires less time than it would take for a message to pass through the processing node if no fault had been registered. This mechanism ensures that, at maximum, only one message may have been lost due to a single node failure. This can be allowed for at the application level by either timing out on an acknowledgement or simply by always sending two messages.

The advantage of using the blocking token mechanism for a distributed parallel system can be appreciated by considering the effect of a single node failure. No central control exists to determine if the node that failed contained the blocking token. Therefore, the detection of a message on the alternative channel should always be acknowledged with a blocking token. This ensures that if the failed node had the blocking token it will be regenerated. Alternatively, if the failed node did not contain this token then having additional tokens presents no problem since each node represents the token as a Boolean flag and as such they will be absorbed as they pass round the ring.

High Reliability

If these algorithms are to be used for safety-critical applications then the question that still needs to be answered is how much reliance can be put on the above designs? The use of a formal development method can increase the confidence that might be placed on any such design. Formal mathematical methods tend to be very time consuming and many programmers lack the necessary proficiency. The approach taken was to use a formal method that could prove the specific properties that were considered applicable to these algorithms. That is, that deadlock is prevented and livelock avoided under all possible program states. The tedious exhaustive search required for safe state analysis was performed by integration of a Petri-net tool into a Computer Assisted Software Engineering (CASE) package. The problem of *state explosion* often experienced with Petri net models was addressed by restricting the model to deal only with communications and by enhancing the model to include colours and timing attributes.

Proving behavioural properties

The first question to be raised in deriving an equivalent Petri net model for a given design is how to do it? It is possible to transform the occam source code to a Petri net but this would soon suffer complexity problems introduced by the flexibility of a programming language. The inverse would be to design a net and transform this into the occam communication harness. The resultant net design was achieved by starting with a state transition diagram which represented all the communications. Occam communications are synchronous and can be represented as state changes in each of the parallel processes, for example, figure 5 shows an occam communication primitive and its equivalent Petri net model. A state transition diagram was developed for both the router and its interactions with the application tasks, then transformed into a Petri net model.

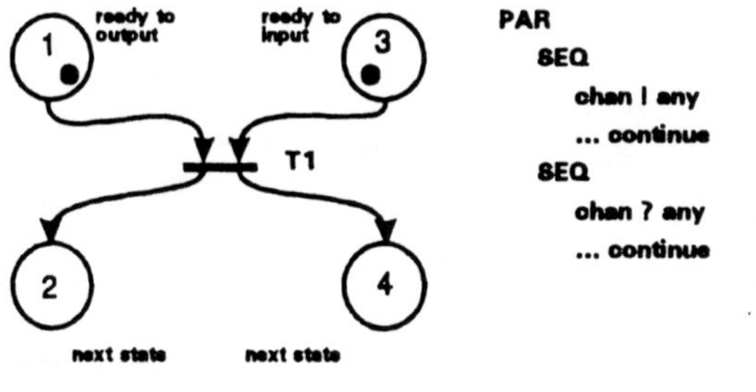

Figure 5 Petri net of an occam channel based on states

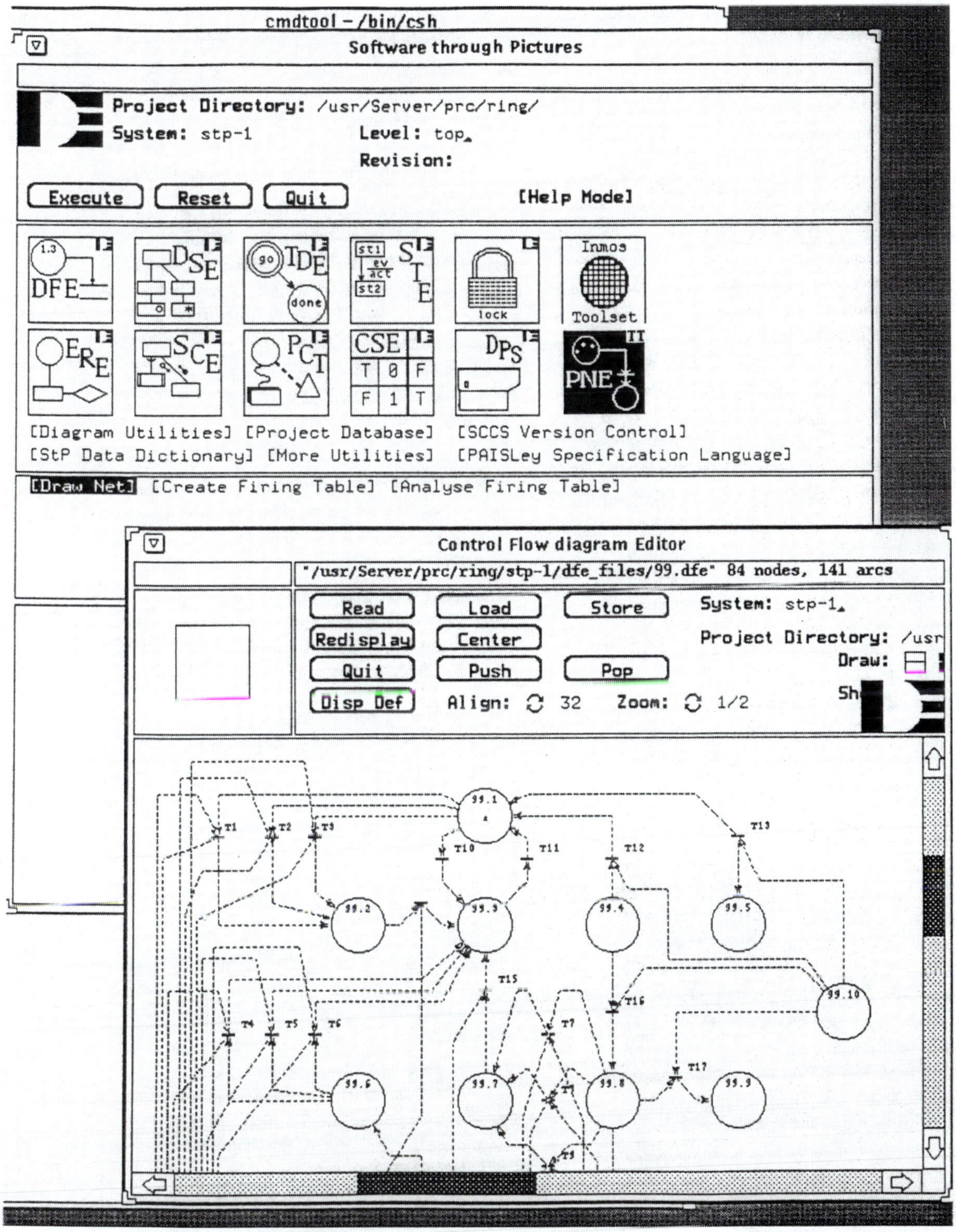

Figure 6 Software through Pictures

Having developed a model it was tested for deadlock by generating all the reachable states. The net is live provided that from any state reached, it is still possible to fire a transitions on the Petri net.

The analysis of livelock requires the addition of timing attributes to each transition. Since each transition represents the transfer of a message then these timing figures can be taken from the message latencies described above. The net can now be analysed to ensure that whenever the choice is given, i.e. more than one transition can fire, these timings can not exclude any choice for too long. That is, will the timings ensure that the control flow will select each available choice before the latency time elapses on that choice. The initial analysis of this net produced a scenario whereby the blocking token could livelock out a message. This would occur whenever the application task tried to pre-empt the blocking token and before it could get access to the ring the token had passed completely round the ring and blocked it again. The probability of this happening was extremely rare and never showed up in testing but the design fault could have remained if a formal analysis had not been pursued. Such faults often remain undetected by testing and only show up after later modifications to a design are made, for example, optimising to achieve performance maximisation.

The use of CASE

The CASE package used was Software though Pictures (StP) developed by IDE. This was chosen as it provides the necessary open environment for user tool integration, see figure 6. The motivation for using CASE is to provide an integrated environment for program development. That is, the CASE tools are used to develop the design, e.g. data flow, state transition diagrams, Petri net model with as much automatic implementation of source code from these designs that can be achieved.

Conclusions

Parallel processing systems can offer many advantages to control applications. It has been shown that the conventional approaches do not necessarily apply to parallel systems. In particular, the timing calculations normally carried out will now have to include message passing. It is possible to provide a virtual environment to the programmer which includes timing attributes by means of a communication harness. Special considerations were needed to ensure that hard real-time constraints can be met under all operating conditions. This, it has been shown, is achievable by defining latencies for the message transfers and designing the system to ensure these latencies are met in practice.

If the reliability of the design is crucial then this can be achieved by using a suitable formal method in the analysis. This uncovered hidden design faults which could have easily remain undetected. The inclusion of these techniques into a CASE tool gave two major benefits: the analysis could be automated; it allowed an integrated systems approach to the problem.

References

[1] Stankovic, J.A.:"Misconceptions about Real-Time Systems - a serious problem for next-generation systems" IEE Computer, October 1988, pp.10-19.

[2] Welch, P H. "Managing Hard Real-Time Demands on Transputers", Applying Transputer Based Parallel Machines, Ed. Andre Bakkers, IOS Press, Amsterdam, September 1987, pp.198-212.

[3] Croll, P.R. & Wilson, G.A.:"Peripheral Handling Techniques for the Transputer", Microprocessors and Microsystems, vol.13, (2), March, 1989, pp.124-128.

[4] Fleming, P.J. & Jones, D.I.:"Parallel Processing in Real-Time control: Applications, Architectures and Implementations" Proc CONPAR 88, Manchester, Sept 88, pp.A235-A336.

[5] Irwin, G.W. & Fleming, P.J.:Editorial, Special Issue, IEE Proc. on "Parallel Processing for real-time control", Pt. D., Vol 137, (4), pp.177-178, July 1990.

[6] Sagoo, J.S. & Holding, D.J.:"The Specification and Design of Hard Real-Time Systems using Timed and Temporal Petri-nets", Microprocessor and Microprogramming, 30, (1990) pp.386-397.

[7] Croll, P.R.:"Deterministic Message Routing for Safety Critical Applications", Occam and the Transputer - Current Developments, Ed. Janet Edwards, IOS Press, Amsterdam, September 1991, pp.234-246.

[8] Croll, P.R. & Nixon, P.A.:"Developing safety-critical software within a CASE environment", IEE colloquium on 'Computer Aided Software Engineering tools for Real-Time Control', No. 1991/087, 24 April, 1991,

Voting Software for Fault-Tolerant Aircraft Flight Control Systems

Julian, M. Bass, Peter J. Fleming

Dept. of Automatic Control and Systems Engineering,
University of Sheffield, P.O. Box 600, Mappin Street, Sheffield, S1 4DU, U.K.

Keywords - Real-Time Systems, Distributed Systems, Fault-Tolerance, Voting, Software Fault-Tolerance.

Abstract

An intelligent and robust voting service is proposed. The novel voting service, is capable of forming a consensus based on performance of redundant applications in the value, time and reliability domains. The voter is distributed using a hot backup approach to ensure timely recovery in the event of voter failure. Specification diversity is advocated in order to improve upon the effectiveness of design diversity. The voter is designed to be a system building block, embedded in a framework of design tools which will allow the developer to express control concepts, such as deadlines, priorities, replication and voting.

1 Introduction

In many modern real-time distributed control systems the costs of system failure, in terms of life and property, are unacceptable. Commercial flight control applications are a good example of systems in which even temporary failure could be disastrous.

Fault-tolerant flight control systems are made more complex by the need to respond to events in the outside environment in real-time. The central issue in the design of such a system is not only "what" happens but "when" it happens [Harrison91].

Parallel systems with multiplicated processing resources can provide the redundancy required to protect against node failure. In striving to avoid a single point of failure, algorithms and strategies are distributed throughout the system. Such distributed systems offer many potential advantages in providing good fault coverage but introduce many problems of their own. These problems include:- concurrency control [e.g., Yuan91]; distributed consensus [e.g., Luan90]; and scheduling, [e.g., Tindell90].

Designers are obliged to provide systems that perform correctly even in unexpected circumstances. The difficulty of obtaining completely reliable hardware and software components has created considerable interest in a systems design approach. Such a unified systems approach, taking full advantage of both fault-avoidance and fault-tolerance, is required to ensure sufficient reliability. Developers using a systems design approach avoid decisions that constrain the design early in its life. For example, the exact mix of hardware and software need not be specified until the module is implemented.

A brief overview of real-time distributed systems is presented in Section 2. The next Section discusses majority voting in the traditional sense, i.e. finding a consensus among application result values:- the "value domain". A discussion follows on the implications of first, timing faults and then, application reliability in a voting system. A unified approach is proposed in which the voting service must be capable of monitoring application performance not only in the value domain but also in the time and reliability domains. Section 4 reports experimental work on the development of such an intelligent voting service.

2 Background

Design diversity is a reality in many applications [Voges88]. Such diversity is useful for minimising common errors introduced during the design and implementation phases. However, design diversity does not guarantee the elimination of specification errors although the existence of competing design teams can subject specifications to closer scrutiny. Traditionally, there are two approaches to implementing design fault-tolerance:- N-Version Programming [Avizienis85] and the Recovery Block [Anderson81]. These strategies use redundant modules of code to protect against faults. Recent work seeks to take advantage of the strong features of both schemes by executing the alternative modules of a recovery block in parallel [Kim89] or using acceptance tests with voters [Kant90].

Currently there is inadequate support for the implementation of real-time control software for reliable systems [Croll91]. Designers need to assess such mechanisms as checkpointing, voting and acceptance testing in order to determine the appropriate strategy to realise specified reliability levels. Modern software engineering methods must be used where systems are complex and have evolving requirements. Our research is concerned with developing a framework of design tools which enable the developer to express control concepts, such as deadlines, priorities, replication and voting. Designers of complex systems need tools to allow the early stages of design to proceed without prejudice to the type of systems, technologies or methodologies that will be used to provide functionality. The tools must enable the designer to select appropriate methods to achieve accuracy, timeliness and reliability of results. This paper explores one aspect of such a framework: voting in a real-time distributed system.

3 Voting Algorithms

The voter is viewed as a system service that can be utilised by a wide range of applications. An intelligent voter can provide greater functionality and provide greater coverage to faults leading to improved reliability. The voter must be designed using formal methods and subjected to much greater verification and validation than would be cost effective for application code. Reuse in a variety of application environments would increase confidence that the voter behaves as specified. Such a "hardened" voter should be seen as a basic system building block in the same way as, say, a floating-point arithmetic library.

A unified approach to voting in the value, time and reliability domains is required. Traditionally, a voter has been seen as a device for masking out result values that do not form a majority. This paper asserts that a voter should also be used to mask out missed deadlines and unreliable application results.

In control system software errors cause discontinuities in the application results. In order to tolerate such errors it is beneficial to know the region of the correct result.

3-1 Voting In The Value Domain

In many voting systems the application results are checked one bit at a time to ensure that the majority of the bits is identical. Bit voting is very susceptible to common errors and requires tight synchronisation of applications systems. It is also possible to check for identical values when the application systems produce integer values.

3-1-1 Distinct Results

It is not possible to vote on identical application system results when floating-point arithmetic is used [Knight86]. Instead a "consensus threshold" is used to eliminate rounding or truncation differences. The voter can then accept application results that lie within the threshold as forming a majority. Several algorithms have been documented that either perform an arbitrary selection of the consensual results, or calculate some average value to provide the voter output [e.g., Lorczak89]. Many applications have numerical limits on the acceptable results that can be produced. It is easy and useful to provide these reasonableness limits to the voter although they are application-specific.

3-1-2 Disageement

In cases where no agreement can be found between application results a majority of the applications are likely to have suffered faults. The application faults may be permanent, transient or due to some intermittent fault. In the case of disagreement it may be necessary to output a safe result or avoid a discontinuity by calculating an approximate result. The approximate results can be used for a few cycles, while some attempt is made to restart the software or reconfigure the system.

3-2 Voting In The Time Domain

The voter must be rigourously designed to ensure that it does not miss deadlines. Problems associated with synchronising application systems that have been developed using diverse design are well documented. A voter must be capable of arbitrating between application systems which are subject to timing as well as value domain faults. This assertion suggests a time domain dual of the consensus threshold. Thus, the temporal behaviour of voting systems can be separated into two areas:-

- temporal behaviour of the voter itself - voter timeliness,
- temporal behaviour of application systems - application timeliness.

3-2-1 Voter Timeliness

The voter must run on separate hardware from the application to provide acceptable fault coverage in all but the smallest of systems. It is therefore much easier to guarantee the timeliness and schedulability of the voter than that of the applications in synchronous systems. The voter is likely to be markedly simpler and smaller than the application systems even though it is intelligent. Thus, it is more difficult to ensure that the application results arrive at the voter on time than it is to gather the majority output.

3-2-2 Application Timeliness

The case where all application results arrive on time, or least do not miss their deadline, is simple and well understood. Voting is performed in the value domain alone. This case is considered no further.

It is necessary to determine if a timing failure is due to a permanent fault or is transient in nature. A transient timing fault will cause a temporary increase in the fault intensity variable for the associated application. This temporary reduction in reliability will be restored when the application has produced correct results, on time, for a specified period. For example, a transient communication fault could cause a communication retry resulting in a missed deadline.

A permanent fault must be masked by the voter but should also cause some attempt to recover from the fault by reconfiguration. The masking of timing faults is required during the period that the system takes to reconfigure and should be regarded as a temporary system condition. Initially, a software reconfiguration may be attempted by reinitialising the software only. If the timing faults persist then a hardware fault must be assumed and a hardware reconfiguration attempted. During this reconfiguration process the voter must mask the timing faults and continue to provide sensible voting results based on the other application results.

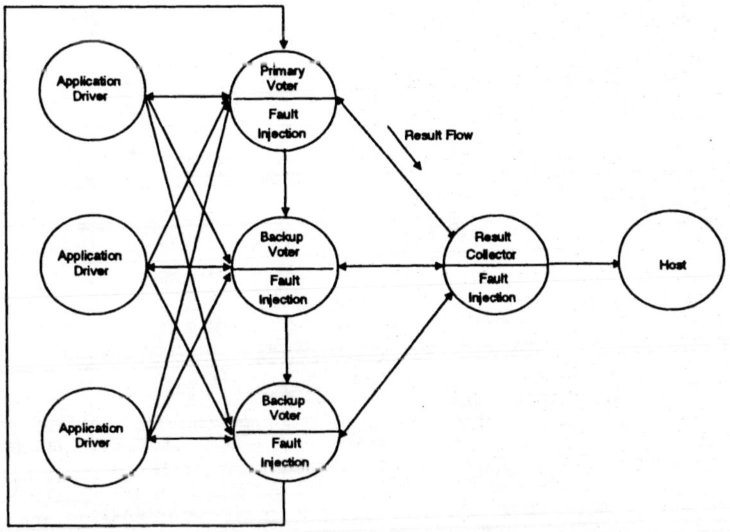

Fig. 1 Logical topology of voter software.

3-3 Voting In The Reliability Domain

The reliability of a voting system depends upon the reliability of:-

- the voter, and
- the application systems.

3-3-1 Voter Reliability

In distributed systems it the voter must not present a single point of failure by being executed on a single piece of hardware. The voter can be algorithmically decomposed to ensure agreement between separately executed modules. Alternatively, the voter can be executed on a single node but provided with hot backups, ready to provide a result in the event of a primary failure.

3-3-2 Application Reliability

The diversely designed applications, within a system, are traditionally intended to achieve the same specified reliability level. We propose that an improvement in overall system reliability can be achieved by designing a "high reliability application". Such an application can be achieved by relaxing the specified result accuracy. A further benefit of a relaxed specification in the value domain is faster execution. The high reliability of the result produced provides information that can be used to prevent common errors forming a majority at the expense of a single correct result. The "high reliability application" result indicates the region in which the correct value lies with a high degree of confidence. This "diverse specification" approach reduces the likelihood of ambiguities and contradictions remaining undetected in a specification. The "high reliability application" is a model for the application used to produce approximately correct results with extreme confidence. A difficulty with this approach will be ensuring the appropriate consistency of the diverse specifications. Further research is required to explore the benefits of diverse specification weighed against consistency and other difficulties.

Although intended to achieve specified reliability levels, application fault intensities will vary in time. Fault intensity variations are due to design faults manifested by a particular input data set, the external environment or hardware deterioration. Each application system must operate within a specified reliability threshold. Failure of an application system, to meet the reliability threshold should precipitate some system reconfiguration in an attempt to recover the errant component. It is unlikely that the designer will have advance knowledge of the exact reliability of an application system. The voter must keep a record of the fault intensity of each application in order to eliminate those results with high uncertainty.

4 Intelligent Voter Experiments

A distributed voter with the ability to tolerate application system timing faults has been built. The voter and application systems are written in Occam and are executed on a network of eleven Inmos T800 Transputers. The application systems are arranged in a triple modular redundant configuration,

Application Results	Application Fault Intensity	Voter Output: Identical Result Voter	Voter Output: Destinct Result Voter	Voter Output: Intelligent Voter
All Identical	Not Applicable	Correct	Correct	Correct
Majority Identical	Not Applicable	Correct	Correct	Correct
All within Consensus	Not Applicable	No Vote	Correct	Correct
Majority within Consensus	Not Applicable	No Vote	Correct	Correct
Majority within Consensus - One correct - One incorrect - One correct with missed deadline	 Low High Moderate	No Vote *	No Vote *	Correct
Complete Disagreement	All High	No Vote	No Vote	Safe Output

* When deadline is missed.

Fig. 2 Voter result classes.

as shown in Fig. 1. Results are passed, through a fully connected communication network, to the distributed voter. The voter comprises a primary and two hot stand-bys. Figure 2 shows a survey of the classes of application result that can be voted upon using traditional voters and the intelligent voter.

4-1 Value Domain Performance

Application outputs fall into a number of classes. The trivial case is when the application outputs arrive at the same time and have identical values. The voter checks for each specified class of output in order to find the highest quality of agreement. The checks are performed in sequence, as shown in Fig. 3, and the best vote found is output. Each application system can produce the following classes of result;

- identically equal
- distinct and inside the consensus threshold
- distinct and outside the consensus threshold
- outside reasonableness threshold.

Each of these classes can be produced with or without periodic timing faults.

4-1-1 Distinct Results

If no pair of the application results are identical a consensus threshold is used. A check is made for majority agreement within the consensus threshold, the fault intensity variable of an application outside the consensus is increased.

4-1-2 Disagreement

The intelligent voter attempts to find the best of the disagreeing application results based on a history of the reliability of the outputs and their previous values. A high reliability estimation provides a window within which the voter result is expected to lie. The window expands as the rate of change of data increases. The result estimation is derived from simple linear prediction or the previous voter output, and is suitable for use in systems with replicated software. In the case where no agreement can be found a safe result is output.

4-2 Time Domain Performance

4-2-1 Voter Timeliness

Initial timings show that the voter takes between 0.2ms and 0.25ms to execute, depending on the level of agreement between application results. These times do not include the group membership algorithm or timeouts for delayed application results. The execution time and timeouts must be considered when deciding the invocation time of the voter.

4-2-2 Application Timeliness

An application that does not provide a result on time is regarded as having produced an unreasonable result. A missed deadline causes an increase in the fault intensity variable for the application.

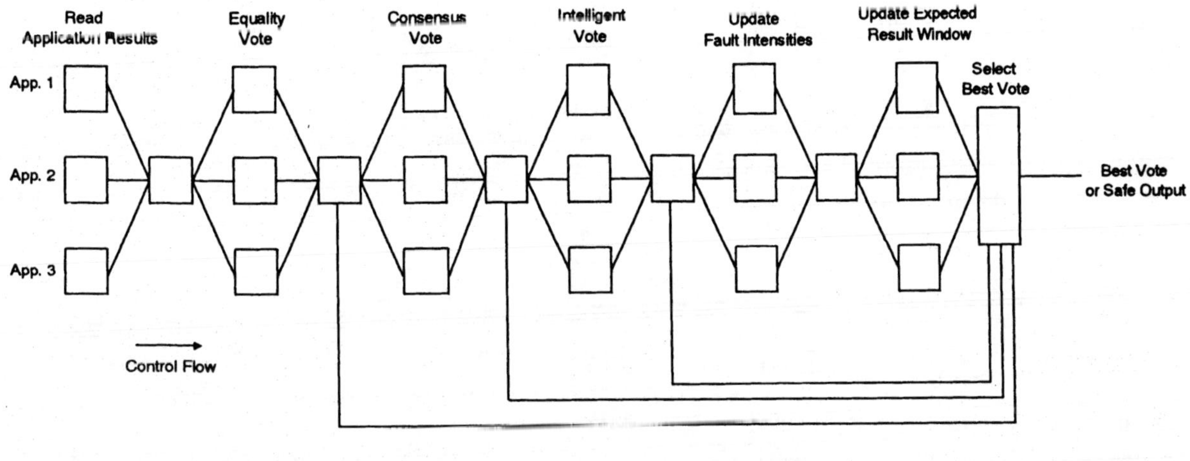

Fig. 3 Sequence of control flow in voter software.

4-3 Reliability Domain Performance

4-3-1 Voter Reliability

The hot backups are used to ensure quick recovery in the event of primary voter failure. A "processor group membership" approach [Cristian88] [Becker92] is used to ensure that one, and only one, voter is primary at all times. A formal design for the distributed voter algorithm will be presented in a future paper.

4-3-2 Application Reliability

The voter keeps a record of faults during recent history of operation for each application. This fault intensity record is used to mask the results of poorly performing applications in the case of disagreement.

5 Future Experimental Work

The voter developed will be applied to an aircraft flight control application utilising a specification diversity approach. A "high reliability" version of the application will be executed in parallel to improve fault coverage, and provide results more quickly, with greater confidence. The reliability is improved by reducing the complexity of the software, and thus the required hardware. The "high reliability" application will, of course, produce results of lower accuracy.

6 Conclusions

Design diversity is required for systems in which failure is unacceptable in terms of loss of life or property. Voting is a fundamental element of such a system. Intelligent voting is required to find a consensus among results of varying performance in the value, time and reliability domains. Intelligent voting with wider coverage is possible if the applications are deliberately specified with different performance goals.

7 Acknowledgements

The authors wish to acknowledge the support and collaboration of George Mann, Steve Ellis and Craig White of Smiths Industries Aerospace and Defence Systems Ltd. We would also like to thank Dr Peter Croll (Department of Computer Science, University of Sheffield) for many constructive comments on the work leading to this paper.

8 References

[Anderson81] Anderson, T. and Lee, P. A., "Fault-Tolerance Principles and Practice", Prentice-Hall International, 1981.

[Avizienis85] Avizienis, A., "The N-Version Approach to Fault-Tolerant Software", IEEE Transactions on Software Engineering, Vol. SE-11, No. 12, December 1985, pp 1491-1501.

[Becker92] Becker, T., "Transparent Service Reconfiguration After Node Failures"Proceedings of the International Workshop on Configurable Distributed Systems, March 1992, pp. 212-23.

[Cristian88] Cristian, F., "Agreeing on Who is Present and Who is Absent in a Synchronous Distributed System", Proceedings of the 18th International Symposium on Fault-Tolerant Computing, 1988, pp 206-211.

[Croll91] Croll, P. R., "Developing Safety-Critical Software within a CASE Environment", IEE Colloqium on "Computer Aided Software Engineering Tools for Real-Time Control", No. 1991/087, April 1991.

[Harrison91] Harrison, D. A., "A Real-Time Language Based on Explicit Timeouts", 1st IFAC Workshop on Algorithms and Architectures for Real-Time Control, September 1991.

[Kant90] Kant, K., "Performance Analysis of Real-Time Software Supporting Fault-Tolerant Operation", IEEE Transactions on Software Engineering, Vol. 39, No. 7, July 1990, pp.906-18.

[Kim89] Kim, K. H. and Welch, H. O., "Distributed Execution of Recovery Blocks: An Approach for Uniform Treatment of Hardware and Software Faults in Real-Time Applications", IEEE Transactions on Computers, Vol. 38, No. 5, May 1989, pp. 626-36.

[Knight86] Knight, J. C. and Leveson, N. G., "An Experimental Evaluation of the Assumption of Independence in Multiversion Programming", IEEE Transactions on Software Engineering, Vol. 12, pp. 96-109, 1986.

[Lorczak89] Lorczak P. R., Caglayan, A. E. and Eckhardt, D. E., "A Theoretical Investigation of Generalized Voters For Redundant Systems", Proceedings of the 19th International Symposium on Fault-Tolerant Computing, 1989.

[Luan90] Luan, S. and Gligor, V. D., "A Fault-Tolerant Protocol for Atomic Broadcast", IEEE Transactions on Parallel and Distributed Systems, Vol. 1, No. 3, July 1990, pp 271-285.

[Tindell90] Tindell, K., "Allocating Real-Time Tasks (An NP-Hard Problem Made Easy), University of York, Dept. of Computer Science Report No. 149, 1990.

[Voges88] Voges, U. (ed.), "Software Diversity in Computerized Control Systems", Springer-Verlag, 1988.

[Yuan91] Yuan, S. and Agrawala, A. K., "Fault-Tolerant Decentralised Commit Protocols", Journal of Parallel and Distributed Computing, Vol. 13, No. 3, November 1991, pp 299-311.

A FAULT-TOLERANT CONTROL SYSTEM FOR A HIGH PERFORMANCE INDUCTION MOTOR DRIVE

Sang-Hoon Kim, Seung Ki Sul, Min-Ho Park

Department of Electrical Engineering, Seoul National University
Seoul, Korea

Abstract. This paper describes a fault-tolerant control system for a high performance induction motor drive. A new fault-tolerant scheme using three microprocessor-based controllers is presented. In the proposed scheme, two of the three processors always compare the results of their computations with each other in order to detect a fault, and when a fault is detected, the third processor is used to locate it. By this method, the proposed scheme guarantees high reliability and provides the flexibility for the utilization of redundancies. Through the comparison of the estimated reliability, it is verified that the proposed scheme gives higher reliablilty than the TMR (Triple Modular Redundancy) to the control system for the induction motor drive.

Keywords. Fault-tolerant control system, reliability, microprocessor-based controller, TMR, fault detection, fault location, retry technique, high performance induction motor drive.

INTRODUCTION

System reliability has been of critical importance in situations where a malfunction of a control system could have catastrophic result[1]. The space shuttle, aircraft flight control systems, and banking systems are such cases. In recent years, the reliability of the techniques become increasingly more important for the general purpose industrial applications[2][3].

Currently, microprocessor-based controllers are commonly used in almost every application area of power electronics. In the past, the progress in this field has been to a great extent related to achive better performance. However little work has been done on methods for improving the reliability of the controller, which is indispensable to the implementation of its algorithms. As the applied systems become larger and the environment gets harsher, the reliability of the control system becomes increasingly important.

The control for a rolling mill machine drive of an iron process is a typical applications in this field where the errorneous outputs of the controller may result in the damage to a plant, the interruption to the process while repairs are carried out, and the cost of repairs. Therefore, it is necessary to keep the overall system failure rate at an acceptable level by ensuring the reliability of such a controller.

The control system reliability improves greatly by the use of fault-tolerant techniques, by which a fault-tolerant control system having high reliability, availability, and safety can be achieved. Here, it is important to select a fault-tolerant technique which can maximize the reliability improvement and system effectiveness, refered as the ratio of utility to cost[4].

TMR (Triple Modular Redundancy) is a well-known fault-tolerant technique[3]. In this technique, the failures in any single module can be tolerated through the majority voting, but the improvement of reliability is restrictive and the utilization of redundancies is not effective. A new technique to overcome problems just mentioned is proposed. With the proposed scheme, a more reliable system than the TMR system is obtainable, and the flexible utilization of a control system is also possible.

The control system for a GTO PWM inverter, which is used to drive high power induction machines such as a traction motor of an electric train and a rolling mill motor of an iron process, is especially required to be highly reliable, available, and safe. For this purpose, the proposed scheme is adopted.

COMPARISON OF FAULT-TOLERANT CONTROLLERS

The overall structure of an applied system is shown in Fig. 1. It consists of an induction motor and a PWM GTO inverter for driving it, a DC-link, several sensors, and a fault tolerant controller which is the main concern in this paper.

The TMR and the proposed fault-tolerant control system are shown in Fig 2 and 3, respectively. In TMR, the three copies of the controller are used, and execute a identical algorithm for control. The three redundant outputs is voted by the majority rule. Afterwards, the correct output is sent to the next level. Compared with TMR, the basic concept of the proposed scheme is as follows: Two of the three identical controllers perform the algorithm for control in parallel and compare the results of those computations for detecting a fault. At the same

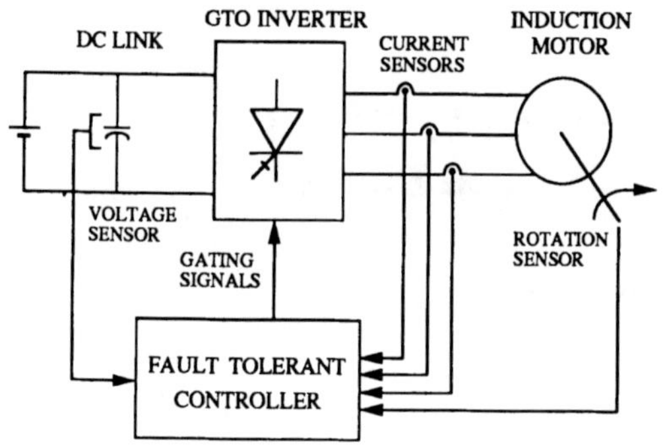

Fig.1 The overall system structure.

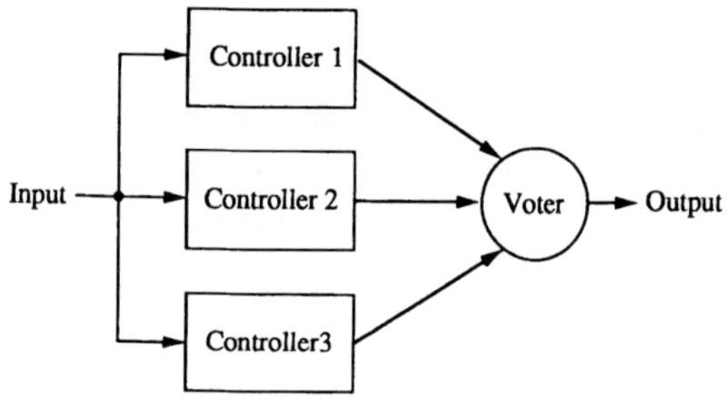

Fig.2 The TMR system.

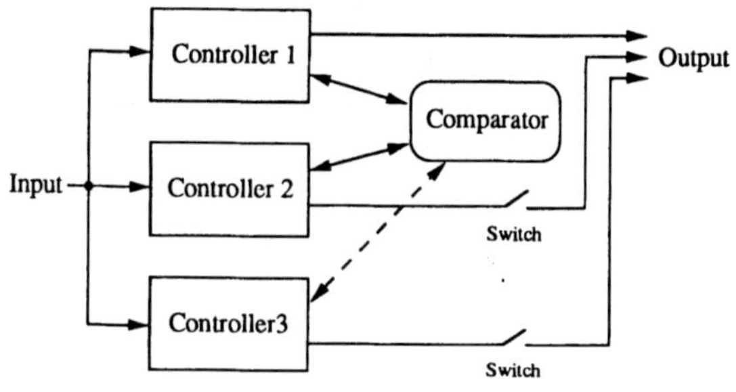

Fig.3 The proposed system.

time, the third one performs the additonal algorithm. In the event of a disagreement, because it is impossible to locate a fault by using the two controllers, the third controller is used to locate the fault. Afterwards, the third controller suspends the task of its own and executes the control algorithm. Through its result, it be determine which of the two controllers are faulty.

In TMR, the three controllers must always perform the whole control algorithms to provide the correct output by the majority voting of the outputs. On the other hand, in the proposed scheme, only two controllers perform the control algorithm to detect a fault in parallel and at the same time the third controller performs the additional algorithm to increase the performance of system. Therefore, in terms of the utilization of redundancies, the proposed scheme is more effective.

THE SYSTEM ARCHITECTURE

The architecture of a fault-tolerant control system using the proposed scheme is shown in Fig. 4 and consists of the three control boards and a common board for the three control boards. Each control board is capable of performing by itself the control algorithm for the induction motor drive.

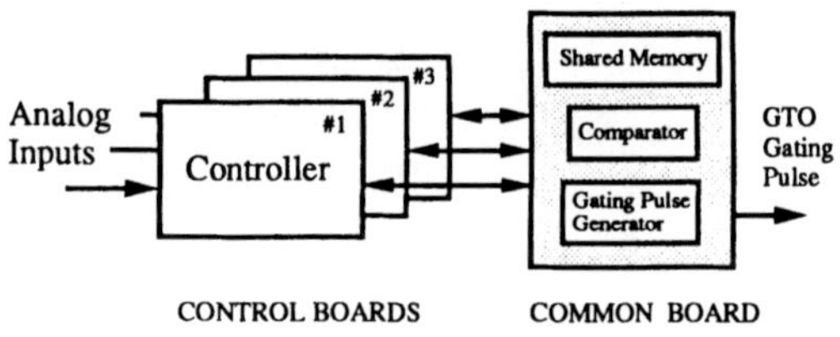

Fig. 4. The architecture of the fault-tolerant control system

Each control board is based on a 16-bit TMS320E14 DSP chip which has a 4K-word program EPROM and 256-word data RAM. It consists of two EPLD chips for the address decoding, a 12-bit A/D converter, a extension slot for the D/A converter, the circuits for the analog inputs, etc. The common board consists of a comparator, a 1K 8-bit triple-port RAM as shared memory, and the circuits for the gate pulse generation. The comparator is implemented by using an EPLD(Erasable Programmable Logic Device). Fig. 5 shows the block diagram of the control board.

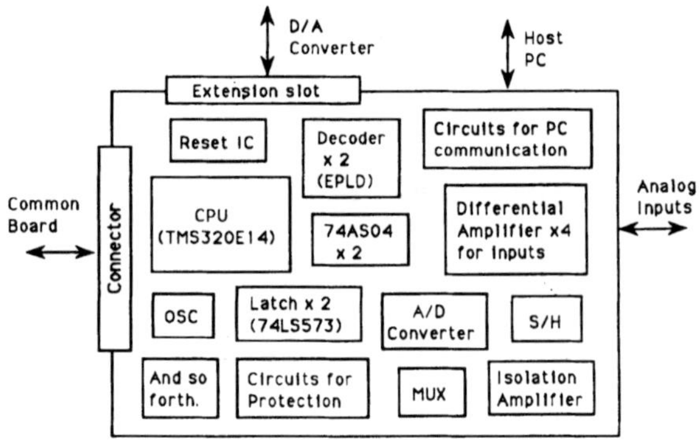

Fig. 5. Block diagram of the control board

Each control board is loosely coupled to remove the disadvantage of perfect synchronization which requires the use of a centralized clock generator and rather sophisticated phase-lock techniques. Therefore, each control board contains its own independent clock generator, where the effect of a clock generator failure can be localized to the board. The controllers communicate only via the shared memory. In this way, the shared parts for the three control boards is reduced at the minimum to prevent the failures of these from downing the whole control system.

THE ANALYSIS AND COMPARISON OF RELIABILITY

For the reliability analysis techniques, Markov modeling is the most commonly used approach[4]. By using it, the effectiveness of the fault tolerance implemented in the system can be analyzed. the Markov model of the proposed system is shown in Fig. 6. Here, it is assumed that only one failure will occur at a time, the system does not contain repair, and the controllers obey the exponential failure law.

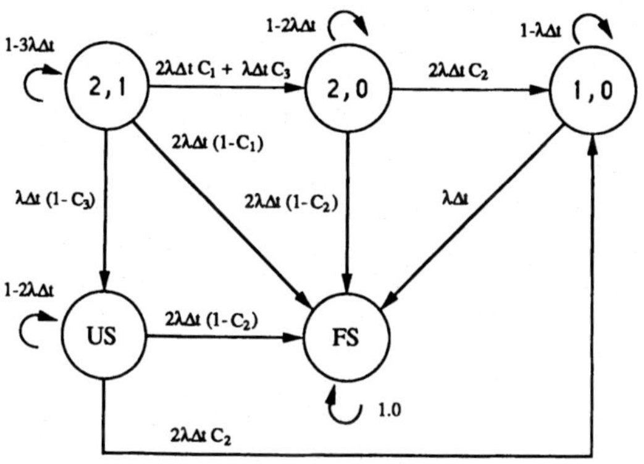

Fig. 6 The Markov model of the proposed scheme.

The states of the proposed model can be partitioned into five major classes: state (2,1) corresponds to the state in which all three controllers are functioning correctly, state (2,0) is the state in which coworker(controller3) has successfully replaced a faulty worker, state (US) is the state in which the fault-free workers did not detect the failure in the coworker, state (1,0) is the state with a online controller, and state (FS) corresponds to the state in which the control system failed.

Here, C_i is the fault coverage factor, that is, the probability that a failure will be correctly handled within the allowable time. The allowable time is the time for detecting, locating, and recovering a failure. It should be much less than the sampling interval of the control system. In the applied system, the maximum switching frequency of the inverter using GTO is limited to about 1KHz. Therefore, the system has a minimum sampling interval of 1msec. The fault coverage must be performed within 1msec, otherwise, the performance of the system will be deteriorated. Because the control algorithm of this system can be executed within 50μsec, this condition is satisfied. C_1 is the probability of transitioning to (2,0) by correctly handling a fault in state (2,1), C_2 is the probability of transitioning from state (2,0) or state (US) to (1,0), and C_3 is the probability of transitioning to (2,0) by correctly detecting faults of coworker in state (2,1).

The discrete-time equation of this model can be written in matrix form as,

$$P(t+\Delta t) = A \cdot P(t) \tag{1}$$

where,

$$P(t+\Delta t) = \begin{bmatrix} P_{(2,1)}(t+\Delta t) \\ P_{(2,0)}(t+\Delta t) \\ P_{(US)}(t+\Delta t) \\ P_{(1,0)}(t+\Delta t) \\ P_{(FS)}(t+\Delta t) \end{bmatrix}, \quad P(t) = \begin{bmatrix} P_{(2,1)}(t) \\ P_{(2,0)}(t) \\ P_{(US)}(t) \\ P_{(1,0)}(t) \\ P_{(FS)}(t) \end{bmatrix}$$

$$A = \begin{bmatrix} 1-3\lambda\Delta t & 0 & 0 & 0 & 0 \\ \lambda\Delta t(2C_1+C_3) & 1-2\lambda\Delta t & 0 & 0 & 0 \\ \lambda\Delta t(1-C_3) & 0 & 0 & 1-2\lambda\Delta t & 0 \\ 0 & 2\lambda\Delta t C_2 & 1-\lambda\Delta t & 2\lambda\Delta t C_2 & 0 \\ 2\lambda\Delta t(1-C_1) & 2\lambda\Delta t(1-C_2) & \lambda\Delta t & 2\lambda\Delta t(1-C_1) & 1 \end{bmatrix}$$

$P(t+\Delta t)$ is the probability state vector at time $t+\Delta t$, $P(t)$ is the probability state vector at time t, and A is the transient matrix. The continuous-time equation can be derived from the discrete-time equation by allowing the time interval Δt to approach zero. The solution of this equation is given by

$$\begin{aligned} P_{(2,1)}(t) &= \exp(-3\lambda t) \\ P_{(2,0)}(t) &= (2C_1+C_3)\cdot[\exp(-2\lambda t)-\exp(-3\lambda t)] \\ P_{(US)}(t) &= (1-C_3)\cdot[\exp(-2\lambda t)-\exp(-3\lambda t)] \\ P_{(1,0)}(t) &= C_2\cdot(2C_1+1)\cdot[\exp(-\lambda t)-2\exp(-2\lambda t) \\ &\quad + \exp(-3\lambda t)] \end{aligned}$$

The reliabilty of the proposed system, $R_P(t)$, is simply the sum of the probabilities that the system is in the operating state. Thus,

$$R_P(t) = P_{(2,1)}(t) + P_{(2,0)}(t) + P_{(US)}(t) + P_{(1,0)}(t) \tag{2}$$

Whereas, those of the TMR and the simplex system (unredundant system) are given by[5],

$$R_{TMR}(t) = 3\cdot\exp(-2\lambda t) - 2\cdot\exp(-3\lambda t) \tag{3}$$

$$R_{sim}(t) = \exp(-\lambda t) \tag{4}$$

For comparison, it is assumed that a voter of the TMR system and a comparator of the proposed system are ideal. The MIL-HDBK-217E formula for estimating individual failure rates of a digital integrated circuit for a control board was used[5]. The estimated failure rate λ for a control board given earlier is 1.66 (failures/10^4hours). Fig. 7 shows the simulated reliabilities of the simplex, the TMR, and the proposed system. The coverage factors C_1, C_2, and C_3 are 0.97, 0.5, and 0.5, respectively.

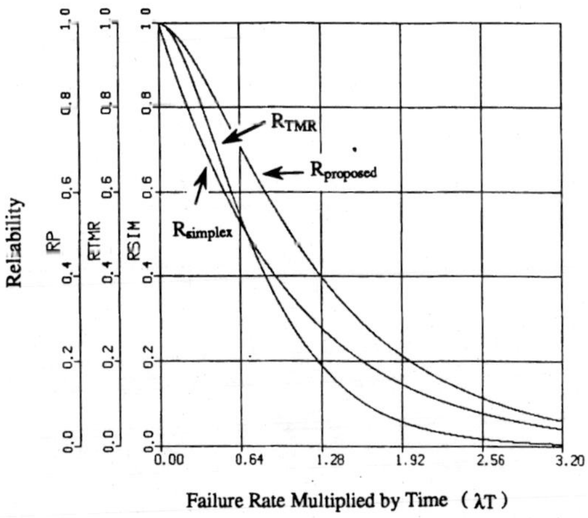

Fig. 7 The comparison of reliabilities.

The reliability of the proposed system is higher than that of the TMR system. Moreover, when a point in time is reached, the TMR system becomes even less reliable than the simplex system. Besides the direct comparison of reliability, there is another comparison method. The parameter for comparison is the mission time, MT[r], which is the time at which the reliability of system falls below level r [4]. The mission times for three systems with a reliability of 0.9 are shown in Fig. 8.:

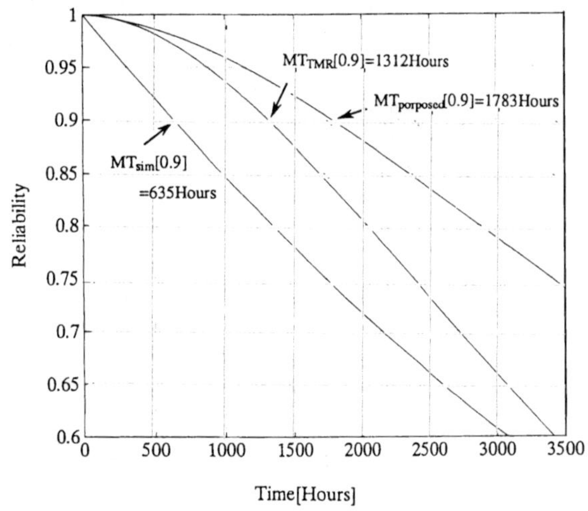

Fig. 8 Mission time comparison

The mission time improvement of the TMR system over the simplex system is approximately double, on the other hand, that of the proposed system is approximately tripled. In other words, the proposed system can operate three times as long as the simplex system and 1.4 times as long as the TMR system while still maintaining a reliability greater than 0.9.

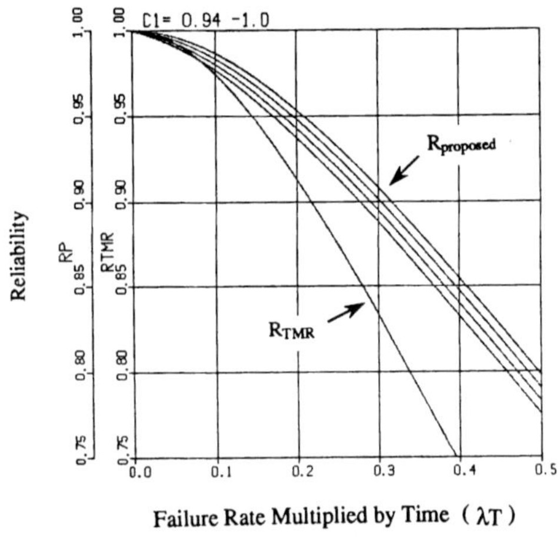

Fig. 9 The reliability of the proposed system for various fault coverage values.

Fig. 9 shows the reliabilities of the proposed system for various values (0.94 ~ 1) of C_1 and for $C_2 = 0.5$, $C_3 = 0.5$. The value of the fault coverage factor C_1 has a significant impact on the reliability of the system.

THE FAULT-TOLERANT TECHNIQUE

A field oriented control is widely used for a high performance ac drive which requires an instantaneous torque response and torque control over a wide speed range[6]. In this paper, it is adopted for the induction motor drive, and PWM based on voltage space vectors[7] is employed as the current regulation method. For the proper control of the whole system under the just mentioned scheme, it is necessary for the controller to be correctly offered the system inputs is required for control algorithm implemention, and sends the correct gating signals to GTO inverter for the induction motor drive. For these purposes, checking for faults is done at this two instances: after the acquisition of the input values, and then after the computation of the results, namely, the gating signals times. All of above mentioned comparisons are performed by a hardware comparator in common board.

The overall control mode is divided into three parts: the fault tolerant mode of three, the fault detection mode of two, and the simplex mode of one healthy controller, respectively.

Fault Tolerant Mode

The state (2,1) of the Markov model corresponds to this mode where the three controllers are fault-free and a faulty controller can be tolerated. Suppose that controller 1 and 2 play a role in the workers, and controller 3 plays a role in the coworker in the beginning. The workers perform the main algorithm, that is, the algorithm of the field oriented control and the current regulation for the induction motor drive. The coworker as a spare always monitors the faults of the workers. In parallel, it performs the algorithm for man-machine interfacing, for improving the performance of the control system, and for detecting the faults of the shared memory and sensors.

The comparison of analog inputs

- At sampling times, all three controllers are offered input values(i_a,i_b,i_c,V_{dc},ω_m) through sensors. The workers convert the analog values into the digital, and send the values to the comparator in order to examine their validity. The workers and the coworker check the signal that indicates the end of comparison of the comparator, and then ascertain whether a disagreement occurrs. In the event of a disagreement, all the controllers write the input values to the shared memory, and the correct values are taken through the coworker's input values. Afterwards, the coworker looks into all the input values of the three controllers to determine whether the disagreement is caused by sensor faults or by the control board faults, and informs the workers of the result. If it is caused by the faults of the current sensors, the corresponding sensor is ignored. The necessary current values (ia,ib,ic) by using two healthy sensors can be obtained, because one current sensor is given as a redundancy to provide a more reliable sensing of the current. However if it is caused by the fault of a voltage sensor or a speed sensor, the system is safely downed. Provided that it is caused by the fault of the input parts in the control board, the controller is offered the necessary input values, via shared memory, by the other controllers.

The comparison of outputs

-The workers calculate the swiching patterns for the GTO inverter, that is, each GTO device's the turn on/ off times. Because the system performance is greatly affected by these values directly, it is very important to compare these values for detecting a faulty result. Therefore, after computing, the workers always compare their results with each other through the comparator. Then, the coworker continuously monitors the result of the comparison. In case of a disagreement, meaning that one of the two workers is faulty, the coworker suspends the task of its own and executes the task of the workers, that is, the control algorithm. Because the frequency of occurrence of transient faults in the digital controllers is much higher than that of permanent ones[8], it is assumed that the disagreement is caused by transient faults. The retry technique is employed to tolerate the transient ones. This technique requires the workers to execute the algorithm once more during coworker's executing for locating a fault. After all three controllers finish computing, each worker examines the probability of the transient fault for its result. In case of transient faults, the control system is still in the fault tolerant mode. If it is the permanent faults, the three controllers compare their computation results for locating a faulty worker. The system mode is degraded to the fault detection mode by isolating the faulty controller.

Fault Detection Mode

This mode corresponds to state (2,0) and state (US). It is assumed that controller 2 is faulty as an example. Then, the coworker (controller 3) takes the place of controller 2 as the worker. A fault is detected by comparing the results of controllers 1 and 3. Here, controller 1 and 3 perform the task which includes the control algorithm and the necessary portion of the coworker's task. In this mode using two controllers, it is impossible to locate the faulty one in case of detecting a fault. Therefore, the time redundancy technique is adopted to each controller for locating a fault. The fundamental concept of this technique is that each controller repeats the computation of the control algorithm once more, and compares the two results to determine if a discrepancy exists. This technique is often suitable when a disagreement is caused by transient faults. If a fault is not located through this technique, the system control is downed with safety. Besides this comparison method, another fault-detecting method exits. This method requires each controller to provide data regarding its status to the other via shared memory periodically. If a controller fails to set the appropriate value in the shared memory, it is faulty. Therefore, the control system is degraded to the simplex mode.

Simplex Mode

State (1,0) of the Markov model corresponds to this mode, where only one fault-free controller takes charge of performing the control algorithm. Of all the three modes, this mode has a fault-detecting method at the lowest level. The faults can only be detected by using a watch dog timer. Each controller has a watch dog timer to permit the detection of runaway conditions, of which time-out signal interrupts itself, and is informed of the other controllers.

CONCLUSION

In the power electronics field, the microprocessor-based controller is widely used to implement advanced control algorithms. However, the its reliability is of little interest. The controller is critical to the system performance, and a controller failure can cause harm to either the system under control or the human. Thus, in this paper, the fault-tolerant control system for a high performance induction motor drive is proposed. Through the comparison of reliabilities, it is verified that the proposed system is more reliable and effective in the utilization of redundancies than the TMR system as fault-tolerant systems using the same number of redundancy.

REFERENCE

[1] A. Avzienis. "Fault-Tolerance : The Survival Attribute of Digital Systems." Proc. IEEE, Vol.66, Oct. 1978.

[2] Steven W. Yates. et.al. "A Fault-Tolerant Multiprocessor Controller for Magnetic Bearings" IEEE Micro, August, 1988.

[3] P. K. Chande. et.al. "Modular TMR Multiprocessor system" IEEE Trans. Ind. Electron., Vol.36, No.1, Feb., 1989.

[4] B. W. Johnson, *Design and Analysis of Fault Tolerant Digital Systems.* Addison-Wesley Publishing Company, Inc. pp. 199-214. 1989.

[5] Reliability Prediction of Electronic Equipment, MIL-HDBK-217E, 1984.

[6] D. W. Novotny, R. D. Lorenz, *Introduction to Field Orientation and High Performance AC Drives.* University of Wisconsin-Madison 1985.

[7] H. Broeck et.al. "Analysis and Realization of a Pulsewidth Modulator Based on Voltage Space Vectors" IEEE Trans. Ind. App., Vol.24, No.1, Jan./Feb., 1988.

[8] J. Sosnowski "Transient Fault Tolerance in Microprocessor Controllers" Proc. the 1st European Workshop on Fault Diagonastics, Vol.1, Aug., 1986.

A Parallel Algorithm for Training Neural Network Based Nonlinear Models

Gordon Lightbody and George W. Irwin

Control Research Group
Department of Electrical and Electronic Engineering
The Queen's University of Belfast
Belfast BT9 5AH

Abstract. This paper describes a parallel algorithm for the training of Multilayer Perceptron based models of nonlinear plant. A training algorithm, based on the Broyden-Fletcher-Goldfharb-Shanno optimisation algorithm is discussed, with test results to show the improvement this algorithm offers over the gradient descent technique. A parallel version of this algorithm is then derived for realisation on an array of transputers. This algorithm was used to train a neural inferential model of a nonlinear chemical plant. It was found that the parallel algorithm provided a highly efficient method of accelerating the neural network training process.

Keywords: Neural Networks, Parallel Training Algorithms, Transputers, Modelling.

INTRODUCTION

Neural Networks such as the Multi-layer Perceptron or Gaussian networks have recently been finding many applications in the field of control engineering, particularly for the modelling of ill-defined plants common in industrial applications (Willis, Di Massimo, Montague, Tham and Morris, 1991). These networks consist of simple processing elements arranged in layers. Each element takes as input the weighed sum of all the outputs of the previous layer and passes this through a nonlinear activation function. The structure is feedforward in nature, with no connection within layers or from outer layers towards the input. It is the nonlinear mapping ability of the feedforward network which has been central to its use in control. Recent theoretical results (Cybenko, 1989; Funahashi, 1989) have rigorously proven that, under certain conditions, a continuous function can be approximated arbitrarily by a feedforward network with only a single hidden layer.

One of the major drawbacks in using the Multilayer Perceptron network, is the extensive periods required for training. Initially the Backpropagation rule was proposed, using gradient descent, (Rumelhart, Hinton and Williams, 1988). However this algorithm is known to have slow convergence properties, as the direction of steepest descent is not the most efficient adaptation direction, (Gill, Murray and Wright, 1981). To help overcome the problems of local minima encountered by gradient descent based training methods, the momentum term was introduced, (Rumelhart, Hinton and Williams, 1986), in which the weight update direction is formed by passing the gradient through what can be thought of as a low pass filter. Here the training gains are kept constant throughout the training process and are chosen to give the fastest convergence whilst ensuring algorithm stability over all of the training set. A number of approaches have been proposed in which the training gains are adaptable, allowing the optimum step to be chosen with each weight update, (Reyneri and Filippi, 1990; Yu and Cheng, 1990).

The training of neural networks in control has been viewed by a number of researchers as a nonlinear system identification problem, in which the weights are to be identified to produce the required response. Sbarbaro and Gawthrop, (1990), applied Stochastic Approximation techniques to the training of Multilayer Perceptron networks, whilst Matthews, (1990), reformulated the network structure and used an Extended Kalman Filter for training. The training process has also been regarded as a nonlinear unconstrained optimisation problem. Sophisticated optimisation algorithms, such as the Levenberg-Marquardt method, (Ruanno, Fleming and Jones, 1991) or the Broyden-Fletcher-Goldfharb-Shanno (BFGS) method, (Battiti and Masulli, 1990), have been found to provide a significant acceleration of the training process over normal Gradient Descent.

Parallel processing offers further possibilities for accelerated training since there are a number of types of concurrency within the neural network training process. Although the network itself consists of many simple parallel elements, during training, the communication required between these processes inhibits the use of coarse grain architectures such as the Inmos transputer. However, by decomposing the network into a number of sub-networks the training problem can be made more suitable for mapping onto coarse grain architectures. Following Oglesby and Mason, (1989), it is also possible to exploit the parallelism of the training data to form a parallel batch training algorithm suitable for mapping onto an array of transputers.

In this paper a BFGS based training algorithm is first proposed. A parallel version of this algorithm is then developed to be mapped onto an array of Inmos transputers. This algorithm is then compared with sequential BFGS methods for the training of an inferential model for a nonlinear CSTR chemical plant.

NEURAL NONLINEAR MODELLING

The following difference equation is widely used to represent a wide class of SISO, discrete-time, nonlinear systems:

$$y_p(k+1)=f(y_p(k),..y_p(k-n),u(k),..u(k-m)) \tag{1}$$

Here the plant output y_p at time $k+1$ is a nonlinear function of past output values and the past inputs. The nonlinear function f(.) is generally unknown, though some idea of its structure may be apparent. Based on the theoretical results of Cybenko (1989) and Funahashi (1989) it has been proposed that neural networks, such as the Multilayer Perceptron can be used to identify this nonlinear vector function and hence form a wide class of realisable nonlinear models, (Narendra and Parthasarathy, 1990; Lightbody and Irwin, 1992; Billings, Jamaluddin and Chen, 1992).

The nonlinear mapping ability of the Multilayer Perceptron network can be best understood by considering the polynomial expansion of each neuron function, (Wray and Green, 1991). The Taylor's expansion of the typical sigmoidal activation function yields an infinite series, whose coefficients fade out with the order of each term. Consider a network with only a single hidden layer of nonlinear neurons. Each node will produce a different infinite polynomial of the inputs to the network, depending on the set of weights connecting that particular hidden node to the input layer. The output layer then combines the polynomials generated by the hidden layer units in a weighted sum, to provide the necessary mapping.

As an example of nonlinear function approximation, using the Multilayer Perceptron, consider the simple single-input, single-output, step function.

$$\begin{aligned} y&=0.3 \quad x<0.5 \\ y&=0.7 \quad x\geq 0.5 \end{aligned} \tag{2}$$

Figure 1 shows the how the accuracy of approximation of a (1,20,1) Multilayer Perceptron model of this function improves as training time increases. Note the similarity to other approximation techniques, such as Fourier series, with the presence of an overshoot which does not disappear with increased training.

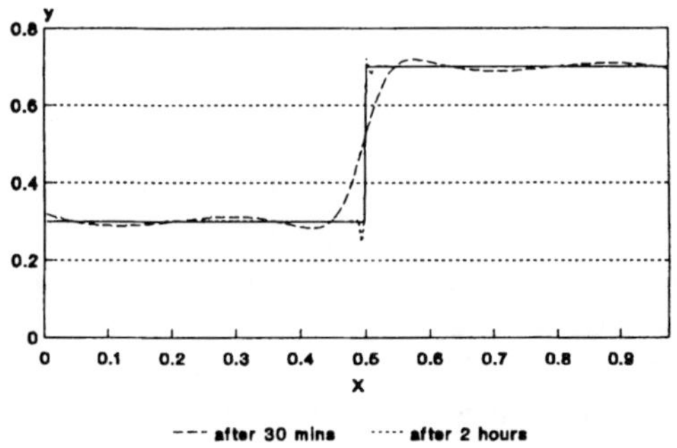

Fig 1. Neural model of a step function

THE BFGS ALGORITHM

The BFGS algorithm is a Quasi-Newton optimisation technique, in which curvature information is used to provide a more accurate descent direction, without actually calculating the second derivatives. Consider the problem of determining the weights to minimise the cost function $J(\bar{w})$. If the present weight set is $\bar{w}=\bar{w}(k)$ then, by Taylor's theorem, a quadratic approximation to the cost function $J(\bar{w})$ about $\bar{w}(k)$ yields:

$$\begin{aligned} J(\bar{w}) \approx J_1(\bar{w}) &= J(\bar{w}(k))+(\bar{w}-\bar{w}(k))^T\bar{g}(k) \\ &+\frac{1}{2}(\bar{w}-\bar{w}(k))^T H(k)(\bar{w}-\bar{w}(k)) \end{aligned} \tag{3}$$

where $\bar{g}(k)$ is the vector of gradients of the cost $J(\bar{w})$ with respect to the weight vector $\bar{w}(k)$, as determined by the batch Backpropagation algorithm and H(k) is the Hessian matrix of second derivatives:

$$\bar{g}(k)=\frac{\partial J(\bar{w}(k))}{\partial \bar{w}(k)} \qquad h_{ij}=\frac{\partial^2 J(\bar{w}(k))}{\partial w_i(k)\,\partial w_j(k)} \tag{4}$$

The next approximation to the minimum of $J(\bar{w})$ is then given by the minimum of $J_1(\bar{w})$ as:

$$\frac{\partial J_1(\bar{w})}{\partial \bar{w}}=\bar{g}(k)+H(k)(\bar{w}-\bar{w}(k))=0 \tag{5}$$

This would be satisfied by the following weight update equation:

$$\bar{w}(k+1)=\bar{w}(k)+\bar{d}(k) \tag{6}$$

where the direction is given from equation 5 as:

$$\bar{d}(k)=-(H(k))^{-1}\bar{g}(k) \tag{7}$$

This update will minimise the quadratic approximation of the cost given in equation (3). In practice, having obtained the direction $\bar{d}(k)$, it is usual to carry out a single line search to determine the step length for the adjustment of the weights along this direction. Hence:

$$\bar{w}(k+1)=\bar{w}(k)+\eta(k)\bar{d}(k) \tag{8}$$

The single line search can be approached in a number of ways. For example, Battiti and Masulli, (1990), used a quadratic approximation to the cost, whose minimum is easily determined. Another quadratic approximation is then produced about this last minimum and the quadratic is again minimised to produce a better solution. This process is repeated until the step that minimises the cost is obtained. In terms of efficiency, it is better that the single line search is inexact, that is, a step which brings the cost close to the minimum along that direction is adequate, (Gill, Murray and Wright, 1981).

The inverse Hessian is difficult to determine and hence Quasi-Newton techniques, such as BFGS, evolved to provide a recursive method to approximate $H^{-1}(k)$ as a positive definite matrix M(k). The matrix M(k) is first set as the identity matrix, giving the initial update direction as the negative gradient. On each iteration through the training set the matrix M(k) is updated, ensuring that it remains positive definite and satisfies the Quasi-Newton condition:

$$M(k)(\bar{g}(k)-\bar{g}(k-1))=\bar{d}(k) \tag{9}$$

The update equation for M(k) is as follows (McKeown, Meegan and Sprevak, 1990):

$$\bar{p}(k+1)=\bar{g}(k+1)-\bar{g}(k)$$

$$\bar{s}(k+1)=\bar{w}(k+1)-\bar{w}(k)$$

$$A(k+1)=(1+\frac{\bar{p}(k+1)^T M(k)\bar{p}(k+1)}{\bar{s}(k+1)^T\bar{p}(k+1)})\frac{\bar{s}(k+1)\bar{s}(k+1)^T}{\bar{s}(k+1)^T\bar{p}(k+1)}$$

$$B(k+1)=-\frac{\bar{s}(k+1)\bar{p}(k+1)^T M(k)+M(k)\bar{p}(k+1)\bar{s}(k+1)^T}{\bar{s}(k+1)^T\bar{p}(k+1)}$$

$$M(k+1)=M(k)+A(k+1)+B(k+1) \tag{10}$$

Battiti and Masulli (1990) suggested that, in order to maintain positive definiteness of the matrix M(k), it should be reinitialised every N iterations to the identity matrix, where N is the number of weights in the network.

The BFGS algorithm has been compared with the batch Backpropagation method for training:

(i) A (2, 15, 1) Multilayer Perceptron to form the simple trigonometric function, $y = \cos(x_1)+\sin(x_2)$
(ii) A (1, 15, 1) Multilayer Perceptron to form the quadratic function $y= x^2$.

A training set of fifty random input-output pairs was produced in each case, with all variables normalised to be in the range [-1, 1]. Both algorithms were executed on a T800 Transputer, both networks used $\tanh(^x/_2)$ neuron nonlinearities and the cost function was as shown in equation (11). Here d_i is the desired output y_i is the network output for the ith pattern and np is the number of patterns in the training set:

$$J=\sum_{i=1}^{np}(d_i-y_i)^2 \tag{11}$$

To compare the two training algorithms, a suitable test was to determine the actual computing time each method required to reduce the training cost of equation (11) to below J=1.0. Note that, since this is effectively a nonlinear optimisation problem, the initial choice of the weights will have a bearing on the evolution of the algorithms. Each algorithm therefore had to be initiated with the same set of weights. Likewise the tests were repeated for a wide range of starting conditions. Table 1 gives the best-case, worst-case and the mean speed ups provided by the BFGS algorithm, as compared with the batch gradient technique with fixed optimal descent step-size. As can be seen from this Table, the BFGS algorithm was at worst over four times faster than the fixed-step, batch Backpropagation algorithm.

TABLE 1 Speed up provided by the BFGS algorithm as compared with batch Backpropagation.

Mapping problem	worst case	mean case	best case
$y=\cos(x_1)+\sin(x_2)$	6.8	9.0	12.0
$y=x^2$	4.4	5.4	7.2

A PARALLEL BFGS ALGORITHM

Two major operations take place on each presentation of the training data. First, the determination of the direction vector, then the choice of the best step along this vector. The determination of the direction requires that the batch gradient be calculated. This process can easily be parallelised, (Oglesby and

Mason, 1989), by splitting the training set data into N subsets, each of which would be assigned to a particular processor. To determine the gradient $\bar{g}_i(k)$ on the ith processor, each data vector from the ith data subset would be applied to a copy of the neural network on that processor. Then, using the backpropagation rule, the instantaneous gradient would be determined. The gradient $\bar{g}_i(k)$ follows from the summation of the instantaneous gradients over that data subset. The complete gradient produced by the full training set would be:

$$\bar{g}(k) = \sum_{i=1}^{N} \bar{g}_i(k) \qquad (12)$$

The direction vector calculation on N processors can be best described as in Figure 2:

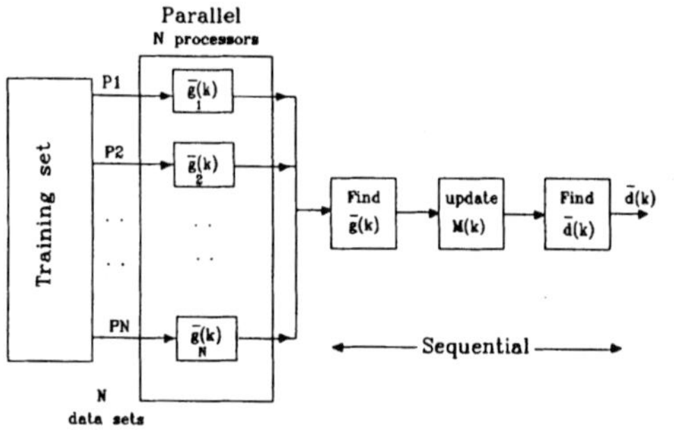

Fig. 2 Parallel direction vector calculation

It was found that, on each presentation of he training set, approximately 20% of the processing time was spent calculating the step size. Hence it was necessary to parallelise the step calculation routine to achieve an adequate efficiency. It should be noted that the single line search proposed by Battiti and Masulli, (1990), cannot be satisfactorily parallelised as it is inherently a sequential process.

A possible alternative method would be to take the weight adjustment direction $\bar{d}(k)$ and to determine the cost over all the patterns, for N_s different step sizes that adequately span the allowable range of steps. A rough cost profile can then be produced as shown in Figure 3. The lowest cost is selected, say for η_i and the range of steps η_{i-1} to η_{i+1} is then examined more closely. An extra N_s steps would be chosen in this range and their associated costs determined to give a more accurate cost profile about the previous minimum point η_i. The step, η_{min}, that now produces the minimum cost would be chosen to adjust the weights according to equation (8).

To determine the cost, for a particular step η, requires that all the weights be adjusted to a dummy weight vector, according to equation (8), then all the data vectors forward propagated and the total cost calculated. With $N_s=6$, twelve cost values would be calculated to provide the desired step-size, as compared to roughly six costs required by the quadratic curve fitting line search method. This technique would therefore be inefficient for use on a single processor. However, this version of the single line search routine allows for ready parallelisation, in which the calculation of a cost or a set of costs can be attributed to a particular processor.

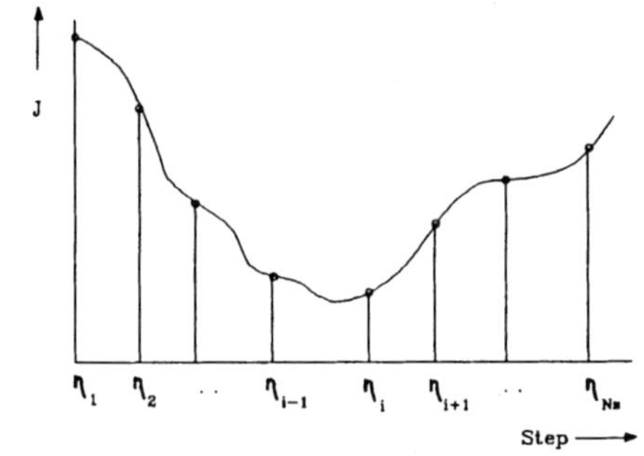

Fig. 3 Approximate $J(\eta)$ cost profile

If the complete algorithm were to be mapped onto N processors, with the number of steps, $N_s=N$, then the structure would be as in Figure 4.

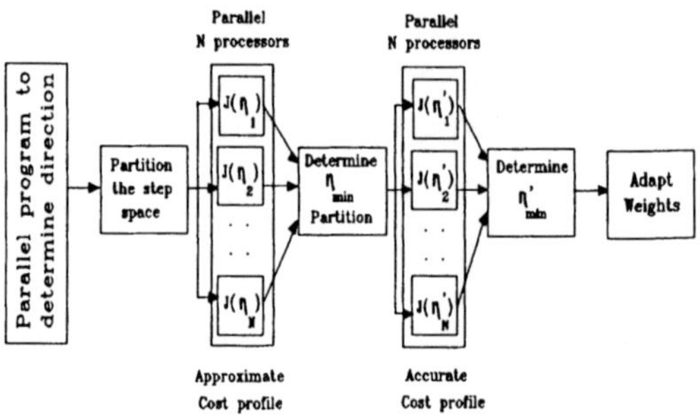

Fig. 4 Parallel single line search

The parallel BFGS algorithm was used to train a neural inferential model of the Continuous-Stirred-Tank-Reactor (CSTR), problem defined by the following equations, (Morningred and others, 1990):

$$\dot{C}_a(t) = \frac{q}{V}\cdot(C_{ao}-C_a(t)) - k_o\cdot C_a(t)\cdot \exp(-\frac{E}{R\cdot T(t)})$$

$$\dot{T}(t) = \frac{q}{V}\cdot(T_o-T(t)) + k_1\cdot C_a(t)\cdot \exp(-\frac{E}{R\cdot T(t)})$$
$$+k_2\cdot q_c(t-d)\cdot(1-\exp(-\frac{k3}{q_c(t-d)}))\cdot(T_{co}-T(t))$$

It suffices to know that, within the CSTR, two chemicals are mixed and react to produce a product compound A at a Concentration $C_a(t)$, with the temperature of the mixture being T(t). The reaction is exothermic, producing heat which acts to slow the reaction down. By introducing a coolant, with flow-rate $q_c(t)$, the temperature can be varied and hence the product concentration controlled. A system time delay, d, of thirty seconds is present. C_{ao} is the inlet feed concentration, q the process flow-rate, T_o and T_{co} the inlet feed and coolant temperatures respectively. All of these are assumed constant at nominal values. Likewise k_o, E/R, v, k_1, k_2 and k_3, are thermodynamic and chemical constants relating to this particular problem.

The measurement of the concentration $C_a(t)$ would be difficult, requiring laboratory analysis or the use of an expensive custom built sensor. In comparison the temperature T(t) would be relatively straight-forward to measure. From knowledge of the approximate order of the plant and the system delay and choosing the sample time as 0.1 minutes, it was decided that a Multilayer Perceptron would be trained to provide the nonlinear inferential estimator of equation (13), relating the present and past temperatures along with the past coolant flow-rates to provide a suitable estimate of the product concentration, $C_a(k)$.

$$\hat{C}_a(k)=\hat{f}(T(k),..T(k-3),q_c(k-1),..q_c(k-8)) \quad (13)$$

The system was excited by a hybrid combination of large step changes and small amplitude PRBS on the coolant flow-rate. A training set of 400 input-output pairs was produced, relating each measured concentration $C_a(k)$ to the present and past three temperatures and the past eight coolant flow rates.

The network to be trained was a (12, 20, 1) Multilayer Perceptron, with $\tanh(^x/_2)$ functions in the hidden layer. The training methods used included the sequential BFGS method, with the quadratic approximation line search on a single T800-25 transputer, the parallel BFGS with the number of steps $N_s=6$, on six T800-25 transputers and the sequential BFGS using the cost profile line search, with $N_s=6$, on a single T800-25 transputer.

A measure of *mapping efficiency* can be obtained by comparing the speed of the parallel algorithm mapped onto the array of transputers, with the same algorithm on one transputer, with the hardware links and processors replaced by software channels and processes. Table 1 gives the times required by the parallel BFGS method on six T800-25 transputers and the same algorithm on a single T800-25 transputer, to complete ten iterations through the training set.

TABLE 2 Time required to complete 10 iterations

Algorithm	time/seconds
Sequential BFGS, new line search, on 1 T800	137.0
Parallel BFGS on 6 T800 processors	27.3

This hence gives a speed-up of 5 times, which results in a mapping efficiency of 84%

To obtain a measure of the *speed* of the parallel BFGS algorithm, it was necessary to compare it with the best sequential method available - the BFGS algorithm with the quadratic approximation line search. Each algorithm was initiated with the same weights and the time required to reduce the cost over the complete training set to below J=0.5 determined. This was repeated for a wide range of starting weights. Table 3 gives the best-case, worst-case and the average training time speed ups provided by the parallel algorithm.

TABLE 3 Speed up provided by the parallel BFGS algorithm over the best sequential method

Mapping problem	worst case	mean	best case
CSTR model J<0.5	5.4	7.7	9.2

In the worst case the parallel algorithm on six transputers was 5.4 times faster than the sequential BFGS method with a quadratic approximation line search. Notice that speed ups in excess of 6 times can be achieved due to the fact that the cost profile method is inherently a more accurate step choice algorithm.

Using the parallel BFGS algorithm on six T800 processors the inferential estimator was trained to give a cost J=0.05, taking approximately five minutes. Figure 5 shows the response of the estimator for a hybrid coolant flow-rate excitation consisting of large step changes with small-signal PRBS superimposed. It should be noted that the model was tested with a signal not encountered in training. As can be seen, the neural network model provides an excellent estimate of the product concentration from temperature and flow-rate information.

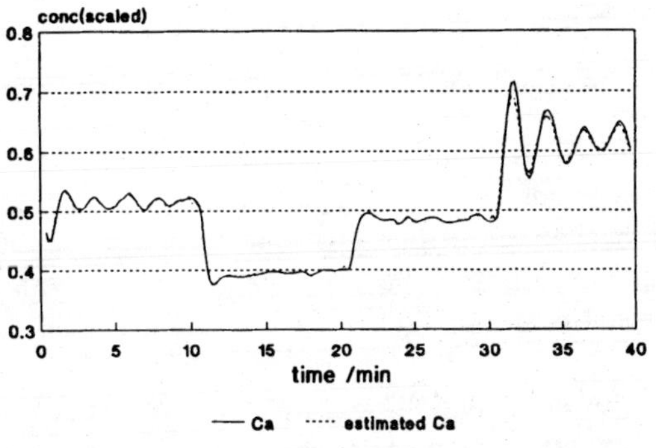

Fig. 5 Response of the neural estimator for the CSTR

CONCLUSIONS

A training algorithm based on the BFGS optimisation technique, was developed and found to perform at least four times faster than the batch Backpropagation training rule. A parallel BFGS training algorithm was then devised,taking advantage of the parallelism present in the training set. To improve the processing efficiency, a parallel single line search routine was proposed. This complete algorithm was mapped onto a pipeline of six T800 transputers with a mapping efficiency of 84%. When compared with the sequential BFGS method with the quadratic approximation single line search, it was found to be at least five times faster. The parallel BFGS algorithm was used to train a neural inferential model of a CSTR chemical plant, to provide estimates of product concentration, from temperature and flow-rate data and was found to give excellent results.

ACKNOWLEDGEMENT

Gordon Lightbody would like to acknowledge the financial support of the Department of Education for Northern Ireland.

REFERENCES

Battiti, R. and F.Masulli, (1990). BFGS optimisation for faster and automated supervised learning. Proc. Int. Neural Net. Conf., Vol. 2, pp. 757-760.

Billings, S.A., H.B. Jamaluddin and S. Chen, (1992). Properties of neural networks with applications to modelling nonlinear dynamical systems. Int. J. Control, Vol. 55, No. 1, pp. 193-224.

Cybenko, G., (1989). Approximations by superpositions of a sigmoidal function. Mathematics of Signals and Systems, Vol. 2, pp. 303-314.

Funahashi, K., (1989). On the approximate realisation of continuous mappings by neural networks. Neural Networks, Vol. 2, pp.183-192.

Gill, P.E. W. Murray and M.H. Wright, (1981). Practical Optimisation, Academic Press, London.

Lightbody, G. and G.W. Irwin, (1992). Neural networks for nonlinear adaptive control. Proc. IFAC Symposium on Algorithms and Architectures for Real-Time Control, Bangor, U.K., IFAC Workshop series, No. 4, pp. 1-13.

McKeown, J.J., D. Meegan and D. Sprevak, (1990). An Introduction to Unconstrained Optimisation, Adam Hilger, Bristol.

Matthews, M.B., (1990). Neural network nonlinear adaptive filtering using the extended Kalman filter algorithm. Proc. Int. Neural Network Conf., INNC90, Paris, Vol. 1 pp.115-118.

Morningred, J.D., B.E. Paden, D.E. Seborg and D.A. Mellichamp, (1990). An adaptive nonlinear predictive controller. Proc. ACC, Vol. 2, pp. 1614-1619.

Narendra, K.S. and K.Parthasarathy, (1990). Identification and control of dynamical systems using neural networks, IEEE Trans. on Neural Networks, Vol.1, No.1, pp. 4-27.

Oglesby, J. and J.S. Mason, (1989). Dynamic scheduling for feedforward neural nets using transputers. Proc. First IEE Conference on Neural Networks.

Reyneri, L.M. and E. Filippi, (1990). Modified backpropagation algorithm for fast learning in neural networks. Electronics Letters, 13th September, Vol. 26, No. 19, pp. 1564-1566.

Ruanno, A.E.B., Fleming, P.J. and D.I. Jones, (1991). A connectionist approach to PID auto tuning. Proc. IEE Control '91 Conf., Vol.2, pp. 762-768.

Rumelhart, D.E., G.E. Hinton and R.J. Williams, (1986). Learning internal representations by error propagation. Parallel Distributed Processing: Explorations in the Microstructure of Cognition, Vol. 1, MIT Press.

Sbarbaro, D. and P.J. Gawthrop, (1990). Learning complex mappings by stochastic approximation. Proc. Int. Joint Conf. on Neural Networks, January 1990.

Willis, M.J., C.D. Massimo, G.A. Montague, M.T. Tham and A.J. Morris, (1991). Artificial neural networks in process engineering, IEE Proc., D,Vol. 138, No. 3, pp. 256-266.

Wray, J and G.G.R. Green, (1991). Analysis of networks that have learnt control problems. Proc. IEE Control '91 Conf., Vol. 1, pp. 261-265.

Yu, X.H. and S.X. Cheng, (1990). Training algorithms for backpropagation neural networks with optimal descent factor. Electronics Letters, 27th September, Vol. 26, No. 20, pp. 1698-1700.

Review of Identification Techniques for Nonlinear Systems Using Neural Networks

Piyush C. Ojha, Robin M. C. De Keyser

Automatic Control Laboratory, University of Ghent, Grotesteenweg Noord 6, B-9052 Zwijnaarde, Belgium

Abstract

Most practical systems exhibit nonlinear characteristics of varying extent e.g from a simple valve to that of pH - processes in a chemical industry. In this paper we briefly review some of the Neural Network methods with a view to identify such systems. This is due to the fact that identification of the unknown system parameters form the backbone of most adaptive control strategies. In this paper the neural net techniques are reviewed from the point of view of approximating the hypersurface which maps the Input-Output space.
Keywords : Identification, Nonlinear Dynamical Systems, Neural Networks

1 Introduction

This paper reviews the possible applications of Neural Networks for *online identification* of dynamical systems exhibiting nonlinear characteristics. It is our intention to use Neural Networks in order to extend the existing model based control techniques to such systems. Narendra and Parthasarthy(1990) explored the possibility of using Backpropagation type neural nets for modelling nonlinear systems. Bhat and coworkers(1990) have used this type of neural nets to model complex nonlinear chemical processes. On the other hand Chen and Billings(1990) used Radial Basis Function type of networks(1990) for the above mentioned purpose. This paper has been reviewed in light of such developments.

2 Theoretical Background

This section provides the theoretical background in light of which the various types of neural networks are discussed later in the paper. In particular the method of orthogonalisation for linearisation of nonlinear systems and the classical identification techniques are discussed and their relevance to neural networks pointed out.

Briefly speaking the concept of orthogonality(Golub and Van Loan, 1989) involves the problem of expressing a vector in an N dimensional space as a linear combination of N independent vectors. In other words we approximate a nonlinear function f by a linear combination of $\Sigma_{i=1}^{N} C_i\phi_i$ in an N dimensional space with constants C_i such that average error $\int(f - \Sigma_{i=1}^{N} C_i\phi_i)^2 dx$ is as small as possible i.e the function f is approximated by a system of linear space formed by the basis vectors, ϕ_i. Alternately we can look at the above mentioned problem from the point of view of classification that is if the subspace of a particular class is spanned by the vector say l_i, then the distance of the input vector of the nonlinear dynamical system in consideration say g, from the subspace l_i can be obtained by decomposing the input vector g as

$$g = \hat{g}_i + \tilde{g}_i \tag{1}$$

where $\hat{g}_i$ is the linear regression or the best linear combination of the prototype that approximate g in the sense of least squares. The best approximation of g can be chosen by the norm $\| \tilde{g}_i \|$, the euclidean distance of g from l_i, which is minimum. The decomposition in (1) may be done by Gram - Schmidt orthogonalisation, Singular Value Decompostion(Golub and Van Loan, 1989) or Orthogonal Least Squares(Chen and Billings,1989). The classification is performed by choosing the subspace for which the linear regression $\hat{g}_i$ is largest or the residual $\tilde{g}_i$ is smallest. Also it may be noted that the decision surface in this case

$$\| \hat{g}_i \|^2 = \| g \|^2 - \| \tilde{g}_i \|^2 \tag{2}$$

is quadratic.

The subspace method is relevant to Kohonen and Radial Basis Function type of neural networks which are discussed in detail in the next section. As pointed out by Kohonen the subspace method can be made adaptive in order to reduce the error e.g to account for coloured noise which occurs frequently in the industry, by rotating the subspace. The concept of rotation is explored from this viewpoint in two dimension. A 2 by 2 orthogonal matrix Q is a rotation if it has the form

$$Q = \begin{bmatrix} \cos\theta & \sin\theta \\ -\sin\theta & \cos\theta \end{bmatrix} \quad (3)$$

If $y = Q^T x$ where

$$x = \begin{bmatrix} \cos\theta_1 \\ \sin\theta_1 \end{bmatrix} \quad (4)$$

then y is obtained by rotating the vector x counterclockwise through an angle θ. As an example if g is an input vector which has been wrongly classified into the subspace spanned by the vector l_i then the multiplication of all the basis vectors of l_i by the scalar $P = 1 - \frac{gg^T}{\| g \|^2}$ would result in a new rotated space l_i' in which all vectors are orthogonal to g i.e the projection of g on l_i' would be zero. This can be carried out iteratively by $P = 1 - \alpha\frac{gg^T}{\| g \|^2}$ where $0 < \alpha < 1$. If α is negative then the projection on g would increase. The idea behind training being that the projection on the wrong subspace is decreased while that on the correct subspace is increased.

Alternately we can discuss the relevance of the basis vectors from the viewpoint of approximating the hypersurface which maps the input - output space of the nonlinear dynamical system in details below. It would be appropriate to consider the kind/shape of decision surface placed by various types of neural networks. Generally speaking the equation of the hypersurface can be given as

$$\begin{aligned} d(x) &= w_1x_1 + w_2x_2 + \ldots + w_nx_n + w_{n+1} \\ &= W^TX + w_{n+1} = 0 \end{aligned}$$

where **X** is the input vector and **W** is the weight or parameter vector. The above equation represents a line when $n = 2$ and a plane when $n = 3$. When $n > 3$ it represents a hyperplane. This is the classification carried out in case of backpropagation type of neural nets. w_{n+1} is the d.c offset speaking in terms of a ***least squares estimator*** terminology and a change in this value amounts to shifting the plane. Similiarly the decision surface can be quadratic or of second order.

$$d^2(x) = \Sigma_{j=1}^{n} w_{jj}x_j^2 + \Sigma_{j=1}^{n-1}\Sigma_{k=j+1}^{n} w_{jk}x_jx_k + \Sigma_{j=1}^{n} w_jx_j + w_{n+1} \quad (5)$$

This equation can be generalised to an r order decision function.

$$d^r(x) = \Sigma_{p_1=1}^{n}\Sigma_{p_2=p_1}^{n} \ldots \Sigma_{p_r=p_{r-1}}^{n} w_{p_1p_2\ldots p_r}x_{p_1}x_{p_2}\ldots x_{p_r} + d^{r-1}(x)$$

This generalised equation can be verified for a quadratic surface (r=2) and a two dimensional data space (n=2).

$$d^2(x) = \Sigma_{p_1=1}^{2}\Sigma_{p_2=p_1}^{2} w_{p_1p_2}x_{p_1}x_{p_2} + d^1(x) \quad (6)$$

where

$$\begin{aligned} d^1(x) &= \Sigma_{p_1=1}^{2} w_{p_1}x_{p_1} + d^0(x) \\ &= w_1x_1 + w_2x_2 + w_3 \end{aligned} \quad (7)$$

Therefore

$$d^2(x) = w_{11}x_1^2 + w_{12}x_1x_2 + w_{22}x_2^2 + w_1x_1 + w_2x_2 + w_3 \quad (8)$$

We can write the generalised equation in a more compact matrix form as shown here. Referring to equation (5) we have

$$\begin{aligned} w_{jj} &= a_{jj} \quad j = 1,2,\ldots,n \\ w_{jk} &= 2a_{jk} \quad j,k = 1,2,\ldots,n; j \neq k \\ w_j &= b_j \quad j = 1,2,\ldots,n \\ w_{n+1} &= c \end{aligned}$$

Then the generalised equation can be written as

$$d(x) = \acute{x}Ax + \acute{x}b + c$$

where

$$A = (a_{jk}) \qquad b = \begin{bmatrix} b_1 \\ b_2 \\ \vdots \\ b_n \end{bmatrix}$$

The matrix **A** determines the shape of the decision surface. When **A** is an identity matrix, the decision surface describes a hypersphere. When **A** is positive definite the decision surface describes a hyperellipsoid with axis in the direction of the eigenvalues of **A**. When **A** is positive semidefinite the decision surface is hyperellipsoid cylinder. When **A** is negative definite the decision surface is hyperboloid. It may be noted that higher order orthogonal and orthonormal polynomial functions may be used to define and generate these decision surface.

Having discussed the shape of the decision surface which divides the input - output space of the dynamical system under consideration we discuss the ***Bayesian type of estimation*** with particular reference to the shape of the decision surface. Bayesian estimation treats the problem of classification of the input output space as a statistical game and has been reviewed here in order to analyse the decision surface. In particular Bayesian type of estimation is relevant for networks using unsupervised learning of which Kohonen network is an example. The Bayes formula can be given as

$$p(\omega_i/x) = \frac{p(\omega_i)p(x/\omega_i)}{p(x)}$$

where $p(\omega_i/x)$ is the probability that vector **x** belongs to class ω_i, $p(x/\omega_i)$ is the likelihood function and indicates the probability that x comes from ω_i. Due to its analytical tractability we now analyse a special case of Bayes classifier for a process with normal frequency distribution. The univariate normal density function can be given as

$$p(x) = \frac{1}{\sqrt{2\pi}\sigma} exp[-\frac{1}{2}(\frac{x-m}{\sigma})^2] \quad (9)$$

where $p(x)$ is the probability density function, m is the mean, σ^2 the variance. Now if we consider M subspaces or classes then we have a multivariate density function as

$$p(x/\omega_i) = \frac{1}{\sqrt{(2\pi)^n}|C_i|^{\frac{1}{2}}}$$

$$exp[-\frac{1}{2}(x-m_i)C_i^{-1}(x-m_i)] \quad (10)$$

where

$$m_i = E_i[x]$$
$$C_i = E_i[(x-m_i)(x-m_i)]$$

where E_i denotes the expectation operator over the input vector of class ω_i
n denotes the dimensionality of the input vector where the covariance matrix C_i is symmetric and positive semidefinite.
C_{kk} is the variance of the kth element of the input vector of a particular class
If $C_{jk} = 0$ then the elements x_i and x_j are statistically independent.
Now the Bayes rule implements the decision function for class ω_i as

$$d_i(x) = p(x/\omega_i)p(\omega_i) \quad (11)$$

We represent the decision function in the logarithmic form as

$$d_i(x) = ln[p(x/\omega_i)] + ln[p(\omega_i)] \quad (12)$$

Substituting for $p(x/\omega_i)$ we have

$$d_i(x) = ln[p(\omega_i)] - \frac{n}{2}ln(2\pi) -$$

$$\frac{1}{2}ln|C_i| - \frac{1}{2}[(x-m_i)C_i^{-1}(x-m_i)] \quad (13)$$

Equation 13 shows that the decision surface is hyperquadratic since there is no term greater than that of second order. Thus at best a Bayes classifier places a quadratic decision surface between two classes and their shape is determined by the covariance matrix C_i. As a special case if all the covariance matrices are equal then the decision surface is linear describing a hyperplane.

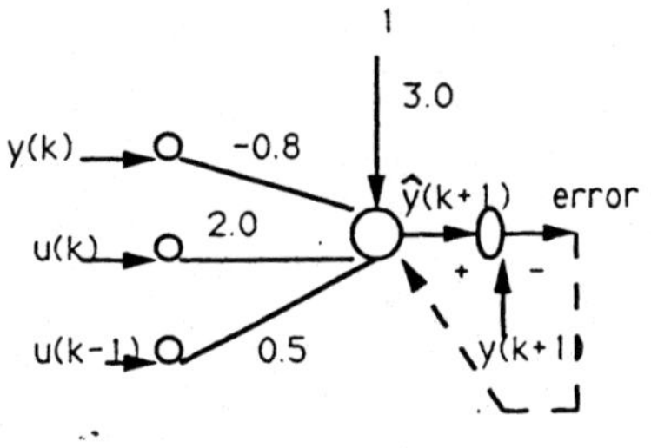

Figure 1: Figure shows the experiment carried out to show the least square nature of performance learning.

Figure 2: Figure shows the convergence of estimated zero position as referred to in figure 1.

3 Neural Networks Approach

This section carries out a comparative study of various types of networks from the point of view of identification of nonlinear systems. The theoretical background from the previous section which discussed the Least Mean Square and Bayesian type of estimation will be used to look at the various types of neural networks in a unified manner.

As is well known the Hebb's learning law carries out the input - output mapping without any error only if the input vector is orthonormal. This problem is sorted out in performance learning. In particular the weights are optimised as shown below

$$F(W) = E[(y_k - \hat{y}_k)]^2$$

where y_k is the output of the process under consideration and $\hat{y}_k$ is the output of the network. In other words the weights are optimised in the least squares sense. Simulation 1 has been carried out to bring out the similarity between the least squares estimator and the perceptron learning rule.
Simulation 1: The known system is taken to be

$$y(k+1) = 2.0u(k) + 0.5u(k-1) - 0.8y(k) + 3.0$$

The perceptron learning methodology used is explained in figure 1. Figure 2 shows the convergence of only one of the parameters (2.0) in order to bring out the similarity with least square estimation. On the contrary in case of Grossberg's learning the weights are adjusted by equating them to the average value of the desired output and not in least squares sense. In particular the processing element receives multiple input signal $x_1, x_2, \ldots, x_n$ and the desired output of the process y of which only one of the $x_i's$ may be active at a time. The idea is that when a particular x_i is positive, its corresponding weight w_i will learn the average of the concurrent y input multiplicatively weighted by x_i. Mathematically the Grossberg's learning law can be expressed as

$$w_i^{new} = w_i^{old} + a[x_i y - w_i^{old}]U(x_i)$$

where $0 \leq a \leq 1$ and

$$U(s) = \begin{cases} 1 & s > 0 \\ 0 & otherwise \end{cases} \tag{14}$$

It may be noted that the above equation adjusts its weights by equating them to the average value of the desired/teaching output and not in the least squares sense.

Having discussed two major learning methodologies we now discuss the various types of existing neural networks. The backpropagation type of neural network typically consists of two or more layers, each layer of which may consist of one or more elements. In a static backpropagation neural net the transfer function of each node is typically a sigmoid or a hyperbolic tan. As explained earlier in section 2 the weights attached to each node are optimised in the least square sense. Nevertheless relatively large number of nodes are required if the backpropagation type neural network is to approximate a relatively complex nonlinear system, thus making it difficult to find a relationship between the parameters/weights and the model in order to use them for existing model based control strategies. Research in this kind of neural network for the purpose of modelling has centred on the idea that the network is trained using a large training set prior to being used for control of a given plant. Besides these networks require to be trained for a very large number of iterations, typically in the range of thousands before they can be used for control purposes.

Simulation 2: This simulation example (Narendra and Parthasarthy, 1990) has been carried out using backpropagation type neural network. The system to be identified is given as $y_s(k+1) = f[y_s(k)] + g[u(k)]$ where $f[y_s(k)] = \dfrac{y_s(k)}{1 + y_s(k)^2}$ and $g[u(k)] = u^3(k)$

The functions f and g are identified separately using Neural networks N_f and N_g which correspond to the functions f and g. The test signal is given as $u(k) = \sin(2\pi k)/25 + \sin(2\pi k)/10$. Figure 4 shows the output of the network N_f where the dotted line indicate the output of the system N_f and the continous line that of the network. Each of the backpropagation networks N_f and N_g have two hidden layers

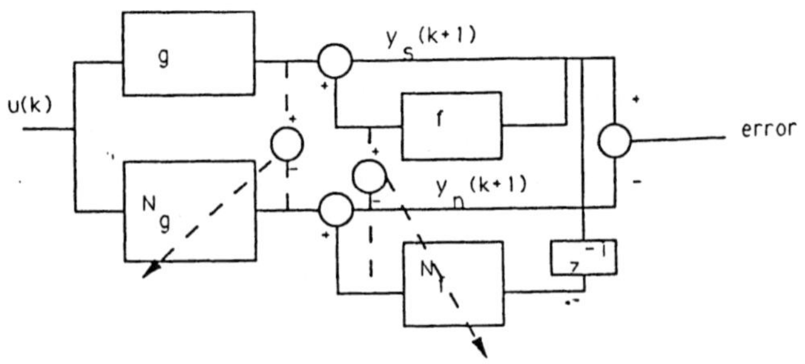

Figure 3: Schematic Diagram of the scheme used for simulation 2

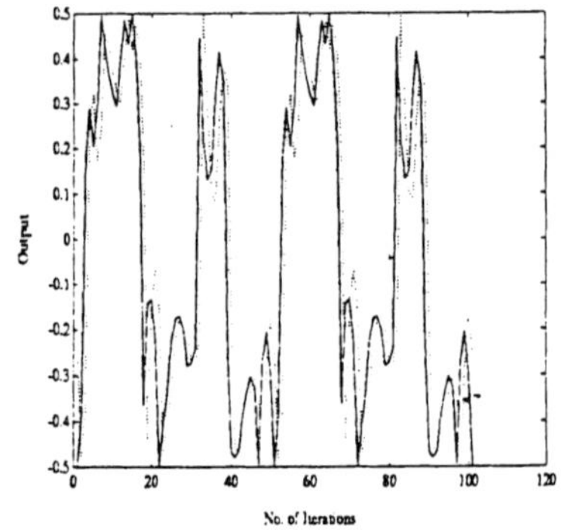

Figure 4: The output of the network N_f as explained in simulation 2.

of 20 and 10 neurons. The networks are trained with learning and momentum term 0.1 and required approximately forty thousand iterations to learn the model. It may be noted that the output of the subsystems g and f are assumed to be measurable.

Kohonen layer exhibits probability density estimation property. This property can be understood as linearising the Input/Output space via probability density estimation. It uses the principle of competition between the nodes based on the concept of unsupervised learning. Thus as new training vectors are entered into the network, the Kohonen vectors are drawn towards them and form a cloud near where the input vectors actually appear as determined by the probability density distribution. Counterpropagation network is an extension of Kohonen network introduced by Hecht Nielsen(Neurocomputing, 1988) to approximate continous functions, where the hidden layer uses the Kohonen structure while the output layer consists of Grossberg layer, explained in section 2.

Simulation 3: This simulation was carried out to bring out the advantages and disadvantages of the counterpropagation type network. It also gives us a good insight into the working of the Kohonen layer. The dotted lines in figure 5 show the output of the system, which is the same as simulation 2, while the continous line shows the output of the network. As explained in case of Grossberg's learning only one of the nodes of the output layer is activated at a time, the selection of which is determined by the hidden

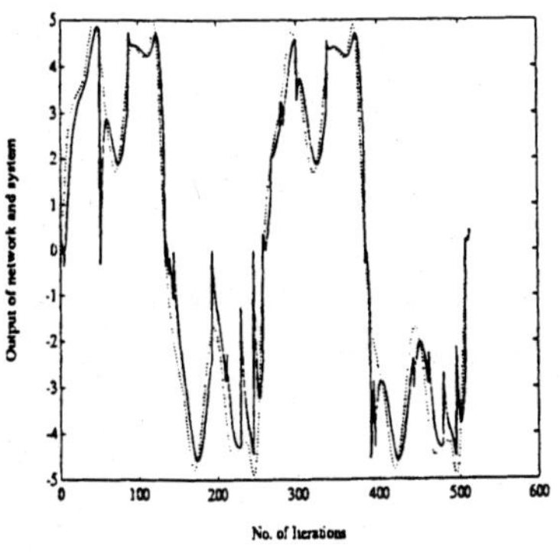

Figure 5: Figure shows the output of Counterpropagation Network.

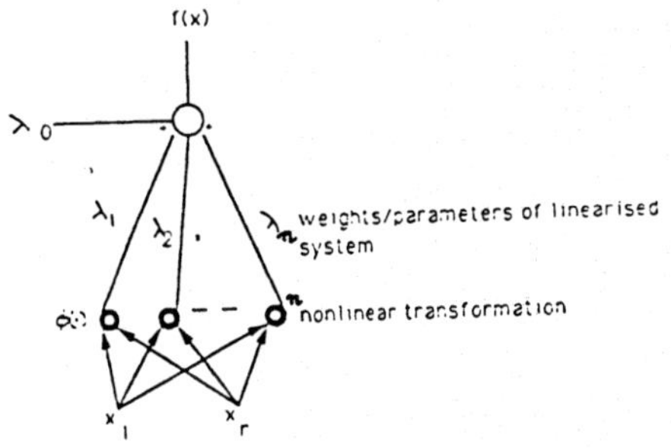

Figure 6: Schematic Diagram of Gaussian Network

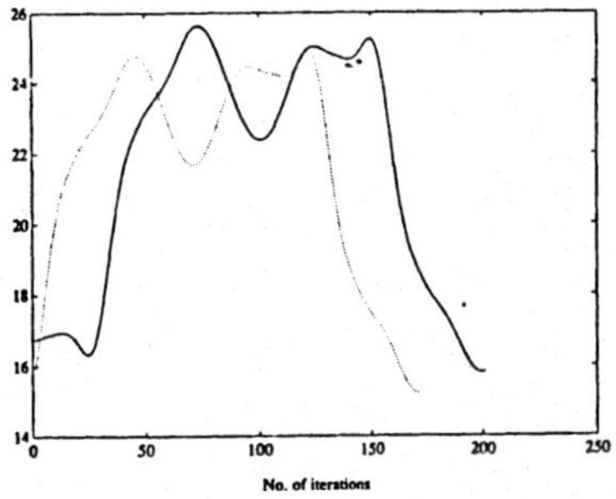

Figure 7: Figure shows the output of the Gaussian network

layer, which uses Kohonen learning, thus exhibiting probability estimation property. The sharp deviations observed in figure 5 are due to activation of different nodes in the hidden layer. Also unlike simulation 2 only one network was used to map the function $y_s(k+1)$ (refer simulation 2). The advantage of this type of neural network was that took only approximately five thousand iterations to track the output. The difference in frequency observed in figure 4 and 5 is due to the fact that input signals of different frequenies were used in the two cases.

As noted earlier the counterpropagation network does not estimate the parameters of the system in least square sense. This however lead us to replace the Grossberg's layer with that of the perceptron. This structure turns out to be similar to that of Radial Basis Function which has been reviewed in detail as it is found to be one of the most useful network from the viewpoint of identification. In particular Chen and Billings(1990) method for using such types of network has been reviewed in the next section.

4 Radial Basis Function Network

Among the various type of existing networks we found Radial Basis Function Network one of the most promising to our need. The system identified is the same as that of Simulation 2. Briefly the RBF network can be regarded as a two layered network which is linear in parameter by fixing all RBF centres and nonlinearity in the hidden layer. In other words the hidden layer performs a fixed nonlinear transformation and maps the input space onto a new space. The output layer then implements a linear combiner and the only adjustable parameters are the weights of this linear combiner. The nonlinearity within the hidden layer can be chosen from a few nonlinear function

- Gaussian function $\phi(v) = \exp(-\frac{v^2}{\beta^2})$
- Thin plate spline function $\phi(v) = v^2 log(v)$
- Multiquadratic function $\phi(v) = (v^2 + \beta^2)^{\frac{1}{2}}$

Referring to the figure 6 the network implements a mapping f_r according as

$$f(x) = \lambda_0 + \Sigma_{i=1}^{n} \lambda_i \phi(\| x - C_i \|)$$

where x is the input vector in the r dimensional data space

$\phi(.)$ is the given nonlinear function
$\| . \|$ is the Euclidean Norm
λ_i are the weights or parameters
n is the number of centres
C_i are the RBF centres

A commonly used method for choosing the centres is to choose arbitrarily some data points as centres. However such a method, in order to achieve a given performance, requires unnecessarily large RBF network, which adds to computational complexity. The Orthogonal Least Squares method is used to obtain an adequate and parsimonious RBF network , that is to select a suitable subset in order to reduce the error to a given limit, which will depend on the pratical application at hand. We can view the RBF network as a special case of the linear regression model.

$$d(t) = \Sigma_{i=1}^{M} p_i(t)\theta_i + \epsilon(t) \tag{15}$$

where $d(t)$ is the desired output
$\theta_i(t)$ are the parameters
$p_i(t)$ are the regressors and are some fixed function of x(t)

$$p_i(t) = p_i[(x(t)]$$

where x(t) is the vector which consists of lagged input and output of the dynamical process under consideration where $y(t)$ is the output and $u(t)$ is the input. Thus

$$x(t) = [y(t-1), \ldots, y(t-n_y), u(t-1), \ldots, u(t-n_u)]$$

Specifically the OLS method is used for selecting a suitable set of RBF centres i.e to select a subset of significant regressors from a given candidate set. We can write equation (15) in a matrix form as

$$d = P\theta + E \qquad (16)$$

where

$$\begin{aligned} d &= [d(1), \ldots, d(N)]^T \\ P &= [p_1, \ldots, p_M] \\ p_i &= [p_i(1), \ldots, p_i(N)]^T, 1 \le i \le M \\ \theta &= [\theta_1, \ldots, \theta_M]^T \\ E &= [\epsilon(1), \ldots, \epsilon(N)]^T \end{aligned}$$

where N is the number of iterations and p_i forms a set of basis vectors, M is the number of centres. Also $P\hat{\theta}$ is the projection of d onto the space spanned by these basis vectors. The Orthogonal Least Square method involves transformation of the set of p_i vectors into a set of orthogonal basis vectors. This is done by decomposing the regression matrix P into.

$$P = WA$$

where A is an $M \times M$ triangular matrix
W is an $N \times M$ matrix such that $W^T W = H$ where H is diagonal with elements h_i. Also the space spanned by the set of orthogonal basis vectors of matrix W is the same space spanned by the set of p_i.

Equation (15) can be rewritten as

$$d = Wg + E$$

Thus the orthogonal LS solution is given by

$$\hat{g} = H^{-1} W^T d$$

The quantities $\hat{g}$ and $\hat{\theta}$ satisfy the triangular system.

$$A\hat{\theta} = \hat{g}$$

This equation can be solved using Gram Schmidt orthogonalisation or OLS, where θ determines the weights associated with the output layer.

Thus referring to the subspace method discussed theoretically in section 2 it can be concluded that the vectors p_i form a set of basis vectors by orthogonalisation of the nonlinear space. Also the dc offset can be estimated by setting one of the p_i terms equal to 1, where the weights λ_0 indicates the estimated dc value.

Simulation 4: This simulation was carried out using the same system as that of simulation 2 with an offset of 2.0. The network used was a Radial Basis Fuction network with nonlinearity in the hidden layer given by gaussian function. Figure 7 shows the output of the network, where the dotted line is the output of the network and the continous line is the output of the system. Though the network was able to track the output in relatively few iterations a shift in the output of the network was observed.

5 Conclusion

Various types of Neural Networks were analysed from the viewpoint of online identification of nonlinear systems. It can be concluded that the Radial Basis type of Neural Networks appear to be one of the most promising for this purpose.

Acknowledgements: We thank our collegues Frits Dumortier and Patrick Van Renterghem for their cooperation.

References

[1] Courant and Hilbert (1970). Methods in Mathematical Physics , Volume 1, First Edition, Interscience Publisher Inc., New York.

[2] Golub G.H and Van Loan C.F(1989), Matrix Computation, second edition, John Hopkins University press.

[3] Narendra K.S and Parthasarty K(1990), Identification and Control of dynamical systems using Neural Networks, IEEE Transaction on Neural Networks, Vol. 1, No. 1(March).

[4] Chen S, Billings S.A and Lou W(1989), Orthogonal least squares methods and their application to nonlinear system identification, International Journal of Control, Vol. 50, no. 5, page 1873 - 1896

[5] Chen S, Billings S.A and Grant P.M(1990), Practical identification of NARMAX models using radial basis functions, International Journal of Control, Vol. 52, no. 6, page 1327 - 1350.

[6] Kohonen T, Self Organisation and Associative memory, second edition, Springer Verlag

[7] Hecht - Nielsen R(1988), Neurocomputing, Addison Wesley Publishing Company.

OPTIMAL EDGE SELECTION BY THE MODIFIED GENETIC ALGORITHM

C. H. Chung and K. S. Lee

Dept. of Control and Instrumentation Eng.
Kwangwoon University
Seoul, 130-701, KOREA

Abstract: Path planning is an important task for optimal motion of a robot in structured or unstructured environment. The goal of this paper is to plan the shortest collision-free path in 3D, when a robot is navigated to pick up some tools or to repair some parts from various locations. To accomplish the goal, the *Path Coordinator* [1] is proposed to have the capabilities of an *obstacle avoidance strategy* [2] and a *traveling salesman problem strategy* (TSP) [3]. The *obstacle avoidance strategy* is to plan the shortest collision-free path between each pair of n locations in 2D or in 3D. The *TSP strategy* [4] is to compute a minimal system cost of a tour that is defined as a closed path navigating each location exactly once. The *TSP strategy* can be implemented by the *Neural Network* [4] [3]. The *obstacle avoidance strategy* in 2D can be implemented by the *VGraph Algorithm* [5]. However, the *VGraph Algorithm* is not useful in 3D, because it can't compute the global optimality in 3D [6]. Thus, the *Path Coordinator* [7] is proposed to solve this problem, having the capabilities of selecting the optimal edges by the *modified Genetic Algorithm* [8] and computing the optimal nodes along the optimal edges by the *Recursive Compensation Algorithm* [9].

The *Neural Network* is applied to a case of a mapping process to solve the TSP with the 3D obstacles. the *Recursive Compensation Algorithm* was also applied to a case where no information on the optimal edges is available. However, the *Recursive Compensation Algorithm* requires information on the optimal edges to get the optimal solution. The exhaustive search algorithm guarantees the optimal edges for the shortest path. It may be applied to find the globally optimal edges for the shortest path in the workspace that has less than 4 obstacles. However, it is not an adequate algorithm to find the optimal edges for the shortest path in the workspace that has many obstacles, since it needs the exponential search time. For ($4 \sim 10$) obstacles, the Modified VGraph algorithm is suggested. This algorithm generates some additional vertices along the edges of the grown obstacles so that no edge is longer than a prespecified maximum length. However, it is not easy to decide how many vertices should be added along the edges of the grown obstacles and the additional vertices need much more computation time. If there are a lot of obstacles in a workspace, this algorithm is not adequate, because it needs a lot of memory space to represent the additional vertices which will result in more computation time. The modified Genetic Algorithm is useful to find the optimal edges for the shortest path in the workspace that has many obstacles. While the modified Genetic Algorithm finds the optimal edges, the *Recursive Compensation Algorithm* calculates the optimal vertices along the optimal edges. The *Recursive Compensation Algorithm* works to compute the cost function for the modified Genetic Algorithm.

1 Introduction

In many path planning algorithms, attempts are made to optimize the path between the start and the goal in terms of a Euclidean distance. The goal of this paper is to plan the shortest collision-free path in 3D, when a robot is navigated to pick up some tools or to repair some parts from various locations. In this paper, a *Planning Coordinator* is proposed to do this goal. To develop theories on the *Planning Coordinator* [10], the following assumptions are made:

- The obstacles are stationary polyhedral objects.
- A navigating robot is a polyhedral object.
- A system's cost function to be optimized is a Euclidean distance.

A navigating robot is shrunk to a configuration point in the *Configuration Space* [11], while the stationary obstacles are expanded to fill all space whenever the presence of a configuration point would imply a collision of the navigating robot with obstacles. Therefore, the path planning algorithm [12] can be formulated as a graph searching problem. The graph is formed by connecting all pairs of visible vertices. This is the well known *VGraph Algorithm* for calculating the shortest collision-free path.

To accomplish this goal, a *Planning Coordinator* must calculate the shortest collision-free path between each pair of the various locations. The *VGraph Algorithm* is useful since it can compute the global optimal path between each pair of the n locations in 2D.

Because the *VGraph Algorithm* cannot compute the global optimal path in 3D, the *VGraph Algorithm* cannot be directly applied for the previous subtasks. The *VGraph Algorithm* is not effective in view of the fact that the *VGraph Algorithm* cannot compute the global optimality in 3D. Instead, a robust algorithm [5] is needed to solve the optimal path between each pair of the n locations in 3D. Since the optimal path in 3D may involve going through points on the optimal edges, a robust algorithm should have the capability of solving the following two problems:

- The first problem is how to select the optimal edges.
- The second problem is how to find the optimal nodes on the optimal edges.

The global optimality of the *Recursive Compensation Algorithm* [1] is proved for 3D convex objects and the *Planning Coordinator* is implemented by using the *Recursive Compensation Algorithm.* While the modified Genetic Algorithm finds the optimal edges, the *Recursive Compensation Algorithm* calculates the optimal vertices along the optimal edges. The *Recursive Compensation Algorithm* can be applied to compute the cost function for the modified Genetic Algorithm [8].

The Genetic Algorithm (GA) [13] may be used to find the optimal edges. However, the Genetic Algorithm does not always guarantee global optimality, since the Genetic Algorithm may converge to the local minima [13]. Therefore, a robust algorithm is needed to guarantee the global optimality. The modified Genetic Algorithm (MGA) used to find the optimal edges in the *Planning Coordinator* has been proved to converge in probability to the global minimum of the cost function [8].

Finally, the *Neural Network* is applied to a case of a mapping process to solve the TSP with the 3D obstacles [7]. Most of the TSP problems have been solved in 2D. However, a robot could visit some locations to pick up some tools or repair some parts in deep sea, in space or on the earth. To solve the TSP with the 3D obstacles, a mapping mechanism is needed to compute the shortest path between two locations. If a location is visible to another location, then a straight line between the two locations is the shortest distance. If a location is not visible to another location, then obstacle-avoidance strategy is needed to calculate the shortest distance between the two locations.

2 Neural Network Application

[Problem Statement] Solve the TSP of 10 locations with the polyhedral obstacles, assuming that $A = B = 50$, $C = 20$, $D = 50$, $u_0 = 0.02$, $n = 15$, and $\tau = 1$ [3]. Calculate the shortest path from the L_i node to the L_j node, assuming that the obstacles are polyhedrons and no optimal edges are known. Suppose that the polyhedral, L_i node and L_j node are represented by the following vertices; $P_1(8,1,-3)$, $P_2(2,1,-3)$, $P_3(2,3,-3)$, $P_4(8,3,-3)$, $P_5(8,1,5)$, $P_6(2,1,5)$, $P_7(2,3,5)$, $P_8(8,3,5)$, $L_i(3,0.5,0.5)$, $L_j(3.5,4,3)$.
(Solution) Most of the TSP problems have been solved in 2D. However, a robot could visit some locations to pick up some tools or repair some parts in deep sea, in space or on the earth. To solve the TSP with the 3D obstacles, a mapping mechanism is needed to compute the shortest path between two locations. If a location is visible to another location, then the straight line between two locations is the shortest distance. If a location is not visible to another location, then obstacle-avoidance strategy is needed to calculate the shortest distance between two locations. Suppose that the polyhedral and Locations are represented by the following vertices; $P_1(8,1,-3)$, $P_2(2,1,-3)$, $P_3(2,3,-3)$, $P_4(8,3,-3)$, $P_5(8,1,5)$, $P_6(2,1,5)$, $P_7(2,3,5)$, $P_8(8,3,5)$, $L_i(3,0.5,0.5)$, $L_j(3.5,4,3)$. The *modified Genetic Algorithm* [8] updates the information on the optimal edges at each generation, according to the fitness of the cost function. Since the *modified Genetic Algorithm* reproduces its population for each generation, the *modified Genetic Algorithm* needs a fast function in order to evaluate the fitness. So the Recursive Compensation Algorithm [1] is applied to compute the fitness function. We set the parameters as follows; size of population = 20, length of chromosome = 12, maximum # of generation = 10, probability of crossover = 0.7, probability of mutation = 0.02, fitness function = $D_{fitness}/D_{RCA} \times 100\%$, where $D_{fitness} = 0.5519 \times 10$ and D_{RCA} = distance computed by the Recursive Compensation Algorithm according to information on the chromosome. To construct the chromosome, We set e_i, $i = 1, 2, \cdots, 12$ as the edges of a polyhedral; $e_1 = \overline{P_1P_2}$, $e_2 = \overline{P_2P_3}$, $e_3 = \overline{P_3P_4}$, $e_4 = \overline{P_1P_4}$, $e_5 = \overline{P_1P_5}$, $e_6 = \overline{P_2P_6}$, $e_7 = \overline{P_3P_7}$, $e_8 = \overline{P_4P_8}$, $e_9 = \overline{P_5P_6}$, $e_{10} = \overline{P_6P_7}$, $e_{11} = \overline{P_2P_7}$, $e_{12} = \overline{P_5P_8}$.

Table 1: Generation 0 by the modified Genetic Algorithm

#	parents	loc.	edges	distance	o.p.r.
1	(0, 0)	0	111100000000	1.599E+01	35%
2	(0, 0)	0	000001100010	1.078E+01	51%
3	(0, 0)	0	100010011000	1.939E+01	28%
4	(0, 0)	0	100010010000	1.337E+01	41%
5	(0, 0)	0	000010010000	1.190E+01	46%
6	(0, 0)	0	010100000000	8.778E+00	63%
7	(0, 0)	0	000010010000	1.190E+01	46%
8	(0, 0)	0	000010011000	1.505E+01	37%
9	(0, 0)	0	000000001101	1.503E+01	37%
10	(0, 0)	0	000000000101	1.163E+01	47%
11	(0, 0)	0	001001100010	1.461E+01	38%
12	(0, 0)	0	010100000000	8.778E+00	63%
13	(0, 0)	0	000000001111	1.757E+01	31%
14	(0, 0)	0	001001100010	1.461E+01	38%
15	(0, 0)	0	110100000000	1.325E+01	42%
16	(0, 0)	0	000001100010	1.078E+01	51%
17	(0, 0)	0	000010010000	1.190E+01	46%
18	(0, 0)	0	010100000000	8.778E+00	63%
19	(0, 0)	0	100010011000	1.939E+01	28%
20	(0, 0)	0	000010011000	1.505E+01	37%

Table 2: Generation 10 by the modified Genetic Algorithm

#	parents	loc.	edges	distance	o.p.r.
1	(10, 9)	1	000001100000	5.519E+00	100%
2	(10, 9)	1	000001100000	5.519E+00	100%
3	(7,15)	12	000001100000	5.519E+00	100%
4	(7,15)	12	000001100000	5.519E+00	100%
5	(14,11)	12	000001100000	5.519E+00	100%
6	(14,11)	12	000001100000	5.519E+00	100%
7	(7,16)	7	000001100000	5.519E+00	100%
8	(7,16)	7	000001100000	5.519E+00	100%
9	(14,15)	12	000001100000	5.519E+00	100%
10	(14,15)	12	000001100000	5.519E+00	100%
11	(11, 1)	10	000001100000	5.519E+00	100%
12	(11, 1)	10	000001100000	5.519E+00	100%
13	(14,19)	12	000001100000	5.519E+00	100%
14	(14,19)	12	000001100000	5.519E+00	100%
15	(14, 8)	12	000001100000	5.519E+00	100%
16	(14, 8)	12	000001100000	5.519E+00	100%
17	(11, 8)	12	000001100000	5.519E+00	100%
18	(1, 1)	12	000001100000	5.519E+00	100%
19	(13, 4)	12	000001100000	5.519E+00	100%
20	(13, 4)	12	000001100000	5.519E+00	100%

Table 1 and Table 2 show that the optimal edges are e_6 and e_7. The Recursive Compensation Algorithm shows that the optimal vertices are $N_1(2.0, 1.0, 1.0682)$ along e_6 and $N_2(2.0, 3.0, 2.0842)$ along e_7 and the shortest path between L_i and L_j is 0.5519453549×10.

The *optimal population ratio* (opr) is defined by

$$\frac{\textit{number of optimal population in a generation}}{\textit{population size in a generation}} \times 100\%,$$

Figure 1 shows the initial state in the Neural TSP. Figure 2 shows the intermediate state ($t = 1.0 \times 10^{-3}$) in the Neural TSP. Figure 3 shows the intermediate state ($t = 2.0 \times 10^{-3}$) in the Neural TSP. Figure 4 shows the intermediate state ($t = 3.0 \times 10^{-3}$) in the Neural TSP. Figure 5 shows the intermediate state ($t = 4.0 \times 10^{-3}$) in the Neural TSP. Figure 6 shows the intermediate state ($t = 5.0 \times 10^{-3}$) in the Neural TSP. Figure 7 shows the intermediate state ($t = 7.0 \times 10^{-3}$) in the Neural TSP. Figure 8 shows the final state in the Neural TSP. Figure 9 shows the comparison between MGA and GA.

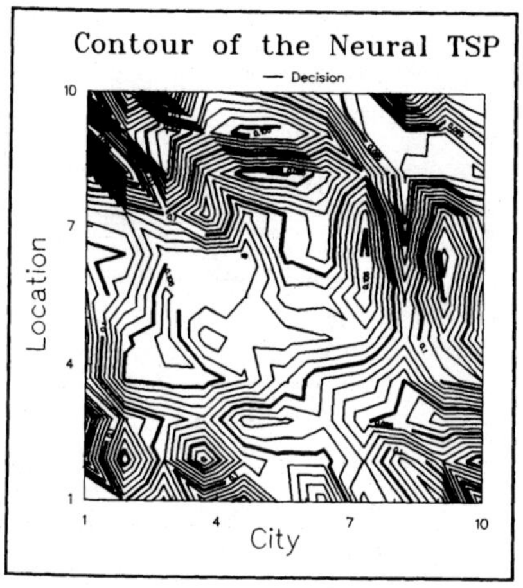

Figure 1: Initial State in the Neural TSP

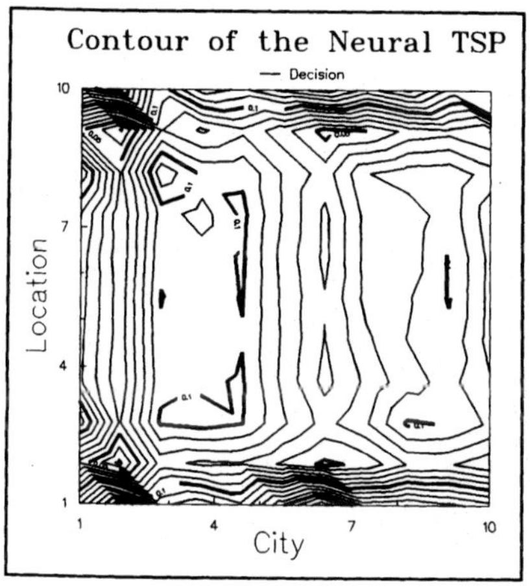

Figure 2: Intermediate State ($t = 1.0 \times 10^{-3}$)

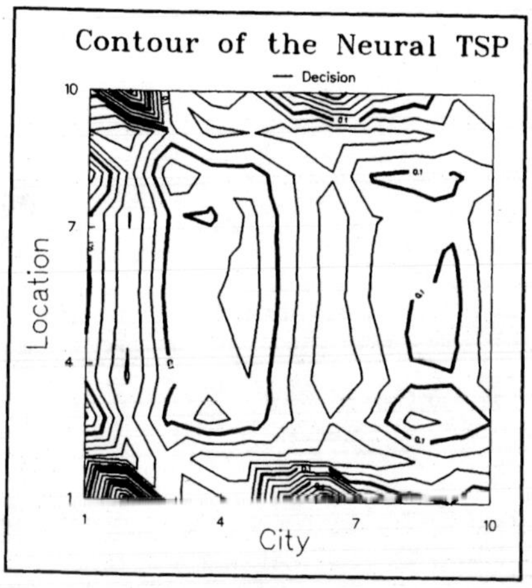

Figure 3: Intermediate State ($t = 2.0 \times 10^{-3}$)

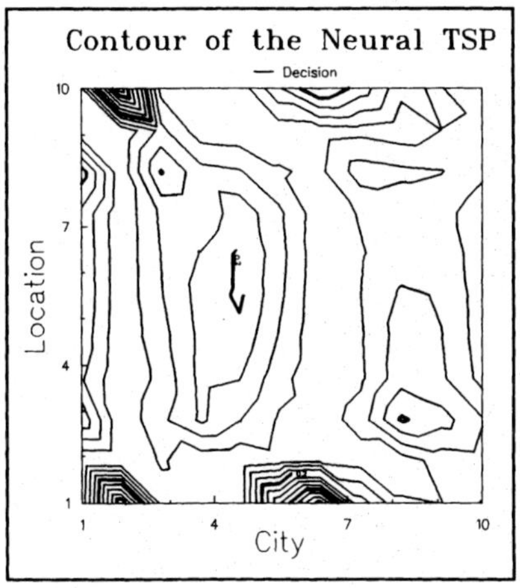

Figure 4: Intermediate State ($t = 3.0 \times 10^{-3}$)

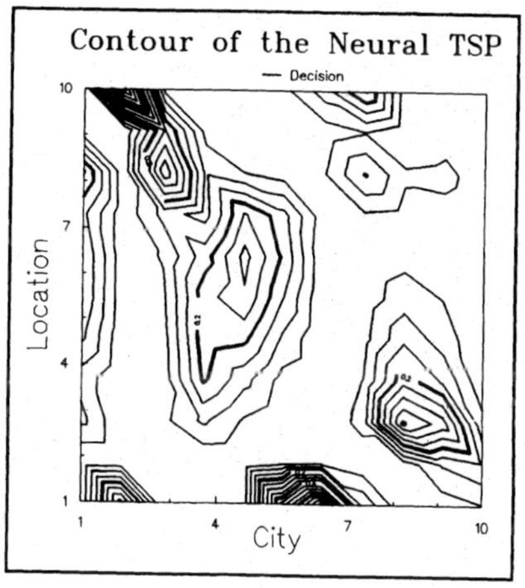

Figure 5: Intermediate State ($t = 4.0 \times 10^{-3}$)

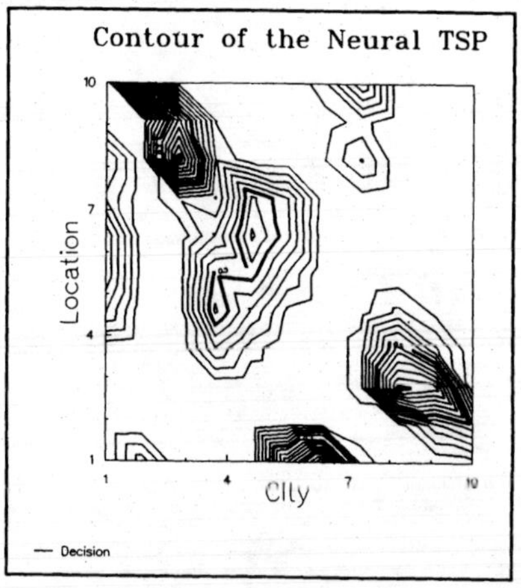

Figure 6: Intermediate State ($t = 5.0 \times 10^{-3}$)

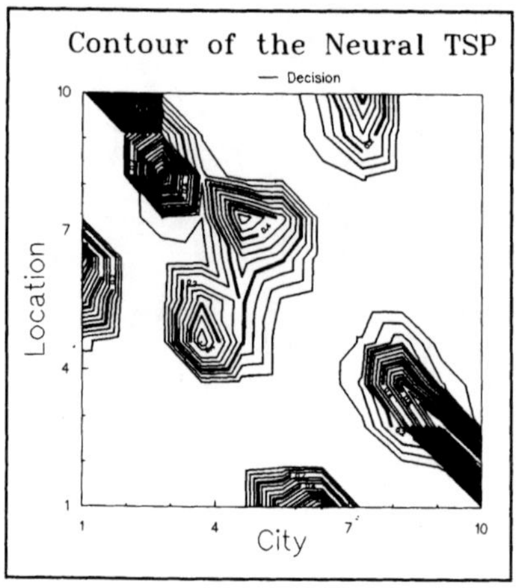

Figure 7: Intermediate State ($t = 6.0 \times 10^{-3}$)

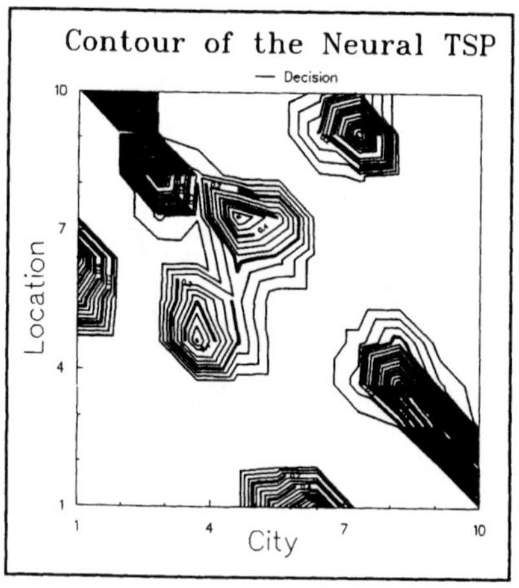

Figure 8: Final State in the Neural TSP

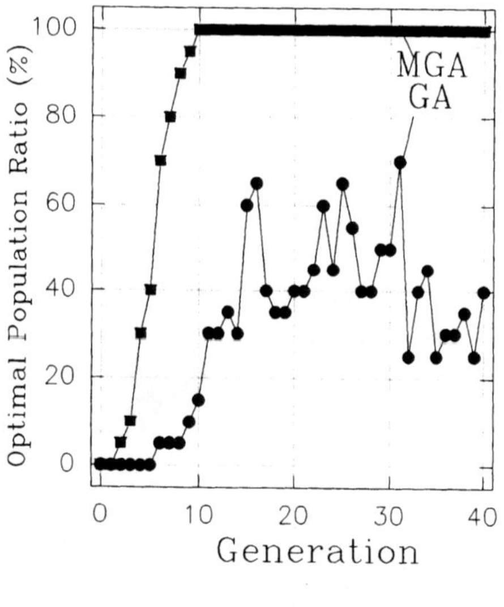

Figure 9: A Comparison between MGA and GA

3 Conclusions

The *Neural Network* is applied to a case of a mapping process to solve the TSP with the 3D obstacles. the *Recursive Compensation Algorithm* was also applied to a case where no information on the optimal edges is available. However, the *Recursive Compensation Algorithm* requires information on the optimal edges to get the optimal solution. The exhaustive search algorithm guarantees the optimal edges for the shortest path. It may be applied to find the globally optimal edges for the shortest path in the workspace that has less than 4 obstacles. However, it is not an adequate algorithm to find the optimal edges for the shortest path in the workspace that has many obstacles, since it needs the exponential search time. For ($4 \sim 10$) obstacles, the Modified VGraph algorithm is suggested. This algorithm generates some additional vertices along the edges of the grown obstacles so that no edge is longer than a prespecified maximum length. However, it is not easy to decide how many vertices should be added along the edges of the grown obstacles and the additional vertices need much more computation time. If there are a lot of obstacles in a workspace, this algorithm is not adequate, because it needs a lot of memory space to represent the additional vertices which will result in more computation time. The modified Genetic Algorithm is useful to find the optimal edges for the shortest path in the workspace that has many obstacles. While the modified Genetic Algorithm finds the optimal edges, the *Recursive Compensation Algorithm* calculates the optimal vertices along the optimal edges. The *Recursive Compensation Algorithm* works to compute the cost function for the modified Genetic Algorithm.

4 Future Work

There are many interesting problems deserving further research for the path planning. Below some of them are listed in no particular order.

- Solve the collision-free path planning for the dynamic obstacles.
- Exchange the knowledge on the path planning with other coordinators for an intelligent robot.
- Consider the general polyhedral objects for the collision-free shortest path.
- Consider not only a Euclidean distance but also time or energy as a system cost function.

References

[1] C. H. Chung and G. N. Saridis, "The recursive compensation algorithm for obstacle avoidance path planning," in *IEEE International Workshops on Intelligent Robots and Systems*, (Tsukuba, Japan), September 1989.

[2] C. H. Chung and G. N. Saridis, "An obstacle avoidance motion organizer for an intelligent robot," Tech. Rep. RAL-TR-88-117, Robotics and Automation Laboratory, Rensselaer Polytechnic Institute, Troy, NY, 1988.

[3] G. V. Wilson and G. S. Pawley, "On the stability of the travelling salesman problem algorithm of hopfield and tank," *Biological Cybernetics*, vol. 58, pp. 63–70, 1988.

[4] J. J. Hopfield, "Neural networks and physical systems with emergent collective computational abilities," in *Proc. Natl. Acad. Sci. USA*, pp. 2554–2558, April 1982.

[5] C. H. Chung and G. N. Saridis, "Path planning for an intelligent robot by the extended vgraph algorithm," in *IEEE International Symposium on Intelligent Control*, (Albany, NY), September 1989.

[6] T. Lozano-Pérez and M. A. Wesley, "An algorithm for planning collision-free paths among polyhedral obstacles," *Communications of the ACM*, vol. 22, pp. 560–570, October 1979.

[7] C. H. Chung and K. S. Lee, "Neural network application to the obstacle avoidance path planning for cim (computer integrated manufacturing)," in *IEEE International Workshops on Intelligent Robots and Systems*, (Osaka, Japan), November 1991.

[8] M. C. Moed and G. N. Saridis, "A boltzmann machine for the organization of intelligent machines," in *Proceedings of 2 Telerobotics Conference*, (Pasadena, CA), January 1989.

[9] C. H. Chung and K. S. Lee, "Hopfield network application to optimal edge selection," in *IEEE International Joint Conference on Neural Networks*, (Singapore), November 1991.

[10] C. H. Chung and G. N. Saridis, "Obstacle avoidance path planning by the extended vgraph algorithm," Tech. Rep. CIRSSE-TR-89-12, Center for Intelligent Robotic Systems for Space Exploration, Rensselaer Polytechnic Institute, January 1989.

[11] R. A. Brooks, "Planning collision-free motions for pick-and-place operations," *The International Journal of Robotics Research*, vol. 2, no. 4, pp. 19–44, 1983.

[12] T. Lozano-Pérez, "Automatic planning of manipulator transfer movements," *IEEE Trans. Systems, Man, and Cybernetics*, vol. 11, pp. 681–698, October 1981.

[13] D. E. Goldberg, *Genetic Algorithms in Search, Optimization, and Machine Learning.* Reading, MA: Addison-Wesley Publishing Company, INC., 1989.

REAL-TIME FAULTS DIAGNOSIS IN A LARGE-SCALE SYSTEM USING NEURAL NETWORKS BASED ON SIGNED GAIN DIGRAPH WITH APPLICATION

Hak-Yeong Jung* and Zeungnam Bien**

*Automatic & Control Department, Research Center KEPCO, Taejeon, SOUTH KOREA
**Department of Electrical Engineering, Korea Advanced Institute of Science and Technology, Taejeon, SOUTH KOREA

Abstract. In a large-scale system such as a power plant which is strongly nonlinear, it is very difficult to locate faults origins when enormous abnormal signals come into exitence in the limited time due to the plant complexity. In this paper, a new approach is proposed in which hierarchically structured neural networks are utilized based on the system directed graph(digraph) with signed gain branches, and a simple method of training data extraction is also given for faults detection, which is crucial aspect for real-time diagnosis in a complex system. This method also comprises the fault propagation probability, fault propagation time, and the fault rates of devices for more accurate diagnosis, and provides auto-tuning of the states of SDG nodes according to the various operating conditions through the plant structure identifier. For each subsystem, there corresponds neural network that performs the faults detection, and the estimation of their faults size.
The plant-wise faults diagnosis is performed for removing the spurious faults using the information from the lower level neural networks, and provides the correponding treatment guideline to the operator at the top level neural network.
The information on the faults size can give the plant operators the plant-wise control strategy.
In a case study of applicaion to a pump system, the proposed scheme diagnoses multi-fault as well as a single fault including sensor fault itself in real-time during quasi-steady state condition.

Keywords. Faults, Diagnosis, Neural Network, Digraph.

INTRODUCTION

For a large-scale system such as a power plant which is composed of many components and strongly nonlinear, faults detection and diagnosis are more and more important when the plant undergoes an abnormal transition.
A good example occurred in Three Mile Island(TMI) accident in America in 1979 shows how immense the economic loss and the resulting influence to the related industry are where the reactor core of the nuclear power plant was severely melted down because of the operator's misoperation from the vast influx of plant information.
For this reason, the automated faults diagnosis and operator guiding system become essential to enhance the safty and the productivity in the power plant and in the production processes.
There are lots of efforts to locate the faults origin from the vast abnormal signals resulting from some troubles in the process.
The technique for automated faults diagnosis can be classified into two approaches, quantitative and qualitative methods depending on how rigorous the used model is.

The quantitative fault diagnosis adopts the thermo-hydraulic mathematical model, which requires extensive engineering manpower and computational horse power , and also has the limited detection capability.
The parameter estimation based fault diagnosis is the typical example (Isermann, 1984).
The qualitative approach, on the other hand, should be able to identify the core knowledge of the process (e.g., the signs of coefficients in all governing equations, and faults and their effects vectors etc.).
The signed directed graph(SDG) is one of the most commonly used qualitative model which is used to represent the causal effects between process variables, and is drawn using nodes corresponding to the process variables or the process devices and branches representing the influence between two adjacent nodes. Kokawa et al.,(1983) have developed fault diagnosis method using digraph and inverse direction search, and D.Q.Qian(1990) improved the Kokawa's method using frame structured failure propagation model and fuzzy concept for the calculation of the failure propagation time, and the failure propagation probability. But, the Kokawa's and D.Q. Qian's methods can not diagnoses multiple faults.
Kramer and Palowitch(1987) derived a rule-based forward reasoning using SDG, which provides efficient method in identifying the possible fault orgin, and Chung-Chien Chang and Cheng-Ching Yu(1990) advanced the kramer and palowitch's apporach by improving the diagnostic resolution.
However, Kramer and palowitch's approach and Chang Yu's appoarch also have the possibility of non-unique diagonstic solution and erronous diagnostic result.
Artificial neural network, recently, has been utilized to fault diagnosis(M.A.Kramer and J.A.Leonard(1990), V.Venkat-asubramanian et al.,(1990), J.C.Hoskins, et al.,(1991))
In the application to the real plant, however, what is important is how to find a good real-time model for fault location so that not the developed, but the incipient abnormalities should be detected as early as possible, while the model describes the plant behavior quite well.
In this study, a new approach is proposed using artificial neural networks based on the system directed graph with the signed gain branches.
The large-scale plant is divided into subsystems for real-time faults detection and each nerual network takes charge of each subsystem, performs the fault diagnosis, calculates the faults size with the knowledge on the granch gain, and gives the corresponding proper treatment guidline to the operator.

The top level neural network performs the plant-wise diagnosis which removes the spurious faults origins by way of backward inference using the declared faults information of the lower level nerual networks.
Several factors, fault propagation propability, fault propagation time, and fault rates of devices are used for more accurate faults diagnosis and the approximate estimation of faults size .
The paper begins with the assumptions, and drawing the digraph with signed gain branches, and next, plant data preprocessing method is described for classifying the various operating conditions. The faults diagnosis procedures and the faults size estimation method are given after that. Finally, simulation and results, and conclusion are followed in sequence.

ASSUMPTIONS FOR FAULTS DIAGONSIS

The operators in a plant can recognize the condition of the plant from the symptoms of the faults propagated devices rather than from the symptoms of the faults origin devices.
Thus, the automated processing of the abnormal data influx is very important for the reliable plant operation and for enhancing the plant availability.
In this section, faults diagnosis method is described under the following assumptions using the fault propagation probability, approximate fault propagation time in a subsystem, and fault rates of devices.
(1) All the sensed data are available throughout the recording, indicating or CRT display.
(2) there exists no negative feedback control loop in the SDG model, but the positive or negative feed-forward loop is allowed here with the fault propagation probability on each path.
(3) The faults diagnosis is performed under the steady-state or quasi-steady state, that is, near steady-state.

DESCRIPTION OF THE FAULT PROPAGATION MODEL

The SDG is used as the fault propagation model that depicts the cause and effect relationship among process variables. In this study, the SDG is reshaped as the following :

$$X_i \overset{\alpha_{ij} \quad p_{ij} \quad t_{ij}}{\circ\!\!\longrightarrow\!\!\circ} X_j \qquad (1)$$

where $\alpha_{ij} = G_{ij}\ \mathrm{sgn}(X_i - X_j)$
P_{ij}= fault propagation probability from node X_i to node X_j
t_{ij}= approximate fault propagration time from node X_i to node X_j
G_{ij}= influence gain from node X_i to X_j, and the $\mathrm{sgn}(X_i - X_j)$ represents the direction of influence, the branch sign.

In the qualitative sense, the precise measurements of nodes are transformed into qualitative states of high, normal, or low which are expressed in terms of deviation variables as follows:
for a typical node X,

$$X = \frac{\text{measured } X(X_m) - \text{normal operating value of } X(\overline{X})}{\text{threshold of } X(X_{ST})} \qquad (2)$$

where the X is denoted as 'normal' for $-1<X<1$, 'high' for $X>1$, and 'low' for $X<-1$.
The value of node X also can be recognized as a continuous value ranging from 'high' level to 'low' level.
The term $\mathrm{sgn}(X_i - X_j)$ also takes the value of '+' or '-'.
If $\mathrm{sgn}(X_i - X_j)$ = '+', it means a positive (or negative) deviation in X_i leads to a positive (or negative) deviation in X_j .
The values of α_{ij}, P_{ij}, t_{ij} can be obtained from the experienced operators, plant data or mathematical model. The fault rates of nodes are also obtained througout the operating data, expertise, or production companies. Once the fault propagation model is available, we may derive the behavior of a given disturbance on node X. For example, in Eq.(1), The state of X_j normalized by it's threshold value can be estimated after t_{ij} as:

$$X_j = \alpha_{ij} \cdot X_i \qquad (3)$$

or reversly estimation of the disturbance can be made from the deviated value of X_i as :

$$X_i = X_j \text{ devided by } \alpha_{ij} \qquad (4)$$

We can approximately measure the size of faults for the faults origin using the qualitative resoning concept as in Eq.(3) and Eq.(4) in this study.

PLANT STRUCTURE IDENTIFIER

From the viewpoint of faults diagnosis and the corresponding compensatory control, there is the fundamental difficulty when the complex plant undergoes its structural changes like a transition to the various operating conditions. The plant characteristics can be identified by the several key parameters, not all the parameters so that the plant structure is classified into several patterns from the readings of the selected key parameters.
The states of SDG nodes can be rearranged automatically according to the classfied patterns with their operating points and thereshold values.
For example, the state X in Eq.(2) can be changed according to the corresponding classified values of $\overline{X}$ and X_{ST}.
In this study, unsupervised learning neural networks are used for the plant structure identifier such as competative learning method and Kohonen's self-organizing feature map(Fig.1).

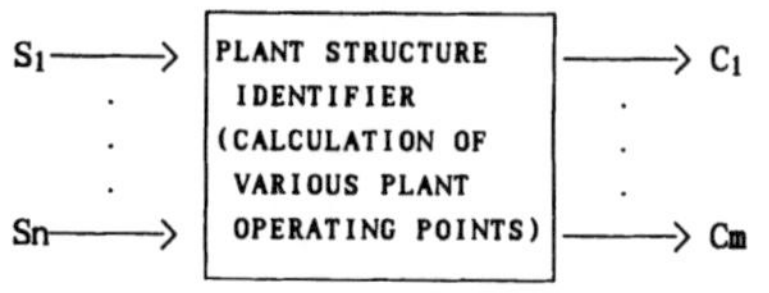

S_i: Plant Key Parameter i, $1 < i \leq n$
C_j: Classified Pattern j for Plant Operating Point and It's Threthold Value, $1 \leq j \leq m$

Fig.1. The Diagram of Plant Structure Identifier

PLANT DESCRIPTION

Division into Subsystem.

Once the fault propagation model, the SDG with signed gain branches here and the plant structure identifier are established for faults diagonsis of the plant, the plant may be devided into several smaller subsystems so that each neutal network takes the charge of each subsystem for faults diagonsis(Fig.2).

Thus, the whole plant is hierarchically composed of modular neural networks, where the top level neural networks identifies the plant-wise faults origins using the declared information from each lower level neural network. The size of a subsystem can be determined depending on the characteristics of the whole system and physical relationship among subsystems considering the real-time diagonsis of the entire system.

A Typical SDG for a Subsystem.

As mentioned above, all the terms, that is, α_{ij}, P_{ij}, and t_{ij} for all branches are not required, rather, only P_{ij}s are employed for faults detection.
In case of feedforward loop with multiple paths from the failed node and the affected node which is conversing node(X_B) from the failed node(X_A) as shown in Fig.3. the resulting fault propagation probablity P_{AB} is the largest one among i P_{AB}s between the two nodes X_A and X_B.
A path may contain more than one node between the failed node and the influenced converging node.
For example, there is no node along the path1 but there are two nodes along the path2 in Fig.3.

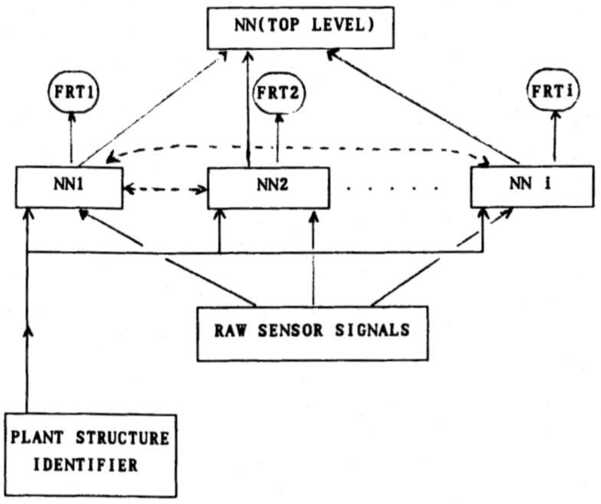

------>:Faults Propagating Direction.
NNi:Neural Network i for Diagnosing Subsystem i.
FRT:Report of Faults Origins and The Corresponding Treatment Guideline.

Fig.2. Hierarchical Structure for Faults Diagnosis in a Large-Scale system.

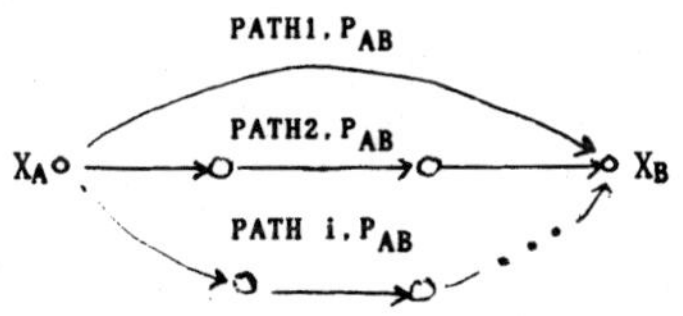

Fig.3 Feed-forward Loop with Multi-Path between Failed Node X_A and Converging Node X_B

The resulting influence on node X_B from the node X_A in Fig.3 can be calculated using the below Eq.(5) with the largest P_{AB} among i P_{AB}s and each path gain α_{AB}, for each path, the path gain α_{AB} and it's corresponding path propagation probability P_{AB} are calculated by multiplying each branch gain and each branch fault propagation probablity between the two nodes.
If the gain on each branch along the path is given in a fuzzy value, Eq.(5) can be calculated using the fuzzy theory.

$$X_B = (\alpha_{AB}(PATH1) + \alpha_{AB}(PATH2) + \cdots + \alpha_{AB}(PATHi)) \cdot X_A \qquad (5)$$

Influence from external subsystem. Consider Fig.4 in which the node X_{B2} in subsystem B has the three external inputs from a subsystem A. and one internal input from node X_{B3} within the subsystem B.
The influence on node X_{B2} from subsystem A can propogate to the subsystem B which is fault-free originally, so that faults are declared not only at subsystem A but also at subsystem B in which the node X_{B2} is suspected of the faults origin in the subsystem B.

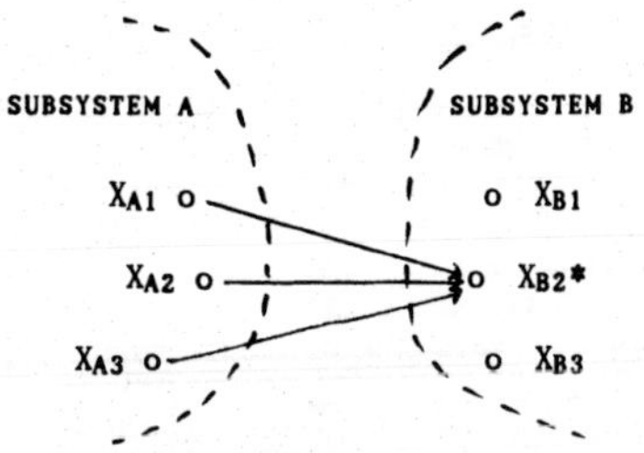

Fig.4 External Influence on Node X^*_{B2}.

In general, if there are k nodes marked with superscript * which are influenced by external inputs from any failed subsystem and internal inputs within the subsystem, then, total influence should be considered at each subsystem which has no internal fault , relevant to this influence for the plant-wise faults diagnosis later.

TRAINING DATA EXTRACTION FOR SUBSYSTEM

It is very important to get the reliable training data which also describe the plant behavior quite well, not based on expert knowledge but based on a systematic model. It is more desirable to explain the extracted data flow, namely, the route of the faults propagation .
The SDG model with only fault propagation probability on each branch as a systematic plant model, as stated at the early section, is used here for generating training data for neural network for faults dectection.
The procedure to get the training data for a typical subsystem is derived below :
1)mark the nodes with superscript * which have the external inputs outside the subsystem that the asterisk marked nodes belong to.
2)select sensors as the sources for providing input training patterns within the subsystem.
The readings from the sensors are transformed to the normalized form by their threshold values using Eq.(2).
3) convert the feedforward loop with multiple paths starting from the failed node as in Fig.3 to a single path with the selected highest fault propagation probability.
4)select the every node including nodes with sensors and sensors themselves, if the selected node is assumed failed and affects any sensor, then mark 'the corresponding fault propagation probability' at the point between the failed node and each fault-affected sensor or mark '0' if not, finally, collect all the fault-assumed nodes and the fault-affected sensors correspondingly forming a relationship table between them.
For example, considering the process in Fig.5-(a) which will be used as a sample subsystem for simulation, Table in Fig.5-(c) derived using above step(4) shows that considering the 4th row, the (fault-assumed)node X_4 affect the sensor nodes S_9 and S_{19} (marked as '1.0' respectively) but does not affect the nodes S_{17} and S_{15} (marked as '0.0' respectively).
5)group the same patterns into one and re-establish the reduced input-output training patterns as in Fig.5-(d) which is the final input-output patterns used for faults diagnosis at lower level neural networks.

FAULTS DIAGONSIS

The faults detection and diagnosis is achieved through the two-stage approach, the incipient faults are detected at subsystem stage, and the plant-wise faults detection is made using the faults information from the subsystems for more accurate multi-fault detection such as removing the spurious faults among susystems.

Faults Detection in a Subsystem The faults diagnosis procedure for a typical subsystem is as follows :

(1) draw the digraph with only fault propagation probability on every branch as stated in the previous section.

(2) get the reduced input-output training data from the degraph as also mentioned earlier.

(3) if there exist several candidates for a fault origin, ordering the candidates for selecting the most suspected candidate is performed based on the calculated fault propagation propability , P_{mn} which is from the candidate m to a fault-affected sensors pattern n , from neural network and fault rate F_m of each candidate m. that is, the product values, P_{mn} x F_m , are compared each other for all candidates, then the candidates are ranked according to these product values, or in a alternative way, the ordering can be done based on the calculated total influence on the nearest affected sensor for each candidate using the path gain which is the total gain from each suspected candidate to the nearest fault-affected sensor with the same concept as in Eq.(3) and Eq.(5) , and thus the candidates are ranked according to deviation contribution of each candidate to the nearest fault-affected sensor.

Faults detection time. The faults detection starts at

least at the indicated time of abnormal state at the nearest sensor added to the longest fault propagation time among sensors in which only paths are selected among the sensors with non-zero fault propagation probability .

Plant-wise Faults-diagonsis.

The plant-wise faults diagonsis is performed at the top level neural network as follows :
(1) construct the fault influence table among subsystems as in Fig.6-(b) in which the fault influnce among subsystems are represented with '1' or '0' at the corresponding element.
For example, considering first row in Fig.6-(b) the fault assumed subsystem A affects the subsystem B as well as the subsystem A itself (denoted as '1') ,but does not affect the subsystem C (denoted as '0') etc., with the similar way as in Fig.5-(c).
In this way, the subsystem of the faults origin is identified from which the faults are propagated into the related subsystems, namely the faults affected subsystems.
(2) there is also the possibility of faults within the fault-affected subsystems except for the propagated faults from the failed subsystems.
If the fault-affected subsystem contains the first fault-influenced nodes marked with asterisk as candidates of faults origin at the fault-affected subsystem , then those nodes are removed from the candiates of the faults origin at the fault-affected subsystem ,which can be identified by back-tracing from the first fault-influenced nodes to the candiates of faults origin of each failed subsystem along the paths with non-zero faults propagation probability.
Thus, the spurious faults are removed through the plant-wise faults diagnosis phase.

Estimation of faults size. If we can get the signed gain on each branch of the SDG , the estimation of each fault size can be made by dividing the amount of the deviation value of the nearest affected sensor with the corresponding path gain as the same concept in Eq.(4).
When the gain on each branch is given in a fuzzy value. the fault size is calculated using the fuzzy theory.

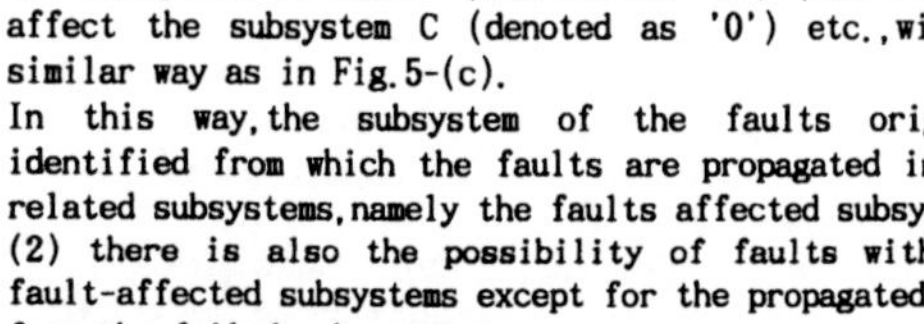

(a)

FAULT AFFECTED SUBSYSTEM					FAULT ASSUMED SUBSYSTEM
A	B	C	.	SUBSYSTEM	
1	1	0	.	A	
0	1	1	.	B	
1	1	1	.	C	
. .	.	. .	.	.	

(b)
Fig.6. (a) A plant Composed of Subsystems.
(b) The Corresponding Fault Influence Table among Subsystems.

SIMULATION RESULTS

The simulation is made on the pump system.
the Fig.(5)-(a),(b),(c),and (d) show the pump system, the coressponding SDG with fault propagation probability on each branch, the training data extraction table, and the reduced input-output training data patterns respectively.
So that, the Fig.(5)-(d) is trained using the error-back propagation four-layer perceptron with 4 input layer neurons, 10 and 22 neurons for two hidden layers respectively ,and 19 output layer neurons.
The resulting output is shown in table 1. in which the five test cases are shown ,that is, the trained pattern, noisy sensor inputs, multi-fault, sensor fault, and the reading pattern of sensors in a quasi-steady state.
The results indicate that the proposed algorithm provides a good diagnosis capability in case of multy-fault as well as a single fault, sensor fault, and noisy inputs.
What is pointed out again here is that the last of the test cases shows that the pattern of the sensors undergoses a transient where some sensors were fully affected while other sensors were less influenced due to the delay of the faults propagation.
In this quasi-steady state case, the faults diagonsis was successfully performed .

CONCLUSION

A new apporach is proposed for faults diagonsis in a large-scale system in real-time, which is achieved by using hierachically structured neural networks, and the appropriate faults detection procedure, and simple data extration method from the SDG.
The SDG model as the plant model is used for getting the reliable plant behavior not based on the human mental knowledge which can be lack of completeness and consistency.
Using the proposed scheme, let me summarize the main characteristics as follows :
(1) the incipient faults detection is made through each lower level neural network while more reliable and accurate faults diagnosis is also made through the plant-wise diagonsis, which enables the real-time faults diagonsis in a large-scale, complex system.
(2) multi-fault as well as a single fault can be detected, and the noise from the sensors is also filtered out due to the inherent characteristics of neural networks, that is, noise-tolerance and non-linear interpolating capabilities.
(3) the detection of sensor fault is also possible through the proper data extraction as in the 16th row of Fig.5-(c).
(4) the faults detection is possible during the weak transient, that is, quasi-steady state condition.
(5) this method also can tell the faults propagation routes and how the faults detection is performed because the SDG model is adopted and the information on the plant behavior is abtained from it.
(6) on the condition that the signed gain on each branch of the digraph can be obtained in any way, the faults size can be approximately estimated by the reverse calculation of the sensor deviation with the total path gain, which may give the plant operators the stability information for the plant-wise control system with the subsystems having their own hi-low disturbance limits.
In addition, the slight change in fault propagation probability in the SDG model can be accommodated because the fault propagation probability is trained through the neural networks.
In this study, the incoming signals is preprocessed and the plant operating conditions are classified so that the state of each node on the SDG is tuned automatically according to the various plant conditions.
The proposed method, however, does not contain the feedback loop in the SDG, and so requires further study to solve this.

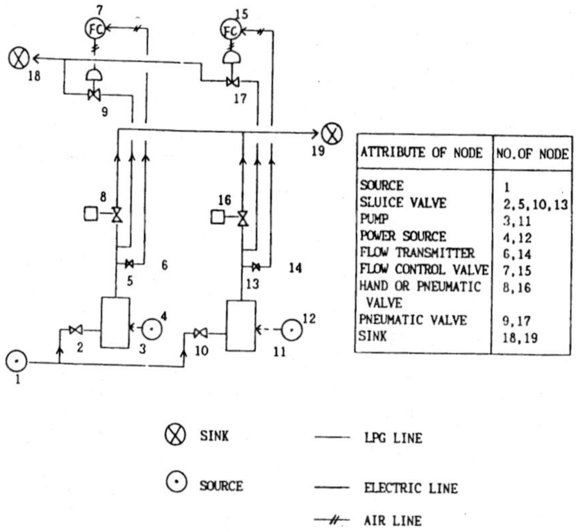

ATTRIBUTE OF NODE	NO.OF NODE
SOURCE	1
SLUICE VALVE	2,5,10,13
PUMP	3,11
POWER SOURCE	4,12
FLOW TRANSMITTER	6,14
FLOW CONTROL VALVE	7,15
HAND OR PNEUMATIC VALVE	8,16
PNEUMATIC VALVE	9,17
SINK	18,19

(A)

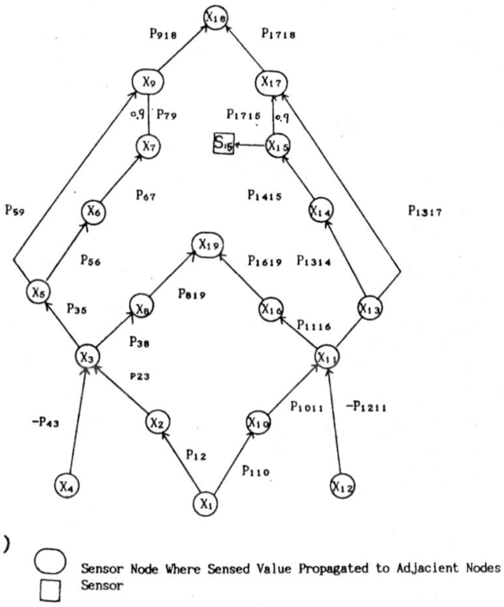

(B)

	SENSORS				
	S_9	S_5	S_{17}	S_{19}	SENSOR & NODES WITHIN THE SUBSYSTEM
1	1.0	1.0	1.0	1.0	X_1
2	1.0	0.0	0.0	1.0	X_2
3	1.0	0.0	0.0	1.0	X_3
4	1.0	0.0	0.0	1.0	X_4
5	1.0	0.0	0.0	0.0	X_5
6	0.9	0.0	0.0	0.0	X_6
7	0.9	0.0	0.0	0.0	X_7
8	0.0	0.0	0.0	1.0	X_8
9	1.0	0.0	0.0	0.0	S_9
10	0.0	1.0	1.0	1.0	X_{10}
11	0.0	1.0	1.0	1.0	X_{11}
12	0.0	1.0	1.0	1.0	X_{12}
13	0.0	1.0	1.0	0.0	X_{13}
14	0.0	1.0	0.9	0.0	X_{14}
15	0.0	1.0	0.9	0.0	X_{15}
16	0.0	1.0	0.0	0.0	S_{15}
17	0.0	0.0	0.0	1.0	X_{16}
18	0.0	0.0	1.0	0.0	S_{17}
19	0.0	0.0	0.0	1.0	S_{19}

(C)

SENSORS				SENSOR & NODE COLLECTION AFFECTING SENSORS																		
S_9	S_{15}	S_{17}	S_{19}	X_1	X_2	X_3	X_4	X_5	X_6	X_7	X_8	S_9	X_{10}	X_{11}	X_{12}	X_{13}	X_{14}	X_{15}	S_{15}	X_{16}	S_{17}	S_{19}
1	1	1	1	1.0	0	0	0	0	0	0	0	0	0	0	0	0	0	0	0	0	0	0
1	0	0	1	0	1.0	1.0	1.0	0	0	0	0	0	0	0	0	0	0	0	0	0	0	0
1	0	0	0	0	0	0	0	1.0	0.9	0.9	0	1.0	0	0	0	0	0	0	0	0	0	0
0	0	0	1	0	0	0	0	0	0	0	1.0	0	0	0	0	0	0	0	0	1.0	0	1.0
0	1	1	1	0	0	0	0	0	0	0	0	0	1.0	1.0	1.0	0	0	0	0	0	0	0
0	1	1	0	0	0	0	0	0	0	0	0	0	0	0	0	1.0	0.9	0.9	0	0	0	0
0	1	0	0	0	0	0	0	0	0	0	0	0	0	0	0	0	0	0	1.0	0	0	0
0	0	1	0	0	0	0	0	0	0	0	0	0	0	0	0	0	0	0	0	0	1.0	0
0	0	0	0	0	0	0	0	0	0	0	0	0	0	0	0	0	0	0	0	0	0	0

(D)

Fig.5 . Training Data Extraction of a Typical Subsystem(Pump).

(a) Pump System in Liquefied Petroleum Gas Plant

(b) Corresponding Signed Gain Digraph

(c) The Input-Ouput Training Data Extraction

(d) The Reduced Input-Output Training Data Table

TABLE 1 Simulation Results

																			INPUT PATTERN			
	X_1	X_2	X_3	X_4	X_5	X_6	X_7	X_8	S_9	X_{10}	X_{11}	X_{12}	X_{13}	X_{14}	X_{15}	S_{15}	X_{16}	S_{17}	S_{19}	S_9	S_{15}	S_{17}	S_{19}
1	.004	.993	.993	.993	.004	.003	.003	.004	.004	.000	.000	.000	.000	.000	.000	.000	.004	.000	.004	1.0	0.0	0.0	1.0
2	.004	.993	.993	.993	.004	.003	.003	.004	.004	.000	.000	.000	.000	.000	.000	.000	.004	.000	.004	1.01	0.01	0.0	0.99
3	.001	.005	.005	.004	.717	.220	.218	.000	.732	.000	.000	.000	.000	.000	.000	.000	.000	.015	.000	1.0	0.0	1.0	0.0
4	.000	.000	.000	.000	.000	.000	.000	.003	.000	.002	.002	.002	.004	.002	.003	.993	.003	.000	.003	0.0	1.0	0.0	0.0
5	.960	.011	.011	.011	.001	.001	.001	.000	.001	.000	.000	.000	.000	.000	.000	.000	.000	.017	.000	1.1	0.9	1.1	0.8

Case1 :The Trained Pattern
Case2 :The Noisy Input Pattern
Case3 :The Multiple Faults(2nd and 8th rows of Fig.5-(D))
Case4 :The Sensor Fault (S15 sensor)
Case5 :The Quasi-Steady State Pattern(1st row of Fig.5-(D))

REFERENCES

Chang, Chung-Chien, and Cheng-Ching Yu (1990), On-line fault diganosis using the signed directed graph, Ind. Engng. Chem. Res. 29, 1290-1299.

Hoskins, J.C., K.M.Kaliyur, and David M. Himmelblau (1991), Fault diagnosis in complex chemical plants using artificial neural networks, AIChE J., Vol.37, 137-141.

Isermann, R. (1984), Process fault detection based on modeling and estimation method-a survey, Automatica, 20, 387-.

Kokawa, M., S. Miyazaki, and S.Shingai-(1983), Fault location using digraph and inverse direction search with application, Automatica, Vol.19, No.6, 729-735.

Kramer, M,A., and B.L.Palowitch, Jr(1987)., A rule-based approach to fault diagnosis using the signed directed graph, AIChE J., Vol.33, No.7, 1067-1078.

Kramer, M.A., and J.A.Leonard (1990), Diagnosis using backpropagation neural networks-Analysis and criticism, Computers Chem. Engng. Vol.14, No.12, 1323-1338.

Qian, D.Q. (1990), An improved method for fault location of chemical plants, Computers Chem. Engng. Vol.14, No.1, 41~48.

Sorsa, T., Heikki N., No Koivo, and H.Koivisto(1991), Neural networks in process fault diagnosis, IEEE. Trans. on systems, man, and cybernetics, Vol.21, No.4, 815~825.

Venkatasubramanian, V., R.Vaidyanathan, and Y.Yamamoto(1990) Process fault detection and diagnosis using neural networks- steady-state process, Computers Chem. Engng. Vol.14, No.7, 699-712.

NEURAL NETWORK IMPLEMENTATION FOR REAL-TIME CLOSED-LOOP MOTION CONTROL OF REDUNDANT ROBOTS

Qin Chen* and J. Y. S. Luh**

*College of Engineering University of California Riverside, California 92521 U.S.A.

**Department of Electrical & Computer Engineering Clemson University Clemson, South Carolina 29634 U.S.A.

ABSTRACT

This paper describes an implementation of closed-loop motion control of redundant robots using a multi-layer neural network. The control scheme is developed based on the well known Hamilton-Jacobi-Bellman algorithm of optimal control. The algorithm requires solving the algebraic matrix Riccati equation, which is v computationally time consuming, and not pract r real time control applications. This pa ustrates that the computing time can b cally reduced by replacing solving the Ri ation with a neural network. The effectiver e neural controller is demonstrated by apply a PUMA 650 robot with a simulated prismati serted between the elbow and the wrist to extra degree of freedom.

I. INTRODUCTION

To increase the de nd versatility in performing various task otic manipulation, robots with kinematic dancy are often employed. The redundant obot provides a self-motion of the mechanism which does not alter the motion of the end-effecto This motion is often referred to as the "solution of null space of the munipulator Jacobian". Thus the complete kinematic solution of the joint velocity constitutes the sum of the contributions from the velocity of the end-effector and the solution c e null space. The former is often derived from optimizing the joint kinetic energy or momentum, while the latter is utilized to achieve a secondary performance such as avoiding kinematic singularity along the motion path, improving manipulator efficiency and flexibility, etc. Because of the additional capability and advantages, kinematically redundant robots attract increasing interests in research community [1].

In [2], Seraji gives a comprehensive review of the published work of this topic over the past decades. Redundancy may be resolved at joint velocity or acceleration level by optimizing some performance criteria. The kinematic properties of redundant robots have been studied by many researchers. Redundancy can also be resolved using the robot dynamics. At either level, one has to deal with the problem of computing the pseudoinverse of a non-square Jacobian matrix. Various approaches to resolving the problem have been proposed in many literature. So far, most of the redundant robots are controlled using open-loop scheme with no feedback of the state of the system. This paper adopts a closed-loop motion control for redundant robots [3]. The difference between the actual position and orientation of the end-effector and the desired trajectory is fed back to generate the control input. The redundancy is resolved at the joint acceleration level, and the self motion of the linkage is utilized to minimize the joint kinetic energy and the control effort. The drawback, however, is the requirement of excessively long computing time during each sampling period which causes the real-time operation impractical.

In the past decade, there has been a dramatic increase in interest in the artificial neural networks. Artificial neural networks are data processing systems which attempt to capture the essence of the electrochemical processing that takes place in human brains. Neural networks are attractive because it is believed that they are closely associated with human brain. The advantages of neural nets are attributed to their structure which is characterized as simple operation and massive parallelism. A neural network consists of many simple processing elements (or neurons) operating in parallel aimed at the dissection of a problem yet to be solved. The recent research in neural networks has shown that the massive parallelism in neural networks makes them computationally powerful and be fault tolerant [4]. Another property of neural nets is their adaptivity and ability to learn. Neural networks are not so much programmed as they are trained with data. Moreover, they improve their performance with experience - the more data they are fed, the more accurate or complete their response.

This paper presents an implementation of a closed-loop control system of a redundant robot using neural network. The preliminary results intend to demonstrate how the computational powerfulness of the neural networks may benefit the control of redundant robots. In the next section, a closed-loop control scheme developed in [3] for redundant robots is described. It will be shown that this scheme requires solving some algebraic matrix Riccati equation which is quite computationally demanding. In Section III, the structure and the training algorithm of a multi-layer neural network which is to be used for redundant robot control are discussed. Section IV describes an experiment of driving a PUMA 560 robot along some given trajectories. Note that PUMA 560 becomes a redundant robot by inserting a simulated prismatic joint between the elbow and the wrist to achieve seven degrees of freedom. The experimental results obtained using neural network are discussed, and compared against those obtained by solving the Riccati equation. A real-time training algorithm is presented in Section VI.

II. CONTROL SCHEME

For generality, the redundant robot considered here consists of linkages of rigid bodies with n revolute and/or prismatic joints. In this paper, the vectors and matrices are represented by bold face letters. Otherwise they are scalars. Let θ be an $n \times 1$ joint rotation or translation vector with respect to joint coordinates; $\mathbf{p}$ an $m \times 1$ position and orientation vector for the end-effector with reference to base coordinates described in Cartesian space. Let $\dot{\theta}$ and $\dot{\mathbf{p}}$ be the velocity vector of the joints and the end-effector described in their respective coordinates. They are related by the manipulator Jacobian matrix $\mathbf{J}$ as

$$\dot{\mathbf{p}} = \mathbf{J}\,\dot{\theta}. \qquad (1)$$

The relationship between the end-effector acceleration and the joint acceleration may be derived by differentiating (1) with respect to time as

$$\ddot{\mathbf{p}} = \mathbf{J}\,\ddot{\theta} + \dot{\mathbf{J}}\,\dot{\theta}. \qquad (2)$$

For a redundant robot, Jacobian matrix $\mathbf{J}$ and its time derivative in (1) and (2) has a dimension of $m \times n$, where $m < n$. Equations (1) and (2) may be used to solve redundancy at the joint velocity and acceleration level, respectively.

By discretizing the time interval of operation, set-points along the Cartesian path corresponding to sampling instants are defined. Now let $\mathbf{x} = \dot{\theta}$. Gupta and Luh proposed a linearized dynamic relationship [3] between every pair of consecutive set-points provided the length is sufficiently small:

$$\dot{\mathbf{x}}(t) = \mathbf{A}\,\mathbf{x}(t) + \mathbf{B}\,\mathbf{u} + \mathbf{M}(t) \qquad (3)$$

where

$$\mathbf{A} = [\delta(-\mathbf{J}^{+}\dot{\mathbf{J}}\,\mathbf{x}) / \delta\mathbf{x}]_{\mathbf{x}_0} - \mathbf{J}^{+}\mathbf{K}_v\,\mathbf{J} \qquad (4)$$

$$\mathbf{B} = (\mathbf{I} - \mathbf{J}^{+}\mathbf{J}) \qquad (5)$$

$$\mathbf{M}(t) = \mathbf{J}^{+}[\mathbf{p}_d + \mathbf{K}_v\,\mathbf{p}_d + \mathbf{K}_p\,(\mathbf{p}_d - \mathbf{p}) - \dot{\mathbf{J}}\,\mathbf{x}_0] - [\delta(-\mathbf{J}^{+}\mathbf{J}\,\mathbf{x}) / \delta\mathbf{x}]_{\mathbf{x}_0}\,\mathbf{x}_0 \qquad (6)$$

in which $\mathbf{x}$ is the joint velocity vector, $\mathbf{x}_0$ is the operating point for linearization, $\mathbf{J}^{+}$ is generalized pseudoinverse of the Jacobian matrix $\mathbf{J}$, $\mathbf{I}$ is an identity matrix, $\mathbf{p}_d$ and $\mathbf{p}$ are the desired and actual end-effector position/orientation vector, $\mathbf{K}_v$ and $\mathbf{K}_p$ are positive definite diagonal matrices satisfying the relation:

$$(\mathbf{p}_d - \mathbf{p}) + \mathbf{K}_v\,(\mathbf{p}_d - \mathbf{p}) + \mathbf{K}_p\,(\mathbf{p}_d - \mathbf{p}) = \mathbf{0}, \qquad (7)$$

and $\mathbf{M}$ is the closed-loop control vector. The optimum control that minimizes the quadratic performance measure

$$\int_0^t [\,\|\mathbf{x}\|_Q^2 + \|\mathbf{u}\|_R^2\,]\,dt$$

where $\mathbf{Q}$ is symmetric positive semi-definite matrices and $\mathbf{R}$ is symmetric positive definite matrix, is known to be:

$$\mathbf{u}(\mathbf{x}) = -\mathbf{R}^{-1}\,\mathbf{B}\,[\mathbf{K}(t)\,\mathbf{x}(t) + \mathbf{s}(t)\,] \qquad (8)$$

where $\mathbf{K}$ and $\mathbf{s}$ are a matrix and a vector, respectively. When the system is discretized in time with sufficiently small time interval within which $\mathbf{K}$ and $\mathbf{s}$ are assumed constant, they satisfy respectively,

$$\mathbf{K}\,\mathbf{B}\,\mathbf{R}^{-1}\,\mathbf{B}\,\mathbf{K} - \mathbf{K}\,\mathbf{A} - \mathbf{A}^{T}\,\mathbf{K} - \mathbf{Q} = \mathbf{0} \qquad (9)$$

and

$$\mathbf{K}\,\mathbf{B}\,\mathbf{R}^{-1}\,\mathbf{B}\,\mathbf{s} - \mathbf{A}^{T}\,\mathbf{s} - \mathbf{K}\,\mathbf{M} = \mathbf{0}. \qquad (10)$$

Thus $\mathbf{K}$ and $\mathbf{s}$ are solved for each small time interval by solving the algebraic Riccati equation (9), the first-order vector algebraic equation (10), and then substituted into (8) to form the closed-loop control function. A block diagram of this control system is shown in Figure 1.

III. NEURAL NET IMPLEMENTATION

3.1 Motivation

When the above control scheme is applied to a PUMA 560 robot, the extra degree of freedom is obtained by "inserting" a prismatic joint between the elbow and the wrist. Note that this prismatic joint is only simulated. The control scheme is implemented on a VAX/750. The time interval of each control input update is 28 msec, which is required by the PUMA robot and specified by its mamufacturer. However, this time interval of 28 msec is too short for the VAX computer to complete the computation of the control input. Consequently, the motion of PUMA robot is not smooth. Based on the simulation result, 800 seconds are needed for the VAX to complete the computation of the control input which drives the robot to travel a trajectory of 1 meter in length. The trajectory is divided into 100 segments. Thus it takes approximately 8 seconds for the computer to calculate the control input for the robot to move through one segment on a given trajectory, not mentioning the time which is required to complete the communication between the computer and the robot. In fact, this communication time is quite long due to the fact that the decoder on PUMA operates at a low speed. But this problem is not addressed here. It can be observed from Figure 1 that computaticn involved is quite intensive, especially the part of solving the Riccati equation to determine matrix $\mathbf{K}$ and then vector $\mathbf{s}$. This observation motivated the investigation of employing a multi-layer neural network to replace the operation of solving the Riccati equations (9) and (10) in the control loop for the purpose of reducing the computing time required to update the control input at each small time interval.

The closed-loop control system diagram with the neural controller is shown in Figure 2. The input to the neural network is the actual joint position vector θ_A, and the output of the neural net is the **u** vector. Thus the most time consuming part of the control scheme is replaced by the neural network. The central task is thus to teach the neural network about the mapping between a given joint position vector θ_A and the desired **u** vector. It has been discussed in the existing literature that a multi-layer neural network may be trained to realize such a function. The training may be performed off-line by exposing the neural network to a set of data obtained when driving the robot through a pre-specified trajectory. The basics of training and using a multi-layer neural network can be found in [4]. Here, back propagation is employed to train the network to implement the θ-to-**u** mapping. Prior to actually incorporating the neural network into the control loop, a number of simulations are conducted to select a proper number of layers and the number of neurons on each layer.

3.2. Neural Network Structure

Figure 3 illustrates a general purpose multi-layer neural network. In addition to the input and the output layers, there are L hidden layers. Let the number of neurons on the input and output layer be K and N, respectively. Let the number of neurons on the i-th hidden layer be M_i, where i = 1, 2, ..., L. Let the output of the neurons on the i-th hidden layer be $\mathbf{q}_i$, where i = 1, 2, ..., L and $\mathbf{q}_i$ has a dimension of $M_i \times 1$. Let $\mathbf{q}_0$ and $\mathbf{q}_{L+1}$ denote the input and the actual (as opposed to the desired) output vector of the neural network, respectively. The output of the j-th neuron on i-th hidden layer is represented by $q_i(j)$, and it is calculated as

$$q_i(j) = f(\mathbf{e}_j^T (\mathbf{W}_{i-1,i}\, \mathbf{q}_{i-1}) - \beta_i(j)) \tag{11}$$

where $\mathbf{W}_{i-1,i}$ is an $M_i \times M_{i-1}$ weight matrix, $\beta_i(j)$ is the offset attached to the j-th neuron on i-th layer, $\mathbf{e}_j^T$ is an $M_i \times 1$ row vector whose j-th component is equal to 1 while the others zero, i = 1, 2, ..., L+1, and function $f(\cdot)$ is a sigmoid logistic non-linearity with the form

$$f(x) = \frac{1}{1 + e^{-x}}. \tag{12}$$

Note that f(x) is bounded below and above by 0 and 1, respectively.

3.3. Neural Network Training

The purpose of neural network training is to determine the weight matrices. The neural network used in our control loop is trained using the well-known back propagation algorithm. Back propagation is an iterative gradient algorithm designed to minimize the mean square error between the actual output vector or scalar of the multi-layer neural network and the desired output. In our case, the input is the 7×1 joint rotation or translation vectotr θ. The output is the 7×1 **u** vector.

The training process is described as follows:

1. Set the components of all the weight matrices $\mathbf{W}_{i-1,i}$ and the offsets β_i to random values on the range of [0,1], where i = 1, 2, ..., L + 1.

2. Present the continuously valued input vector $\mathbf{q}_0 = \theta_A$ and specify the desired output $\mathbf{u}_d = \mathbf{u}_{Ric}$. Note that the training data is generated on the computer by equation (8) with respect to a given trajectory.

3. Use the formulas in Figure 3 to calculate the actual output by forward propagating the input θ_A through the network.

4. At t-th iteration, use the following recursive algorithm starting at the output layer and working back to the first hidden layer. The weights are adjusted by

$$w_{ab}(t+1) = w_{ab}(t) + \eta\, \delta_b\, i_a + \alpha\, (w_{ab}(t) - w_{ab}(t-1)) \tag{13}$$

In this equation $w_{ab}(t)$ is the weight from the a-th neuron on a hidden layer or an input to the b-th neuron on the layer above at time t, i_a is either the output of neuron a on a hidden layer or is the a-th component of the input vector, η is a gain term which is on the range of [0,1], and δ_b is an error term for neuron b. The third term in (13) is the momentum term, which makes training faster sometimes, where $0 < \alpha < 1$. If neuron b is on the output layer, then

$$\delta_b = q_{L+1}(b)(1 - q_{L+1}(b))\,(u_d(b) - q_{L+1}(b)) \tag{14}$$

where $u_d(b)$ and $q_{L+1}(b)$ are the desired and the actual output of b-th neuron, respectively.

If, however, neuron b is on an internal hidden layer, then

$$\delta_b = i_a(1 - i_a) \sum_k (\delta_k\, w_{bk})$$

where k is over all the neurons in the layer above neuron b. The offsets β_b are adapted by

$$\beta_b(t+1) = \beta_b(t) - \rho\, \delta_b$$

where $0 < \rho < 1$.

5. Terminate the training process if the weights do not change anymore. Otherwise, go to Step 2.

V. EXPERIMENTAL RESULTS

Experiments of driving PUMA robot along different trajectories are conducted. Two of them are described here.

Case 1: A straight line parallel to x-axis for a distance of 1 meter. The trajectory is divided into 100 segments.

Case 2: A circular arc trajectory whose radias is 5 meters and length is also 1 meter. The trajectory is divided into 100 segments.

Even thought the number of neurons on the input and output layer is fixed to be 7 in the above cases, the number of hidden layers is undetermined, neither is the number of neurons on each hidden layer. The simulation indicates that the accuracy of the neural network as far as the θ_A to **u** mapping is concerned, and the speed of training, depends to a large degree on the above two parameters. A number of simulations are conducted in seeking for a proper number of layers and the number of neurons on each layer. As a result, a neural network, with 2 hidden layers and 10 and 12 neurons on each hidden layer, is selected. The mean square error (MSE) between the neural network output and the desired output is defined as $\|\mathbf{u}_{nn} - \mathbf{u}_{Ric}\|$. It is within 1% after 4000 iterations during training.

The timing data obtained in the two cases is shown below.

Total Elapsed Time (including comm. time)

	Case 1	Case 2
NN	104.59sec	112.18 sec
Riccati	826.06sec	857.04 sec

It is measured that the time required for sending and receiving data between the robot and the VAX for 100 segments is 45 seconds. Subtracting this amount of time from the above time durations yields:

Total Elapsed Time (excluding comm. time)

	Case 1	Case 2
NN	59.59sec	67.18 sec
Riccati	776.06sec	812.04 sec

Thus neural networks runs 12 times faster than solving the Riccati equation.

VI. REAL-TIME TRAINING ALGORITHM

The control scheme presented above requires the off-line training of the neural network. To enhance the capability of the controller, a real-time training algorithm is sought.

Let $\mathbf{p}_d$ and θ_d be the desired postion/orientation vector of the end-effector and its corresponding joint displacement vector, respectively. Let $\mathbf{e} = \theta - \theta_d$. Choose the control

$$\mathbf{B}\ \mathbf{u} = \ddot{\theta}_d - \mathbf{K}_v \dot{\mathbf{e}} - \mathbf{K}_p \mathbf{e} - \mathbf{A}\ \dot{\theta} - \mathbf{M}(t) \tag{14}$$

where $\mathbf{K}_v$ and $\mathbf{K}_p$ are diagonal positive definite matrices. Since $\dot{\mathbf{x}} = \ddot{\theta}$, then substituting (14) into (3) yield

$$\ddot{\mathbf{e}} + \mathbf{K}_v \dot{\mathbf{e}} - \mathbf{K}_p \mathbf{e} = \mathbf{0}$$

so that $\theta \rightarrow \theta_d$ asymptotically. The real-time training algorithm for the neural network in the controller is as follows:

Input Data: $\theta_d(k)$ for k = 1, 2, ..., L; $\theta(0)$, $\dot{\theta}(0)$, $\ddot{\theta}(0)$.

L = total number of sampling periods for the trajectory.

Γ = maximum number of iterations for training the weight matrix of the neural network for each sampling period of the trajectory.

Ω = total number of repeated trials of neural network's weight matrix training for the entire trajectory.

δ = allowable tolerance for trained input **B u** from the neural network.

Step 0: Precalculate $\dot{\theta}_d(k)$ and $\ddot{\theta}_d(k)$ by definite differences. Arbitrarily assign numerical values to the initial weight matrix **W**(0) of the neural network.

Step 1: Set $\omega = 1$ (Initiate the first trial)

Step 2: Set k = 1 (First sampling period)
Set $\gamma = 1$ (First iteration)

Step 3: Use (4) to compute **A**(k-1).
Use (5) to compute **B**(k-1).
Use (6) to compute **M**(k-1).
Use (14) to compute **B**(k-1)**u**(k-1).

Step 4: Feed θ(k-1), $\dot{\theta}$(k-1), $\ddot{\theta}$(k-1) and **W**(k-1) into the neural network to obtain $[\mathbf{B}(k-1)\mathbf{u}(k-1)]_n$.

Step 5: Train the neural network using data:

B(k-1) **u**(k-1), θ(k-1), $\dot{\theta}$(k-1), $\ddot{\theta}$(k-1) to obtain **W**(k).

Step 6: If $\|[\mathbf{B}(k-1)\mathbf{u}(k-1)]_n - [\mathbf{B}(k-1)\mathbf{u}(k-1)]\| < \delta$, then go to Step 8. Otherwise continue.

Step 7: If $\gamma < \Gamma$, then go to Step 4. Otherwise continue.

Step 8: If k < L, then measure θ(k) and compute $\dot{\theta}$(k) and $\ddot{\theta}$(k) set k = k + 1, set $\gamma = 1$, and go to Step 3. Otherwise continue.

Step 9: If $\omega < \Omega$, then set $\omega = \omega + 1$ and go to Step 2. Otherwise continue.

VII. CONCLUSION

A closed-loop motion-control scheme for redundant robots may be implemented using a multi-layered neural network. The utilization of neural network reduces the time, which is required to compute the control input, by over 10 times. The preliminary experimental results demonstrate that the neural network has a potential for redundant robot control in real-time. This algorithm was test on a simulated planar robot with satisfactory results. The experiment on PUMA robot using this algorithm is in progress.

Acknowledgement - The authors acknowledge Mr. Sunil Gupta for providing the experimental data, and the Center for Advanced Manufacturing at Clemson University for their support.

REFERENCES

1. Special Section on Kinematic Redundancy, Edited by J.Y.S. Luh, IEEE Transactions on Robotics and Automation, Vol.7, No.5, October, 1991, pp.577-606.

2. Seraji, H. "Configuration Control of Redundant Manipulators: Theory and Implementation," IEEE Transactions on Robotics and Automation, August, Vol.5, No.4, 1989.

3. Gupta, S. and J.Y.S. Luh, "Closed-loop Control of Manipulators with Redundant Joints Using the Hamilton - Jacobi - Bellman Equation," Proceedings of IEEE International Conference on Robotics and Automation, april 9-11, 1991, Sacramento, California, pp.472-277.

4. Lippman, R. P, "An Introduction to Computing with Neural Nets," IEEE ASSP Magazine, April, 1987.

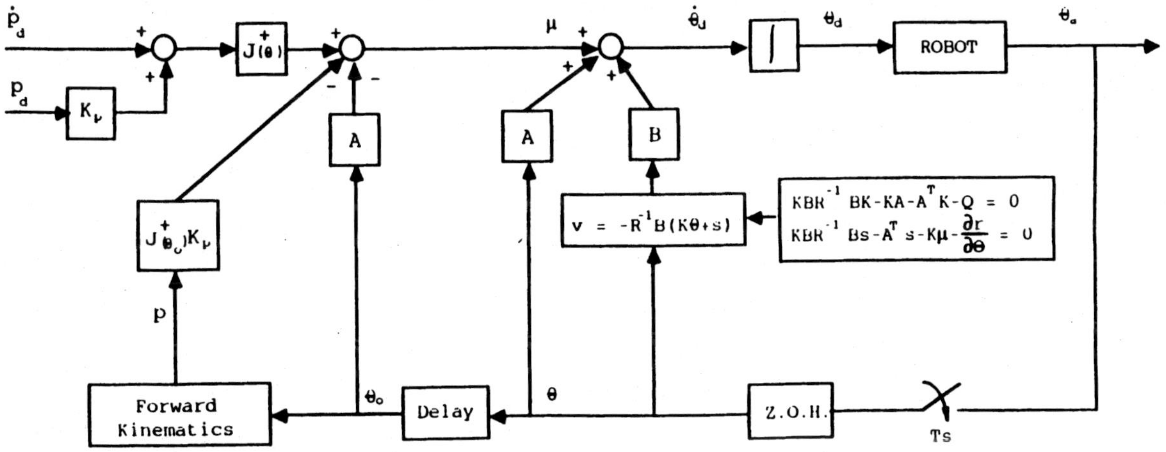

θ_d = Desired joint position generated by controller

θ_a = Actual joint position reached by robot

Figure 1 Closed-loop Resolved Rate Control Scheme

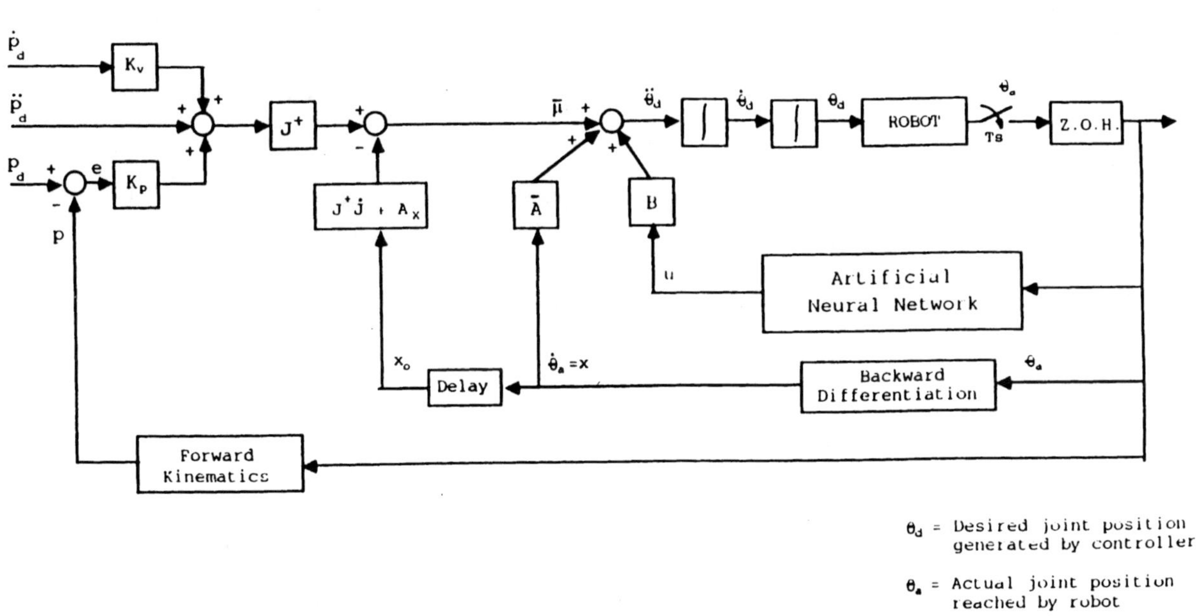

Figure 2 Block Diagram of Neural Network Controller

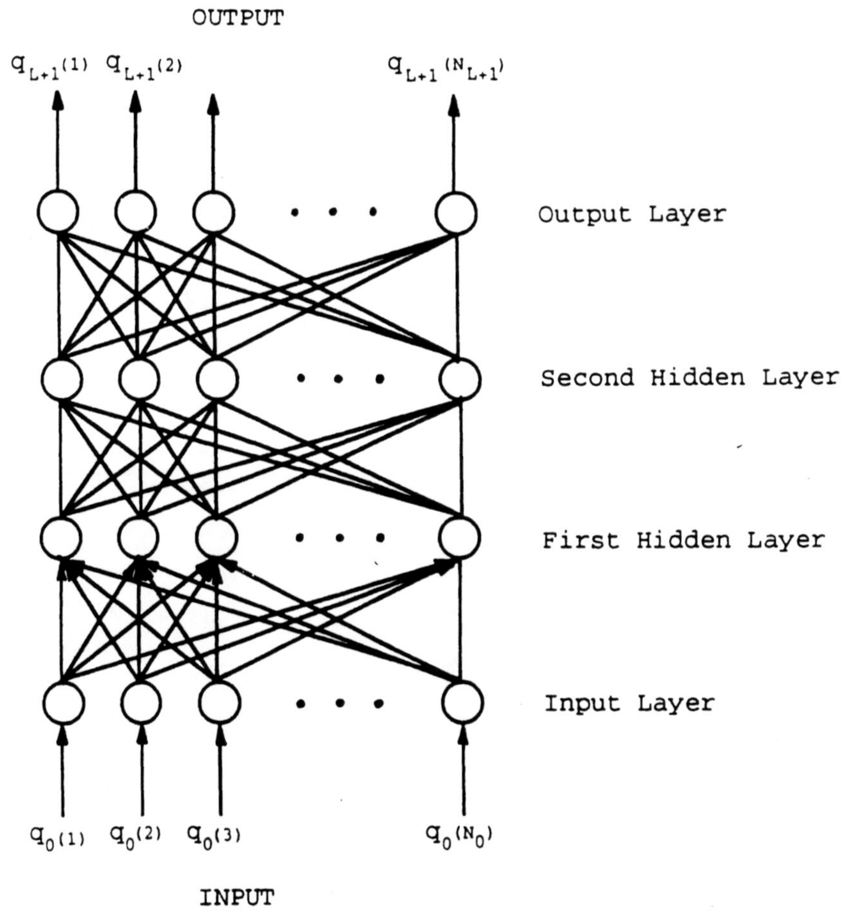

Figure 3 A General Purpose Multi-layered Neural Network

SYSTEM IDENTIFICATION USING NEURAL NETWORKS

K. Endo and Y. Ishida

Department of Electronics and Communication, Meiji University, Kawasaki, Japan

Abstract. This paper presents an identification method of dynamical systems using neural networks. The network learning algorithm described here is based on the conjugate gradient method. The proposed algorithm is applied to the identification problem of a linear dynamical system. Simulation results show that our algorithm is superior to the standard back-propagation algorithm. Furthermore this paper describes a gain adjustment method of I-PD control systems using neural networks. The weights of a neural network are considered compensation elements of an I-PD controller. These weights are given by learning the neural network. The conjugate gradient method is then utilized as the learning algorithm. Simulation results show that our method is effective to deal with unknown dynamical systems.

Keywords. Neural nets (networks); conjugate gradient method; system identification; I-PD controllers; gain adjustment.

INTRODUCTION

In order to design controllers for a dynamical system, it is necessary to have a model that will adequately describe the system's motion. Recently, many studies have been performed to apply neural networks to identification and control of dynamical systems. Neural networks have the potential to treat many problems, which can not be handled by conventional approaches, and the advantage to be directly used as controllers. However, a globally optimal solution is not guaranteed because the slope of the optimization surface can have many local optimal. The purpose of this paper is to introduce a new network learning algorithm which is superior to the standard back-propagation algorithm. The proposed algorithm is based on the conjugate gradient method. Furthermore, in this paper, a means to identify dynamical systems is described. A new method to determine compensation elements of I-PD control systems is also presented. Both these methods are constructed by the neural network.

Firstly, we apply the neural network based on the conjugate gradient method to identification of dynamical systems. In simulation experiments, the transfer function of the controlled object is identified by giving the input and output signals of the object as an input pattern to the neural network. Output signals of the identified model are compared with those of the controlled object. Furthermore the proposed method, that is, the learning algorithm based on the conjugate gradient method is compared with the standard back-propagation algorithm for the number of iterations to minimize the error between the output of the neural network and the output of the controlled object. According to simulation results, we can see that the proposed algorithm is effective to identification of dynamical systems.

Secondly, in this paper we propose a new method that regards the weights of a neural network as compensation elements of the I-PD controller. These weights can be determined by the learning of the neural network. The neural network consists of three layers, and the conjugate gradient method is utilized as the learning algorithm. The integral result of the error obtained by subtracting the controlled value from the reference value, the controlled value and its differential result are used as an input pattern to the neural network. The weights of the neural network are adjusted to minimize the error between the controlled value and the target value, that is, the output value of a model. Simulation results show that our method is effective to deal with unknown dynamical systems.

SYSTEM IDENTIFICATION USING NEURAL NETWORKS

Structure of System Identification Using Neural Networks

If the controlled object is the n-th order system, the pulse transfer function G(z) is generally shown as follows:

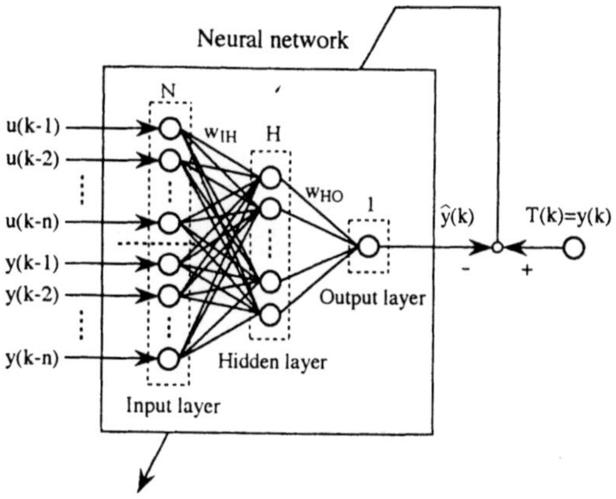

Fig. 1. Structure of a neural network for system identification

$$G(z) = \frac{Y(z)}{U(z)} = \frac{a_1 z^{-1} + \dots + a_n z^{-n}}{b_0 + b_1 z^{-1} + \dots + b_n z^{-n}} \quad (1)$$

where U(z) and Y(z) are the z-transforms of input signals u(i) and output signals y(i) of the controlled object, and $i=1,2,\cdots,k,\cdots$. Therefore the k-th output signal y(k) becomes

$$y(k) = a_1 u(k-1) + \dots + a_n u(k-n) - b_1 y(k-1) - \dots - b_n y(k-n) \quad (2)$$

Note that $b_0=1$ in Eq. (1). The y(k) is obtained by the input and output signals from $(k-1)$-th to $(k-n)$-th.

The architecture of a neural network for system identification is illustrated schematically in Fig. 1. This network has three layers that are an input layer, one hidden layer and an output layer. There are N nodes in the input layer, H nodes in the hidden layer and one node in the output layer. However N equals $2n$ where n is the order of the controlled object. In Fig. 1, w_{IH} is the coupling coefficient between an input layer node and a hidden layer node, and w_{HO} is the coupling coefficient between a hidden layer node and an output layer node. These coupling coefficients are called weight coefficients or weights. The input and output signals from $(k-1)$-th to $(k-n)$-th are given as an input pattern to the neural network for estimation of the k-th output signal. The output signal y(k) is used as the target signal T(k). The estimated k-th output signal $\hat{y}(k)$ is obtained from the output layer.

The outputs of H hidden layer nodes are determined by summing inputs multiplied by each weights w_{IH} and passing through a nonlinear function called the sigmoid function F:

$$F(x) = \frac{1-e^{-x}}{1+e^{-x}} \quad (3)$$

The output of one output layer node is determined by summing outputs of H hidden layer nodes multiplied by each weights w_{HO}. However the sigmoid function is not used at the output layer.

Network Learning Algorithm Based on the Conjugate Gradient Method

The learning procedure at the m-th iteration requires to compare the output vector $\mathbf{Y}_i$ of the network due to an input pattern vector with a target vector $\mathbf{T}_i$. The resulting error is

$$f_i(\mathbf{W}) = \frac{1}{2}|\mathbf{T}_i - \mathbf{Y}_i|^2 \quad (4)$$

where $\mathbf{W}$ is the weight vector. We will try to minimize the least squared function made up from the outputs of all the input patterns:

$$\phi(\mathbf{W}) = \sum_{i=1}^{p} f_i(\mathbf{W}) \quad (5)$$

where p is the number of input pattern vectors. The gradient vector of $\phi(\mathbf{W})$ at point $\mathbf{W}$ is given by

$$\nabla\phi(\mathbf{W}) = \sum_{i=1}^{p} \nabla f_i(\mathbf{W}) \quad (6)$$

We assume the following single variable function (Charalambous, 1990):

$$\Psi(\alpha) = \phi(\mathbf{W}^{(m)} + \alpha\mathbf{D}^{(m)}) \quad (7)$$

where $\mathbf{W}^{(m)}$ is the weight vector at the m-th iteration and $\mathbf{D}^{(m)}$ is the direction based on the conjugate gradient method as follows (Polak, 1971):

$$\mathbf{D}^{(m)} = -\nabla\phi(\mathbf{W}^{(m)}) + \beta_m \mathbf{D}^{(m-1)} \quad (8)$$

where

$$\beta_m = \frac{(\nabla\phi(\mathbf{W}^{(m)}) - \nabla\phi(\mathbf{W}^{(m-1)}))^T \nabla\phi(\mathbf{W}^{(m)})}{\nabla\phi(\mathbf{W}^{(m-1)})^T \nabla\phi(\mathbf{W}^{(m-1)})} \quad (9)$$

Then we can update the weight vector as follows:

$$\mathbf{W}^{(m+1)} = \mathbf{W}^{(m)} + \alpha_m \mathbf{D}^{(m)} \quad (10)$$

where α_m is the value of a that will sufficiently decrease the function value of $\Psi(\alpha)$. The flow chart of the conjugate gradient learning algorithm is shown in Fig. 2.

Simulation Results of System Identification

In this section, simulation results of system identification using the neural network are presented. Let us assume that the transfer function of the controlled object is the 2nd order system as follows:

$$G(s) = \frac{\omega_n^2}{s^2 + 2\zeta\omega_n s + \omega_n^2} \quad (11)$$

with $\zeta=1$ and $\omega_n=0.7$. Then the pulse transfer function G(z) becomes

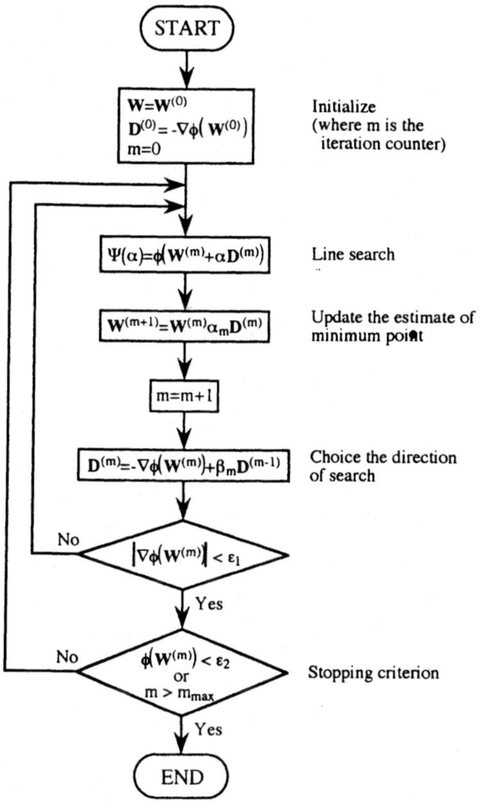

Fig. 2. Conjugate gradient learning algorithm

$$G(z) = \frac{0.0192z^{-1}+0.0167z^{-2}}{1-1.621z^{-1}+0.657z^{-2}} \qquad (12)$$

where the sampling period is 0.3(sec). Random signals uniformly distributed in the interval [-1,1] are given as the input signals u(i) to the controlled object. The output signals y(i) are obtained by inputting random signals into the pulse transfer function G(z) of Eq. (12). These input signals u(i) and output signals y(i) are utilized as input patterns of the neural network shown in Fig. 1.

Fig. 3 shows the output of the controlled object shown by Eq. (11) or Eq. (12) and the model obtained by identification using the neural network after the identification procedure has been carried out for 80 iterations. The number of input patterns is 100. The numbers of input layer nodes and hidden layer nodes are 4 and 3 respectively. Fig. 3 (a) is the result obtained by the standard back-propagation algorithm and (b) is the result by the proposed algorithm mentioned above, that is, the network learning algorithm based on the conjugate gradient method. From this figure, it is evident that (b) is superior to (a) for the convergence of output error. Fig. 4 shows the output signals of the identified model for the unknown input patterns, that is, the testing patterns. These output signals nearly equal those of the controlled object. Therefore we can see that the transfer function of the controlled object is identified by the neural network.

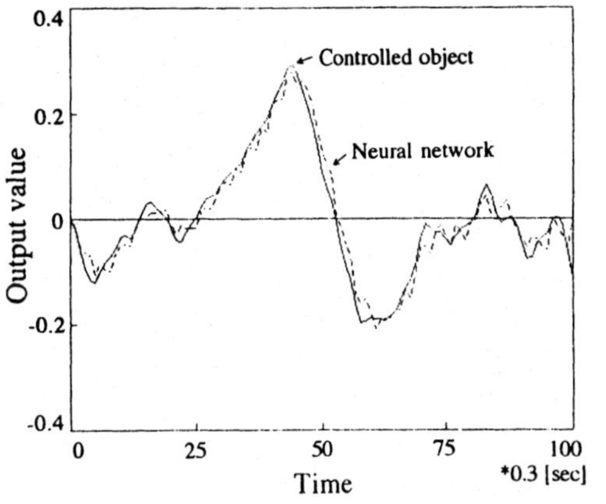

(a) The standard back-propagation algorithm

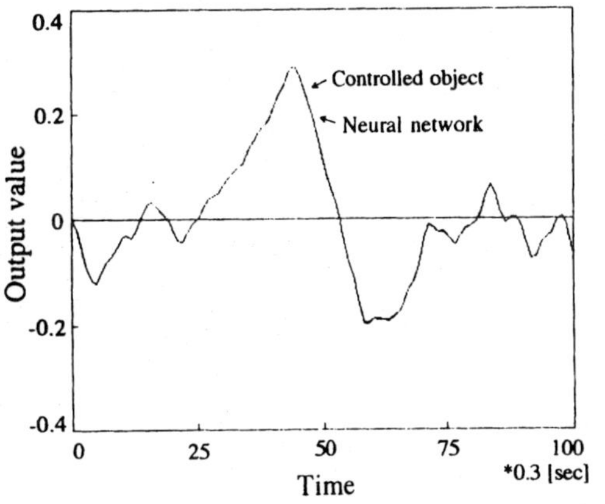

(b) The proposed algorithm

Fig. 3. Identification results

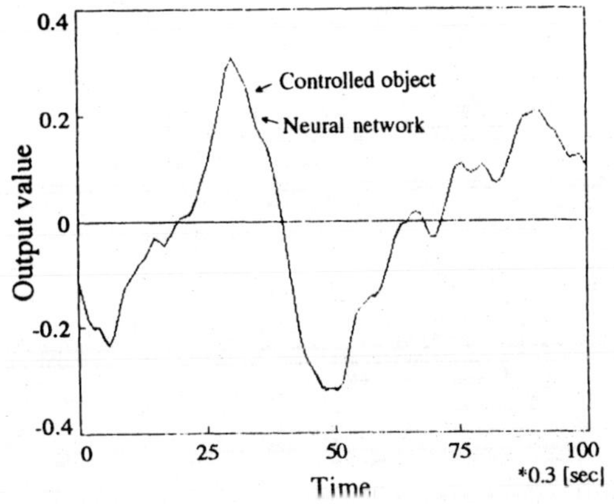

Fig. 4. Identification result for unknown input patterns

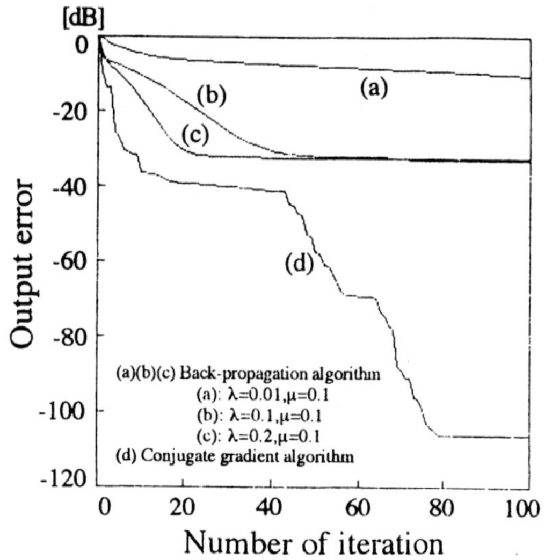

Fig. 5. Convergence of output errors

Fig. 5 shows the convergence of output error with the conjugate gradient algorithm and the one with the standard back-propagation algorithm for three sets of the learning coefficient λ and the momentum coefficient μ. From this figure, the convergence depends on the value of the coefficients λ and μ for the standard back-propagation algorithm and the superiority for convergence of the conjugate gradient algorithm over the standard back-propagation algorithm is apparent.

In Eq. (2), if a series of the input and output signals is $\{u,y\}=\{1,0,\cdots,0\}$, the k-th output signal becomes $y(k)=a_1$ where $\{u,y\}=\{u(k-1),\cdots,u(k-n),y(k-1),\cdots,y(k-n)\}$. In the same way, if $\{u,y\}=\{0,1,0,\cdots,0\}$, then $y(k)=a_2$. Therefore, if $\{u,y\}=\{1,0,\cdots,0\}$ is given as the input pattern to the neural network, the coefficient a_1 of the pulse transfer function in Eq. (1) can be determined. If $\{u,y\}=\{0,1,0,\cdots,0\}$ is given, we can get the coefficient a_2. However, in the case that $\{u,y\}=\{1,0,\cdots,0\},\{u,y\}=\{0,1,0,\cdots,0\},\cdots,\{u,y\}=\{0,\cdots,0,1\}$ are given as the input patterns, the average error of estimated coefficients is approximately 1.7%. So we give some values from -1 to 1 such as $\{u,y\}=\{0.3,0,\cdots,0\}$ to the neural network and take the average values of the outputs obtained by them. According to the experimental result, the average error of estimated coefficients is approximately 0.5%.

GAIN ADJUSTMENT OF I-PD CONTROL SYSTEM

Structure of I-PD Control System Using Neural Networks

Fig. 6 shows the structure of the control system with I-PD compensation elements. G(z) is the pulse transfer function of the controlled object. K_i denotes the integral element which is multiplied by the integral of the error obtained by subtracting the controlled value y(k) from the reference value r(k). K_p is the proportional element which is multiplied by the controlled value. K_d is the derivative element which is multiplied by differential of the controlled value. The manipulated variable u(k) is obtained by summing the operation results of the I-PD compensation, and transferred to the controlled object.

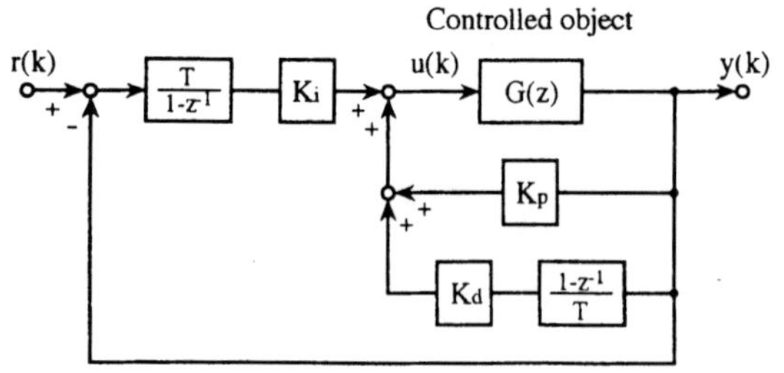

Fig. 6. Block diagram of I-PD control system

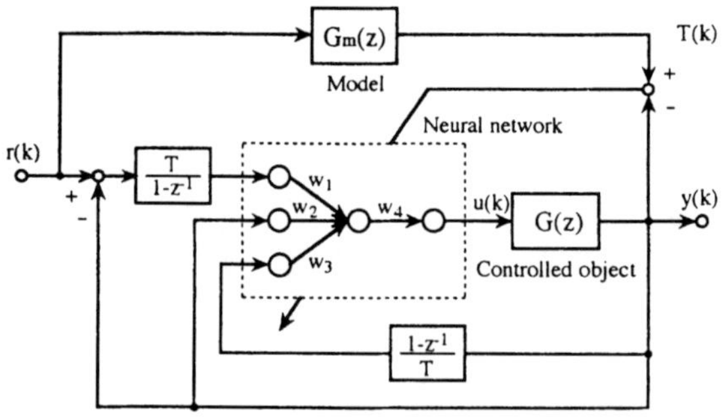

Fig. 7. I-PD control system using a neural network

In this paper, a new method to obtain each coefficient of I-PD compensation elements is suggested. This method is a means to regard weights of a neural network as coefficients of I-PD compensation elements. The weights are obtained by a learning algorithm of the neural network.

Fig. 7 illustrates the structure of the I-PD control system composed by a neural network. Inside of a dotted line is the neural network which has three layers that are an input layer, one hidden layer and an output layer. In this research, the input layer has 3 nodes, the hidden layer has 1 node and the output layer has 1 node. The integral result of the error obtained by subtracting the controlled value y(k) from the reference value r(k), the controlled value and its differential result are used as the input pattern to the input layer. Therefore the weights (w_1,w_2,w_3) from the input layer to the hidden layer in the neural network are considered compensation elements of the I-PD control system. Also the weight (w_4) from the hidden layer to the output layer is considered an amplifier gain. These weights (w_1,w_2,w_3,w_4) of the neural network can be obtained by the learning algorithm based on the conjugate gradient method. However, the output value from a model is used as the target value T(k). The weights are adjusted to minimize the error between the controlled value y(k) and the target value T(k).

Simulation Results of Gain Adjustment

In this section, simulation results on gain adjustment of an I-PD control system using the neural network are shown. The transfer function G(s) of the controlled object is assumed to be the 2nd order system as follows:

$$G(s) = \frac{k}{s(s+\omega)} \qquad (13)$$

with ω=50 and k=45. Because controlled objects such as DC servo motor are often identified as the 2nd order system shown in Eq. (13). Then the pulse transfer function G(z) becomes

$$G(z) = \frac{0.00192z^{-1}+0.00162z^{-2}}{1-1.607z^{-1}+0.607z^{-2}} \qquad (14)$$

where the sampling period is 0.01(sec). On the other hand, the transfer function $G_m(s)$ of the reference model is assumed to be the 3rd order system as follows:

$$G_m(s) = \frac{\omega_1\omega_2^2}{(s+\omega_1)(s^2+2\zeta\omega_2 s+\omega_2^2)} \qquad (15)$$

with ω_1=10, ω_2=10 and ζ=1. The pulse transfer function $G_m(z)$ of this model is produced by sampling $G_m(s)$ at 100Hz. The initial weights of the neural network are determined by random signals uniformly distributed in the interval [-1,1]. The number of input patterns is 100.

Fig. 8 shows step responses after the learning of the neural network. One is the step response of the reference model. The other is the step response of the plant, that is, the controlled object. The learning procedure has been carried out for 200 iterations. From this figure, we can see that the controlled value nearly equals the target value. Therefore appropriate weights of the neural network, that are compensation elements for the I-PD control system, can be obtained by the learning of the neural network.

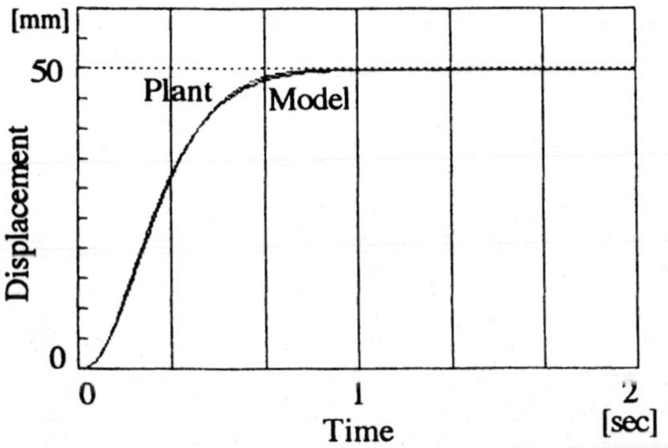

Fig. 8. Step responses after learning

Although the simulation experiment in this section has been performed for the transfer function of Eq. (13), the same results are also obtained for other transfer function, e.g., Eq. (11).

CONCLUSIONS

In this paper, we have proposed a new method for identification of dynamical systems. This is a method to utilize a neural network. The conjugate gradient method has been used as the learning algorithm of the neural network. Input signals to the controlled object and the corresponding output signals are given to the input layer of the neural network. Simulation experiments have been carried out for the 2nd order system. In the simulation experiments, outputs of the identified model have been compared with outputs of the controlled object. The errors between them have been shown for the number of iterations. The transfer function of the controlled object assumed for the simulation experiments has been determined from the weights of the neural network after learning. The estimated error was approximately 0.5%. According to the experimental results, we have found out that the learning algorithm based on the conjugate gradient method is superior to the standard back-propagation algorithm for identification of unknown systems.

Furthermore we have suggested a new method to determine compensation elements of an I-PD control system. This is a method to regard the weights of a neural network as compensation elements of the I-PD control system. The integral result of the error obtained by subtracting the controlled value from the reference value, the controlled value and its differential result are used as the input pattern. The weights of the neural network have been obtained by the learning algorithm based on the conjugate gradient method. Simulation experiment has been performed on a transfer function which is the 2nd order system. The 3rd order system has been utilized as the reference model. Simulation result has shown that the controlled value nearly equals the target value. Therefore we consider that the proposed method is effective to deal with unknown dynamical systems.

ACKNOWLEDGMENTS

The authors wish to acknowledge Dr. T. Honda and Dr. H. Kamata of Meiji University, Dr. C. Charalambous of Kuwait University, and Dr. Y. Ogawa for their helpful advises. This work was supported by SMC Corporation.

REFERENCES

Charalambous, C. (1990). A conjugate gradient algorithm for the efficient training of artificial neural networks. *C.R.S.T Technical Reports 90/06*. Available from the University of Western Australia, Department of Electrical and Electronic Engineering.

Ishida, Y., K. Endo, H. Kamata and **C. Charalambous** (1991). Vowel recognition by a neural network using vocal tract area function. *Proceedings of the 1991 IEEE International Joint Conference on Neural Networks (IJCNN-91)*, vol.3, 2144-2149.

Pattichis, C.S., C. Charalambous, C.N. Schizas and **L.T. Middelton** (1991). EMG diagnosis using the conjugate gradient backpropagation neural network algorithm. *Proceedings of the 1991 International Conference on Artificial Neural Networks (ICANN-91)*, vol.2. North Holland, Amsterdam. pp.1621-1624.

Polak, E. (1971). Computational methods in optimization: a unified approach. Academic Press, New York.

ARTIFICIAL NEURAL NETWORK APPROACH TO INFERENTIAL CONTROL OF VOLATILITY IN REFINERY PLANTS

Heon Chang, Xuemin Shen and Ming Rao

Department of Chemical Engineering, University of Alberta
Edmonton, Alberta Canada T6G 2G6

ABSTRACT In this paper, Artificial Neural Network (ANN) has been applied to predict the volatility of the product in refinery plants. To maximize the product yield, ANN model can then be used for inferred volatility control. Two case studies have been carried out for different processes using the Backpropagation (BP) network. In the first case study, to increase learning speed, the Modified Backpropagation (MBP) network was used to train ANN with Generalized Descent (GD) method. ANN model and Regression Analysis (RA) model were compared in representing the relation between the plant stream data and the product volatility. Results show ANN model to be capable at predicting the product volatility and superior to RA model. In the second case study, Conjugated Gradient (CG) method were used in training ANN. ANN model is also shown to be able to fit even nosy plant data.

Keywords artificial neural network, backpropagation, generalized descent method, conjugate gradient method, inferential control, volatility, refinery

INTRODUCTION

In recent years, Artificial Neural Network (ANN) technology has begun being introduced in the control community because there are systems, for which rigid mathematical models do not exist or are not sufficient to meet a full-scale identification requirement, such as highly nonlinear chemical processes or very complex industrial plants. In these cases, ANN may be an effective tool to cope with these problems, especially for systems where the characteristics are totally unknown. In fact, there exist very complex systems with unknown characteristics and uncertainties that are difficult to identify using mathematical models. To model such systems, ANN can provide some promising solutions.

ANN has proven effective at solving pattern recognition problems in a wide variety of areas including image processing, speech recognition, sensor interpretation and system identification (Haesloop and Holt, 1990). ANN can learn and adapt themselves to inputs from actual processes, thus allowing representation of complex engineering systems that are difficult to model with traditional physical engineering relations. Several factors motivate the use of ANN:

1. general mapping capabilities
2. reliance on the knowledge specified through learning instead of a pre specified algorithm
3. ability to use vast amounts of sensory information
4. capability to respond with high speed to sensory inputs
5. greater degree of robustness and fault tolerance due to their distributed representation

Many areas of chemical engineering such as fault detection and diagnosis (Watanabe and co-workers, 1989), process control (Donat and co-workers, 1990; Cooper and co-workers, 1992), process design, and process modeling and simulation (Hernandez and Arkun, 1990) can take advantage of these above factors to organize and detect features from unpredictable and/or imprecise process data.

Establishing a dynamic process model is often the first step toward implementing a modern control algorithm. Because of the inherent complexity and interactions in chemical process systems, most models for chemical process control are typically determined from a known input and output sequence. These identified models are usually linear approximations about a chosen mean operating point. Therefore, as the process moves away from the linearization point, or if the operating point is changed, the discrepancy between the linear model and true nonlinear system behavior becomes more significant. ANN offers the opportunity to directly model nonlinear processes. With a more accurate nonlinear model, the plant-model divergence mentioned above may be reduced.

ANN modeling is such a concerned area with a non-programmed adaptive approach to processing engineering data systems. ANN can adaptively develop transformations in response to their environment. In contrast to programmed computing, ANN is able to develop a mapping function from examples of that function's operation. For many chemical engineering systems that are difficult to model, there exists a significant amount of data charting the system. Thus, these stored examples can be used to develop an ANN model that describes the input/output behavior of the process. In addition to self-organizing capability of ANN, another advantage is the parallelism inherent in the neurocomputing architecture. This allows ANN system to be implemented using highly parallel hardware to achieve real-time performance.

ARTIFICIAL NEURAL NETWORK

From an engineering point of view, a new definition for ANN modeling involves unknown parameters (the weight values) and given network topology (units, layers and connections). It can be referred to as a non-parametric model that is totally different from conventional mathematical models. By this definitions, ANN can be used to represent the input-output data and relationships of any physical system, as long as sufficient input-output system data are collected. It is also assumed that ANN modeling would only be used for systems whose mathematical models cannot be obtained. Otherwise, conventional identification methods are probably superior.

The new concept of non-parametric identification is based on the properties of ANN using the given topology and adjustable weights. The main advantages from introducing ANN are derived from the utilization of their learning ability, associative memory, pattern classification and other such features.

Architecture

The three-layer perceptron neural network shown in Fig. 1 is composed of many interconnected processing units or neurons (N_0 neurons in the input layer, N_1 neurons in the hidden layer and N_2 neurons in the output layer) organized in successive layers. The analysis and design of any multilayer neural networks can be done similarly. The network is fully connected because each neuron distributes its output to each neuron in the next layer. Neuron i in layer j receives N_{j-1} inputs from layer j-1 with connection strengths or associated weights. The neuron i first computes the weighted sum of the N_{j-1} inputs,

$$S_{i,j} = \sum_{k=1}^{N_{j-1}} W_{i,k,j} X_j \qquad (1)$$

where X's and W's represent the given input and its associated weight, respectively. Then it outputs a nonlinear function of the sum in (1), which serves as an input to the next layer.

$$Y_{i,j} = F(S_{i,j}) \qquad (2)$$

The activation function F is often chosen to be a sigmoid. In order to use the backpropagation algorithm, the sigmoidal function must be continuous and strictly monotonic. Typical choices for the activation function are $F(x) = 1/(1+e^{-x})$ and the hyperbolic tangent, $F(x) = \tanh(x)$.

The first or input layer is composed of fan-out units. These neurons do not perform any computation but simply distribute their inputs to all neurons in the next layer.

Learning Paradigm

Although a broad range of ANN architectures and learning paradigms are available, the BackPropagation (BP) algorithm for multilayer feedforward networks is the most common approach for current engineering applications because it is a simple and powerful method which first appeared as an answer to the problem of training feedforward ANN with hidden units and was later established as applicable to the more general class of recurrent networks.

BP is a supervised synchronous learning paradigm. It is supervised because for every input presented for learning, the expected or correct corresponding output is available so the network can modify its weights accordingly by applying Generalized Descent (GD) method or Conjugate Gradient (CG) method. It is synchronous because all the weights are modified at each learning step. The network is expected to learn a number of data patterns composed of inputs and associated outputs. The objective of using the BP algorithm is to minimize the overall error, E, between the desired and actual outputs:

$$E = \sum_{l=1}^{p} \sum_{k=1}^{N_2} e_{k,2} \qquad (3)$$

$$e_{k,2} = \frac{1}{2} (D_k - Y_{k,2})^2 \qquad (4)$$

where D's, Y's and p represents the desired output, the actual output and the number of data set used to train the network, respectively. The error of the desired output and actual output is required to be smaller than the specified error tolerance for the ANN to identify the system using the input-output data.

Before learning begins, all the weights are initialized at small random values between -1 and 1. The first pattern is then presented to the network. Learning takes place in two successive steps: forward and backward passes. During the forward pass, each neuron i in layer j computes the weighted sum of its inputs S_{ij} and outputs Y_{ij} as in (1) and (2), respectively. This process is called forward because it takes place in layer j-1 before layer j and therefore propagates from the input layer to the output layer. At this point the activations or outputs of all the neurons of the network are available. During the backward pass, the output error is backpropagated in the reverse direction. Each neuron computes an associated error and the calculation propagates from the output layer to the input layer.

Generalized descent method. The basic strategy behind GD method is to minimize the overall error by performing gradient descent in parameter space based on errors between a set of network mappings and the corresponding set of desired mapping examples (Rumelhart, 1986; Caudill, 1988).

Once $e_{k,2}$ in (4) has been computed for one pattern, the weights are modified as,

$$W_{i,k,j} = W_{i,k,j} + \eta \Delta W_{i,k,j} \qquad (5)$$

Hidden to output layer:

$$\Delta W_{i,k,2} = (Y_{k,2} - D_k) Y_{k,2} (1 - Y_{k,2}) Y_{i,1} \qquad (6)$$

Input to hidden layer:

$$\Delta W_{i,k,1} = \sum_{j=1}^{N_2} \Delta W_{k,j,2}(1-Y_{k,2})W_{i,k,1}X_i \qquad (7)$$

The next pattern is then presented and the calculations are repeated.

This standard GD method is known to have several major drawbacks (Leonard and Kramer, 1990; Donat and co-workers, 1990; Copper and co-workers, 1992). First, it adapts the weights based on the error from a single pattern representation instead of the overall error E which is the true objective function to be minimized. Secondly, the rates of convergence of this algorithm are extremely slow and the improvement per iteration falls sharply. Third drawback is that this method often lead to a local minimum.

Conjugate Gradient method. To overcome the above deficiencies of GD method, CG method can be used to train the network. This method adapts the weights based on the overall error which is the true objective function to be minimize. Therefore it is more robust and has better convergence properties than GD method (Cooper and coworkers, 1992).

For the overall error, E in (3), the CG method updates the weights at kth iteration as:

$$\mathbf{w}^{k+1} = \mathbf{w}^k + \alpha^k \mathbf{s}^k \qquad (8)$$

where **w** is the vector of weights and α is found by carrying out a unidimensional line minimization in the given search direction **s**. The new search direction at each iteration is found by (Reklaitis and co-workers, 1983)

$$\mathbf{s}^k = -\mathbf{g}^k + \left(\frac{\|\mathbf{g}^k\|^2}{\|\mathbf{g}^{k-1}\|^2} \right) \mathbf{s}^{k-1} \qquad (9)$$

with $\mathbf{s}^0 = -\mathbf{g}^0$. The steepest descent vector, **g**, is calculated by extension from GD method. The elements of **g** are given by

Hidden to output layer:

$$g_{i,k} = \sum_{l=1}^{p} \Delta W_{i,k,2}^{(l)} \qquad (10)$$

Input to hidden layer:

$$g_{i,k} = \sum_{l=1}^{p}\sum_{j=1}^{N_2} \Delta W_{k,j,2}^{(l)}(1-Y_{k,2}^{(l)})W_{i,k,1}X_i^{(l)} \qquad (11)$$

The training procedures are iteratively carried out until the ANN can represent the relationships of these data within the specified error tolerance. The non-parametric model is obtained by the fully connected structure and values of the weights. If well trained, the neural net can identify the system in terms of the input-output data.

PREDICTION OF VOLATILITY

Case Study I

At the 'refinery A', the 'unit 1' processes a sweet mixed blend crude, a combination of several crudes, to produce the 'product A' and the 'product B'. The 'variable 1', which has its primary impact on the volatility of the 'product A', is adjusted on a basis of the 'quality A' of the 'product A'; on the other hand, laboratory measurement of the 'product A' volatility is used as a basis to adjust the drawoff rate (yield) of the 'product A', which primarily affects its 'quality A'.

In order to maximize the yield of the 'product A', its volatility has been controlled by the following control strategy:

1. fixing the yield of the 'product A' by operator entered targer
2. controlling the 'quality A' of the 'product A' by manipulating the 'product B' draw rate
3. resetting the yield target on the basis of laboratory measurement of the 'product A' volatility

Volatility "below specification" indicates a lost opportunity for higher yield. Volatility "above specification" is off-specification product: hence a deterrent to raising the yield target. The controller then tended to operate with the conservative yield target of the 'product A', which resulted in volatility giveaway. The yield target was usually reset on the basis of laboratory measurement of the 'product A' volatility once per day (more often with large giveaway or when "above specification"). Therefore, maximization of the 'product A' yield could not entirely satisfied by this control strategy.

To reduce the volatility giveaway, an on-line volatility analyzer can be used. The use of the volatility analyzer, however, requires a high maintenance: frequent recalibration and regular decoking. An alternative is to control the volatility on the basis of an volatility inferred from tower data. The operator can then enter a volatility target directly. While satisfactory performance can be expected from a very well maintained volatility analyzer, the inferred volatility scheme is easily justified based on the low development cost. Therefore, use of a volatility analyzer could be avoided if the volatility can be predicted reliably from stream quality data.

From the observation of the 'product A' distillation curve, it was empirically known that the volatility of the 'product A' depends on its 'quality A', the 'variable 1' and the 'variable 2'

Regression Analysis model. The correlation of laboratory measurement of the 'product A' volatility to stream quality data has been done by the linear regression. Because of difficulty in correlating a time-stamped laboratory results to noisy plant data (error in the dependent variable plays havoc with traditional statistical methods), the simulated data for 'unit 1' were used to determine the functional relationship. The real data were then served to adjust the parameters of the regression equation. The final correlation was established between the volatility of the 'product A', the 'quality A' of the 'product A', the 'variable 1' and the 'variable 2'. The RA model is

$$Y=a_0+a_1X_1+a_2X_1^2+a_3X_1^3+a_4X_2+a_5X_3+a_6X_3^2 \qquad (12)$$

where

Y = the volatility of the 'product A'
X_1 = the 'quality A' of the 'product A'
$X_2 = (\alpha_1-\beta_1)/\beta_1$
$X_3 = (\alpha_2-\beta_2)/\beta_2$
α_1 = the 'variable 1'
α_2 = the 'variable 2'
$\beta_1 = a_7+a_8X_1$
$\beta_2 = a_9+a_{10}X_1$
$a_1,a_2,a_3,a_4,a_5,a_6,a_7,a_8,a_9,a_{10}$ = constants

Artificial Neural Network model. The GD method was used to train the network. In order to increase the learning speed of the ANN when the GD method is used, a new learning rate function $\eta(t)$ is introduced instead of the constant η in the standard BP algorithm. It is designated and used in Modified Backpropagation (MBP) as follows:

$$\eta(t) = \begin{cases} \lambda_0 & \text{for } \|\Delta W_{i,j,k}\|^2 > \delta \\ \lambda + \lambda_0 e^{-\mu t} & \text{for } \|\Delta W_{i,j,k}\|^2 \leq \delta \end{cases} \quad (13)$$

where λ and λ_0 are parameters to be chosen, and are very small ($\lambda_0 > \lambda > 0$, μ and $\delta > 0$). They are system dependent and usually determined by experience and knowledge on neural networks.

In the initial stage, $\eta(t)$ has a relatively larger value because λ_0 is chosen bigger, $\lambda_0 > \lambda$, in order to promote a fast learning convergence of the MBP. When the weight matrix is convergent to a certain degree ($\|\Delta W_{i,j,k}\|^2 \leq \delta$), $\eta(t)$ starts decreasing until $\eta(t) = \lambda$, and λ is selected relatively smaller for a better convergence in the final stage.

The ANN selected for the prediction of 'product A' volatility from the stream data has 3 units in the input layer (N_0=3) with 5 neurons in the single hidden layer (N_1=5) and 1 neuron in the output layer (N_2=1). The 70 data set used for learning was directly taken from the data used for regression analysis at the 'refinery A'. The MBP algorithm was used with λ_0=5, λ=0.5, μ=0.09. After learning, the overall error was 0.00007.

In this case study, a comparison of the performance between the RA model and the ANN model for fitting the data is shown in Fig. 2. The results show that the ANN model clearly outperforms the regression model for all points. With the data set not used for ANN learning, the ability of the ANN model to predict the volatility of the 'product A' was also investigated. Figure 3 shows a comparison of RA model and ANN model predictions for the volatility of the 'product A' from the 'quality A' of the 'product A', the 'variable 1' and the 'variable 2'. The results show that the ANN model also accommodates all 9 points much better than the RA model.

Long term changes in operating conditions or change crude composition will affect the accuracy of the ANN model as well as the RA model. Some maintenance of the correlation is expected (i.e. periodically retraining the network for more accurate prediction of volatility or including crude composition changes in ANN learning input variables).

Although ANN has the capability to cope with the nonlinearity, complexity and uncertainty of dynamic systems in terms of their numerical characteristics in identifying the relationships and features of the system input-output data. ,some of the crucial ANN problems that need to be overcome are

1. ANN needs too much learning time (usually, the smaller acceptable error, the longer the training time). Therefore, ANN may not be a good choice for time-varying systems.
2. Convergence and stability cannot be guaranteed.
3. Due to a lack of robustness, some initial parameters(weights, learning rate, etc.) may strongly influence the performance of ANN, thus global assurance is needed.

Case Study II

At the 'refinery B', the 'unit 2' also processes several other kinds of crudes. According to demands, the 'unit 2' produces the 'product C' or the 'product D' along with the 'product C'. When the 'product D' is produced, a similar volatility control strategy is used to maximize the 'product D' yield.

The ANN selected for the prediction of the 'product D' volatility from the stream data has 11 units in the input layer (N_0=11) with 13 neurons in the single hidden layer (N_1=13) and 1 neuron in the output layer (N_2=1). The 97 data set used for learning was collected from the real operating data used for two years at the 'refinery B'. In this case study, the CG method was used to train the network. After learning, the overall error of 0.00325 was obtained.

In this case study, the ability of the ANN model for fitting the real plant data is shown in Fig. 4. The results show that, to a degree, the ANN model is able to fit even nosy plant data. Figure 5 shows the prediction of the 'product D' volatility from the plant stream data by the ANN model.

CONCLUSIONS

In this paper, we have demonstrated that ANN can model the product volatility at refinery plants. The ANN model matches the data much better than the RA model. In addition, the ANN model has the ability to fit nosy plant data and to predict the product volatility. In conclusion, the ANN shows promise for dealing with more complex multidimensional mapping and inherent parallel architecture should allow real-time performance even for very complex mapping.

ACKNOWLEDGMENTS

This project is financially supported by Imperial Oil Limited University Research Grants. The authors wish to acknowledge Dr. S. Woo, Dr. Mahalec and A. McIntosh for their significant advice and support.

REFERENCES

Caudill, M. (1988), Neural networks primer. AI

Expert, March, 53

Cooper, D. J., L. Megan, and R. F. Hinde, Jr. (1992), Comparing two neural networks for pattern based adaptive process control. AIChE Journal, 38, 41.

Donat, J. S., N. Bhat and T. J. McAvoy (1990). Optimizing neural net based predictive control. Proceedings of American Control Conference, San Diego, 2466.

Hernandez, E. and Y. Arkun (1990). Neural network modeling and an extended DMC algorithm to control nonlinear systems. Proceedings of American Control Conference, San Diego, 2454.

Haesloop, D. and B. R. Holt (1990). A neural network structure for system identification. Proceedings of American control Conference, San Diego, 2460.

Leonard, J. and M. A. Kramer (1990). Improvement of the back-propagation algorithm for training neural networks. Computer and Chemical Engineering, 14, 337.

Reklaitis, G. V., A. Ravindran and K. M. Ragsdell (1983). Engineering Optimization Methods and Application. John Wiley and Sons, New York.

Rumelhart, D. E. and McClelland, J. L. (1986). Parallel Distributed Processing: Explorations in the Microstructure of Cognition: 1. Foundations, MIT Press, Cambridge.

Tveter, D. (1991), Getting a fast break with backprop. AI Expert, July, 36.

Vemuri, V. (1988). Artificial Neural Networks : Theoretical concepts. Computer Society press of the IEEE, Washington D. C.

Wasserman, P. D. (1989) Neural Computing Theory and Practice. Van Nostrand Reinhold, New York.

Watanabe, K., I. Matsuura, M. Abe, M. Kubota and D. M. Himmeblau (1989). Incipient fault diagnosis of chemical processes via artificial neural networks. AIChE Journal, 35, 1903.

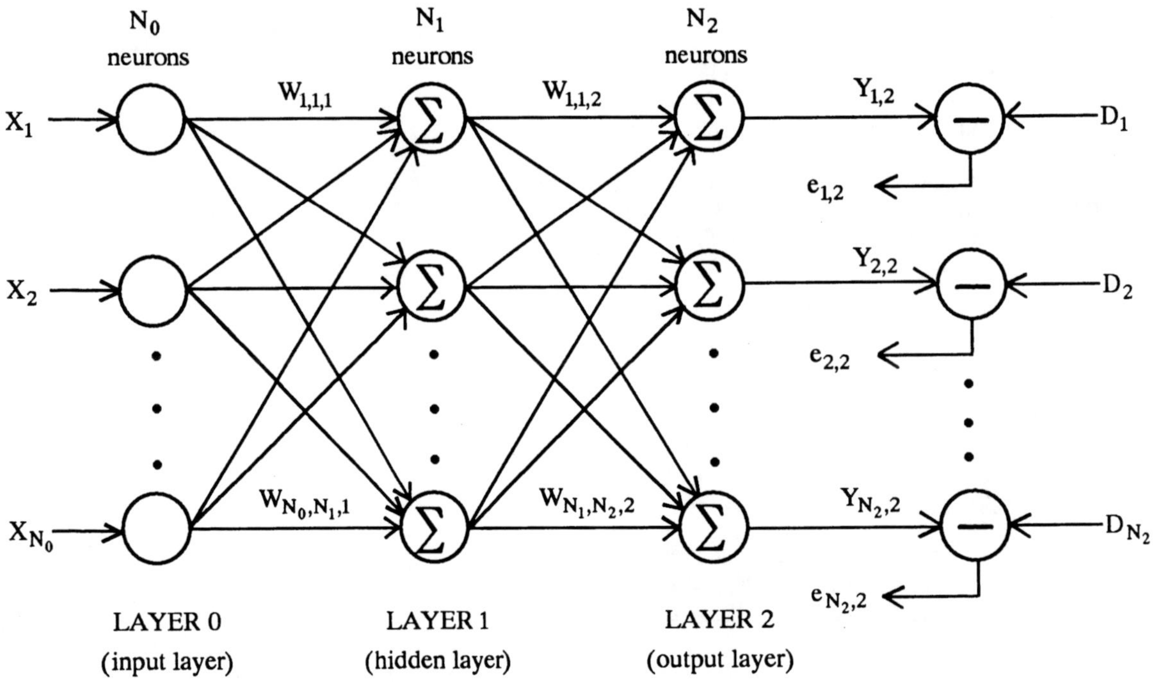

Fig. 1. Three-layer Backpropagation Network

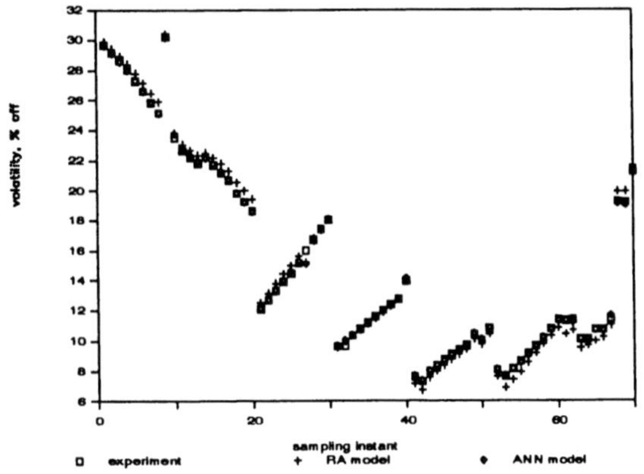

Fig. 2. Comparison of ANN Model and RA model on fitting the data

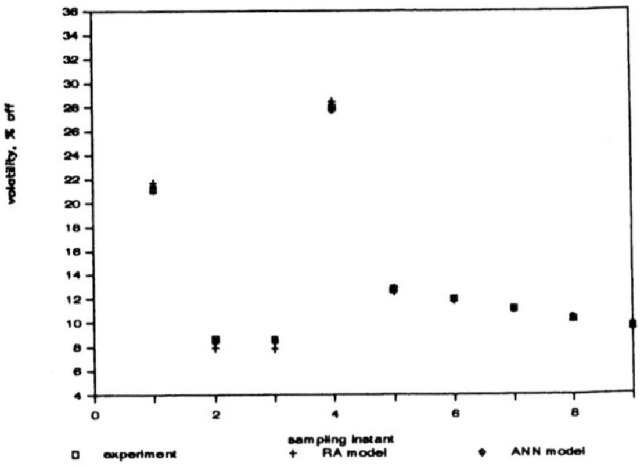

Fig.3. Comparison of ANN Model and RA Model on predicting Volatility from 'quality A', 'variable 1' and 'variable 2'

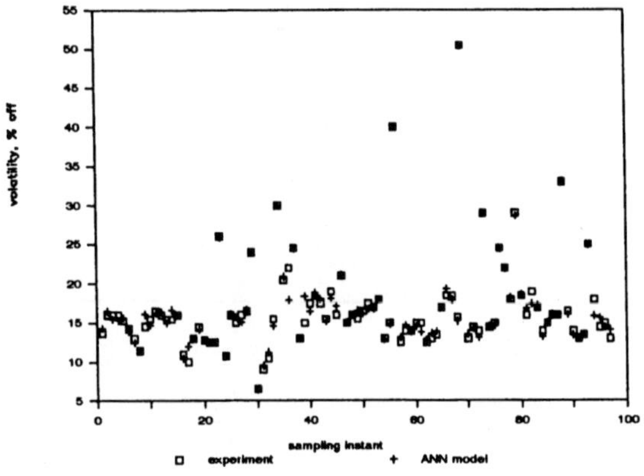

Fig. 4. Fitting noisy Plant Stream data

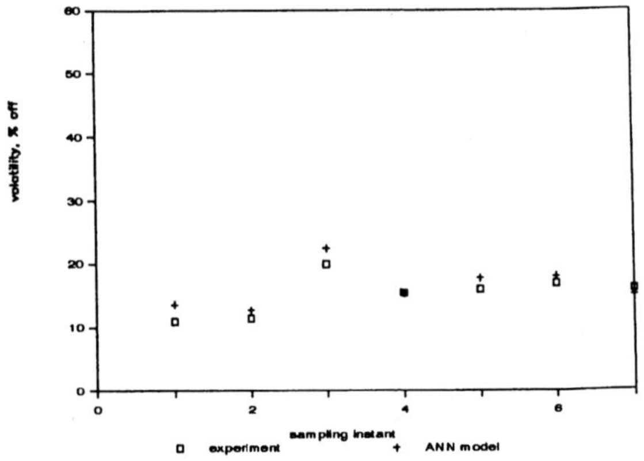

Fig. 5. Prediction of Volatility from real Plant Stream data

On-Line Identification of State-Space Models via Exploitation of Displacement Structure

Young Man Cho, Guanghan Xu, Thomas Kailath and R. Roy

Information Systems Lab. Stanford University Stanford, CA 94305
E-mail: young@isl. stanford. edu,phone: [415]725-5698, FAX: [415]723-8473

Abstract. The computational burden of state space model identification has limited its real-time application though it offers some important advantages over other methods based on input/output transfer functions. A recently proposed state space identification method [1] uses ideas from the theory of displacement structure and significantly reduces the computational burden. For an $M \times N$ data matrix ($N > M$), this batch method requires only $O(MN)$ flops while the existing method takes $O(MN^2)$ flops. However, for slowly changing systems, it is more desirable to update the model as time goes on with the minimal computational burden. In this paper, we extend our results of the batch processing algorithm to allow updating of the identified state space model with $O(M^2)$ flops. Again, the theory of displacement structure and of the fast subspace decomposition (FSD) technique are essential to the derivation of this fast algorithm. Some computer simulation results are also presented.

Keywords. system identification, state space models, displacement structure, fast subspace decomposition, update and downdate

1 Introduction

The direct identification of state space models has recently attracted much attention [2; 3; 4; 5; 6; 7]. This recent surge of interest stems from some unique advantages of state space identification over the input-output (I/O) based identification:

1. State space descriptions are numerically better conditioned than I/O descriptions [7; 8].
2. Estimation of the number of states (or the model order) forms an integral part of the state space identification process [7].
3. State space model identification can be more readily extended to multi-input, multi-output (MIMO) systems [7].

However, most of the past contributions deal with nonrecursive batch processing algorithms, which are appropriate for off-line identification of time-invariant systems. On the other hand, on-line identification of state space models has drawn less attention mainly due to its high computation and storage costs even though it can be applied to a wider class of systems (*e.g.*, time-variant or nonlinear systems). We recently proposed [1] a faster batch identification method for state space models that exploits the displacement structure of Toeplitz-like matrices arising in the identification procedure and requires only $O(MN)$ flops when the size of the data matrix is $M \times N$. In some cases, the computational burden of our algorithm is acceptable even for on-line identification. However, since the data window N is usually much larger than the sliding window size M, the resulting heavy computational cost may preclude the possibility of on-line identification. If the identification results can be updated in a recursive fashion, the computational complexity per update will only be a function of M but not N. In this paper, we shall extend our fast algorithm for batch processing and present a new recursive method to update the identification results provided a few new data samples. Given d new data samples, by exploiting the Hankel (or Toeplitz) structure of the data matrix, the new method reduces computational complexity to $O(M^2 + dM)$ flops, $d \ll N$. As in the batch algorithm, this recursive method also update the order of the system based on the new data samples.

In Section 2, we introduce the existing state space identification algorithms of De Moor *et al.* [3] and review our batch identification method. In Section 3, we adapt the identification technique for batch processing to allow recursive processing. In Section 4,

*This research was supported by the Advanced Research Projects Agency of the Department of Defense and was monitored by the Air Force Office of Scientific Research under Contract No. F49620-90-C-0014. This manuscript is submitted for publication with the understanding that the US Government is authorized to reproduce and distribute reprints for Government purposes notwithstanding any copyright notation hereon. The views and conclusions contained in this document are those of the authors and should not be interpreted as necessarily representing the official policies or endorsements, either express or implied, of the Advanced Research Projects Agency or the U.S. Government.

we apply the newly developed algorithm to identify a simulated system.

2 A Conventional State Space Identification Method

A state space model for discrete-time lumped systems is considered,

$$\mathbf{x}_{k+1} = \mathbf{A}\mathbf{x}_k + \mathbf{B}\mathbf{u}_k, \qquad \mathbf{y}_k = \mathbf{C}\mathbf{x}_k + \mathbf{D}\mathbf{u}_k, \qquad (1)$$

where $\mathbf{x}_k$ ($\in \Re^n$) is the state of the linear system at time k, $\mathbf{u}_k$ ($\in \Re^{n_i}$) is the input to the linear system, and $\mathbf{y}_k$ ($\in \Re^{n_o}$) is the observed output. The objective[1] is to estimate $\mathbf{A}, \mathbf{B}, \mathbf{C}$, and $\mathbf{D}$ from $\{\mathbf{u}_k\}$ and $\{\mathbf{y}_k\}$.

Following the notation of De Moor *et al.* [3], we start with the easily obtained equation

$$\mathbf{Y} = \mathbf{\Gamma}_M\mathbf{X} + \mathbf{H}\mathbf{U}\ , \qquad (2)$$

where

$$\mathbf{Y} = \begin{bmatrix} \mathbf{y}_k & \mathbf{y}_{k+1} & \cdots & \mathbf{y}_{k+N-1} \\ \mathbf{y}_{k+1} & \mathbf{y}_{k+2} & \cdots & \mathbf{y}_{k+N} \\ \cdots & \cdots & \cdots & \cdots \\ \mathbf{y}_{k+M-1} & \mathbf{y}_{k+M} & \cdots & \mathbf{y}_{k+N+M-2} \end{bmatrix}$$

$$\mathbf{\Gamma}_M = \begin{bmatrix} \mathbf{C}^T & \mathbf{A}^T\mathbf{C}^T & \ldots & (\mathbf{A}^{M-1})^T\mathbf{C}^T \end{bmatrix}^T$$

$$\mathbf{X} = [\mathbf{x}_k\ \mathbf{x}_{k+1}\ \cdots\ \mathbf{x}_{k+N-1}]$$

$$\mathbf{H} = \begin{bmatrix} \mathbf{D} & 0 & \cdots & 0 \\ \mathbf{CB} & \mathbf{D} & \cdots & 0 \\ \mathbf{CAB} & \mathbf{CB} & \cdots & 0 \\ \cdots & \cdots & \cdots & \cdots \\ \mathbf{CA}^{M-2}\mathbf{B} & \mathbf{CA}^{M-3}\mathbf{B} & \cdots & \mathbf{D} \end{bmatrix},$$

where $\mathbf{Y} \in \Re^{M_o \times N}$, $\mathbf{\Gamma}_M \in \Re^{M_o \times n}$, $\mathbf{X} \in \Re^{n \times N}$, $\mathbf{H} \in \Re^{M_o \times M_i}$, $\mathbf{U} \in \Re^{M_i \times N}$ (constructed in the same fashion as $\mathbf{Y}$), and $M_o = n_o M$, $M_i = n_i M$. Next, we compute the complete SVD of $\mathbf{U}$,

$$\mathbf{U} = \mathbf{\Phi}\begin{bmatrix} \mathbf{S} & \mathbf{O} \end{bmatrix}\begin{bmatrix} \mathbf{\Psi}_1 \\ \mathbf{\Psi}_2 \end{bmatrix}, \quad \mathbf{U} = \mathbf{\Phi}\mathbf{S}\mathbf{\Psi}_1.$$

We can define $\mathbf{U}^\perp = \mathbf{\Psi}_2^T$ because of the fact that $\mathbf{U}\mathbf{U}^\perp = \mathbf{0}$. Applying $\mathbf{U}^\perp$ to (2) yields

$$\mathbf{Y}\mathbf{U}^\perp = \mathbf{\Gamma}_M\mathbf{X}\mathbf{U}^\perp. \qquad (3)$$

This relation shows that the column span of $\mathbf{\Gamma}_M$ can be determined from $\mathbf{Y}\mathbf{U}^\perp$. Exploiting the special structure of $\mathbf{\Gamma}_M$, we can estimate $\mathbf{A}$ and $\mathbf{C}$ by using the ESPRIT algorithm [1; 9; 10] originated from sensor array processing.

Knowing $\mathbf{\Gamma}_M$, we can form a matrix $(M_o - n) \times M_o$ matrix $\mathbf{Q}$ such that $\mathbf{Q}\mathbf{\Gamma}_M = \mathbf{O}$. Then premultiplying $\mathbf{Q}$ and postmultiplying $\mathbf{U}^\dagger$, we obtain

$$\mathbf{Q}\mathbf{Y}\mathbf{U}^\dagger = \mathbf{Q}\mathbf{H} \triangleq \mathbf{K}.$$

Since $\mathbf{H}$ is a linear function of $\mathbf{B}$ and $\mathbf{D}$ given $\mathbf{A}$ and $\mathbf{C}$, $\mathbf{B}$ and $\mathbf{D}$ can be obtained by solving a set of linear equations.

[1] Of course, since the states are not directly observed, the system matrices are identifiable only to within an arbitrary non-singular state (similarity) transformation.

Additional treatment of noise

De Moor *et al.* [3] did not specifically discuss how to handle the additive noise. Here we shall show how to do this so that the obtained estimates are asymptotically unbiased. With the additive noise, (2) becomes

$$\mathbf{Y} = \mathbf{\Gamma}_M\mathbf{X} + \mathbf{H}\mathbf{U} + \nu, \qquad (4)$$

where ν denotes the noise Hankel matrix constructed in a manner similar to $\mathbf{Y}$. Postmultiplying $\mathbf{U}^\perp$ yields

$$\mathbf{Y}\mathbf{U}^\perp = \mathbf{\Gamma}_M\mathbf{X}\mathbf{U}^\perp + \nu\mathbf{U}^\perp. \qquad (5)$$

The noise ν is assumed to be a zero-mean Gaussian random vectors with variance $\sigma^2\mathbf{I}$. Therefore, the covariance matrix of $\mathbf{Y}\mathbf{U}^\perp$ is

$$\begin{aligned} \mathbf{C}_y &= E\{\mathbf{Y}\mathbf{U}^\perp(\mathbf{Y}\mathbf{U}^\perp)^T\} \\ &= \mathbf{\Gamma}_M\mathbf{X}\mathbf{X}^T\mathbf{\Gamma}_M^T + \sigma^2\mathbf{\Sigma}_n \end{aligned}$$

where the new noise covariance matrix $\mathbf{\Sigma}_n$ can be explicitly evaluated given $\mathbf{U}^\perp$ (see [4] for the details). With $\mathbf{\Sigma}_n$, we can pre- and post-multiply both sides of (5) by $\mathbf{\Sigma}_n^{-1/2}$ and $\mathbf{\Sigma}_n^{-T/2}$, respectively, to prewhiten the noise. Thus,

$$\begin{aligned} \widetilde{\mathbf{C}}_y &= \mathbf{\Sigma}_n^{-1/2}\mathbf{C}_y\mathbf{\Sigma}_n^{-T/2} \\ &= \mathbf{\Sigma}_n^{-1/2}\mathbf{\Gamma}_M\mathbf{X}\mathbf{X}^T\mathbf{\Gamma}_M^T\mathbf{\Sigma}_n^{-T/2} + \sigma^2\mathbf{I}. \end{aligned} \qquad (6)$$

Since $\mathbf{\Gamma}_M$ is a rank $n(\ll M_o)$ matrix, $\widetilde{\mathbf{C}}_y$ can be expressed as a low-rank matrix plus a shift ($\sigma^2\mathbf{I}$). Such matrix structure falls into a standard subspace fitting framework and various subspace decomposition methods (*e.g.*, eigendecomposition) can be applied to estimate the range space of the low-rank matrix, *i.e.*, $span\{\mathbf{\Sigma}_n^{-1/2}\mathbf{\Gamma}_M\}$.

Let $(\lambda_i, \mathbf{e}_i)_{i=1}^{M_o}$ denote all the eigenpairs of $\widetilde{\mathbf{C}}_y$. It is not difficult to see that the eigenvalues have the following distribution: $\lambda_1 \geq \cdots \geq \lambda_n > \lambda_{n+1} = \cdots = \lambda_{M_o}$ and the desired subspace $span\{\mathbf{\Sigma}_n^{-1/2}\mathbf{\Gamma}_M\}$ equals $span\{\mathbf{e}_1, \ldots, \mathbf{e}_n\}$. By checking the number of smallest repeated eigenvalues, we can determine n and then the desired subspace $span\{\mathbf{\Gamma}_M\}$ is just $span\{\mathbf{\Sigma}_n^{1/2}[\mathbf{e}_1, \ldots, \mathbf{e}_n]\}$. The above shows a way of exactly determining $span\{\mathbf{\Gamma}_M\}$ in the presence of noise.

A Fast State Space Model Identification: Batch Processing

In the existing methods, the SVD (singular value decomposition) of $\mathbf{U}$ requires $O(MN^2)$ flops, while the ED (eigendecomposition) of $\mathbf{Y}\mathbf{U}^\perp\mathbf{U}^{\perp T}\mathbf{Y}^T$ costs $O(M^3)$ flops. As M and N become large, these computational costs are very burdensome. In addition, the existing identification scheme requires too much storage. Storing $\mathbf{Y}$, $\mathbf{U}$, $\mathbf{U}^\perp$, $\mathbf{Y}\mathbf{U}^\perp$ and $\mathbf{U}^\dagger$ demands $O(M^2 + MN + N^2)$ bytes. These computational and storage requirements may often preclude the real-time implementation of the state space model identification. This motivates the search for faster methods that require less storage. The basic ideas behind the fast identification method arise from the following observations.

1. Since the Hankel matrices $\mathbf{Y}, \mathbf{U}$ have the so-called displacement structure (about which more is said in Appendix A), these matrices can be determined uniquely by a few (*i.e.*, displacement rank $\alpha \leq M, N$) vectors, comprising a so-called generator matrix. The product, inverse and pseudo-inverse of Hankel (or Toeplitz) matrices also preserve the displacement structure and can be computed through generator matrices, which results in a significant computational reduction.

2. Given the covariance matrix either explicitly or implicitly (*e.g.*, as a product of several matrices), we can exploit its low-rank + shift structure by using the recently developed Fast Subspace Decomposition (FSD) algorithm [12] to replace the more costly ED or SVD.

The resulting fast identification technique for batch processing reduces the computational load from $O(MN^2 + M^2N + M^3)$ to $O(MN + M^2)$ flops For details, refer to [1].

3 A Fast State Space Model Identification: Recursive Processing

In the following, we focus our discussion on on-line *recursive* identification techniques to cope with time-varying scenarios. In recursive identification techniques, update/downdate and exponential weighting are commonly used. We adopt the former in our new recursive method for the identification of state space models.

Since the fast batch technique for state space identification reduces the computational load to $O(MN + M^2)$ flops, we may be able to use it in several on-line applications. However, the computational load $O(MN)$ may not be acceptable for certain on-line applications when the number of data samples, N, is quite large. Since only one measurement comes into and only one goes out of the data window N, most of the computations involved are redundant. Judicious use of such redundancy can reduce the computational load further. It is reasonable to conjecture that the computational cost is only a function of M but not N.

The inherent difficulty in making the fast batch algorithm recursive stems from the nonrecursive nature of the Schur algorithm and FSD. These two problems seem to be invincible and are not dealt with in this paper. As a result, the recursive algorithm requires at least $O(M^2)$ flops which is the computational load of the Schur algorithm and FSD. Another major difficulty in the recursive identification technique arises from the noise covariance matrix update. Since the noise covariance matrix update in the fast identification method for batch processing involves the computation of a projection matrix of size $N \times N$ and the summation of $O(NM)$ elements, this step should be modified significantly (see Section 3.1). Except for updating of noise covariance matrix, the basic structure of the batch algorithm is maintained. In fact, the recursive identification technique is an extension of the fast batch technique. Updating of the so-called generator matrices with sizes much smaller than the original matrices, constitutes a basic idea in the recursive algorithm. Of course, the displacement structure of the data matrices $\mathbf{U}, \mathbf{Y}$ and the low rank + shift structure of the covariance matrix play crucial roles. Before proceeding, the reader is suggested to refer to Appendix A for reviews of displacement structure and the generalized Schur algorithm, because they are essential to the derivation of the new identification algorithm. For simplicity of exposition, we transform all the Hankel matrices in (4) to Toeplitz matrices by reversing the order of their columns or equivalently by post-multiplying by $\tilde{\mathbf{I}}$, a matrix with ones on the antidiagonal and zeros elsewhere. We explain the algorithm step by step.

3.1 Step 1 and 2: The Noise Covariance Matrix Update

This subsection covers the method of updating the noise covariance. The basic assumption is that sensor noises are uncorrelated and temporally white. Then, the noise covariance matrix has a form $\mathbf{\Sigma}_n = \mathbf{\Sigma} \otimes \mathbf{I}_{n_o \times n_o}$, where $\mathbf{\Sigma}$ is the noise covariance matrix of one sensor only, and $\otimes$ denotes the Kronecker product. Then $\mathbf{\Sigma}_{p,q}$ turns out to be $\sum_{m=1}^{N-p+q} \mathbf{P}_{m-p+q,m} \sigma^2$ (see [1]). Since $\mathbf{\Sigma}_{p,q}$ is a function of $p-q$, $\mathbf{\Sigma}$ is a Toeplitz matrix and we have only to compute the first column or row of $\mathbf{\Sigma}$ to construct $\mathbf{\Sigma}$. In the fast identification technique for batch processing, we compute $\mathbf{P}_u^\perp$ explicitly and then find the first column or row of $\mathbf{\Sigma}$ just by summing the diagonal $(\mathbf{\Sigma}_{1,1})$ and subdiagonal elements $(\mathbf{\Sigma}_{p,1},\ 2 \leq p \leq M)$ of $\mathbf{P}_u^\perp$. The resulting computational load is $O(MN)$. However, in order to compute the noise covariance matrix with $O(M^2)$ or less, we should avoid direct computation of $\mathbf{P}_u^\perp$ because of the inevitable $O(MN)$ additions required to compute the elements of $\mathbf{\Sigma}$ from $\mathbf{P}_u^\perp$. As a result, we must develop an algorithm that computes $\mathbf{\Sigma}$ without forming $\mathbf{P}_u^\perp$ directly.

Updating $\mathbf{\Sigma}_n$[2] without forming $\mathbf{P}_u^\perp$

Our goal here is to update the noise covariance matrix $\mathbf{\Sigma}$ with $O(M^2)$ or less flops. As shown earlier, $\mathbf{\Sigma}_{p,q} = \sum_{m=1}^{N-p+q} \mathbf{P}_{m-p+q} \sigma^2$. In other words, $\mathbf{\Sigma}$ is a Toeplitz matrix and it's $(1,j)$-th element $\mathbf{\Sigma}_j$ can be found as sums of $(j-1)$-th subdiagonal elements of $\mathbf{P}_u^\perp$.

$$\mathbf{\Sigma}_j = \mathrm{Tr}(\mathbf{Z}^{j-1}\mathbf{P}_u^\perp), \quad \mathbf{P}_u^\perp = \mathbf{I} - \mathbf{U}^T(\mathbf{U}\mathbf{U}^\perp)^{-1}\mathbf{U}$$

where $1 \leq j \leq M$ and $\mathbf{Z}$ is a nilpotent matrix whose elements are zero except the unity elements on the first lower diagonal. When $j = 1$,

$$\mathbf{\Sigma}_1 = \mathrm{Tr}(\mathbf{P}_u^\perp) = N - M > 0$$

For $2 \leq j \leq M$,

$$\mathbf{\Sigma}_j = \mathrm{Tr}(\mathbf{Z}^{j-1}\mathbf{P}_u^\perp) = -\mathrm{Tr}[(\mathbf{U}\mathbf{U}^T)^{-1}\mathbf{U}\mathbf{Z}^{j-1}\mathbf{U}^T].$$

Note that we reduce the dimension of the matrices from $N \times N$ to $M \times M$. However, since $\mathbf{Z}$ lies between

[2] In the following derivation, we deal with the single input case. Extension to the multi-input case is straightforward.

$\mathbf{U}$ and $\mathbf{U}^T$, it is not easy to relate the computation of Σ_j to that of Σ_{j+1}. Since $\mathbf{U}$ is a Toeplitz (with displacement structure),

$$\mathbf{Z}_N\mathbf{U}^T - \mathbf{U}^T\mathbf{Z}_M = [\ g_1 \quad g_2\ ][\ h_1 \quad h_2\ ]^T, \quad (7)$$

where

$$g_1 = \left[1/2 \quad u_{N+M-1} \quad u_{N+M-2} \quad \ldots \quad u_{M+1}^T\right]^T$$

$$g_2 = \left[1/2 - u_{N+M-1} \quad - u_{N+M-2} \quad \ldots \quad - u_{M+1}\right]^T$$

$$h_1 = \left[-u_{N+1} \quad - u_{N+2} \quad \ldots \quad - u_{N+M-1} \quad 1/2\right]^T$$

$$h_2 = \left[-u_{N+1} \quad - u_{N+2} \quad \ldots \quad - u_{N+M-1} \quad - 1/2\right]^T$$

Also, note the dimensions of two shift-down matrices $\mathbf{Z}_N$ and $\mathbf{Z}_M$ are N and M respectively. In the following derivation, the subscripts will be dropped unless there is any ambiguity. By (7), it can be shown that the Σ_j is the sum along the j-th subdiagonal of the following matrices Ω_1 and Ω_2:

$$\Omega_l = \begin{bmatrix} h_l^T\mathbf{Z}^{M-2} \\ h_l^T\mathbf{Z}^{M-3} \\ \vdots \\ h_l^T \end{bmatrix} (\mathbf{U}\mathbf{U}^T)^{-1}\mathbf{U} \begin{bmatrix} g_l^T \\ g_l^T\mathbf{Z}^T \\ \vdots \\ g_l^T(\mathbf{Z}^{M-2})^T \end{bmatrix}^T$$

$$= \mathbf{H}_l(\mathbf{U}\mathbf{U}^T)^{-1}\mathbf{U}\mathbf{G}_l$$

where $l = 1,\ 2$.
Specifically,

$$-\Sigma_2 = \mathrm{Tr}[\mathbf{Z}^{M-2}(\Omega_1 + \Omega_2)]$$

$$\vdots$$

$$-\Sigma_M = \mathrm{Tr}[\Omega_1 + \Omega_2]$$

Now, the question is whether we can compute Ω_l with $O(M^2)$ flops. We use the theory of displacement to solve this problem [11]. We observe that Ω_l is a product of Toeplitz matrices,

$$\Omega_l = \mathbf{H}_l(\mathbf{U}\mathbf{U}^T)^{-1}\mathbf{U}\mathbf{G}_l \in \Re^{(M-1)\times(M-1)}$$

where $\mathbf{H}_l, \mathbf{U}, \mathbf{G}_l$ are all Toeplitz matrices. We can compute Ω_l with $O(M^2)$ flops using the generalized Schur Algorithm.
Form an embedded matrix $\mathbf{W}_l$:

$$\mathbf{W}_l = \begin{bmatrix} \mathbf{U}\mathbf{U}^T & \mathbf{U}\mathbf{G}_l \\ \mathbf{H}_l & \mathbf{O} \end{bmatrix} \in \Re^{(M+M-1)\times(M+M-1)}.$$

Note that the Schur complement of the (1,1) block is $-\Omega_l$ from which Σ can be found with $O(M^2)$ flops. The generator of $\mathbf{W}_l$ can be updated with $O(M)$ flops even though it takes $O(MN)$ flops to compute it (see [13]). We apply the generalized Schur algorithm M-times to $\mathbf{W}_l$ to find the generator of Ω_l, which requires $O(M^2)$ flops. Ω_l can be computed with $O(M^2)$ flops from its generator since most multiplications are repeated. The subdiagonals of Ω_l are summed to yield the Σ_j, which costs $O(M^2)$ flops. At this point, we are done with the $O(M^2)$ update of the noise covariance matrix Σ.
The next step is to compute $\Sigma^{1/2}$ and $\Sigma^{-1/2}$. Again using the Schur algorithm, we can compute them with $O(M^2)$ flops. The details can be found in [1].

3.2 Step 3: FSD Signal Subspace Estimation

Once the covariance matrix is formed either explicitly or implicitly (a product of several matrix factors), we can estimate the signal subspace including its dimension. Recently, Xu and Kailath proposed a Fast Subspace Decomposition (FSD) method [12] using the well-known Lanczos algorithm. This new method can estimate the signal subspace including its dimension in $O(nM^2)$ flops for an $M \times M$ symmetric matrix with n-dimensional signal subspace, where n is usually much smaller than M. For a brief introduction to FSD, refer to [12].
Here, we propose to apply FSD to find the signal subspace $span\{\mathbf{E}_s\}$ of the symmetric matrix $\Sigma_n^{-1/2}\mathbf{Y}\mathbf{P}_u^{\perp}\mathbf{Y}^T\Sigma_n^{-T/2}$. The trace and determinant of $\Sigma_n^{-1/2}\mathbf{Y}\mathbf{P}_u^{\perp}\mathbf{Y}^T\Sigma_n^{-T/2}$ should be computed as prerequisites for the FSD detection scheme to estimate the system order n.
We start by performing triangular factorization of $\mathbf{Y}\mathbf{P}_u^{\perp}\mathbf{Y}^T$. Since $\mathbf{U}$ and $\mathbf{Y}$ are block-Toeplitz matrices and have displacement structure, $\mathbf{Y}\mathbf{P}_u^{\perp}\mathbf{Y}^T$ also possesses displacement structure. With the same trick used in the previous subsection, we form an embedded matrix whose Schur complement is $\mathbf{Y}\mathbf{P}_u^{\perp}\mathbf{Y}^T$.

$$\mathbf{W} = \begin{bmatrix} \mathbf{U}\mathbf{U}^T & \mathbf{U}\mathbf{Y}^T \\ \mathbf{Y}\mathbf{U}^T & \mathbf{Y}\mathbf{Y}^T \end{bmatrix}, \qquad \mathbf{F} = \mathbf{Z}^{n_i} \oplus \mathbf{Z}^{n_o}$$

After M_i steps of the Schur algorithm, we obtain a generator of $\mathbf{Y}\mathbf{P}_u^{\perp}\mathbf{Y}^T$. After another M_o steps, we obtain the Cholesky factor of $\mathbf{Y}\mathbf{P}_u^{\perp}\mathbf{Y}^T$. These two iterations require $O((M_i + M_o)^2)$ flops. The determinant and trace of $\Sigma_n^{-1/2}\mathbf{Y}\mathbf{P}_u^{\perp}\mathbf{Y}^T\Sigma_n^{-T/2}$ can be computed with $O(M_o^2)$ flops as in the batch processing. This completes the FSD prerequisite.
Now we apply FSD to $\Sigma_n^{-1/2}\mathbf{Y}\mathbf{P}_u^{\perp}\mathbf{Y}^T\Sigma_n^{-T/2}$ to find the signal subspace. Since the Cholesky factor of $\mathbf{Y}\mathbf{P}_u^{\perp}\mathbf{Y}^T$ has already been computed, we can work on the product of four triangular factors. If we perform the ED of $\Sigma_n^{-1/2}\mathbf{Y}\mathbf{P}_u^{\perp}\mathbf{Y}^T\Sigma_n^{-T/2}$, we have to compute the product of four triangular factors explicitly, which requires an additional $O(M_o^3)$ flops. However, the key step in the Lanczos algorithm is the matrix-vector product which can be achieved by four sequential matrix-vector multiplications if we only have the factors of the matrix. In other words, the explicit computation of the four matrix products can be avoided and the computational cost of each step is $O(M_o^2)$ flops. As shown in [12], only $O(n)$ Lanczos steps is required to estimate the signal subspace. Thus, the total computational cost for FSD is $O(nM_o^2)$ flops.
At Step 4, $\mathbf{A}$ and $\mathbf{C}$ are estimated through the TLS ESPRIT algorithm mentioned in Section 2. At Step 5, we compute $\mathbf{K} = \mathbf{Q}\mathbf{Y}\mathbf{U}^{\dagger}$ to estimate $\mathbf{B}$ and $\mathbf{D}$ while exploiting the displacement structure of $\mathbf{U}$ to find $\mathbf{U}^{\dagger}$. $\mathbf{Q}$ is replaced by $\mathbf{E}_s^{\perp} = \mathbf{I} - \mathbf{E}_s(\mathbf{E}_s^T\mathbf{E}_s)^{-1}\mathbf{E}_s^T$ since the computation of $\Gamma_M^{\perp}$ requires more flops.

3.3 Block Update and Downdate

For a large data window N, the system characteristic usually does not change much within a sample period. As a result, updating the system parameters at every sample instant may not be necessary. Instead, updating the system parameters at every, say d, samples may be more reasonable. Besides the fact that the update at every sample is not necessary, a block update may also have computational advantage. The reason is as follows. The Schur and FSD algorithms cannot be implemented recursively. If the estimates are updated at each data sample, the total updating cost for d data samples will be $O(dM^2)$ flops. Now since the generators of the embedded matrices $\mathbf{W}$ can be updated with $O(M)$ flops for each data sample, the generator update for d samples costs $O(dM)$ flops. With the generators of $\mathbf{W}$, we can apply the generalized Schur algorithm to find the generators of $\mathbf{\Omega}, \mathbf{Y}\mathbf{P}_{\mathbf{u}}^{\perp}\mathbf{Y}^T, \mathbf{Y}\mathbf{U}^{\dagger}$ and then apply the FSD method to find the signal subspace. Since the FSD and Schur algorithms require $O(M^2)$ flops, the total computational complexity of a block update is $O(M^2 + dM)$ flops.

Finally, we compare the computational load of our batch and recursive algorithms in Table 1.

Steps	Find	*Batch*	*Recursive*
1	$\mathbf{U}^{\perp}$	$O(MN)$	$O(M^2 + dM)$
2	$\mathbf{\Sigma}_n^{1/2}$, $\mathbf{\Sigma}_n^{-1/2}$	$O(M^2)$	$O(M^2 + dM)$
3	$\mathbf{E}_s$, n	$O(MN)$	$O(M^2 + dM)$
5	$\mathbf{QYU}^{\perp}$	$O(MN)$	$O(M^2 + dM)$

TABLE 1 Comparative flop counts of the batch and recursive methods. In the recursive processing, the flop counts are tabulated for the block update of block size d.

4 Simulations

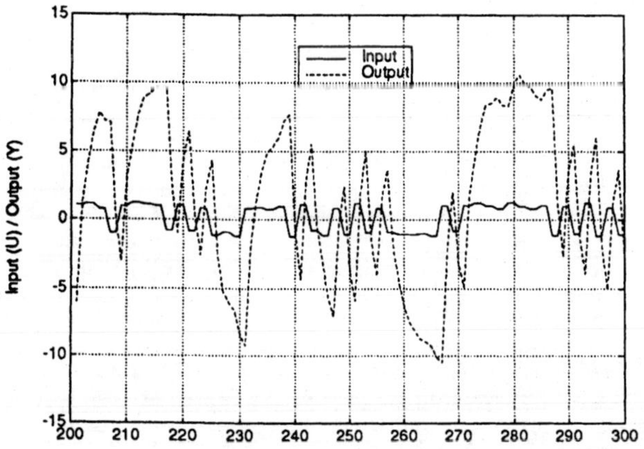

Fig. 1. A snapshot of input/output data used in the simulation.

We applied the new identification technique to the following single input, single output and two state-simulated system,

$$\mathbf{A}_k = \begin{bmatrix} 0.7 + 0.2\frac{\exp(\frac{-k}{2000})-1}{\exp(-1)-1} & 0 \\ 0 & 0.2 - 0.1\frac{\exp(\frac{-k}{2000})-1}{\exp(-1)-1} \end{bmatrix}$$

$$\mathbf{B}_k = \begin{bmatrix} 2 & 1 \end{bmatrix}^T, \quad \mathbf{C}_k = \begin{bmatrix} 1 & 2 \end{bmatrix}, \quad \mathbf{D}_k = 0.05$$

where k is a time index starting from 0. The eigenvalues of the system matrix $\mathbf{A}_k$ (the system "poles") drift from $(0.7, 0.2)$ to $(0.9, 0.1)$ during the simulation as shown in Fig. 2. The exponentially changing eigenvalues can arise, for example in a spring-mass system where the mass is exponentially decreasing due to leakage, say. A PRBS (pseudo random binary sequence) input (see Fig. 1) was designed and applied to the system to obtain a system model. No input process noise was added. The total time span of the simulation is 2100 samples. White Gaussian noise of $\sigma = 0.5$ was added to the output. Fig. 1 shows the resulting output. The signal to noise ratio for each state output was computed under the assumption that noise was added to each state output and turns out to be 28 dB for state 1 and 13 dB for state 2.

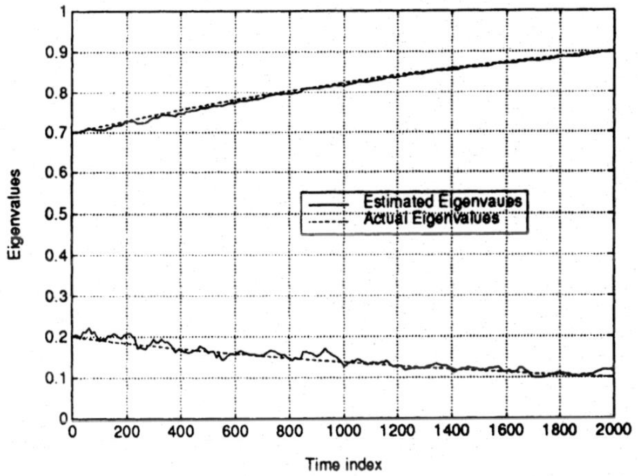

Fig. 2. Trajectory of the actual and estimated eigenvalues of the simulated system.

The new identification technique was applied to the input/output data in Fig. 1 with data window size $N = 100$ and sliding window size $M = 20$. The identification technique could successfully detect the number of states, 2, throughout the simulation (this is mainly due to the high signal to noise ratio). Fig. 2 shows the identified eigenvalues together with the actual eigenvalues. The algorithm clearly tracks the actual eigenvalues. There are two interesting things to note in Fig. 2. First, we can observe that the estimates lag behind the true values. This is due to the relatively big data window size. Since we assume that the system is time-invariant within the data window, the estimated parameters should be such that they best explain the data within the data window. Then, roughly speaking, the average eigenvalues within the data window would be proper estimates and the estimates would lag behind the true values in our example where the eigenvalues are exponentially decaying or monotonically changing. When the simulation was performed with a smaller data window size, the extent of lag became smaller. Second, the eigenvalue estimate corresponding to the first state is better than the eigenvalue estimate of the second state. This is mainly due to the higher signal to noise ratio of the first state output than the second. Overall, we can conclude the proposed identification scheme performs well for the given example.

5 Conclusions

A new recursive algorithm for identification of state space models has been presented. It exploits the displacement and low-rank + shift structures of the matrices that arise in the identification procedure as in the batch processing case [1]. The key idea in the new recursive approach is to update the so-called generator vectors instead of matrices, which reduces the computational cost down to $O(M^2 + dM)$ flops when we update the system parameters at every d samples. Such reductions of computational burden may allow on-line identification of state space models in many potential applications. The results of computer simulation have given some evidence of the viability of the fast recursive identification technique.

A Brief Description of Displacement Structure and Generalized Schur Algorithm

Let $\mathbf{A} \in \Re^{m\times n}$ and let $\{\mathbf{F}_f, \mathbf{F}_b\}$ be strictly lower triangular matrices. If $\mathbf{A} - \mathbf{F}_f\mathbf{A}\mathbf{F}_b^T$ is a rank $\alpha(< m, n)$ matrix, which can be expressed as $\mathbf{X}\mathbf{Y}^T$, where $\mathbf{X} \in \Re^{m\times\alpha}$ and $\mathbf{Y} \in \Re^{n\times\alpha}$, then $\mathbf{A}$ has displacement structure with a generator $\{\mathbf{X}, \mathbf{Y}\}$.

It can be easily shown that a generator of a matrix with displacement structure carries sufficient information of a matrix. Therefore, operation of such a structured matrix can be achieved by manipulating a small number (2α) of generator vectors. Since for $\alpha \ll m, n$, the size of a generator is much smaller than a matrix, manipulating the generator can result in a significant computational reduction.

The well-known generalized Schur algorithm is an important method to partially or completely triangularize a structured matrix [11]. Since many matrix operations can be eventually viewed as a triangular factorization problem, the generalized Schur algorithm can be applied to exploit displacement structure to reduce the computational load.

Before presenting the generalized Schur algorithm, we assume that we have a procedure called $MakeProper$ that can covert a given generator $\{\mathbf{X}, \mathbf{Y}\}$ into a proper generator $\{\mathbf{X}_p, \mathbf{Y}_p\}$. We say that a generator is proper with respect to the j-th column if for a certain i, all the elements in the i-th row of $\mathbf{X}$ and above, except for the element $[X]_{i,j}$, are zeros, and the same is true for $\mathbf{Y}$. Now, we present the generalized Schur algorithm.

The Generalized Schur Algorithm

Input: A generator $\{\mathbf{X}, \mathbf{Y}\}$ of $\mathbf{A} \in \Re^{m\times n}$ with respect to $\{\mathbf{F}_f, \mathbf{F}_b\}$.

Output: (i) Partial triangular factors $\mathbf{L} \in \Re^{m\times r}$ and $\mathbf{U} \in \Re^{r\times n}$ of $\mathbf{A}$.
(ii) A generator $\{\mathbf{X}, \mathbf{Y}\}$ of the Schur complement of the $r \times r$ leading principal submatrix of $\mathbf{A}$.

Procedure GeneralizedSchur
    for $k := 1$ to r
        $MakeProper$
        The kth column of $\mathbf{L} := \mathbf{x}_{pvt}$
        The kth row of $\mathbf{U} := \mathbf{y}_{pvt}^T$
        Replace $\mathbf{x}_{pvt}$ with $\mathbf{F}_f\mathbf{x}_{pvt}$
        Replace $\mathbf{y}_{pvt}$ with $\mathbf{F}_b\mathbf{y}_{pvt}$
    end
    return(L,U,{X,Y})

At the k-th step of the above generalized Schur algorithm, $\{\mathbf{X}, \mathbf{Y}\}$ is assumed to be proper with respect to the pvt-th column and $\mathbf{x}_{pvt}, \mathbf{y}_{pvt}$ are the pvt-th columns of $\mathbf{X}, \mathbf{Y}$. Note that the above procedure needs $O(\alpha(m+n)r)$ flops.

REFERENCES

[1] Y.M. Cho, G. Xu, and T. Kailath, Fast identification of state-space models via exploitation of displacement structure, *IEEE Trans. Aut. Control*, February 1992, Submitted.

[2] A. van der Veen, E. F. Deprettere, and A. L. Swindlehurst, SVD-Based Estimation of Low-Rank System Parameters, In E. F. Deprettere and A. J. van der Veen, editors, *Algorithms and Parallel VLSI Archetecture*. Elsevier Science Publishers, 1991.

[3] B. De Moor, M. Moonen, L. Vandenberghe, and J. Vandewalle, A geometrical approach for the identification of state space models with singular value decomposition, In *Proc. IEEE ICASSP*, volume 4, pages 2244–2247, New York, NY, 1988.

[4] A. Swindlehurst, R. Roy, B. Ottersten, and T. Kailath, A subspace fitting approach for identification of linear state space models, *IEEE Trans. Aut. Control*, 1991, Submitted.

[5] M. H. Verhaegen and P. DeWilde, Subspace model identification. part I: The Output-Error State Space Model Identification class of Algorithms, *Int. J. Control*, 1991, accepted for publication.

[6] W. E. Larimore, System identification, reduced-order filtering and modeling via canonical variate analysis, In *Proc. CDC*, pages 445–451, June 1983.

[7] W. E. Larimore, Canonical variate analysis in identification, filtering and adaptive control, In *Proc. CDC*, pages 596–604, December 1990.

[8] L. Ljung, *System Identification : Theory for the User*, Prentice Hall,Inc., Englewood Cliffs, New Jersey 07632, 1987.

[9] R. Roy, A. Paulraj, and T. Kailath, Esprit- a subspace rotation approach to estimation of parameters of cisoids in noise, *IEEE Trans. ASSP*, **34**(5):1340–1342, Oct. 1986.

[10] R. Roy and T. Kailath, **ESPRIT** – Estimation of Signal Parameters via Rotational Invariance Techniques, *IEEE Trans. on ASSP*, **37**(7):984–995, July 1989.

[11] J. Chun, T. Kailath, and H. Lev-Ari, Fast parallel algorithms for QR and triangular factorization, *SIAM J. Sci. Stat. Comput.*, **8**(6):899–913, November 1987.

[12] G. Xu and T. Kailath, "A Fast Algorithm for Signal Subspace Decomposition and Its Performance Analysis", In *Proc. ICASSP 91 Conf.*, volume 5, pages 3069–3072, Toronto, Canada, May 1991.

[13] Y.M. Cho, G. Xu, and T. Kailath, Fast recursive identification of state-space models via exploitation of displacement structure, *Automatica*, July 1992, Submitted.

Index Rule Scheduling Policies Applied To Identification

C. Olivier and O. Dessoude

ETCA/CREA, Laboratoire "Système de Perception", 94114 Arcueil, France

Abstract. The optimal scheduling of probabilistic tests in order to identify N independent stationary objects in parallel obeys an index rule. The practical implementation of this decision scheme calls for the discretization of an equivalent information state space and for the efficient computation of the allocation indices. Different cost functions are proposed and numerical results show the improvement with respect to a myopic policy.

Keywords. Stochastic scheduling, parameter identification, dynamic allocation indices.

1 Introduction

1.1 Problem statement

We suppose that each object O_n, $n \in \{1, \ldots, N\}$, is characterized by a static parameter to be identified, Θ_n, belonging to the finite set of values $\{1, \ldots, P\}$, and that this parameter will be observed in discrete time through a single sensor, characterized by a set of P stationary probability density functions $p_i(z_k) = p(z_k \mid \Theta_n = i)$, $i \in \{1, \ldots, P\}$. The simultaneous identification of these N objects, to be performed by a single sensor, states a classical scheduling problem, *i.e.*, the allocation of the sensor at each time instant to one of these objects, in order to optimize some predefined cost function.

1.2 Equivalent full state information formulation

As usual for this kind of problem with incomplete observations (2), the trick is to change the state space and to introduce a new framework with complete information. This is done by defining an *information state* for each object O_n, as the vector of probabilities for its unknown parameter as it is known at time t_k, *i.e.*, $p_n^k = \{(p(\Theta_n = i))\}_{i \in \{1,\ldots,P\}}$. This vector will change each time the parameter Θ_n is observed, *i.e.*, each time the identification process chooses object O_n. Depending on the actual result of the observation z_k, an *a posteriori* probability vector will be computed, using Bayes formula :

$$p(\Theta_n = i \mid z_k) = \frac{p(\Theta_n = i)\, p(z_k \mid \Theta_n = i)}{p(z_k)}, \qquad (1)$$

with the normalization constant

$$p(z_k) = \sum_{i=1}^{P} p(\Theta_n = i)\, p(z_k \mid \Theta_n = i). \qquad (2)$$

This posterior thus defines the new probabilities vector as $p_n^{k+1} = \{p(\Theta_n = i \mid z_k)\}_{i \in \{1,\ldots,P\}}$ for the observed object.

The current information state at time t_k will therefore be a set of N probabilities vectors $(p_1^k, \ldots, p_N^k)$ corresponding to the N objects. The dimensionality of this state space may be thus understood as "S_P^N", where S_P is the simplex of order P defined by

$$S_P = \{\{p_i\} \mid \sum_{i=1}^{P} p_i = 1, p_i \geq 0\}.$$

1.3 State space discretization

1.3.1 Unordered version

The problem will now be discretized in order to deal with a finite number of information states. Be Δ a divisor of the interval $[0, 1]$, *i.e.*, a real number such that $\frac{1}{\Delta}$ is an integer N_Δ. The probabilities vectors $p_{n,i}$ will be discretized at the levels α_i, $i \in \{1, \ldots, P\}$ such that :

1. $\forall i, \exists$ an integer k_i s.t.$\alpha_i = k_i \Delta$;
2. $\sum_{i=1}^{P} \alpha_i = 1$.

This constant pace framework can be improved in a second stage with an adaptive pace discretization. The number of states $B(N_\Delta, P)$ obtained in the discretization of the simplex S_P at Δ, *i.e.*, the number of P-tuples α_i, $i \in \{1, \ldots, P\}$ satisfying these two conditions is given by the following recursive formula :

Lemma 1

$$B(N_\Delta, P) = \sum_{n=0}^{N_\Delta} B(n, P-1), P > 1;$$
$$B(N_\Delta, 1) = 1.$$

1.3.2 Ordered version

A way to contain the combinatorial explosion is to look for an ordered version, with an additional requirement on the α_i :

$\alpha_1 \leq \alpha_2 \leq \ldots \leq \alpha_P$.

The new number of ordered states $B^*(N_\Delta, P)$ is given by another recursive formula :

Lemma 2

$$\begin{aligned} B^*(N_\Delta, P) &= B^*(N_\Delta, P, 0) \\ B^*(N_\Delta, P, q) &= 0, \; if\, n < 0; \\ &= 1, \; if\, p = 1; \\ &= B^*(N_\Delta - P * q, P, 0) \; if\, q > 0; \\ &= \sum_{n=0}^{N_\Delta} B^*(N_\Delta - n, P-1, n) \\ & \textit{in other cases.} \end{aligned}$$

The following array gives an idea of the number of states obtained in both versions, the figures for the ordered case being written between brackets :

$B(B^*)$	$N_\Delta = 100$	$N_\Delta = 200$
$P = 2$	101(51)	201(101)
$P = 3$	5151(884)	20301(3434)
$P = 5$	4598126(46262)	70058751(643287)

The ratio $\frac{B}{B^*}$ is roughly comparable to $P!$. Even in ordered versions, there still is a heavy combinatorial factor, and an important fact to notice is that *the ordering implies a stringent symmetry constraint on the sensor output*. It means that given a state $x = (p_1, \ldots, p_N)$, all the states $x_\sigma = (p_{\sigma(1)}, \ldots, p_{\sigma(N)})$ are equivalent, for all the permutations σ in $\{1, \ldots, N\}$. Be x_0 the specific state $(\alpha, 1-\alpha, 0, \ldots, 0)$, and a permuted version $x_1 = (1-\alpha, \alpha, 0, \ldots, 0)$. For x_0 and x_1 to be equivalent, we assume that there must be the same number of possible transitions from x_0 and x_1 to other equivalent states with the same probabilities. In fact, these probabilities are the observation result probabilities, $\{p_0(z = z_i)\}_{i=1,\ldots,P}$ or $\{p_1(z = z_i)\}_{i=1,\ldots,P}$, depending what the initial state is, x_0 or x_1,and meaning that $\forall\alpha$, there exists a permutation σ' such that $p_0(z = z_i) = p_1(z = z_{\sigma'(i)})$. But be $Z_{ij} = p(z = z_i \mid \Theta = \Theta_j)$:

$$\begin{aligned} p_0(z = z_i) &= \alpha Z_{i1} + (1-\alpha) Z_{i2}; \\ p_1(z = z_{\sigma'(i)}) &= (1-\alpha) Z_{\sigma'(i)1} + \alpha Z_{\sigma'(i)2}, \end{aligned}$$

so that :

$$\forall i, \alpha \underbrace{\left(Z_{i1} + Z_{\sigma'(i)1} - (Z_{i2} + Z_{\sigma'(i)2})\right)}_{\lambda_{\sigma'(i),i}} = \underbrace{Z_{\sigma'(i)1} - Z_{i2}}_{\mu_{\sigma'(i)i}}. \qquad (3)$$

We know that α will be discretized on N_Δ values. Assume $N_\Delta > P$. For each i, the ratio $\frac{\mu_{\sigma'(i)i}}{\lambda_{\sigma'(i),i}}$ may only take P different values corresponding to the different possible values of $\sigma'(i)$. Be then $r_{ji} = \frac{\mu_{j,i}}{\lambda_{j,i}}$. If we pick an i and a specific α_i such that $\alpha_i \neq r_{ji}, \forall j$, there is no permutation σ' satisfying the ith equation 3. Then, for these equations to be still true, there must exist a permutation σ' such that $\lambda_{\sigma'(i),i} = \mu_{\sigma'(i)i} = 0, \forall i$, or $Z_{i1} = Z_{\sigma'(i)2}$ and $Z_{\sigma'(i)1} = Z_{i2}, \forall i$. This is true for the couple of indices $(1,2)$ but we could obtain the same with any couple $(m,n), m \neq n$ so that in fact, $\forall (m,n), m \neq n, \exists \sigma' \mid \forall i, Z_{im} = Z_{\sigma'(i)n}$ and $Z_{in} = Z_{\sigma'(i)m}$. This means that each column of matrix Z will have the same elements but in a different order. Proceeding further, it is in fact possible to show that the only possible ordered cases occur with an observer of the form $Z_{ij} = \delta_{\sigma(i),j}\lambda + (1 - \delta_{\sigma(i),j})\mu$ for some permutation σ in $\{1, \ldots, P\}$ and two real numbers $0 < \lambda, \mu < 1$.

1.4 Transition probabilities matrix

Be E_d the departure state corresponding to a probabilities vector $p_{E_d} = (\alpha_{1,d}, \ldots, \alpha_{P,d})$ Being given an observation result z_k, the output vector $p_{n,i}$ is given by equation 1,*i.e.* :

$$p_{n,i} = p(\Theta_n = i \mid z_k) = \frac{\alpha_{i,d} p(z_k \mid \Theta_n = i)}{\sum_{p=1}^{P} \alpha_{p,d} p(z_k \mid \Theta_n = i)}.$$

It will be approximated by the closest discretized state under a chosen distance, say E_f. The observation result z_k therefore corresponds to a *transition* in the state space from E_d to E_f. Define then the transition probability of the equivalent Markov Chain

$$T_{d,f} = Pr(E^{k+1} = E_f \mid E^k = E_d).$$

Be now $Z_{d,f}$ the set of observation results z for which the posterior state $p_{|z}^{k+1}$ precisely admits E_f as the closest discretized state :

$$\{z \mid d(p_{|z}^{k+1}, p(E_f)) = \min_j d(p_{|z}^{k+1}, p(E_j)) \mid p^k = p_{E_d}\}.$$

The transition from E_d to E_f will then occur with probability

$$T_{d,f} = \int_\Omega 1_{Z_{d,f}}(z) p(z) dz,$$

where 1_A is the indicator function of set A.

As a matter of fact, there is then no complexity increase in considering sensors with continuous outputs. This being kept in mind, sensors with a finite number of outputs $z_k = a_j, j = 1, \ldots, Q$ with probabilities given by a stationary observation matrix $p_{i,j} = p(z_k = a_j \mid \Theta_i)$ will be considered in the sequel without loss of generality.

2 Cost functions and sensor control policies

2.1 Cost functions

Since the final goal is to identify the real values of the parameters $\Theta_n, n \in \{1, \ldots, N\}$, the cost functions of interest have to be related to the discrepancy, for each object, between the actual information state and the final state which would correspond to the identification of the actual parameter value for this object. They are then functions $c_1(p_n^k)$ of the probability vectors of each object.

The following elementary costs may then be considered (among others) :

1. $\left\{ \begin{array}{ll} c_1^\varepsilon(p_n^k) = & 0 \text{ is } \max_{i=1,\ldots,P}(1 - p_{n,i}^k) > 1 - \varepsilon \\ c_1^\varepsilon(p_n^k) = & 1 \text{ else} \end{array} \right\}$,
(identification at confidence level $1 - \epsilon$, with $\epsilon \geq \Delta$).

2. $c_2(p_n^k) = 1 - (\max_{i\in\{1,\ldots,P\}} p_{n,i}^k)$ (probability of the event " the maximum likelihood state does not correspond to the real state, this is shown in the appendix)

3. $c_3(p_n^k) = \sum_{i=1}^{P} p_{n,i}^k(1 - p_{n,i}^k)$ (expectation of the distance between 1 and the current probability of the real state)

4. $c_4(p_n^k) = \sum_{i=1}^{P} p_{n,i}^k \log(p_{n,i}^k)$ (probability vector entropy).

The considered cost functions will be discounted version of these, of the form :

$$J_q = \sum_{k=0}^{\infty} \beta^k \sum_{n=1}^{N} c_q(p_n^k), 0 \leq \beta < 1. \quad (4)$$

The significance of these corresponding discounted criteria are then explained below :

1. J_1 : sum of the discounted times to identify the true values of the parameters at confidence levels $1 - \epsilon$;
2. J_2 : discounted integral over time of the sum of the probabilities of making a wrong decision at the different time instants regarding what the true value is;
3. J_3 : discounted integral over time of the discrepancy between the fully identified state 1 and the current probability of the real state;
4. J_4 : discounted integral over time of the sum of the probabilities vector entropies.

2.2 Control policies

Consider some state-dependent scheduling policy g for our sensor, leading to the observation of object O_{u_k} at t_k if $u_k = g^k(p_1^k, \ldots, p_n^k)$. If the admissible policies have to be stationary, the decision space, without any further simplification, will be B^N, where B is the cardinal of the discretized simplex S_P.

Index scheduling policies are of particular interest as a sub-family of these policies, since they allow a very simple sensor control, based upon off-line computation. Such policies u_γ are characterized by an index γ, actually a scalar function on the current information state of each object :

$$\gamma \;:\; p_n^k \longmapsto \gamma_n^k = \gamma(p_n^k),$$

and consist in the selection of the largest current index among all objects :

$$u_{\gamma k} = \arg \max_{n\in\{1,\ldots,n\}} \gamma_n^k.$$

In our case, it turns out that index policies are optimal, allowing a tremendous shrink of the complexity to B from B^N.

3 Multi-armed bandit formulation and optimal solution.

3.1 Gittins' fundamental result.

We now recall the optimality result due to Gittins (1).

Theorem 1 *Be the following optimization problem :*

$$\max_{u\in U} E\left(\sum_{l=0}^{\infty} \beta^l r(p_{u_{k+l}}^{k+l}) \mid p_n^k, n \in \{1,\ldots,n\}\right), \quad (5)$$

and the index family :

$$\gamma_g(p_i^k) = \max_{\tau\geq 1} \frac{E\left\{\sum_{l=0}^{\tau-1} \beta^l r(p_i^{k+l})\right\}}{E\left\{\sum_{l=0}^{\tau-1} \beta^l\right\}} \quad (6)$$

where τ is a stopping time. Then the optimal strategy for 5 is u_{γ_g}.

3.2 A cost-equivalency lemma.

Before proceeding further, we shall introduce the following technical lemma (4) (3) :

Lemma 3 *Minimizing criterion $J_1 = E\{\sum_{k=0}^{\infty} \beta^k \sum_{n=1}^{N} c(p_n^k)\}$, is equivalent to maximizing the criterion $J_2 = E\{\sum_{l=0}^{\infty} r(p_{u(l)}^l)\beta^l - (1-\beta)r(p_{u(0)}^0)\}$, where $r(p_{u_k}^k) = c(p_{u_k}^k) - c(p_{u_k}^{k+1})$.*

A proof is given in the appendix.

3.3 The optimal solution

Using theorem 1 and lemma 3, we define, for each possible criterion J_q, an equivalent reward $r_q(p^k)$ by

$$r_q(p_n^k) \;=\; c_q(p_n^k \mid z)_d - (c_q(p_n^k)),$$

where z is the observation result and $p_n^{k+1} \mid z$ the posterior probability vector. $p_d(x)$ stands for the closest discretized vector in the state space to x, and the corresponding index family is

$$\gamma_q(p_n^k) = \max_{\tau\geq 1} \frac{E\left\{\sum_{l=0}^{\tau-1} \beta^l r_q(p_n^{k+l}) \mid p_n^k\right\}}{E\left\{\sum_{l=0}^{\tau-1} \beta^l\right\}}.$$

Then the optimal strategy for minimizing $J_q' = J_q - \beta r_q(p_{u(0)}^0)$ is the optimal strategy for maximizing $\sum_{l=0}^{\infty} r_q(p_{u(l)}^l)\beta^l$, *i.e.*, u_{γ_q}. Before turning to the actual efficient computation of these indices, a question to be answered is to know if these policies actually yield to the identification of the true parameter values.

4 Convergence aspects.

4.1 Theoretical convergence condition

Note $\Theta_t(i)$ the true value of the parameter for object O_i. Before any discretization of the state space, the simultaneous identification would require that $\forall i \in 1, \ldots, N, \forall j \neq t(i)$,

$$\frac{p(\Theta(O_i)) = \Theta_j) \mid z_1, \ldots z_k)}{p(\Theta(O_i) = \Theta_{t(i)}) \mid z_1, \ldots z_k)} \rightarrow 0$$

when $k \rightarrow \infty$.

Elementary results from probability theory then show that to be identified, each object must be observed infinitely often, meaning that $\forall i \in 1, \ldots, N$,

$$p(\forall m, \exists k > m \mid O_i \text{ is observed at time } t_k) = 1.$$

4.2 Misidentification probability

In a discretized state space with pace Δ, convergence actually means to reach an identification level -*i.e.*, a value of the maximum probability- greater that $1 - \frac{\Delta}{2}$-, since then the most likely state will then be of the form $p_{n,i}^k = \delta_{i,i_0}$ for some i_0. Let us call this state I_{i_0}. Then there still is a small probability that, at some point in the identification process of object O_n, a state I_{i_0} may be reached first, Θ_{i_0} *not* being the true value of the parameter Θ_n. Once in state I_i, Bayes formula lets the state invariant, whatever the observation result. This is an absorbing state and then, there is no chance to correctly identify the true parameter value, whatever the scheduling policy used in the future. In fact, to be more accurate, let τ_i the first hitting time of state I_i. The misidentification probability for object O_n, p_n^{mis} is then given by, if Θ_t is the true value of the parameter :

$$p_n^{mis} = \sum_{i \neq t} p(\tau_i = \arg\min_j \tau_j).$$

Be i_0 the initial state and

$$h_{i,j}(i_0) = \sum_{i \neq t} p(\tau_i = \arg\min_j \tau_j \mid i_0).$$

Be $h_{i,j}$ the vector of these probabilities for $i_0 \neq i, j$.

Lemma 4 (Misidentification probability) *Be $T_{(i,j)^c}$ the restriction of T obtained by removing states i et j from the equivalent Markov Chain. Be T_i the probabilities vector such that : $T_i(x) = p(x_{k+1} = i \mid x_k = x), \forall x \neq i, j$. Then :*

$$h_{i,j} = (I - T_{(i,j)^c})^{-1} T_i, \quad (7)$$

and if the initial probabilities vector for Θ is p^0, the expected misidentification probability $\hat{p_{mis}}$ is :

$$\hat{p_{mis}} = \sum_{n=1}^{N} p_n^0 \sum_{i \neq n} h_{I_i, I_n}(p^0). \quad (8)$$

The proof is given in the appendix.

As a conclusion, the discretization of the state space implies the existence of a misidentification probability, and especially under index rules scheduling policies.

5 Efficient computation of the indices.

The practical computation of the indices over a finite state space is given by a classical algorithm due to Varaiya, Walrand and Buyukkoc (4) which we shall first summarize. We consider a convenient discretization of the state space, as described above, leading to $B(N_\Delta, P)$ states $\{1, \ldots, I = B(N_\Delta, P)\}$. The idea is to define, for any subset A of our state space, its associated stoppping time $\tau^A = \min\{k \geq 1 \mid x_k \in I - A\}$ and the two expressions :

$$W^A(i) = E\left\{ \sum_{k=0}^{\tau^A - 1} \beta^k r(x_k) \mid x_0 = i \right\},$$

$$V^A(i) = E\left\{ \sum_{k=0}^{\tau^A - 1} \beta^k \mid x_0 = i \right\}.$$

For any set A, these are the solutions of two linear systems

$$W^A = (I - \beta P^A)^{-1} R,$$
$$V^A = (I - \beta P^A)^{-1} Q,$$

using the following vectorial notations :

$$W^A = (W^A(1) \ldots W^A(I))^t,$$
$$V^A = (V^A(1) \ldots V^A(I))^t,$$
$$R = (r^A(1) \ldots r^A(I))^t,$$
$$Q = (1 \ldots 1)^t.$$

What is more, with $A = \{j \in E \mid \gamma(j) \leq \gamma(i)\}$, we have $\gamma(i) = \frac{W^A(i)}{V^A(i)}$.

Let us define $I_0 = \emptyset$, $I_1, \ldots, I_J$, a partition of the state space with decreasing indices, and $A_0, A_1, \ldots, A_n$ the family of subsets such that $A_n = \bigcup_{m=0}^{n} I_m$, $n \in \{0, \ldots, J\}$. We may now express the optimal stopping time using these definitions :

$$\forall i \in I_n \quad \tau^*(i) = \tau^{A_n}.$$

Hence a recursive algorithm for the construction of I_n and its Gittins index :

$$I_0 = \emptyset,$$
$$I_{n+1} = \{i \in E - A_n \mid \frac{W^{A_n}(i)}{V^{A_n}(i)} = \max_{j \in E - A_n} \frac{W^{A_n}(j)}{V^{A_n}(j)}\}.$$

Let us now explain how this algorithm can be made much more efficient using simple algebra. Knowing the inverse of a given matrix, Sherman-Morrison formula allows us to derive the inverse of this matrix after changing one row or column :

$$(M + u \otimes v)^{-1} = M^{-1} - \frac{(M^{-1}u) \otimes ((M^{-1})^t v)}{1 + v^t M^{-1} u}.$$

As each step of the previous algorithm modifies only a few rows of $I - \beta P^A$, generally not more than one. The recursive algorithm can be reformulated very naturally :

$$M_0 = I$$

Fig. 1. Construction of Gittins indices.

$$\begin{aligned} I_0 &= \emptyset \\ M_{n+1} &= M_n - \frac{\beta(M_n^{-1}(P)_{*i}) \otimes (M_n^{-1})_{i*}}{1+\beta(P)^t_{*i}(M_n^{-1})_{i*}}, \\ I_{n+1} &= \{i \in E - A_n \mid \\ & \frac{(M_n R)_i}{(M_n Q)_i} = \max_{j\in E-A_n} \frac{(M_n R)_j}{(M_n Q)_j}\}, \end{aligned}$$

where $(M)_{i*}$, resp.$(M)_{*i}$, denotes the ith row, resp. column, of matrix M. Its complexity is reduced from I^4 to I^3, as each iteration involves only one matrix-vector product, one scalar product and one tensorial product, instead of a full matrix inversion.

6 Numerical simulations.

We consider the identification of $N = 4$ objects, the parameter taking $P = 2$ values Θ_1, Θ_2, with an observer matrix $M_{i,j} = p(z = i/\Theta = \Theta_j)$ given by :

$$M = \begin{pmatrix} 0.70 & 0.30 \\ 0.30 & 0.70 \end{pmatrix}.$$

The discretization pace is $\Delta = 0.01$, so that in an ordered version there are $B^* = 51$ states. The discount factor is $\beta = 0.999$, and the cost function is c_1^ϵ with $\epsilon = 0.015$. The initial state is $p(\Theta_n = \Theta_1) = p(\Theta_n = \Theta_2) = \frac{1}{2}$, for $n = 1, \ldots, 4$. The following results have been obtained over 500 simulation runs. Misidentification occurred 18 times.

Results	Myopic policy	Index policy
Average cost	150.7	119.0
Cost variance	58.5	52.0

The myopic policy is obtained by observing the object which yields the maximum expected immediate reward. Note that, in the case of cost c_1^ϵ, most of the states are equivalent with respect to this policy, *i.e.*, of expected immediate reward 0, except if there is a probability, from these states, to reach a state I_i after a single observation. The improvement provided by an index policy with respect to this myopic one is around 21%. When $\beta \longrightarrow 1$, the criterion corresponds to the sum of the identification times for the different objects and the improvement is stabilized around 25%. And when $\beta \longrightarrow 0$, it of course tends to 0 since the contribution of the immediate time step becomes infinitely important with respect to the other time steps, yielding the optimality of the myopic policy.

Finally, figure 1 displays the first steps of the construction of the indices for another discretization ($\Delta = 1000$). Each line describes the state space after each iteration of the algorithm, the black pixels corresponding to states whose index is already known, *i.e.*, belonging to the A_n sets. In other words, this figure shows the growing of the A_n sets and gets an insight into the structural complexity of this problem. With some imagination and a great deal of intuition, one could claim that the contribution of the dynamic allocation indices lies here in the fringe of this fractal triangle...

7 Conclusion

The problem of sequentially identifying several objects with one sensor admits an optimal index policy. In particular, an adequate choice of the cost function leads to the minimum discounted time identification of the objects. The practical implementation of this type of policies supposes the discretization of a probabilistic state space. Numerical simulations show the improvement one can expect by using index policies. The misidentification probability drawback could be reduced by an adaptive discretization scheme of the state space, characterized by a refinement of the pace in the vicinity of absorbing states I_i.

REFERENCES

[1] J.C. Gittins. *Multi-armed bandit allocation indices.* John Wiley & Sons, Chichester, 1989.

[2] P.R. Kumar and P.P. Varaiya. *Stochastic systems : estimation, identification and adaptive control.* Prentice-Hall, Englewood Cliffs, 1986.

[3] X. Merlo. *Techniques probabilistes d'intégration et de contrôle de la perception en vue de son exploitation par le système de décision d'un robot.* Thèse de Doctorat, Institut Polytechnique de Lorraine, 1988.

[4] P. P. Varaiya, J. C. Walrand, and C. Buyukkoc. Extensions of the multiarmed bandit problem : the discounted case. *IEEE Transactions on Automatic Control*, AC-30(5):426–439, 1985.

Appendix

Lemma 5 *The event " the maximum likelihood state does not correspond to the real state" has probability* $1 - (\max_{i\in\{1,\ldots,P\}} p^k_{n,i})$.

Proof : Be t the real value of the parameter. Then, where $\delta_{\{i,j\}}$ stands for the Kronecker index :

$$\begin{aligned} & p(\arg\max_{i=1,\ldots,N} p_i^k = t) \\ &= \sum_{j=1}^{N} \delta_{\{\arg\max_{i=1,\ldots,N} p_i^k, j\}} p(\Theta_j = \Theta_t) \end{aligned}$$

$$= \sum_{j=1}^{N} \delta_{\{\arg \max_{i=1,\dots,N} p_i^k, j\}} p_j^k$$

$$= \max_{i=1,\dots,N} p_i^k,$$

and this yields the result. □

Proof of lemma 3 : Be n and k respectively the object and time indices. First note that :

$$\begin{aligned} c(p_n^k) &= c(p_n^0) + \sum_{l=0}^{k-1} (c(p_n^{l+1}) - c(p_n^l)) \\ &= c(p_n^0) - \sum_{l=0}^{k-1} r(p_n^l). \end{aligned}$$

Then, with the convention that the sum cancels when $k = 0$:

$$\begin{aligned} &\sum_{k=0}^{\infty} \beta^k \sum_{n=1}^{N} c(p_n^k) \\ &= \sum_{k=0}^{\infty} \beta^k \sum_{n=1}^{N} \left(c(p_n^0) - \sum_{l=0}^{k-1} r(p_n^l) \right) \\ &= \underbrace{\frac{\sum_{n=1}^{N} c(p_n^0)}{1-\beta}}_{C_0} - \underbrace{\sum_{k=0}^{\infty} \beta^k \sum_{n=1}^{N} \sum_{l=0}^{k-1} r(p_n^l)}_{C_1}. \end{aligned}$$

Then :

$$\begin{aligned} &C_1 \\ &= \sum_{k=2}^{\infty} \left(\sum_{n=1}^{N} r(p_n^l) \sum_{l=0}^{k-1} \beta^l \right) \\ &= \sum_{l=0}^{\infty} \left(\sum_{n=1}^{N} r(p_n^l) \sum_{k=\max(2,l+1)}^{\infty} \beta^k \right) \\ &= \sum_{l=0}^{\infty} \left(\sum_{n=1}^{N} r(p_n^l) \sum_{k=l+1}^{\infty} \beta^k \right) - \beta \sum_{n=1}^{N} r(p_n^0) \\ &= -\beta \sum_{n=1}^{N} r(p_n^0) + \sum_{l=0}^{\infty} \left(\sum_{n=1}^{N} r(p_n^l) \frac{\beta^{l+1}}{1-\beta} \right). \end{aligned}$$

But

$$\sum_{n=1}^{N} r(p_n^l) = r(p_{u(l)}^l),$$

since the cost will not be modified at time t_l except for the object $u(l)$ chosen at this specific instant. Finally :

$$\begin{aligned} &\sum_{k=0}^{\infty} \beta^k \sum_{n=1}^{N} c(p_n^k) \\ &= \frac{\sum_{n=1}^{N} c(p_n^0)}{1-\beta} - \sum_{l=0}^{\infty} r(p_{u(l)}^l) \frac{\beta^{l+1}}{1-\beta} + \beta r(p_{u(0)}^0) \\ &= \frac{\sum_{n=1}^{N} c(p_n^0) + \beta \left((1-\beta) r(p_{u(0)}^0) - \sum_{l=0}^{\infty} r(p_{u(l)}^l) \beta^l \right)}{1-\beta}, \end{aligned}$$

and this yields the lemma. □

Proof of lemma 4 : It is easy to see that, for $x \neq i, j$, $y \neq i, j$:

$$h_{i,j}(x) = \sum_{y \neq i,j} T_{x,y} h_{i,j}(y) + T_{x,i}.1 + T_{x,j}.0,$$

because $h_{i,j}(i) = 1$ and $h_{i,j}(j) = 0$ by definition. This yields the result. □

A CURVE-BROKEN-LINE METHOD FOR SYSTEM IDENTIFICATION

Fanjin KONG and Robin M. C. DE KEYSER

Automatic Control Laboratory, University of Gent
Grotesteenweg Noord 6, B-9052 Gent, Belgium

Abstract. The methods used for system identification are divided into two classes, the nonparametric methods (IIR Infinite Impulse Response) and the parametric methods (such as ARMAX, etc.). For a slowly time-variant and/or a slightly non-linear and/or a noisy system, the IIR method requires too much computational effort to be used efficiently.

This paper discusses the relationship between the IIR model and the parametric model, and introduces a new method, so called Curve-Broken-Line (CBL) method, to estimate the IIR coefficients of the system identified from the input and output signals. Then the parameters of the parametric model could be calculated from the IIR coefficients by using the relationship between the IIR model and the parametric model. The advantage of the CBL method is that it doesn't need much a priori knowledge of the process.

The CBL method can be used in the case of on-line parameter identification, when the parameters of the system vary slowly, or the system is slightly nonlinear or noisy, so that it is necessary to update the parameters regularly.

Keywords. System identification, Impulse response, Parameter estimation, Recursive least squares, Models

INTRODUCTION

The methods used for system identification are divided into two classes, the nonparametric methods and the parametric methods. The literature referring to this area is very extensive (Gustavsson, 1972; Eykhoff, 1974; Rake, 1980; Graupe and co-workers, 1980; Åström, 1980; Strejc, 1980). Some authors also paid attention to the relationship between the coefficients estimated by using IIR method and parametric methods (Ljung and Söderström, 1983; Söderström and Stoica, 1989).

Consider a linear time invariant discrete system with time delay d as follows

$$A(q^{-1})y(t)=B(q^{-1})u(t-d) \qquad (1)$$

or

$$A(q^{-1})y(t)=\mathcal{B}(q^{-1})u(t) \qquad (2)$$

where

$$A(q^{-1})=1+a_1q^{-1}+\ldots+a_{na}q^{-na} \qquad (3)$$

$$B(q^{-1})=\underline{b}_0+\underline{b}_1q^{-1}+\ldots+\underline{b}_\beta q^{-\beta} \qquad (4)$$

$$\mathcal{B}(q^{-1})=b_0+b_1q^{-1}+\ldots+b_{nb}q^{-nb}$$

$$\text{with } b_i=0 \;\; i=0,\ldots,d-1; \quad b_{d+i}=\underline{b}_i \;\; i=0,\ldots,\beta \;\; \text{and } nb=\beta+d \qquad (5)$$

u(t) is the input signal and y(t) is the output signal. Throughout the paper, it is assumed that the given system is asymptotically stable.

By long division, we have

$$y(t)=\frac{\mathcal{B}(q^{-1})}{A(q^{-1})}u(t)$$

$$=\sum_{i=0}^{\infty}h_b(i)q^{-i}u(t)=\sum_{i=0}^{\infty}h_b(i)u(t-i) \qquad (6)$$

where $h_b(i)$ $i=0, \ldots, \infty$ are the coefficients of the infinite impulse response (IIR) of the system with

$$h_b(i)=0 \qquad i=0,\ldots,d-1 \qquad (7)$$

$$h_b(i)=b_i-a_1h_b(i-1)-\ldots-a_jh_b(i-j)-\ldots-a_{na}h_b(i-na)$$

$$i=d,\ldots,nb \;\; \text{and} \;\; h_b(i-j)=0 \;\; \text{if} \;\; i-j\le d-1 \qquad (8)$$

$$h_b(i)=-a_1h_b(i-1)-\ldots-a_jh_b(i-j)-\ldots-a_{na}h_b(i-na)$$

$$i\ge nb+1 \;\; \text{and} \;\; h_b(i-j)=0 \;\; \text{if} \;\; i-j\le d-1 \qquad (9)$$

Looking at Eqs. (6) and (9), if the IIR coefficients of the system are known exactly, we can evaluate the parameters a_i $i = 1, \ldots, na$ by using na linear independent equations like Eq. (9) with indices $i \ge nb+1$. Otherwise, we must use the Least Squares (LS) algorithm or Recursive Least Squares (RLS) algorithm to compute the parameters a_i $i = 1,\ldots,na$ from ψ linear independent equations with $\psi > na$. Then the values of the parameters b_i $i = 0,\ldots,nb$, or equivalently the parameters $\underline{b}_i$ $i = 0,\ldots,\beta$ and the time delay d can be obtained by using Eqs. (7) and (8).

In Eq. (6), the sum $\Sigma h_b(i)u(t-i)$ contains infinite terms. But as pointed out by Ljung and Söderström (1983), we generally assume that the data collection is initialized at time t=0, and this problem is usually resolved in practice by summing up only to i=t in Eq. (6). Since $h_b(i)$ decays exponentially, this approximation is reasonable, except possibly for small t. Thus it is permissible to assume that $|h_b(i)| \approx 0$ if index i is larger than a sufficiently large positive integer N_b. The infinite sum in Eq. (6) can then be replaced by a finite sum and a small value $o_1(t)$,

$$y(t) = \sum_{i=0}^{N_b} h_b(t)\,u(t-i) + o_1(t) \qquad (10)$$

where

$$o_1(t) = \sum_{i=N_b+1}^{\infty} h_b(i)\,u(t-i) << \sum_{i=0}^{N_b} h_b(i)\,u(t-i) \qquad (11)$$

Although Eq. (10) contains a finite sum, it is not practical to be used into an on-line identification program because of too much calculation needed (N_b being of the order of 100).

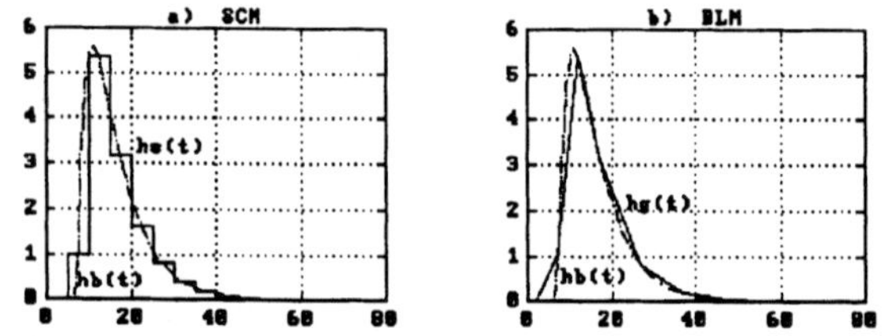

Figure 1 a) Staircase Method.
b) Broken-Line Method

Several simplifying methods can be applied to reduce the computational effort. The simpler one is the Staircase Method (SCM). In this method, a staircase curve $h_s(t)$ is used to approximate the real IIR curve $h_b(t)$ of a given plant, as shown in Fig. 1 (a). The number of the SCM coefficients is reduced a great deal, but it is subjected to low precision. Another one is the Broken-Line Method (BLM), recently introduced by Gu and co-workers (1990). BLM uses a broken-line $h_g(t)$ as the approximation of $h_b(t)$, as shown in Fig. 1 (b). The BLM possesses a higher precision than SCM, but the authors did not refer to the relationship between BLM and parametric models.

As discussed above, since the first nb+na linear independent equations of Eqs. (7) - (9) contain the main information of a given parametric model, the IIR coefficients included in these nb+na equations ought to be estimated as exactly as possible. Now, let's introduce our Curve-Broken-Line (CBL) method. Looking at the curves in Fig. 2, suppose that the curve of the IIR expression is $h_b(t)$ and the curve of the CBLM is $h_1(t)$ respectively. Notice that, in Fig. 2, due to the time delay of the system, $h_1(t) = h_b(t) = 0$ from the initial time 0 to the time d-1, and $h_1(t) = h_b(t)$ $t = d, \ldots, l_0$, here l_0 is a sufficiently large positive integer (>> nb+1) selected to ensure the precision requirement when evaluating the parameters a_i $i = 1, \ldots, na$, b_i $i = 0, \ldots, \beta$, and d. Other time indices l_1, l_2, ..., l_n are positive integers selected to make the broken-line part of $h_1(t)$ compared to $h_b(t)$ $t>l_0$, under the given precision requirement, as close as possible.

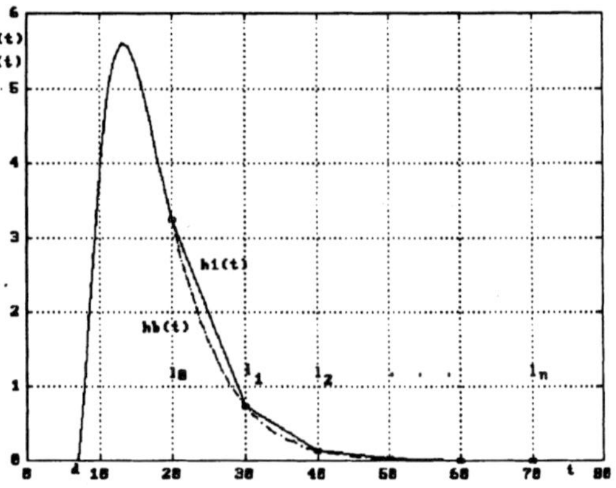

Figure 2 The IIR Curve $h_b(t)$ and the CBL $h_1(t)$

The first feature of CBLM, different from BLM, is that the parameters of a given plant are obtained from its IIR estimations. This identified parametric model can be used for further control purpose in which a parametric model is required. Secondly, the error o(t) due to the truncation and due to the approximation between $h_1(t)$ and $h_b(t)$ is also estimated, and is used as a pointer to show the identification precision. Thirdly, in the case of long-range predictive control, such as the Extended Horizon Predictive Control and a simple version of the Generalized Predictive Control, the IIR coefficients of a given plant are required (Scattolini and Bittanti, 1990). Although the CBL curve is not a smooth curve, a smooth IIR curve also can be calculated from the identified parametric model.

This paper discusses, in Section 2, the CBL method and the estimation for $h_b(t)$. Then in Section 3, an algorithm indicates how to calculate the parameters of the parametric model from the coefficients of the CBL model. Section 4 is an example.

THE CURVE-BROKEN-LINE (CBL) METHOD AND THE ESTIMATION FOR $h_b(t)$

By using CBL method, from Eq. (10) we obtain

$$y(t) = \sum_{k=0}^{l_0} h_b(k)\,u(t-k) + \sum_{k=1}^{l_1-l_0} [h_b(l_0) + k\frac{h_b(l_1)-h_b(l_0)}{l_1-l_0}]\,u(t-l_0-k) + \ldots + \sum_{k=1}^{l_n-l_{n-1}} [h_b(l_{n-1}) + k\frac{h_b(l_n)-h_b(l_{n-1})}{l_n-l_{n-1}}]\,u(t-l_{n-1}-k) + o_2(t) + o_1(t) \qquad (12)$$

$$= \sum_{k=0}^{l_0} h_b(k)\,u(t-k) + h_b(l_0)\sum_{k=1}^{l_1-l_0-1}(1-\frac{k}{l_1-l_0})\,u(t-l_0-k) + h_b(l_1)[\sum_{k=1}^{l_1-l_0}\frac{k}{l_1-l_0}u(t-l_0-k) + \sum_{k=1}^{l_2-l_1-1}(1-\frac{k}{l_2-l_1})u(t-l_1-k)]$$

$$+\dots+h_b(l_{n-1})\Big[\sum_{k=1}^{l_{n-1}-l_{n-2}}\frac{k}{l_{n-1}-l_{n-2}}u(t-l_{n-2}-k)+$$

$$+\sum_{k=1}^{l_n-l_{n-1}-1}\Big(1-\frac{k}{l_n-l_{n-1}}\Big)u(t-l_{n-1}-k)\Big]+$$

$$+h_b(l_n)\sum_{k=1}^{l_n-l_{n-1}}\frac{k}{l_n-l_{n-1}}u(t-l_{n-1}-k)+o(t) \tag{13}$$

where

$$o_2(t)=\sum_{i=l_0+1}^{N_b}h_b(i)\,u(t-i)-$$

$$-\sum_{i=1}^{n}\sum_{k=1}^{l_i-l_{i-1}}\Big[h_b(l_{i-1})+k\frac{h_b(l_i)-h_b(i_{i-1})}{l_i-l_{i-1}}\Big]u(t-l_{i-1}-k) \tag{14}$$

$$o(t)=o_1(t)+o_2(t) \tag{15}$$

and

$$l_n=N \tag{16}$$

Define

$$L_i=l_i-l_{i-1} \qquad i=1,\dots,n \tag{17}$$

$$h_1(i)=h_b(i) \qquad i=0,\dots,l_0 \tag{18}$$

and

$$h_1(l_0+i)=h_b(l_i) \qquad i=1,\dots,n \tag{19}$$

Furthermore, define the new input w(.) as follows

$$w(t-i)=u(t-i) \qquad i=0,\dots,l_0-1 \tag{20}$$

$$w(t-l_0)=u(t-l_0)+\sum_{k=1}^{L_1-1}\Big(1-\frac{k}{L_1}\Big)u(t-l_0-k) \tag{21}$$

$$w(t-l_0-i)=\sum_{k=1}^{L_i}\frac{k}{L_i}u(t-l_{i-1}-k)+$$

$$+\sum_{k=1}^{L_{i+1}-1}\Big(1-\frac{k}{L_{i+1}}\Big)u(t-l_i-k) \qquad i=1,\dots,n-1 \tag{22}$$

$$w(t-l_0-n)=\sum_{k=1}^{L_n}\frac{k}{L_n}u(t-l_{n-1}-k) \tag{23}$$

By using above definitions, Eq. (12) can be re-written as follows

$$y(t)=\sum_{i=0}^{l_0+n}h_1(i)\,w(t-i)+o(t) \tag{24}$$

Eq. (24) is a variant of Eq.(10). Obviously, the number of coefficients of $h_1(t)$ is much less than that of $h_b(t)$ in Eq. (10).

Usually, there are disturbances in the system. A frequently used parametric model is the following ARMAX-model

$$A(q^{-1})y(t)=B(q^{-1})u(t-d)+C(q^{-1})e(t) \tag{25}$$

or

$$y(t)=\frac{B(q^{-1})}{A(q^{-1})}u(t)+\frac{C(q^{-1})}{A(q^{-1})}e(t) \tag{26}$$

where $A(q^{-1})$, $\underline{B}(q^{-1})$ or $B(q^{-1})$, u(t), y(t) and time delay d are expressed just the same as in Eq. (5) or Eq. (6). {e(t)} is a sequence of independent random variables with zero mean and variance σ^2.

$$C(q^{-1})=1+c_1q^{-1}+\dots+c_{nc}q^{-nc} \tag{27}$$

If necessary, Eq. (25) or Eq. (26) can be extended by $D(q^{-1})z(t)$, where z(.) means a measurable disturbance.

Using the same method described above, we obtain

$$y(t)=\sum_{i=0}^{\infty}h_b(i)\,u(t-i)+\sum_{i=1}^{\infty}h_e(i)\,e(t-i)+e(t) \tag{28}$$

where $h_b(i)$ i = 0, ..., ∞ are the same as in Eqs. (7) - (9), and

$$h_e(i)=c_i-a_1h_e(i-1)-\dots-a_jh_e(i-j)-\dots-a_{na}h_e(i-na)$$

$$i=1,\dots,nc \tag{29}$$

$$h_e(i)=-a_1h_e(i-1)-\dots-a_jh_e(i-j)-\dots-a_{na}h_e(i-na)$$

$$i\ge nc+1 \tag{30}$$

In Eqs. (29) and (30), $h_e(0)=1$ and $h_e(i-j)=0$ if $i-j < 0$. These sequences $\{h_b(i)\}$ and $\{h_e(i)\}$ will tend to zero exponentially in practice. Similarly, as stated before, we can express Eq. (28) as follows,

$$y(t)=\sum_{i=0}^{N_b}h_b(i)u(t-i)+\sum_{i=1}^{N_e}h_e(i)e(t-i)+e(t)+o_1(t) \tag{31}$$

where N_b and N_e are two sufficiently large positive integers with $N_b \gg na+nb$ and $N_e \gg na+nc$.

$$o_1(t)=\sum_{i=N_b+1}^{\infty}h_b(i)\,u(t-i)+\sum_{i=N_e+1}^{\infty}h_e(i)\,e(t-i) \tag{32}$$

Then, by using CBLM, we obtain

$$y(t)=\sum_{i=0}^{l_0+n}h_1(i)\,w(t-i)+\sum_{i=1}^{p_0+m}h_2(i)\,v(t-i)+e(t)+o(t) \tag{33}$$

where L_i, $h_1(.)$ and w(.) are defined as in Eqs. (17) - (23). Besides, P_i, $h_2(.)$, v(.) and o(t) are listed in the Appendix.

Define

$$\theta(t)=[h_1(0)\dots h_1(l_0+n)\ h_2(1)\dots h_2(p_0+m)\ o(t)]^T \tag{34}$$

and

$$x(t)=[w(t)\dots w(t-l_0-n)\ v(t-1)\dots v(t-p_0-m)\ 1]^T \tag{35}$$

thus

$$y(t)=x^T(t)\theta(t)+e(t) \tag{36}$$

Define the criterion function with fixed forgetting factor μ as follows [Goodwin and Payne (1977), Strejc (1980), Graupe and coworkers (1980), Landau (1990)]

$$J(t)=\sum_{i=1}^{t}\mu^{(t-i)}[y(i)-x^T(i)\hat{\theta}(t)]^2 \tag{37}$$

Then use the recursive least squares algorithm to compute the estimation of $\hat{\theta}$

$$\hat{\theta}(t+1)=\hat{\theta}(t)+K(t+1)[y(t+1)-x^T(t+1)\hat{\theta}(t)] \tag{38}$$

$$K(t+1)=\frac{P(t)x(t+1)}{\mu+x^T(t+1)P(t)x(t+1)} \tag{39}$$

$$P(t+1)=\frac{[I-K(t+1)x^T(t+1)]P(t)}{\mu} \tag{40}$$

where

$$P(0)=I/\delta \qquad 0<\delta<<1 \tag{41}$$

Since the sequence of $\{e(t)\}$ is unmeasurable in practice, the posterior prediction error (residual) $\hat{e}(t)$ will be used to replace the real error $e(t)$, here

$$\hat{e}(t)=y(t)-x^T(t)\hat{\theta}(t) \tag{42}$$

On the other hand, for an ill-conditioned matrix $P(t)$ and/or a matrix $P(t)$ with a large dimension, numerical stability problem can take place. Thus a popular factorization known as U-D factorization in which the matrix $P(t)$ is written as (Bierman (1977), Ljung and Söderström (1983))

$$P(t)=U(t)D(t)U^T(t) \tag{43}$$

is used to guarantee that $P(t)$ remains positive definite, and that the rounding errors do not affect the solution significantly.

ALGORITHM

In above sections, we have discussed the relationship between the parameters of the ARMAX-model of Eq. (25) or Eq. (26) and the values of $h_b(i)$ and $h_e(i)$ of the IIR expression in Eq. (28). In this section, we will establish the algorithm required in the case of Eq. (25).

Since only the variations of the output signal with respect to variations in the input signals are used for the estimation algorithm, the d.c. values y_{dc} and u_{dc} of the output and input signals must be cancelled before being used in the algorithm below. y_{dc} and u_{dc} can be estimated just by averaging [Isermann (1980)]

$$y_{dc}=\frac{1}{N_y}\sum_{k=1}^{N_y}y(k) \tag{44}$$

and

$$u_{dc}=\frac{1}{N_u}\sum_{k=1}^{N_u}u(k) \tag{45}$$

before the identification of the dynamics starts. Here, N_y and N_u are two sufficiently large positive integers.

1. Select p_i $i=0,\dots,m$ and l_i $i=0,\dots,n$, where $p_0 >>$ nc+na and $l_0 >>$ nb+na. Since the orders of the system are usually low (≤ 4) in practice, p_0 and l_0 will not be too large (usually, the larger the values of l_0, p_0, n and m, the better the identified results).

2. For every sample instant t, use Eqs. (20) – (23) to calculate $w(t-i)$ and use Eqs. (71) – (74) to calculate $v(t-i)$ under the assumptions that the data collection is initialized at time t=0.

After every estimate of the parameters $\theta(t)$, use Eq. (42) to calculate the posterior prediction error $\hat{e}(t)$ which will be used to replace the real error $e(t)$ in the next estimation period when calculating $v(t-i)$.

3. Use the recursive least squares algorithm Eqs. (38) – (40) with forgetting factor μ, combined with the U-D factorization algorithm, to compute the estimation of θ.

4. Use Eqs. (7) – (9) to calculate the parameters a_i $i = 1, \dots,$ na, b_i $i = 0, \dots,$ nb, or equivalently the parameters $\underline{b}_i$ $i = 0, \dots, \beta$ and the time delay d. And use Eq. (29) to calculate the parameters c_i $i = 1, \dots,$ nc.

Define

$$H_\tau=\begin{bmatrix} -\hat{h}_1(j-1) & \cdots & -\hat{h}_1(j-i) & \cdots & -\hat{h}_1(j-\tau+1) & \vdots & -\hat{h}_1(j-\tau) \\ -\hat{h}_1(j) & \cdots & -\hat{h}_1(j-i+1) & \cdots & -\hat{h}_1(j-\tau+2) & \vdots & -\hat{h}_1(j-\tau+1) \\ \vdots & & \vdots & & \vdots & \vdots & \vdots \\ -\hat{h}_1(l_0-2) & \cdots & -\hat{h}_1(l_0-i-1) & \cdots & -\hat{h}_1(l_0-\tau) & \vdots & -\hat{h}_1(l_0-\tau-1) \end{bmatrix}$$

$$=[H_{\tau-1}\vdots h_\tau] \qquad 1\le\tau\le l_0-j-1 \tag{46}$$

where j is a pre-selected number with $nb+1 \le j \le l_0-na-2$, and $\hat{h}_1(j-i) = 0$ if $j-i < 0$.

1). Calculate the determinants of the matrices $(H_\tau^T H_\tau)$ $\tau = 1, \dots, l_0-j-1$. According to the matrix theory, if $(H_\tau^T H_\tau)$ is a regular matrix, $\det(H_{\tau+1}^T H_{\tau+1})$ can be computed recursively, as follows

$$\det(H_{\tau+1}^T H_{\tau+1})=\det\left(\begin{bmatrix} H_\tau^T \\ \text{---} \\ h_{\tau+1}^T \end{bmatrix}[H_\tau\vdots h_{\tau+1}]\right)$$

$$=\det\left(\begin{bmatrix} H_\tau^T H_\tau & \vdots & H_\tau^T h_{\tau+1} \\ \text{---} & + & \text{---} \\ h_{\tau+1}^T H_\tau & \vdots & h_{\tau+1}^T h_{\tau+1} \end{bmatrix}\right)$$

$$=\det(H_\tau^T H_\tau)\det(h_{\tau+1}^T h_{\tau+1}-h_{\tau+1}^T H_\tau(H_\tau^T H_\tau)^{-1}H_\tau^T h_{\tau+1}) \tag{47}$$

Thus, according to Eq. (9), for $\tau = \tau 1 =$ na, since the rank of the matrix H_τ is equal to na, the inequality below is available, i.e.

$$|\det(H_{\tau+1}^T H_{\tau+1})| << |\det(H_\tau^T H_\tau)| \tag{48}$$

or equivalently,

$$|\det(H_{\tau1+1}^T H_{\tau1+1})| / |\det(H_{\tau1}^T H_{\tau1})| =$$

$$|\det(h_{\tau1+1}^T h_{\tau1+1}-h_{\tau1+1}^T H_{\tau1}(H_{\tau1}^T H_{\tau1})^{-1}H_{\tau1}^T h_{\tau1+1})| \le \phi << 1 \tag{49}$$

where ϕ is a given estimation and calculation error. Define

$$H_b=H_{\tau1} \tag{50}$$

$$h_b=[\hat{h}_1(j)\ \hat{h}_1(j+1) \dots \hat{h}_1(l_0-1)]^T \tag{51}$$

$$\hat{A}=[\hat{a}_1\ \hat{a}_2 \dots \hat{a}_{na}]^T \tag{52}$$

According to Eq. (9), we have

$$\hat{A}=(H_b^T H_b)^{-1} H_b^T h_b \tag{53}$$

2). Using Eqs. (7) and (8), we obtain

$$\begin{bmatrix} \hat{b}_0 \\ \hat{b}_1 \\ \vdots \\ \hat{b}_\alpha \end{bmatrix}=\begin{bmatrix} \hat{h}_1(0) & 0 & \cdots & 0 \\ \hat{h}_1(1) & \hat{h}_1(0) & \cdots & 0 \\ \vdots & \vdots & & \vdots \\ \hat{h}_1(\alpha) & \hat{h}_1(\alpha-1) & \cdots & \hat{h}_1(\alpha-na) \end{bmatrix}\begin{bmatrix} 1 \\ \hat{a}_1 \\ \vdots \\ \hat{a}_{na} \end{bmatrix} \tag{54}$$

where $\hat{h}_1(\alpha-j) = 0$ if $\alpha-j < 0$. Here, α is an appropriately large positive integer ($\geq nb+d$).

The time delay d can be simply estimated by using the following inequalities (Isermann, 1980). If

$$|\hat{b}_0|,\ |\hat{b}_1|, \ldots,\ |\hat{b}_{r-1}| \ll \sum_{i=0} \hat{b}_i \tag{55}$$

and

$$|\hat{b}_r| \gg |\hat{b}_{r-1}| \tag{56}$$

then

$$\hat{d} = r \tag{57}$$

By the same way, the order nb also can be estimated as follows. If

$$|\hat{b}_{k+1}|,\ |\hat{b}_{k+2}|, \ldots,\ |\hat{b}_{\alpha}| \ll \sum_{i=0}^{\alpha} \hat{b}_i \tag{58}$$

and

$$|\hat{b}_k| \gg |\hat{b}_{k+1}| \tag{59}$$

then

$$nb = k - \hat{d} \tag{60}$$

3). Compute the parameters ci i=1, ..., nc.
a). Using Eq. (29), we obtain

$$\begin{bmatrix} \hat{c}_1 \\ \hat{c}_2 \\ \vdots \\ \hat{c}_{na} \end{bmatrix} = \begin{bmatrix} \hat{h}_2(1) & 1 & 0 & \cdots & 0 \\ \hat{h}_2(2) & \hat{h}_2(1) & 1 & \cdots & 0 \\ \vdots & \vdots & \vdots & & \vdots \\ \hat{h}_2(na) & \hat{h}_2(na-1) & \hat{h}_2(na-2) & \cdots & 1 \end{bmatrix} \begin{bmatrix} 1 \\ \hat{a}_1 \\ \vdots \\ \hat{a}_{na} \end{bmatrix} \tag{61}$$

b). Calculate the zeros z_i i=1, ..., na of the polynomial

$$C^*(q) = q^{na} + \hat{c}_1 q^{na-1} + \ldots + \hat{c}_{na-1} q + \hat{c}_{na} \tag{62}$$

c). Since the zeros of $C(q^{-1})$ in Eq. (27) and those of $C^*(q)$ in Eq. (62) are each others' mirror images in the unit circle (i.e. if z_i is a zero of $C(q^{-1})$, then $1/z_i$ is a zero of $C^*(q)$), according to the spectral factorization theorem, $C(q^{-1})$ can always be chosen so that $C^*(q)$ has all its zeros inside or on the unit circle without affecting the second-order properties of the disturbance term. Thus, if the norm $\| z_i \| > 1$ for some i $\in$ [1 na], let $z_i^* = 1/z_i$, otherwise, let $z_i^* = z_i$. Then re-calculate the parameters c_i i=1, ..., na in Eq. (62) as follows,

$$\hat{c}_1 = (-1)\sum_{i=1}^{na} z_i^*,\ \hat{c}_2 = (-1)^2 \sum_{i=1}^{na} \sum_{j=i+1}^{na} z_i^* z_j^*, \ldots, \hat{c}_{na} = (-1)^{na} \prod_{i=1}^{na} z_i^* \tag{63}$$

d). If

$$|\hat{c}_{k+1}|,\ |\hat{c}_{k+2}|, \ldots,\ |\hat{c}_{na}| \ll \sum_{i=1}^{na} \hat{c}_i \tag{64}$$

and

$$|\hat{c}_k| \gg |\hat{c}_{k+1}| \tag{65}$$

then

$$nc = k \tag{66}$$

Finally, we have

$$C(q^{-1}) = 1 + \hat{c}_1 q^{-1} + \ldots + \hat{c}_{nc} q^{-nc} \tag{67}$$

In practice the orders of the system are usually less than or equal to 4, so that the evaluation of Eqs. (44) - (67) is not too difficult.

Notice that since it is impossible to obtain a single zero of a polynomial with real negative value by sampling a continuous time system, all such kind of zeros should be rejected in Step 4.

5. Repeat Steps 2 - 4 in the case of on-line identification. If $\hat{\sigma}(t)/y(t) > \gamma$, the given precision requirement, after certain cycles of calculation, it means that maybe either the original selection points for l_i i=0, ..., n and/or p_i i=1, ..., m must be adjusted, or there are offsets between the input and output signals, or the given precision requirement needs revision. Then restart the algorithm.

EXAMPLE

[Åström (1980)]. The data for the example was generated by the model of Eq. (25) with

$$A(q^{-1}) = 1 - 1.5\ q^{-1} + 0.7\ q^{-2}$$

$$B(q^{-1}) = q^{-1} + 0.5\ q^{-2}$$

$$C(q^{-1}) = 1 - 0.7\ q^{-1}$$

The disturbances $\{e(t)\}$ were gaussian with zero mean and standard deviation 0.5. The input signal u is a PRBS signal switching between -1 to 1 generated by a eleven-bit shift register with initial value 200.

In Step 1, the p_i and l_i are selected as those being listed in following Table 1.

Table 1 The Positions of the Points of $h_1(t)$ and $h_2(t)$

i	0	1	2	3	4	5	6	7
l_i	18	25	32	39	46	53	63	
p_i	15	22	29	36	43	50	57	67

In the recursive least squares algorithm Eqs.(38) – (40), the forgetting factor μ is selected as 0.98.

In Step 4, the dimension of matrix H_τ is selected as 8 x 5, i.e. j = 10 and τ = 5. The estimates of parameters in $A(q^{-1})$ and $B(q^{-1})$ are shown in Fig. 3 below. The parameters were calculated when the sample instant t is larger than 30.

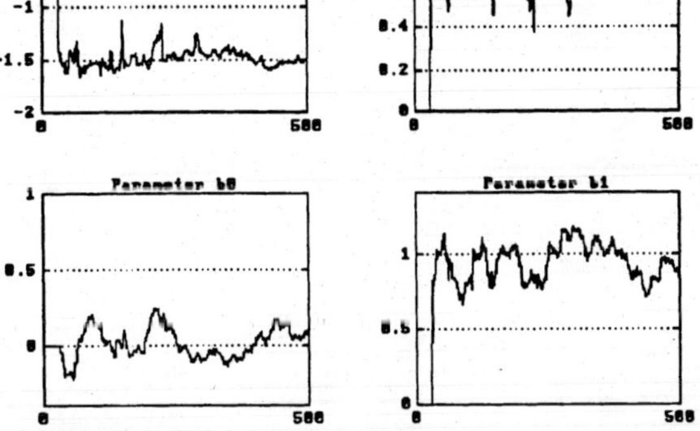

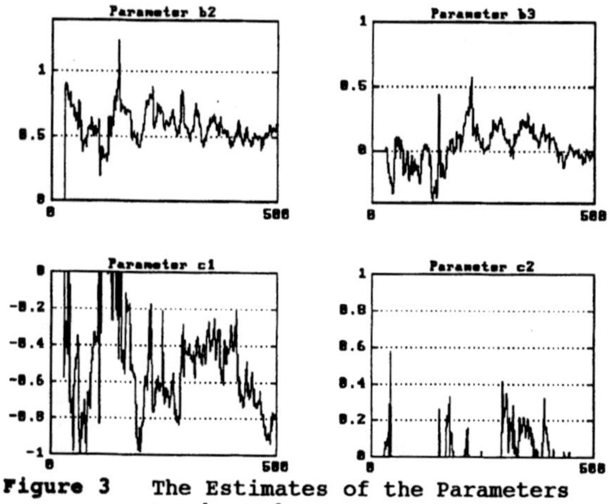

Figure 3 The Estimates of the Parameters a_i, b_i and c_i

In Fig. 3, the orders of na and nb are equal to 2, and the time delay is 1. The mean values and the standard deviations of the parameters are listed in Table 2.

Table 2 The Mean Values and the Standard Deviation of the Parameters a_i, b_i and c_i

	Mean	Deviation	Real Value
$\hat{a}_1$	-1.4837	0.0780	-1.5
$\hat{a}_2$	0.6901	0.0786	0.7
$\hat{b}_0$	0.0151	0.1012	0
$\hat{b}_1$	0.9418	0.1186	1
$\hat{b}_2$	0.5847	0.1288	0.5
$\hat{b}_3$	0.0320	0.1498	0
$\hat{c}_1$	-0.5183	0.2507	-0.7
$\hat{c}_2$	0.0397	0.0829	0

CONCLUSION

The CBL method calculates and estimates the parameters of the parametric model from the coefficients of the IIR model. It is not necessary to know much a priori knowledge about the system. This method can thus be applied in the case of on-line parameter identification, where the parameters of the system may vary slowly, or when the system is slightly nonlinear or noisy, so that it is necessary to update the parameters regularly.

APPENDIX

$$P_i = p_i - p_{i-1} \qquad i=1,\ldots,m \tag{68}$$

$$h_2(i) = h_e(i) \qquad i=1,\ldots,p_0 \tag{69}$$

$$h_2(p_0+i) = h_e(p_i) \qquad i=1,\ldots,m \tag{70}$$

$$v(t-i) = e(t-i) \qquad i=1,\ldots,p_0-1 \tag{71}$$

$$v(t-p_0) = e(t-p_0) + \sum_{k=1}^{P_1-1}\left(1-\frac{k}{P_1}\right)e(t-p_0-k) \tag{72}$$

$$v(t-p_0-i) = \sum_{k=1}^{P_i}\frac{k}{P_i}e(t-p_{i-1}-k) + \sum_{k=1}^{P_{i+1}-1}\left(1-\frac{k}{P_{i+1}}\right)e(t-p_i-k) \qquad i=1,\ldots,m-1 \tag{73}$$

$$v(t-p_0-m) = \sum_{k=1}^{P_m}\frac{k}{P_m}e(t-p_{m-1}-k) \tag{74}$$

$$o(t) = o_1(t) + \sum_{i=l_0+1}^{N_b} h_b(i)u(t-i) - \sum_{i=1}^{n}\sum_{k=1}^{L_i}\left[h_b(l_{i-1}) + k\frac{h_b(l_i)-h_b(l_{i-1})}{L_i}\right]u(t-l_{i-1}-k) + \sum_{i=p_0+1}^{N_e} h_e(i)e(t-i) - \sum_{i=1}^{m}\sum_{k=1}^{P_i}\left[h_e(p_{i-1}) + k\frac{h_e(p_i)-h_e(p_{i-1})}{P_i}\right]e(t-p_{i-1}-k) \tag{75}$$

REFERENCES

Åström, K. J. (1980). Maximum Likelihood and Prediction Error Method. Automatica 16, 551-574.

Bierman, G. J. (1977). Factorization Methods for Discrete Sequential Estimation. Academic Press, New York.

Eykhoff, P. (1974). System Identification : Parameter and State Estimation. Wiley, London.

Goodwin, G. C. and R. L. Payne (1977). Dynamic System Identification : Experiment Design and Data Analysis. Academic Press, New York.

Graupe, D., V. K. Jain and J. Salahi (1980). A Comparative Analysis of Various Least-Squares Identification Algorithms. Automatica 16, 663-681.

Gu X. Y. and Jin Y. Y. and Qu B. C. (1990). Identification Method for Parsimonious Impulse Response Model and Its Application to DMC. Northeast University of Technology, China. Internal Report.

Gustavsson, I. (1972). Comparison of Different Methods for Identification of Industrial Processes. Automatica 8, 127-142.

Isermann, R. (1980). Practical Aspects of Process Identification. Automatica 16, 575-587.

Landau, I. D. (1990). System Identification and Control Design Using P. I. M. + Software. Prentice-Hall, (INPG/CNRS), France.

Ljung, L. and T. Söderström (1983). Theory and Practice of Recursive Identification. MIT Press, Cambridge, Mass.

Rake, H. (1980). Step Response and Frequency Response Methods. Automatica 16, 519-526.

Scattolini, R. and S. Bittanti (1990). On the Choice of the Horizon in Long-range Predictive Control —— Some Simple Criteria. Atuomatica 26, 915-917.

Söderström, T. and P. Stoica (1989). System Identification. Prentice Hall, London.

Strejc, V. (1980). Least Squares Parameter Estimation. Automatica 16, 535-550.

ROBUST FAILURE DETECTION AND ISOLATION IN NON-LINEAR CONTROL SYSTEMS

Alexey Ye. Shumsky

Department of Radio Electronics, Far Eastern Polytechnical Institute, Vladivostok, Russia

Abstract. Using the concept of redundancy relations a solution is given to the problem of robust failure detection and isolation in non-linear discrete-time control systems. To obtain robust redundancy relations the method of non-linear transformation of system and interval analysis is used. The method for optimal choosing redundancy relations is developed.

Keywords. Failure detection; non-linear systems; robustness; non-linear transformations; optimization; interval analysis.

INTRODUCTION

An active approach to the design of fault tolerant control systems is to first detect and isolate a failed component and next reconfigurate the control system to accommodate the failure, hence, an integral part of such a system is failure detection and isolation (FDI). A wide variety of techniques has been proposed in recent years to solve FDI problem (see, for example, surveys by Iserman, 1984, and Frank, 1990). In one way or another, the procedure of FDI essentially consists of two stages. The first stage is the generation of signals, called residuals, whose values are nominally zero or close to zero when no failure in system is presented and must be distinguishably different from zero under failure conditions. The second stage involves using residuals to make the appropriate decisions. The procedures for residual generation depend in turn on models relating the measured variables and can use full or reduced-order observers, detection filters, Kalman filters or follow from the unified so-called parity space approach. As model errors and noise disturbance also influence on the residuals it leads in the issue to robust FDI, i.e. the design of diagnostic procedure that is maximally sensitive to the effects of failures and minimally sensitive to model errors and noise disturbance. The well known solutions of such a problem primarily concern the stage of residual generation and deal with linear systems. In Chow and Willsky (1984) the design of a robust residual generation process was formulated as a minimax optimization problem. Proposed criterion specifies robustness with respect to a particular operating point, thereby allowing the possibility of adaptive choosing the best redundancy relations. However, a drawback of this approach is that it leads to an extremely complex on-line optimization problem. In Low and others (1986) an alternative measure of optimum for redundancy relations, leading to far simpler optimization procedure involving a single singular value decomposition, was developed. The criterion taking into account relative contribution of the effects of disturbances due to the model errors, noise and the failures was described in Frank (1990). It is different the criterion used in Lou and others (1986) which does not guarantee independence of the residual of each system input. The optimization problem, following from this criterion, reduces to a generalized eigenvalue-eigenvector problem. Solutions of robust residual generation problem described in Low and others (1986) and in Frank (1990) may be considered only as approximation to optimal due to their independence of operating points. In this work an optimal solution is achieved by using non-linear redundancy relations (even if initial system is linear) which are adaptive to behavior of inputs of system. As rule, realization of such solution does not need a complex on-line optimization procedure.

PROBLEM FORMULATION

Let system subjected to diagnosis is described by discrete-time model

$$x(t+1)=f(x(t),u(t),p)$$
$$y(t)=h(x(t),p) \qquad (1)$$

where $u\in R^l$ is the input vector, $x\in R^n$ is the directly unobservable state vector, $y\in R^s$ is the output vector, f and h in common case are non-linear vector functions. During the normal operation the vector of uncertain parameters p takes values in a specified subset P of R^m. Failures result in moving parameters from their tolerances. It is supposed that measured vectors of input and output are

$$\hat{u}(t)=u(t)+\zeta(t), \quad \hat{y}(t)=y(t)+\eta(t) \quad (2)$$

where $\zeta(t)$ and $\eta(t)$ are zero-mean Gaussian noise vectors with famous covariance matrices. Under this conditions the spectrum of the next problems is developed
(1) defining robust redundancy relations for noise-free conditions;
(2) optimal choosing redundancy relations which have selective sensitivity to distortions of definite parameters;
(3) expansion of robust redundancy relations for noise conditions.
Because to obtain robust redundancy relations for non-linear systems the special nonequivalent non-linear transformation of initial model is required the next questions are discussed in addition to ones mentioned above
(4) special mathematical techniques;
(5) special kind of nonequivalent non-linear transformation of model (1).

ROBUST REDUNDANCY RELATIONS

In order to develop a clear picture of using redundancy for residual generation in non-linear systems, consider the case of following deterministic model

$$x_1(t+1)=f_1(u(t),y(t),p),$$

$$x_i(t+1)=f_i(u(t),y(t),x_1(t),\ldots x_{i-1}(t),p) \quad (3)$$

$$i=2,3,\ldots k$$

$$y_*(t)=h_*(x_k(t),p) \quad (4)$$

where x_i one may consider as the subvector of x, $y_*(t)=y(t)$. The simple way of residual generation for some nominal value of p is the next

$$r(t)=y(t)-y_*(t)$$

according which residual $r(t)$ is generated as the difference between output of system $y(t)$ and its estimation $y_*(t)$ obtained by open-loop generation based on the model (3), (4). Due to the structure of this model behavior of the residual is independent of initial state of system for all $t\geq k$. To obtain such property for common case of model (1) it must be transformed to the form (3), (4). Because the equivalent (i.e. keeping the input-output map) transformation of model (1) to model (3), (4) in non-linear case may be impossible, we assume vectors $x_1,\ldots,x_k$, y_* depend on vectors x and y accordingly

$$x_i(t)=\phi_i(x(t)) \quad (5)$$

$$y_*(t)=\psi(y(t)) \quad (6)$$

where ϕ_i, $i=1,2,\ldots k$, and ψ are vector functions. As result the rule of residual generation is modified as follows

$$r(t)=\psi(y(t))-y_*(t) \quad (7)$$

Clearly, during the normal operation of system when value of vector p is equal to nominal the right-hand side of relation (7) is equal to zero. With the exception of x_i, $i=1,2,\ldots,k$, from (3), (4) we turn to input-output relation for estimation $y_*(t)$ in the next form

$$y_*(t)=F(Y^k(t-1),U^k(t-1),p)$$

where $Y^k(t-1)=[y'(t-1),\ y'(t-2),\ldots,\ y'(t-k)]'$, $U^k(t-1)=[u'(t-1),u'(t-2),\ldots,u'(t-k)]'$ and mark ' denote the transposition.

Let now $p\in P$ and real vector functions $F_{min}(Y^k(t-1),\ U^k(t-1))$ and $F_{max}(Y^k(t-1),\ U^k(t-1))$ be the lower and upper boundary vector functions for interval vector function $F(Y^k(t-1),U^k(t-1),p)$, i.e. functions $F_{min}(Y^k(t-1),U^k(t-1))$ and $F_{max}(Y^k(t-1),U^k(t-1))$ give the best approximation for all functions $F(Y^k(t-1),U^k(t-1),p)$, $p\in P$, such that $F_{min}(Y^k(t-1),U^k(t-1))\leq F(U^k(t-1),Y^k(t-1),p))\leq F_{max}(Y^k(t-1),U^k(t-1))$. Therefore, during the normal operation of system (under $p\in P$) $F_{min}(Y^k(t-1),\ U^k(t-1))\leq \leq\psi(y(t))\leq F_{max}(Y^k(t-1),U^k(t-1))$. The last inequality leads to the pair of robust redundancy relations in the form of inequalities

$$r_{max}(t)=\psi^*(t)-F_{min}(Y^k(t-1),U^k(t-1))\geq 0$$

$$r_{min}(t)=\psi^*(t)-F_{max}(Y^k(t-1),U^*(t-1))\leq 0$$

where function $\psi^*(t)$ is determined as follows $\psi^*(t)=\psi(y(t))$.
Checking these relations composes the maintenance of robust diagnostic procedure. Violation of one of these relations results from the moving at least one of the parameters from its tolerance. Functions $\psi^*(t)-F_{min}(Y^k(t-1),U^k(t-1))$, $\psi^*(t)-F_{max}(Y^k(t-1),U^k(t-1))$ we call upper redundancy function (URF) and lower redundancy function (LRF) accordingly. To resume mentioned above we formulate robust diagnostic procedure for noise-free conditions as one of calculating URF and LRF of measured inputs and outputs taken over a finite time window and then comparison obtained values with zero threshold. The stage of transition from interval function to lower and upper boundary functions is essential for obtaining LRF and UHF . To get such functions for concrete interval function the well known procedures may be used . The most simple solution of this problem one can have when considered interval function is polynomial one.

MATHEMATICAL TECHNIQUES

Now we give the brief introduction into special mathematical techniques, used for determination of functions ϕ_i, ψ (Zhirabok, Shumsky 1987, 1989). Let G denote a set of vector functions on R^n. A pair of vector functions $\alpha(x)$, $\beta(x)\in G$ is said to be connected by partial order relation $\leq$ ($\alpha\leq\beta$) iff there exists vector function γ determined on the set of

values of function α such that $\beta=\gamma\alpha$. To examine the fulfillment of this relation the rank criterion of functional dependence may be used. If $\alpha\leq\beta$ and $\beta\leq\alpha$ then α and β are equivalent, denoted $\alpha\cong\beta$, because relation $\cong$ is reflexive, symmetric and transitive. The equivalent transformations of vector functions may be used for their simplification.
Let binary operation $\times$ denote

$$\alpha \times \beta \triangleq \begin{bmatrix} \alpha \\ \beta \end{bmatrix}$$

The binary operation $\square$ we introduce as follows:

$\alpha\square\beta\in G$ and if $\alpha\leq\gamma$, $\beta\leq\gamma$, then $\alpha\square\beta\leq\gamma$.

It follows from definition of partial order relation that expelling functional dependent components is the equivalent transformation for vector functions. As result to find the vector function $\gamma\cong\alpha\square\beta$ it is sufficient to determine full set of its functional independent components γ_j such the next equalities are fulfiled

$$\text{Rank}\begin{bmatrix} \partial\alpha/\partial x \\ \partial\gamma_j/\partial x \end{bmatrix}=\text{Rank}\begin{bmatrix} \partial\alpha/\partial x \end{bmatrix}$$

$$\text{Rank}\begin{bmatrix} \partial\beta/\partial x \\ \partial\gamma_j/\partial x \end{bmatrix}=\text{Rank}\begin{bmatrix} \partial\beta/\partial x \end{bmatrix}$$

where $\partial\alpha/\partial x$, $\partial\beta/\partial x$ and $\partial\gamma_j/\partial x$ are the Jacobian matrices. These equalities leads to partial differential equations concerning component γ_j. Its integration in common case may be difficult. The essential simplification of this problem may be achieved by restrictions which are put on set G. Thus, in linear case we have

$$\alpha \square \beta = A\,x$$

where A is the matrix with maximum rank whose each row is linear dependent on rows as both Jacobian matrix $d\alpha/dx$ and Jacobian matrix $d\beta/dx$.
We define the binary relation Δ, operators M and m as follows

$$(\alpha,\beta)\in\Delta \quad \text{iff} \quad \Pi_u \times \alpha\Pi_x \leq\beta f$$

$$(M(\beta),\beta)\in\Delta \text{ and if } (\alpha,\beta)\in\Delta \text{ then } \alpha\leq M(\beta)$$

$$(\alpha,\ m(\alpha))\in\Delta \text{ and if } (\alpha,\beta)\in\Delta \text{ then } m(\alpha)\leq\beta$$

where Π_u, $\Pi_u(x,u)=u$, and Π_x, $\Pi_x(x,u)=x$, are projections, f is the vector function from model (1).

Theorem 1: For arbitrary functions $\alpha,\beta\in G$ it is true:

(1) $\alpha\times\beta\leq\alpha$, $\alpha\times\beta\leq\beta$

(2) if $\alpha\leq\beta$ then $\alpha\times\beta\cong\alpha$

(3) if $\alpha\leq\beta$ then $m(\alpha)\leq m(\beta)$

(4) if $\alpha\leq\beta$ then $M(\alpha)\leq M(\beta)$

(5) $M(\alpha\times\beta)\cong M(\alpha)\times M(\beta)$

(6) $\alpha\leq M[m(\alpha)]$

Proof: see Zhirabok and Shumsky (1987).
Let us consider the common way of operator M calculation (Zhirabok and Shumsky, 1989). Let the function βf be transformed to the next form

$$\beta f(x,u,p)=g(a_0(x),a_1(x),\ldots,a_s(x),u,p)$$

where a_0, $a_1,\ldots,a_s$ are arbitrary vector functions, for which there are such values of input $u=c_1$, $u=c_2,\ldots,u=c_r$ that the next conditions are fulfilled

$$\beta f(x,c_1,p)\times\beta f(x,c_2,p)\times\ldots\times\beta f(x,c_r,p)\leq a_i$$
$$i=1,2,\ldots,s.$$

Then $M(\beta)\cong a_0\times a_1\times\ldots\times a_s$. Since the examination of last conditions may be difficult, the next rule of operator M calculation can be used in series of cases.

Theorem 2: Let the function $\beta\in G$ have one component and the function $\beta f(x,u,p)$ be written as

$$\beta f(x,u,p)=a_0(x,p)+b_0(u)+\sum_{j=1}^{s} a_j(x,p)b_j(u)$$

where the system of functions 1, $b_1(u),\ldots,b_s(u)$ is linear independent. Then $M(\beta)\cong a_0\times a_1\times\ldots\times a_s$.
Proof: see Zhirabok and Shumsky (1987).

By using the property (5) of operator M the result of last theorem may be spread for a multicomponent case.
The next theorem gives the rule for calculating operator m.

Theorem 3: Let $(\beta\Pi_x\times\Pi_u)\square f\cong\alpha f$.
Then $m(\beta)\cong\alpha$

TRANSFORMATION OF INITIAL MODEL

Let us find the necessary and sufficient conditions in which nonequivalent non-linear transformation of model (1) to model (3),(4) exists. Substituting right-hand side of relation (3) to left-hand side of (5) and right-hand side of the first relation (1) to right-hand side of (5) and replacing vector y with function h (according second relation (1)) we come to equations

$$\begin{aligned} &f_1(h(x),u,p)=\phi_1 f(x,u,p) \\ &f_i(h(x),u,\phi_1(x),\ldots,\phi_{i-1}(x),p)= \\ &=\phi_i f(x,u,p), \quad i=1,2,\ldots,k \end{aligned} \tag{8}$$

From (1),(4) and (6) we get

$$h_*(\phi_k(x),p)=\phi\, h(x\ ,p) \tag{9}$$

To expel unknown functions f_i, $i=1,2,\ldots,k$, let us puss from (8),(9) to the functional inequalities concerning vector functions ϕ_i, ϕ. Using the definition of partial order relation for vector functions, operation $\times$ and operator m we obtain from (8) and(9) respectively

$$m(h)\leq\phi_1, \quad m(h\times\phi_1\times\phi_2\times\ldots\times\phi_{i-1})\leq\phi_i \tag{10}$$
$$i=2,3,\ldots,k$$

$$\phi_k\leq\phi h \tag{11}$$

Taking into account the property (1) of operation $\times$ and property (3) of operator m, we obtain from (10) $\phi_k \le \phi_{k-1} \le \dots \le \phi_1$ At last, referring to the property (2) of operation $\times$, we come from (10) to the next inequalties

$$m(h) \le \phi_1,\ m(h \times \phi_{i-1}),\quad i=2,\dots,k \quad (12)$$

Thus, fulfilling the functional inequalities (12),(11) is the necessary condition for the existence of model (1) transformation. The sufficiency of this condition follows from the equivalence of (10), (11) and (8), (9) and possibility to pass from last equations to (3), (4) by replacing functions h, ϕ_i with corresponding vectors y, x_i. Among possible transformations of (1) the most interested is such that allows to get the minimum vector function ψ, denoted as ψ_o. The components of ψ_o make form the basis for all functions ψ, satisfying inequality (11). To determine function ψ_o, the next algorithm, based on resolution of (11), (12) is proposed

<u>Algorithm</u> 1

1. Take i:=2, $\phi_{1,o} := m(h)$.
2. Calculate $\phi_{i,o} := m(\phi_{i-1,o} \times h)$.
3. If function $\phi_{i,o}$ is not equal to $\phi_{i-1,o}$, then increment i and go to the step 2, otherwise go to the next step.
4. Determine function ψ_o as a minimum vector function, satisfying inequality $\phi_{i,o} \le \psi_o h$. End.

The number of iterations, required for fulfillment of Algorithm 1, can not exceed the number of functional independent components of function $\phi_{i,o}$, i.e n. To obtain redundancy relations one may choose any scalar function ψ as a function depended on basis functions. To obtain functions f_i, i=1,2,...,k, h_* it is necessary to solve equations (8),(9) under famous function ψ. But to solve these equations functions ϕ_i, i=1,2,...,k, must be determined first from the next algorithm.

<u>Algorithm</u> 2.

1. Take j:=1, k:= number of iterations, required for fulfillment of Algorithm 1, $\psi_1(X)$:= vector function with minimum number of components such that, firstly, each of its components is depended on components of $\phi_{k,o}$ and, secondly, function ψh depends on function ψ_1.
2. Calculate function $M(\psi_j)$.
3. If function $M(\psi_j)$ depends on h, take k:=i and go to the step 5, otherwise to the next step.
4. Determine vector function ψ_{j+1} with minimum number of components such that, firstly, each of its components is depending on components of $\phi_{k-j,o}$ and, secondly, depends on functions ψ_{j+1} and h. Take j:=j+1 and go to the step 2.
5. For each j=k,k-1,...,1 take $\phi_{k-j+1} := \psi_{j*}$, where ψ_{j*} is obtained from ψ_j by excepting those components, which are depending on components of functions $\psi_k, \psi_{k-1}, \dots, \psi_{j+1}$. End.

The procedure of determining functions ψ_j is the base of Algorithm 2 (step 1 and step 2). The possibility to determine them for each j follows immediately from properties (4) and (6) of operators. The convergence of Algorithm 2 follows from the final value of k.

OPTIMAL REDUNDANCY RELATIONS

We have seen earlier, one may take scalar function ψ as a function of basis functions and to obtain the pair of robust redundancy relations. The useful expansion of the set of possible redundancy relations may be achieved if to construct function ψ as a function of basis functions taken in different instants of time $\psi_o(t)$, $\psi_o(t-1)$, ..., $\psi_o(t-m)$ i.e $\psi(t)=\Phi(\psi_o(t),\psi_o(t-1),\dots,\psi_o(t-m)))$. In this case corresponding URF and LRF may be determined under interval function

$$F(Y^k(t-1),U^k(t-1),p) = \Phi(F_1(Y^{k1}(t-1),U^{k1}(t-1),p),\dots,F_s(Y^{ks}(t-1),U^{ks}(t-1),p),\dots,F_1(Y^{k1}(t-m),U^{k1}(t-m),p),\dots,F_s(Y^{ks}(t-m),U^{ks}(t-m),p))$$

where $F_j(Y^{kj}(t-\tau),U^{kj}(t-\tau),p)$ is the interval function that corresponds to basis function $\psi_{oj}(y(t-\tau+1))$, $\tau=1,\dots,m$, $j=1,\dots,s$, $k=\max\{k_1,k_2,\dots,k_s\}+m$.

Clearly, there are many candidate redundancy relations for a given system. The maximum sensitivity of redundancy relation to distortions of definite parameter we consider as the base for choosing concrete version. The minimum length ω_i of interval such that the moving parameter from its range under other parameters in their tolerances results in violation of redundancy relations we use as the criterion to characterize the sensitivity of redundancy relation to parameter p_i. Let $\Gamma_j=[\gamma_{jmin},\gamma_{jmax}]$ be the tolerance for parameter p_j, j=1,2,...,m. The next theorem gives the rule for determining ω_i as a function of $U^k(t-1)$, $Y^k(t-1)$.

<u>Theorem</u> 4: Let the next conditions are fulfilled

(1) $\gamma^*_{i\,min} = \gamma^*_{min}(Y^k(t-1),U^k(t-1))$ and $\gamma_i^*{}_{max} = \gamma_i^*{}_{max}(Y^k(t-1),U^k(t-1))$ are the minimum and maximum solutions of equations

$$F(Y^k(t-1),U^k(t-1),p)-F_{min}(Y^k(t-1),U^k(t-1))=0$$
$$F(Y^k(t-1),U^k(t-1),p)-F_{max}(Y^k(t-1),U^k(t-1))=0 \qquad (13)$$

concerning parameter p_i under $p_j \in \Gamma_j$, $i \neq j$;
(2) function $F(Y^k(t-1),U^k(t-1),p)$ is the monotonous one for p_i under $p_j \in \Gamma_j$, $i \neq j$, $p_i \in [\gamma^*_{imin}, \gamma_{imin}]$, $p_i \in [\gamma_{i\,max}, \gamma^*_{i\,max}]$ and in the neighborhood of $\gamma^*_{i\,min}$, $\gamma^*_{i\,max}$.

Then $\omega_i = \gamma^*_{i\,max} - \gamma^*_{i\,min}$.

The proof of this theorem is omitted due to the limited volume of the paper.

To reduce problem of choosing redundancy relations to optimization one it is proposed to use the next structure of function $\psi^*(t)$

$$\psi^*(t)=\sum_j \sum_\tau \alpha_{j\tau}(t)\psi_{oj}(y(t-\tau+1)) \qquad (14)$$

under coefficients $\alpha_{j\tau}(t)$ determined from demand on minimum of ω_i at each time step. As result we come to the following minimax problem

$$\min_{\alpha_{j\tau}} \max_{p_j \in \Gamma_j} \omega_i(U^k(t-1),Y^k(t-1),\alpha_{11},\dots,\alpha_{sm},p)$$

Since it has a trivial solution ($\alpha_{j\tau}=0$), without loss of generality we can restrict $\alpha_{j\tau}$

$$\sum_j \sum_\tau \alpha^2_{j\tau}(t)=1$$

If equations (13) admit a closed form solution one may obtain relations for calculating optimal coefficients $\alpha_{j\tau}$ at the stage of synthesis of diagnostic procedure in the form of functions

$$\alpha_{j\tau}(t)=\alpha_{j\tau}(Y^k(t-1),U^k(t-1)) \qquad (15)$$

In this case optimal function $\psi^*(t)$ is obtained by substituting (15) to (14). Otherwise, determining optimal coefficients needs on-line solution of optimization problem and may lead to extensive computation under diagnostic process.
Thus, to develop the optimal robust diagnostic procedure one can carry out the next actions
(1) to obtain function ψ_o (from algorithm 1) and to realize transformation of initial model (using algorithm 2 and relations (8),(9)) for each basis function;
(2) to get input-output relations for estimations y_{*i}, corresponded to basis functions $\psi_{oi}(y(t))$, i=1,2,...,m, in the form of interval functions;
(3) to find optimal functions $\psi^*(t)$ by solving minimax problem for some fixed index m and for each parameter of system and then to obtain corresponding URF and LRF.

DIAGNOSIS UNDER NOISE CONDITIONS

Let measured input and output vectors satisfy (2). Substituting expressions for vectors u(t), y(t) to interval function $F(Y^k(t-1),U^k(t-1),p)$ one obtain function

$$F(\hat{Y}^k(t-1),\hat{U}^k(t-1),\zeta(t-1),\eta(t-1),p)$$

depended on measured vectors $\hat{Y}^k(t-1)=[\hat{y}'(t-1),\hat{y}'(t-2),\dots,\hat{y}'(t-k)]'$, $\hat{U}^k(t-1)=[\hat{u}'(t-1),\hat{u}'(t-2),\dots,\hat{u}'(t-k)]'$ and noise vectors $\zeta^k(t-1)=[\zeta'(t-1),\zeta'(t-2),\dots,\zeta'(t-k)]'$, $\eta^k(t-1)=[\eta'(t-1),\eta'(t-2),\dots,\eta'(t-k)]'$. Its linearization concerning vectors $\zeta^k(t-1)$, $\eta^k(t-1)$ about zero values of these vectors gives

$$F(\hat{U}^k(t-1),\hat{Y}^k(t-1),\zeta^k(t-1),\eta^k(t-1),p)\approx$$
$$\approx F(\hat{U}^k(t-1),\hat{Y}^k(t-1),p)+$$
$$+D(\hat{U}^k(t-1),\hat{Y}^k(t-1),p)[\zeta^k(t-1),\eta^k(t-1)]'$$

where
$$D(\hat{U}^k(t-1),\hat{Y}^k(t-1),p)=\frac{dF(\hat{U}^k(t-1),\hat{Y}^k(t-1),\zeta^k(t-1),\eta^k(t-1),p)}{d(\zeta^k(t-1),\eta^k(t-1))}$$
under $\zeta^k(t-1)=0$, $\eta^k(t-1)=0$.
Also, function $\phi^*(t)$ may be perform as a function of measured vectors $\hat{y}(t)$, $\hat{y}(t-1),\dots,\hat{y}(t-m)$ and noise vectors $\eta(t)$, $\eta(t-1),\dots,\eta(t-m)$. Its linearization about zero values of $\eta(t)$, $\eta(t-1),\dots,\eta(t-m)$ gives

$$\psi(\hat{Y}^m(t-1),\hat{y}(t),\eta^m(t-1),\eta(t))\approx$$
$$\approx\psi(\hat{Y}^m(t-1),\hat{y}(t))+$$
$$+\Phi(\hat{Y}^m(t-1),\hat{y}(t))[\eta^{m\prime}(t-1),\eta'(t)]'$$

where
$$\Phi(\hat{Y}^m(t-1),\hat{y}(t-1))=\frac{d\,\psi(\hat{Y}^m(t-1),\hat{y}(t),\eta^m(t-1),\eta(t))}{d\,(\eta^m(t-1),\eta(t))}$$
under $\eta^m(t-1)=0$, $\eta(t)=0$.

Define
$$T_\beta(t)=\max\{T_\beta(t,p),\ p\in P\}$$
where positive real number $T_\beta(t,p)$ determines the boundary of interval $[-T_\beta(t,p),T_\beta(t,p)]$ such that the value of random function
$$\Omega(\hat{U}^k(t-1),\hat{Y}^k(t-1),\hat{y}(t),\zeta^k(t-1),\eta^k(t-1),\eta(t),p)=$$
$$=D(\hat{Y}^k(t-1),\hat{U}^k(t-1),p)[\zeta^k(t-1),\eta^k(t-1)]'+$$
$$+\Phi(\hat{Y}^m(t-1),\hat{y}(t))[\eta^{m\prime}(t-1),\eta'(t)]',\ k\geq m,$$
belongs to this interval with probability β.
Under low noise conditions when linearization gives a good approximation of functions $F(\hat{Y}^k(t-1),\hat{U}^k(t-1),\zeta^k(t-1),\eta^k(t-1),p)$ and $\psi(\hat{Y}^m(t-1),\hat{y}(t),\eta^m(t-1),\eta(t))$ the next inequality will be fulfilled for all $p\in P$ with probability not less β

$$F_{min}(\hat{Y}^k(t-1),\hat{U}^k(t-1))-T_\beta(t)\le$$
$$\le\psi^*(t)\le$$
$$\le F_{max}(\hat{Y}^k(t-1),\hat{U}^k(t-1))+T_\beta(t)$$

It gives the robust redundancy relations in the form

$$\Gamma_{max}(t)=\psi^*(t)-F_{min}(\hat{Y}^k(t-1),\hat{U}^k(t-1))\ge \ge T_\beta(t)$$
$$\Gamma_{min}(t)=\psi^*(t)-F_{max}(\hat{Y}^k(t-1),\hat{U}^k(t-1))\le \le -T_\beta(t) \qquad (16)$$

Thus, robust diagnostic procedure for noise conditions consists in calculating URF and LRF and comparison obtained values with nonzero thresholds $T_\beta(t)$ and $-T_\beta(t)$ correspondingly.

Calculating threshold $T_\beta(t)$ at each time step one may realize by variance

$$\sigma^2(t)=\max\{\sigma^2(t,p),\ p\in P\}$$

where $\sigma^2(t,p)$ is the variance of variate $\Omega(\hat{Y}^k(t-1),\hat{U}^k(t-1),\hat{y}(t),\zeta^k(t-1),\eta^k(t-1),\eta(t),p)$ determined by the next relation

$$\sigma^2(t,P)=\sum_{i,j}\frac{d\Omega(\hat{Y}^k(t-1),\hat{U}^k(t-1),\rho^k(t),\hat{y}(t),p)}{d\rho_i^k(t)}\cdot\frac{d\Omega(\hat{Y}^k(t-1),\hat{U}^k(t-1),\rho^k(t),\hat{y}(t),p)}{d\rho_j^k(t)}K_{ij}$$

where the noise vector $\rho^k(t)=[\eta'(t),\eta_k'(t-1),\dots,\eta'(t-k),\zeta'(t-1),\dots,\zeta'(t-k)]$, $\rho_i^k(t)$ is the corresponding component of vector $\rho^k(t)$, K_{ij} is the element of covariance matrix for noise vector $\rho^k(t)$. Other words, the expression for calculating $\sigma^2(t)$ may be obtain in the form of upper boundary function for interval function $\sigma^2(t,p)$. Therefore one may calculate the rating of the variance $\sigma^2(t)$ at each time step by the measured vectors $\hat{U}^k(t-1)$, $\hat{Y}^k(t-1)$, $\hat{y}(t)$ under famous covariance matrix.

To get optimal redundancy relations under noise conditions one must take into account the influence of threshold on ω_i. To determine optimal coefficients $\alpha_{j\tau}(t)$ in this case the iterative procedure is proposed. At the first its step the determining coefficients $\alpha_{j\tau}(t)$ is fulfilled by resolving minimax problem under noise-free conditions. At the each next step (1) the new value of threshold is calculated under values of coefficients $\alpha_{j\tau}(t)$ determined at the preceding step and (2) the new values of coefficients are obtained under equations (13) whose right-hand sides are corrected by substituting thresholds $T_\beta(t)$ and $-T_\beta(t)$ correspondingly. Procedure is finished if value of ω_i determined at the some step is not less then one obtained at the preceding step. Because such optimization procedure must be realized on-line, it leads to increasing necessary computations under diagnostic process.

CONCLUSION

This work has developed methods for determining robust redundancy relations for FDI in non-linear dynamic systems. Two cases of noise-free and low noise conditions were considered. In both cases diagnostic procedure includes calculating non-linear redundancy functions of measured inputs and outputs of system taken over a finite time window and threshold testing obtained values. Under noise-free conditions threshold is equal to zero. Under low noise conditions threshold is determined at each time step by variance that may be calculated analogous to redundancy relations as the function of measured inputs and outputs. Method guarantees the fulfilling redundancy relations (with set probability for low noise case) during normal operation of system. Calculation of covariance of noise vectors in the expression for variance allows to make diagnostic procedure adaptive to alternation of noise characteristics. The optimization procedure was developed to obtain redundancy relations with selective (maximally high) sensitivity to definite failures. It allows to organize not only detection but isolation of failures in system too. Using optimal redundancy relations for noise case needs in resolving on-line optimization problem. Without significant loosing optimality one may overcome this drawback by using redundancy relations which are optimal for noise-free case.

REFERENCES

Chow, E.Y. and A.S. Willsky. (1984). Analytical redundancy and the design of robust failure detection systems. IEEE Trans.Aut.Control. AC-29, 603-614.

Frank, P.M. (1990). Fault diagnosis in dynamic systems using analytical and knowledge-based redundancy - a survey and some new results. Automatica, 26, 453-474.

Isermann, R. (1984). Process fault detection based on modeling and estimation methods - a survey. Automatica, 20, 387-404.

Lou, X.C., A.S. Willsky and G.L. Verghese. (1986). Optimally robust redundancy relations for failure detection in uncertain systems. Automatica, 22 , 333-344.

Zhirabok, A.N. and A.Ye. Shumsky. (1987) Functional diagnosis of continuous dynamic systems described by equations whose right-hand side is polynomial. Automation and remote control, No 8, p.p. 154-164.

Zhirabok, A.N. and A.Ye. Shumsky. (1989) Functional diagnosis of nonstationary dynamic systems. Automation and remote control, No11, p.p. 146-154.

A NEW DYNAMIC NEUROCONTROL ARCHITECTURE FOR ROBOT MANIPULATORS

Young Joo Moon and Se Young Oh

Department of Electrical Engineering
Pohang Institute of Science and Technology
Pohang, Korea 790-600

Abstract Neural network control has many innovative potentials for fast, accurate, and intelligent adaptive control. A novel neurocontrol architecture is introduced in which an inverse dynamic neural controller in parallel with a linear neurocontroller cooperates to learn through training to control a robot manipulator accurately under different movement speeds and payloads. Compared with the previous type of neurocontrollers using a fixed PD gain controller, the proposed architecture not only eliminates the painful process of gain tuning but also exhibits a superior control performance because the linear controller constantly adapts its gains while moving. Experimental verification of the controller was carried out for the dynamic control of a PUMA arm demonstrating its superior dynamic control performances compared with a conventional PD or another neurocontrol architecture.
Keywords neural network, neurocontrol, robot manipulators, dynamics, inverse dynamics, system identification, control architecture, linear neurocontroller

INTRODUCTION

Dynamic control of robots is essential to maintain good dynamic performances over a wide range of operating conditions, such as different speeds and payloads. For dynamic control, joint torque computation based on the robot's inverse dynamic model is needed. However, problems with computational dynamics include the difficulty of obtaining a good dynamic model as well as an intense computational burden and hinder its practical use for industrial robots.
Recently, neurocontrol (neural network control) is receiving a great deal of attention due to the following pragmatic advantages:

(Due to the massive parallelism of neural networks)
1) Real-Time Control
2) Fault-Tolerance
3) Robustness Against Noise

(Due to the learning capability of neural networks)
1) No need for an exact model of the plant nor of its environment
2) Automatic compensation of unmodeled effects
3) Adaptation to changing environment
4) Constant performance improvement with learning
5) Easy handling of sensor data fusion
6) No need for knowledge on difficult control laws
7) Simple programming

Various neurocontrol architectures have been developed in [1,2,3,4]. Most of these use a linear controller in parallel with the neural network to provide initial learning data as well as to supplement the feedforward neural controller after learning. In general, choosing a good gain set for the linear controller is not trivial but essential for the stable learning of a given trajectory.
In this paper, a novel neurocontrol architecture has been developed where a single layer perceptron replaces the fixed-gain linear controller. This behaves like a variable-gain linear controller where the optimal gain set is learned through training. Two ways of training the neural net for control is developed here. The first method does not use any plant identifier while the second one uses a neural identifier.

THE PROPOSED NEUROCONTROL ARCHITECTURE

Figure 1 shows the proposed control architecture. The multilayer perceptron neural network provides the dynamic torque required to move the robot from the current state to the desired state (compensating for changing dynamics) while the single layer perceptron acts as the linear feedback controller. The sum of the two neural nets provides the actuator driving signal. The neural network slowly learns the robot's dynamic properties by constantly gathering control data while moving while the linear neurocontroller provides the error-driven torque and adapts fast to changing conditions thereby acting as a reflexive controller. Hereafter, the inverse dynamic neurocontroller portion will be labeled as IDNC and the linear neurocontroller portion as LNC.

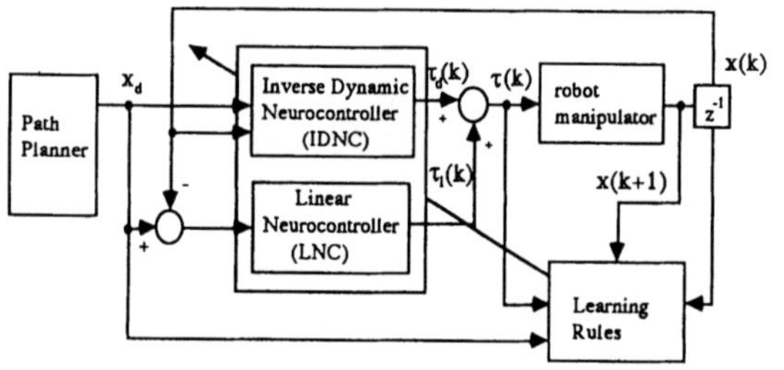

Fig. 1. The Proposed Neurocontrol Architecture

The training cycle consists of the following.

The IDNC weights are initialized randomly while the LNC weights are initialized to zero values. For a given desired state of the robot, the summation of the linear feedback torque and the inverse dynamic torque drives the actuator of the robot arm. The weights of the LNC may be adjusted by the delta rule in a direction to reduce the state error. Alternatively, when the system identification is performed, this result can be used for weight adjustment. The weights of the IDNC are adjusted by backpropagating the linear feedback torque output resembling feedback error learning by Kawato[1]. At first, the IDNC output is very small and the LNC output dominates. However, as learning gradually proceeds, the IDNC torque increases while the LNC torque decreases. Eventually after learning is complete, the IDNC will provide almost most of the torque required and the LNC will just correct minor errors. But when the environment suddenly changes and the IDNC had no time to respond due to its inherently slow learning, the LNC will take over the major role of error correction briefly while the IDNC learns the new environment.

TRAINING THE PROPOSED NEUROCONTROLLER

Training The IDNC

Training of the IDNC weights is very simple since the linear feedback error torque is directly used as the error signal. This error is backpropagated to adjust the weights.

Training The LNC

Two methods of training the LNC are proposed here. The robot state x is defined as the concatenation of joint positions x_p and rates x_v, therefore a 2n dimensional vector where n is the number of degrees of freedom. Method I simply uses the summation of the joint position error and rate error as the training error for the neural net and Method II derives the training error from a more sophisticated system identification process.

Training without system identification :Method I Training LNC requires an error signal, namely the difference between desired and actual output values. Using the 2n state errors (n position errors and n rate errors) directly as the teaching error will not work since the LNC has only n torque outputs. Therefore, the position and rate errors are summed resulting in n errors which serve as the training signals for the n outputs. Then, a simple delta rule is used to train the weights of the LNC. Figure 2 illustrates this training scheme.

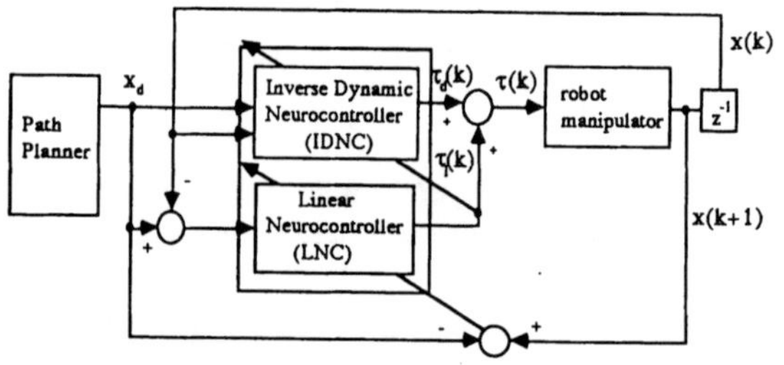

Fig. 2. Neurocontroller Training Scheme : Method I

Training the LNC using system identificaton :Method II A discrete version of a robot's dynamic equations of motion may be written as:

$$x(k+1) = F(x(k),\tau(k)) \tag{1}$$

where x(k+1), x(k) are state vectors of the robot at time k and k+1 respectively and $\tau(k)$ is the joint torque vector applied at time k. This equation basically states that the robot changes its state from x(k) to x(k+1) when a joint torque $\tau(k)$ is applied. This function F is nonlinear in robots and thus makes it very difficult to solve for $\tau(k)$ for control. Therefore, a linear time-varying approximation of the dynamics has been used instead:

$$x(k+1) = A(k)x(k) + B(k)\tau(k) \qquad (2)$$

When neural nets are used to approximate the coefficient matrices A(k) and B(k), the following form is more adequate:

$$x(k+1) = A(x(k)) + B(x(k))\tau(k) \qquad (3)$$

A(x(k) and B(x(k)) are approximated by using two multilayer neural networks of A(x(k),w) and B(x(k),w). These nets are trained by gathering and observing data while moving the arm with a PD controller around a desired trajectory. Hence, the identification is good only in the vicinity of the desired trajectory. Attempt to train it in the entire workspace has not been successful so far, but not deemed necessary for the current application. Some noise is added to the PD controller to obtain some generalization effects. The net for A(x(K)) is trained by setting $\tau(k)=0$ and observing x(k) and x(k+1).

After completing the learning of A(x(k)), learning B(x(k)) starts. Eq. (3) can be rewritten as:

$$x(k+1) - A(x(k),w) \cong B(x(k))\tau(k) \qquad (4)$$

The net for B(x(k)) is trained by moving the arm and observing x(k), x(k+1) and $\tau(k)$. The teaching error signal is then:

$$e = x(k+1) - A(x(k),w) - B(x(k),w)\tau(k) \qquad (5)$$

Unfortunately, this error signal can not be directly used to train B(X(k),w). Rather it is the output of a combined network shown in Fig. 3. The error in Eq. (5) is backpropagated through a linear network with $\tau(k)$ as weights and then through the weights of B(x(k),w).

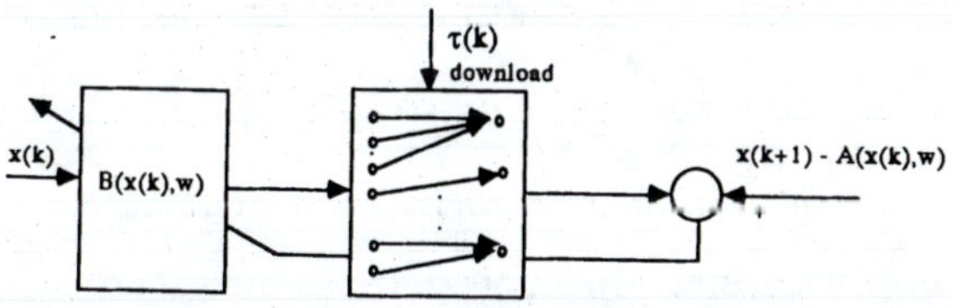

Fig. 3. Weight Learning Scheme for the Network for B(x(k))

Training of the LNC is effected by obtaining a solution $\tau^*_l(k)$ in the following equation:

$$x_d = A(x(k)) + B(x(k))(\tau_d(k) + \tau^*_l(k)) \qquad (6)$$

But normally, error exists in finding $\tau^*_l(k)$ and the following will hold:

$$x(k+1) = A(x(k)) + B(x(k))(\tau_d(k) + \tau_l(k)) \qquad (7)$$

Hence, the error to be minimized is:

$$\begin{aligned} e &= x_d - x(k+1) \\ &\cong [x_d - A(x(k),w)] - B(x(k),w)(\tau_d(k) + \tau_l(k)) \end{aligned} \qquad (8)$$

Here, A and B are used as identified in the previous subsection and τ_d are given. Again this signal can not directly be used to train the LNC and a scheme similar to identifying B in the previous subsection is used. This is shown in Fig. 4.

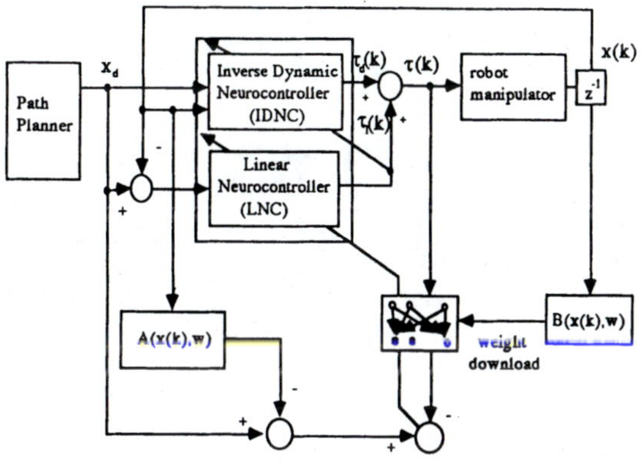

Fig. 4. Neurocontroller Training Scheme : Method II

DYNAMIC COMPUTER SIMULATION

The tracking control performances of the PUMA 560 robot arm are compared herein using the proposed control architecture, the conventional PD controller, and a previous neurocontrol architecture. Results indicate the superior tracking accuracies of the proposed neurocontrol architectures at various speeds and payloads.

Tracking Performance at Various Movement Speeds

In the first experiment, the robot was driven to follow a circle of circumference 1.256 m in 3 seconds with the average velocity of 0.41m/sec. Both methods as discussed were used. The system identification neural networks were trained using 10,000 training points randomly generated around the desired state space trajectory. Figure 5 shows the evolution of the average RMS error as learning iterations continue. The proposed control architecture using either method shows a superior performance to the PD controller (PDC) or the other neurocontroller using a fixed PD controller in parallel. Between two methods proposed, Method II using sytem identification exhibits more stable convergence characteristic.

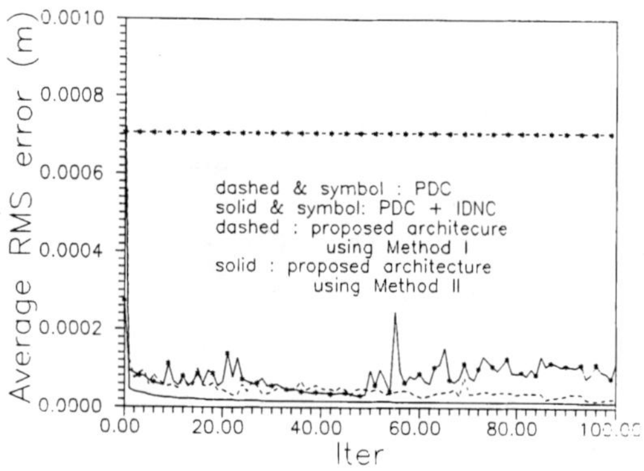

Fig. 5. Comparision of Learning curves Using Different Architectures

For a vivid comparison of the tracking performances in Cartesian workspace, the plots of the actual trajectories for the proposed architecture II after sufficient training at the lower speed and the PDC are given in Fig. 6. The proposed architecture shows an almost perfect tracking performance while the PDC shows some error. It clearly demonstrates the necessity for dynamic control which will be even more pronounced at higher speeds.

In order to examine how each controller reacts to a higher speed command, the robot was sped up by a factor of three after 20 such training iterations and was run for 180 additional iterations. The evolution of the average RMS error for the lower and higher speeds is shown in Fig. 7. It demonstrates a clear advantage at high speeds for the proposed controller compared with the PD controller or the previous type of neurocontroller using a fixed-gain PD. The proposed controller appears clearly effective in adapting to fast changing dynamics. The cooperation of the IDNC and LNC seems ideal in that the IDNC slowly learns the required dynamic torques and the LNC quickly learns the optimal gains needed to correct against minor errors resulting from differences between the neural dynamic model and the true dynamics.

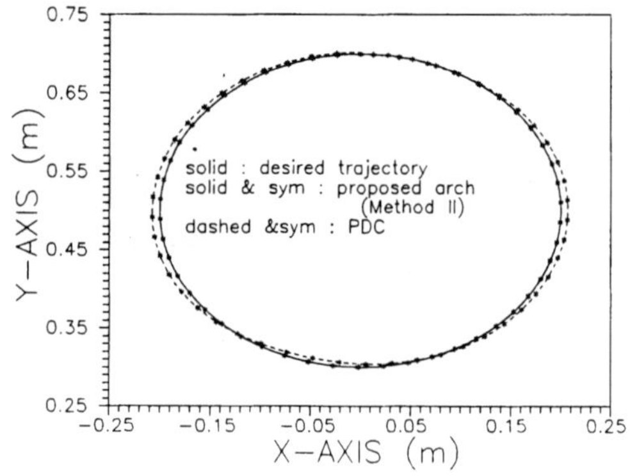

a) Plot of the Trajectory's X-Y Plane Projection

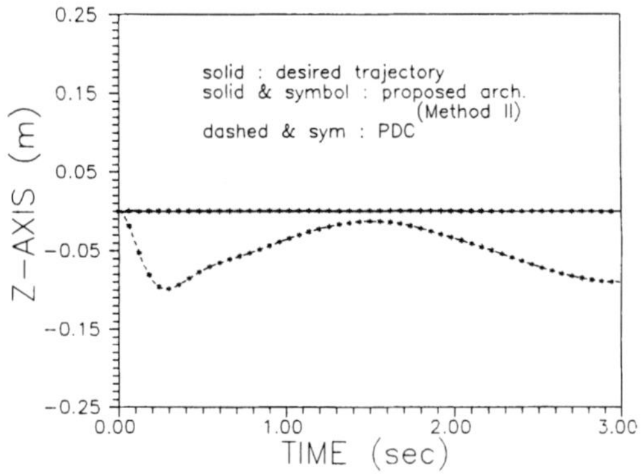

b) Plot of the Trajectory's Z Position

Fig. 6. Cartesian Tracking Performance for the Proposed and PD Controllers

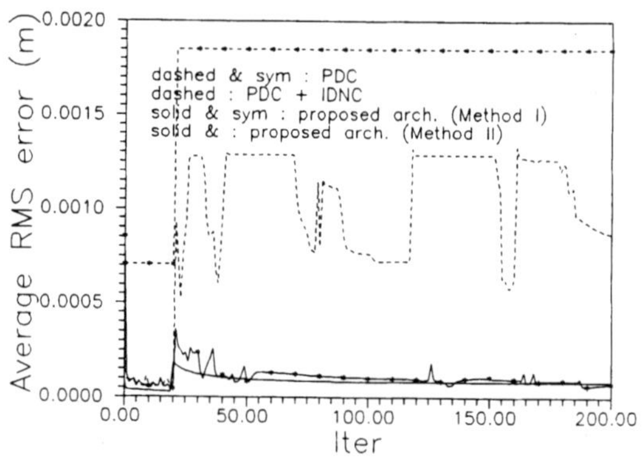

Fig. 7. Comparison of Capabilities of Different Controllers to Adapt to a Faster Motion Command (A factor of three speed increase is requested after 20 iterations of movement.)

Tracking Experiment With A Sudden Change in Payload

For demonstrating the controller's ability to adapt to changing payloads, after the robot drew the circle with an average velocity of 0.41m/sec without load up to 20 iterations, it suddenly picked up a 3 kg load and continued to run for 180 more iterations. Figure 8 shows the learning curves for different controllers. The proposed neurocontroller clearly works better than the PD controller with no learning capability, but not significantly better than the previously researched type of neurocontrollers. Apparently, dynamic change for increasing the speed by a factor of three for the PUMA arm seems to be more intense than for picking up a small load.

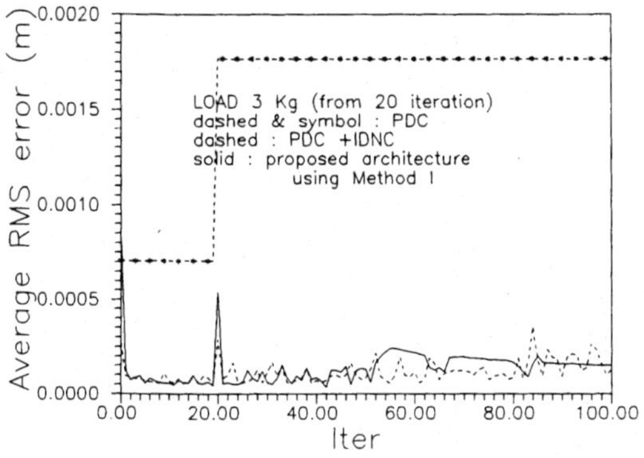

Fig. 8. Comparison of Capabilities of Different Controllers to Adapt to a 3 kg Load Change

CONCLUSION

A novel neurocontrol architecture is introduced in which an inverse dynamic neural controller in parallel with a linear neurocontroller cooperate to learn to control a robot manipulator accurately under different movement speeds and payloads. Compared with the previous types of neurocontrollers which use an inverse dynamic neurocontroller and a fixed PD gain controller, the proposed architecture not only eliminates the painful process of gain tuning but also exhibits superior dynamic performances because the linear controller constantly adapts its gains while moving. It is to be noted that in most cases Method I requiring no system identification will suffice. However, when the dynamic change is so great that system identification is needed before control, Method II must be used. Currently, system identification is carried out off-line in the vicinity of desired trajectory only using the neural net. It is turned off in actual control action but it may be turned on as needed even during control movement. Finally, although experimental verification of the controller is made with robots, it can be used to control any dynamic systems. As such, like any neural controller, it acts as a universal controller.

REFERENCE

[1] H.Miyamoto, M Kawato, T. Setoyama and Suzuki " Feedback-Error Learning Neural Network for Trajectory Control of a Robotic Manipulator," Neural Networks, Vol. 1, No.3,1988

[2] W. T. Miller,R. P. Hewes, F. H. Glanz, and L. G. Kraft, "Real Time Dynamic Control of an Industrial Manipulator Using a Neural-Network Based Learniing Controller", IEEE Transactions on Robotics and Automation, vol. 6, no. 1, pp 1-8,Feb. 1990.

[3] G.J. Wang, H.M.S Yeh, "Unsupervising Adaption Neural Networks," International Joint Conference on Neural Networks,June,1990.

[4] K. S. Narendra and K. Parthasarathy, "Identification and Control of Dynamical Systems Using Neural Networks," IEEE Transactions on Neural Networks,vol. 1,no.1,pp.4-27,1990.

[5] S.D Wang, H.M.S Yeh, "Self-Adaptive Neural Arichitecture for Control Applications," International Joint Conference on Neural Networks,June,1990.

[6] A. Guez, J.L. Eilbert, and M. Kam, "Neural Network Architecture for Control," IEEE Control Systems Magagine, pp.22-24, April 1988.

[7] Y.H. Pao, "Adaptive Pattern Recognition and Neural Networks," Addison Wesley, 1989.

A SIGNAL PROCESSOR BASED REAL TIME CONTROL FOR A POSITION AND ORIENTATION MEASUREMENT SYSTEM FOR INDUSTRIAL ROBOTS

Helmut GANDER, Johann Peter PRENNINGER, Markus VINCZE

Institute of Flexible Automation, Vienna University of technology

Abstract. This paper presents a real time controller for a position and orientation measurement system for industrial robots. The measurement system uses the beam of a laser interferometer and a mirror mounted on a cardan joint to follow a retroreflector which is mounted to the robot's end effector. The position of the end effector is measured with an accuracy in the range of micrometers and is computed in real time. The orientation is determined by analyzing the intensity profile of the reflected laser beam. The tracking controller drives the motors of the cardan joint in a way that enables the measurement system to track the position of highly dynamic robots. Since the control loops for the two axes are coupled and time-variant, a real time controller is used to achieve optimal tracking performance.

Keywords. Automatic Control, Computer Control, Control Theory, Digital Control, Predictive Control, Position and Orientation Measurement of Industrial Robots, Real Time Control;

INTRODUCTION

The application of robots in assembly grew rapidly in recent years and is still growing. Accompanying came the demand and first installments of off-line progamming systems. One of the critical items in off-line programming is the accuracy of the robot. Contrary to repeatability this characteristic measure is not necessary in teaching the movements of a robot. Therefore developers basically worked on improving repeatability. Though this also increases accuracy it can be said that the accuracy of an industrial robot is generally at least a magnitude lower than its repeatability.

Another factor for this bias in research and development was the absence of appropriate measurement systems. Repeatability can be easily metered with local position and orientation measurement equipment. Accuracy needs a system that reaches the entire workspace of the robot and works in a well established reference coordinate system. Since recent applications such as lasercutting or metrology robotics also need high path accuracy the measurement system must be able to track the robot's movement.

Due to this challenging conditions dynamic accuracy measurement systems are rare and limited to either straight lines, slow motion or low accuracy (Jiang, Black, Duraisamy, 1988).

The subsequently introduced Laser Tracking System (LTS) overcomes most of the shortcomings of other systems. One critical point in obtaining the required tracking capability is the control of the mirror mounted to the cardan joint. It can be shown that the control loops for the two axes are coupled and time-variant. Thus conventional control algorithms must be robust and do not perform satisfactorily. It will be shown that a real time controller can be used to increase the tracking performance considerably. The real time controller is combined with the real time position measurement system and is based on a digital signal processor.

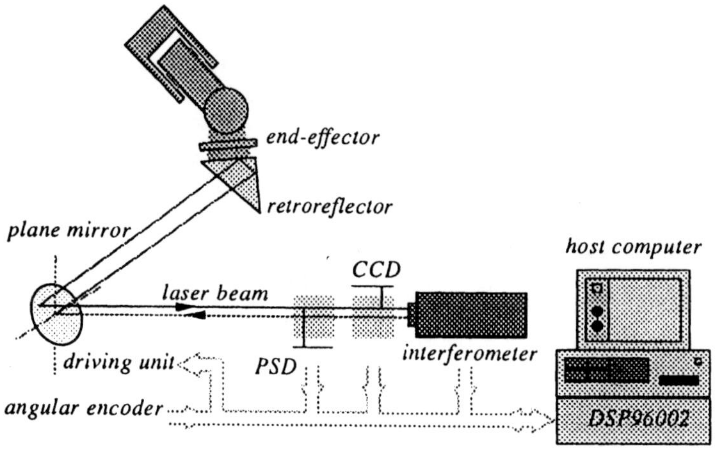

Fig. 1. Functional principle of the LTS.

Functional Principle of the Laser Tracking System (LTS)

Figure 1 shows the functional principle of the LTS. The beam of a high speed HeNe-laser interferometer is deflected by a plane mirror mounted on a cardan joint and hits the retroreflector which is mounted to the robot's end effector. The beam is reflected parallelly, is again deflected by the plane mirror and reaches the interferometer.

The parallel displacement of the incoming laser beam that is twice the distance between the laser beam and the centerpoint of the retroreflector, is measured with a position sensitive diode (PSD). Ideally, the laser beam hits the centerpoint of the retroreflector and there is no parallel displacement. When the robot starts moving the laser beam does not hit the centerpoint, thus there is a displacement of the laser beam. This displacement can be regarded as the tracking error and is minimized by the tracking controller through turning the axes of the cardan joint.

Two facts limit the maximum tracking error. First the reference and measurement beam must overlap to ensure correct measurements of the interferometer. Second the tracking error must not exceed 2 mm, otherwise the incoming laser beam does not hit the PSD and track is lost (larger PSD's should not be used due to insufficient dynamics and accuracy). These facts directly restrict the maximum allowable dynamic potential of the robot during measurement, especially the maximum acceleration. Thus a controller has to be developed that keeps the tracking error as small as possible.

The data from the interferometer, the angular encoders and the PSD are used to determine the position of the robot's end effector in real time with sampling rates up to 10 kHz (Gander and colleagues, 1991). The specifications of the LTS regarding to position measurement shows Table 1.

TABLE 1 Specifications of Position Measurement

measurement accuracy	50 μm
repeatability	10 μm
sampling rate	max. 10 kHz
measurement volume	>> 1 m^3

The orientation of the retroreflector can be extracted by analyzing the reflected laser beam with a vision system (Prenninger, Gander, Vincze, 1991). The specifications of orientation measurement are presented in Table 2.

TABLE 2 Specifications of Orientation Measurement

measurement range (Roll)	unlimited
measurement range (P/Y)	± 30 degree
measurement accuracy	better than 10 arcsec
resolution	1 arcsec

The tracking controller ensures that the laser beam follows arbitrary movements of the robot. The specifications that are considered to be sufficient for high dynamic operation are shown in Table 3.

TABLE 3 Specifications of the Target Tracking Controller

maximum velocity of the robot	6 m/s
maximum acceleration of the robot	100 m/s^2

Control Loop

Ideally the laser beam hits exactly the centerpoint of the retroreflector (see dashed line in fig. 2). When the robot starts moving the beam does not hit the centerpoint, thus there is a parallel displacement of the laserbeam, that is measured by the PSD. Figure 2 illustrates the optical path for one dimension.

$$\begin{pmatrix} x_R \\ y_R \\ z_R \end{pmatrix} = \begin{pmatrix} -\frac{1}{2}+\cos^2\beta_m\cos^2\gamma_m & -\cos^2\beta_m\sin\gamma_m\cos\gamma_m & \sin\beta_m\cos\beta_m\cos\gamma_m \\ \cos^2\beta_m\sin\gamma_m\cos\gamma_m & \frac{1}{2}-\cos^2\beta_m\sin^2\gamma_m & \sin\beta_m\cos\beta_m\sin\gamma_m \\ -\sin\beta_m\cos\beta_m\cos\gamma_m & \sin\beta_m\cos\beta_m\sin\gamma_m & \frac{1}{2}-\sin^2\beta_m \end{pmatrix} \cdot \begin{pmatrix} l \\ \Delta y \\ \Delta z \end{pmatrix} \quad (1)$$

x_R, y_R, z_R	coordinates of the retroreflector	$\Delta y, \Delta z$	beam displacement
β_m	measured angle of the horizontal axis	l	beam length
γ_m	measure angle of the vertical axis		

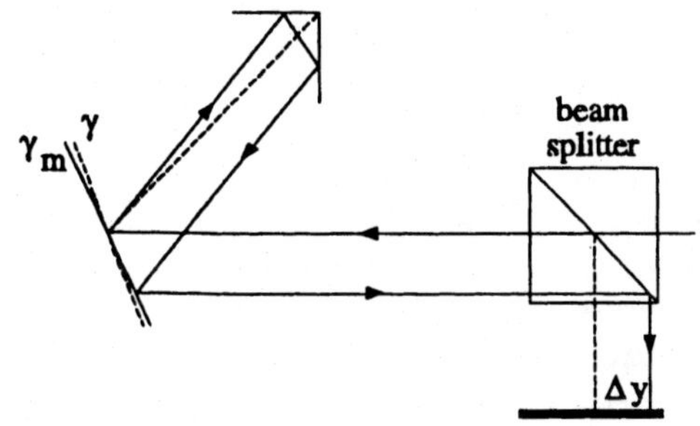

Fig. 2. Tracking Error in Detail.

The position of a beam striking a position sensitive diode can be determined by measuring the photocurrents on each side of the PSD and applying Eq. (2).

$$\Delta y = \frac{I_2 - I_1}{I_2 + I_1 - I_d}$$
$$\Delta z = \frac{I_4 - I_3}{I_4 + I_3 - I_d} \quad (2)$$

where

$\Delta y, \Delta z$	is the beam displacement,
I_1, I_2, I_3, I_4	are the photocurrents and
I_d	is the dark current.

The displacement of the laser beam is regarded the tracking error and is minimized by the tracking controller through turning the axes of the cardan joint. Figure 3 shows the closed control loop. The set point angles are the angles that are necessary for zero beam displacement and are thus defined by the position of the robot.

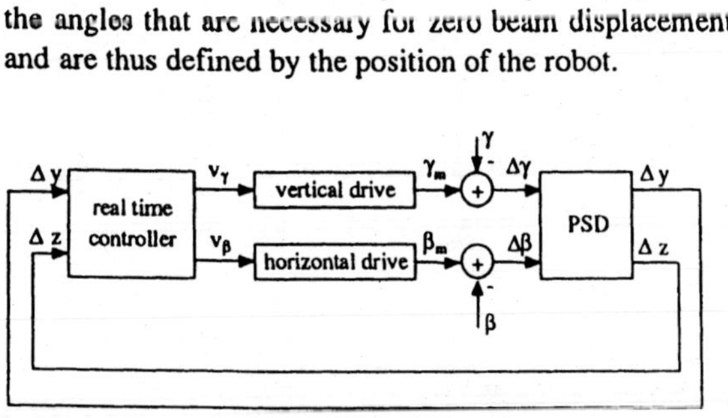

Δy	displacement of the beam	γ_m	angle vertical axis
Δz	displacement of the beam	β	set point angle horizontal axis
v_β	angular velocity horizontal axis	γ	set point angle vertical axis
v_γ	angular velocity vertical axis	Δβ	angle error horizontal axis
β_m	angle horizontal axis	Δγ	angle error vertical axis

Fig. 3. Control Loop.

The open loop consists of the horizontal and the vertical drive and the PSD. The inputs of the system are the angular velocities of the drives, the outputs are the signals from the PSD (the beam displacement). The transfer function of the system is determined by the drives and by the PSD. Due to the geometric arrangement there is a coupling between the two axes. This coupling depends on the angles of the axes. Additionally the length of the beam is a factor for both axes.

The angle errors can be computed using the beam displacement, the length of the beam and the angle of the axes of the cardan joint. Eq. 1 is the transformation of the sensor data to the position of the endeffector without correction of systematic errors.

The set point angles are the angles that are necessary for no beam displacement (Δy = Δz = 0) and can be computed according to Eq. 3,

$$\gamma_s = \arctan\left(\frac{y_R}{x_R + \frac{l}{2}}\right)$$
$$\beta_s = \arctan\left(\frac{-z_R}{\sqrt{y_R^2 + (\frac{l}{2} + x_R)^2}}\right) \quad (3)$$

where

β_s	is the set point angle of the horizontal axis,
γ_s	is the set point angle of the vertical axis, and
x_R, y_R, z_R	are the coordinates of the retroreflector.

These calculations are very time consuming and cause a dead-time in the control loop. Thus the equations are linearized under the assumption that the tracking error is very small (in the range of some millimeters).

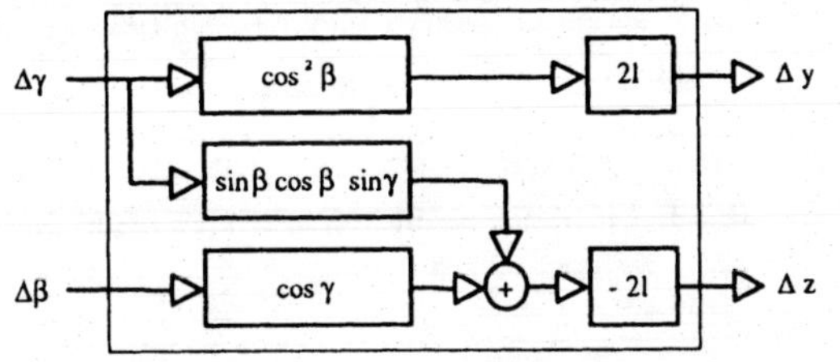

β	angle horizontal axis	Δγ	angle error vertical axis
γ	angle vertical axis	Δy	displacement of the beam
Δβ	angle error horizontal axis	Δz	displacement of the beam

Fig. 4. Functional Equivalent of the PSD.

Figure 4 shows the functional equivalent for the PSD that is valid only for small beam displacements.

During a movement of the robot the length of the beam and the angles of the axes change, thus the open loop is somehow time-variant.

REAL TIME CONTROLLER

Conventional control algorithms (PID) that are applied to time-variant systems must be robust enough to avoid instabilities, but this in turn decreases the tracking performance considerably.

Thus a real time controller is used to account for the changes in the controlled system. The hardware architecture of the tracking controller is shown in figure 5. The same system is used for measurement and real time position computation.

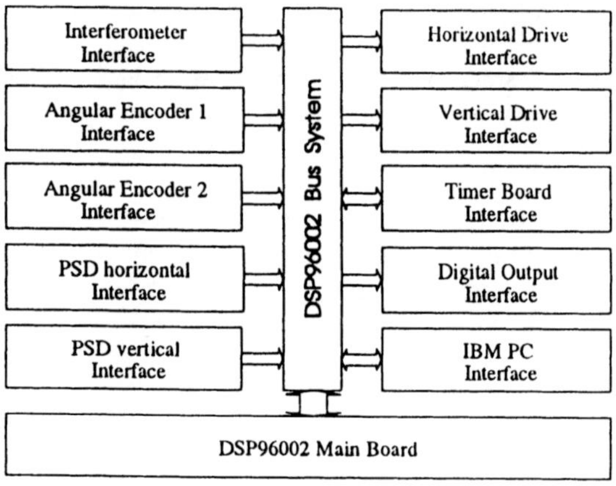

Fig. 5. Hardware Architecture for the Real Time Controller.

The hardware is based on a Digital Signal Processor (DSP) DSP96002 from Motorola. This DSP offers an arithmetic speed of 50 MFLOPS at 33 MHz (60 MFLOPS at 40 MHz). Via interfaces the DSP has access to the data from the interferometer, the PSD and the angular encoders. Additionally there are interfaces to the host computer (IBM - PC), to the power electronics of the motor drives and to the digital output of the retroreflector's position. The timer board generates programmable sampling times for the real time position computation and the target tracking controller. The sampling time currently used is 100 μs.

The coupling that is introduced by the PSD and the factor that is introduced by the beam length is eliminated using Eq. 4.

$$\Delta\gamma = \frac{\Delta y}{2l\cos^2\beta}$$
$$\Delta\beta = -\frac{1}{2l}\left(\tan\gamma\tan\beta\Delta y + \frac{\Delta z}{\cos\gamma}\right) \qquad (4)$$

where

γ is the angle of the vertical axis,
β is the angle of the horizontal axis,
$\Delta\gamma$ is the angle error of the vertical axes,
$\Delta\beta$ is the angle error of the horizontal axes,
Δy, Δz is the beam displacement and
l is the beam length.

The resulting transfer function of the control loops are linear and time invariant. The drives of both axes can be characterized by an IT_2 transfer function. The resonant frequencies are 160 Hz for the horizontal and 80 Hz for the vertical axis.

Using a conventional PID controller in combination with Eq. 4 a maximum acceleration of 20 rad/s^2 around the vertical axis is possible (the horizontal axis is better). This has been proven experimentally. Since the specifications regarding to the maximum acceleration could not be reached, a new tracking algorithm has been developed.

The new algorithm is based on the fact, that the movement of the retroreflector can not change abruptly. This is due to the inertia of the robot's end effector.

Thus the future position of the retroreflector can be estimated and is regarded as the set point for the tracking control. A predictive controller minimizes the future tracking errors and considerably increases the tracking performance. Simulations have shown that accelerations of more than 100 rad/s^2 around the vertical axis are possible.

Figure 6 shows the software modules of the tracking algorithm.

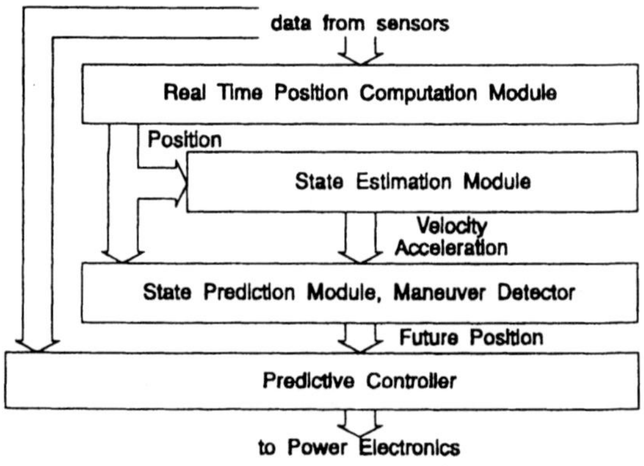

Fig. 6. Software Modules of the Tracking Controllers.

State Estimator

The state estimator uses the actual and the past positions of the robot's end effector to estimate the velocity and the acceleration of the robot. This is only possible since the position is computed in real time.

Due to the small values of the velocity and the acceleration (for example: an acceleration of 10 g = 1µm/sampling interval^2) the effects of measurement noise (partly due to quantization) on the state have to be eliminated by low-pass filtering.

Predictor

The predictor estimates the future position of the robot's end effector using the current state (measured position, estimated velocity and acceleration) of each coordinate. Eq. 5 shows how the future position of the robot's end effector is computed.

$$x_{n+1} = \Phi \cdot x_n \qquad (5)$$

with

$$x = (x \quad \dot{x} \quad \ddot{x} \quad y \quad \dot{y} \quad \ddot{y} \quad z \quad \dot{z} \quad \ddot{z})^T$$

and

$$\Phi = \begin{pmatrix} 1 & T & T^2 & & & & & & \\ 0 & 1 & T & & \ddots & & & \ddots & \\ 0 & 0 & 1 & & & & & & \\ & & & 1 & T & T^2 & & & \\ & \ddots & & 0 & 1 & T & & \ddots & \\ & & & 0 & 0 & 1 & & & \\ & & & & & & 1 & T & T^2 \\ & \ddots & & & \ddots & & 0 & 1 & T \\ & & & & & & 0 & 0 & 1 \end{pmatrix}$$

where

- x,y,z are the coordinates of the robot,
- $\cdot$ stands for time derivative,
- T stands for the transpose and
- T is the sampling period.

Assuming constant acceleration of the robot T can also be regarded as the look ahead time for the prediction of the position and velocity of the robot's end effector.

Predictive Controller

In contrast to conventional control algorithms predictive controllers minimize the actual *and* the future tracking error using Eq. (6).

$$J = \sum_{i=1}^{m} [w(k+i) - y(k+i)]^2 + \sum_{j=1}^{n} \Delta u^2 (k+j-1) = Min \qquad (6)$$

where

- J is the cost function that is minimized,
- w is the set point value,
- y is the real value,
- Δu is the change of the controller output,
- m is the cost horizon and
- n is the control horizon (Tsang, 1988).

Since the future tracking errors can not be measured, a model of the controlled system is used for the prediction of the future output signals.

In the case of the LTS there are no exact future set values, but the predicted future set point angles that are required for no tracking error. This predicted future angles are regarded as the set point angles for the predictive controller of each axis. Thus the difference between the predicted angles and the real angles of the axes are minimized. The predicted values are updated every sampling period using the measured actual position of the robot's end effector.

Consequently the tracking performance that can be reached by the predictive controller mainly depends on the state predictor of the target tracking controller.

CONCLUSION

The LTS described is used for the determination of robot position and path accuracy characteristics. The measurement procedure requires the precise tracking of the moving robot. Due to high acceleration and speed capabilities of modern robots this task needs an advanced control algorithm. We propose a real time controller that works similar to a predictive controller. With a state estimator and a predictor the future position of the robot is estimated. The predictive controller then minimizes the actual and future tracking error. First results with a slightly simplified controller show the feasibility of the approach.

ACKNOWLEDGEMENTS

The research was supported by the Austrian Science Foundation, projects P6796PHY and P8190TEC.

REFERENCES

Gander, H., J. P. Prenninger, M. Vincze, G. Zeichen (1991). A new measurement system for advanced modelling and identification for robot control. *Symposium on Robot Control*, 16.-18.9.1991, Vienna, Austria.

Jiang, B. C., J. T. Black, R. Duraisamy (1988). A review of recent developments in robot metrology. *Journal of Manufacturing Systems*, Vol. 7, Nr. 4, pp. 339-357.

Prenninger, J. P., H. Gander, M. Vincze (1991). Real time 6DOF measurement of robot end effectors. *International Robots & Vision Conference*, 22.-24.10.1991, Detroit, Michigan USA.

Tsang, T. T. C., D. W. Clarke (1988). Generalised predicitve control with input constraints. *IEE proceedings*, Vol. 135, No. 6.

ACCURACY INCREASING OF ROBOT REAL-TIME CONTROL

E. P. Kukareko, A. P. Pashkevich, D. E. Khmel, A. N. Korzun, Y. L. Yurkewich

Minsk Radioengineering Institute, 6 P. Brovka, 220600, Belarus

Abstract The paper presents an algorithm for robot control accuracy increasing through precise definition of locations in a technological program. The proposed approach is based on computing homogeneous transformation matrix which describes the relation between locations in real robot environment and their images in world model maintained by robot off-line programming system. Experimental verification of the algorithm is presented. Some computational experiments were carried out to investigate proposed algorithm robustness and reliability. The workcell calibration method was integrated into robot off-line programming system for PUMA and KUKA robots.

Keywords workcell calibration, parameter estimation, industrial robots, iterative methods, convergence of numerical methods

INTRODUCTION

Accuracy of robot real-time control depends on two main factors: accuracy of path description in terms of location sequence and accuracy of trajectory realization by control system. This paper deals with robot accuracy increasing by precise definition of programmed points using off-line programming system. In the last decade robot off-line programming (OLP) systems become an important part of computer-integrated manufacturing. Such systems offer many benefits to robot programmers enabling them to use all advantages of current programming techniques. But real-time application of such technique is limited by inaccuracy of a "world model" (i.e object position and orientation in robot environment), which is used in OLP system (Craig, 1986). During development and further production use its necessary to keep the internal model maintained by the programming system ("world model") in correspondence with actual state of robot's environment. One of the ways to make programs developed on OLP system usable for real-time control is an integration of workcell calibration methods into a robot programming system.

WORKCELL CALIBRATION ALGORITHM

Proposed method can be applied to a large class of applications that involve actions performed relative to rigid objects. For such cases correspondence between "real world" and "world model" can be described by a matrix equation.

$$R = T_{cal} * P, \qquad (1)$$

where T_{cal} is [4x4] homogeneous matrix (offset matrix);
$Ri = (x_ri\ y_ri\ z_ri)$ - Cartesian coordinates in "real world", further named as "real points";
$Pi = (x_pi\ y_pi\ z_pi)$ - Cartesian coordinates in "world model", further named as "program points".

Generally homogeneous matrix is defined by 6 parameters (3 linear and 3 angular). We will use Euler angels (roll-pitch-jaw) for orientation description for which T_{cal} is given by

$$'T_{cal} = \begin{vmatrix} C_aC_b & -C_aS_bS_c-S_aC_c & -C_aS_bC_c+S_aS_c & 'x \\ S_aC_b & -S_aS_bS_c+C_aC_c & -S_aS_bC_c-C_aS_a & 'y \\ S_b & C_bS_c & C_bC_a & 'z \\ 0 & 0 & 0 & 1 \end{vmatrix} \qquad (2)$$

,

where $C_a = \cos(a)$, $C_b = \cos(b)$, $C_c = \cos(c)$,
$S_a = \sin(a)$, $S_b = \sin(b)$, $S_c = \sin(c)$,
(x y z) are linear offset parameters.
(a b c) are angular offset parameters.

Assuming that differences between real points and program points are small, expression (2) can be simplified to

$$T_{cal} = \begin{vmatrix} 1 & -a & -b & x \\ a & 1 & -c & y \\ b & c & 1 & z \\ 0 & 0 & 0 & 1 \end{vmatrix} \quad (3)$$

We can rewrite matrix equation (1) in scalar form as:

$$\begin{aligned} r_x &= p_x - a*p_y - b*p_z + x \\ r_y &= a*p_x + p_y - c*p_z + y \\ r_z &= b*p_z + c*p_y + p_z + z \end{aligned} \quad (4)$$

To obtain offset matrix parameters (x y z a b c) least-square method was used. It was assumed that two sets of point coordinates {Ri} and {Pi} are known which correspond to real points and program points respectively. Let us define an objective in the form,

$$\begin{aligned} f = {} & \Sigma(p_x - a*p_y - b*p_z + x - r_x)^2 + \\ & + \Sigma(a*p_x + p_y - c*p_z + y - r_y)^2 + \\ & + \Sigma(b*p_z + c*p_y + p_z + z - r_z)^2 \rightarrow \min \end{aligned} \quad (5)$$

Setting to zero partial derivatives with respect to x,y,z,a,b,c results in

$$\begin{aligned} {}^1/_2\, {}^{df}/_{dx} &= \Sigma(p_x-a*p_y-b*p_z+x-r_x)=0, \\ {}^1/_2\, {}^{df}/_{dy} &= \Sigma(a*p_x+p_y-c*p_z+y-r_y)=0, \\ {}^1/_2\, {}^{df}/_{dz} &= \Sigma(b*p_z+c*p_y+p_z+z-r_z)=0, \\ {}^1/_2\, {}^{df}/_{da} &= \Sigma[(p_x-a*p_y-b*p_z+x-r_x)*(-p_y) + \\ & + (a*p_x+p_y-c*p_z+y-r_y)*(p_x)] = \\ & = (p^{x2}+p_y^2)*a+(p_y*p_z)*b+(-p_x*p_z)*c + \\ & + (-p_y)*x+(p_x)*y+(r_x*p_y-r_y*p_x)=0, \\ {}^1/_2\, {}^{df}/_{db} &= \Sigma[(p_x-a*p_y-b*p_z+x-r_x)*(-p_z) + \\ & + (b*p_z+c*p_y+p_z+z-r_z)*(p_x)] = \\ & = (p_y*p_z)*a+(p_x^2+p_z^2)*b+(p_x*p_z)*c + \\ & + (-p_z)*x+(p_x)*z+(r_x*p_z-r_z*p_x)=0, \\ {}^1/_2\, {}^{df}/_{dc} &= \Sigma[(a*p_x+p_y-c*p_z+y-r_y)*(-p_z) + \\ & + (b*p_z+c*p_y+p_z+z-r_z)*(p_y)] = \\ & = (-p_y*p_z)*a+(p_x*p_y)*b+(p_y^2+p_z^2)*c + \\ & + (-p_z)*y+(p_y)*z+(r_y*p_z-r_z*p_y) = 0. \end{aligned} \quad (6)$$

The system of linear equations (6) can be written in matrix form,

$$Ax = B,$$

where $B = (\Sigma(x_r-x_p) \quad \Sigma(y_r-y_p) \quad \Sigma(z_r-z_p) \quad \Sigma(y_rx_p-x_ry_p) \; \Sigma(z_rx_p-x_rz_p) \; \Sigma(z_ry_p-y_rz_p))^T$,

$X = (x \; y \; z \; a \; b \; c)^T$,

$$A = \begin{vmatrix} 1 & 0 & 0 & -\Sigma y_p & -\Sigma z_p & 0 \\ 0 & 1 & 0 & \Sigma x_p & 0 & -\Sigma z_p \\ 0 & 0 & 1 & 0 & \Sigma xp & \Sigma y_p \\ -\Sigma y_p & -\Sigma z_p & 0 & \Sigma(x_p^2+y_p^2) & \Sigma ypzp & -\Sigma x_pz_p \\ \Sigma x_p & 0 & -\Sigma z_p & \Sigma y_pz_p & \Sigma(x_p^2+z_p^2) & \Sigma x_py_p \\ 0 & \Sigma x_p & \Sigma y_p & -\Sigma x_pz_p & \Sigma x_py_p & \Sigma(y_p^2+z_p^2) \end{vmatrix} , \quad (7)$$

Index i in Eqs.(4 - 7) is omitted for readability.

To simply (7) the original coordinate system is translated by a vector which components are mean values of all program points {Pi}.

$$\begin{aligned} 'p_x^i &= p_x^i - {}^1/_n\, \Sigma p_x^i, & 'r_x^i &= r_x^i - {}^1/_n\, \Sigma p_x^i, \\ 'p_y^i &= p_y^i - {}^1/_n\, \Sigma p_y^i, & 'r_y^i &= r_y^i - {}^1/_n\, \Sigma p_y^i, \\ 'p_z^i &= p_z^i - {}^1/_n\, \Sigma p_z^i, & 'r_z^i &= r_z^i - {}^1/_n\, \Sigma_z^i, \end{aligned} \quad (8)$$

As a result we obtain matrix A in form

$$\begin{vmatrix} 1 & 0 & 0 & 0 & 0 & 0 \\ 0 & 1 & 0 & 0 & 0 & 0 \\ 0 & 0 & 1 & 0 & 0 & 0 \\ 0 & 0 & 0 & \Sigma(x_p^2+y_p^2) & \Sigma ypzp & -\Sigma x_pz_p \\ 0 & 0 & 0 & \Sigma y_pz_p & \Sigma(x_p^2+z_p^2) & \Sigma x_py_p \\ 0 & 0 & 0 & -\Sigma x_pz_p & \Sigma x_py_p & \Sigma(y_p^2+z_p^2) \end{vmatrix} , \quad (9)$$

and linear Ax=B can be solved separately for ('x 'y z') and ('a 'b 'c).

From the first three equations we have the linear offset

$$\begin{aligned} 'x &= \Sigma\, (r_{xi}-p_{xi}), \\ 'y &= \Sigma\, (r_{yi}-p_{yi}), \\ 'z &= \Sigma\, (r_{zi}-p_{zi}). \end{aligned} \quad (10)$$

To find the angular offset is necessary to solve linear system

$$\Sigma \begin{vmatrix} 'p_x^2+'p_y^2, & 'p_y'p_z & -'p_x'p_z \\ p_y'p_z' & p_x^2+'p_z^2, & p_x'p_y \\ -'p_x'p_z & 'p_x'p_y & 'p_y^2+'p_z^2 \end{vmatrix} \begin{vmatrix} a \\ b \\ c \end{vmatrix} = \Sigma \begin{vmatrix} 'r_x*'p_y-'r_y*'p_x \\ 'r_z*'p_x-'r_x*'p_z \\ 'r_z*'p_y-'r_y*'p_z \end{vmatrix} . \quad (11)$$

Substituting ('x 'y 'z 'a 'b 'c) to (2) we get matrix $'T_{cal}$. Matrix T_{cal} for original coordinate system is given by

$$T_{cal} = D^{-1}*T_{cal}*D,$$

where D is homogeneous matrix of linear displacement

$$D = \begin{vmatrix} 1 & 0 & 0 & {}^1/_n\Sigma p_x^i \\ 0 & 1 & 0 & {}^1/_n\Sigma p_y^i \\ 0 & 0 & 1 & {}^1/_n\Sigma p_z^i \\ 0 & 0 & 0 & 1 \end{vmatrix}$$

Then the set of points $\{^1P_i\}$ is transformed to a new set | | $\{^2P_i\}$. | |

$$^{j+1}P_i = {}^jT_{cal}*{}^jP_i, \,|$$

where j = 1,2,..,n - iteration number

The sets $\{^jP_i\}$ and $\{R_i\}$ are compared with each other. The procedure is repeated several times until desired accuracy is achieved. In practice 4-5 iterations is enough for good convergence. The final calibration matrix T_{cal} used in expression (1) is a product of $^jT_{cal}$

$$^0T_{cal} = {^1T_{cal}} * {^2T_{cal}} * \ldots * {^nT_{cal}}.$$

SOFTWARE FOR ROBOTIC CELL CALIBRATION

Proposed algorithm was integrated into robot off-line programming system. The calibration subsystem of the OLP consists of two parts: program kernel that compute calibration matrix T_{cal} as described before and accessory programs. The accessory programs are intended for file reading and writing, robot language program parsing and modification.

Program points $\{P_i\}$ and real points $\{R_i\}$ are obtained from two programs in robot language: one of them is prepared on OLP, another by teaching in real robot environment. These point sets are used for computing matrix T_{cal} and calibration of a robot technological program.

During parsing the technological program is analyzed and commands to be modified are detected. The commands arguments (point coordinates) are corrected according to expression (1). Sample program in SRCL (Siemens Robot Control Language) before and after calibration is presented on Fig.1.

Before calibration:

```
DEF HP11
PTP X -959.072 Y -1384.66 Z 916.444
  A 167.821 B 55.08 C -90.1566
PTP X -958.994 Y -1351.61 Z 977.672
  A 167.805 B 55.074 C -90.1396
PTP X -956.888 Y -1324.14 Z 915.706
  A 168.5 B 55.0 C -90.0
END HP11
```

After calibration:

```
DEF HP11
PTP X -961.072 Y -1392.61 Z 913.398
  A 167.709 B 55.0158 C -90.0459
PTP X -960.994 Y -1359.67 Z 974.691
  A 167.67 B 55.0065 C -90.0117
PTP X -958.849 Y -1332.14 Z 912.716
  A 167.656 B 55.0021 C -89.9936
END HP11
```

Fig.1. Example of robot program calibration.

The calibration subsystem was written in C and C++ languages for IBM PC and was built in off-line programming systems of Granat RM-01 and VAZ PR 161/15, PR 161/60 robots. These robot kinematics and programming languages are similar to the well-known Unimation PUMA-560 and KUKA IR 161/15, IR 161/60. The algorithm was used on manufacturing floor to increase industrial robot accuracy in arc welding in automotive industry.

EXPERIMENTAL VERIFICATION

The workcell calibration method was approved on experiments with industrial robot PM-01(PUMA-560). There were 11 test points situated on the rigid body arbitrary oriented in the robot workspace. Robot was taught in all of them and their coordinates were obtained using control system as a measuring device. The experiment was carried out twice for different positions of the body. Corresponding sets of coordinates $\{S1_i\}$ and $\{S2_i\}$ are presented in tables 1 and 2 respectively.

Then 8 points from the set $\{S1_i\}$ was used as program points $\{P_i\} = \{S1_i, i=1,..,8\}$ and 8 points from the set $\{S2_i\}$ was treated as real points $\{R_i\} = \{S2_i, i=1,..,8\}$.

TABLE 1 Set of Test Points $\{S1_i\}$

N	X	Y	Z
1	-201.28	523.91	-45.41
2	-229.41	628.03	2.38
3	-335.47	671.50	53.31
4	-455.03	626.56	74.66
5	-519.44	521.81	54.66
6	-492.97	417.03	3.94
7	-389.00	376.25	-45.22
8	-269.59	422.53	-66.03
9	-203.25	580.13	-24.03
10	-276.22	660.00	29.91
11	-447.00	387.00	-23.66

TABLE 2. Set of Test Points $\{S2_i\}$

N	X	Y	Z
1	-213.78	516.88	-44.28
2	-237.25	622.00	3.59
3	-341.13	669.44	54.41
4	-461.91	628.03	77.44
5	-530.31	524.94	58.66
6	-507.56	418.63	8.72
7	-406.88	374.03	-40.00
8	-285.75	416.47	-61.72
9	-213.31	572.47	-22.97
10	-283.13	655.25	30.69
11	-463.19	387.41	-18.13

Using algorithm described above we obtain the following calibration matrix

$$T_{cal} = \begin{vmatrix} 0.999 & 0.032 & 0.009 & -28.777 \\ -0.033 & 0.999 & 0.012 & -13.647 \\ -0.009 & -0.013 & 1.000 & 6.404 \\ 0 & 0 & 0 & 1 \end{vmatrix}$$

Applying Eq.(1) to the set $\{S1_i\}$ we get a new set $\{S2_i'\}$(Table 3), which should be very close to $\{S2_i\}$.

TABLE 3 Set of Test Points $\{S2_i'\}$

N	X	Y	Z
1	-213.37	515.93	-43.87
2	-237.66	621.49	2.83
3	-341.79	669.02	54.13
4	-462.54	628.26	77.10
5	-530.49	525.42	59.00
6	-507.90	419.22	9.38
7	-405.77	374.47	-40.16
8	-285.11	416.58	-62.60
9	-213.31	572.44	-23.19
10	-283.16	655.31	30.36
11	-463.19	387.37	-18.23

An average difference between $\{S2_i\}$ and $\{S2_i'\}$ coordinates is equal 0.98 mm and can be considered as a prove of the proposed algorithm fitness. It should be mentioned that the error of 0.98 mm is mainly a result of manual teaching but its quite enough for arc and spot welding applications.

It was established that the manual teaching error is a random value with distribution function shown on Fig. 2. Its mean value slightly varies from operator to operator and is about 0.2 mm.

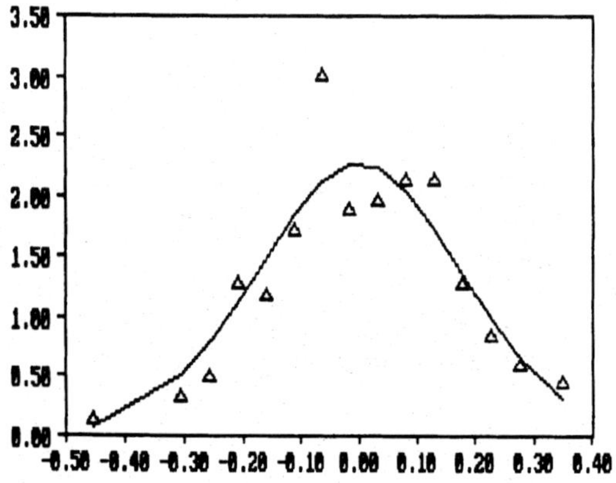

Fig. 2. Noise distribution function.

Some computational experiments were carried out to investigate proposed algorithm robustness and reliability. Accuracy dependence on number of points used for forming linear equations and their layout in robot workspace were investigated.

Program points {Pi} were regular distributed in space. Real points {Ri} were generated in accordance to Eq.(1) and noise component was added to all of them. Algorithm dependence on number of points {Pi} and {Ri} is shown on Fig. 3. Experiment proves that 8-10 points provide acceptable for practical application accuracy.

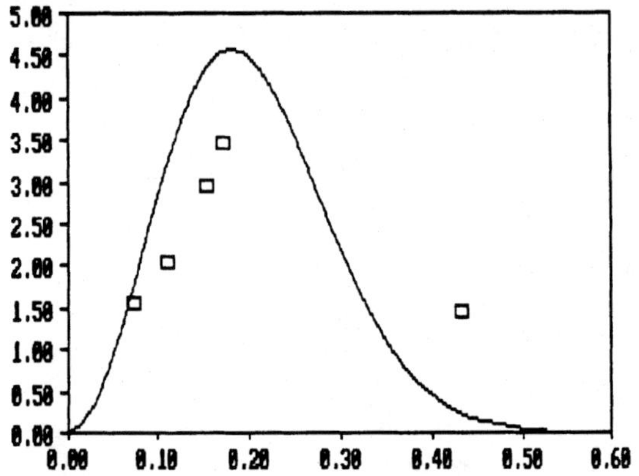

Fig. 3. Accuracy dependence on number of points.

The accuracy dependence on points layout in space was not discovered.

CONCLUSION

In this paper an approach to robot control accuracy increasing through workcell calibration has been proposed. The least square method was used to generate a linear system that leads to an efficient procedure of calibration matrix computing. Such approach allows to avoid many difficulties especially when robot performs irreversible operations (e.g. welding, drilling, etc.) and the same technological programs are applied to several similar robotic cell.

An experimental verification of algorithm has been presented. The proposed workcell calibration method has been successfully implemented and tested using KUKA and PUMA robots.

This work is a part of the main project on robot control accuracy increasing which also includes off-line manipulator and tool geometry identification and on-line control algorithms correction (Kukareko,1989).

REFERENCES

Craig, J.J. Introduction to Robotics. Mechanics and Control. Addison-Wesley, 1986.

Kukareko E.P., Korzun A.N., Pashkevich A.P. Industrial robot and robotic workcell calibration. Minsk, 1989.

DESIGN OF A SIMPLE AND ROBUST CONTROL SYSTEM FOR A SINGLE-LINK FLEXIBLE ROBOT ARM

Kang-Bark Park, Ju Jang Lee

Department of Electrical Engineering, Korea Advanced Institute of Science and Technology, 373-1 Kusong-dong Yusong-gu 305-701, Korea

Abstract. Lightweight robot manipulators have considerable structural flexibility. Hence, the elastic behavior of the robot arms must be considered in control system design. Owing to the complexity of a dynamic model of flexible robots, it is desirable to design a controller simple. Furthermore, since the robot must handle a wide variety of payloads, the robustness of the control system becomes very important. In this paper, a simple and robust control system is proposed. The closed-loop system is shown to be asymptotically stable in the sense of Lyapunov and robust to uncertainty in system parameters. The simulation results are presented to show the robustness of the simple controller against payload variations.

Keywords. Variable structure control; Control system design; Robust Control; Feedback control; Stability; Flexible manipulator.

INTRODUCTION

Robot manipulators have been widely used both in dangerous circumstances and for industrial applications. They have been traditionally modeled as a chain of rigid links with colocated actuators and sensors, to ensure stable and reliable control. Moreover, the robot arms are generally made large and massive in order to remain its rigidity while carrying an assigned payload. Present generation manipulators are limited to a load-carrying capacity of typically 5–10% of their own weight by the need of structural rigidity.

Today, there is an increasing requirement for manipulators with high speed, precision and payload-handling capabilities as a result of demand for higher productivity. These qualities cannot be achieved with existing massive and heavy robot manipulators. For higher operating speeds, the manipulators should be made lightweight, but lighter members are more likely to deform elastically. Therefore, it is necessary to include dynamic effects of the distributed link flexibility in the model of the manipulators.

Such flexible manipulators provide diverse advantages as follows: higher speed, smaller actuators, lower energy consumption, lower overall cost, safer operation due to reduced inertia, less bulky design, lighter overall mass to be transported, and so on. For the flexible manipulators, however, the equations describing the manipulator dynamics become more complex; besides, they complicate the control system design which focuses primarily on the compensation for bending effects. Furthermore, since the robot must handle a wide variety of payloads, the robustness of the control system becomes very important.

Recently, numerous investigators have analyzed several modeling methods of a flexible manipulator. One of them is a modal analysis method. Dominant eigen-modes are identified by experiment, approximated modes are obtained through solving the governing partial differential equations and assumed modes can be chosen from approximation of the actual system dynamics. Finite element method (FEM) has also been used. This method discretizes the actual system into a number of elements, whose elastic and inertial properties are obtained from the actual system. These methods give the approximated static and dynamic properties of the actual system. Because the dynamics of mechanical systems with distributed flexibility are described by infinite-dimensional mathematical models, and a finite-dimensional model of the system is needed for the design of a finite-dimensional controller.

There are also various schemes proposed for the design of controllers. A technique for end-point control corroborated by experimental work has been introduced by Cannon and Schmitz (1984). By using the Linear Quadratic Gaussian (LQG) controller design method, they successfully implement an noncolocated controller for the robot. However, since the controller is sensitive to parameter variations, its performance will be degraded when payload or typical parameters of the robot are varying with time. Rovner and Cannon (1987) used Recursive Least Square (RLS) algorithm to identify the system transfer function with unknown payload on the tip. The scheme requires a learning period which takes about two seconds with a sampling frequency of 50Hz. The experiments have shown good results, but the robustness of the algorithm still depends upon the number of coefficients of the transfer function to be identified. Other approaches are as follows: finite element approach, modal control, Model Reference Adaptive Control (MRAC), conventional control, feedforward control, and combined state space and frequency domain techniques, acceleration feedback control (Balas, 1978; Book and Majette, 1983).

For the design of a flexible manipulator controller, there are two facts to consider: first, the simplicity; second, the robustness. Owing to the complexity of dynamic model of flexible manipulators, it is desirable to design a controller simple. Furthermore, robustness of the controller is also very important for the robot to handle a wide range of payload variations. In general, control systems with state observer have been used. The control systems, however, require heavy computational load. Moreover, observer must be varied in operation to cope with payload variations. Hence, load forecast is needed.

In this paper, a simple and robust controller is proposed. For the simplicity, the control system only uses measurable data such as: hub ratio, tip position and tip ratio. Sliding mode control method, one of robust control methods, is introduced for the robustness. On account of robustness, load forecast is not necessary.

DYNAMIC MODEL OF A FLEXIBLE ROBOT ARM

Consider a uniform, slender beam connected via a rigid hub to the armature of an electric motor and having a payload as shown in Fig. 1. Where x, L, ρ, E, I and M_p are distance along the length of the beam, length of the beam, mass per unit length, Young's modulus of elasticity and cross-sectional area moment of inertia and payload mass. I_h is the hub inertia and I_b is the inertia of the beam about the motor armature ($= \frac{1}{3}\rho L^3$).

The following assumptions are made:

- The deflection w is small ($w \ll L$).
- Euler-Bernoulli beam assumptions are used.
- Beam inertia and flexibility are uniformly distributed over the link length.
- The motion occurs only in the horizontal plane.

Because of the last assumption, the effects of gravity are not considered.

The displacement of points along the deformed profile of the beam is described in terms of radial and circumferential coordinates x and y. From Fig. 1, it is apparent that y is related to the angle of rotation of rigid mode, q_o, and the flexural displacement of the beam, w, as follows:

$$y(x,t) = w(x,t) + x \cdot q_o(t).$$

In terms of this variable a fourth-order partial differential equation of motion for a single-link flexible arm is given in the form (Cannon, 1984)

$$EI\frac{\partial^4 y}{\partial x^4} + \rho\frac{\partial^2 y}{\partial t^2} = 0 \tag{1}$$

with four boundary conditions:

$$\begin{aligned} y(0,t) &= 0, & (2)\\ EIy''(0,t) + \tau - I_h\ddot{\theta}_h &= 0, & (3)\\ EIy''(L,t) &= 0, & (4)\\ EIy'''(L,t) - M_p\ddot{y}(L,t) &= 0, & (5) \end{aligned}$$

where $\dot{}$ and $'$ denote the derivative with respect to time and x. Then the following equation can be obtained.

$$\ddot{q}_i + 2\xi_i\omega_i\dot{q}_i + \omega_i^2 q_i = \frac{\phi_i'(0)}{I_T}\tau,$$

where $I_T = I_h + I_b + M_pL^2$, q_i is a modal function, $\phi_i(x)$ is a mode shape, and $i = 0, 1, 2, \cdots$. The model can be reconfigured in the state-space form

$$\begin{aligned} \dot{\mathbf{x}} &= \mathbf{A}\mathbf{x} + \mathbf{B}\tau, \qquad (6)\\ \mathbf{y} &= \mathbf{C}\mathbf{x}, \end{aligned}$$

where

$$\mathbf{x}^t = \begin{bmatrix} q_o & \dot{q}_o & q_1 & \dot{q}_1 & \cdots & q_n & \dot{q}_n \end{bmatrix}$$

$$\mathbf{A} = \begin{bmatrix} 0\ 1 & & & \\ 0\ 0 & & & \\ & 0 & 1 & & \\ & -\omega_1^2 & -2\xi_1\omega_1 & & \\ & & & \ddots & \\ & & & & 0 & 1\\ & & & & -\omega_n^2 & -2\xi_n\omega_n \end{bmatrix}$$

$$\mathbf{B}^t = \frac{1}{I_T}\left[0 \ 1 \ 0 \ \phi_1'(0) \ \cdots \ 0 \ \phi_n'(0)\right]$$

$$\mathbf{C} = \begin{bmatrix} L & 0 & \phi_1(L) & 0 & \cdots & \phi_n(L) & 0 \\ 0 & L & 0 & \phi_1(L) & \cdots & 0 & \phi_n(L) \\ 0 & 1 & 0 & \phi_1'(0) & \cdots & 0 & \phi_n'(0) \end{bmatrix}$$

where ^{t} denotes transpose and the three rows of C are the tip-position sensor, tip-rate sensor and the hub-rate sensor measurement vectors.

Practically, a model based on as few as four modes may be suitable for simulation and control system design. In this paper, the model based on three modes ($n = 3$) is considered.

DESIGN OF A CONTROLLER

In this paper, one of the control objectives is to introduce additional damping into the flexible motion. This has been done by using additional sensors to measure the flexible link vibrations and feed measurements of sensors back to the controller. Generally, strain gauges have been used in order to observe the states. This method, however, leads to heavy computational load. Hence, a simple control system using only directly measurable outputs is presented in this section.

Theoretically, any flexible system has an infinite number of elastic modes. Due to physical limitations, a limited number of sensors and actuators can be applied, thus restricting the controller design to a few critical modes in case of state utilization. Note that outputs of sensors would contain information about the unmodeled as well as the modeled modes. This is referred to as observation spillover. Similarly, the control action would affect both the modeled and unmodeled modes leading to the control spillover. Based on the work done by Balas (1978), the higher unmodeled modes may have a harmful or destabilizing effect on the system response. To avoid the observation spillover, low pass filtering of the sensor outputs can be used. Control spillover can also be avoided by inserting the Low Pass Filter (LPF) in the controller output stage. In this paper, the controller is designed based on the model equation (6).

For the system (6), consider the following switching surface

$$s_{rigid} = \dot{e}_{rigid} + \lambda_1 e_t, \tag{7}$$

where $e_{rigid} = \theta_d - q_o$, $e_t = \theta_d - \theta_t$, θ_d denotes the desired tip angular position and θ_t denotes the tip angular position. In the sliding mode, $\dot{s}_{rigid} = 0$. Applying this condition to equation (7) leads to the following equation:

$$\begin{aligned} \dot{s}_{rigid} &= \ddot{e}_{rigid} + \lambda_1 \dot{e}_t \\ &= -\ddot{q}_o + \lambda_1 \dot{e}_t \\ &= -\frac{1}{I_T}\tau + \lambda_1 \dot{e}_t \\ &= 0. \end{aligned}$$

Therefore, equivalent iuput torque is

$$\tau_{eq} = \hat{I}_T \lambda_1 \dot{e}_t.$$

Now, consider the following control law

$$\tau = \hat{I}_T\left(\lambda_1 \dot{e}_t + \lambda_2 \dot{e}_h + K sgn(s)\right), \tag{8}$$

where $\lambda_1 > 0$, $\lambda_2 > 0$, $e_h = \theta_d - \theta_h$, θ_h denotes the hub angle, and

$$\begin{aligned} \hat{I}_T &= \sqrt{I_{T_{max}} I_{T_{min}}}, \\ s &= \dot{e}_t + \lambda_1 e_t, \\ sgn(s) &= \begin{cases} 1 & s > 0 \\ 0 & s = 0 \\ -1 & s < 0, \end{cases} \\ K &= |(\beta - 1)\lambda_1 \dot{e}_t - \lambda_2 \dot{e}_h| + \beta M + \eta, \\ M &> 0, \\ \beta &= \sqrt{\frac{I_{T_{max}}}{I_{T_{min}}}}, \\ \eta &> 0. \end{aligned}$$

With this controller, following lemma is obtained.

lemma 1 *For the system (6) and the proposed controller (8), the sliding mode exists provided that*

$$\sup_t \left| \sum_{i=1}^{\infty} \frac{\phi_i(L)}{L} \ddot{q}_i(t) \right| < M.$$

proof
It is sufficient to show that $s\dot{s} \le 0$, and $s\dot{s} = 0$ if and only if $s = 0$.

$$\begin{aligned} s\dot{s} &= s\left[\ddot{e}_t + \lambda_1 \dot{e}_t\right] \\ &= s\left[-\ddot{q}_o - \sum_{i=1}^{\infty} \frac{\phi_i(L)}{L}\ddot{q}_i(t) + \lambda_1 \dot{e}_t\right] \\ &= s\left[\lambda_1 \dot{e}_t - \sum_{i=1}^{\infty} \frac{\phi_i(L)}{L}\ddot{q}_i(t) - \frac{1}{I_T}\tau\right] \\ &= s\left[\lambda_1 \dot{e}_t - \sum_{i=1}^{\infty} \frac{\phi_i(L)}{L}\ddot{q}_i(t)\right. \\ &\qquad \left. - \frac{\hat{I}_T}{I_T}\left(\lambda_1 \dot{e}_t + \lambda_2 \dot{e}_h + K sgn(s)\right)\right] \\ &= s\left[\left(1 - \frac{\hat{I}_T}{I_T}\right)\lambda_1 \dot{e}_t - \sum_{i=1}^{\infty} \frac{\phi_i(L)}{L}\ddot{q}_i(t)\right. \\ &\qquad \left. - \frac{\hat{I}_T}{I_T}\left(\lambda_2 \dot{e}_h + K sgn(s)\right)\right] \end{aligned}$$

$$
\begin{aligned}
&= s\frac{\hat{I}_T}{I_T}\left[\left(\frac{I_T}{\hat{I}_T}-1\right)\lambda_1\dot{e}_t-\lambda_2\dot{e}_h-Ksgn(s)\right.\\
&\quad\left.-\frac{I_T}{\hat{I}_T}\sum_{i=1}^{\infty}\frac{\phi_i(L)}{L}\ddot{q}_i(t)\right]\\
&< -\eta|s|.
\end{aligned}
$$

□

Furthermore, following theorem is also obtained for the stability.

theorem 1 *For the system (6) and the proposed controller (8), the overall system is asymptotically stable in the sliding mode.*

Proof

Let us choose the Lyapunov function candidate as

$$V=\frac{1}{2}\left[\sum_{i=0}^{n}\left(\omega_i^2q_i^2+\dot{q}_i^2\right)+\frac{\lambda_1^3}{4\lambda_2}\frac{\hat{I}_T}{I_T}e_t^2\right].$$

Then V is a p.d.f. And in the sliding mode,

$$s=\dot{e}_t+\lambda_1e_t=0.$$

Therefore,

$$
\begin{aligned}
\dot{V} &= \sum_{i=0}^{n}\left[\omega_i^2q_i\dot{q}_i+\dot{q}_i\left(-\omega_i^2q_i-2\xi_i\omega_i\dot{q}_i\right.\right.\\
&\quad\left.\left.+\frac{\hat{I}_T}{I_T}\phi_i'(0)\tau\right)\right]+\frac{\lambda_1^3}{4\lambda_2}\frac{\hat{I}_T}{I_T}e_t\dot{e}_t\\
&= \frac{\hat{I}_T}{I_T}\sum_{i=0}^{n}\phi_i'(0)\dot{q}_i\left(\lambda_1\dot{e}_t+\lambda_2\dot{e}_h\right)\\
&\quad-\frac{\lambda_1^2}{4\lambda_2}\frac{\hat{I}_T}{I_T}\dot{e}_t^{\,2}-\sum_{i=0}^{n}2\xi_i\omega_i\dot{q}_i^{\,2}\\
&= -\frac{\hat{I}_T}{I_T}\dot{e}_h\left(\lambda_1\dot{e}_t+\lambda_2\dot{e}_h\right)-\frac{\lambda_1^2}{4\lambda_2}\frac{\hat{I}_T}{I_T}\dot{e}_t^2\\
&\quad-\sum_{i=0}^{n}2\xi_i\omega_i\dot{q}_i^{\,2}\\
&= -\frac{1}{4\lambda_2}\frac{\hat{I}_T}{I_T}\left(2\lambda_2\dot{e}_h+\lambda_1\dot{e}_t\right)^2-\sum_{i=0}^{n}2\xi_i\omega_i\dot{q}_i^{\,2}\\
&\leq 0.
\end{aligned}
$$

Therefore, $\dot{V}=0$ if and only if $\dot{q}_i\equiv0\ \forall\ i$. This fact implies that $\dot{e}_t=\dot{e}_h=0$ and $\ddot{q}_i=0\ \forall\ i$ and $q_i=0\ \forall\ i\geq1$. Furthermore, $s=\dot{e}_t+e_t=0$. Hence, $e_t=0$ and for the rigid mode,

$$
\begin{aligned}
e_t &= \theta_d-\sum_{i=0}^{n}\frac{\phi_i(L)}{L}q_i\\
&= \theta_d-q_o\\
&= 0.
\end{aligned}
$$

Therefore, $q_o=\theta_d$, $q_i=0\ \forall i\geq1$ point in the state space is asymptotically stable by La Salle's theorem (Vidyasagar, 1978). □

Now, consider the meaning of the term $\lambda_2\dot{e}_h$ (= hub error rate) in the controller (8). Assume that the exact value of I_T is known and K is zero for the simplicity of analysis. Then the control law is rewritten in the form:

$$\tau=I_T\left(\lambda_1\dot{e}_t+\lambda_2\dot{e}_h\right). \tag{9}$$

And the system (6) is rewritten as follows:

$$\ddot{q}_i+2\xi_i\omega_i\dot{q}_i+\omega_i^2q_i=\frac{\phi_i'(0)}{I_T}\tau.$$

Inserting equation (9) into this equation gives:

$$
\begin{aligned}
&\ddot{q}_i+\left[2\xi_i\omega_i+\phi_i'(0)\left(\lambda_1\frac{\phi_i(L)}{L}+\lambda_2\phi_i'(0)\right)\right]\dot{q}_i\\
&+\left[\omega_i^2+\phi_i'(0)K_i\frac{\phi_i(L)}{L}\right]q_i\\
&=\phi_i'(0)\left(\lambda_1\left(\dot{e}_t+\frac{\phi_i(L)}{L}\dot{q}_i\right)\right.\\
&\left.+\lambda_2\left(\dot{e}_h+\phi_i'(0)\dot{q}_i\right)\right).
\end{aligned}
$$

In this equation, the right-hand side does not contain the ith mode and the sign of $\phi_i'(0)\phi_i(L)$ are not same for all modes. Therefore, it is desirable to use the $\dot{e}_h$ factor in the feedback mechanism. Since the factor $\lambda_2\dot{e}_h$ leads to the term $\lambda_2\left(\phi_i'(0)\right)^2\dot{q}_i$, additional damping is introduced. Because dominant vibrational mode is generally the first one and there is an uncertainty in payload, design of λ_2 value must be based on these factors.

SIMULATION RESULTS

In order to demonstrate the effectiveness of the controller developed in the previous section, numerical simulation has been performed on a single-link three-mode flexible manipulator.

TABLE 1 Modal Parameters (0.0 *kg* Payload)

i	ω_i (rad/sec)	$\phi_i'(0)$	$\phi_i(L)$	ξ_i
0	0	1	1	0
1	42.0	5.478	-0.359	0.0015
2	108.2	5.929	0.217	0.002
3	261.0	2.387	-0.178	0.002

TABLE 2 Modal Parameters (0.2 *kg* Payload)

i	ω_i (rad/sec)	$\phi_i'(0)$	$\phi_i(L)$	ξ_i
0	0	1	1	0
1	41.1	6.731	-0.286	0.0015
2	107.4	7.367	0.167	0.002
3	259.7	2.965	-0.135	0.002

TABLE 3 Modal Parameters (0.4 kg Payload)

i	ω_i (rad/sec)	$\phi_i'(0)$	$\phi_i(L)$	ξ_i
0	0	1	1	0
1	40.6	7.788	−0.244	0.0015
2	107.1	8.565	0.140	0.002
3	259.1	3.446	−0.113	0.002

The model parameters for the first three flexible modes with various payloads are given in TABLE 1, 2 and 3.

In the TABLE 1, 2 and 3, mode number zero means the rigid mode. For the rigid mode, $\phi_0(x)$ is defined as follows:

$$\phi_0(x) = x.$$

From these tables, it is obvious that the system parameters are largely affected by payload variations.

It is assumed that the flexible manipulator is initially at rest. The controller in equation (8) is discontinuous and it is well known that synthesis of such a controller gives rise to chattering of trajectory about sliding surface $s = 0$. In order to avoid the chattering phenomenon, the function $sgn(s)$ in the controller (8) has been replaced by $sat(s)$. The function $sat(s)$ is defined as follows

$$sat(s) = \begin{cases} 1 & s > \delta \\ s/\delta & |s| \leq \delta \\ -1 & s < -\delta \, , \end{cases}$$

where $\delta > 0$.

Figs. 2~5 show the performance of the proposed controller. These figures demonstrate that the tip angular position θ_t is successfully regulated for various payloads. Figure 3 shows that there is no chattering in the control torque τ.

In practice, gripper of the flexible manipulator may drop payload in the course of motion. In this case, the payload is changed abruptly. Figs. 4~5 show the performance of such a case. In Fig. 4, gripper drop the 0.4kg payload at 0.9sec, and Fig. 5 shows the control input.

CONCLUSION

A simple and robust control system of a single-link flexible arm is proposed. This controller does not need to estimate the modal functions of the system. It only uses output measurements such as: tip position, tip position rate and hub rate. Conventional control system uses vibrational modes of the system as states. Hence, it requires heavy computational load because of many matrix and vector operation. In addition, because the conversion matrix that is used in state estimation is affected by payload variation, conventional control system may be sensitive to payload variation and parameter uncertainty. In the proposed controller, these drawbacks are eliminated by using robust control law: variable structure controller. It is clear that the proposed controller has a simple form. In order to verify the robustness of the controller, the simulation results are given in the previous section.

REFERENCES

Balas, M. J. (1978). Feedback control of flexible systems. IEEE Trans. Aut. Control., 23, 673–679.

Book, W. J., and M. Majette (1983). Controller design for flexible, distributed parameter mechanical arms via combined state space and frequency domain techniques. J. Dynamic Syst., Meas, and Contr., 105, 245–254.

Cannon, R. H. Jr., and E. Schmitz (1984). Initial experiments on the endpoint control of a flexible one-link robot. Int. J. Robotics Res., 3, 62–75.

Itkis, U. (1976). (Ed.) *Control Systems of Variable Structure.* Wiley & Sons, New York.

Meirovitch, L. (1967). (Ed.) *Analytical Methods in Vibrations.* MacMillan, New York.

Rovner, D., and R. H. Jr, Cannon (1987). Experiments toward on-line identification and control of a very flexible one-link manipulator. Int. J. Robotics Res., 6, 3–19.

Utkin, V. I. (1977). Variable structure systems with sliding modes. IEEE Trans. Aut. Control., 22, 212–222.

Vidyasagar, M. (1978). (Ed.) *Nonlinear Systems Analysis.* Prentice–Hall, New Jersey.

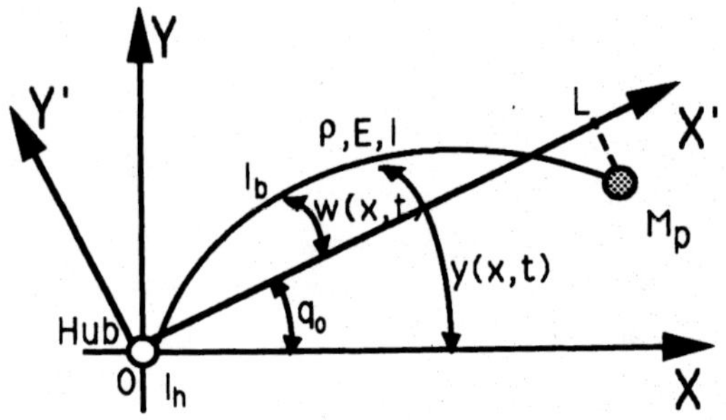

Fig. 1. Geometry of the flexible arm.

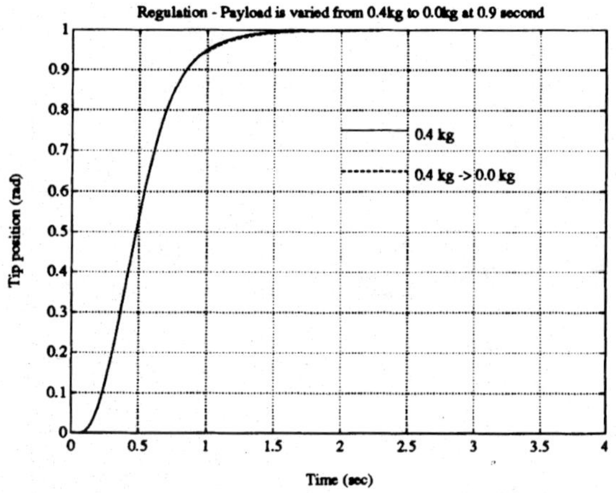

Fig. 4. Tip position; payload variation.

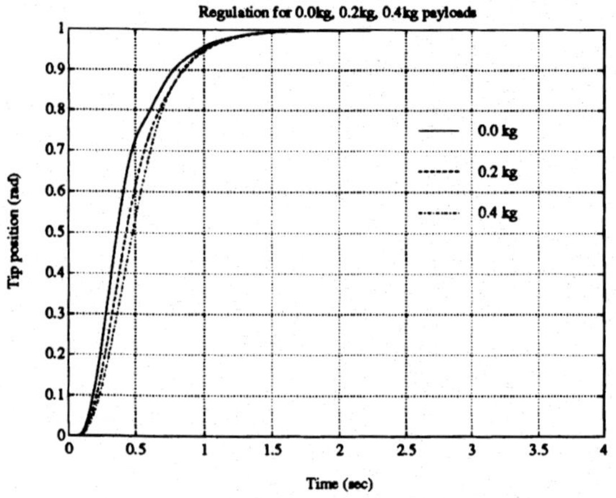

Fig. 2. Tip position.

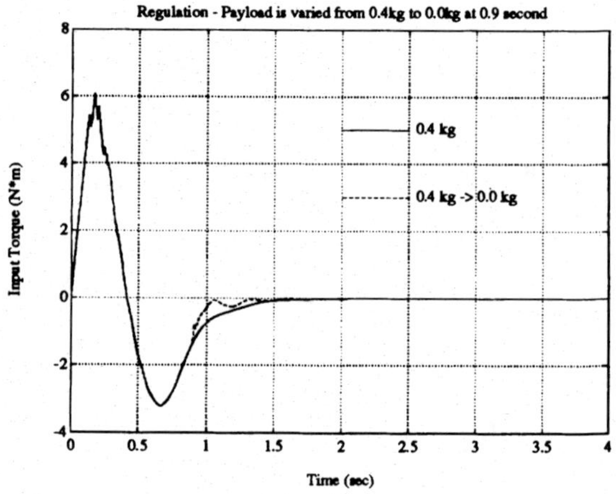

Fig. 5. Control Input; payload variation.

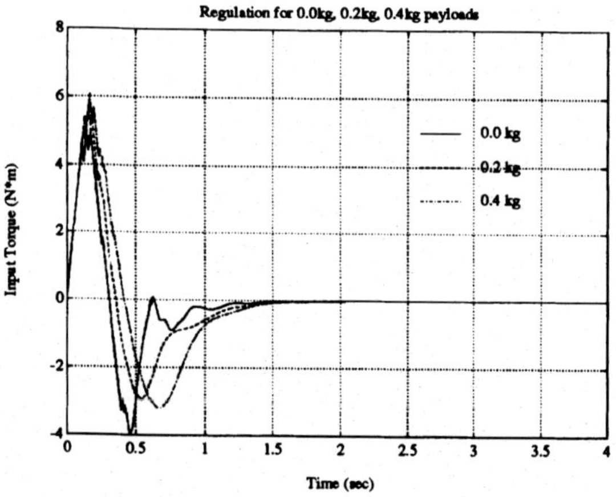

Fig. 3. Control input torque.

SUPPRESSION METHOD OF DISTURBANCE TORQUE DUE TO SYSTEM DRIFT VIA AN OBSERVER IN THE ROBOT CONTROL SYSTEM

Dong-Il Kim, Byeong Hwan Jeon and Sungkwon Kim

Control R/D Team, Production Engineering Division,
Consumer Electronics Business, Samsung Electronics,
416, Maetan-3-Dong, Kwonsun-Ku, Suwon city, Kyungki-Do 441-742, Korea

Abstract The degradation of performances in the robot manipulator control system results from disturbance torques due to the system drift, frictions in the gears, Coriolis, gravity loading, and centrifugal effects of the link. The torque due to the system drift is closely related to the difference between ground levels and the variation of characteristics of the control system, while the others depend on the manipulator's physical parameters, instantaneous joint configuration, joint velocity and acceleration, and the load it is carrying.

In this paper, an estimation algorithm of the system drift in the robot manipulator control system is proposed, which is based on the gradient search method. This algorithm is executed in the steady state for the elimination of the system drift because the variation of the characteristic of the control system due to the change in temperature, supply voltage, and etc. are slow compared with the motion of the robot manipulator.

The proposed algorithm is implemented on the control system for a robot manipulator with six degrees of freedom, which has an 80386/80387 as a main processor and a TMS 320C30 as a processor for motion control. Experimental results show that the proposed suppression method of the system drift via an observer can be effectively used in controlling robot manipulators with high dynamic performance at both the transient and steady states.
Keywords robot manipulator, disturbance torques, system drift, gradient search method, observer, velocity and acceleration feedforward

INTRODUCTION

The performance of the robot manipulator control system is evaluated by repeatability, accuracy, transient response and settling time for the commanded motion, system robustness for power on/off, and the characteristic of acceleration/deceleration under large load change. The performance degradation of the robot manipulator control system arises from disturbance torques due to the system drift, frictions in the gears, Coriolis, gravity loading, and centrifugal effects of the link. Torques due to Coriolis, gravity loading, centrifugal effects and frictions depend on the physical parameters of the robot manipulator, instantaneous joint configuration, joint velocity and acceleration, and the load it is carrying. On the other hand, the torque due to the system drift is caused by the variations in temperature and supply voltage in the robot manipulator control system which change the characteristic values of analog components such as capacitors, resistors, and linear ICs. As a result, control parameters are greatly changed and the physical quantities such as commanded inputs and feedback values are deflected by offset values. Aforementioned disturbance torques have the time-varying characteristics and act on the actuators of the robot manipulator, so accuracy, repeatability, acceleration/deceleration, and settling time of the commanded motion are greatly deteriorated. Furthermore the cycle time of the robot manipulator which is the measure of productivity is damaged.

Thanks to the inherent advantages, the recent digital control system of the robot manipulator is sure to eliminate disturbance torques due to both the physical structure of the robot manipulator and the system drift. However, the robot manipulator control systems which are widely used in industry up to date mostly consist of digital position controllers and analog velocity controllers. The disturbance torques due to Coriolis, gravity loading, and centrifugal effects can be compensated by precomputing the magnitudes of them and feeding the computed values into the digital position controller. By the similar way, the disturbance torque due to friction in the gear can be compensated in the velocity control loop independently of the digital or analog control system [1]. On the contrary, this compensation method does not apply to the elimination of the disturbance torque due to the system drift because it appears as the uncertain offset in the desired behavior of the analog circuit. To overcome this difficulty, it is required to obtain the information of the system drift through the indirect method. If combined with the compensation methods for the disturbance torques due to physical structure of the robot manipulator, the compensation method for the system drift will enable the robot manipulator to carry out diverse and complex jobs with high dynamic performance.

This paper discusses the observer which makes it possible to estimate the system drift in the robot manipulator control system. The characteristic of the system drift which varies much more slowly than that of the motion of the robot manipulator has motivated the on-line estimation by an observer based on the gradient search method. The estimated value is feedforwarded into the output of the position controller to compensate the disturbance torque due to the system drift. In the actual implementation, the compensation algorithm for elimination of the disturbance torques in the control of the robot manipulator consists of the observer and the velocity and acceleration terms. It has been executed on the robot manipulator control system with Intel 80386/80387 as a main CPU and TMS320C30 as a motion control CPU.

(7), it is assumed that the position, that is the integral of the desired trajectory, is controlled by a digital computer theory while the velocity by the system consisting of analog circuits. If the control gains are appropriately adjusted and there exists no system drift in the robot manipulator control system, the proposed controller of (7) can compensate the disturbance torques due to Coriolis, gravity loading, and centrifugal effects so efficiently as to produce desirable tracking errors and robustness even in the presence of the model inaccuracies. The friction torques due to gears can be easily suppressed by the compensation in the velocity control loop, which is not mentioned in this paper.

The system drift, which results from the variations in temperature and supply voltage in the robot manipulator control system, is one of the main factors which cause both great tracking errors and mechanical vibrations at the goal position. Accordingly, it is also required to eliminate the system drift in order to eliminate such undesirable phenomena perfectly, as shown in (7).

Now, the estimation method of the system drift is described. When performing tasks, the robot manipulator repeats the motion according to the trajectory which consists of acceleration, constant velocity, and deceleration. The physical quantities of actuators, which drive the axes of motion of the robot manipulator, are in the steady state both under constant speed and at a standstill. The disturbance torque due to the system drift can be efficiently compensated in the steady state of the robot manipulator. Since the system drift varies much more slowly compared with the motion of the robot manipulator, the only compensation at a standstill is sufficient to eliminate it. When the robot manipulator arrives at the goal position after deceleration, the output of the integral controller in the position control loop of joint j is expressed as

$$E_j(t) = f(\theta_j^* - q_{jf},\ D_j(t)-X_j(t)), \qquad (8)$$

where θ_j^* is the commanded reference position, q_{jf} is the feedback position, $D_j(t)$ is the offset due to the system drift, and $X_j(t)$ is the estimated value of $D_j(t)$. Because $D_j(t)$ can not be expressed in the explicit form, $X_j(t)$ should be found indirectly in order to eliminate the effect of $D_j(t)$ on control performance. From (8), we see that iterative numerical search procedures can be very effective in estimating $D_j(t)$ [7]. As a consequence, the following observer, which has a negative search direction of the first order differentiation of $E_j(t)$, is used for the estimation of $D_j(t)$:

$$X_j(k+1) = X_j(k) + \mu_j(-\Delta k), \qquad (9)$$

where

$$\Delta k = \left.\frac{\partial E_j}{\partial X_j}\right|_{E=E_{jk}}, \qquad (10)$$

and μ_j is the adaptation constant. The adaptation constant should be appropriately chosen to guarantee the fast convergence time of the observer. Practically, the most optimal value of μ_j is obtained by the trial and error method. Fig.1 shows the iterative search procedures of the observer given by (9) and (10).

Now, the trajectory planning method which is used in experiments is described. First, let us define the variables:

f_{ij}	position displacement of joint j,
f_{oj}	desired trajectory of joint j,
V_{jmax}	maximum speed of joint j,
Q(a,b)	quotient of a divided by the integer b,
R(a,b)	remainder of a divided by the integer b,
K_{ap}	number of sampling periods needed for the completion of acceleration/deceleration.

In order to obtain the desired trajectory, the position displacement of joint j is divided by V_{jmax} to generate the quotient $Q(f_{ij}, V_{jmax})$ and the remainder $R(f_{ij}, V_{jmax})$. Based on these values, the position displacement except $R(f_{ij}, V_{jmax})$, which is denoted as $f_q(k)$, is distributed over $Q(f_{ij}, V_{jmax})$ sampling periods as follows:

$$f_q(k) = V_{jmax}[u(k)-u(k-Q(f_{ij}, V_{jmax}))], \qquad (11)$$

where u(k) is the unit step function. The convolution of $f_q(k)$ and a_k yields

$$\begin{aligned} f_{oj}(k) &= \sum_{n=0}^{k} f_q(n)a_{k-n}, \\ &= \sum_{n=0}^{k} V_{jmax}(a_{k-n}-a_{k-n-g}), \quad g = Q(f_{ij}, V_{jmax}) \qquad (12) \\ &= f_{oj}(k-1)+V_{jmax}(a_k-a_{k-g}). \end{aligned}$$

In (12), a_k and $f_{oj}(k)$ represent the profiles of acceleration, that is torque, and the instantaneous velocity of joint j, respectively. As a consequence, the desired trajectory of joint j for $f_q(k)$ is given by

$$\begin{aligned} P_{oj}(k) &= \sum_{n=0}^{k} f_{oj}(n), \\ &= P_{oj}(k-1)+f_{oj}(k), \quad f_{oj}(0) = 0. \qquad (13) \end{aligned}$$

Finally, the remained $R(f_{ij}, V_{jmax})$ is divided by $Q(f_{ij}, V_{jmax})+K_{ap}$ to generate $Q(R(f_{ij}, V_{jmax}),\ Q(f_{ij}, V_{jmax})+K_{ap})$ and these resultant values are distributed evenly over the whole range of the desired trajectory in the discrete time domain. As a consequence, fij is transformed into the desired trajectory, whose integral is equal to the position displacement of joint j. The simultaneous control of multiple axes of motion in industrial robots can be easily obtained on the basis of the above procedure.

EXPERIMENTAL RESULT

A robot manipulator with six degrees of freedom is driven by the control algorithm including the proposed compensation algorithm, which is implemented on the control system consisting of microprocessors (Intel 80386/80387, TMS

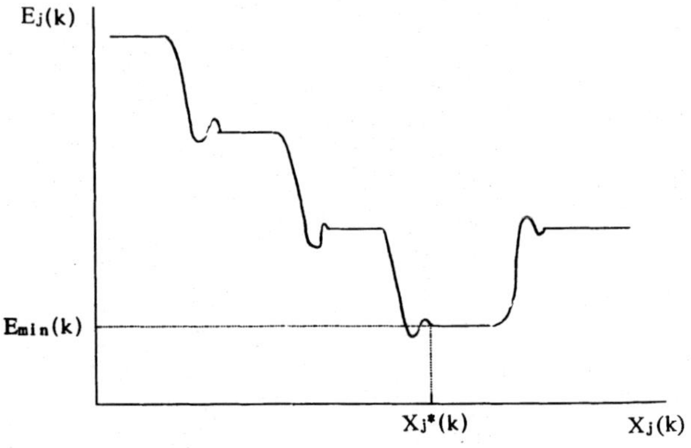

Fig.1 The iterative search procedures of the observer.

The experimental results demonstrate that the proposed compensation algorithm effectively can eliminate the disturbance torques due to both the system drift and the physical structure of the robot manipulator.

ESTIMATION OF THE SYSTEM DRIFT

Let's consider the robot manipulator driven by AC servo motors with impressed current control. The voltage equations of the permanent magnet AC servo motor are expressed as

$$\begin{bmatrix} V_{as} \\ V_{bs} \\ V_{cs} \end{bmatrix} = \begin{bmatrix} R_s + PL_s & PM & PM \\ PM & R_s + PL_s & PM \\ PM & PM & R_s + PL_s \end{bmatrix} \begin{bmatrix} i_{as} \\ i_{bs} \\ i_{cs} \end{bmatrix} + \Phi_m \omega_r \begin{bmatrix} \sin\theta_r \\ \sin(\theta_r + 2\pi/3) \\ \sin(\theta_r + 4\pi/3) \end{bmatrix}, \quad (1)$$

and

$$T_e = K_T \Phi_m (i_{as} \sin\theta_r + i_{bs} \sin(\theta_r + 2\pi/3) + i_{cs} \sin(\theta_r + 4\pi/3)), \quad (2)$$

where

V_{as}, V_{bs}, V_{cs} stator phase voltages
i_{as}, i_{bs}, i_{cs} stator phase currents
Φ_m rotor flux
ω_r rotor speed
R_s stator resistance
L_s stator self-inductance
M mutual inductance between the stator and the rotor
K_T torque constant
P $=d/dt$
θ_r angular displacement in radians between the magnetic axis of stator phase A and the rotor

If the AC servo motor is fed by the following stator current commands under impressed current control, where motor currents are essentially sinusoidal and are regulated within fixed bands [2], [3]:

$$[u_1 \; u_2 \; u_3]^T = [I_m \sin\theta_r \;\; I_m \sin(\theta_r + 2\pi/3) \;\; I_m \sin(\theta_r + 4\pi/3)]^T, \quad (3)$$

the generated motor torque is given by

$$T_e = 1.5 K_T \Phi_m I_m, \quad (4)$$

where I_m is the amplitude of stator current commands. In order to combine the dynamics of the AC servo motors into those of the robot manipulator, it is necessary to draw relations between the quantities at the end of the gear box. From (4), the j-th joint torque Tej of the robot manipulator is given by

$$T_{ej} = N_j K^j{}_T \Phi^j{}_m I_{mj}, \quad j = 1, \ldots, n. \quad (5)$$

Then, the dynamic behavior of a rigid-link robot manipulator is governed by the following second-order differential equation [4], [5]:

$$\sum_{j=1}^{n} D_{ij} \ddot{q}_j + I_{ai} \ddot{q}_i + \sum_{j=1}^{n} \sum_{k=1}^{n} D_{ijk} \dot{q}_j \dot{q}_k + D_i = T_{ei}, \quad i = 1, \ldots, n, \quad (6)$$

where

$q_i, \dot{q}_i, \ddot{q}_i$ the position, velocity and acceleration of joint i,
T_{ei} the torque (or force) acting at joint i,
D_{ii}, D_{ij} the effective and coupling inertias,
I_{ai} the reflected actuator inertia of joint i,
D_{ijj}, D_{ijk} the coefficients of centripetal and Coriolis forces,
D_i the gravity force.

Although the dynamics of the AC servo motor are reduced to almost linear system by impressed current control, (6) shows that the dynamic behavior of the robot manipulator is highly coupled. Because of complexity and nonlinearities between the dynamic equations of manipulator links, it is difficult to directly design control laws. This difficulty can be overcome by compensating the disturbance torques due to friction, Coriolis, gravity loading, and centrifugal effects by precomputing the magnitudes of them and feeding the computed values into the appropriate nodes of the robot manipulator control system. As the more efficient method, the differential geometric control theory can be applied to produce a linearly decoupled dynamic equations of manipulator links [6]. However, the precise information of the physical parameters of the robot manipulator is required in order to obtain the desired performance of the velocity and position tracking of each joint in both cases. The errors are inevitable in measuring the parameters of the robot manipulator, so the trade-off between control performance and difficulty in implementation should be taken into consideration.

In this paper, the following control inputs are designed in order to suppress the disturbance torques as well as to obtain good control performance in tracking the commanded velocity and position of each joint:

$$I_{mj} = K_{WPj}(\omega_j^* - \dot{q}_{jf}) + K_{WIj} \int (\omega_j^* - \dot{q}_{jf}) + K_{WAj} \omega_j^*,$$

$$\omega_j^* = K_{\theta Pj}(\theta_j^* - q_{jf}) + K_{\theta Ij} \int (\theta_j^* - q_{jf}) + K_{\theta Dj} d(\theta_j^* - q_{jf})/dt + K_{Aj} \theta_j^* + K_{AAj} d\theta_j^*/dt + X_j(t), \quad j = 1, \ldots, n, \quad (7)$$

where

q_{jf}, q_{jf} the position and velocity feedback values of joint j
θ_j^* the desired position command of joint j,
K_{WPj} the velocity proportional gain of joint j,
K_{WIj} the velocity integral gain of joint j,
K_{WAj} the first order velocity feedforward gain of joint j,
$K_{\theta Pj}$ the position proportional gain of joint j,
$K_{\theta Ij}$ the position integral gain of joint j,
$K_{\theta Dj}$ the position derivative gain of joint j,
K_{Aj} the first order position feedforward gain of joint j,
K_{AAj} the second order position feedforward gain of joint j,
$X_j(t)$ the estimated value of the system drift of joint j,

in the velocity and position control loops for each joint of the robot manipulator control system. In the implementation of

320C30)), an embedded PLC with 1024 I/O ports, an embedded vision system, and a network controller. The Intel 80386/80387 executes the main control algorithm which is composed of inverse kinematics, trajectory planning, external communications, and service routines for peripheral devices while the TMS 320C30 the motion control algorithm for axes of motion of the robot manipulator and the compensation algorithm for the disturbance torques. The robot manipulator with six degree of freedom, which has the specification of Table 1, is used as the controlled object [8]. Fig.2 and 3 show the block diagram representation and the photograph of the entire control system of the robot manipulator, respectively.

Table 1. Body specifications of the robot manipulator.

	Link coordinate parameters					
Joint i	θ_i	α_i	a_i	d_i	Joint Range	Speed
1	180°	90°	0	0	±160°	150°/s
2	-90°	0°	00mm	165mm	±160°	150°/s
3	0°	90°	0	0	±135°	150°/s
4	0°	90°	0	300mm	±180°	163°/s
5	0°	90°	0	0	±105°	163°/s
6	0°	0°	0	86mm	±280°	204°/s

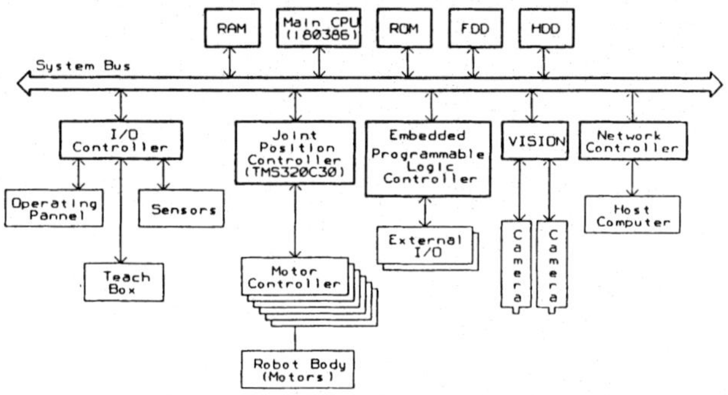

Fig.2 The hardware block diagram of the control system for the robot manipulator.

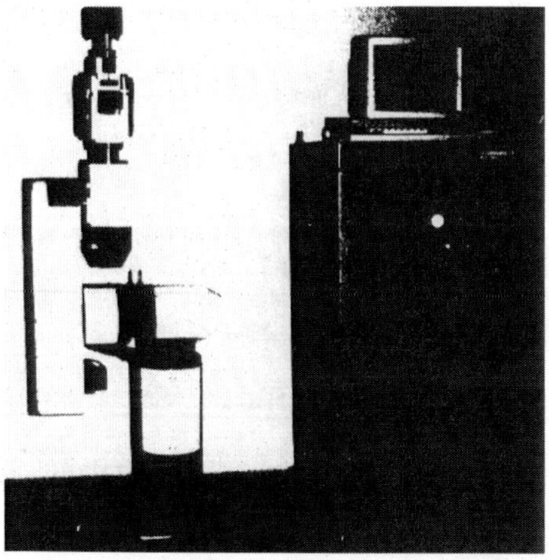

Fig.3 The photograph of the robot manipulator and control system.

The experimental results for two cases are discussed. In experiments, the PTP(point to point) motion of the robot manipulator is only considered and the experimental results for joint 1 is presented. The first case concentrates on the suppression of the disturbance torques due to the physical structure of the robot manipulator with velocity and acceleration feedforward compensation terms. In this case, the system drift is assumed not to exist. Fig.4(a) shows the experimental result for case where velocity and acceleration feedforward compensation terms are incorporated. On the contrary, Fig.4(b) demonstrates the experimental result for case of the control algorithm of no compensation. In Fig.4, P_{o1}^* and P_{o1} are the desired trajectory and the actual trajectory of joint 1, respectively and the difference between them is the tracking errors. Experimental results show that the control algorithm with velocity and acceleration feedforward compensation terms can control the robot manipulator with better control performance than that without them.

In the second experiment, the suppression performance of the disturbance torque due to the system drift via an observer is investigated. In this experiment, the robot manipulator is driven under the state where the optimal velocity and acceleration feedforward compensation terms are incorporated into the position control loop for elimination of the disturbance torque due to the physical structure of the robot manipulator. Fig.5(a) shows the actual trajectory of joint 1 of the robot manipulator in the presence of the system drift with the magnitude of 0.98 V, which is equivalent to the 200 pulse offset in the desired trajectory. As is shown in Fig.5(a), undesirable oscillation lasts till the robot manipulator comes to stop in the presence of the system drift. On the other hand, Fig.5(b) shows the actual trajectory when the observer is included in the control algorithm. Comparison between these two experimental results in Fig.5 demonstrates that the proposed compensation method based on the observer can eliminate the disturbance torque due to system drift to the degree where the desirable dynamic performance of the robot manipulator is obtained.

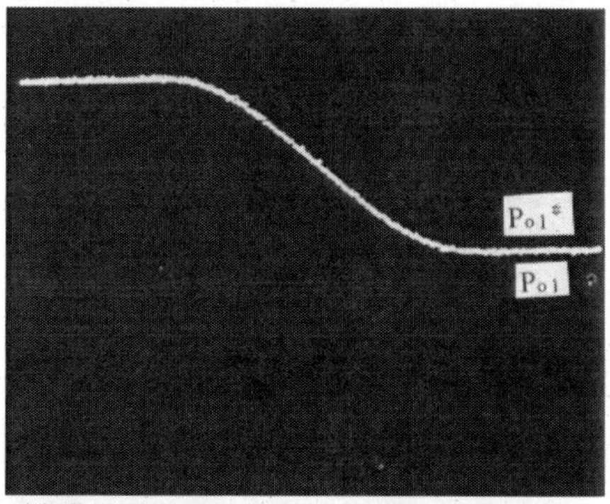

(a) Experimental result for compensation.

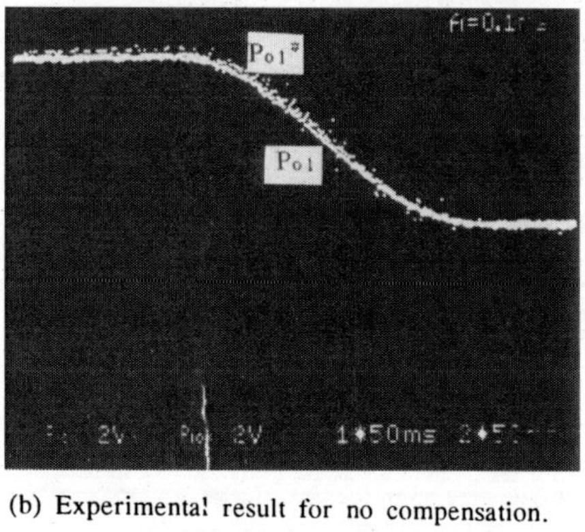

(b) Experimental result for no compensation.

Fig.4 Experimental results for velocity and acceleration feedforward compensation.

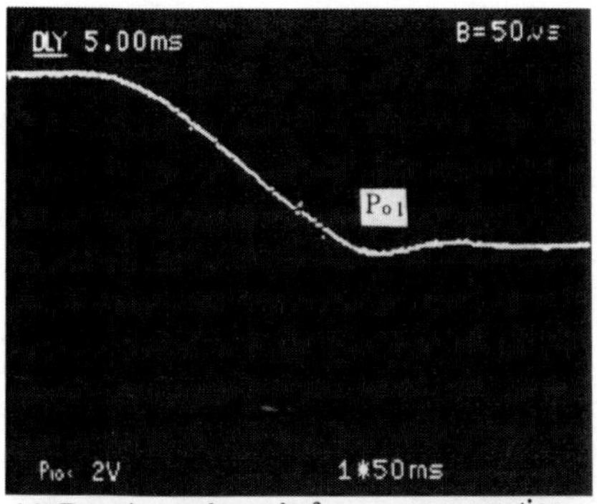

(a) Experimental result for no compensation.

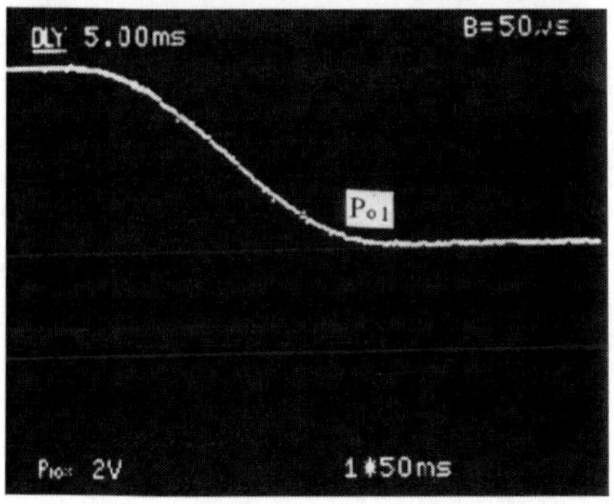

(b) Experimental result for compensation.

Fig.5 Experimental results for compensation of the system drift.

CONCLUSION

A suppression method of the disturbance torque due to the system drift is proposed on the basis of the observer which is derived from the gradient search method. This algorithm combined with the acceleration and velocity feedforward control terms can effectively eliminate the disturbance torque due to both physical structure of the robot manipulator and the system drift in the robot manipulator control system.

The proposed algorithm, which is implemented on the control system with an 80386 /80387 as a main processor and a TMS 320C30 as a processor is applied to the control of a robot manipulator with six degrees of freedom. Experimental results show that the proposed suppression method of the system drift via an observer can be effectively used in controlling robot manipulators with high dynamic performance at both the transient and steady states.

REFERENCES

[1] K. S. Fu, R. C. Gonzalez, and C. S. Lee, Robotics: Control, Sensing, Vision, and Intelligence, Ch.5, McGraw -Hill, 1987.

[2] Brod, D. M. and Novotny, D. W., "Current Control of VSI-PWM Inverter," IEEE Trans. Ind. Appl., vol. 21,pp.562-570, 1985.

[3] Kim, D. I., "Control of Induction Motors via Singular Perturbation Technique and Nonlinear Feedback Control," .IEEE IECON'90, pp.909-914, Asilomar, U.S.A.,1990.

[4] A. K. Bejcy, "Robot Arm Dynamics and Control," JPL, Tech. Memo. 33-669, 1974.

[5] Paul, R., Robot Manipulator: Mathematics, Programming, and Control, Cambridge, MA: MIT Press, 191.

[6] Tarn, T. J., Bejczy, A. K., Yung, X., and Li, Zuofeng, "Effect of Motor Dynamics on Nonlinear Feedback Robot Control Arm Control," IEEE Trans. on Robotics and Automat., vol.7, no.1, pp.114-122, 1991.

[7] Goodwin and Sin, Adaptive Filtering Prediction and Control, Ch.8, Prentice Hall, 1984.

[8] Samsung Electronics, User's Manual - A1-U, 1991.

AN EXTENDED WORKSPACE MAPPING ALGORITHM AND ITS IMPLEMENTATION IN A NUCLEAR TELE-ROBOTIC CONTROL SYSTEM

Young Soo Park*, Woo Tae Jeong*, Ji Sup Yoon*, Jae Sol Lee*,
Hyun Soo Park*, Hyung Suck Cho**

**Remote Technology Department, Korea Atomic Energy Research Institute Daejon, Republic of Korea*
***Precision Engineering Department, Korea Advanced Institute of Science and Technology Daejon, Republic of Korea*

Abstract. When a universal master control device is used for the operation of master-slave telerobots, the accessible workspaces of slave manipulators are reduced. To complement such a problem, a new kinematic mapping strategy is introduced so that the slave manipulator's workspace is fully utilized. By expressing the slave's joint angles in complex numbers, an extended workspace is defined. The slave end effector's location in the extended workspace is then mapped to a location within its reachable workspace. A few complex to real joint angle mapping schemes are suggested and implemented in a master-slave manipulator system consisting of a universal master control mechanism built and modified in the laboratory. It is observed that the proposed mapping provides uninterrupted motion over the entire workspace of the slave manipulator.

Keywords. *Teleoperation, extended workspace, workspace mapping, universal master controller, master-slave system*

INTRODUCTION

For performing delicate tasks in nuclear facilities and other hazardous environments, telerobotics may provide safe and effective solutions. A typical form of telerobotic system is the master-slave type in which a remotely located slave robot replicates the spatial motion of a human operated master mechanism. Unlike the replica master mechanisms of early days, universal master controllers adopt kinematically unsymmetric master and slave manipulators. This type of input device is versatile in that it makes it feasible to apply one device for the control of various slave robots. Here, the differences in their kinematic and dynamic structures are adequately transformed and compensated in the control computer. Added to the versatility, it is also possible to impose more sophisticated control actions using such control systems and useful results are frequently reported.

Despite its many merits, a significant drawback in using universal master mechanism is that it limits the effective workspace of the slave manipulator. In a master-slave type telemanipulator system, the slave's end effector location assumes the spatial location pointed by the master mechanism. Therefore, if the location happens to lie outside the slave manipulator's workspace limit, the slave robot's inverse position kinematics produces one or more of the joint angles in complex numbers. Since complex joint angle can not be assumed by a physical manipulator, robot controller flags an error and usually terminates in such a case. Therefore, universal master mechanisms are currently designed to operate within small work volumes and used only for force reflecting purposes. Obviously small work space of the master reduces the effective workspace of the slave robot, therefore, its manipulation capability is significantly lost. A scaling transformation between the two workspaces may improve the situation to a certain extent. Nevertheless, since typical robotic manipulators have multi degrees-of-freedom articulated structures, their workspaces take complicated shapes and a simple scaling transformation in cartesian space can not provide effective recovery of the lost work volume. For such reasons, rate controlled devices such as switch box, joystick and force ball are still in wide use in spite of their brute inconveniences in use and inaccurate positioning capabilities.

In the realm of manipulator kinematics it is theoretically possible to represent the end effector locations outside a manipulator's workspace boundary using complex valued joint angles. Clearly, joint angles with imaginary parts cannot be implemented physically. However, if complex joint angles are mapped to real numbers by some suitable mapping, the robot approximates the configuration required to produce the desired location within the limits of its work space and the task may continue. Therefore, rather than adopting a master mechanism of limited work volume, it is suggested in this paper to use a workspace mapping method for the control of kinematically unsymmetric master-slave system. Since the precise nature of the motion approximation depends on the kinematic properties of the robot and the complex-to-real mapping used, it is important to investigate various mapping processes and develop an appropriate mapping method.

Very few research efforts are, however, reported in the use of complex numbers in manipulator's inverse kinematics solutions except Lipkin and Park (1988), and Pohl and Lipkin (1990). To this end, this paper presents a few complex to real workspace mapping algorithms and their behaviors are investigated by implementing them in a unilateral master-slave manipulator system, which consists of a universal master input device and a 3 d.o.f. electromechanical manipulator. To implement the algorithm, the control system hardware is developed using a multi axes' motor controller and a computer system in VME bus.

EXTENDED WORKSPACE MAPPING

In a physical robotic manipulator, its work space is limited by three regions: the region corresponding to complex joint angles, the region limited by joint limits and the region that flags kinematic singularities. Furthermore, for a kinematically unsymmetric master-slave telerobotic system, the slave manipulators work space is further limited by the mismatched workspace between the master and the slave manipulators. In this study, it is assumed that the manipulators are capable of making full revolutions at all the joints and the master mechanism has workspace large enough to cover the entire workspace of the slave manipulator. Fig. 1 shows the master and slave's workspaces in conceptual form. This ideal assumption makes it possible to represent all the locations specified in the cartesian reference frame with complex numbered sets of joint angles.

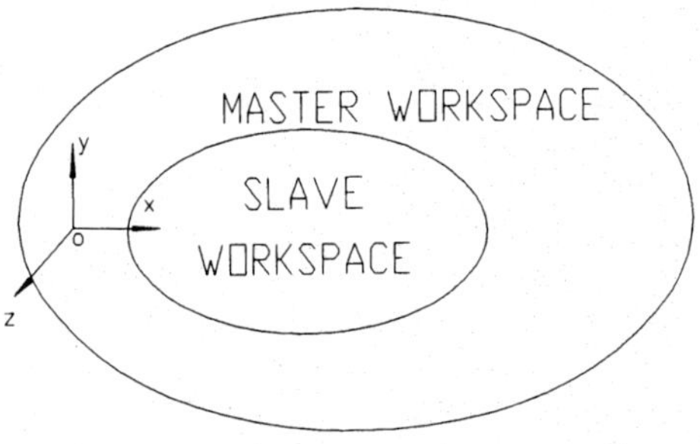

Fig. 1. Workspace of master-slave system

For a manipulator with joints, the forward and inverse kinematics problem is represented as followings:

Forward kinematics: $\underline{X} = f(\underline{\theta})$,

Inverse kinematics: $\underline{\theta} = f^{-1}(\underline{X})$,

where $\underline{X}$ is the forward kinematics function and $\underline{X}$ and $\underline{\theta}$ are n-dimensional vectors of cartesian locations and joint angles, respectively.

Referring to Fig 2, the locus of $\underline{\theta}$ values that produce real $\underline{X}$ composes a manifold **M** in complex joint space $\mathbf{C}^N$. The set **S** is a real valued subset of **M** defined by $\mathbf{Im}(\underline{\theta}) = \underline{\mathbf{O}}$ and is the physically achievable joint configurations. **M** and **S** are given as

$M = \{\underline{\theta}: \mathrm{Im}(f(\underline{\theta})) = \underline{0}\}$ or $\{\underline{\theta}: f(\underline{\theta})\ \mathrm{Real}\}$,

$S = \{\underline{\theta}: \mathrm{Im}(f(\underline{\theta})) = \underline{0}\}$ or $\{\underline{\theta}: \underline{\theta}\ \mathrm{Real}\}$

The workspace of manipulator **W** is then defined as the image of **S** over the transformation f. The extended workspace $\mathbf{W}_e$ is similarly defined as the image of **M** over f. Thus,

$\mathbf{W} = f(\mathbf{S})$ and $\mathbf{W}_e = f(\mathbf{M})$.

Therefore, in complex inverse kinematics, for any real valued $\underline{X}$, there is a complex valued $\underline{\theta}$ such that $\underline{\theta} = f^{-1}(\underline{X})$, and if $\underline{X} \in \mathbf{W}$ then $\underline{\theta}$ has non-zero imaginary part. Then the complex to real joint angle mapping is described as

$\underline{\theta}' = g(\underline{\theta})$ or $\underline{X}' = h(\underline{X})$,

where $\underline{\theta}' \in$ **S** and $\underline{X} \in \mathbf{W}$. Also, the transformations g and h are the mappings from extended workspace to real workspace in joint coordinate system and cartesian coordinate system, respectively. The operation of master-slave manipulator system requires a successive set of transformations as shown in fig 3. Here, the subscript M indicates the master and S indicates the slave, and the single quote ' indicates the value obtained after the extended workspace mapping. As indicated in the figure, the complex-to-real mapping can be implemented both in the joint coordinate system and in the cartesian coordinate system. However, mapping in cartesian domain is quite difficult to realize because generally a manipulator's work space takes complicated shapes and its determination requires much computational effort which can only be accomplished by off-line computation. Thus, in this study, it is proposed to perform the mapping in joint coordinate system.

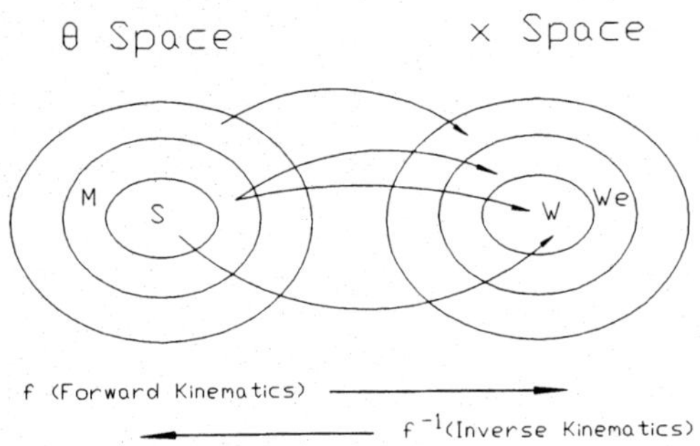

Fig. 2. Extended workspace

MASTER-SLAVE MANIPULATOR SYSTEM HARDWARE

The hardware of the master-slave manipulator system for this research project consists of a universal master input device, a slave manipulator, and a VME based control system. The master input device is designed to be operated with two hands: a spatial position input box and a 3 d.o.f. joystick to specify the three orientations of the wrist. The spatial position input box is shown in Fig. 4 where the spatial position of the handle is determined by the lengths of the four retractable strings attached to it. The other ends of the strings are attached to potentiometers so that the lengths of the string can be measured. The hand orientation is obtained directly from the three tilt angles of the joystick.

To be used as the slave manipulator, an electromechanical manipulator shown in Fig. 5 is modified. This manipulator is originally designed for handling heavy materials in nuclear facilities. Including the mobile trolley and the telescopic mast, it

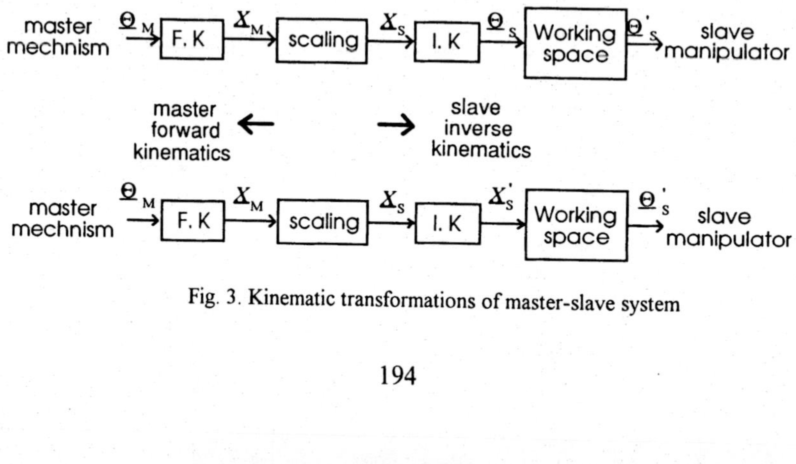

Fig. 3. Kinematic transformations of master-slave system

has total of 10 degrees-of-freedom motion, each of which is rate controlled using a switch box as the input device. For the research purpose, optical position encoders are mounted to the last three joints so that they can be servo controlled.

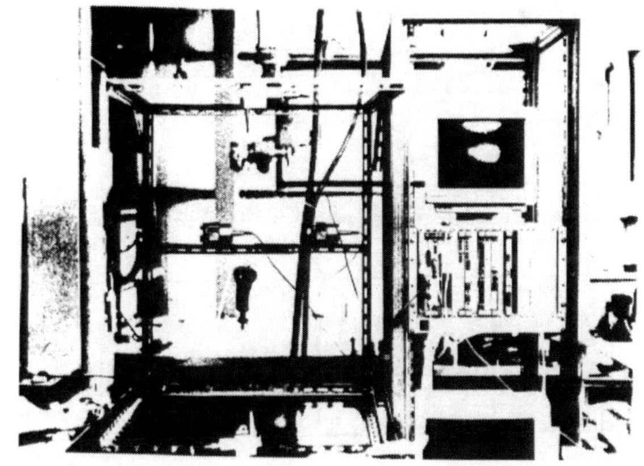

Fig. 4. Master input device and control computer

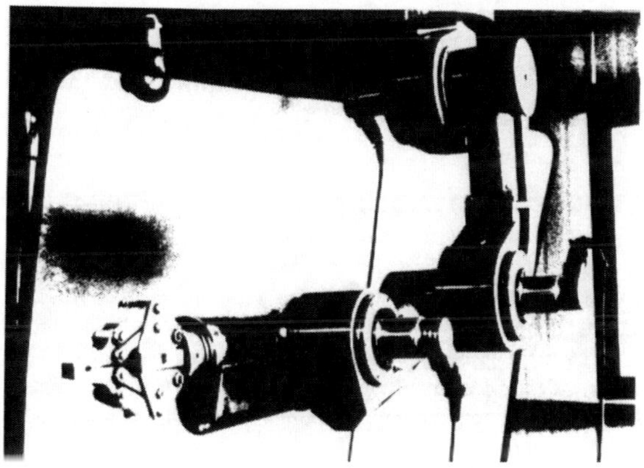

Fig. 5. Slave manipulator

The control computer system is assembled in VME bus environment as shown in Fig. 6. It includes a multitasking VME embedded PC module, an 8-axes motor controller module(PMAC), and an A/D converter module. Both axes controller and the A/D converter module include dual ported rams, which are address mapped through the VME bus to be accessed by the system controller computer module. PMAC controller is equipped with variable gain PID controller and notch filter. Figure 7 shows the control loop of the PMAC controller. To obtain the best control performance, gain tuning is performed. To do this, a step reference input is given and the set point tracking response of the joint actuation motor is observed. The gains are selected to give critically damped response. Figure 8 shows the first joint motor's step response after the gain tuning is done.

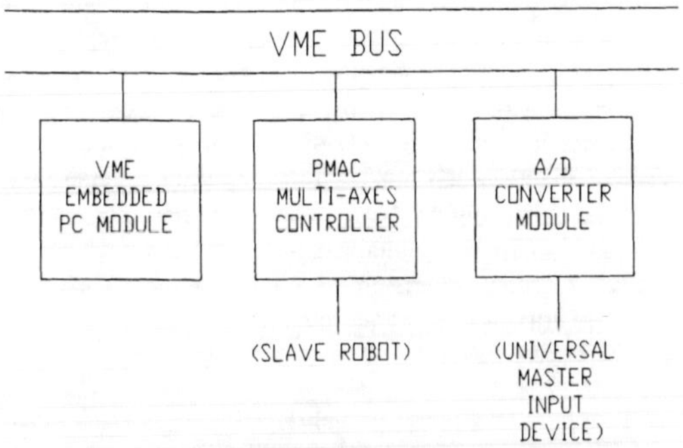

Fig. 6. Structure of control computer system

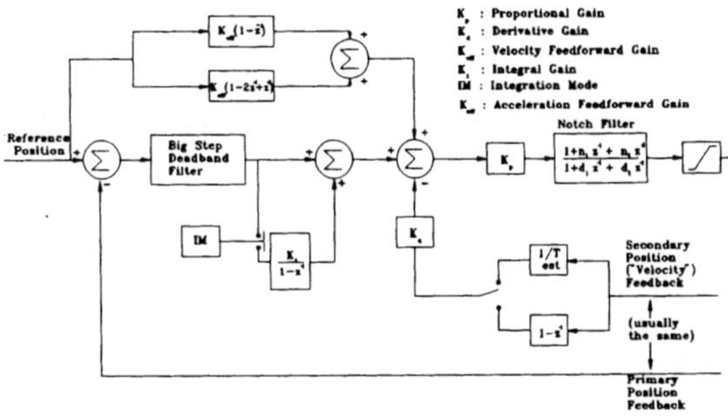

Fig. 7. Block diagram of multi axes controller (PMAC)

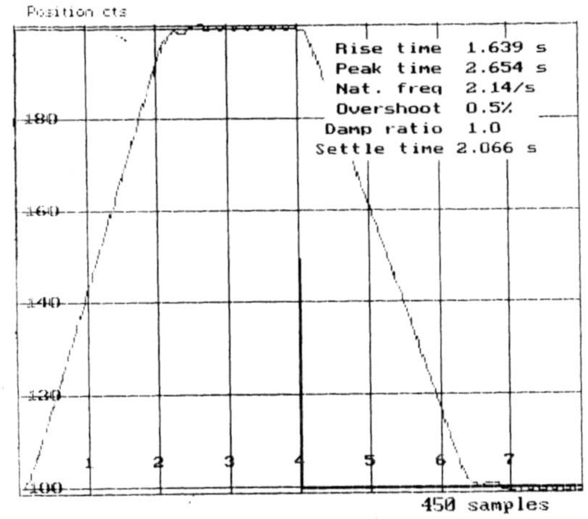

Fig. 8. Step response of the joint actuating motor

KINEMATIC ANALYSIS OF THE MASTER-SLAVE SYSTEM

With the coordinate frame specified in Fig. 9 and the lengths of the string are given as l_1, l_2, l_3, and l_4, the forward kinematics equations to compute the spatial position (x, y, z) are as follows,

$$y = \frac{Y}{2} - \frac{l_1^2 - l_2^2 - Y^2}{2Y}, \tag{1}$$

$$x = X - (h^2 - y^2)^2, \tag{2}$$

$$z = \frac{z}{2} - \frac{l_3^2 - l_4^2 + Z^2}{2Z}, \tag{3}$$

where X, Y and Z are the dimensions of the input box in x, y and z directions, respectively.

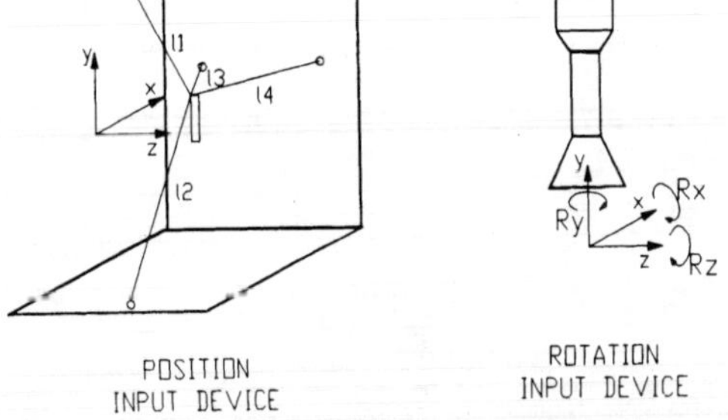

Fig. 9. Coordinate definitions of master input device

The slave manipulator is a 3 d.o.f. articulated type. Kinematically it is considered as a 3 d.o.f. planar manipulator. Its inverse kinematics problem is then solved by closing the loop by adding a hypothetical link $\underline{a}_{41}$ in the direction $\underline{a}_{41}$. Fig. 10 shows a four bar mechanism which is kinematically equivalent to the spatial manipulator. Here, a_{12}, a_{23}, a_{34} and a_{41} are the link lengths of the equivalent four bar mechanism and $\underline{a}_{12}$, $\underline{a}_{23}$, $\underline{a}_{34}$ and $\underline{a}_{41}$ are the unit vectors in the direction of the respective links. The joint angles θ_i s are defined as the angles between the unit vectors a_{ki} and a_{ij} measured in clockwise direction. ϕ_1 is the actuator angle measured from the x-axis of the reference frame.

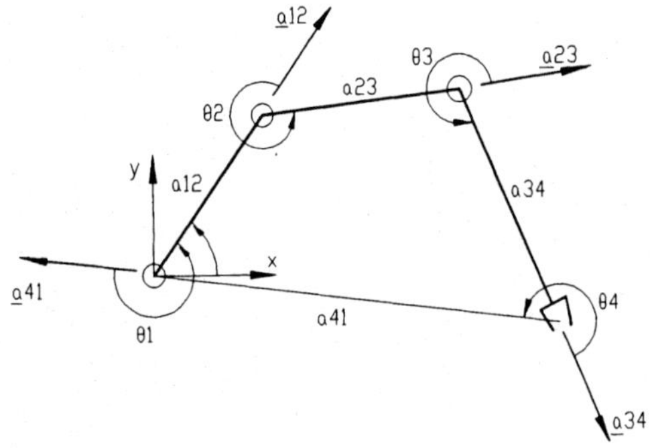

Fig. 10. Planar manipulator represented as a four bar mechanism

Given the planar position, (x, y), and the wrist orientation θ_4, the inverse kinematics equations are derived from the vector loop closure equation of the form.

$$a_{12}a_{12} + a_{23}a_{23} + a_{41}a_{34} + a_{41}a_{41} = 0. \tag{4}$$

Then the inverse kinematics routines are given as the followings. The first joint angle, θ_1, is obtained by solving the input-output equation:

$$AC_1 + BS_1 - D = 0, \tag{5}$$

where $A = 2a_{12}(a_{34}C_4 + a_{41})$,

$B = -2a_{12}a_{34}S_4$,

$D = a_{23}^2 - a_{12}^2 - a_{34}^2 - a_{41}^2 - 2a_{34}a_{41}C_4$.

$$\text{Thus,}\quad C_i^{(\pm)} = \frac{AD\mp B(A^2+B^2-D^2)^{1/2}}{A^2+B^2}, \tag{6}$$

$$S_i^{(\pm)} = \frac{BD\pm A(A^2+B^2-D^2)^{1/2}}{A^2+B^2}, \tag{7}$$

Then the joint angles are determined as

$$\theta_1^{(\pm)} = \mathrm{ATAN}(S_1^{\pm}, C^{\pm}).$$

Here and after, it is denoted as $C_i = \cos\theta_i$ and $S_i = \sin\theta_i$. The second and the third joint angles, 2 and 3, are calculated from sets of simultaneous trigonometric equations. For joint 2,

$$A_1C_2 + B_1S_2 - D_1 = 0,$$
$$A_2C_2 + B_2S_2 - D_2 = 0, \tag{8}$$

where $A_1 = a_{23}C_1$, $B_1 = -a_{23}S_1$,

$D_1 = a_{41} - a_{34}C_4 - a_{12}C1$,

and $A_2 = a_{23}C_1$, $B_2 = a_{23}C_1$,

$D_2 = -a_{12}S_1 + a_{34}S_4$.

Similarily, for joint 3,

$$A_1C_3 + B_1S_3 - D_1 = 0,$$
$$A_2C_3 + B_2S_3 - D_2 = 0, \tag{9}$$

where $A_1 = a_{23}C_4$, $B_1 = -a_{23}S_4$,

$D_1 = a_{12}C_1 - a_{34}C_4 - a_{41}$,

and $A_2 = -a_{23}S_4$, $B_2 = -a_{23}C_4$,

$D_2 = -a_{12}S_1 + a_{34}S_4$.

These equations are solved using Cramer's rule

$$C_i = \frac{\begin{vmatrix} D_1 & B_1 \\ D_2 & B_2 \end{vmatrix}}{\begin{vmatrix} A_1 & B_1 \\ A_2 & B_2 \end{vmatrix}}, \quad S_i = \frac{\begin{vmatrix} A_1 & D_1 \\ A_2 & D_2 \end{vmatrix}}{\begin{vmatrix} A_1 & B_1 \\ A_2 & B_2 \end{vmatrix}} \tag{10}$$

for i = 2, 3. Then, the joint angles are computed by

$$\theta_i = \mathrm{ATAN}(S_i, C_i).$$

As it can be seen from the above procedures, two sets of solution exist for a given end effector location.

PROPOSED EXTENDED WORKSPACE MAPPING METHODS

Mapping in joint angle space can be achieved in two ways: serial mapping and parallel mapping. Serial mapping is accomplished by mapping the complex joint angles into real angles as soon as they are found. Whereas, in parallel mapping the joint angles are calculated in complex mummers until all the joint angles are determined, then these are mapped to real values. In the presented work, both serial and parallel mapping methods are investigated.

A most simple method of serial mapping is proposed to simply discard the imaginary parts of the joint angles as soon as they appear. This scheme is computationally simple and as the simulation in Fig. 11(a) shows it successfully maps into reachable workspace. However, this method tends to propagate the error in joint angle calculations to the remaining joints.

Another proposed method is a parallel mapping in which the joint angles are preserved in complex domain until all joint angles are computed, then the imaginary parts are discarded. However, in actual implementation, parallel mapping requires different formulation of inverse kinematics equations. As it can be seen in the previous section, in solving inverse kinematics problem, sine and cosine values of the joint angles are obtained. When these values are real, the joint angles are determined with proper recognition of their quadrant information with a special version of inverse tangent function, ATAN2. In the actual implementation of this function in standard computer programming languages, proper quadrants are determined by comparing the real values of the sine and cosine. However, complex numbers can not be compared in this manner and the logic of ATAN2 function can not be for computing inverse tangent value of complex angles. Since in parallel mapping, complex joint angles has to be maintained until all the joint angles are calculated, a complex version of ATAN2 function should be devised. In complex algebra, inverse tangent of a complex number Z is given as

$$\tan^{-1}(Z) = \tfrac{1}{2}\log\left|\frac{i+Z}{i-Z}\right| i - \tfrac{1}{2}[\mathrm{Arg}(\frac{i+Z}{i-Z}) - 2n\pi]. \tag{11}$$

For a given complex value $Z = C_i / S_{ii}$, direct application of this identity for the computation of joint angle gives multiple solutions: two numbers which are radians apart. The proper identification of correct quadrant is difficult, if at all possible. To avoid this problem, the inverse kinematic routines are reformulated by making the following tan-half angle substitutions,

$$\cos\theta = {(1-t^2)} \Big/ {(1+t^2)}, \tag{12}$$

$$\sin\theta = {2t} \Big/ {(1+t^2)}, \tag{13}$$

where $t = \tan(\theta/2)$. Then the trigonometric equations, (5), (8), (9) are converted to polynomial equations. Solving these equation results in values of $t = \tan(\theta/2)$. This tan-half angle can be directly applied to equation (11) to give unique values of joint angle, θ.

$$\theta = 2\tan^{-1}(t) = i \log\left|\frac{i+Z}{i-Z}\right| - \mathrm{Arg}\left(\frac{\mathrm{i}+\mathrm{Z}}{\mathrm{i}-\mathrm{Z}}\right). \tag{14}$$

Thus, the parallel mapping method is restructured by making the tan-half angle substitution to uniquely determine the complex joint angles. Figure 11(b) shows a simulation result.

Besides the proposed simple mapping methods, in-depth analyses on the workspace and manipulator's dynamic capacities may lead to more effective mapping methods. It is suggested to consider inertia matrix (Asada, 1983), payload capacity and velocity and acceleration capacities (Vertut and Coiffet, 1984) in developing a heuristic mapping algorithm of the following form

$$\underline{\theta'} = \alpha I(\theta) + \beta P(\theta) + \gamma V(\theta) + \varepsilon A(\theta),$$

where I, P, V and A are some index function representing the inertia, payload capacity, velocity and acceleration capacities, respectively. Also, α, β, γ and ε are weighting factors. At this point, detailed investigations on the various manipulator capacities are not performed, partly because they require precise measurement on dynamic parameters of the slave manipulator.

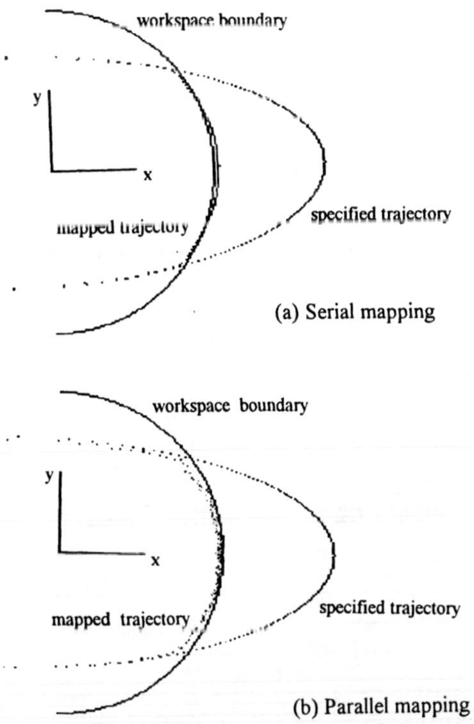

Fig. 11. Simulation of workspace mapping algorithms

PHYSICAL IMPLEMENTATIONS OF THE MAPPING ALGORITHM

The proposed serial and parallel mapping algorithms are implemented in the controller of the master-slave manipulator system to test its functionality. This experimental process is focused on verifying the system's capability on providing continuous kinematic coordination between the master and slave sides over the entire workspace assumed by the master mechanism. Therefore, the effectiveness of the mapping algorithms is evaluated based on their functionality; whether it can provide continuous mapping in the extended workspace of the slave manipulator. This is partly due to the difficulties encountered in devising experimental set up for measuring the manipulator's spatial location. Consequently, the manipulator configurations are identified with visual observations. Also, the dynamic characteristics of the manipulator is left for further studies. In the test operation, using the master input device, the slave manipulator is guided along a trajectory similar to that adopted in the simulation: the trajectory starts from a position inside the reachable workspace, leads out of the workspace boundary and reenters into the workspace. Both serial and parallel methods are tested for complex to real mapping.

Observation on the actual motion trajectory of the slave manipulator reveals that the manipulator follows different trajectories in the two cases. Serial mapping results in fully extended arm configuration over the manipulators in complex joint angle space. Such behaviors are consistent with those predicted by the simulation result. On the other hand, parallel mapping accompanies lateral movement close to vibrations in the direction of extended arm. In both cases, the manipulator maintains continuous motion over the extended workspace, and it may be stated that the proposed mapping methods are both capable of providing suitable mapping and consequently maximum utilization of slave manipulator is achieved.

CONCLUSION

To enhance the workspace capability of the unilateral master-slave type telemanipulator system using universal master input device, the use of workspace mapping methods is proposed and simple mapping algorithms are implemented in an actual master-slave system. The proposed mapping is performed in complex joint space, and physically it corresponds to a spatial mapping in extended workspace whose concept is introduced in this paper. As a related work, a 6 d.o.f. universal master mechanism is designed in such a way to possess large workspace. Also, to be used as a slave manipulator, an originally non servo driven electromechanical manipulator is redesigned to include servo control loop, and VME bus based master-slave control system is developed. Currently, only the mapping algorithms of elementary characteristics are tested. However, it is identified that simple mapping algorithms in complex joint angle space can provide uninterrupted motions over the entire workspace of the slave manipulator.

At this point, numerous areas of further researches seem to be in order. Although elementary mapping methods are introduced in this paper, there are many other ways of mapping in extended workspace with in-depth considerations on various dynamic characteristics of the manipulator. Also, an index to properly evaluate the performance of mapping methods is needed. Since the algorithm's performance is determined by the complex mapping process, a detailed study into the physical interpretation of complex joint angles is also required.

REFERENCES

Lipkin, H. (1983). *Kinematic Control of a Robotic Manipulator with a Unilateral Manual Controllers.* M.S. Thesis,

University of Florida.

Park, Y.S. (1988). *Practical Application of Complex Joint Angle Solutions for Robotic Manipulators*, M.S.Thesis, Georgia Institute of Technology

Lipkin, H., and Duffy, J. (1985). *A Vector Analysis of Robot Manipulators*, Recent Advances in Robotics, John Wiley & Sons, New York.

Lipkin, H., and Park, Y.S. (1988). *Root-Locus analysis of Robot Kinematics*, to appear in ASME journal of Machine Design.

Pohl, E.D., and Lipkin, H. (1990). *Kinematics of Complex Joint Angles in Robotics*, IEEE Robotics and Automation, PP. 86-91.

Churchill, R.V., and Brown, J.W. (1985). Complex Variables and Applications, McGraw-Hill, New York.

Fu, K.S., et. al. (1987). Robotics: Control, Sensing,, Vision, and Intelligence, McGraw-Hill, New York.

Pieper, D.L., and Roth, B. (1969). *The Kinematics of Manipulators Under Computer Control*, Proceedings of the 2nd International Congress on the Theory of Machines and Mechanisms, Vol. 2, PP.159-168

Siciliano, B. (1990). *Kinematic Control of Redundant Robot Manipulators: A Tutorial*, Journal of Intelligence and Robotic Systems, Vol. 3, PP. 201-212.

Acai, T., and Nakano, E. (1985). *Bilateral Control System for Manipulators with Completely Different Configurations*, ICAR, PP. 303-309.

Asada, H. (1983). *A Geometrical Representation of Manipulator Dynamics and Its Application to Arm Design*, ASME J. of DSMC 104-3, PP. 131-142.

Vertut, J., and Coiffet, P. (1984). Robot Technology: Teleoperation and Robotics Evolution and Development, Prentice-Hall, New Jersey, PP. 163-177.

Kumar, A. (1981). *The Workspace of a Mechanical Manipulator*, ASME J. of Mechanical Design, Vol. 103, PP. 665-672.

REAL-TIME ALGORITHM OF THE INVERSE DYNAMICS OF FLEXIBLE ARMS

WANG Guoli, LU Guizhang

Artificial Intelligence & Robotic Research Laboratory
Department of Computer and System Sciences
Nankai University, Tiajin 300071, P.R. China

Abstract. In this paper, the problem of solving the noncausal stable inverse dynamics of the single-link flexible arms in real-time is studied. The state-space form of the noncausal stable inverse dynamics is derived, and an effective algorithm of finding the noncausal stable inverse dynamics in real-time is proposed. Computer simulation studies are given to illustrate our results in this paper.

Keywords. Robots, Inverse Systems, Feedforward Control

INTRODUCTION

When compared to feedback control strategies, the forwardfeed compensation technique based on the inverse dynamics is preferable and attractive for most end-effector high-speed, precision tracking tasks of the flexible arms because complex sensor systems are not necessary. But the rigid inverse dynamic approach fails to be extended to flexible robot arms, and the traditional causal inverse method can not be used to find the stable inverse dynamic solution. These difficulties are largely due to the fact that the inverse of the system with the nonminimum phase properties resulted from the flexibility is unstable.

Instead of the traditional unbounded causal inverse, new bounded noncausal inverse solutions have been found based on Fourier transform technique both in frequency domain and in time domain by E. Bayo and co-workers (Bayo 1987, Bayo,etal. 1989, Bayo and Moulin 1989, Moulin and Bayo 1991), which validity has been proved by several successful experiments (Bayo etal. 1988, 1989). These valuable works make it possible that the forwardfeed compensation method can be applied to the flexible robot systems. Unfortunately, the computational cost of the algorithms given by the previous approaches can still pose significant challenges for their real-time implementations. In the frequency approach (Bayo 1987, Bayo etal. 1988, 1989) the following numerical evaluations are necessary: FFT, solution of linear system of complex equations, inverse FFT, which limit the application of this approach to the tasks of the trajectory tracking control off-line. For the purpose of solving the inverse dynamic solution on-line, an algorithm in time domain was proposed (Bayo and Moulin 1989), in which only evaluation of convolution integral is performed necessarily in real-time while the calculations of the impulse response function can be done off-line based on the frequency method.

In this paper, an alternative algorithm of finding the stable noncausal inverse dynamic solution is presented for the single-link flexible arm, in which the state-space form of the noncausal stable inverse system is constructed based on the causal unstable inverse system instead of the forward dynamics. The generation of the state-space form of the noncausal inverse system in time domain can be easily performed off-line, as its output the noncausal stable inverse dynamic solution can be obtained efficiently on-line, which provides the possibility of developing various real-time control laws based on inverse dynamics for the end-point high-speed trajectory tracking.

Firstly, preliminaries on the formulations of the forward dynamics and the inverse dynamics of the flexible arm is presented in section 2. Then the state-space formulation of the noncausal stable inverse dynamics is followed in section3, and its capablity of the real-time implementation is discussed in section 4. Finally, computer simulation studies are given to illustrate our results in this papers.

PRELIMINARIES

In this paper consider a single-link flexible robot which rotates in horizontal plane and its configuration is shown in Fig.1, where the inertia frame is denoted by OXY, and the rotating frame by oxy. The flexible link is modeled by a slender uniform beam which has bending deflection only in oxy plane. Let l be the length of the link, ρ the mass per length, and EI is the stiffness of the flexible beam. Then the net displacement of the flexible link is given by

$$y(t,x) = x\theta(t) + w(t,x) \tag{1}$$

where $w(t,x)$ denotes the deflection at the point x and time t in the rotating frame, and $\theta(t)$ is the rotation angle of the hub.

Neglecting rotatory inertia and shear deformation and according to the Bernoulli-Euler theory, the motion equation is

$$\rho A\ddot{y} + EID^4y = 0 \tag{2}$$

with the boundary conditions

$$\begin{aligned} &y(t,0) = 0 \quad Dy(t,0) = \theta(t) \\ &EID^2y(t,0) + \tau(t) = J_hD\ddot{y}(t,0) \\ &D^2y(t,l) = 0 \quad D^3y(t,l) = 0 \end{aligned} \tag{3}$$

where D is used here and below to denote the spatial differential operator d/dx, J_h is the moment of inertia of the hub, and τ is the torque applied to the link.
Motion equations (2)(3) can be written into the following compact evolution equation

$$\ddot{u} + Au = \Omega\tau \tag{4}$$

where $u(t,x) = [y(t,x)\ Dy(t,0)]^\tau$, and $\Omega = [0\ J_h^{-1}]^\tau$, A is the elasticity operator

$$Au(t,x) = [\frac{EI}{\rho}D^4y(t,x)\ -\frac{EI}{J_h}D^3y(t,0)]^\tau \tag{5}$$

which is defined on $H = L^2(0,l) \times R^1$ with the product

$$\langle u,v\rangle = \rho\int_0^l u_1v_1dx + J_hu_2v_2$$

It can be verified that A is a closed operator with purely discrete spectrum on H, and has the eigenvalues $\{\lambda_n\}_{n=0}^{\infty}$ and normalized eigenvectors $\{\Psi_n(x) = [\phi_n(x)\ D\phi_n(0)]^\tau\}_{n=0}^{\infty}$, which satisfying

$$\lambda_0 = 0,\ \ 0 < \lambda_1 < \lambda_2 < \ldots < \infty,\ \lim_{n\to\infty}\lambda_n = \infty$$

Then $u(t,x)$ has the following modal expansion:

$$u(t,x) = \sum_{n=0}^{\infty} u_n(t)\Psi_n(x) \tag{6}$$

where $u_n, n = 0,1,\ldots$ are the solutions of

$$\ddot{u}_n + \lambda_n u_n = \psi_n\tau,\ \ \psi_n = D\phi_n(0) \tag{7}$$

Now the evolution equation can be written into the modal state-space form:

$$\dot{x} = Ax + B\tau \tag{8}$$

where

$$A = blockdiag[A_0\ A_1\ A_2\ \ldots]\ \ A_n = \begin{bmatrix} 0 & 1 \\ -\lambda_n & 0 \end{bmatrix} \tag{9}$$

$$B = [B_0^\tau\ B_1^\tau\ B_2^\tau\ \ldots]^\tau \qquad B_n = \begin{bmatrix} 0 \\ \psi_n \end{bmatrix} \tag{10}$$

Let y_o denote the output variable of the forward dynamics, and

$$y_o = Cx,\ \ C = [C_0\ C_1\ C_2\ \ldots],\ \ C_n = [c_{n1}\ c_{n2}] \tag{11}$$

Applying Laplace transformation to (8) and (11) gives the transfer function from the torque τ to the output y_o

$$G(s) = \sum_{n=0}^{\infty}\frac{(c_{n1}+c_{n2}s)\psi_n}{s^2+\lambda_n} \tag{12}$$

The purpose of the inverse dynamics is to find the torque that must be applied by the actuator in order to achieve the desired tip motion prescribed by the trajectory planing. The tip displacement of the flexible link is naturally used as the output variable in extensive literatures, i.e. $y_o = y_{tip}$, named by tip-output. Recently, the well-posedness problem of the inverse dynamics of the flexible robot arms is investigated in Wang (1992), which shows that the inverse dynamics with tip-output is ill-posed due to the fact that the forward dynamics is not invertible. The ill-posedness of the inverse dynamics with tip-output may result in the fact that the inverse dynamics can work well only when the poor accurate forward dynamics must be used. In order to obtain the well-posed inverse dynamics without loss the accuracy of the end-point tracking control, the pseudo tip-output $\hat{y}_{tip}$ is redefined, and integrated output-variable/inverse-dynamic-control design is carried out, in which the following requirements should be met:

(1) the forward dynamics must be invertible;

(2) the inverse dynamics is stable;

(3) $\|\hat{y}_{tip} - y_{tip}\|_{L^2(0,T)}$ should be small as could as possible while

$$\|y_o - y_{tip}\|_{L^2(0,T)} \to 0\ \ as\ \|w(t,l)\|_{L^2(0,T)} \to 0$$

Thus, in what follows we assume that the output variable y_o satifies the following constraints imposed by (1)(2)(3):

(A1) y_o is pseudo tip-output, i.e. $y_o = \hat{y}_{tip}$ satisfying (3) posed above;

(A2) the relative degree of the forward dynamics is 2;

(A3) the L^∞-stable inverse dynamic system exists, i.e. $\inf_{Res=0}|s^2G(s)| > 0$.

By (A1)(A2) we see that

(c1) $c_{n2} = 0$,

(c2) $\sum_{n=0}^{\infty} c_{n1}\psi_n \neq 0$

For a given L^∞-stable noncausal system $G(s)$ with the set of its poles $\sigma(G) \subset C^+ \cup C^-$, the impulse response function $g(t)$ can be obtained by

$$g(t) = \begin{cases} \frac{1}{2\pi i}\int_{-\infty}^{\infty} G_+(j\omega)e^{j\omega t}d\omega & t < 0 \\ \frac{1}{2\pi i}\int_{-\infty}^{\infty} G_-(j\omega)e^{j\omega t}d\omega & t > 0 \end{cases} \tag{13}$$

where $G(s) = G_-(s) + G_+(s)$, $\sigma(G_+) \subset C^+$ while $\sigma(G_-) \subset C^-$, and $C^+ = \{s \in C|Res > 0\}$, $C^- = \{s \in C|Res < 0\}$. The details of the noncausal stable inverse dynamics can be found in Wang (1992).

STATE-SPACE FORMULATION OF NONCAUSAL STABLE INVERSE DYNAMICS

The practical implementation requires that the finite dimensional approximation to the inverse dynamics be made. Here N-modal truncation approxamation to (8)(11) gives:

$$\dot{x}^N = A^Nx^N + B^N\tau \tag{14}$$

$$y_o^N = C^Nx^N \tag{15}$$

where

$$A^N = blockdiag[A_0; A_1; \ldots; A_N]$$
$$B^N = [B_0^\tau\ B_N^\tau\ \ldots\ B_N^\tau]^\tau$$
$$C^N = [C_0\ C_1\ \ldots\ C_N]$$

and the transfer function of the N-modal state-space form is expressed by

$$G^N(s) = \sum_{n=0}^{N}\frac{c_{n1}\psi_n}{s^2+\lambda_n} \tag{16}$$

The causal inverse system can be obtained by

$$\Sigma_m^c \begin{cases} \dot{x}_m = W_mx_m + U_m\bar{y} \\ \tau^N = V_mx_m + q^N\bar{y} \end{cases} \tag{17}$$

where $\Sigma_m(V_m, W_m, U_m, q^N)$ is the minimum realization form of the quartet (V, W, U, q^N), and

$$W = A^N - B^Nq^NC^N(A^N)^2\ \ U = B^Nq^N \tag{18}$$

$$V = -q^NC^N(A^N)^2\ \ q^N = C^NA^NB^N \tag{19}$$

and $\bar{y}$ is defined by

$$\bar{y} = \begin{cases} (y_o^N)^{(2)} & 0 \le t \le T \\ 0 & others \end{cases} \quad (20)$$

By (A3) we see that W has the eigenvalue spectrum $\sigma(W_m) \subset C^+ \bigcup C^-$. Thus there exists the appropriate modal matrix $M = [M_1 \ M_2]$ such that

$$\Lambda = M^{-1} W_m M = blockdiag[\Lambda_+ \ \Lambda_-] \quad (21)$$

or

$$\Lambda_+ = M_1^+ W_m M_1 \quad \Lambda_- = M_2^+ W_m M_2 \quad (22)$$

where $\sigma(\Lambda_+) = \sigma(W_m) \bigcap C^+$, $\sigma(\Lambda_-) = \sigma(W_m) \bigcap C^-$ and the matrix S^+ is the left inverse of the matrix S, defined $S^+ = (S^\# S)^{-1} S^\#$, $\#$ designating conjugate transpose. Consequently, Σ_m is equivalent to the following form

$$\Sigma^c \begin{cases} \dot{x}_+ = \Lambda_+ x_+ + U_+ \bar{y} \\ \dot{x}_- = \Lambda_- x_- + U_- \bar{y} \\ \tau^N = V_+ x_+ + V_- x_- + q^N \bar{y} \end{cases} \quad (23)$$

where

$$U_+ = M_1^+ U_m \qquad U_- = M_2^+ U_m \quad (24)$$
$$V_+ = V_m M_1 \qquad V_- = V_m M_2 \quad (25)$$

And the transfer function $H(s)$ of Σ^c can be decomposed into two parts, causal stable part and causal unstable part, i.e.

$$H(s) = (s^2 G^N(s))^{-1} = H^+(s) + H^-(s) \quad (26)$$

where

$$H^+(s) = V_+(sI - \Lambda_+)^{-1} U_+ \quad (27)$$
$$H^-(s) = V_-(sI - \Lambda_-)^{-1} U_- \quad (28)$$

Applying (13) to (26) gives the explicit expression of the impulse response function of the noncausal stable inverse dynamics as follows

$$h(t) = \begin{cases} h_+(t) = V_+ exp(\Lambda_+ t) + U_+ + q^N \delta(t) & t \ge 0 \\ h_-(t) = V_- exp(\Lambda_- t) + U_- + q^N \delta(t) & t \le 0 \end{cases} \quad (29)$$

Thus the input-output formulation of the noncausal stable inverse dynamics is given by

$$\tau^N(t) = \begin{cases} \int_0^T h_-(t-\tau)\bar{y}(\tau)d\tau & t > T \\ \int_0^t h_-(t-\tau)\bar{y}(\tau)d\tau + \\ + \int_t^T h_+(t-\tau)\bar{y}(\tau)d\tau & 0 \le t \le T \\ \int_0^T h_+(t-\tau)\bar{y}(\tau)d\tau & t < 0 \end{cases} \quad (30)$$

and its state-space formulation is expressed by

$$\Sigma_-^c \begin{cases} \dot{x}_- = \Lambda_- x_- + U_- \bar{y} & 0 < t < \infty \\ \tau_- = V_- x_- & x_-(0) = 0 \end{cases} \quad (31)$$

$$\Sigma_+^c \begin{cases} \dot{x}_+ = \Lambda_+ x_+ + U_+ \bar{y} & -\infty < t < T \\ \tau_+ = V_+ x_+ & x_+(T) = 0 \end{cases} \quad (32)$$

$$\tau = \bar{\tau} + q^N \bar{y} \quad (33)$$

where

$$\bar{\tau}(t) = \begin{cases} \tau_-(t) & t > T \\ \tau_+(t) + \tau_-(t) & 0 \le t \le T \\ \tau_+(t) & t < 0 \end{cases} \quad (34)$$

In summary, the state-space form of the noncausal stable inverse dynamics can be generated by the following algorithm off-line:

1 To estabilish the state-space form of the forward dynamics (14)(15);

2 The causal inverse system can be obtained by (18)(19).

3 The state-space form of the noncausal stable inverse dynamics can be generated by (31)(32)(33) which are based on two-block-diagonlization procedure (22)(24)(25)

It should be noted that the modal matrix construction plays an important role in the two-block-diagonlization procedure. Recently, the matrix sign function provides an effectively computational approach to find the modal matrix (Shieh etal. 1987). The matrix sign function of W_m is defined by

$$\begin{aligned} sign(W_m) &= M^{-1}[I_{n1} \bigoplus I_{n2}]M \\ sign^+(W_m) &= 0.5(I_{nw_m} + sign(W_m)) \\ sign^-(W_m) &= 0.5(I_{nw_m} - sign(W_m)) \end{aligned}$$

where $nw_m = dim(W_m)$, $n1 + n2 = nw_m$, and $n1$ and $n2$ are the dimensions of the matrices Λ_+ and Λ_-, respectively. The matrices M_1 and M_2 in (24) and (25) can be obtained by

$$\begin{aligned} M_1 &= ind[sign^+(W_m)] = [m_1^+ \ \dots \ m_{n1}^+] \\ M_2 &= ind[sign^-(W_m)] = [m_1^- \ \dots \ m_{n2}^-] \end{aligned}$$

where $\{m_i^+\}_{i=1}^{n1}$ and $\{m_i^-\}_{i=1}^{n2}$ are the dependent column vectors of $sign^+(W_m)$ and $sign^-(W_m)$, respectively. The approximation of matrix sign function can be obtained by the following fast recursive algorithm (Shieh etal 1983)

(i) $k = 1$

$$sign_1(W_m) = W_m$$

(ii) $k = k + 1$

$$sign_{mk}(W_m) = sign_{fk}[sign_{nk}(W_m)]$$

where $mk = fk \times nk$

$$\begin{aligned} sign_j[W_m] = {} & [(I_{nw_m} + W_m)^j - (I_{nw_m} - W_m)^j] \\ & [(I_{nw_m} + W_m)^j + (I_{nw_m} - W_m)^j]^{-1} \end{aligned}$$

(iii) if the accuracy is desired, then stop; otherwise, goto step (ii).

To illustrate the off-line algorithm of generating the state-space form of the noncausal stable inverse dynamics, consider the following two degree of freedom system given in Bayo and Moulin(1989):

$$\begin{cases} \begin{bmatrix} 2 & 1 \\ 1 & 2 \end{bmatrix} \begin{bmatrix} \ddot{u} \\ \ddot{v} \end{bmatrix} + \begin{bmatrix} 1 & -1 \\ -1 & 1 \end{bmatrix} \begin{bmatrix} u \\ v \end{bmatrix} = \begin{bmatrix} 1 \\ 0 \end{bmatrix} \tau \\ y = v \end{cases} \quad (35)$$

We assume that $\ddot{y}(t)$ is given by

$$\ddot{y} = \begin{cases} 1 & 0 < t < 1 \\ -1 & 1 < t < 2 \end{cases} \quad (36)$$

which is prescribed by the trajectory planning. Now the problem is to estiblish the state-space form of the noncausal stable inverse dynamics. Applying the off-line algorithm posed above to (35) gives:

- *step 1.* Let $x = [u \ \dot{u} \ v \ \dot{v}]^\tau$, then the state-space form of the forward dynamics (35) can be expressed by

$$\begin{cases} \dot{x} = Ax + B\tau \\ y = Cx \end{cases} \quad (37)$$

where

$$A = \begin{bmatrix} 0 & 1 & 0 & 0 \\ -1 & 0 & 1 & 0 \\ 0 & 0 & 0 & 1 \\ 1 & 0 & -1 & 0 \end{bmatrix}, \quad B = \begin{bmatrix} 0 \\ 2/3 \\ 0 \\ -1/3 \end{bmatrix}$$

$$C = [0 \ 0 \ 1 \ 0]$$

It is easily to be verified that $CB = 0$ and $CAB = -1/3$, thus the relative degree of the forward dynamics is two, and the forward dynamics is invertible.

- *step 2* (18) and (19) estiblish the causal inverse system of (35), which is given by

$$\begin{cases} \dot{x} = Wx + Uy^{(3)} \\ \tau = Vx + qy^{(3)} \end{cases}$$

where

$$W = \begin{bmatrix} 0 & 1 & 0 & 0 \\ 1 & 0 & -1 & 0 \\ 0 & 0 & 0 & 1 \\ 1 & 0 & 0 & 0 \end{bmatrix}, \quad U = \begin{bmatrix} 0 \\ -2 \\ 0 \\ 1 \end{bmatrix}$$

$$V = [0\ 0\ 1\ 0] \quad q = -3$$

and its minimum relization form is

$$W_m = \begin{bmatrix} 0 & 1 \\ 1 & 0 \end{bmatrix}, \quad U_m = \begin{bmatrix} -2.1213 \\ 0 \end{bmatrix}$$

$$V_m = [0\ \ 4.2426]$$

- *step 3* Applying the fast algorithm of finding the matrix sign function to W_m gives

$$sign(W_m) = \begin{bmatrix} 0 & 1 \\ 1 & 0 \end{bmatrix}$$

$$sign^+(W_m) = 0.5 \begin{bmatrix} 1 & 1 \\ 1 & 1 \end{bmatrix}$$

$$sign^-(W_m) = 0.5 \begin{bmatrix} 1 & -1 \\ -1 & 1 \end{bmatrix}$$

then we have $M_1 = [1\ 1]^\tau$ and $M_2 = [-1\ 1]^\tau$. (22)(24)and (25) give

$$\Lambda_+ = 1 \quad U_+ = -1.0607 \quad V_+ = 4.2426$$
$$\Lambda_- = -1 \quad U_- = -1.0607 \quad V_- = -4.2426$$

then the state-space form of the noncausal stable inverse dynamics can be expressed by

$$\Sigma^c_- \begin{cases} \dot{x}_- = -x_- - 1.0607\bar{y} & 0 < t < \infty \\ \tau_- = -4.2426x_- & x_-(0) = 0 \end{cases}$$

$$\Sigma^c_+ \begin{cases} \dot{x}_+ = x_+ - 1.0607\bar{y} & -\infty < t < T \\ \tau_+ = 4.2426x_+ & x_+(2) = 0 \end{cases}$$

$$\tau = \bar{\tau} - 3\bar{y}$$

where $\bar{\tau}$ and $\bar{y}$ are defined by (34) and (20), respectively.

REAL-TIME IMPLEMENTATION

It is interesting to note that although the desired motion time T is finite the domain of the torque produced by the noncausal inverse dynamics is $[-\infty, \infty]$, thus the finite time appoximation for the solution of the noncausal inverse dynamics is necessary for the practical implementation. In what follows, we assume that the values of $\tau(t)$ over finite interval $[T_i, T_f]$ is needed, where $T_i < 0$ and $T_f > T$ are the begining and terminal time of the torque working, respectively, and the trajectory planning provides the values of the tip displacement $y^{(3)}_{tip}(t)$ over the interval $[0, T]$ at ny instants seperated by Δt, which can be expressed by

$$y_{tip}(t) = \sum_{k=1}^{ny} y(k)\phi_k(t), \tag{38}$$

$$\phi_k = \begin{cases} 1 & t \in [(k-1)\Delta t, k\Delta t] \\ 0 & others \end{cases} \tag{39}$$

The discrete-time model corresponding to the state-space form of the noncausal inverse dynamics with the sampling period Δt can be expressed by

$$\Sigma^d_- \begin{cases} x^d_-(k+1) = \Lambda^d_- x^d_-(k) + U^d_- \bar{y}(k) & k > 0 \\ \tau_-(k) = V_- x^d_-(k) \quad x^d_-(0) = 0 \end{cases} \tag{40}$$

$$\Sigma^d_+ \begin{cases} x^d_+(k-1) = \Lambda^d_+ x^d_+(k) + U^d_+ \bar{y}(k) & k < ny \\ \tau_+(k) = V_+ x^d_+(k) \quad x^d_+(ny) = 0 \end{cases} \tag{41}$$

$$\tau(k) = \bar{\tau}(k) + q^N \bar{y}(k) \tag{42}$$

where $f(k) \equiv f(k\Delta t)$,

$$\Lambda^d_- = e^{\Lambda_- \Delta t} \qquad U^d_- = (\Lambda^d_- - I_{n1})\Lambda^{-1}_- U_- \tag{43}$$

$$\Lambda^d_+ = e^{\Lambda_+ \Delta t} \qquad U^d_+ = (\Lambda^d_+ - I_{n2})\Lambda^{-1}_+ U_+ \tag{44}$$

Let $ni = T_i \Delta t^{-1}$ and $nf = T_f \delta t^{-1}$, combining (40)(41) with (42) gives

$$\tau(k) = \begin{cases} V_-(\Lambda^d_-)^{k-ny-1} x^d_-(ny), & ny < k \le nf \\ \sum_{j=0}^{k-1} V_-(\Lambda^d_-)^{k-j-1} U^d_- y(j) & \\ + \sum_{j=k+1}^{ny} V_-(\Lambda^d_-)^{j-k-1} U^d_- y(j) & \\ + q^n y(k) \qquad 0 \le k \le ny & \\ V_+(\Lambda^d_+)^{ny-k-1} x^d_-(ny), & ni \le k < 0 \end{cases} \tag{45}$$

Now we denote the sequence $\{V_-(\Lambda^d_-)^k U^d_-\}^{nf}_{k=0}$ by $\{h_-(k)\}^{nf}_{k=0}$, and $\{V_+(\Lambda^d_+)^k U^d_+\}^{ny-ni}_{k=0}$ by $\{h_+(k)\}^{ny-ni}_{k=0}$. and,

$$H^k_- = \begin{cases} [h_-(0) \ \dots \ h_-(k-1)] & 0 < k \le ny \\ [h_-(k-ny) \ \dots \ h_-(k)] & k > ny \end{cases} \tag{46}$$

$$H^k_+ = \begin{cases} [h_+(0) \ \dots \ h_+(ny-k-1)] & 0 \le k < ny \\ [h_+(-k) \ \dots \ h_+(ny-k)] & k < 0 \end{cases} \tag{47}$$

$$Y^k_- = \begin{cases} [y(k-1) \ \dots \ y(0)]^\tau & 0 < k \le ny \\ [y(ny) \ \dots \ y(0)]^\tau & k > ny \end{cases} \tag{48}$$

$$Y^k_+ = \begin{cases} [y(k+1) \dots \ y(ny)]^\tau & 0 \le k < ny \\ [y(0) \ \dots \ y(ny)]^\tau & k < 0 \end{cases} \tag{49}$$

Thus the solution of noncausal inverse dynamics can be expressed by

$$\tau(k) = \begin{cases} H^k_- Y^k_- & k > ny \\ H^k_- Y^k_- + q^N y(k) & k = ny \\ H^k_- Y^k_- + H^k_+ Y^k_+ + q^N y(k) & 0 < k < ny \\ H^k_+ Y^k_+ + q^N y(k) & k = 0 \\ H^k_+ Y^k_+ & k < 0 \end{cases} \tag{50}$$

In practical applications, $\{y(k)\}^{ny}_{k=0}$ is provided by the trajectory planning, thus Y^k_- for $0 < k \le nf$ and Y^k_+ for $ni \le k < ny$ are known. H^k_- for $0 < k \le nf$ and H^k_+ for $ni \le k < ny$ can be obtain from the state-space form of the noncausal inverse dynamics off-line. For the purpose of solving the inverse dynamics on-line, only numerical calculations in (50) are needed, which involve for every k, $ni \le k \le nf$

- $ny + 1$ real multiplications;
- ny real additions;

and can be easily performed in real-time by commom microprocessors.

NUMERICAL EXAMPLES

In what follows the real-time algorithm proposed based on the elegant state-space form of noncauasl stable inverse dynamics is applied to solve the stable inverse dynamics. Here the basic parameters used for the simulation are

$$l = 1[M] \qquad EI = 5.74[N.M^2]$$
$$\rho = 0.2332[KG/M] \quad J_h = 0.005[KG.M^2]$$

The pseudo tip-output y_o is defined by

$$\hat{y}_{tip} = lDy(t,0) + 0.9w(t,l)$$

The desired tip acceleration is shown in Fig.2. The torques provided by the inverse dynamics for $N = 3$ and $N = 4$ are shown in Fig.3, while the desired tip motion and the tip motions when the torques shown in Fig.3 are applied to the flexible link are shown in Fig.4.

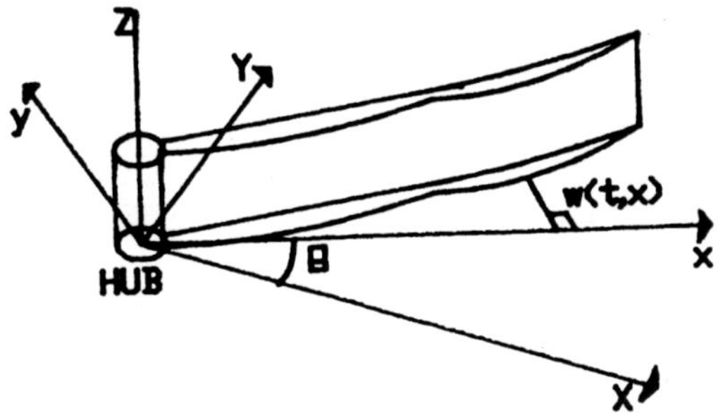

Fig.1 The structure of a planar single-link flexible arm

CONCLUSIONS

For the purpose of solving the noncausal stable inverse dynamics on-line, the elegant state-space form of the noncuasal stable inverse dynamics is derived, and an effective algorithm in real-time is proposed. Compared to Bayo's algorithm in time domain, the algorithm proposed in this paper has the following advantages:

- the matrix operations are only needed in the generation procedure of the state-space form of the noncausal stable inverse dynamics;
- the state-space form of the noncausal stable inverse dynamics is prefered to the input-output formulation when the inverse dynamic analysis and the feedforward controller design are carried out;
- more effective real-time implenmentation may be possible.

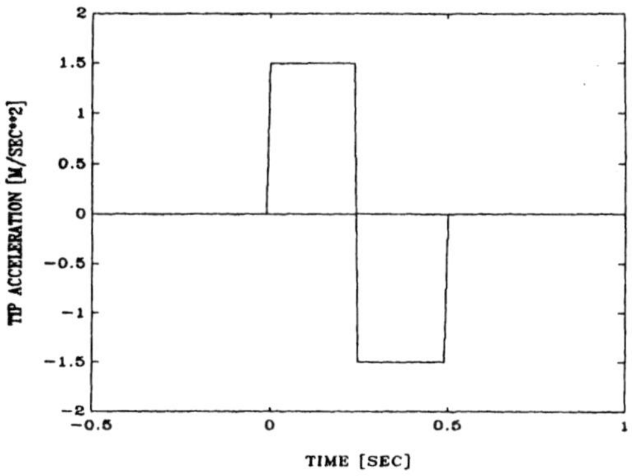

Fig.2 The desired tip acceleration

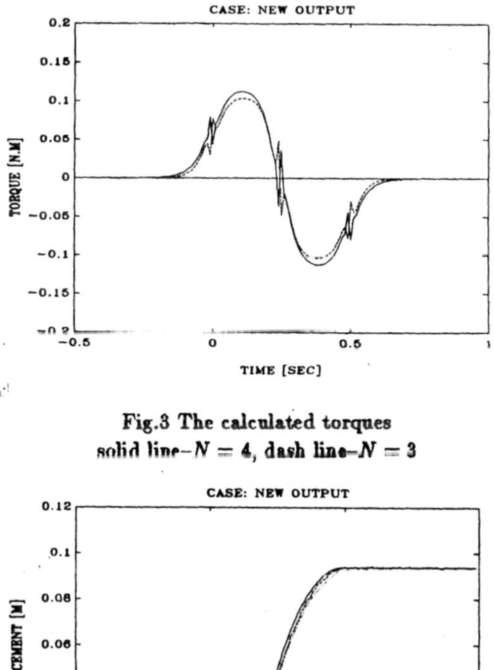

Fig.3 The calculated torques solid line–$N = 4$, dash line–$N = 3$

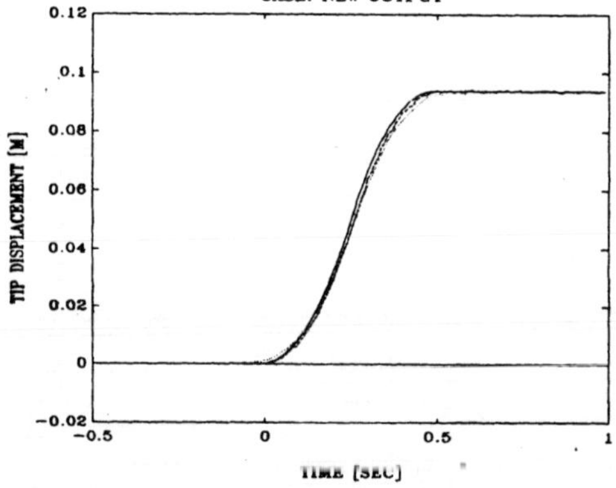

Fig.4 The tip motion solid line–desired tip motion dash line–$N = 4$, dot line–$N = 3$

REFERENCES

Bayo, E. (1987). J. Robotics systems, 4(1), 63-75.

Bayo, E. et al., (1988). Int. J. of Robotics and Automation, 3(3), 150-156.

Bayo, E. et al., (1989). Int. J. of Robotics Research, 8(6), 49-62.

Bayo, E and Moulin, H. (1989). Proc. of the IEEE Int. Confere. on Robotics and Automation, 710-715.

Moulin, H. and Bayo, E. (1991). ASME J. of Dyna. sys., Meas., and Contr., 113, 320-324.

Shieh, L. S. etal.(1984). Int. J. Systems Sci., 15(11), 1203-1220.

Wang, G. (1992). Doctoral Thesis, Nankai University, Tianjin, PRC, Chapter1-2.

REAL-TIME NAVIGATION AND OBSTACLE AVOIDANCE BASED ON GRIDS METHOD FOR FAST MOBILE ROBOT

Zhao-Qing Ma, Zeng Ren Yuan

Department of Computer Science and Technology, Tsinghua University, Beijing, PRC

ABSTRACT We have developed and implemented a real–time navigation and obstacle avoidance method based on grids on our mobile robot THMR–2. This method permits the detection of unknown obstacles and avoidance collision based on the information of ultrasonic sensors while autonomously steering the mobile robot toward the given target in smooth and continuous motion. Experimental results are given in some typical environment. The strong power of the method has been demonstrated.

Keywords: Mobile robot, Real–time Navigation and Obstacle Avoidance.

INTRODUCTION

The autonomous mobile robot must have the capabilities of perception, decision making, planning, navigation and control in order to carry out intelligent tasks. Real–time obstacle avoidance and navigation is one of the key issues to successful application of mobile robot system. The difficulties lie in that a lot of information about the environment is required while the information must be processed quickly. Robotics researchers have done much work in this field. During the past few years, potential field methods (PFMs) and grids methods have gained increased popularity in the field of robots and mobile robots. But there are some significant problems that are inherent to PFMs[3]. Borenstein[1] developed a method (VFH) which used grids for the representation of obstacles. The robot controlled by VFH performs very well. However, a disadvantage of the methods based on grids is that the information of environment representation will become heavier and heavier in order to satisfy high resolution of environment representation when the robot moves. So, the grids method is limited in practical use.

In this paper, based on [1], we introduce a real–time navigation and obstacles avoidance method based on grids for mobile robot. This method contributes a significant improvement on the grids method and has implemented on our mobile robot THMR–2. We present our work in the following: (1) Selection of the size of the grid. (2) Determination of the movable window for the rectangular mobile robot in shape. (3) Determination of weight parameters. (4) Optimization of the environment information.

CONTROL ALGORITHM

The representation of environment

The representation of obstacles depends strongly on the characteristies of the sensors that provide the raw data. In many mobile robots, ultrasonic sensors are used because of their low cost and easy operation. Around our mobile robot THMR–2, 15 ultrasonic sensors are mounted on a horizontal semicircle covering a 180 ° panorama. The sensor can detect the distance 0.7–7m in its direction.

We use a two–dimensional Cartesian historgam grid for the representation of obstacles, each cell in the histogram grid holds a certainty value CV that represents the confidence of the algorithm in the existance of an obstacle at that location. A probability profile is projected onto a cell in the histogram grid by a rang reading. For ultrasonic sensors, this cell corresponds to the measured distance and lies on the acoustic axis of the sensor. A probabilistic distribution is actually obtained by continuously and rapidly sampling each sensor while the robot is moving. Thus, the same cell and its neighboring cells are repeatedly incremented. This results in a histogramic probability distribution in which high CVs are obtained in cells close to the actual location of the obstacle.

The size of a cell has some effect on the performance of the control algorithm. If the size of a cell is small, the resolution of environment representation is high, and the power of finding a path in cluttered environment is strong, but the capability of resistance to disturbance is week, and the information of environment representation is heavy, thus the procedure of making a decision is computationally intensive and would impose a heavy time penalty. The size of a cell is also depended on the property of sensors. If sensors are quick and accurate at sampling environment, the size of a cell can be small. If not, the size of a cell will not be too small. In our implementation, the size of a cell is 0.3m * 0.3m. We set up a environment coordinate system to hold the detailed description of the robot's environment. In this coordinate system, the two–dimensional Cartesian histogram grid is continuously updated in real–time with range data sampled by the onboard range sensors. The histogram grid is absolute and does not change with the robot's momentary location.

By the following equation (2–1), the measured point on corresponding to obstacles can be transformed into the environment coordinate system. (see Fig.1).

$$\begin{cases} x'_e = X_r + d_o * cos(\alpha_r + \alpha_i) \\ y'_e = Y_r + d_o * sin(\alpha_r + \alpha_i) \end{cases} \qquad (2-1)$$

By the following equation (2–2), the coordinate (x'_e, y'_e) can be mapped into a cell(i,j) at that location in the environment coordinate system.

$$\begin{cases} x_e = INT(x'_e / w) * W + INT(w/2) \\ y_e = INT(y'_e / w) * W + INT(w/2) \end{cases} \qquad (2-2)$$

where

(x_e, y_e) Coordinate of the cell(i,j) in the environment coordinate system

W Width of the cell(i,j).

<u>The determination of the direction</u>

Along with the robot moves a notional window overlaying a semicircle region with the radius R in the environment coordinate system. We will call this region "movable window", and the cell that momentarily belongs to the movable window will be called "window cell". In our implementation, the radius R is 3.6m. Only the window cells have immediate influence on the robot control.

For symmetrically shaped mobile robot, the centre of the movable window is defined at the geometric centre of the robot. For rectangular shaped mobile robots, the centre of the movable window can be defined at the point from which the distance to the robot's front edge equals to its distance to both sides of the robot. We set up a window coordinate system at the movable window. The centre of the window coordinate system is defined at the centre of the window. The direction of the X–axis is defined in the direction from the window centre to the front of the robot right ahead (see Fig.2).

By the following equation (2–3), the cell(i,j) in the environment coordinate can be transformed into the movable window.

$$\begin{cases} x_w = (x_e - x_r) * cos(\alpha_r) \\ \qquad + (y_e - y_r) * sin(\alpha_r) - D_w \\ y_w = (y_e - y_r) * cos(\alpha_r) \\ \qquad + (x_e - x_r) * sin(\alpha_r) \end{cases} \qquad (2-3)$$

where

(x_w, y_w) Coordinate of the cell (i,j) in window coordinate system.

(x_r, y_r, α_r) The position and the orientation of the robot in the environment coordinate system.

(x_e, y_e) Coordinate of the cell(i,j) in the environment coordinate system.

D_w The distance between the window centre to the centre of the robot.

Around the centre of the movable window, we separete the movable window to n angular sectors of width θ (see Fig.2). θ may be chosen arbitrarily but must be such that $n = 180 / \theta$ is an integer. Each sector k corrsponds to a discrete angle quantized to multiples of θ. In each sector k, there is a value H_k which represents the obstacle density in the direction k. We will treat the content of each window cell as an obstacle vector, the direction of which is determined by the direction β from the cell to the window centre.

$$\beta_{i,j} = tan^{-1}(y_w / x_w) \qquad (2-4)$$

and the magnitude is given by

$$M_{i,j} = CV_{i,j}^2 * [a - b * D_{i,j}] \qquad (2-5)$$

where

$\beta_{i,j}$ Direction from the window cell(i,j) to the window centre.

$M_{i,j}$ Magnitude of the obstacle vector at the window cell(i,j).

$CV_{i,j}$ Certainty value of the window cell(i,j).

$D_{i,j}$ Distance between the window cell(i.j) and the window centre.

(x_w, y_w) Coordinate of the window cell(i,j).

a,b Positive constants.

specifically, a and b are chosen such that $a - b * R = 0$ where R is the radius of the movable window. When $D_{i,j} = R, M_{i,j}$ is equal to zero for the farthest window cell and increases lineraly for closer cells. If a and b are large, the difference there is in $M_{i,j}$ is great with different $D_{i,j}$, this is favourable to determining the direction, but if a and b are too large, this is not favourable to smoothing sectors by the following smoothing function.

Sector k is established through

$$k = INT(\beta_{i,j} / \theta) \qquad (2-6)$$

For each sector k, the obstacle density H'_k is calculated by

$$H'_k = \sum_{i,j} M_{i,j} \qquad (2-7)$$

Because of the discrete nature of the grid, the result of the mapping may appear ragged and cause errors in the selection of the steering direction. Therefore a smoothing function is applied to H'_k which is defined by

$$H_k = (H'_{k-L+1} + 2H'_{k-L+2} + \cdots + L \cdot H'_k + \cdots + 2H'_{k+L-2} + H'_{k+L-1}) / (2L - 1) \qquad (2-8)$$

In our current implementation, $L = 3$ yields satisfactory smoothing results. Any valley comprising of several neighboring sectors with obstacle densities below a threshold is a candidate–valley, the algorithm selects the one that most closely matches the direction to the target.

Optimal stratagem

As sensors continuously sample environment, the information the computer gained will be greater and greater, This would not only overflow the memory but also impose a heavy time penalty. This will not satisfy the requirement of real–time response. For this reason, the representation of environment must be optimal. We found the information of the region that is in back of a distance to the robot is usefulless. So, if these information would be deleted, there was not much effect on making a determination. Optimal step is as following

(1). We get a cell from the environment coordinate system.

(2). By equation (2–2) transform it into the window coordinate system.
(3). If the cell is in back of the robot and the distance between the robot and the cell is farther than a constant TH, we will delete this cell from the environment coordinate system.
In this way, the information of the environment will be light, and the work space of a robot will not be limited.

EXPERIMENTAL RESULTS

Experimental system

This control algorithm has been implementated on our mobile robot THMR–2 which is developed by the department of Computer Science and Technology in Tsinghua University. The THMR–2 is rectangular (1.5m × 0.8m)in shape with three wheels, two real wheels are driven separatly, the front one moves freely. A run's maximum speed can be set at the beginning of a run. The robot tries to maintain this speed during the run unless there is a drastic change in direction. A reduction in speed is necessary to allow for the steering wheels to turn into the new direction, this reduction in speed is implemented by the following function

$$speed' = speed_m * (1 - c_1 * k - c_2 * |k - k_o|) \quad (3-1)$$

where

c_1, c_2	Positive constants.
k_o	the right line direction of the robot
k	the determined travel direction.
$speed_m$	the maximum travel speed of a run.

Then, we can calcalute the actual speeds of both real wheels

$$speed_L = speed' * (1 - c_3 * |k - k_o|)$$
$$speed_R = speed' * (1 + c_3 * |k - k_o|) \quad (3-2)$$

where

c_3	Positive constants.
$speed_L$	The speed of the left wheel.
$speed_R$	The speed of the right wheel.

THMR–2 has a two–layer control system, two on–board 8031 computers serve as the ultrasonic sensors subsystem and the location subsystem of the robot. The control algorithm proposed runs in a 16MHZ, IBM PC 80386 with co–processor 80387 computer.

Experiment discussion

In extensive tests, we run the algorithm to control our mobile robot THMR–2 through difficult courses in which obstacles are unmarked. Our mobile robot THMR–2's maximum speed can reach 1.10m / sec. In most experiments the robot runs at its run's maximum speed (0.28m / sec). This speed was reduced only when an obstacle was approached frontally.
In Fig.3, if we place obstacle A, THMR–2 will travel along the path 1 to reach the target. When obstacle A is taken away, test again, THMR–2 will travel along the path 2 automically. That expresses THMR–2 has the capability of real–time local planning and usually selects the direction that most closely matches the direction to the target.
Fig.4 shows the path that THMR–2 travels along a obstacle and seeks the target automically. The test shows the navigation power of THMR–2. The result expresses that the path is smooth, continuous and the travel is stable.
THMR–2 has the capability of travelling throught a narrow door. In this case, some present ultrasonic sensors are momentary failture, THMR–2 can use the environment coordinate system to decide the direction. THMR–2 can also travel from wide corridor to narrow corridor without oscillation.

CONCLUSIONS

The control algorithm proposed in this paper has been implemented on our mobile robot THMR–2. This algorithm is computationally efficient, very robust, insensitive to misreadings of ultrasonic sensors and allows continuous and fast motion of the mobile robot without stopping in front of obstacles. When present ultrasonic sensors are momentary failture, THMR-2 can also find the right path. We use optimal stratagem to prevent the informention of environment representation from getting too heavy.

The experiment shows that THMR-2 controlled by our control algorithm performs very well in real-time obstacle avoidance, navigation and planning. THMR-2 also appears some intelligence.

The idea of the control algorithm has perspective in practice. That can easily combines the local planning with the global planning, and can also be used in real-time building map.

REFERENCES

〔1〕 Borenstein, J. and Koren, Y., "Real-time Obstacle Avoidance for Fast Mobile Robots in Cluttered Environments" The 1990 IEEE International Conference on Robotics and Automation, Cincinnati, Ohio, May 13-18, 1990, pp.572-577.

〔2〕 Koren, Y., and Borenstein, J., "Potential Field Methods and Their Inherent Limitations for Mobile Robot Navigation", Proceedings of the 1991 IEEE International Conference on Robotics and Automation, Sacramento, California-April, 1991, pp.1398-1404.

〔3〕 Borenstein, J. and Koren, Y., "Real-time Obstacle Avoidance for Fast Mobile Robots", IEEE Transactions on System, Man, and Cybernetics, Vol.19, No.5, Sept / Oct, 1989, pp.1179-1187.

〔4〕 Moravec, H.P., "Sensor Fusion in Certainty Grids for Mobile Robots", Al Magazine, Summer 1988, pp.61-74.

〔5〕 Arkin, R.C., "Motor Schema-Based Mobile Robot Navigation". The International Journal of Robotics Research, August 1989, pp.92-112.

〔6〕 Elfes, A., "Sonar-based Real-world Mapping and Navigation", IEEE J.Robotics Automat., Vol. RA-3, No.3., 1987, pp.249-265.

〔7〕 Zhao-Qing Ma, "Research on Real-time Navigation and Dynamic Obstacle Avoidance for Mobile Robot", M.S. Dissertation, Computer Science and Technology, Tsinghua University, Beijing, March, 1992.

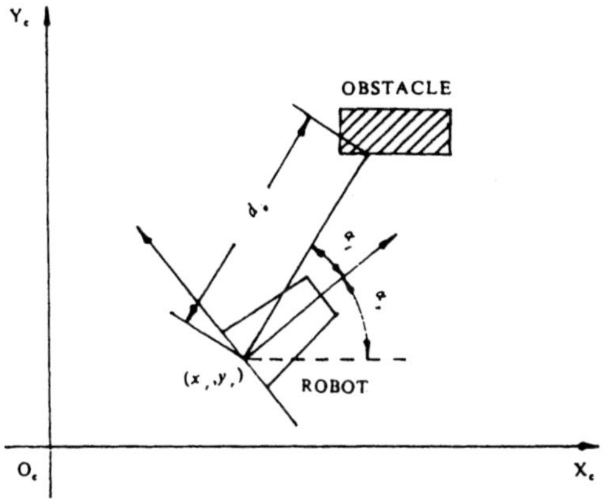

Fig.1. Mapping of the measured point on corresponding to obstacles onto the environment coordinate system.

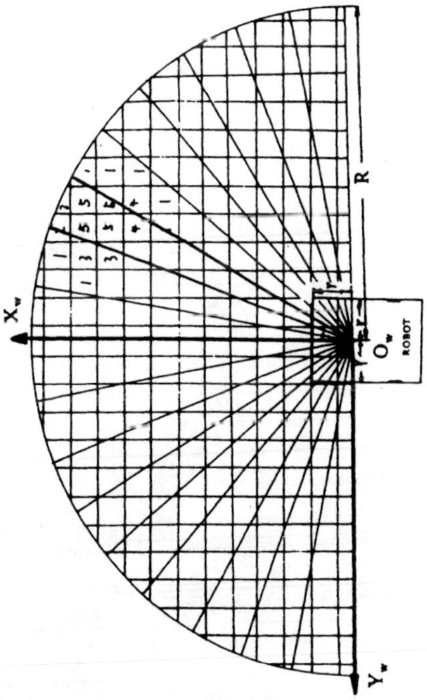

Fig.2. Grid representation of the environment and the movable window.

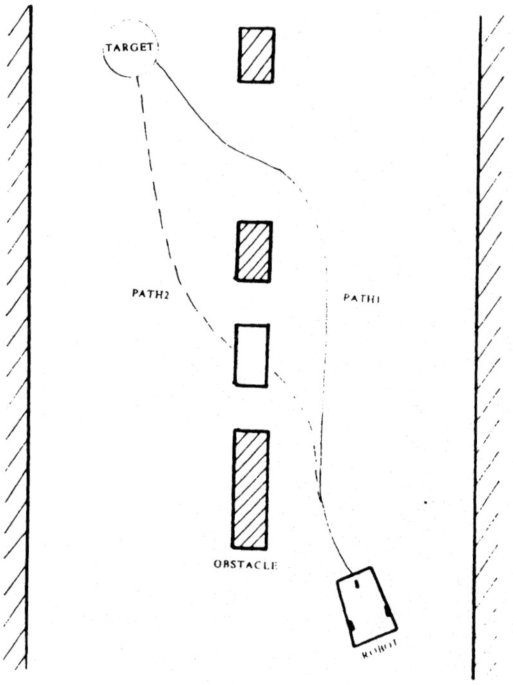

Fig.3. Robot always selects the optimal direction to the target

Fig.4(a)

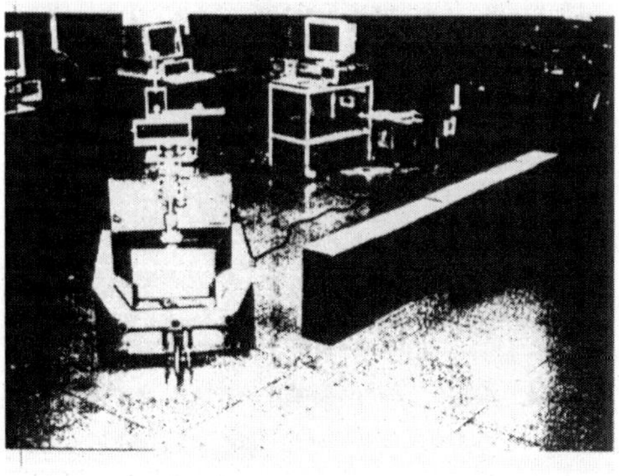

Fig.4(b)

Fig.4(c)

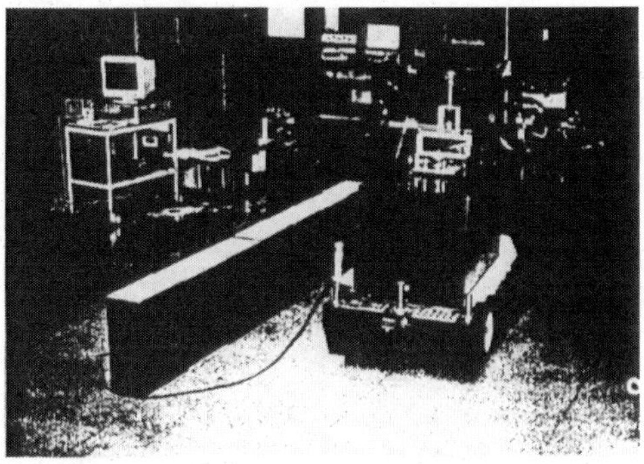

Fig.4 (d)

Fig.4. Robot travels along a obstacle and seek the target automically

STRUCTURED MATRICES AND FAST RLS ADAPTIVE FILTERING

Ali H. Sayed and Thomas Kailath

Information Systems Laboratory, Stanford University, Stanford, CA 94305

Abstract. We present a new approach to the discrete-time Chandrasekhar recursions and some generalizations thereof. We extend the recursions to a class of structured time-variant state-space models, and discuss connections with the (generalized) Schur algorithm. We also apply the extended recursions to the adaptive filtering problem and give a transparent derivation of fast recursive least-squares algorithms.

Keywords. Recursive least squares, Adaptive systems, State-space methods, Kalman filters, Chandrasekhar recursions, Schur algorithm, Time-variant models.

INTRODUCTION

Adaptive filtering is widely used to cope with time-variations of system parameters and to compensate for the lack of a priori knowledge in the statistical properties of the input data. For a variety of reasons, the least-squares criterion is used to derive linear estimators for the desired parameters (Kailath, 1981; Haykin, 1991) A wide range of algorithms and schemes has been developed to exploit the data structure and to reduce the computational complexity. These basically fall into three main groups of adaptive filtering algorithms: Recursive Least Squares (RLS) and the corresponding fast versions, Lattice Least-Squares (LLS), and QR-based least-squares (see *e.g.*, (Haykin, 1991) and the references therein for more details on the subject). The derivation of the different schemes may be put into a unified framework (Sayed and Kailath, 1992) by relating the filtering problem to convenient structured (covariance) matrices. We focus here on the RLS approach and show that the associated fast versions follow by applying the (generalized) Schur algorithm to a Toeplitz-like covariance matrix. Fast RLS algorithms were originally derived in (Carayannis, Manolakis and Kalouptsidis, 1983; Cioffi and Kailath, 1984). It was later shown (Slock, 1989; Houacine, 1991) that these fast versions also follow by considering the Chandrasekhar equations (Kailath, Vieira and Morf, 1978) associated with the linear least-squares estimation problem. We present a new approach to the discrete-time Chandrasekhar recursions and some generalizations thereof. The approach is a simplification and extension of the derivation in (Kailath, Sayed and Lev-Ari, 1992). We extend the Chandrasekhar recursions to a class of *structured time-variant* state-space models, and show that within this framework, fast versions of the recursive least-squares algorithm in the single channel, multichannel, nonlinear or multidimensional contexts, also follow naturally.

* This work was supported in part by the Air Force Office of Scientific Research, Air Force Systems Command under Contract AFOSR91-0060. The work of the first author is also supported by a fellowship from Fundação de Amparo à Pesquisa do Estado de São Paulo and by Escola Politécnica da Universidade de São Paulo, Brazil.

First, a brief review of known results. Consider a $p \times 1$ process $\{\mathbf{y}_i\}$ with an n-dimensional state-space model

$$\begin{aligned} \mathbf{x}_{i+1} &= F_i \mathbf{x}_i + G_i \mathbf{u}_i \\ \mathbf{y}_i &= H_i \mathbf{x}_i + \mathbf{v}_i \qquad \text{for } i \geq 0 \end{aligned} \tag{1}$$

where $\{F_i, G_i, H_i\}$ are known matrices with dimensions $n \times n$, $n \times m$ and $p \times n$ respectively. We assume that $\mathbf{x}_0, \mathbf{u}_i$ and $\mathbf{v}_i$ are stochastic variables that satisfy

$$E\mathbf{x}_0 = \bar{\mathbf{x}}_0, \quad E(\mathbf{x}_0 - \bar{\mathbf{x}}_0)(\mathbf{x}_0 - \bar{\mathbf{x}}_0)^* = \Pi_0$$

$$E\mathbf{u}_i \mathbf{x}_0^* = E\mathbf{v}_i \mathbf{x}_0^* = 0, \quad E\mathbf{v}_i = E\mathbf{u}_i = 0$$

$$E \begin{bmatrix} \mathbf{u}_i \\ \mathbf{v}_i \end{bmatrix} \begin{bmatrix} \mathbf{u}_j^* & \mathbf{v}_j^* \end{bmatrix} = \begin{bmatrix} Q_i & C_i \\ C_i^* & R_i \end{bmatrix} \delta_{ij}$$

with R_i positive-definite. The symbol $*$ denotes complex conjugation and the letter E denotes expected value. Let $\hat{\mathbf{x}}_{i|i-1}$ be the linear least-squares estimate of $\mathbf{x}_i$ given $\{\mathbf{y}_0, \ldots, \mathbf{y}_{i-1}\}$. The Kalman filter (Kalman, 1960) computes this quantity via the recursion: $\hat{\mathbf{x}}_{0/-1} = \bar{\mathbf{x}}_0$,

$$\hat{\mathbf{x}}_{i+1|i} = F_i \hat{\mathbf{x}}_{i|i-1} + K_i R_{\epsilon,i}^{-1} \epsilon_i \tag{2}$$

where $\epsilon_i = \mathbf{y}_i - H_i \hat{\mathbf{x}}_{i|i-1}$, $R_{\epsilon,i} = E(\epsilon_i \epsilon_i^*)$ and $K_i = E(\mathbf{x}_{i+1} \epsilon_i^*)$. Kalman showed that K_i and $R_{\epsilon,i}$ can be computed by the expressions

$$K_i = F_i P_{i|i-1} H_i^* + G_i C_i, \quad R_{\epsilon,i} = H_i P_{i|i-1} H_i^* + R_i$$

where $P_{i|i-1}$ is the error covariance in the one-step prediction of $\mathbf{x}_i$: $P_{i|i-1} = E(\mathbf{x}_i - \hat{\mathbf{x}}_{i|i-1})(\mathbf{x}_i - \hat{\mathbf{x}}_{i|i-1})^*$, and satisfies the Riccati difference recursion: $P_{0|-1} = \Pi_0$,

$$P_{i+1|i} = F_i P_{i|i-1} F_i^* - K_i R_{\epsilon,i}^{-1} K_i^* + G_i Q_i G_i^*$$

We can check that the number of operations (*i.e.* multiplications and additions) needed in going from index i to index $(i+1)$ in the Riccati recursion is $O(n^3)$, and this is true whether or not the state-space model has constant parameters. However, one expects a computationally more efficient procedure in the case of time-invariant (also called constant-parameter) models $\{F, G, H, Q, R, C\}$. Indeed, it has been shown (see *e.g.*, (Kailath, Vieira and Morf, 1978)) that in the constant-parameter case the complexity can be reduced to $O(n^2\alpha)$ per iteration, where

$$\begin{aligned} \alpha &= \text{rank } (F\Pi_0 F^* + GQG^* - \bar{K}_{p,0}\bar{K}_{p,0}^* - \Pi_0) \\ &= \text{rank } (P_{1|0} - P_{0|-1}) \end{aligned}$$

and $\bar{K}_{p,i} = K_i R_{e,i}^{-*/2}$. This is achieved by using the so-called Chandrasekhar recursions to compute $\{K_i,\ R_{e,i}\}$ for use in the formulas (2). There are many forms for the Chandrasekhar recursions, but we shall give here perhaps the simplest (so-called square-root) version.

SQUARE-ROOT CHANDRASEKHAR ALGORITHM

Let $\delta P_i = P_{i+1|i} - P_{i|i-1}$. It turns out that for constant-parameter models, the quantity δP_i often has low rank, much less than n, and this fact can be exploited to find a lower complexity algorithm. Observe that δP_i is a Hermitian matrix, so that it has only real eigenvalues. We can factor it (nonuniquely) as

$$\delta P_i = P_{i+1|i} - P_{i|i-1} = L_i S_i L_i^* \tag{3}$$

where S_i is an $\alpha \times \alpha$ signature matrix, viz., a diagonal matrix with as many ± 1 on the diagonal as δP_i has positive and negative eigenvalues. In fact we can assume, as shown ahead, that S_i is the same for all i (that is, $S_i = S\ \forall i$). We now form the pre-array (the reason for choosing such pre-arrays can be justified geometrically as in (Morf and Kailath, 1975). They also arise naturally in the context of the Schur algorithm (Kailath, Sayed and Lev-Ari, 1992), as shown later in this paper. In this section, we shall give a simple algebraic verification.)

$$A_i = \begin{bmatrix} R_{e,i}^{1/2} & HL_i \\ \bar{K}_{p,i} & FL_i \end{bmatrix}$$

Let Θ_i be any $J = (I \oplus S_i)$−unitary matrix ($\Theta_i J \Theta_i^* = J$) that triangularizes A_i. That is,

$$A_i \Theta_i = \begin{bmatrix} X & 0 \\ Y & Z \end{bmatrix}$$

Comparing the entries on both sides of the equality $A_i J A_i^* = A_i \Theta_i J \Theta_i^* A_i^*$, we get

$$\begin{aligned} XX^* &= R_{e,i} + HL_i S_i L_i^* H^* \\ &\overset{(3)}{=} R_{e,i} + H(P_{i+1|i} - P_{i|i-1})H^* \\ &= R_{e,i} + R_{e,i+1} - R_{e,i} = R_{e,i+1} \end{aligned}$$

So we can choose $X = R_{e,i+1}^{1/2}$. Moreover,

$$\begin{aligned} YX^* &= K_i + FL_i S_i L_i^* H^* \\ &= K_i + F(P_{i+1|i} - P_{i|i-1})H^* = K_{i+1} \end{aligned}$$

and hence we can identify $Y = \bar{K}_{p,i+1}$. Finally, $YY^* + ZS_iZ^* =$

$$\begin{aligned} &= K_i R_{e,i}^{-1} K_i^* + FL_i S_i L_i^* F^* \\ &= K_i R_{e,i}^{-1} K_i^* + F(P_{i+1|i} - P_{i|i-1})F^* \end{aligned}$$

Therefore, $ZS_iZ^* = P_{i+2|i+1} - P_{i+1|i}$. Comparing with (3) we see that we can choose $S_{i+1} = S_i$ and $Z = L_{i+1}$. So we are led to the following recursion

$$\begin{bmatrix} R_{e,i}^{1/2} & HL_i \\ \bar{K}_{p,i} & FL_i \end{bmatrix} \Theta_i = \begin{bmatrix} R_{e,i+1}^{1/2} & 0 \\ \bar{K}_{p,i+1} & L_{i+1} \end{bmatrix} \tag{4}$$

where Θ_i is *any* $J = (I \oplus S)$−unitary matrix that produces a block zero in the (1,2) entry on the right-hand side of (4). These are called the *square-root Chandrasekhar recursions*. We can verify that each such iteration takes only $O((n+p)(n+\alpha)) = O(n^2\alpha)$ computations when $n \geq p$, as is often the case. We remark that the relation among the (generalized) Levinson algorithm, the (generalized) Schur algorithm and the Chandrasekhar recursions has been pointed out already in (Kailath, Vieira and Morf, 1978; Kailath, Sayed and Lev-Ari, 1992).

STRUCTURED TIME-VARIANT MODELS

The derivation of the Chandrasekhar recursions (4) is based on the fact that δP_i has low rank for constant-parameter systems, as shown in expression (3). We now show how to extend (4) to a class of *time-variant* state-space models that exhibit certain structure in their time-variation. For this purpose, we consider again the state-space model given by (1), and we shall say that it is a *structured* time-variant model if there exist $n \times n$ matrices Ψ_i such that F_i, G_i, and H_i vary according to the following rules:

$$H_i = H_{i+1}\Psi_i,\quad F_{i+1}\Psi_i = \Psi_{i+1}F_i,\quad G_{i+1} = \Psi_{i+1}G_i \tag{5}$$

It is clear that constant-parameter systems satisfy (5) with $\Psi_i = I$. We shall show later that another special case of (5) arises in the recursive least-squares problem. We further assume that the covariance matrices R_i, C_i and Q_i are time-invariant, even though the restrictions on Q_i, R_i and G_i can be dropped (Sayed and Kailath, 1992) .

It turns out that in order to extend the Chandrasekhar recursions to time-variant structured models, we need to consider the difference $\delta_{\Psi_i} P_i = P_{i+1|i} - \Psi_i P_{i|i-1}\Psi_i^*$, which we factor (nonuniquely) as follows

$$P_{i+1|i} - \Psi_i P_{i|i-1}\Psi_i^* \doteq L_i S_i L_i^* \tag{6}$$

where S_i is an $\alpha \times \alpha$ signature matrix. In fact, it also turns out that S_i is the same for all i ($S_i = S$). The reason for considering (5) and the difference (6) will become clear as soon as we give a simple algebraic verification of the proposed recursions, as well as when we discuss the connections with the (generalized) Schur algorithm. We now form the pre-array

$$A_i = \begin{bmatrix} R_{\epsilon,i}^{1/2} & H_{i+1}L_i \\ \Psi_{i+1}\bar{K}_{p,i} & F_{i+1}L_i \end{bmatrix}$$

and choose an arbitrary $J = (I \oplus S)-$unitary matrix, Θ_i, that triangularizes A_i. That is,

$$A_i\Theta_i = \begin{bmatrix} X & 0 \\ Y & Z \end{bmatrix}$$

Comparing the entries on both sides of the equality $A_iJA_i^* = A_i\Theta_iJ\Theta_i^*A_i^*$, we can check easily, as done in the previous section, that $X = R_{\epsilon,i+1}^{1/2}$, $Y = \bar{K}_{p,i+1}$, and $Z = L_{i+1}$. For example, we have

$$\begin{aligned} XX^* &= R_{\epsilon,i} + H_{i+1}L_iS_iL_i^*H_{i+1}^* \\ &= R_{\epsilon,i} + H_{i+1}(P_{i+1|i} - \Psi_iP_{i|i-1}\Psi_i^*)H_{i+1}^* \\ &= R_{\epsilon,i} + H_{i+1}P_{i+1|i}H_{i+1}^* - H_iP_{i|i-1}H_i^* \\ &= R_{\epsilon,i+1} \end{aligned}$$

which shows that we can choose $X = R_{\epsilon,i+1}^{1/2}$. A similar argument holds for Y and Z. Therefore, we are led to the following (square-root) *extended Chandrasekhar recursion*

$$\begin{bmatrix} R_{\epsilon,i}^{1/2} & H_{i+1}L_i \\ \Psi_{i+1}\bar{K}_{p,i} & F_{i+1}L_i \end{bmatrix}\Theta_i = \begin{bmatrix} R_{\epsilon,i+1}^{1/2} & 0 \\ \bar{K}_{p,i+1} & L_{i+1} \end{bmatrix}$$

where Θ_i is *any* $J = (I \oplus S)-$unitary matrix that produces a block zero in the (1,2) entry on the right-hand side of the above expression.

CONNECTION TO THE SCHUR ALGORITHM

Let $\mathcal{R} = E(\mathbf{y}_i\mathbf{y}_j^*)_{i,j=0}^{N}$ denote the covariance matrix of the output process $\{\mathbf{y}_i\}$, and define $\mathcal{Z}$ to be the lower triangular shift matrix with ones on the pth subdiagonal and zeros elsewhere. Clearly $\mathcal{R}$ is a Hermitian positive-definite block-matrix with $p \times p$ block-entries. We now verify that, for a structured time-variant state-space model as in (5), the covariance matrix $\mathcal{R}$ exhibits (displacement) structure, in the sense that $\mathcal{R} - \mathcal{Z}\mathcal{R}\mathcal{Z}^*$ has low rank. We then show that the (generalized) Schur algorithm reduces to the extended Chandrasekhar recursions when the extra structure provided by the assumed state-space model is properly incorporated into the Schur recursions. Let $\hat{\mathbf{x}}_i = E\mathbf{x}_i$, and define $\Pi_i = E(\mathbf{x}_i - \hat{\mathbf{x}}_i)(\mathbf{x}_i - \hat{\mathbf{x}}_i)^*$ to be the state covariance matrix. From the state equation (1) it follows that Π_i obeys the recursion $\Pi_{i+1} = F_i\Pi_iF_i^* + G_iQG_i^*$. Moreover, the following simple identities are easy to check

$$\begin{aligned} \Pi_{i+1} - \Psi_i\Pi_i\Psi_i^* &= F^{[i]}\Delta F^{*[i]} \\ E(\mathbf{y}_i\mathbf{y}_i^* - \mathbf{y}_{i-1}\mathbf{y}_{i-1}^*) &= H_iF^{[i-1]}\Delta F^{*[i-1]}H_i^* \\ E(\mathbf{y}_i\mathbf{y}_{i+1}^* - \mathbf{y}_{i-1}\mathbf{y}_i^*) &= H_iF^{[i-1]}\Delta F^{*[i]}H_{i+1}^* \end{aligned}$$

where $F^{[i]} = F_iF_{i-1}\dots F_1$, $F^{[0]} = I$ and $\Delta = \Pi_1 - \Psi_0\Pi_0\Psi_0^*$. These expressions lead to $\mathcal{R} - \mathcal{Z}\mathcal{R}\mathcal{Z}^* =$

$$\begin{bmatrix} R_{\epsilon,0} & K_0^*H_1^* & K_0^*F^{*[1]}H_2^* & \dots \\ H_1K_0 & H_1\Delta H_1^* & H_1\Delta F^{*[1]}H_2^* & \\ H_2F^{[1]}K_0 & H_2F^{[1]}\Delta H_1^* & H_2F^{[1]}\Delta F^{*[1]}H_2^* & \\ H_3F^{[2]}K_0 & H_3F^{[2]}\Delta H_1^* & H_3F^{[2]}\Delta F^{*[1]}H_2^* & \\ \vdots & & & \ddots \end{bmatrix}$$

There is significant redundancy in the elements of $\mathcal{R} - \mathcal{Z}\mathcal{R}\mathcal{Z}^*$. To explore this, we factor the leading $2p \times 2p$ principal submatrix into the (rank and inertia revealing) form

$$\begin{bmatrix} R_{\epsilon,0} & K_0^*H_1^* \\ H_1K_0 & H_1\Delta H_1^* \end{bmatrix} =$$

$$\begin{bmatrix} R_{\epsilon,0}^{1/2} & 0 \\ A & B \end{bmatrix}\begin{bmatrix} I & 0 \\ 0 & S \end{bmatrix}\begin{bmatrix} R_{\epsilon,0}^{1/2} & 0 \\ A & B \end{bmatrix}^*$$

for some signature matrix S and matrix entries A and B to be determined. Comparing both sides of the above expression we conclude that $A = H_1\bar{K}_{p,0}$ and $BSB^* = H_1(P_{1|0} - \Psi_0P_{0|-1}\Psi_0^*)H_1^*$. This suggests that we introduce the (nonunique) factorization $P_{1|0} - \Psi_0P_{0|-1}\Psi_0^* = L_0SL_0^*$, where L_0 is $n \times \alpha$ and S is an $\alpha \times \alpha$ signature matrix (notice that this derivation motivates the introduction of (6)). Hence we can take $B = H_1L_0$. Moreover, we can now check that $\mathcal{R} - \mathcal{Z}\mathcal{R}\mathcal{Z}^* = \mathcal{G}J\mathcal{G}^*$, where $J = I \oplus S$ and

$$\mathcal{G} = \begin{bmatrix} R_{\epsilon,0}^{1/2} & 0 \\ H_1\bar{K}_{p,0} & H_1L_0 \\ H_2F^{[1]}\bar{K}_{p,0} & H_2F^{[1]}L_0 \\ H_3F^{[2]}\bar{K}_{p,0} & H_3F^{[2]}L_0 \\ \vdots & \vdots \end{bmatrix}$$

We say that $\mathcal{R}$ is a Hermitian close-to-Toeplitz matrix with respect to $(\mathcal{Z}, J)$, and $\mathcal{G}$ is called a *generator* matrix. Notice that the rows of $\mathcal{G}$ are closely related, which is a consequence of the underlying state-space model. The block-triangular factorization of close-to-Toeplitz matrices, such as $\mathcal{R}$, can be computed efficiently (and recursively) by using the Schur reduction procedure (Lev-Ari and Kailath, 1986), which reduces to the following generator recursion: $\mathcal{G}_0 = \mathcal{G}$,

$$\begin{bmatrix} 0_p \\ \mathcal{G}_{i+1} \end{bmatrix} = \mathcal{G}_i\Theta_i\begin{bmatrix} 0_p & \\ & I_\alpha \end{bmatrix} + \mathcal{Z}\mathcal{G}_i\Theta_i\begin{bmatrix} I_p & \\ & 0_\alpha \end{bmatrix}$$

where Θ_i is a J-unitary matrix chosen such that the top p rows of $\mathcal{G}_i$ (denoted by g_i) are reduced to the form $g_i\Theta_i = \begin{bmatrix} r_i & 0 \end{bmatrix}$, where r_i is a $p \times p$ matrix. Therefore, the generator $\mathcal{G}_{i+1}$ is obtained by multiplying $\mathcal{G}_i$ by Θ_i and then shifting down the first p columns of $\mathcal{G}_i\Theta_i$ by p steps. We now apply this algorithm to the generator of $\mathcal{R}$. The first step involves multiplying by Θ_0, which is the identity matrix since the first block-row of $\mathcal{G}_0$ already has a $p \times \alpha$ block zero, and shifting down the first block-column:

$$\mathcal{G}_1 = \begin{bmatrix} R_{\epsilon,0}^{1/2} & H_1L_0 \\ H_1\bar{K}_{p,0} & H_2F^{[1]}L_0 \\ H_2F^{[1]}\bar{K}_{p,0} & H_3F^{[2]}L_0 \\ \vdots & \vdots \end{bmatrix}$$

Let Θ_1 be a $J-$unitary matrix such that

$$\begin{bmatrix} R_{\epsilon,0}^{1/2} & H_1L_0 \end{bmatrix}\Theta_1 = \begin{bmatrix} X & 0 \end{bmatrix}$$

Then we may check easily (as done before) that

$$\begin{bmatrix} R_{e,0}^{1/2} & H_1 L_0 \\ H_1 \bar{K}_{p,0} & H_2 F^{[1]} L_0 \end{bmatrix} \Theta_1 = \begin{bmatrix} R_{e,1}^{1/2} & 0 \\ H_2 \bar{K}_{p,1} & H_2 L_1 \end{bmatrix}$$

where $P_{2|1} - \Psi_1 P_{1|0} \Psi_1^* = L_1 S L_1^*$. Therefore $\mathcal{G}_1 \Theta_1$ is equal to

$$\begin{bmatrix} R_{e,1}^{1/2} & 0 \\ H_2 \bar{K}_{p,1} & H_2 L_1 \\ H_3 F^{[2,2]} \bar{K}_{p,1} & H_3 F^{[2,2]} L_1 \\ H_4 F^{[3,2]} \bar{K}_{p,1} & H_4 F^{[3,2]} L_1 \\ \vdots & \vdots \end{bmatrix}$$

where $F^{[i,j]} = F_i F_{i-1} \dots F_j$. Next we shift down the first p columns, form Θ_2 and so on. We see that because of the special state-space structure of the elements of the generator of $\mathcal{R}$, there is significant redundancy in the generator array: the first two nonzero rows tell enough to fill out all other rows. That is exactly the (Chandrasekhar algorithm) simplification provided by the assumption of an underlying state-space model. So the basic recursion is just the extended Chandrasekhar algorithm:

$$\begin{bmatrix} R_{e,i}^{1/2} & H_{i+1} L_i \\ \Psi_{i+1} \bar{K}_{p,i} & F_{i+1} L_i \end{bmatrix} \Theta_i = \begin{bmatrix} R_{e,i+1}^{1/2} & 0 \\ \bar{K}_{p,i+1} & L_{i+1} \end{bmatrix} \quad (7)$$

RECURSIVE LEAST-SQUARES

We now consider an important special case of (5) that arises in the recursive least-squares problem in adaptive filtering. The basic problem reads as follows: Given pairs of data points, $\{\mathbf{u}_i, d(i)\}_{i=0}^{N}$, where $\mathbf{u}_i$ is a $1 \times M$ row vector that consists of the values of M input channels at time i,

$$\mathbf{u}_i = \begin{bmatrix} u_1(i) & u_2(i) & \dots & u_M(i) \end{bmatrix} \quad (8)$$

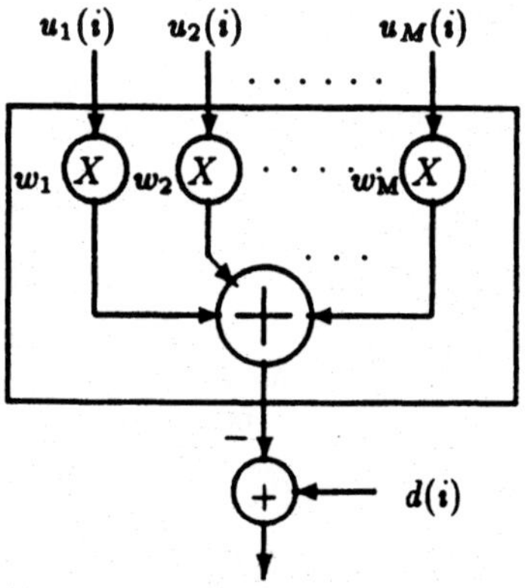

Fig. 1. **Linear least-squares estimator.**

($d(i)$ and $u_j(i)$, $j = 1, \dots, M$, are assumed scalar for simplicity), we are required to determine (recursively) the linear least-squares estimate of an $M \times 1$ column vector of unknown tap weights (see Fig. 1),

$$\mathbf{w} = \begin{bmatrix} w_1 & w_2 & \dots & w_M \end{bmatrix}^T,$$

so as to minimize the exponentially weighted error sum:

$$\min_{\mathbf{w}} \mathcal{E} = \min_{\mathbf{w}} \sum_{i=0}^{N} \lambda^{N-i} |d(i) - \mathbf{u}_i \mathbf{w}|^2, \quad 0 < \lambda \le 1 \quad (9)$$

The parameter λ is often called the forgetting factor, since past inputs are (exponentially) weighted less than the more recent values. We may rewrite $\mathcal{E}$ as follows

$$\mathcal{E} = \lambda^N \sum_{i=0}^{N} \left| \frac{d(i)}{(\sqrt{\lambda})^i} - \frac{\mathbf{u}_i \mathbf{w}}{(\sqrt{\lambda})^i} \right|^2,$$

which shows that minimizing $\mathcal{E}$ is equivalent to the following minimization problem:

$$\min_{\mathbf{x}_0 = \mathbf{w}} \sum_{i=0}^{N} |y(i) - \mathbf{u}_i \mathbf{x}_i|^2 \quad (10)$$

where we defined the normalized quantities $y(i) = d(i)/(\sqrt{\lambda})^i$ and $\mathbf{x}_i = \mathbf{w}/(\sqrt{\lambda})^i$. This simple manipulation of the expression for the total error shows that we can always rescale the original problem (9) (with an exponential factor λ) to another problem (10) with $\lambda = 1$. Now, the minimization in (10) may be easily recast into a Kalman filtering problem by considering the following M-dimensional state-space model

$$\begin{aligned} \mathbf{x}_{i+1} &= \lambda^{-1/2} \mathbf{x}_i, \quad \mathbf{x}_0 = \mathbf{w}, \quad \Pi_0 = \infty I \\ y(i) &= \mathbf{u}_i \mathbf{x}_i + v(i), \quad E v(i) v^*(j) = \delta_{ij} \quad (11) \end{aligned}$$

Let $\mathbf{w}_i$ denote the weight vector estimate given the input data up to time i. This is clearly related to the state-estimate $\hat{\mathbf{x}}_{i+1|i} = \mathbf{w}_i/(\sqrt{\lambda})^{i+1}$, which is given by $\hat{\mathbf{x}}_{0|-1} = 0$,

$$\hat{\mathbf{x}}_{i+1|i} = \lambda^{-1/2} \hat{\mathbf{x}}_{i|i-1} + \mathbf{k}_i r_{e,i}^{-1} \epsilon(i) \quad (12)$$

where $\epsilon(i) = y(i) - \mathbf{u}_i \hat{\mathbf{x}}_{i|i-1}$, $\mathbf{k}_i = \lambda^{-1/2} P_{i|i-1} \mathbf{u}_i^*$, $r_{e,i} = 1 + \mathbf{u}_i P_{i|i-1} \mathbf{u}_i^*$, and $P_{i|i-1}$ is the error covariance matrix that satisfies the Riccati difference equation

$$P_{i+1|i} = \lambda^{-1} \left[P_{i|i-1} - P_{i|i-1} \mathbf{u}_i^* r_{e,i}^{-1} \mathbf{u}_i P_{i|i-1} \right] \quad (13)$$

Recursions (12) and (13), along with the expressions for $\mathbf{k}_i$ and $r_{e,i}$, constitute, apart from normalization, the well known *RLS algorithm*. Indeed, the Kalman filter variables $\{\mathbf{k}_i, r_{e,i}, P_{i+1|i}\}$ are essentially scaled versions of the RLS variables as usually described in the literature. To clarify this point, if we apply the matrix inversion lemma (see *e.g.*, (Kailath, 1980)) to (13), then we readily conclude that

$$\begin{aligned} P_{i+1|i}^{-1} &= \lambda P_{i|i-1}^{-1} + \lambda \mathbf{u}_i^* \mathbf{u}_i, \quad P_{0|-1}^{-1} = 0 \\ &= \lambda \sum_{j=0}^{i} \lambda^{i-j} \mathbf{u}_j^* \mathbf{u}_j \equiv \lambda \Phi_i \end{aligned}$$

where Φ_i is usually referred to as the weighted autocorrelation matrix (Haykin, 1991). If we define the RLS variable $P_i = \Phi_i^{-1}$, then we get $P_i = \lambda P_{i+1|i}$. Moreover, the *a priori* error $e(i) = d(i) - \mathbf{u}_i \mathbf{w}_{i-1}$, is clearly related to the innovation by $\epsilon(i) = e(i)/(\sqrt{\lambda})^i$.

The *a posteriori* error $e^p(i) = d(i) - \mathbf{u}_i\mathbf{w}_i$, may be written as

$$\begin{aligned} e^p(i) &= d(i) - (\sqrt{\lambda})^{i+1}\mathbf{u}_i\hat{\mathbf{x}}_{i+1|i} \\ &= e(i)\left[1 - \sqrt{\lambda}\mathbf{u}_i\mathbf{k}_i r_{e,i}^{-1}\right] = e(i) r_{e,i}^{-1} \end{aligned}$$

That is, the conversion factor γ_i, which converts the a priori error $e(i)$ into the a posteriori error $e^p(i)$, is given by $\gamma_i = r_{e,i}^{-1}$. We may finally, rewrite the Kalman filter equations in terms of the original RLS variables:

$$\begin{aligned} \mathbf{w}_i &= \mathbf{w}_{i-1} + \mathbf{g}_i e(i), \quad \mathbf{w}_{-1} = \mathbf{0} \\ \mathbf{g}_i &= \frac{\lambda^{-1} P_{i-1}\mathbf{u}_i^*}{1 + \lambda^{-1}\mathbf{u}_i P_{i-1}\mathbf{u}_i^*} \qquad (14) \\ P_i &= \lambda^{-1}\left[P_{i-1} - \mathbf{g}_i\mathbf{u}_i P_{i-1}\right], \quad P_{-1} = \sigma I \end{aligned}$$

for $\sigma \gg 1$, and where we defined $\mathbf{g}_i = \sqrt{\lambda}\mathbf{k}_i r_{e,i}^{-1}$. In summary, the Kalman variables and the RLS variables are related as in TABLE 1. We remark that we do not really need to rewrite the filter equations in terms of the original RLS variables, since computing $\hat{x}_{i+1|i}$ is enough for determining $\mathbf{w}_i$. We established the correspondence between the Kalman and the RLS variables for the sake of comparison with the expressions that are usually described in the literature.

TABLE 1 Correspondence between the Kalman and RLS variables.

KF	RLS
$y(i)$	$d(i)/(\sqrt{\lambda})^i$
$\mathbf{x}_i$	$\mathbf{w}/(\sqrt{\lambda})^i$
$\hat{\mathbf{x}}_{i+1\|i}$	$\mathbf{w}_i/(\sqrt{\lambda})^{i+1}$
$\lambda P_{i+1\|i}$	$P_i = \Phi_i^{-1}$
$\sqrt{\lambda}\mathbf{k}_i r_{e,i}^{-1}$	$\mathbf{g}_i$
$e(i)$	$e(i)/(\sqrt{\lambda})^i$
$r_{e,i}^{-1}$	γ_i

For a variety of reasons, it is sometimes preferable to start the recursive algorithm with a *non-zero* initial value $\mathbf{w}_{-1}$. This allows for instance, to change the tracking capability of the algorithm. It also allows to start the adaptive scheme with an initial guess obtained from a gradient-based method, such as the LMS algorithm (see (Haykin, 1991) for a related discussion). This change may be incorporated into our derivation by redefining the total error criterion (9):

$$\mathcal{E} = (\mathbf{w} - \bar{\mathbf{w}})^*\Pi_0^{-1}(\mathbf{w} - \bar{\mathbf{w}}) + \sum_{i=0}^{N} \lambda^{N-i}|d(i) - \mathbf{u}_i\mathbf{w}|^2$$

with $\bar{\mathbf{w}} = E\mathbf{w}$, $E(\mathbf{w} - \bar{\mathbf{w}})(\mathbf{w} - \bar{\mathbf{w}})^* = \Pi_0$. The state-space model (11) remains unchanged except for the new value of Π_0. Moreover, we now use the filter equations (12–13) with the new initial conditions $P_{0|-1} = \Pi_0$ and $\hat{\mathbf{x}}_{0|-1} = \mathbf{w}_{-1} = \bar{\mathbf{w}}$ (we now have $P_{i+1|i}^{-1} = \lambda\Phi_i + \lambda^{i+1}\Pi_0$).

FAST RECURSIVE LEAST SQUARES

We showed in the previous section that the RLS adaptive algorithm can be obtained by setting up a suitable state-space model (11), and by using the Riccati-based Kalman filter recursions (12–13). The state-space model we set up has special structure: $F = \lambda^{-1/2}I, G = 0, Q = 0$, and $R = 1$ are constant, while $H_i = \mathbf{u}_i$ is not. We now further assume that the input channels $\{u_1(.), u_2(.), \ldots, u_M(.)\}$ have a shift structure in the following sense: $u_j(i) = u_{j-1}(i-1)$. If we denote the value of the first channel at time i by $u(i)$, then this corresponds to having an input row $\mathbf{u}_i$ of the form

$$\mathbf{u}_i = \left[\begin{array}{cccc} u(i) & u(i-1) & \ldots & u(i-M+1) \end{array}\right] \qquad (15)$$

The shift structure in $\mathbf{u}_i$ suggests that we might be able to get fast RLS algorithms by using the extended Chandrasekhar recursions in place of the Riccati recursions. In fact this is true, and many results in the literature can be obtained in a more transparent (square-root array) form, and many variations and extensions derived in this way. To exploit the shift structure in $\mathbf{u}_i$, we consider the following $(N+1)-$ (not $M-$) dimensional state-space model

$$\begin{aligned} \mathbf{x}_{i+1} &= \lambda^{-1/2}\mathbf{x}_i, \quad \mathbf{x}_0 = \begin{bmatrix} \mathbf{w} \\ \mathbf{0} \end{bmatrix} \\ y(i) &= \mathbf{h}_i\mathbf{x}_i + v(i), \quad Ev(i)v^*(j) = \delta_{ij} \qquad (16) \end{aligned}$$

where $\mathbf{x}_i$ is now an $(N+1)\times 1$ state-vector with trailing zeros (added for convenience), and

$$\mathbf{h}_i = \left[\begin{array}{ccccc} u(i) & u(i-1) & \ldots & u(0) & 0_{N-i} \end{array}\right]$$

is a $1\times(N+1)$ row vector. An initial state covariance matrix (with trailing zeros) is assumed, viz.,

$$E(\mathbf{x}_0 - \bar{\mathbf{x}}_0)(\mathbf{x}_0 - \bar{\mathbf{x}}_0)^* = \begin{bmatrix} \Pi_0 & \mathbf{0} \\ \mathbf{0} & \mathbf{0} \end{bmatrix} \equiv \Pi_0 \oplus \mathbf{0}$$

where Π_0 is an $M \times M$ positive definite matrix. The Kalman equations of the previous section can now be rewritten as

$$\hat{\mathbf{x}}_{i+1|i} = \lambda^{-1/2}\hat{\mathbf{x}}_{i|i-1} + \mathbf{k}_i r_{e,i}^{-1}\left[y(i) - \mathbf{h}_i\hat{\mathbf{x}}_{i|i-1}\right]$$

$$r_{e,i} = 1 + \mathbf{h}_i P_{i|i-1}\mathbf{h}_i^*, \quad \mathbf{k}_i = \lambda^{-1/2}P_{i|i-1}\mathbf{h}_i^*$$

$$P_{i+1|i} = \lambda^{-1}\left[P_{i|i-1} - P_{i|i-1}\mathbf{h}_i^* r_{e,i}^{-1}\mathbf{h}_i P_{i|i-1}\right] \qquad (17)$$

with $P_{0|-1} = \Pi_0 \oplus \mathbf{0}$. The gain vector $\bar{\mathbf{k}}_{p,i} = \mathbf{k}_i r_{e,i}^{-*/2}$ also has trailing zeros, $\bar{\mathbf{k}}_{p,i}^* \equiv \left[\begin{array}{cc} \mathbf{c}_i^* & \mathbf{0} \end{array}\right]$, say. The computational complexity of the RLS algorithm is $O(M^2)$ operations (multiplications and additions) per time step. However, *though time-variant*, the special structure of $\mathbf{h}_i$, viz., $\mathbf{h}_i = \mathbf{h}_{i+1}Z$, may be further exploited to reduce the operation count to $O(M)$, where Z denotes the shift matrix with ones on the first subdiagonal. Observe that this relation (along with $F_{i+1}Z = ZF_i$, since $F_i = \lambda^{-1/2}I$) shows that the state-space model (16) is indeed structured. The reduction in operation count can be now achieved by using the extended Chandrasekhar recursions (7) with $\Psi_i = Z$, $F_i = \lambda^{-1/2}I$, $G_i = 0$. To apply these recursions, we first introduce the (nonunique) factorization $L_0 S L_0^* = P_{1|0} - ZP_{0|-1}Z^* =$

$$\lambda^{-1}\left(\begin{bmatrix}\Pi_0 & 0\\ 0 & 0\end{bmatrix} - \bar{\mathbf{k}}_{p,0}\bar{\mathbf{k}}_{p,0}^*\right) - Z\begin{bmatrix}\Pi_0 & 0\\ 0 & 0\end{bmatrix}Z^*$$

where L_0 and S are $(N+1)\times\alpha$ and $\alpha\times\alpha$ matrices respectively. The factor L_0 is of the form $L_0^* = \begin{bmatrix}\tilde{L}_0^* & 0\end{bmatrix}$, where $\tilde{L}_0$ is $(M+1)\times\alpha$. Let $\tilde{\mathbf{h}}_i$ be the row vector of the first $M+1$ coefficients of $\mathbf{h}_i$. Writing down the extended Chandrasekhar recursions (7) we obtain

$$\begin{bmatrix} r_{e,i}^{1/2} & \tilde{\mathbf{h}}_{i+1}\tilde{L}_i \\ \begin{bmatrix}0\\ \mathbf{c}_i\end{bmatrix} & \lambda^{-1/2}\tilde{L}_i\end{bmatrix}\Theta_i = \begin{bmatrix} r_{e,i+1}^{1/2} & 0 \\ \begin{bmatrix}\mathbf{c}_{i+1}\\ 0\end{bmatrix} & \tilde{L}_{i+1}\end{bmatrix}$$

where Θ_i is *any* $J = (1\oplus S)$—unitary matrix that produces the zero entry on the right hand-side of the above expression. The computational complexity of each step is $O(\alpha M)$, where the value of α depends on the choice of Π_0. This recursion is a square-root version of fast RLS algorithms discussed in the literature (Carayannis, Manolakis and Kalouptsidis, 1983; Cioffi and Kailath, 1984). We note that the rotation Θ_i may be implemented in a variety of ways such as: elementary rotations, Householder transformations, square-root and/or division-free rotations, etc.. These may differ in computational complexity, numerical behaviour, and ease of hardware (VLSI) implementation. We also remark that explicit expressions for Θ_i can be written down in each case, leading to various explicit sets of equations.

In the prewindowed case $u(i) = 0$ for $i \le 0$. This implies $\bar{\mathbf{k}}_{p,0} = 0$ and $P_{1/0} - ZP_{0/-1}Z^* =$

$$\lambda^{-1}\begin{bmatrix}\Pi_0 & 0\\ 0 & 0\end{bmatrix} - Z\begin{bmatrix}\Pi_0 & 0\\ 0 & 0\end{bmatrix}Z^* \qquad (18)$$

A convenient choice is $\Pi_0 = \text{diag}\ \{\lambda, \lambda^2, \ldots, \lambda^M\}$, which leads to $\alpha = 2$, and we may choose $S = 1\oplus -1$, and

$$\tilde{L}_0 = \begin{bmatrix} 1 & 0 & \ldots & 0 & 0\\ 0 & 0 & \ldots & 0 & \lambda^{M/2}\end{bmatrix}^T$$

In the covariance case (where the past data $\{u(-M+1), \ldots, u(-1), u(0)\}$ is assumed available), and for the same choice of Π_0, we get $\alpha = 3$ (see (Sayed and Kailath, 1992) for more details). We also remark that other choices of Π_0 give other values of α.

EXTENSIONS AND CONCLUDING REMARKS

Recall that at the beginning of the last section we assumed that the channel inputs obey a shift structure, viz., $\mathbf{u}_j(i) = \mathbf{u}_{j-1}(i-1)$. This reflected into a special structured state-space model (16) with $\mathbf{h}_i = \mathbf{h}_{i+1}Z$. Our derivation, however, makes it clear that we can also obtain fast algorithms for other cases where the input channels do not necessarily exhibit a shift structure, the case usually considered in the literature. For example, if the input vector $\mathbf{u}_i$ in (8) satisfies a relation of the form $\mathbf{u}_i = \mathbf{u}_{i+1}\Psi$, for some constant matrix Ψ, then the state-space model (11) is also structured, and we may write down the corresponding extended Chandrasekhar recursions. For matrices Ψ that are relatively sparse, in the sense that $\Psi\bar{k}_{p,i}$ requires $O(M)$ operations, and for appropriate choices of Π_0, we are also led to a fast RLS algorithm.

Moreover, the state-space model for the RLS problem ((11) or (16)) has a *row* H matrix, whereas our derivation allows for models with more general matrices H: Such models arise for example, in the multichannel, multidimensional and/or nonlinear adaptive problems, where in many instances, choice of Ψ with a block shift structure is convenient, such as: $\Psi = Z\oplus Z\oplus\ldots\oplus Z$. The details of all these connections will be discussed elsewhere.

REFERENCES

Kailath, T. (1981). *Lectures on Wiener and Kalman Filtering*, 2nd ed., Springer-Verlag, NY. 187 pp.

Haykin, S. (1991). *Adaptive Filter Theory*, 2nd ed., Prentice Hall, Englewood Cliffs, NJ. 854 pp.

Sayed, A. H. and T. Kailath (1992). A unified square-root derivation of adaptive filtering algorithms. Tentative title, to be submitted.

Carayannis, G., D. Manolakis and N. Kalouptsidis (1983). A fast sequential algorithm for least squares filtering and prediction. *IEEE ASSP*, 31, 1394–1402.

Cioffi, J. and T. Kailath (1984). Fast recursive-least-squares transversal filters for adaptive filtering. *IEEE ASSP*, 32, 304–337.

Slock, D. T. M. (1989). *Fast Algorithms for Fixed-Order Recursive Least-Squares Parameter Estimation.* PhD thesis, Stanford University, CA.

Houacine, A. (1991). Regularized fast recursive least squares algorithms for adaptive filtering. *IEEE SP*, 39, 860–870.

Kailath, T., A. C. Vieira and M. Morf (1978). Orthogonal transformation (square–root) implementations of the generalized Chandrasekhar and generalized Levinson algorithms. *Lecture Notes in Control and Information Sciences*, 32, 81–91, Springer-Verlag, NY.

Kailath, T., A. H. Sayed and H. Lev-Ari (1992). Generalized Chandrasekhar recursions from the generalized Schur algorithm. *IEEE AC*, submitted for publication (see also *Proc. ICASSP*, 4, 549–552, 1992).

Kalman, R. E. (1960). A new approach to linear filtering and prediction problems. *Trans. ASME J. Basic Eng.* 82, 34–45.

Morf, M. and T. Kailath (1975). Square root algorithms for least squares estimation. *IEEE AC* 20, 487–497.

Lev-Ari, H. and T. Kailath (1986). Triangular factorization of structured Hermitian matrices. *Operator Theory: Advances and Applications* 18, 301–324, Birkhäuser, Boston.

Kailath, T. (1980). *Linear Systems.* Prentice Hall Englewood Cliffs, NJ. 682 pp.

DEVELOPMENT OF A TRACK-WHILE-SCAN FILTER WITH PSEUDOMEASUREMENTS

Taek Lyul Song, Jo Young Ahn, and Tae Yoon Um

Agency for Defense Development
P.O. Box 35 (4-2-2)
Daejeon, Korea

Abstract In this paper, a track-while-scan filter algorithm developed by using pseudomeasurements is proposed, and its structure is compared with other filter algorithms such as those of a constant gain filter, and the extended Kalman filter (EKF). The proposed filter is based on pseudo-linear structure of measurements obtained by transforming the original nonlinear measurements of target range, azimuth, and radial velocity from a search radar. The proposed algorithm is computationally efficient and bias-free while it provides satisfactory performance. Unbiased characteristics of the proposed filter is analyzed theoretically.

Monte Carlo simulation results are included to demonstrate effectiveness of proposed algorithm compared with other filter algorithms.

Keywords pseudomeasurement, real-time fire control, unbiasedness, constant gain filter, extended Kalman filter.

1. INTRODUCTION

A track-while-scan filter algorithm is required for antiaircraft gun fire control systems or missile fire control systems for real-time threat evaluation, target designation, and control of gun or launcher turrets from information gathered by a search radar. The information is corrupted by system noises and fixed and moving clutters result in creating false targets. Therefore, it is important to associate this information with existing tracks or initiate new tracks.

The association task is carried out with aid of track-while-scan filters which generate predicted positions and sizes of acceptable correlation gates from state estimates and covariances. The filters also generate accurate target positions which are to be transferred to tracking radars or fire control systems as required by a target designation scheme. This paper concentrates on constructing a computationally efficient track-while-scan filter with satisfactory performance while logics and algorithms for data association and track management are resorted to Farina and Studer (1985).

Target state prediction and estimation belong to nonlinear estimation area where optimal state estimate does not exist, hence suboptimal approaches are commonly used. The extended Kalman filter (EKF) is applied to the problem quite frequently, however stability of state estimates is not guaranteed by its ad hoc nature. A Kalman filter type structured filter called pseudo-measurement filter (PMF) utilizing pseudo-linear measurements obtained by transforming nonlinear measurements of target position, is introduced in Aidala (1977), however it produces biased state estimates. Aidala and Nardone (1982) analyzed the biasedness results from correlation between filter gain and residual of the PMF. This paper utilizes pseudomeasurements based on a rotating coordinate system generated by state prediction to accomodate track-while-scan radar applications. In order to generate bias-free estimates, filter gain structure similar to the modified gain extended Kalman filter (MGEKF) of Song and Speyer (1985), which generates seemingly unbiased estimates in most of simulations, is employed and bias-free characteristics of state estimates are theoretically analyzed in this paper. Filter structure in this paper is quite different from the one in Song et al. (1988), where state estimates and covariances for filter gain calculation are processed in different coordinate systems. The proposed filter is computationally efficient due to its decoupled structure.

2. TARGET DYNAMICS AND PSEUDOMEASUREMENTS

2.1 Target dynamics

In order to establish a mathematical model of the system dynamics, coordinate systems are defined first. An inertial reference coordinate system (IRCS) on which all the fire control calculations are based is defined as (N,E) where N and E stand for the north and the east. A predicted line of sight coordinate system (PLCS) is defined as (X,Y) where X is along the predicted range of target and Y is chosen such that the PLCS is right handed. Angular relation between the IRCS and the PLCS is shown in Fig. 1.

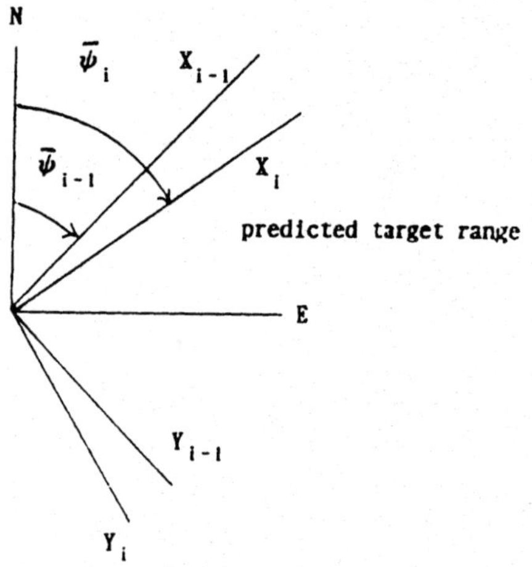

$\bar{\psi}_i$: predicted azimuth angle of target at time t_i

Fig.1. Angular relation between the IRCS and PLCS

The PLCS is not an inertial coordinate system since the predicted line of sight is rotating during radar search mode. However, the PLCS is regarded mathematically as an inertial coordinate system for the time interval between radar measurements since the PLCS is rotating incrementally with update rate corresponding to radar search rate.

Target dynamics in the PLCS are assumed linear and written in the following Langevin form

$$\dot{x}(t) = A\, x(t) + B\, \omega(t) \qquad \text{— (1)}$$

where $x(t) = (X,\ Y,\ \dot{X},\ \dot{Y})^T$, $x(0)$ is a Gaussian vector, $\omega(t)$ is a two-element zero-mean white Gaussian noise vector with covariance $E[\omega(t)\omega^T(\tau)] = q\, I\, \delta(t-\tau)$, the 4×4 matrix A with 2×2 partition matrices satisfies

$$A = \begin{bmatrix} 0 & I \\ 0 & 0 \end{bmatrix} \qquad \text{— (2)}$$

and $B = [0,\ I]^T$ is a 4×2 coefficient matrix.

The target dynamics described in (1) are discretized with time interval Δt to be implemented in a digital computer. At the ith time stage $t = t_i$, the target dynamics can be written in a discretized form as

$$x_i = \phi(\Delta t)\, x_{i-1} + W_{i-1} \qquad \text{— (3)}$$

where $\phi(\Delta t)$ is the 4×4 state transition matrix given by

$$\phi = \begin{bmatrix} I & \Delta t I \\ 0 & I \end{bmatrix}, \quad \Delta t = t_i - t_{i-1}. \qquad \text{— (4)}$$

A zero-mean white Gaussian noise vector sequence W_{i-1} is defined as

$$W_{i-1} = \int_{t_{i-1}}^{t_i} \phi(t_i - \lambda)\, B\, \omega(\lambda)\, d\lambda, \qquad \text{— (5)}$$

and its covariance Q_{i-1} can be calculated by

$$Q_{i-1} = q \int_{t_{i-1}}^{t_i} \phi(t_i - \lambda)\, B\, B^T\, \phi^T(t_i - \lambda)\, d\lambda,$$

$$= q \begin{bmatrix} \frac{\Delta t^3}{3} I & \frac{\Delta t^2}{2} I \\ \frac{\Delta t^2}{2} I & \Delta t I \end{bmatrix} \qquad \text{— (6)}$$

where the power spectral density q of $\omega(t)$ is chosen according to target maneuverability.

2.2 Pseudomeasurement formulation

The original nonlinear measurements from a doppler search radar are target range z_R, target azimuth from the north Z_ψ, and target radial velocity $z_{\dot{R}}$ which are corrupted by noises. The original measurements at the ith time stage can be expressed in the PLCS as follows by referring Fig.2.

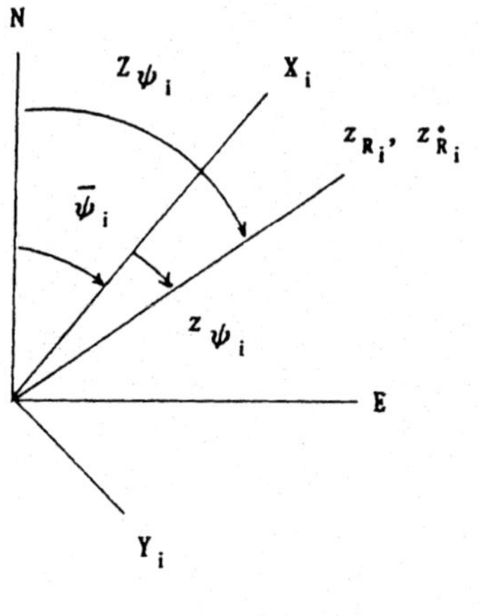

Fig.2. Radar measurements

$$z_i = \begin{bmatrix} z_R \\ z_\psi \\ z_{\dot{R}} \end{bmatrix}_i = \begin{bmatrix} \sqrt{X^2+Y^2} + v_R \\ \tan^{-1}\dfrac{Y}{X} + v_\psi \\ \dfrac{X\dot{X}+Y\dot{Y}}{\sqrt{X^2+Y^2}} + v_{\dot{R}} \end{bmatrix}_i ,$$

$$\text{and } z_{\psi_i} = Z_{\psi_i} - \bar{\psi}_i \qquad \text{— (7)}$$

where the measurement noises v_R, v_ψ and $v_{\dot{R}}$ are uncorrelated zero-mean white Gaussian with covariances σ_R^2, σ_ψ^2 and $\sigma_{\dot{R}}^2$. The predicited azimuth angle $\bar{\psi}_i$ is obtained by using incremental azimuth angle $\Delta\bar{\psi}_i$ such that $\bar{\psi}_i = \bar{\psi}_{i-1} + \Delta\bar{\psi}_i$.

By applying simple trigonometric relation to (7), the original nonlinear measurements in (7) are transformed to pseudolinear structured measurements with state dependent non-Gaussian noise sequences.

The pseudomeasurements are obtained as

$$y_i \triangleq \begin{bmatrix} z_R \\ 0 \\ z_{\dot{R}} \end{bmatrix}_i = \begin{bmatrix} \cos z_\psi & \sin z_\psi & 0 & 0 \\ -\sin z_\psi & \cos z_\psi & 0 & 0 \\ 0 & 0 & \cos z_\psi & \sin z_\psi \end{bmatrix}_i \cdot \begin{bmatrix} X \\ Y \\ \dot{X} \\ \dot{Y} \end{bmatrix}_i + \begin{bmatrix} v_R \\ \sqrt{X^2+Y^2}\,\sin v_\psi \\ v_{\dot{R}} \end{bmatrix}_i$$

$$\triangleq H(z_{\psi_i})\, x_i + \upsilon_i(x_i, v_i) \qquad \text{— (8)}$$

where $\upsilon(x_i, v_i)$ is the state dependent noise. Since the state variable x_i and noise v_i are independent, and v_i is a zero-mean white Gaussian noise, υ_i is also a zero-mean noise. The covariance of υ_i is obtained in the following section.

3. DEVELOPMENT OF A SUBOPTIMAL FILTER

3.1. Filter structure

The state variable x_i expressed in the PLCS at the i-1th time stage is defined as x_i^{i-1}, then the target dynamics expressed in this PLCS is

$$x_i^{i-1} = \phi(\Delta t)\, x_{i-1}^{i-1} + W_{i-1} \qquad \text{— (9)}$$

The predicted state estimate $\bar{x}_i^{i-1}$ satisfies

$$\bar{x}_i^{i-1} = \phi(\Delta t)\, \hat{x}_{i-1}^{i-1} \qquad \text{— (10)}$$

where $\hat{x}_{i-1}^{i-1}$ is the measurement updated estimates at the i-1th time stage, and $\Delta t = T\,(1 + (Z_{\psi_i} - Z_{\psi_{i-1}})/2\pi)$ with search radar scan period T. For the state prediction in (10), Δt of the previous scan is used. Note that if we assume $\hat{x}_{i-1}^{i-1}$ as the conditional mean of x_{i-1}^{i-1} given information $Z_{i-1} = \{z_1, z_2, \ldots, z_{i-1}\}$ then $\bar{x}_i^{i-1} = E[\,x_i^{i-1} | Z_{i-1}]$. Note also that the conditional probability density of x_{i-1}^{i-1} given Z_{i-1} for arbitrary i is assumed Gaussian to derive a suboptimal filter in this section. From (10), the incremental azimuth angle $\Delta\bar{\psi}_i = \tan^{-1}(\bar{Y}_i^{i-1} / \bar{X}_i^{i-1})$ is obtained to set up the PLCS at the ith time stage. The state estimate $\bar{x}_i^{i-1}$ expressed in the new PLCS is obtained as

$$\bar{x}_i^i = C(\Delta\bar{\psi}_i^{i-1})\, \bar{x}_i^{i-1} \qquad \text{— (11)}$$

where

$$C(\Delta\bar{\psi}_i^{i-1}) = \begin{bmatrix} C_1(\Delta\bar{\psi}_i^{i-1}) & 0 \\ 0 & C_1(\Delta\bar{\psi}_i^{i-1}) \end{bmatrix}$$

$$\text{with } C_1(\cdot) = \begin{bmatrix} \cos(\cdot) & \sin(\cdot) \\ -\sin(\cdot) & \cos(\cdot) \end{bmatrix}. \qquad \text{— (12)}$$

In the PLCS at the ith time stage, the pesudo-measurement y_i^i can be written as

$$y_i^i = H(z_{\psi_i})\, x_i^i + \upsilon_i(x_i^i, v_i) \qquad \text{— (13)}$$

where $x_i^i = C(\Delta\bar{\psi}_i^{i-1})\, x_i^{i-1}$ The meacurement updated estimate $\hat{x}_i^i$ is obtained from the following structure.

$$\hat{x}_i^i = \bar{x}_i^i + K_i\,(\,y_i^i - H(z_{\psi_i})\,\bar{x}_i^i\,). \qquad \text{— (14)}$$

If we subtract $\hat{x}_i^i$ from x_i^i and define $e_i^i = x_i^i - \hat{x}_i^i$, $\bar{e}_i^{i-1} = x_i^{i-1} - \bar{x}_i^{i-1}$ and $\bar{e}_i^i = x_i^i - \bar{x}_i^i$, then we get from (14) and (13)

$$e_i^i = (I - K_i H(z_{\psi_i}))\bar{e}_i^i + K_i\,\upsilon_i(x_i^i, v_i) \qquad \text{— (15)}$$

where from (11) $\bar{e}_i^i = C(\Delta\bar{\psi}_i^{i-1})\,\bar{e}_i^{i-1}$, and from (9) and (10)

$$\bar{e}_i^{i-1} = \phi(\Delta t)\, e_{i-1}^{i-1} + W_{i-1}. \qquad \text{— (16)}$$

3.2. Filter gain for unbiased estimates

A sufficient condition that the estimate $\hat{x}_i^i$ is unbiased is considered first. For the unbiased estimate, the unconditional expectation of e_i^i defined as $E[e_i^i]$ should be zero. The unconditional expectation can also be obtained from

$$E[e_i^i] = E^{Z_{i-1}}[E[e_i^i|Z_{i-1}]] \qquad (17)$$

where $Z_{i-1} = \{ Z_{i-2} \cup z_{i-1} \}$, and $E[\cdot\ |Z_{i-1}]$ is the conditional expectation of $(\cdot)$ given the information Z_{i-1}.

Therefore, if $E[e_i^i|Z_{i-1}]$ vanishes, the estimate $\hat{x}_i^i$ is unbiased. The conditional expectation of e_i^i given Z_{i-1} is

$$E[e_i^i|Z_{i-1}] = E[\bar{e}_i^i|Z_{i-1}] - E[K_iH_i(z_{\psi_i})x_i^i|Z_{i-1}]$$
$$+ E[K_iH_i(z_{\psi_i})\bar{x}_i^i|Z_{i-1}] + E[K_i\upsilon_i(x_i^i, v_i)|Z_{i-1}]. \qquad (18)$$

The first term of the RHS of (18) is zero since $E[\bar{e}_i^i|Z_{i-1}] = E[x_i^i|Z_{i-1}] - \bar{x}_i^i = 0$. If $K_i = K_i(Z_{i-1})$ is chosen, then the second term calculated with the assumption that $X \gg Y$ becomes

$$K_iE[H(z_{\psi_i})x_i^i|Z_{i-1}] = K_i\begin{bmatrix} e^{-\sigma_\psi^2/2}\ \bar{X}_i^i \\ 0 \\ e^{-\sigma_\psi^2/2}\ \bar{\dot{X}}_i^i \end{bmatrix} \qquad (19)$$

where the facts $\bar{Y}_i^i = 0$, $E[\cos v_\psi] = e^{-\sigma_\psi^2/2}$, and $E[\sin v_\psi] = 0$ are used. Similarly, the third term is calculated as

$$K_iE[H(z_{\psi_i})|Z_{i-1}]\,\bar{x}_i^i$$
$$= K_i\begin{bmatrix} e^{-\sigma_\psi^2/2} & 0 & 0 & 0 \\ 0 & e^{-\sigma_\psi^2/2} & 0 & 0 \\ 0 & 0 & e^{-\sigma_\psi^2/2} & 0 \end{bmatrix}\begin{bmatrix} \bar{X}_i^i \\ 0 \\ \bar{\dot{X}}_i^i \\ \bar{\dot{Y}}_i^i \end{bmatrix} \qquad (20)$$

where the assumptions $X \gg Y$ and independence of X and Y are used. From (19) and (20), the second and the third terms of (18) vanish. The last term of (18) also vanishes since v_i is zero-mean and independent of x_i^i and Z_{i-1}, i.e.,

$$K_iE[\upsilon_i|Z_{i-1}] = K_iE[F(x_i^i)f(v_i)|Z_{i-1}] = 0, \qquad (21)$$

where

$$F(x_i^i) = \begin{bmatrix} 1 & 0 & 0 \\ 0 & \sqrt{X_i^2 + Y_i^2} & 0 \\ 0 & 0 & 1 \end{bmatrix}^i,$$

and $f(v_i) = (v_{R_i},\ \sin v_{\psi_i},\ v_{\dot{R}_i})^T$. $\qquad$ (22)

From the above discussion, we know that if the filter gain K_i is chosen as a function of past measurements Z_{i-1}, the estimate is unbiased.

Next, the algorithm for filter gain is obtained. The covariance of $\bar{e}_i^{i-1}$, $\bar{P}_i^{i-1}$ is obtained as

$$\bar{P}_i^{i-1} = \phi\, P_{i-1}^{i-1}\, \phi^T + Q_{i-1}, \qquad (23)$$

where P_{i-1}^{i-1} is the conditional covariance of e_{i-1}^{i-1} given Z_{i-1}. The conditional covariance of $\bar{e}_i^i$ given Z_{i-1} is obtained from (11) as

$$\bar{P}_i^i \triangleq E[\bar{e}_i^i\bar{e}_i^{iT}|Z_{i-1}] = C(\Delta\bar{\psi}_i^{i-1})\bar{P}_i^{i-1}C(\Delta\bar{\psi}_i^{i-1})^T. \qquad (24)$$

In this paper, in order to preserve decoupled structure of X and Y channels, and for computationally efficient filter algorithms, $C(\Delta\bar{\psi}_i^{i-1})$ in (24) is assumed to be I since $\Delta\bar{\psi}_i^{i-1}$ is a small quantity. Therefore,

$$\bar{P}_i^i \simeq \bar{P}_i^{i-1}. \qquad (25)$$

The covariance of e_i^i is considered next.

$$e_i^ie_i^{iT} = (I - K_iH(z_{\psi_i}))\bar{e}_i^i\bar{e}_i^{iT}(I - K_iH(z_{\psi_i}))^T$$
$$- (I - K_iH(z_{\psi_i}))\,\bar{e}_i^if^T(v_i)F^T(x_i^i)K_i^T$$
$$- K_iF(x_i^i)f(v_i)\bar{e}_i^{iT}(I - K_iH(z_{\psi_i}))^T$$
$$+ K_iF(x_i^i)f(v_i)f^T(v_i)F^T(x_i^i)K_i^T \qquad (26)$$

where $\upsilon_i = F(x_i^i)\,f(v_i)$ in (21) is used. Calculation of the conditional expectation of (26) given Z_{i-1}, i.e., P_i^i is very complicate because of correlation between $\bar{e}_i^i$ and z_{ψ_i} for given Z_{i-1}. In order to simplify the calculation, $H(z_\psi)$ is approximated under the assumption that z_ψ is a small quantity such that

$$H(z_{\psi_i}) \simeq \begin{bmatrix} 1 & 0 & 0 & 0 \\ 0 & 1 & 0 & 0 \\ 0 & 0 & 1 & 0 \end{bmatrix} + \begin{bmatrix} 0 & z_{\psi_i} & 0 & 0 \\ -z_{\psi_i} & 0 & 0 & 0 \\ 0 & 0 & 0 & z_{\psi_i} \end{bmatrix}. \quad — (27)$$

The first term of the RHS of (27) is H(0). Then,

$$P_i^i \triangleq E[e_i^i e_i^{iT} | Z_{i-1}] = (I - K_i H(0))\bar{P}_i^i (I - K_i H(0))^T + K_i E[F(x_i^i) R_i F^T(x_i^i) | Z_{i-1}] K_i^T + \text{correlation term}, \quad — (28)$$

where $R_i = E[f(v_i) f^T(v_i)] = \text{diag}(\sigma_R^2, (1 - e^{2\sigma_\psi^2})/2, \sigma_{\dot{R}}^2)$. Further approximation is employed such that $E[F(x_i^i) | Z_{i-1}] \simeq F(\bar{x}_i^i) = \text{diag}(1, \bar{X}_i^i, 1)$ and neglecting the correlation term, P_i^i becomes

$$P_i^i = (I - K_i H(0))\, \bar{P}_i^i (I - K_i H(0))^T + K_i\, F(\bar{x}_i^i)\, R_i\, F^T(\bar{x}_i^i)\, K_i^T. \quad — (29)$$

The filter gain K_i is chosen to minimize the trace of P_i^i (Gelb, 1974), then K_i becomes

$$K_i = \bar{P}_i^i H^T(0)(H(0)\bar{P}_i^i H^T(0) + F(\bar{x}_i^i) R_i F^T(\bar{x}_i^i))^{-1}, \quad — (30)$$

which is a function of the past measurements Z_{i-1}.

For filter initialization, $x_0 = (z_{R_0}, 0, z_{\dot{R}_0}, 0)^T$, and $P_0 = \text{diag}(\sigma_R^2, z_{R_0}^2 \sigma_\psi^2, \sigma_{\dot{R}}^2, \sigma_{\dot{Y}}^2)$ are used. Approximation of $\bar{P}_i^i$ in (25) and the diagonal forms of $F(\bar{x}_i^i)$ and R_i make a decoupled structure of X and Y channels such that the computation of covariances and gains is algebraically simple. Moreover, computation of covariance matrices requires 6 storages instead of 10 storages for coupled cases. Stability of the proposed filter algorithm can be analyzed by a method similar to Song and Speyer (1985).

A newly developed constant gain filter and the extended Kalman filter are briefly introduced in the next section for filter performance comparison.

4. A CONSTANT GAIN FILTER AND THE EKF

A constant gain filter considered in this section utilizes a more simplified target dynamics and measurements. That is to say, as target dynamics in the PLCS, the equations

$$x_i^{i-1} = \phi\, x_{i-1}^{i-1}, \text{ and } x_i^i = C\,(\Delta \bar{\psi}_i^{i-1})\, x_i^{i-1} \quad — (31)$$

are used, and as measurements the following equation is used.

$$\begin{bmatrix} z_R \\ z_R z_\psi \\ z_{\dot{R}} \end{bmatrix}_i = \begin{bmatrix} 1 & 0 & 0 & 0 \\ 0 & 1 & 0 & 0 \\ 0 & 0 & 1 & 0 \end{bmatrix} \begin{bmatrix} X \\ Y \\ \dot{X} \\ \dot{Y} \end{bmatrix}_i + \begin{bmatrix} v_R \\ X v_\psi \\ v_{\dot{R}} \end{bmatrix}_i. \quad — (32)$$

The above is derived from (8) with approximations of negligibly small z_ψ and $z_\psi \dot{Y}$.

With the above dynamics and measurements, a Kalman filter type filter structure is formulated first, however the filter gain is calculated by using covariance obtained as if $C(\Delta \bar{\psi}_i^{i-1}) \simeq I$ as the suboptimal filter in the previous section to ensure decoupling of X and Y channels. For the decoupled X channel, the filter gain is obtained with an assumption of using a constant Δt value in the filter structure. The covariance of the measurement updated estimate of the X channel of (31) and (32) can be written as (Jazwinski, 1970)

$$P_i = \phi(i\Delta t)[P_0^{-1} + \sum_{j=1}^{i} \phi^T(j\Delta t) H_x^T R_x^{-1} H_x \phi(j\Delta t)]^{-1} \cdot \phi^T(i\Delta t) \quad — (33)$$

where

$$P_0 = R_x = \begin{bmatrix} \sigma_R^2 & 0 \\ 0 & \sigma_{\dot{R}}^2 \end{bmatrix}, \text{ and } H_x = I. \quad — (34)$$

After some algebraic manipulations, P_i is obtained

$$P_i = \frac{\sigma_R^2}{(i+1)(1+i(i+2)F/12)} \begin{pmatrix} \frac{6+Fi(2i+1)}{6} & \frac{iF}{2\Delta t} \\ \frac{iF}{2\Delta t} & \frac{F}{\Delta t^2} \end{pmatrix} \quad — (35)$$

with a dimensionless parameter $F = \sigma_{\dot{R}}^2 \Delta t^2 / \sigma_R^2$. Therefore, the filter gain K_{x_i} for the X channel is obtained as

$$K_{x_i} = P_i H_x^T R_x^{-1} = \frac{1}{(i+1)(1+i(i+2)F/12)}$$

$$\times \begin{bmatrix} \dfrac{6+Fi(2i+1)}{6} & \dfrac{i\Delta t}{2} \\ \dfrac{iF}{2\Delta t} & 1 \end{bmatrix}. \quad \text{— (36)}$$

The filter gain is used with increasing i upto a certain predetermined value of i to ensure filter convergence and a fixed value for i is used thereafter.

Similarly for the decoupled Y channel, the filter gain K_{Y_i} is obtained as

$$K_{Y_i} = \frac{1}{1 + i(i+1)(i+2)G/12} \begin{bmatrix} 1 + \dfrac{i(i+1)(2i+1)G}{6(i+1)} \\ \dfrac{iG}{2\Delta t} \end{bmatrix} \quad \text{— (37)}$$

where $P_0 = \mathrm{diag}\,(z_{R_0}^2 \sigma_\psi^2, \sigma_{\dot{Y}}^2)$, the covariance of measurement noise $R_y = \bar{X}_i^{i\,2} \sigma_\psi^2 \simeq z_{R_0}^2 \sigma_\psi^2$, and a dimensionless parameter $G = \sigma_{\dot{Y}}^2 \Delta t^2 / (z_{R_0}^2 \sigma_\psi^2)$ are used in the derivation. As in the X channel, i in (37) is increased to a predetermined value and a fixed value is used thereafter.

The extended Kalman filter (EKF) formulation is obtained in the IRCS with linear target dynamics and nonlinear measurement equations. A detailed derivation is omitted here. Interested reades should refer Jazwinski (1970).

5. SIMULATION RESULTS AND CONCLUSIONS

Monte Carlo simulation results are presented to demonstrate the performance of the proposed filter in comparison with the constant gain filter and the EKF. A target trajectory coordinatized in the IRCS is shown in Fig. 3, where the target maneuvers with 3g turn acceleration 25 seconds after tracking initiation. Target velocity is 300 m/sec with constant height of 700 m.

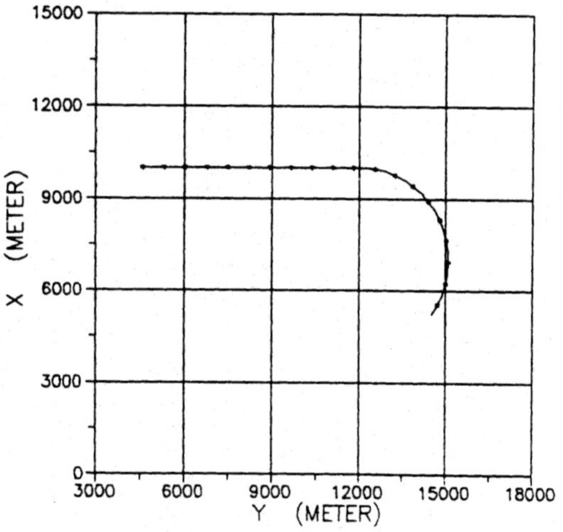

Fig.3. Target trajectory

The simulation program is executed in the IBM PC 386 with 25 MHz clock cycle, and the results of 100 runs of Monte Carlo simulation are transformed to the IRCS for comparison and presented in Figs. 4 ~ 5. The errors in the position estimate of the proposed filter, at a specific time $t = t_i$, in Fig.4, for example, are plotted by taking averaged value of errors in the predicted state estimate X_i^i transformed to the IRCS.

The scan period T of a search radar is 1.2 seconds. The power spectral density q for the proposed filter and the EKF is chosen to be always equivalent to 2g maneuver such that the filters become mismatched, and as measurement noise variances $\sigma_R^2 = 1600\ m^2$, $\sigma_\psi^2 = 25\ (mrad)^2$, and $\sigma_{\dot{R}}^2 = (7.5)^2 (m/sec)^2$ are used. Initial values for $\hat{x}_0$ and P_0 are chosen according to the filter initialization described in Section 3 with $\sigma_{\dot{Y}}^2 = 5000\ (m/sec)^2$. Since the EKF is formulated in the IRCS, a proper coordinate transformation is used for the initial values of the EKF. After 5 measurements, the gains are fixed for the constant gain filter.

As shown in Figs. 4 ~ 5, the proposed filter shows the best performance among the three filters with rapid convergence and stability with respect to target maneuvers so that the proposed filter can be used to hand over accurate target data to a tracking radar or fire control systems after 3 or 4 scan periods. If the filter parameter q is adjusted according to correlation gate size management, the performance of the proposed filter could be more improved. For 1.2 seconds of scan period, the central processing unit (CPU) time of 0.695 msec, 0.335 msec, and 6.22 msec are required for a single run of the proposed filter, the constant gain filter, and the EKF, respectively. Even taking into account of time required for data association and correlation gate size management, the proposed filter algorithm can be used to track several targets in real-time.

The results indicate that the proposed filter has computational effectiveness while providing accurate target data in a track-while-scan radar application. The computational effectiveness results from the decoupled structure of X and Y channels, and the bias-free characteristics of the filter described in Section 3 assist to provide accurate target state estimates.

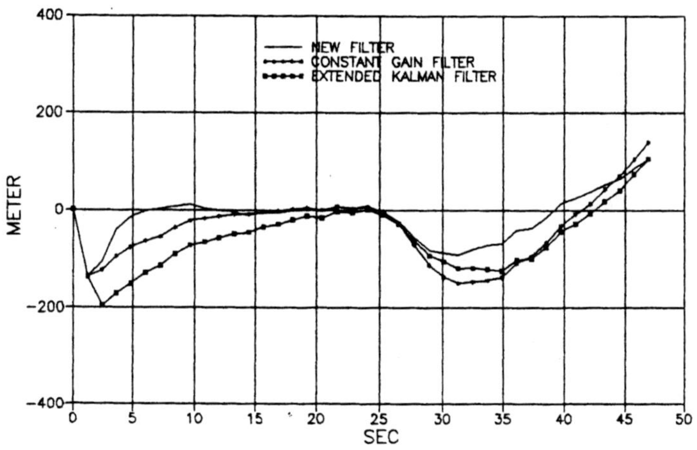

Fig.4. X-position errors

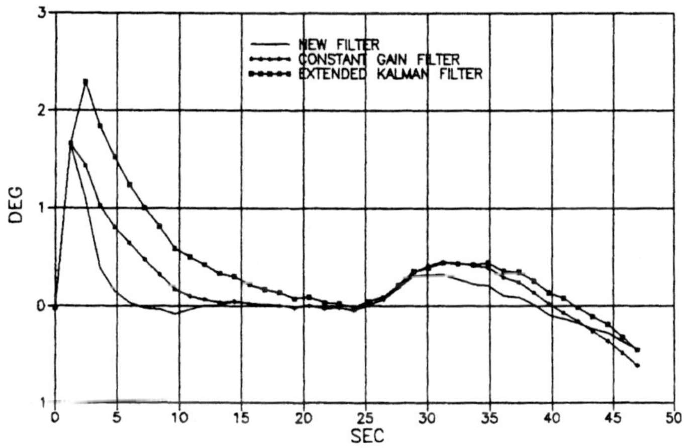

Fig.5. Azimuth angle errors

REFERENCES

Aidala, V.J. (1977). Behavior of the Kalman filter applied to bearings-only target motion analysis. Advances in Passive Tracking, Vol.1, Naval Postgraduate School.

Aidala, V.J. and S.C. Nardone (1982). Biased estimation properties of the pseudolinear tracking filter. IEEE trans. Aerospace and Electronic Syst.,AES-18(4).

Farina, A. and F.A. Studer (1985). Radar Data Processing Vol.1 - Introduction and Tracking. John Wiley & Sons Inc, New York

Gelb, A. (1974). Applied Optimal Estimation. The MIT Press, Cambridge.

Jazwinski, A.H. (1970). Stochastic Processes and Filtering Theory. Academic Press, New York and London.

Song, T.L. and J.L. Speyer (1985). A stochastic analysis of a modified gain extended Kalman filter with applications to estimation with bearings only measurements. IEEE Trans. Automatic Control, AC-30(10).

Song, T.L., J.Y. Ahn, and C.B. Park (1988). Suboptimal Filter Design with pseudomeasurements for target tracking. IEEE Trans. Aerospace and Electronic Syst., AES-24(1).

A REDUCED ORDER EXTENDED KALMAN FILTER ALGORITHM FOR PARAMETER AND STATE ESTIMATION OF AN INDUCTION MOTOR

Pietro Muraca, Ciro Picardi

Dipartimento di Elettronica, Informatica e Sistemistica - Universitá della Calabria
87030 RENDE - Italy

Abstract It is known that the application of the field-oriented control strategy to induction motors needs the knowledge of the rotor flux components and/or the rotor time constant. The aim of this paper is to give contributions to the solution of the above problem by means of real-time algorithms. The first part of the paper gives a short introduction about Kalman Filter theory. The second part is dedicated to the development of a suitable deterministic reduced order model with only the unmeasurable rotor fluxes as states, from which two reduced-order Kalman filters are obtained. The former one is called Reduced Order Linear Kalman Filter (RKF) and represents an effective algorithm for the estimation of the rotor flux components. The second filter, called Reduced Order Extended Kalman Filter (REKF) is obtained first developing a non linear model where the three state variables are the rotor flux components and the rotor time constant and then applying the extended Kalman Filter theory. The last part of the paper is related to the simulations results and the implementation tests both for RKF and REKF algorithms to verify their effectiveness in the problems of real-time applications.

Keywords Induction Motor, Field Oriented Control, Parameter and State Estimation, Kalman Filter

INTRODUCTION

In the last years the increasing availability of low cost digital processing hardware has produced much interest in the employement of field oriented control to induction motors, which thus behave like direct current motors. However the full advantages of field orientation are obtained only if the instantaneous magnitude and orientation of the rotor flux vector is defined as accurate as possible. This information can be obtained by a direct or an indirect measurement procedure. The first solution has the disadvantage that it needs an auxiliary flux sensor, like Hall probes or search coils; this prevents the method of being applied to drives that are already in use and makes the manufacturing of new drives more expensive. The indirect measurement procedure consists in a real-time estimation of the rotor flux components from easily measurable quantities as the stator currents and voltages and the mechanical rotor speed.

Different authors presented either full or reduced order observers, characterized by both a rapid convergence to zero of the estimation error and a good robustness with respect to variations of the machine parameters (Bellini et al., 1988; La Cava et al., 1990). Nevertheless the measurement noise, the system disturbances and the model mismatching limits the performance of the observer-based systems and cannot be compensated with a deterministic approach. The problem can be solved by using a Kalman Filter (KF), which is a linear, recursive optimal filter capable of estimating unknown states of a linear system in a noisy environment (Lima et al., 1989; Ammar et al., 1991). Nevertheless, the performance of the KF depends largely on the quality of the motor model; unfortunately, some parameters, in particular the rotor resistance, cannot be exactly known, because they are strongly influenced by machine operating conditions. One possibility is to view the parameters as fictitious states and use the Extended Kalman Filter (EKF) for the ensuing non linear filtering problem.

Recently different combinated state-parameter estimators based on the EKF have been proposed. The most of them estimate not only the stator and rotor flux components but also the rotor resistance , which is fundamental for a good performance of the indirect field oriented control (Lipo and Zai 1987; La Cava et al., 1989; Acarnley et al., 1989; Capolino and Du 1991; Dell'Aquila et al., 1991) The EKF is known to be computationally expensive, this cost being a function of the state dimension in the underlying

This work is partially supported by the M.U.R.S.T. (fondi 40%)

model. In order to obtain an efficient algorithm for a real-time estimation of the rotor flux components and the rotor time constant, this paper proposes a reduced-order EKF .

The first part of the paper is dedicated to a short presentation of KF theory. After recalling the continuous and discrete time models of the induction motor, the second part is related to the development of a suitable deterministic reduced order model with only the unmeasurable rotor fluxes as states, from which two reduced-order Kalman filters are obtained. The former is called Reduced Order Linear Kalman Filter (RKF) and represents an effective algorithm for the estimation of the rotor flux components, also when the stator current measurements are not completely noiseless, but however of good quality. The second algorithm, called Reduced Order Extended Kalman Filter (REKF) is obtained first developing a non linear model where the three state variables are the rotor flux components and the rotor time constant and then applying the EKF.

Simulations have been carried out to verify the feasibility of estimating the rotor flux components and the rotor time constant with the proposed algorithms. Finally the paper gives the results about two implementation tests both for RKF and REKF algorithms to verify their effectiveness in the problems of real-time applications.

KALMAN FILTERING

This section gives a short introduction about Kalman filter theory in order to provide notations for the reader and to recall the basic formulas both for the basic linear Kalman Kilter (KF) and its Extended version (EKF). Moreover, this presentation will simplify the descriptions of the reduced-order filters proposed by the authors .

Linear Kalman Filter (KF)

The KF can be introduced by considering the linear discrete time system described by the state space equations:

$$x(k+1)=F(k)x(k)+G(k)u(k)+w(k) \tag{1}$$

$$y(k)=H(k)x(k)+v(k) \tag{2}$$

where $x(k)$ is the state vector, $y(k)$ is the output or measurement vector and $u(k)$ is the known input. Also $F(k)$, $G(k)$ and $H(k)$ are the system, input and output or measurement matrices respectively. Furthermore, $w(k)$ and $v(k)$ represent the system and measurement noises characterized by:

$$\begin{aligned} &E[w(k)]=0, \quad E[v(k)]=0, \ \forall\, k \\ &E[w(k)w^T(j)]=Q(k)\delta_{k,j} \quad Q\geq 0, \\ &E[v(k)v^T(j)]=R(k)\delta_{k,j} \quad R>0 \\ &E[w(k)v^T(j)]=0, \ \forall\, k,j \end{aligned} \tag{3}$$

where $\delta_{k,j}$ is the Kronecker delta. The initial state $x(0)$ is an independent random vector with known statistics

$$E[x(0)]=x_0, \quad E[(x(0)-x_0)(x(0)-x_0)^T]=P_0 \tag{4}$$

The problem is to find an estimate $x(k/k)$ of the state $x(k)$ at time k with its error covariance $P(k/k)=E[(x(k)-x(k/k))(x(k)-x(k/k))^T]$. The notation $x(k/k)$ implies the estimate is to be based on all the measurements obtained up to and including $y(k)$ obtained at time k. The equations for the Kalman filter are:

$$x(k+1/k)=F(k)x(k/k)+G(k)u(k) \tag{5}$$

$$P(k+1/k)=F(k)P(k/k)F^T(k)+Q(k) \tag{6}$$

$$x(k+1/k+1)=x(k+1/k)+ K(k+1)[y(k+1)-H(k+1)x(k+1/k)] \tag{7}$$

$$K(k+1)=P(k+1/k)H^T(k+1) [H(k+1)P(k+1/k)H^T(k+1)+R(k+1)]^{-1} \tag{8}$$

$$P(k+1/k+1)=[I-K(k+1)H(k+1)]P(k+1/k) \tag{9}$$

$$x(0/0)=x_0; \qquad P(0/0)=P_0 \tag{10}$$

Extended Kalman Filter (EKF)

To use KF with nonlinear system models it is necessary to first linearise the model about a nominal or auxiliary state trajectory to produce a linear perturbation model. The basic KF is used to estimate the perturbation states and then these are combined with the auxiliary states to produce the state estimates of the non linear model. When the auxiliary state is made equal to the most recent KF estimate the procedure is known as the Extended Kalman Filter (EFK). The EKF can be applied as a state estimator for non linear systems. It can carry out combined state and parameter estimation treating selected parameters as extra states and forming an augmented state vector. The result is that whether the original state space model was linear or not the augmented model will be non linear because of multiplication of states.

Let the non linear model of the system be described by the following equations:

$$x(k+1)=f(x(k),u(k))+w(k) \tag{11}$$

$$y(k)=H(k)x(k)+v(k) \tag{12}$$

where $x(k)$, $y(k)$ and $u(k)$ are respectively the state, the measurement and the input vectors; $f(x(k),u(k))$ is a vector state function of state vector and $H(k)$ is the measurement matrix. Furthermore, $w(k)$ and $v(k)$ represent the system and measurement noises with statistic described by Eq. (3). The equations of EKF are :

$$x(k+1/k)=f(x(k),u(k)) \tag{13}$$

$$\Phi(k)=\left[\frac{\partial f(x(k),u(k))}{\partial x(k)}\right]_{x(k)=x(k/k)} \tag{14}$$

$$P(k+1/k)=\Phi(k)P(k/k)\Phi^T(k)+Q(k) \tag{15}$$

$$x(k+1/k+1)=x(k+1/k)+ K(k+1)[y(k+1)-H(k+1)x(k+1/k)] \quad (16)$$

$$K(k+1)=P(k+1/k)H^T(k+1) [R(k)+H(k+1)P(k+1/k)H^T(k+1)]^{-1} \quad (17)$$

$$P(k+1/k+1)=[I-K(k+1)H(k+1)]P(k+1/k) \quad (18)$$

$$x(0/0)=x_0; \quad P(0/0)=P_0 \quad (19)$$

In general the EKF is not optimal and, due to the linear approximation, it is possible for the estimates to diverge so that care must be taken in its application. In some case, it is convenient to use the following rather than Eq. (18) :

$$P(k+1/k+1)=[I-K(k+1)H(k+1)]P(k+1/k) [I-K(k+1)H(k+1)]^T+K(k+1)R(k+1)K^T(k+1) \quad (20)$$

INDUCTION MOTOR MODELS

Continuous Time Model

Under the usual hypotheses of anisotropic structure of the machine, of sinusoidal spatial distribution of the m.m.f. and of magnetic circuit linearity, neglecting the iron losses and assuming as state variables the components of the stator current i_s and those of the rotor flux ϕ_r, according to a rotating reference frame (d and q), the model of the electromagnetic circuits of the induction motor is:

$$\dot{i}_s = A_{11} i_s + A_{12}\phi_r + B_1 v_s \quad (21)$$

$$\dot{\phi}_r = A_{21} i_s + A_{22}\phi_r \quad (22)$$

where

$$i_s=[i_{sd}, i_{sq}]^T; \quad \phi_r=[\phi_{rd}, \phi_{rq}]^T; \quad v_s=[v_{sd}, v_{sq}]^T$$

The quantities v_{sd} and v_{sq} are the components of the motor supply voltage v_s according to the reference frame. The two-dimensional matrices A_{ij} $(i,j=1,2)$ and B_1 can be expressed in the following particular way:

$$A_{ij}=a_{ij}^{(1)}I+a_{ij}^{(2)}J; \quad B_1=b_1^{(1)}I \quad (23)$$

where

$$I=\begin{bmatrix}1 & 0\\ 0 & 1\end{bmatrix}; \quad J=\begin{bmatrix}0 & 1\\ -1 & 0\end{bmatrix}$$

$$a_{11}^{(1)}=-\frac{1}{\sigma}\left[\frac{1}{Ts}+\frac{1-\sigma}{T_r}\right]; \quad a_{11}^{(2)}=\omega_a; \quad a_{12}^{(1)}=\frac{K_r}{\sigma L_s T_r};$$

$$a_{12}^{(2)}=-\frac{K_r}{\sigma L_s}; \quad a_{21}^{(1)}=\frac{L_M}{T_r}; \quad a_{21}^{(2)}=0$$

$$a_{22}^{(1)}=-\frac{1}{T_r}; \quad a_{22}^{(2)}=\omega-\omega_a; \quad b_1^{(1)}=\frac{1}{\sigma L_s};$$

$$T_s=\frac{L_s}{R_s}; \quad T_r=\frac{L_r}{R_r}; \quad \sigma=1-\frac{L_M^2}{L_s L_r}; \quad K_r=\frac{L_M}{L_r}$$

R_s and R_r being the stator and rotor resistances, L_s and L_r the stator and rotor self-inductances and L_M the mutual inductance; in addition ω and ω_a are the angular speed of the motor and that of the reference frame (expressed in electrical radians per second) equal to the supply angular frequency. The above formulation of the model of the induction motor gives state variables with constant non-zero magnitudes in the steady- state which is the best way in order to use statistical methods in conjunction with the model.

Discrete Time Model

Assuming as constant the values of the motor angular speed and the supply angular frequency during each sampling interval and neglecting the effects of the modulation introduced by the inverter (considering during the sampling interval only the mean values of the components of the stator voltage), the following discrete-time model can be obtained by Eqs. (21,22):

$$i_s(k+1)=F_{11}(k)i_s(k)+F_{12}(k)\phi_r(k)+G_1(k)v_s(k) \quad (24)$$

$$\phi_r(k+1)=F_{21}(k)i_s(k)+F_{22}(k)\phi_r(k)+G_2(k)v_s(k) \quad (25)$$

in which matrices F_{ij} $(i,j=1,2)$ and G_i $(i=1,2)$, obtained by the discretization of the continuous-time model, have a structure similar to that of matrices A_{ij} and B_1, i.e.

$$F_{ij}=f_{ij}^{(1)}I+f_{ij}^{(2)}J; \quad G_i=g_i^{(1)}I+g_i^{(2)}J \quad (26)$$

A full-order discrete-time observer of the induction motor (with either deterministic or stochastic approach) can be carried out from Eqs. (24,25), but with matrices F_{ij} and G_i calculated by approximated methods. In fact, the exact determination of the above matrices requires an expensive computational effort and prohibit their on-line calculation. Unfortunately it is not possible to carry out these calculations off-line before starting the observer since the system matrices A_{ij} and therefore F_{ij} and G_i depend on the varying rotor speed and/or the supply angular frequency. A mostly used approximated method is the integration algorithm of Euler, which is characteristic of quasi-continuous approaches. In this case, taking into account Eq. (26), one obtains:

$$f_{ij}^{(1)}=1+a_{ij}^{(1)}T \quad \forall i=j; \qquad f_{ij}^{(1)}=a_{ij}^{(1)}T \quad \forall i\neq j,$$

$$f_{ij}^{(2)}=a_{ij}^{(2)}T; \quad g_1^{(1)}=b_1^{(1)}T; \quad g_1^{(2)}=g_2^{(2)}=g_2^{(1)}=0 \quad (27)$$

where T represents the length of the sampling interval. Obviously, the smaller T is better the Euler approximation is; otherwise this requires a fast microprocessor system for the on-line implementation of the filter. The employment of a stochastic approach overcomes in part these difficulties, because it intrinsically takes into account the model mismatchings and therefore a not very small length T can be chosen.

REDUCED ORDER LINEAR KALMAN FILTER (RKF)

As the stator currents are measurable it is possible to reformulate the motor model with only the unmeasurable rotor fluxes as states.

By suitable manipulations of Eqs. (24,25), the following 2th order model can be obtained:

$$\phi_r(k+1)=F_{22}(k)\,\phi_r(k)+u_r(k) \tag{28}$$

$$y_r(k)=\phi_r(k) \tag{29}$$

where

$$u_r(k)=F_{21}(k)\,i_s(k)+G_2(k)\,v_s(k) \tag{30}$$

$$y_r(k)=F_{22}(k-1)\,\phi_r^m(k-1)+u_r(k-1) \tag{31}$$

$$\phi_r^m(k)=F_{12}^{-1}(k)\,[\,i_s(k+1)-F_{11}(k)\,i_s(k)-G_1(k)\,v_s(k)\,] \tag{32}$$

The quantity ϕ_r^m, obtained as combination of stator voltage and current vectors at time $k+1$ represents an "indirect measure" of the rotor flux vector at previous time k. It is important to note that the possibility to use an efficient reduced-order estimator is related to the good quality of the stator current measurements. Nevertheless, even if these were completely free of noise, a stochastic approach is necessary to take into account the approximations of the model and the fact that the input modulated voltages can be considered as the addition of sinusoidal deterministic signals plus white Gaussian noises. The reduced order model given by Eqs. (28, 29) is deterministic and no model or measurement uncertainties are included. To extend it to a stochastic state model requires the addition of two noise vectors: a 2×1 system noise vector $w(k)$ and a 2×1 mesurement noise vector $v(k)$, with the characteristics given by Eq. (3). Then, to the so obtained stochastic discrete model it is possible to apply the current KF algorithm described by Eqs. (5-10), where :

$x(k/k)=\phi_r(k/k)$; $u(k)=u_r(k)$; $y(k)=y_r(k)$;

$F(k)=F_{22}(k)$; $G(k)=H(k+1)=I$

It is easy to observe that the correction vector, which is added to the prediction vector in Eq. (7), can be expressed as:

$$\phi_{rc}(k+1)=K(k+1)\,F_{22}(k)\,[\,\phi_r^m(k)-\phi_r(k/k)\,] \tag{33}$$

Therefore the correction of the prediction at time $k+1$ to obtain the estimation at time $k+1$ is based on the comparison between the indirect measure of the rotor flux and the estimation evaluated at the previous time k. The diagram given in Fig.1 illustates the structure of the described RKF.

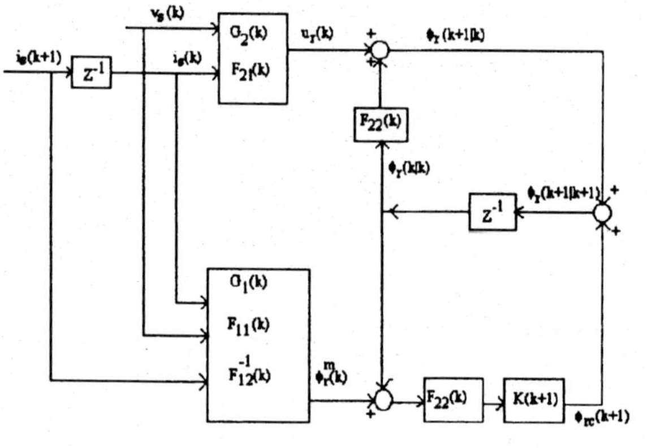

Fig.1 Reduced Linear Kalman Filter (RKF)

REDUCED ORDER EXTENDED KALMAN FILTER (REKF)

In order to obtain a Reduced order Extended Kalman Filter (REKF) capable of providing the joint estimation of the rotor flux componentes and the rotor time constant, to the model given by Eqs. (28,29) the following discrete equation must be added:

$$T_r(k+1)=T_r(k) \tag{34}$$

where the parameter T_r has been defined as state variable. Then introducing the new state vector :

$$x_e(k)=[\,i_s^T(k),T_r(k)\,]^T \tag{35}$$

the discrete state equations for the augmented system are the following:

$$x_e(k+1)=f\,(x_e(k),u_r(k)) \tag{36}$$

$$y_r(k)=H_e\,x_e(k) \tag{37}$$

with

$$f\,(x_e(k),u_r(k)=\begin{bmatrix}F_{22}(T_r(k),k\,)\,\phi_r(k)+u_r(k)\\ T_r(k)\end{bmatrix}$$

$$H_e=\begin{bmatrix}1&0&0\\0&1&0\end{bmatrix}$$

$$u_r(k)=F_{21}(T_r(k),k\,)\,i_s(k)+G_2(T_r(k),k\,)\,v_s(k)$$

$$y_r(k)=F_{22}(T_r(k),k\,)\,\phi_r^m(k-1)+u_r(k-1)$$

$$\phi_r^m(k)=F_{12}^{-1}(T_r(k),k\,)i_s(k+1)-F_{11}(T_r(k),k\,)\,i_s(k)-$$
$$G_1(T_r(k),k\,)\,v_s(k)$$

Also in this case the extension to a stochastic state model requires the addition of the noise vectors: a 3×1 system noise vector $w(k)$ and a 2×1 mesurement noise vector $v(k)$ with statistics described in Eq. (3). Then the EKF algorithm summarized through Eqs. (13-20) can be applied with the following equivalences:

$x(k/k)=x_e(k/k)$; $u(k)=u_r(k)$; $y(k)=y_r(k)$.

$f\,(x(k),u(k))=f\,(x_e(k),u_r(k))$; $H(k+1)=H_e$

The diagram given in Fig. 2 illustrates the structure of the proposed REKF.

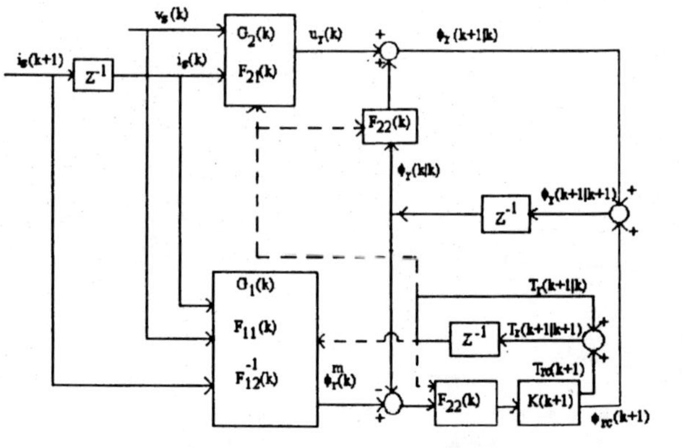

Fig. 2 Reduced Order Extended Kalman Filter (REKF)

SIMULATION RESULTS

In order to verify the effectiveness of the proposed algorithms, some simulations have been carried out using the actual parameters of an induction motor characterized by the following rated values:

Power = $280\,KW$; Voltage = $2300\,V$; Frequency = $50\,Hz$; Pole pairs = 3

$R_s = 0.312\,\Omega$; $R_r = 0.203\,\Omega$;

$L_r = L_s = 0.179\,H$; $L_M = 0.174\,H$

The rated values of the stator current and rotor flux have been assumed equal to $150\,A$ and $6.88\,V\,sec.$. Various tests have been effected considering different operating conditions of the motor; this section shows the most significative results in order to clearly point out the good quality of the estimations provided by the two algorithms. In both cases the results refers to the condition in which the motor operates at constant angular speed ($\omega = 310\,el.rad./sec$, $\omega_a = 314\,el.rad./sec$) and starts with the steady-state values of the state variables corresponding to a sinusoidal supply voltage characterized by $v_{sd} = 2100\,V$ and $v_{sq} = 0$. This last choice has not to be considered restrictive, because the random noise, which has been added to the model used for the simulation of the motor, can be thought as comprehensive of the harmonics present in the modulated voltages given by the inverter supplying the motor.

The current measurements, which represent input signals for the filters, have not be assumed completely noiseless, that is a noise with very small standard variation has been added to the stator current components given by the simulation of the motor. However, the simulation results has shown that this does not affect the good quality of the estimations provided by the two types of Kalman filters.

In all the tests the initial estimates of the rotor flux components have been set to zero so as to verify the quality of the convergence towards the exact values. The initial estimation of the rotor time constant, for the REKF, has been set to the double of the actual value . The shown results refers to the employment of the Euler approximation method for the calculation of the estimated values. A length of the sampling interval T equal to $0.5\,ms$ has been selected for the examined operating conditions. All covariance matrices $Q(k)$, $R(k)$ have been chosen constant and diagonal; also matrix P_0 has been chosen diagonal.

Estimation with RKF

Fig. 3 shows the results obtained by the proposed RKF. The continuous curve concerns the actual value of $\phi_{rd}(k)$, the dashed curve is the estimated value $\phi_{r_d}(k/k)$ and the dotted curve shows the quantity $\phi_{rd}^{m}(k)$. It is important to note that the times scale on the horizontal axis, shown in Fig. 3 starts with a value equal to 0.02 sec. In such a way, rather than the convergence phase (which, however is very fast), the figure points out that the estimation given by the RKF is a filtered version of the actual value.

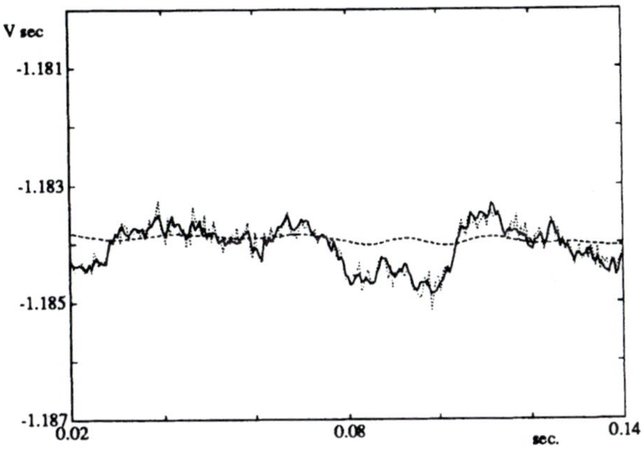

Fig. 3 - Estimation of ϕ_{rd} by RKF.

For sake of simplicity the results for ϕ_{rq}, wich are analogous to those for ϕ_{rd} are not shown.

The computational effort of the RKF is much reduced in comparison with that required by a similar full order KF for two reasons : the reduced order of the underlying motor model and a gain matrix $K(k+1)$ proportional the identity matrix as consequence of the particular structure of matrices F_{ij} , G_i, P_0 and of the output equation for the RKF. To test the effectiveness of the above algorithm in the problems of real-time applications, it is implemented in Borland C++ language with floating point over 32 bits on a 80486-based machine with a main clock at 25 MHz. This test has given a time of about $3\,ms$ for one calculation step; therefore it is obvious to expect that by using a more powerful machine such as DSP-based with high speed floating-point option the computation time for one step will be much lower than $3\,ms$ and hence lower than the length of the sampling interval.

Estimation with ERKF

Figs. 4 and 5 show the most significative results given by the REKF.

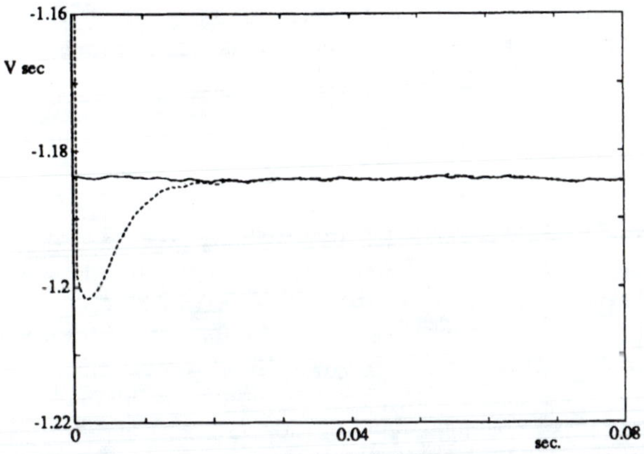

Fig. 4 - Estimation of ϕ_{rd} by REKF.

In particular Fig. 4 reports the shapes of the actual and the estimate values (continous and dashed lines respectively) of ϕ_{rd} beginning from the initial time instant in which the estimated value is set to zero. There is evidence that the estimation is very fast and accurate.
The estimation of the rotor time constant, beginning from a value equal to the double of the actual one (i.e. an error of 100 %) is illustrated in Fig. 5.

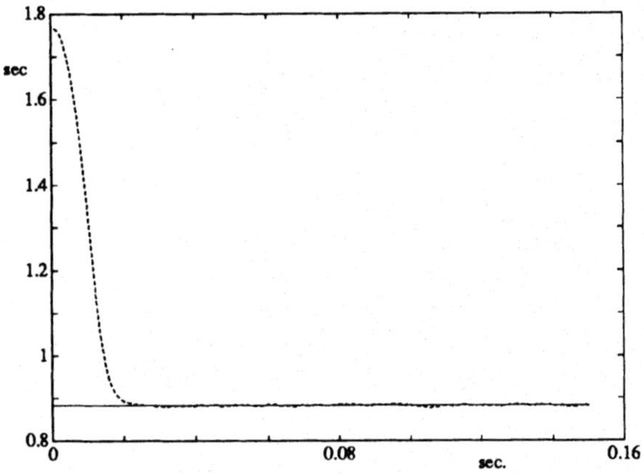

Fig. 5 - Estimation of T_r by REKF.

The shape of the estimation transient shows clearly the fast convergence phase of the algorithm; after an interval equal to about 0.02 sec. the estimated rotor time constant practically reaches the actual value. The parameter sensitivity of REKF algorithm has been tested in order to verify its robustness. The different tests performed have shown that the identification of the rotor time constant is very robust in front of the variations of the stator resistance. In the case of the mutual inductance, the absolute sensitivity is enough low but algorithm instability occurs when this parameter is varied.

Also for the REKF algorithm an implementation test of the same type described for the RKF algorithm has been carried out. This test has given a time of about $16\,ms$ for one calculation step; this considerable increase of the time for the REKF compared with that for RKF, besides to the higher order of the filter is due to the fact that it is not possible completely to take advantage of the particular structure of the dynamic and input matrices of the system model.

CONCLUSIONS

This paper has presented two reduced order Kalman Filters, one of linear type (RKF) and the other one of extended type (REKF), which provide two real-time algorithms to be used for the determination of the rotor flux components and/or the rotor time constant, which is necessary for implementation of the direct and/or indirect field-oriented control strategy to induction motor drives. Simulations results have shown a fast convergence phase and a good accuracy for the estimations given by both the algorithms. Some implementation tests have shown the effectiveness of the two algorithms, because the times necessary for one calculation step have resulted low enough to think that they represent an acceptable computation effort for powerful machines such as DSP-based. The authors are presently engaged in experimentally implementing the proposed algorithms on DSP-based machines.

AKNOWLEDGEMENTS

The authors sincerely thank Eng. Domenico Famularo and Eng. Massimo Riccio (Univ. della Calabria) for their contributions in the numerical tests.

REFERENCES

Acarnley P.P., Atkinson D.J. and Finch J.W., (1989) *Parameter Identification Techniques for Induction Motor Drives*, European Power Electronics Conf., Aachen .

Ammar F.B., de Fornel B., Mirzaian A. and Pietrzak-David M., (1991) *Field- Oriented Control of High-Power Induction Motor Drives by Kalman Filter Flux Observation*, European Power Electronics Conf., Firenze .

Bellini A., Figalli G. and Ulivi G., (1988) *Analysis and Design of a Microcomputer-based Observer for an Induction Machine*, Automatica , Vol. 24, No. 4.

Capolino G.A., Du B., (1991) *Extended Kalman Observer for Induction Machine Rotor Currents*,European Power Electronics Conf., Firenze .

Dell'Aquila A., Lovecchio F.S., Salvatore L. and Stasi S., (1991) *Induction Motor Parameter Estimation via EKF*, European Power Electronics Conf., Firenze .

La Cava M., Muraca P. and Picardi C., (1990) *Estimation of the Flux Components of an Induction Motor with a Sliding Mode Observer*, IEEE Int. Workshop on Intelligent Motion Control, Instanbul.

La Cava M., Picardi C. and Ranieri F.,(1989) *Application of the Extended Kalman Filter to Parameter and State Estimation of Induction Motors*, Int. Journal of Modelling & Simulation, Vol. 9, No. 3.

Lima A.M.N., de Fornel B. and Pietrzak-David M., (1989) *On Stochastic Filtering Techniques and its Applications to AC Numerical Drive Systems* , European Power Electronics Conf., Aachen.

Lipo T. A., Zai L.C. (1987) *An Extended Kalman Filter Approach to Rotor Time Constant Measurement in PWM Induction Motor Drives* IEEE IAS Annual Meeting, Atlanta.

ON STOCHASTIC LYAPUNOV FUNCTION METHOD IN OPTIMAL LINEAR FILTERING PROBLEM

Hoang H. S. and O. Talagrand,

Laboratoire de Meteorologie Dynamique, Ecole Normale Superieure,
24 rue Lhomond, 75231 PARIS Cedex, FRANCE

Abstract. The stable filters (SF) designed by inversion of direct Lyapunov function (LF) method are proposed. Some properties of SF are studied. The stability of SF can be easily established by using the equation constructed for given stochastic LF. The existence and uniqueness conditions of design problem are investigated. It is shown that in particular when the statistics of initial state are given not perfectly, a performance of estimates produced by SF may be better than a performance of usual Kalman filter (KF) estimates. Some numerical examples are given.

Keywords. Filtering; stability; Lyapunov methods ; time-varying parameters; real-time data processing.

INTRODUCTION. PROBLEM STATEMENT

The KF technique is most important tool in real-time data processing. The optimality of filters, however, does not imply its stability (Jazwinski, 1970). The first results related to stability of optimal filter are developed in (Kalman,1963; Sorenson,1967) from which we know that if the linear system is uniformly completely observable and uniformly completely controllable then the Kalman filter is uniformly asymptotically stable,

That property implies that the effect of the initial statistics - error covariance matrix (ECM) - $P(0)$ is forgotten as more and more data are processed (Jazwinski, 1970). This fact is very important since in practice $P(0)$ is often poorly known and sometimes rather arbitrary set. Furthermore this implies that the computation of $P(t)$ is stable. To prove this result one often uses the so-called direct Lyapunov function method (LF-method) which consists in selecting a LF for considered system. It means that to solve the problem of stability of filter we have to demonstrate an existence of LF possessing the necessary properties. In general that is a difficult task (Kalman, 1963; Jazwinski, 1970; Lypsher and Shiryaev, 1974).

We will follow in this paper the "inversion" approach. We are seeking the "optimal" filter (Def.1.1) such that this filter possesses a LF of given form which (i.e. LF) changes along the trajectory of estimates produced by the filter by prescribed law. Thus, according to the inversion approach our task is not to establish an existence of LF , but inversely, the structure and parameters of LF are given a-priori and our problem is to design a filter possessing this LF.

The idea on possible inversion of direct LF-method was first suggested by Zubov (1963) for optimal control problem. Some procedures of inverse LF-method and design of control algorithms are given in (Crutko, 1983, 1984, 1989). In (Hoang et al, 1991) this idea is proposed to design the optimal nonlinear filters for general class of nonlinear systems with dependent noises.

The key idea underlying in an inversion of LF- method for filter design consists of the following: let $\{x(t), z(t)\}$, $t \geq 0$ be partially observed random vector-process, described by some difference equations, where $x(t) \in R^n$ is the state, $z(t) \in R^p$ is the observation part. To estimate $x(t)$ given by $z(\tau), 0 \leq \tau \leq t$ let $F(\delta,\lambda) = \{x'(t) : x'(t) = \delta\xi + \lambda\}$ be some parameterized class of filters which is assumed to be selected a-priori, where ξ is a known function measurable with respect to $\{x'(t), z(t), 0 \leq t < t\}$.

Definition 1.1. The estimator $\hat{x}(t) \in F(\delta,\lambda)$ is called ***optimal*** if it satisfies the following conditions

$$\hat{x}(t) \text{ is unbiased: } E\{\hat{x}(t)\} = E\{x(t)\} \quad (1a)$$

$$J(\delta) = Sp\{\delta^T\delta\} \rightarrow \min_\delta \quad (1b)$$

$$\Delta v(x,t) + \beta v(x,t) = 0, \; \beta > 0 \quad (1c)$$

here $v(x,t) \hat{=} Sp\{P(t)\}$, $\Delta v(x,t) \hat{=} v(x,t+1) - v(x,t)$

$P(t) \hat{=} E\{\tilde{x}(t)\tilde{x}^T(t)\}$, $\tilde{x}(t) \hat{=} x(t) - \hat{x}(t)$

The function $v(x,t)$, thus, may be considered as a stochastic LF (SLF) and the problem is to choose the "control" variable $\delta = \delta(t)$ to drive the state $\hat{x}(t)$ of filtering equation so that along its trajectory the function $v(x,t)$ should satisfy Eq.(1c) ($\lambda = \lambda(t)$ often is chosen to satisfy the condition (1a)) . In (Hoang et al., 1991) under some reasonable conditions one can prove that a solution of formulated problem exists, unique and $\hat{x}(t)$ is stable, i.e. $\lim_{t\rightarrow\infty} P(t) < \infty$ provided $Sp\{P(0)\} < \infty$.

In this paper we apply the results in (Hoang et al., 1991) to linear filtering problem

$$x(t+1) = \varphi(t)x(t) + G(t)w(t), \; t = 0,1,... \quad (2a)$$

$$z(t) = H(t)x(t) + v(t), \; t = 1,2,... \quad (2b)$$

where $x(0), \{w(t)\}, \{v(t)\}$ in general are correlated and their covariance matrices may be singular (Nguyen and Hoang, 1983) $\{w(t)\}, \{v(t)\}$ are zero-mean sequences. The following special class of recursive filters, commonly used in real-time data processing, is proposed

$$\hat{x}(t+1) = \varphi(t)\hat{x}(t) + K[z(t+1) - H(t+1)\varphi(t)\hat{x}(t)]$$
$$t = 0,1,..., \hat{x}(0) = E(x(0)) \quad (3a)$$

Thus, Eq. (3a) may be considered as a recursive algorithm for estimating the time-varying vector $x(t)$. For $F(\delta,\lambda)$ we have $\delta = (I_n, \delta'), \xi = [(\varphi.x')^T, (z - H\varphi x')^T]^T, \lambda = 0$, $\delta' \hat{=} K \in R^{nxp}$.

II. PRELIMINERY RESULTS

2.1. Without loss of generality, let $z(t)$ be scalar ($p=1$)

$$\hat{x}(t+1/t) \hat{=} \varphi(t)\hat{x}(t), \; \tilde{x}(t+1/t) \hat{=} x(t+1) - \hat{x}(t+1/t)$$
$$\tilde{z}(t+1/t) \hat{=} z(t+1) - H(t+1)\hat{x}(t+1/t) \quad (3b)$$

and for any two random vectors $\xi, \eta, \bar{\xi} \hat{=} E(\xi), K\xi \hat{=} E(\xi\xi^T)$, $K\xi\eta \hat{=} E(\xi\eta^T)$. First for simplicity let in (2)

$$\varphi(t) = I_n, \; G(t) = 0 \tag{4}$$

The algorithm (3) now takes the form

$$\widehat{x}(t+1) = \widehat{x}(t) + K(t+1)[z(t+1) - H(t+1)\widehat{x}(t)]$$
$$t = 0,1,\ldots, \widehat{x}(0) = E(x(0)) \tag{5}$$

By Def. 1.1 the matrix K is proposed to be chosen from criterion (1b) with $\delta \hat{=} K$. Let $|\cdot|$ denote the euclidian norm .

Lemma 2.1. Let $\chi \in R^n, \zeta \in R^1, \chi, \zeta$ be random values. Let further

$$K_{\chi\zeta} \neq 0 \tag{6a}$$
$$K_\lambda = \lambda K_{\chi\zeta}[1+\lambda K_\zeta]^{-1} \tag{6b}$$
$$\lambda_1 = -\alpha' - \beta, \; \lambda_2 = -\alpha' + \beta, \tag{6c}$$
$$\alpha' = -[K_\zeta]^{-1} > 0, \; \beta = [\Delta]^{1/2}[bK_\zeta]^{-1}$$
$$\Delta = b^2 - ac > 0, \; a = bK_\zeta, \; b < 0, \; c > 0 \tag{6d}$$

Under condition (6),

$$|K_{\lambda_1}| < |K_{\lambda_2}| \tag{7}$$

Proof. To prove (7) let us show the equivalent version

$$e = |K_{\lambda_1}|^2 - |K_{\lambda_2}|^2 < 0 \tag{8}$$

Since $|K_\lambda|^2 = \lambda^2 |K_{\chi\zeta}|^2 [1+\lambda K_\zeta]^{-2}$ then for $K_{\chi\zeta} \neq 0$ the inequality (8) is equivalent to $e_1 = \lambda_1^2[1+\lambda_1 K_\zeta]^{-2} - \lambda_2^2[1+\lambda_2 K_\zeta]^{-2} < 0$. One sees that $e_1 < 0$ iff (if and only if) $[\lambda_1 - \lambda_2][\lambda_1 + \lambda_2 + 2\lambda_1\lambda_2 K_\zeta] < 0$ but as $\lambda_1 - \lambda_2 = -2\beta > 0$ then the last inequality is valid iff $e_2 = \lambda_1 + \lambda_2 + 2\lambda_1\lambda_2 K_\zeta < 0$. For $e_3 = \lambda_1 + \lambda_2 = -2\alpha' < 0$, $e_4 = \lambda_1\lambda_2 = \alpha'^2 - \beta^2 == ac(bK_\zeta)^{-2} < 0$ under condition (6b) $\Delta > 0$. Hence $e_2 = e_3 + 2e_4 K_\zeta < 0$ which proves the lemma ⊗

<u>*Comment 2.1.*</u>

- In (6) we implicitly assumed $K_\zeta \neq 0$. Indeed this condition always holds by applying Lemma A2 (Appendix) to (6a) .
- The Eq. (6b) is correct if $\lambda \neq -[K_\zeta]^{-1}$. For λ_1, λ_2 defined in Lemma 2.1 this condition is fulfilled since $\alpha' = -[K_\zeta]^{-1}, \beta < 0$.
- In the form

$$K_\mu = K_{\chi\zeta}[\mu + K_\zeta]^{-1}, \; \mu = 1/\lambda \tag{9}$$

a parameter μ plays the role of "regularizing" parameter (Nguyen and Hoang, 1983; Hoang and Talagrand, 1992).

Theorem 2.1. Hypothesis in Lemma 2.1 , $\chi \hat{=} x - \widehat{x}(t)$, $\zeta \hat{=} z(t+1) - H(t+1)\widehat{x}(t)$. Then $\forall \beta$ satisfying

$$0 < \beta < \rho^2, \; \rho^2 = (K_{\chi\zeta}^T K_{\chi\zeta})[K_\zeta Sp(K_\chi)]^{-1} \tag{10}$$

the estimator $\widehat{x}(t+1)$ (Def. 1.1) for problem (2)(3)(4) always exists, unique and is defined by (6b) subject to $\lambda = \lambda_1$.

Proof. Unbiassness of $\widehat{x}(t+1)$ is provided by structure of (5). The formulated filter design now reduces to minimization of (1b) subject to constraint (1c). The use of multipliers Lagrange method leads to

$$\tilde{J}[K,\lambda] \rightarrow \min\{K,\lambda\}$$
$$\tilde{J}[K,\lambda] = J[K] + \lambda\{\Delta v(x,t) + \beta v(x,t)\} \tag{11}$$

From definitions
$\Delta v(x,t) = v(x,t+1) - v(x,t) = E\{(\chi - K\zeta)^T(\chi - K\zeta)\} - E\{\chi^T\chi\} =$
$= -2K^T K_{\chi\zeta}^T + K^T K K_\zeta$ therefore $\tilde{J}[K,\lambda] = Sp(K^T K +$
$+ \lambda[K^T K K_\zeta - 2K^T K_{\chi\zeta}^T + bK_\chi])$.

The condition of minimization for $\tilde{J}[K,\lambda]$ is

$$\partial\tilde{J}[K,\lambda] / \partial K = 0$$
$$\partial\tilde{J}[K,\lambda] / \partial\lambda = \Delta v + \beta v = 0 \tag{12}$$

The first one in (12) leads to (6b). By setting (6b) into the second one we find

$$a\lambda^2 + 2b\lambda + c = 0 \tag{13a}$$
$$a = bK_\zeta, \; b = \beta K_\zeta Sp(K_\chi) - K_{\chi\zeta}^T K_{\chi\zeta}, \; c = bSp(K_\chi) \tag{13b}$$

Eq.(13), by (6b), has two solutions λ_1, λ_2 defined by (6c). Thus, the optimal λ minimizing $J[K]$, as shown by Lemma 2.1, is equal to $\lambda = \lambda_1$. To specify (10), observing that from $\Delta > 0$ subject to (6a) we come to the right of inequalities (10). The condition $\beta > 0$ is posed by (1c) ⊗

<u>*Comment 2.2.*</u>

- Condition for existence of λ should be $\Delta \geq 0$. However, due to (6a) , $\Delta = 0$ iff $\beta = \rho^2$ or by (13b), $b = 0$. The last implies $a = 0$ and (13a) becomes $c = 0$ which is impossible by assumption $c > 0$ (6d). Thus, in this case there is no λ satisfying Eq.(13) .

- Due to Lemma A2 and Lemma 2.1, condition (6a) is most important one to provide an existence of solution of considered problem. From Lemma A4, if $P(0) > 0$, (6a) always takes the place provided $H(t) \neq 0$ .When (6a) holds, using Lemma A3 we conclude $0 < \rho^2 \leq 1$.

- Note that λ_1 may be represented in form

$$\lambda_1 = \alpha'\gamma[1-\alpha]^{-1/2}, \; \gamma = 1 - [1-\alpha]^{1/2}, \; \alpha = \beta\rho^{-2} \tag{14}$$

which leads to interesting interpretation of SF (see next Section).

III. DESIGN OF STABLE FILTER FOR SYSTEM STATE

3.1. As outlined in § I, a structure of filter is selected in the form (3a) and the gain coefficient $K(t)$ is assumed to be chosen so that $\widehat{x}(t)$ should be optimal in the sense of Def. 1.1. Rewrite (3a) as

$$\widehat{x}(t+1) = \widehat{x}(t+1/t) + K(t+1)\widetilde{z}(t+1/t) \tag{15a}$$

Correspondingly we have

$$\widetilde{x}(t+1/t) = \varphi(t)\widetilde{x}(t) + G(t)w(t) \tag{15b}$$
$$\widetilde{x}(t+1) = \widetilde{x}(t+1/t) - K(t+1)\widetilde{z}(t+1/t) \tag{15c}$$

Theorem 3.1. Suppose the hypothesis of Lemma 2.1 holds where $\chi \hat{=} \widetilde{x}(t+1/t), \; \zeta \hat{=} \widetilde{z}(t+1/t)$. Under the condition

$$g = (\rho^2 - 1)\Pi + 1 > 0 \tag{16a}$$
$$\rho^2 = \rho^2(t+1/t) = K_{\chi\zeta}^T K_{\chi\zeta}[K_\zeta Sp(K_\chi)]^{-1} \tag{16b}$$
$$\Pi = \Pi(t+1/t) = Sp(K_\chi)[Sp(P(t)]^{-1}, \; P(t) = E\{\widetilde{x}(t)\widetilde{x}^T(t)\} \tag{16c}$$

for all β , $0 < \beta < g$ the SLF $v(x,t) = Sp\{P(t)\}$ computed along the trajectory of (15a) will change according to Eq. (3c). For fixed value β the optimal gain $K(t) = K(t,\lambda)$ is defined by (6b) subject to $\lambda = \lambda_1$ where λ_1 is of the form (6c) or by (14a) with

$$\alpha = \beta_g \rho^{-2} = (\Pi + \beta - 1)[\Pi\rho^2]^{-1}, \; \beta_g = [1 + \beta - 1]/\Pi$$
(14c)

The proof of Theorem is similar to that of Theorem 2.1 , noticing that the second term in (12), after substitution of $K(\lambda)$, becomes $K_\zeta K^T K - 2K_{\chi\zeta}^T K + Sp(A) = 0$ where

$$A = K_\chi + (\beta - 1)P(t) \tag{17a}$$

and the coefficients in (13a) now are

$$a = bK_\zeta, \; b = K_\zeta Sp(A) - K_{\chi\zeta}^T K_{\chi\zeta}, \; c = Sp(A) \tag{17b}$$

<u>*Comment 3.1.*</u> Rewrite (16a) as

$$\Pi < (1-\rho^2)^{-1} \tag{18}$$

The left of (18) represents a ratio of one-step predicted

estimate error (PEE) to filtered estimate error (FEE). It means that for existence of $\beta > 0$ this ratio must be less than some threshold which is equal to the inverse of $1-\rho^2$. In practice, as the rule, FEE < PEE hence $\Pi \geq 1$. In particular, under condition (4), $\Pi = 1$, and for $0 < \rho^2 < 1$ the inequality (18) automatically holds.

3.2. We show now that under conditions of Theorem 3.1 the SLF $v(x,t)$ indeed changes according to the rule (3c). Really, setting λ_1 given by (14) into (6b), after several transformations we arrive to

$$K(t,\lambda_1) = \gamma K^*(t), K^*(t) = K_{\chi\zeta}K_\zeta^{-1} \qquad (19)$$

here γ is defined by (14). Mention that K^* is of the form of the gain for suboptimal in mean square filter (SOMSF) (Nguyen and Hoang, 1991) (or Kalman filter (KF) in uncorrelated case). From (19),

$$P(t+1) = K_\chi + (\gamma^2 - 2\gamma)K^*(t+1)K_{\chi\zeta}^T = K_\chi - \alpha K^*(t+1)K_{\chi\zeta}^T \qquad (20)$$

since $\gamma^2 - 2\gamma = -\alpha$. The Eq.(20) and (14c) imply

$$P(t+1) = K_\chi - [\Pi+\beta-1][\Pi\rho^2]^{-1}K_\chi^{-1}K_{\chi\zeta}K_{\chi\zeta}^T =$$
$$= K_\chi - [\Pi+\beta-1][\Pi\rho^2]^{-1}\{K_\zeta Sp\{K_\chi\}\}^{-1}K_{\chi\zeta}K_{\chi\zeta}^T Sp\{K_\chi\}$$

or

$$Sp\{P(t+1)\} = Sp\{K_\chi\} - [\Pi+\beta-1][\Pi\rho^2]^{-1}\rho^2 Sp\{K_\chi\} =$$
$$= -[\beta-1][\Pi]^{-1}Sp\{K_\chi\} = (1-\beta)Sp\{P(t)\}$$

IV. SOME PROPERTIES OF STABLE FILTER

4.1. *Stability*. From (1c) for $\Psi(t,0) \hat{=} \Pi_{\tau=0}^{t-1}(1-\beta(\tau))$ we have $v(t) = \Psi(t,0))v(0)$ hence for all $v(0) < \infty$ it implies $v(t) < \infty$ since $0 < (1-\beta) < 1$. Thus, the designed filter is stable. If there exists a positive constant c such that $g(t+1/t) \geq c$, $\forall t$ then SF evidently is asymptotically exponential stable for the choice $\beta(t) = \beta^*$, $\beta^* \leq c$, $\forall t > 0$ where β^* is constant. In this case we can prove a convergence (in mean square sense) of estimate produced by SF to true system state.

4.2. *Relationship to SOMSF*. In (Nguyen and Hoang, 1991) the SOMSF in class of filters (3a) is proposed which is derived by solving the optimization problem

$$J(K) = Sp\{P(t+1)\} \to \min_K \qquad (21)$$

It can be shown that a solution of this problem always exists, unique and is defined by (19). When the system and observation noises in (2) are white and mutually uncorrelated (and with $x(0)$) this SOMSF reduces to KF. It is interesting to note that SOMSF does not belong to class of SF (and this concerns, of course, the KF too). However, as will be shown, the SOMSF can be obtained as a limit of SF.

Theorem 4.1. The SOMSF is a limit of TSF as $\beta \to g$.

Corollary 4.1. The KF is a limit of TSF as $\beta \to g$.

Proof of Theorem 4.1 follows from (19)(20) (14) and $\alpha = [\Pi-1+\beta][\Pi-1+g]^{-1}$ (see (14c)).

Since the KF is widely practiced, let us make some comparison of SF with KF. Assume $\{w(t)\}$, $\{v(t)\}$ are white, mutually uncorrelated and are uncorrelated with $x(0)$. In what follows we shall use the low index for corresponding SF and KF when needed and for two SPD (symmetric positive definitive) $P(1), P(2)$ we write $P(1) < P(2)$ if $P(2) - P(1)$ is SPD.

Theorem 4.2. The ECM $P_{sf}(t,\beta)$ is a strictly monotone decreasing matrix-function of β i.e. $P_{sf}(t,\beta_2) < P_{sf}(t,\beta_1)$, $\beta_1 < \beta_2$ and $\lim_{\beta \to g} P_{sf}(t,\beta) = P_{kf}(t)$.

Theorem can be proved by the method similar to that of Theorem 4.1.

Theorem 4.2 shows that the precision of $\hat{x}_{sf}(t,\beta)$ depends considereblly on the choice of β (see Example 1) and the KF-estimate is the best one.

4.3. *Performance sensitivity of SF and KF*. The results presented above are valid if all the system parameters and statistics of initial state and noises are given correctly. In practice, unfortunately, the situation is not so fine. What is the behavior of SF in such situation ? As to KF, there are many works concerned with this question (Jazwinski, 1970; Sorenson, 1967; Griffin and Sage, 1968; Ljung and Kailath, 1977; Safonov and Athans, 1979).

Let us present here a simple observation related to comparison of sensitivity of SF and KF to error in initial ECM $M = P(0)$. Let θ denote a vector of all unknown parameters. For instance let $\theta = M$ be given instead of true one $\theta^* = M^*$ and $M^* < M$. Assume $P(\theta)$ is evaluated by (20) (and by Riccati equation for KF) subject to θ. Thus, $P(\theta)$ is not the true ECM. We denote the true ECM by $P^*(\theta)$. From Theorem 4.2 for $\beta_1 < \beta_2$

$$P_{sf}(M,\beta_1) > P_{sf}(M,\beta_2) > P_{kf}(M) \qquad (22)$$

$$P_{sf}^*(M^*,\beta_1) > P_{sf}(M^*,\beta_2) > P_{kf}(M^*) \qquad (23)$$

The question we are interested is whether there exists the relationship similar to (22),(23) for $P_{sf}^*(M)$?

Theorem 4.3. Suppose $\beta_1 < \beta_2$ and $\gamma_i = \gamma(\beta_i)$, $i = 1,2$. Then

$$P_{sf}^*(M,\beta_1) > P_{sf}^*(M,\beta_2) > P_{sf}^*(M) \qquad (24)$$

iff the following inequality holds

$$D = (\gamma_1 + \gamma_2)K(M)K^T(M)K_z(M^*) - 2K(M)K_{\chi\zeta}^T(M^*) < 0 \qquad (25)$$

If $D > 0$ then

$$P_{sf}^*(M,\beta_1) < P_{sf}^*(M,\beta_2) < P_{kf}^*(M) \qquad (26)$$

Proof. For fixed β and $\gamma = \gamma(\beta)$,

$$P_{sf}^*(M,\gamma) = M^* + \gamma^2 K(M)K^T(M)K_\zeta(M^*) - 2\gamma K(M)K_{\chi\zeta}^T(M^*) \qquad (27)$$

Calculate $P_{sf}^*(M,\gamma_1) - P_{sf}^*(M,\gamma_2)$ by using (27), one can verify that the first inequality in (24) holds iff $D < 0$ since $\gamma_1 - \gamma_2 < 0$. The second one follows immediately from Theorem 4.2. The formula (2.6) holds by similar argument ⊗

Corollary 4.2. Under condition (4) and $n = 1$, $M^* < M$, the relation (24) is valid iff

$$m_L < M^* < M, m_L = (\gamma_1+\gamma_2)RM[2R+H^2(2-\gamma_1-\gamma_2)M]^{-1} \qquad (28)$$

Let $\gamma_2 = 1$. Then (28) expresses the condition under which the KF-estimate is better than SF-estimate with chosen parameter γ_1. If we keep γ_1 to be close enough to 1, m_L will approach to M and for "almost" all M which differ from M^*, the SF-estimate is better than the KF-estimate. Thus by keeping β close to g we can guarantee a good performance for SF-estimates.

By making the use of methods in proving Lemma A4 and Theorem 7.6 (Jazwinski, 1970) one can estiblish

Theorem 4.4. Let $P^*(0) \leq P(0)$, $Q^*(t) \leq Q(t)$, $R^*(t) \leq R(t)$ and $P(t), M(t)$ be ECM for fitered and predicted estimate respectively. Then for fixed value γ

$$P^*(\gamma,\theta^*,t) \leq P(\gamma,\theta,t); M^*(\gamma,\theta^*,t) \leq M(\gamma,\theta,t) \qquad (29)$$

Theorem 4.4 implies that although the actual values $P(0), Q(t), R(t)$ are not given perfectly we can use the computed "variances", as in KF case (Jazwinski, 1970), as the upper bounds for $P^*(\gamma,\theta^*,t), M^*(\gamma,\theta^*,t)$ and to determine whether the conservative estimates of $P(0), Q(t), R(t)$ give satisfactory filter performance. For further discussion see (Jazwinski, 1970, § 7.8).

V. EXAMPLES

Example 1. Let $z(t) = 2x + v(t)$, $t = 1,2,\ldots$ where x, $v(t)$ are zero-mean uncorrelated, $Var(x) = 1$, $E\{v(t)v(\tau)\} = 2\delta_{t\tau}$. We use three methods to estimate the unknown x : KF-method, SF-method and SA-method (stochastic approximation method). The ECM for three methods are presented in Table 1.

Table 1. Comparison of three methods

obs.	P_{kf}	P_{sa}	P^1_{sf}	P^2_{sf}	P^3_{sf}
1	.3333	3.000	.5000	.4118	.3333
2	.2000	.5000	.3333	.2593	.2001
3	.1429	.2780	.2500	.1892	.1429
4	.1111	.1950	.2000	.1489	.1112
5	.0909	.1500	.1667	.1228	.091
6	.0769	.1220	.1428	.1045	.0777
7	.0667	.1030	.1250	.0909	.0667
8	.0588	.0890	.1110	.0804	.0589
9	.0526	.0827	.1000	.0721	.0527
10	.0476	.0759	.0909	.0654	.0447

In Table 1, $P^i_{sf} = P_{sf}(\beta_i), \beta_1 = f(2), \beta_2 == f(1.4), \beta_3 = f(1.001)$, $f(a) = 2P(t)[a+2P(t)]^{-1}$, $a > 1$, since $\rho^2 = 2P(t)/[1+2P(t)]$; for SAM , $K(t) = 1/t$. Results in Table 1 show that the performances of SF-estimates depend considerablly on the choice of β and these numerical results are in agreement with Theorem 4.2. For $\beta = \beta_1$ after only 3 observations the SAM gives the better estimate and this tendency is unchangable as t increases. For $\beta = \beta_2$ a preference of SF to SAM is quite reasonable. Finally, SF subject to $\beta = \beta_3$ gives the estimates with accuracy comparable with estimates derived by KF (their differences are of order 10^{-4}).

Example 2. Consider the filtering problem for 2-dimensional states of system (2) subject to $\varphi = |\varphi_{ij}|^2_{i,j=1}$, $\varphi_{11} = \varphi_{22} = 0.2$, $\varphi_{12} = -\varphi_{21} = -0.458$, $H = (1,-0.5)$, $E\{x(0)\} == (1,0)^T$, $Var\{x(0)\} = I$, $\{w(t)\}, \{v(t)\}$ are zero-mean uncorrelated with $x(0)$ sequences, $w(t)$ is white with unit variance, $v(t)$ is correlated, $E\{v(t)v(\tau)\} = 1/t$, $t \geq \tau$. In Table 2 - the numerical results calculated for some t. The condition (18) is satisfied for all t and the SLF $v(x,t)$ decreases quite rapidly after 6 steps of data processing.

Table 2. SF for two-dimensional case

ob.	ρ^2	Π	$[1-\rho^2]^{-1}$	β	$v(x,t)$
1	0.381	1.25	1.616	0.221	1.746
2	0.381	1.488	1.616	0.073	1.634
3	0.447	1.474	1.807	0.178	1.286
4	0.461	1.584	1.855	0.139	1.286
5	0.476	1.827	1.907	0.034	1.226
6	0.485	1.882	1.942	0.022	1.198

Example 3. Consider the sensitivity of $P_{KF}(t+1), P_{SF}(t+1,\gamma)$ of KF and SF to change of initial ECM $M = P(t)$. Since the KF is the limit of SF as $\gamma_2 \rightarrow 1$, the value m_L in (28a) now is

$$m_L = (\gamma_1+1)RM[2R+H^2(1-\gamma_1)M]^{-1} \tag{28b}$$

From (28a,b) it is evident that when there is a misspecification in initial ECM, if (28a,b) hold then the KF gives the estimates which are better (i.e. their ECM are smaller) in comparison with SF-estimate. For instance, $M = H = R = 1$. Eq. (28b) yields $m_L = (\gamma_1 + +1)[3-\gamma_1]^{-1}$. For different γ_1 the lower bounds m_L for M^* are presented in Table 3.

Table 3. Lower bounds m_L for M^*

γ_1	m_L	γ_1	m_L
0.1	0.3793	0.6	0.6667
0.2	0.4286	0.7	0.7391
0.3	0.4815	0.8	0.8182
0.4	0.5389	0.9	0.9048
0.5	0.6		

It is important to emphasize that when γ_1 tends to 1, m_L tends to 1 too. It means that by holding γ_1 to be close to 1, a performance of SF-estimate will be better than performance of KF if M is different from M^* . This fact is shown numerically in Table 4 where the performances of KF-estimate $\widehat{x}_{kf}$ and SF-estimate $\widehat{x}_{sf}(\gamma_1)$, $\gamma_1 = 0.5$ and $\gamma_1 = 0.9$ as a function of M^* are presented. The computation is based on formula (27).

Table 4. Performance of KF and SF

M^*	$P_{sf}(0.5)$	$P_{sf}(0.9)$	P_{kf}
0.1	0.1187	0.2327	0.275
0.2	0.175	0.263	0.3
0.3	0.2312	0.2932	0.325
0.4	0.2875	0.3235	0.35
0.5	0.3438	0.3537	0.375
0.6	0.4	0.384	0.4
0.7	0.4563	0.4142	0.425
0.8	0.5125	0.4445	0.45
0.9	0.5687	0.4747	0.475

Thus, for $M^* = 0.6$ the KF and SF with $\gamma_1 = 0.5$ give the estimates of the same performance $P_{kf} = P_{sf}(0.5) = 0.4$ while they are given by different formulas: $\widehat{x}_{kf}(t+1) = \widehat{x}_{kf}(t) + [z(t+1) - \widehat{x}_{kf}(t)]/2$, $\widehat{x}_{sf}(t+1) = \widehat{x}_{sf}(t) + [z(t+1) - \widehat{x}_{sf}(t)]/4$ if the true ECM of $\widehat{x}_{kf}(t), \widehat{x}_{sf}(t)$ are M^* . A performance of SF with $\gamma_1 = 0.9$ in general is better than a performance of KF.

VI. CONCLUSION

In this work a new method of design of stable filters based on SLF-method is suggested. The main feature of this method is lying in idea of inversion of direct LF-method which reduces the filtering problem to constrainted optimization. The conditions for existence and uniqueness of SF are investigated. The proposed SF and KF, as we see, are of the same order of complexity from point of view of their implementation in real-time order . In general the choice of parameter γ (or β, μ ...) depends on the level of uncertainty given a-priori (Nguyen and Hoang,1983; Hoang and Talagrand, 1992). Since in practice, as the rule, the a-priori information on system parameters and statistics of initial state and noises is given not perfectly the stability of constructed filter can guarantee a good performance of estimates. For the designed SF a problem of its stability may be easily solved by the use of Eq. (1c) as shown in § IV. We demonstrated also that in the case of a-priori uncertainty the proposed SF , under some conditions, are preferable in comparison with standard KF. A problem of sensitivity of SF to different sources of uncertainties (system parameters, noise statistics ...), of course, is of importance and requires the futher study.

Acknowledgement. The authors are mostly grateful to Ministere de la Recherche et Technologie for its financial support during visit of first author in Labo de Météorologie Dynamique (LMD, ENS, Paris, France) and to Director of LMD, Prof. R. Sadourny for his valuable encouragement and suggestions in this work.

APPENDIX

Lemma A1. Let $\beta \rightarrow 0$. Then $\beta_g \rightarrow [\Pi-1]/\Pi$ $\alpha \rightarrow [\Pi-1][\Pi\rho^2]^{-1}, \gamma \rightarrow [1-g/\Pi\rho^2]^{1/2}$. For the case $\beta \rightarrow g$ we have $\beta_g \rightarrow \rho^2, \alpha \rightarrow 1, \gamma \rightarrow 1$.

Lemma A1 may be proven by direct calculation.

Due to Lemma A1, the SF is redefined by $\alpha \in (\alpha_L, 1), \alpha_L = \max\{0, (\Pi-1)(\Pi\rho^2)^{-1}\}$ if $g > 0$. For $g < 0$ we can redetermine the SF as the filter of the form (15a)(19)(20) subject to $\alpha \in (0,1)$.

Lemma A2. Let ζ, χ be defined as in Lemma 2.1. Then under condition (6a) we have $K\zeta \neq 0, Sp[K_\chi] \neq 0$.

Proof. Let χ_i be ith component of χ. Lemma now follows from $|E\{\chi_i\zeta\}| \leq |K_\zeta^{1/2}| |K_{\chi_i}^{1/2}|$ $\otimes$

Lemma A3. The following inequalities hold : $0 \leq \rho^2 \leq 1$

Proof. From $E\{\chi_i\zeta\}^2 \leq E\zeta^2 E\chi_i^2$, a summation of both sides of this inequality gives $\Sigma_{i=1}^n E\{\chi_i\zeta\}^2 \leq \Sigma_{i=1}^n E\zeta^2 E\chi_i^2$. The left of this inequality is $\Sigma_{i=1}^n E\{\chi_i\zeta\}^2 = K_{\chi\zeta}^T K_{\chi\zeta}$ whereas the right is $\Sigma_{i=1}^n E\zeta^2 E\chi_i^2 = E\zeta^2 \Sigma_{i=1}^n E\chi_i^2 = E\zeta^2 Sp\{K_\chi\}$ $\otimes$

Lemma A4. Let $P(0) > 0$, $\varphi(t)$ or $Q(t)$ be nonsingular, $R(t)$ be nonsingular. Then for all t we have $P(t,\alpha) > 0, \forall \alpha \in (0,1)$ where

$$P(t+1,\alpha) = M(t+1,\alpha) - \alpha K(t+1) K_{\chi\zeta}^T$$
$$M(t+1,\alpha) = \varphi(t) P(t+1,\alpha) \varphi^T(t) + Q(t)$$

Lemma may be proved by induction as follows. Consider the problem of estimating $x(t+1)$ in

$$x(t+1) = \varphi(t)x(t) + w(t),$$
$$z(t+1) = H(t+1)x(t+1) + v(t+1)$$

with $Ex(t) = \hat{x}_{sf}(t,\alpha), E\{\tilde{x}(t)\tilde{x}^T(t)\} = P_{sf}(t,\alpha)$. From Riccati equation for KF it follows that $P_{kf}(t) > 0$ provided that the conditions of lemma are satisfied (Jazwinski, 1970). We have $\delta P = P_{kf}(t) - P_{sf}(\alpha,1) = (\alpha-1)K^* HM < 0, \alpha \in (0,1)$ or $0 < P_{KF}(1) < < P_{SF}(\alpha,1)$.

REFERENCES

Crutko P.D. (1983) The Lyapunov functions in inverse problems for dynamical controlled systems. Scalar models. J. News of Academy of Sciences of USSR, Technical Cybernetics, N° 4.

Crutko P.D. (1984) The Lyapunov functions in inverse problems for dynamical controlled systems. Multidimensional models. J. News of Academy of Sciences of USSR, Technical Cybernetics, N° 4.

Crutko P.D. (1989) Invesion of direct Lyapunov method in control problems for dynamical systems. J. News of Academy of Sciences of USSR, Technical Cybernetics, N° 3.

Heffes H. (1966) The effect of erroreous models on the Kalman filter response. J. IEEE Trans. Autom. Contr., AC-11, pp. 541-543.

Hoang H.S., and Nguyen T.L. , Pham H.T. (1991) Time-stable non-linear filters: Stochastic Lyapunov function approach. Proc. Int. Conf. on Mathematical Theory of Networks and Systems, June, Kobe, Japan.

Hoang H.S., and Talagrand O. (1992) On regularization of best linear estimator in linear inversion with nonnegative covariance structure. Proc.5th Int. Meeting on Statistical Climatology, June, Ontario, Canada.

Jazwinski A.H. (1970) Stochastic processes and filtering theory. Academic Press, New York.

Kalman R.E. (1963) New methods in Wiener filtering theory. Proc. Symp. Eng. Appl. Random Theory and Probability (Bogdanoff J.L. and Kozin F., eds.),Wiley,NY.

Lipsher R.S., and Shiryaev A.N. (1974) Statistics of random processes. Nauka, Moskow.

Ljung L., and Kailath T. (1977) Efficient change of initial condition, dual Chandrasekhar equations and some applications. J. IEEE Trans. Autom. Contr., June, pp. 443-447.

Nguyen T.L., and Hoang H.S. (1983) On solution of ill-posed optimal linear filtering with correlated noises. J. Automat.& Remote Control, N°4, pp. 58-73.

Nguyen T.L., and Hoang H.S. (1991) Approximate approach to nonlinear filtering with dependent noises. Part I: Nonlinear filter suboptimal in mean square. To appear in Kybernetik.

Safonov S., and Athans M. (1979) On stability theory. Proc. IEEE Conf. on Decision and Control, Jan., San Diego,

Sorenson H.W. (1967) On the error behavior in linear minimum variance estimation problems. J. Trans. Autom. Contr., AC-12, pp. 557-562. CA.

Zubov V.I. (1963) On theory of analytic design of regulators . J. Automat. & Remote Contr., N°8.

Specification of Intelligent Controllers for Discrete Event Systems in A Temporal Logic Framework

Dan Ionescu, Jing-Yue Lin, Hyung Soo Hwang

Department of Electrical Engineering, University of Ottawa, Ottawa, Ontario, Canada K1N 6N5

Keywords. Intelligent control, Real-Time Control, Discrete event systems; Temporal Logic Models.

Abstract: In this paper, a temporal logic model is used for designing controllers. The plant is formalized as a system driven by inputs and events that force a sequence of states in the system state space. Temporal logic models are built for the plant, and controller; the closed loop system specification are also written in the same language. A synthesis method for building an intelligent controller is developed based on the reachability property. An example demonstrates the viability of this approach.

1 Introduction

Intelligent controllers received recently considerable attention from the research community. Despite their belonging to the control and systems field, the intelligent controllers do not benefit from the elegance of a theoretical approach such as their continuous (or discrete) control systems counter-parts, where plant and controller behavior is described by differential (difference) equations. This is due to a heteromorphic knowledge required for a domain implementation, where control and computer engineers, together with computer scientists cooperate for merging in the same system different and sometimes oposite concepts. Other reasons for the amalgamate concepts melted in an intelligent system are the layered structure of an intelligent system and the various sources and types of information which have to be coalesced.

The domain literature reflects, therefore, this situation. Concepts generated by control engineers put the emphasis on adaptivity [Tsypkin 1973] or tuning [Astrom 1992], as a source and domain of intelligence. A layered architecture [Albus 1990] characterizes the theory developed by computer engineers dealing with the design of complex robot tasks [Valavanis 1990]. An attempt for a brief classification of various methods applied in intelligent control is made by Ward [Ward 1990]. However, the information given in the latter is poor and the number of methods cited incomplete, which is expectable for such a domain. From an artificial intelligence (AI) point of view an intelligent system, however, must include a reasoning mechanism, a rule base and a database of facts characterizing the environment in which the intelligent controller interracts. These rules and database are time dependent.

The motivation for this paper consists in seting up a theoretical basis which will allow the designers to abstractly specify the models of intelligent controllers. For this purpose the theory developed by Lin and Ionescu [Lin 1990 and 1991b] is applied to intelligent controllers.

In what follows an attempt to characterize the intelligent controllers through a set of rules written in a temporal logic syntax is provided. An intelligent controller synthesis procedure, based on the above models, and their theoretical implications is developed.

The plant, the controller, and the closed loop system will be described as temporal logic models (TLM). Based on this type of model a synthesis procedure will be developed.

Intelligent controllers have to react to external stimuli (events) and to elaborate a system strategy corresponding to a set of rules validated by the above stimuli, in real-time. Therefore, external events have an essential role to play in controlling such systems. The theory and the desgn procedures developed in this paper will take into account the above aspects.

2 Temporal Logic Models

In order to describe the plant, the controller, and the closed loop dynamics using a set of rules a temporal logic framework is considered. In this framework, a dynamical model for systems reacting to events is developed.

The language and the proof system of the temporal logic used in this paper are same as those in [Thistle 1986]. The temporal logic is an extension of the ordinary logic to include the notion of time, providing five modal operators: $\Box$ (henceforth), $\Diamond$ (eventually), $\bigcirc$ (next), $\mathcal{U}$ (until) and $\mathcal{P}$ (precedes).

According to Lin and Ionescu [Lin 1990 and 1991b] a dynamical formula $l(e,s)$ has the following form:

$$\Box[\delta = e \wedge x = R \Rightarrow (\bigcirc x) = Q]$$

which is related to the transition f by $s = x$, and $s' = f(e,s) = \bigcirc x$, and says that the occurrence of event e changes the value of state x from R to Q.

A temporal logic model (TLM) of a process is defined by a 5-tuple $M = (S, E, s_0, \delta, l)$ where S is a finite set of states; E is a finite set of events, each event being viewed as a transition from a state to another state; s_0 is an initial state; δ is a firing function, a mapping from S into E^* (the set of subsets of E) such that $\forall s \in S, \forall e \in \delta(s), e(s) \stackrel{\text{def}}{=} \delta(s,e)$ is defined where $\delta(s)$ is the set of events firing in the state s; and l is a labelling function from E into F^* where F is the set of logical formulas.

Determined by the firing function δ, $e(s) \stackrel{\text{def}}{=} \delta(s,e)$ is a transition from one state to another; and hence, it may be called as a next state function, which gives a dynamical behavior to real-time program.

A state of a TLM corresponds the execution of exactly one event. To reflect this fact, we systematically add to our specifications for each process in the TLM the following:

$$\Box\{[\vee_{1\le i\le N}(x_i = P)] \wedge [\wedge_{1\le i<j\le N} \neg(x_i = P \wedge x_j = P)]\}$$

where N is the dimension of the state, $x_1, x_2, \cdots, x_N$ are all the components of the state, and $\neg$ is the logical negation.

Consider now an example [Zhou 1990] of a manufacturing process where two machines are used in series, having a buffer between them, for storing the intermediate parts, and a robot for loading and unloading the parts for both machines. The final product is obtained by processing a part first on machine m_1, and then on machine m_2. However, either m_1 or m_2 might fail while processing the part. The part will be processed for completion after the machine which failed gets repaired. The robot can not be interrupted while loading or unloading one of the two machines. The same restriction applies to the processing of any of the parts on any of the two machines.

In order to describe this industrial process as a TLM the following notations will be used.

The five possible states of a process are:

- WP which stands for "waiting for being processed"
- PM_1 which stands for "being processed on machine M_1"
- B which stands for "being buffered"
- PM_2 which stands for "being processed on machine M_2"
- ST which stands for "going to be stored"

For any i $(0 < i < n)$ the events of a process i will be:

- a_i the event of starting the processing on machine M_1, and correspondingly the predicate showing this hapening will be $stprm1(x)$
- b_i the event of finishing processing on M_1, and $fnprm1(x)$ being the corresponding predicate
- c_i and $stprm2(x)$ being the event and the predicate for starting the processing on M_2, respectively
- d_i and $fnprm2(x)$ being the event and the predicate for starting the processing on M_2, respectively
- o_i and $rt(x)$ being the event and the predicate for returning outside the processing cycle for preparing the processing of the next raw material, respectively

The event-condition table and the event-state diagram are given at the end of the paper in Table 1 and Figure 1 respectively.

The plant (process) model is then given by the following TLM:

$$\Box[stprm1(x) \wedge WP(x) \Rightarrow PM1(\bigcirc x)] \qquad (P1)$$

$$\Box[fnprm1(x) \wedge PM1(x) \Rightarrow B(\bigcirc x)] \qquad (P2)$$

$$\Box[stprm2(x) \wedge B(x) \Rightarrow PM2(\bigcirc x)] \qquad (P3)$$

$$\Box[fnprm2(x) \wedge PM2(x) \Rightarrow ST(\bigcirc x)] \qquad (P4)$$

$$\Box[rt(x) \wedge ST(x) \Rightarrow WP(\bigcirc x)] \qquad (P5)$$

Formula $(P1)$ describes the effect of the request of processing the part x on machine M_1: its state changes from $WP(x)$ to $PM1(x)$, while the states of the other processes remain the same. $(P2)-(P6)$ are similar to $(P1)$, describing the effects of the finishing processing ofpart x on machine M_1, the buffering of part x when getting out of machine M_1, the start of processing part x on machine M_2, the finishing of processng on M_2 commencement of writing, and the end of writing, respectively, of process x. The above rules are shown on the plant event-state graph in Figure 2. To these rules an initial condition has to be added.

Initial condition

$$WP(x) \qquad (P6)$$

It means that initially all processes on both machines are idle.

3 Intelligent Controller Specification

Specifying an intelligent controller as a TLM, requires to write the rules that reflect the restrictions imposed on the closed loop system.

Attaching a controller to a system governed by dynamical rules means to merge two sets of rules,- those of the plant and of the controller - into a single set - the rules governing the dynamics of the composite system.

The composite system contains, however, two real-time systems which have to be synchronized.

This implies the development of theoretical tools which will provide the designer with the correct procedure.

To this end the concept of reachability and composition and synchronization of two ore more processes, will be introduced.

For a temporal logic model $M = (\mathbf{S}, \mathbf{E}, \mathbf{F}^*, f, s_0, l)$, a state s is immediately reachable from s_0 if there is an event $e \in E_{s_0}$ such that $f(e, s_0) = s$.

Obviously, the initial states are reachable, therefore, in a TLM $M = (\mathbf{S}, \mathbf{E}, \mathbf{F}^*, f, s_0, l)$, $s' = f(e, s)$ is reachable from s for $e \in E_s$ if and only if the dynamical formula $l(e, s) \in \mathbf{F}^*$ is true.

A reachability graph can therefore be built following the state transitions of the plant encoded in the above rules. The transition graph of the plant is given in Figure 3, and the corresponding reachability graph for the the same plant is given in Figure 4.

For the composition and synchronization of two DES, results from Henessy's process theory are used. A TLM is proved to be a $\Sigma-$algebra [Henessy 1988].

A possible composition operation can be used to build a TLM from its component TLMs, say two TLMs, by putting together of two TLMs with synchronization between selected events.

The dynamic behavior of the composed TLM is now given in terms of the projection mappings from $\bar{M}$ into M and M'.

The next result obtained by Lin and Ionescu [Lin 1991b] gives the theoretical fundations for the design of composite intelligent controllers and plant systems.

Let (Φ, Ψ) and (Φ', Ψ') be the projections from $\bar{M} = M \| M'$ to M and M' respectively, *i.e.* $\Phi : \bar{\mathbf{S}} \rightarrow \mathbf{S}$, $\Psi : \bar{\mathbf{E}} \rightarrow \mathbf{E}$, $\Phi' : \bar{\mathbf{S}} \rightarrow \mathbf{S}'$, and $\Psi' : \bar{\mathbf{E}} \rightarrow \mathbf{E}'$. Then

(1): The projection mappings are $\Sigma-$homomorphisms;

(2): $t = \bar{f}(e, s)$ in $\bar{M}$ if and only if $\Psi(t) = f(\Phi(e), \Psi(s))$ in M and $\Psi'(t) = f'(\Phi'(e), \Psi'(s))$ in M'.

A controller synthesis procedure can be obtained by using reachability and composition mmethods.

(i): Specify the required behavior of the closed-loop system M by temporal logic formulas;

(ii): Find out the set of states which should not be reached by the closed-loop system according to (i). Denote the set by R_u;

(*iii*): Construct the controller M_c such that

$$R(M, s_0) \times R(M_c, q_0) = R(M, s_0) \times Q - R_u$$

(*iv*) Synchronize the controlled events in the plant and in the controller; and compose the plant M and the controller M_c.

Returning to the example given in section 2, the closed loop specifications have to express the fact that the parts are processed in the order of their arrival: if part i precedes part j in its arrival at the processing machine, then part i will be sent before part j to that machine.

If the machine M_1 may fail while processing a job then the robot returns to the sate WP or B.

If the machine M_2 may fail while processing a job then the robot returns to the sate WP.

One has also to add the rule that specifies that two parts can not be processed on the same machine at the same time - the mutual exclusion rule.

With this, the closed loop system specifications can be written as follows:

$$\Box[\wedge_{i=1}^{n}[x_i = PM_1 \Rightarrow \wedge_{i\neq j=1}^{n} x_j \neq PM_2] \qquad (CL1)$$

$$\Box[\wedge_{i=1}^{n}[x_i = PM_2 \Rightarrow \wedge_{i\neq j=1}^{n} x_j \neq PM_1 \qquad (CL2)$$

Here (CL1) and (CL2) represent the mutual exclusion condition.

$$\Box\wedge_{i,j=1,i\neq j}^{n}(x_i \neq PM_1 \Rightarrow (\delta = a_i\mathcal{P}\delta = \dashv_{|} \Rightarrow \delta = \lfloor_{\rangle}\mathcal{P}\delta = \lfloor_{|}))] \qquad (CL3)$$

$$\Box\wedge_{i,j=1,i\neq j}^{n}(x_i \neq PM_2 \Rightarrow (\delta = r_i\mathcal{P}\delta = \nabla_{|} \Rightarrow \delta = \lceil_{\rangle}\mathcal{P}\delta = \lceil_{|}))] \qquad (CL4)$$

Here (CL3) and (CL4) express the condition according to which the parts are processed in their arrival order.

The next rules will encode the fail condition imposed on either machine M_1 or $M - 2$. Tey are:

- If machine M_1 may fail while processing a job, then the robot returns either to the waiting to be processed (WP) or to the buffer (B) state.

- If machine M_2 may fail while processing a job, then the robot returns to the waiting to be processed (WP) state.

Consider now that $EQ(x, y$ representsthe fact "x identical to y", and let Y_1 and Y_2 be assigned the value 1 or 0 for representing the facts "machine M_1 or machine M_2 are functinal or not" respectively. Let also assign nonnegative integer values to C_1 and C_2 which will be used as counters for the raw materials entering the processing stage and for the buffering the parts, respectively. Two more predicates are to be defined for being able now to write the last rules:

- $Add1(n)$ which stands for increasing n by 1.

- $Sub1(n)$ which stands for decreasing n by 1.

With these the rules are:

$$\Box[WP(x) \wedge EQ(Y_1, 1) \wedge \neg EQ(C_1, 0) \Rightarrow Sub1(\bigcirc C_1) \wedge PM_1(x)] \qquad (CL5)$$

$$\Box[PM_1(x) \wedge EQ(Y_1, 0) \wedge EQ(C_2, 0) \Rightarrow WP(x)] \qquad (CL6)$$

$$\Box[PM_1(x) \wedge EQ(Y_1, 0) \wedge \neg EQ(C_2, 0) \Rightarrow B(x)] \qquad (CL7)$$

$$\Box[PM_1(x) \wedge EQ(Y_1, 1) \Rightarrow Add1(\bigcirc C_2) \wedge B(x)] \qquad (CL8)$$

$$\Box[B(x) \wedge EQ(Y_2, 1) \wedge \neg EQ(C_2, 0) \Rightarrow Sub1(\bigcirc C_2) \wedge PM_2(x)] \qquad (CL9)$$

$$\Box[PM_2(x) \wedge EQ(Y_2, 0) \wedge EQ(C_1, 0) \Rightarrow WP(x)] \qquad (CL10)$$

$$\Box[PM_2(x) \wedge EQ(Y_2, 0) \wedge EQ(C_1, 0) \Rightarrow B(x)] \qquad (CL11)$$

$$\Box[PM_2(x) \wedge EQ(Y_2, 1) \Rightarrow DS(x)] \qquad (CL12)$$

$$\Box[DS(x) \wedge \neg EQ(C_1, 0) \Rightarrow WP(x)] \qquad (CL13)$$

The iniatial condition for the system is as follows:

$$EQ(Y_1, 1) \wedge EQ(Y_2, 1) \wedge EQ(C_1, 0) \wedge EQ(C_2, 0) \qquad (CL14)$$

Considering now the two sets of rules (P1) to (P6) and (CL1) to (CL14) together with the corresponding initial conditions, and the reachability graph given in Figure 4, an appropriate intelligent controller is formalized as follows:

$$\Box[stprm1(x) \wedge EQ(Y_1, 0) \wedge \neg EQ(C_1, 0) \Rightarrow Sub1(\bigcirc C_1)] \qquad (C1)$$

$$\Box[stprm1(x) \wedge EQ(Y_1, 1) \wedge \neg EQ(C_1, 0) \Rightarrow EQ(\bigcirc C_1, C_1)] \qquad (C2)$$

$$\Box[fnprm1(x) \wedge EQ(Y_1, 1) \Rightarrow Add1(\bigcirc C_2)] \qquad (C3)$$

$$\Box[stprm1(x) \wedge EQ(Y_2, 1) \wedge \neg EQ(C_2, 0) \Rightarrow Sub1(\bigcirc C_2)] \qquad (C4)$$

$$\Box[stprm1(x) \wedge EQ(Y_2, 0) \Rightarrow EQ(\bigcirc C_2, C_2)] \qquad (C5)$$

The initial condition of the controller is then defined by:

$$\neg EQ(C_1, 0) \wedge EQ(Y_1, 1) \qquad (C6)$$

The controller structure obtained so far is to be implemented as a set of rules which are solved with the help of a reasoning mechanism. This gives the controller the capacity to reason on various situation through the processed has to pass, making it an intelligent tool for process supervision.

4 Conclusions

An abstract model which can be used for designing intelligent controllers following a technique used in control system design, was introduced. the formalisms allows the graph manipulation and the use of the formal logic in its extension: the temporal logic. Temporal logic was considered because of its capacity to combine rules, proper to reasoning processes, with the variable time which is proper to dynamic systems. Expressing the rules for a system charcterization is a simple process; it might, however, have complexity implications when dealing with complex real systems.

References

Ya.Z.Tsypkin and Z.J.Nikolic (1973). Foundations of the theory of learning systems, Academic Press, New York, London, 1973.

Hennessy, M. (1988), *Algebraic Theory of Processes*, The MIT press, Massachusetts.

Lin, J.-Y., and D. Ionescu (1990). A generalized temporal logic approach for control problems of a class of nondeterministic discrete event systems. *Proc. 29th IEEE Conf. Decision & Control*, Honolulu, Hawaii, Dec 5-7, 1990. pp.3440 3445.

Lin, J.-Y., and D. Ionescu (1991b). A temporal logic approach to the control of discrete event systems. *Proc. the 1991 American Control Conference*, Boston, Massachusetts, June 26-28, 1991. pp.2934-2935.

Thistle, J.G., and W.M. Wonham (1986). Control problems in a temporal logic framework. *Int. J. Control*, **44**, 943-976.

Astrom, K. J., Hang, C. C., Persson, P., and HO, W. K., (1992) Towards intelligent PID control Automatica, IFAC ,Vol.28, No.1, pp.1-9.

Cleveland, B. and Meystel, A., (1990) Dynamic predictive planning + fuzzy logic control = intelligent control, Proc. 5th IEEE Int. Symposium on Intelligent Control, Philadelpia, Sep.5-7,1990. pp.1217-1224.

Zhou, M. C., Dicesare, F., and Guo,D.,(1990). Modeling and performance analysis of a resource-sharing manufacturing systems using stochastic Petri nets, Proc. 5th IEEE international symposium on intelligent control, Philadelpia,Sep. 5-7, 199, pp.1005-1010.

Ward, T. L., Ralston, P. A., and Stoll, K. E.,(1990) Intelligent control of machines and processes, Computers and industrial engineering. Vol.19. No.1-4.pp.205-209.

Albus, J. S.,(1990). A theory of intelligent systems. Proc. 5th IEEE international symposium on intelligent control, Philadelpia, Sep.5-7, 1990, pp.866-875.

precondition	event	postcondition
WP(S1)	a (e1)	PM(S2)
PM1(S2)	b (e2)	B(S3)
B(S3)	r (e3)	PM2(S4)
PM2(S4)	d (e4)	DS(S5)
DS(S5)	o (e5)	WP(S1)

Table 1 Event and state table

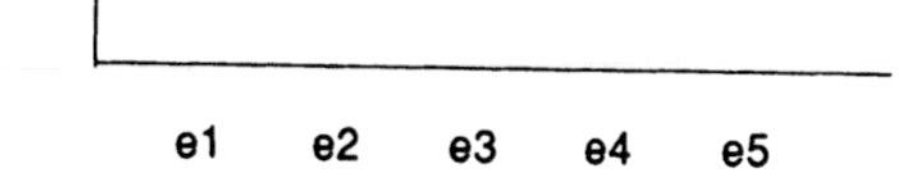

Figure 1 Event and state correspondence

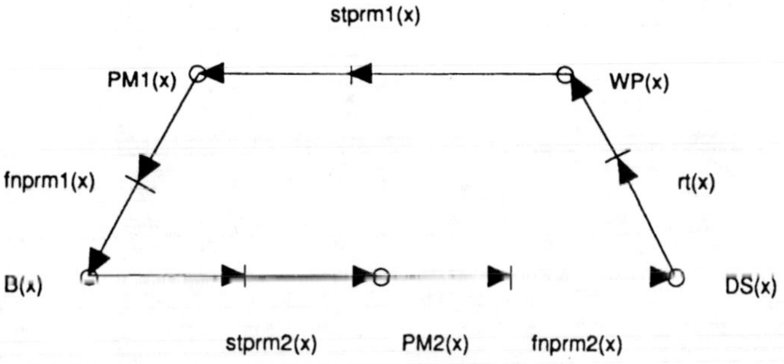

Figure 2 Event and state graph

State	x1	x2	state	x1	x2
S0	WP	WP	S13	WP	B
S1	PM1	WP	S14	WP	PM2
S2	B	WP	S15	PM1	PM2
S3	PM2	WP	S16	B	PM2
S4	DS	WP	S17	DS	PM2
S5	WP	PM1	S18	WP	DS
S6	B	PM1	S19	PM1	DS
S7	PM2	PM1	S20	B	DS
S8	DS	PM	S21	PM2	DS
S9	PM1	B	S22	PM1	PM1
S10	PM2	B	S23	PM2	PM2
S11	B	B	S24	DS	DS
S12	DS	B			

Table2 Reachability table

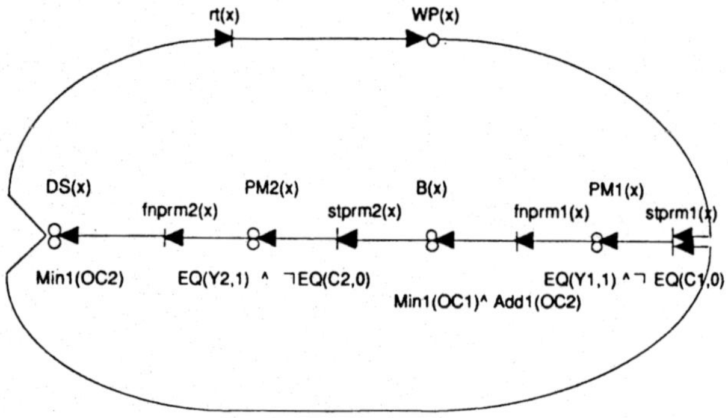

Figure 5 The composite system graph

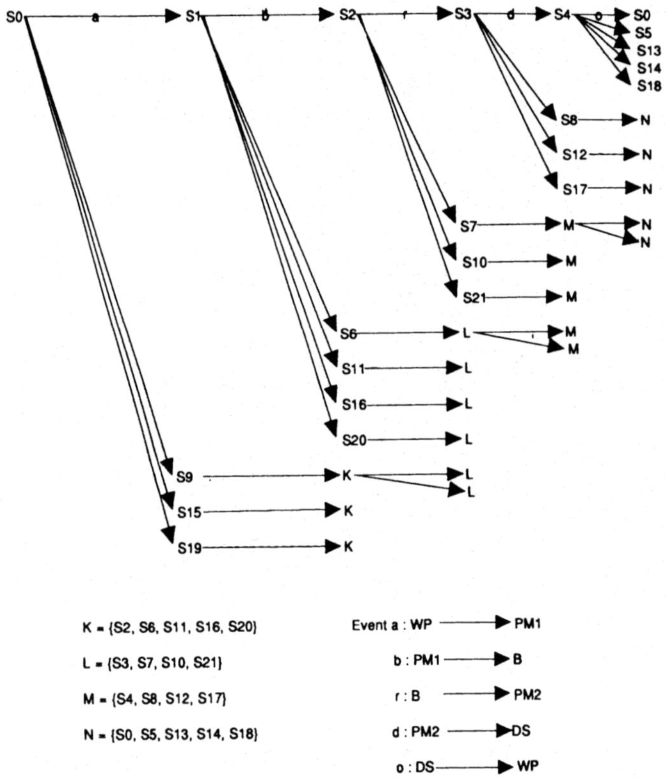

Figure 3 State transition graph

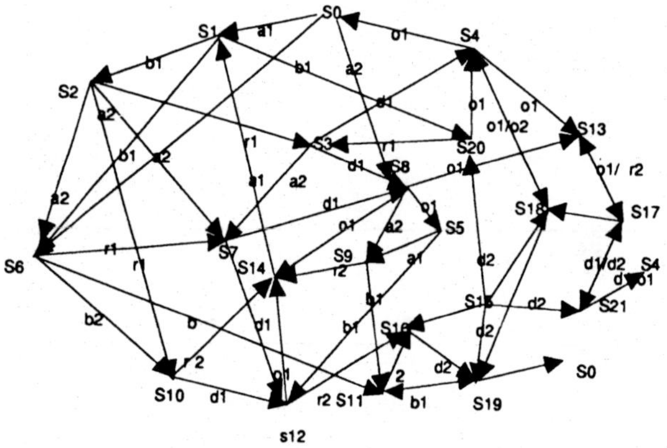

Figure 4 Reachability graph

A NEW DECOMPOSITION METHOD OF RELATION MATRICES OF MULTIVARIABLE FUZZY SYSTEMS

Gi-Joon Jeon, Kyun Kyung Lee, Pyeong Gi Lee

Department of Electronic Engineering, Kyungpook University, Daegu, Korea

Abstract. For the design of multivariable fuzzy control systems decomposition of control rules is preferable since it alleviates the complexity of the problem. In some systems, however, inference error of the Gupta's decomposition method is inevitable because of its approximate nature. In this paper we define an index of applicability which decides whether the decomposition method can be applied to a multivariable fuzzy system or not. We also propose a new decomposition method of relation matrices of multivariable fuzzy systems using the index of applicability. As an illustrative example, the existing decomposition method is applied to an ENOR gate model and the utility of the index of applicability is verified by the output inference error of the model.

Keywords. Multivariable fuzzy system; index of applicability; relation matrix; decomposition; inference error; projection

INTRODUCTION

Since Zadeh(1965) has introduced the fuzzy set theory, numerous fuzzy control strategies have been developed and many examples of applications to diverse systems(e.g. Kickert and Van Nauta Lemke,1976; King and Mamdani,1977) have been reported in a variety of academic and professional journals across the world. Most of the contributions, however, are made on single-input single-output systems contrary to the real world in which multivariable systems prevail.

The analysis and design procedures for multivariable fuzzy systems are very difficult mainly due to multidimensional nature of control rules and relation matrices. To get rid of the dimensionality problem and computational burden of the multivariable systems, Gupta, Kiszka and Trojan(1986) propose a decomposition method of multivariable control rules in which the rules are simplified by max min operations. Consequently, it reduces computational time and storage requirement when the outputs of the systems are inferred from inputs and relation matrices. This decomposition method works well in some systems, but it is not acceptable in other systems because of the large inference error resulted from the ignorance of some control rules during the execution of decomposition.

In this paper defined is an index of applicability which decides whether the decomposition of the relation matrix can be applied to mutivariable fuzzy systems or not. Using the index of applicability a new method of decomposition of relation matrices of multivariable fuzzy systems is proposed. As an illustrative example, the existing decomposition method is applied to an ENOR gate model and the utility of the index of applicability is verified by the inference error of the model.

OUTPUT INFERENCE FROM TWO-DIMENSIONAL RELATION MATRICES

Consider a multivariable fuzzy system with N inputs and M outputs as shown in Fig. 1.

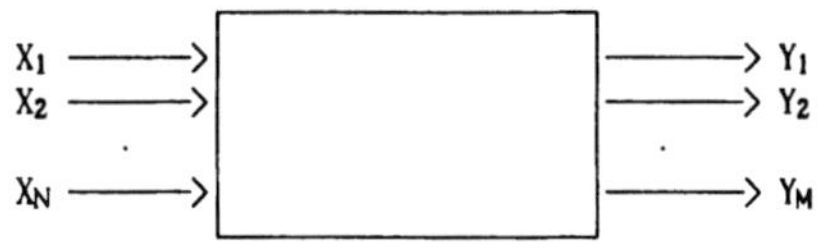

Fig. 1. Block diagram of a Multivariable System.

Suppose that K control rules of the system are given by

$$\begin{array}{ll} \text{IF } X_1 \text{ is } X_{1(1)} \text{ AND} \cdots X_N \text{ is } X_{N(1)} \text{ THEN } Y_1 \text{ is } Y_{1(1)} & \\ \quad \text{AND} \cdots Y_M \text{ is } Y_{M(1)} & \\ \quad \vdots & \\ \text{ALSO} & \\ \quad \vdots & \\ \text{IF } X_1 \text{ is } X_{1(k)} \text{ AND} \cdots X_N \text{ is } X_{N(k)} \text{ THEN } Y_1 \text{ is } Y_{1(k)} & (1) \\ \quad \text{AND} \cdots Y_M \text{ is } Y_{M(k)} & \\ \quad \vdots & \\ \text{ALSO} & \\ \quad \vdots & \\ \text{IF } X_1 \text{ is } X_{1(K)} \text{ AND} \cdots X_N \text{ is } X_{N(K)} \text{ THEN } Y_1 \text{ is } Y_{1(K)} & \\ \quad \text{AND} \cdots Y_M \text{ is } Y_{M(K)}. & \end{array}$$

where the inputs and the outputs are normalized fuzzy sets, $X_{N(k)}$ is the fuzzy value of the N-th process input in the k-th control rule, and $Y_{M(k)}$ is that of the M-th process output in the k-th rule. Let the dimensions of discrete fuzzy inputs and outputs be given by

$$\dim[\, X_n \,] = p_n \quad , \; n=1,\cdots,N$$

$$\dim[\, Y_m \,] = q_m \quad , \; m=1,\cdots,M.$$

Following conditions are satisfied since the elements(i_1, $i_2 \cdots i_N, j_1, j_2 \cdots j_M$) obtained by quantizing physical values are some elements in the universe of discourse.

$$\begin{array}{ll} 1 \le i_n \le p_n \quad , \; n=1,\cdots,N & \\ & (2) \\ 1 \le j_m \le q_m \quad , \; m=1,\cdots,M & \end{array}$$

From control rules established by experts' experience, a relation matrix which is necessary to the inference of the m-th output is defined as

$$R_m = \bigvee_{k=1}^{K} \{ X_{1(k)} \wedge X_{2(k)} \wedge \cdots \wedge X_{N(k)} \wedge Y_{m(k)} \}, \quad (3)$$

$$1 \leq m \leq M$$

where $\vee$ is the max_operator and $\wedge$ the min_operator.

Given the current inputs $X_1, X_2, \cdots, X_N$, the present m-th output Y_m is computed by the following compositional rule of inference.

$$Y_m = X_1 \circ X_2 \circ \cdots \circ X_N \circ R_m , \quad (4)$$

$$1 \leq m \leq M$$

where o is the max-min composition of fuzzy relations. In case of systems with many inputs, the method of inference expressed by Eq.(4) accompanies such difficulties as large computation time and storage space requirement because of the multidimensionality of the relation matrix Eq.(3) and the fuzzy output equation, Eq.(4). Therefore, it is very difficult to infer the outputs of the systems with many inputs by those operations. To get rid of this problem, Gupta, Kiszka and Trojan(1986) proposed a inference method which is composed of N·M two-dimensional relation matrices. The inference method is expressed in a matrix form as follows:

$$\begin{bmatrix} Y_1 \\ Y_2 \\ \cdot \\ \cdot \\ Y_M \end{bmatrix} = [\, X_1 \; X_2 \; \cdots \; X_N \,] * \begin{bmatrix} R_{11} & \cdots & R_{1M} \\ R_{21} & \cdots & R_{2M} \\ \cdot & \cdot & \cdot \\ \cdot & & \cdot \; \cdot \\ R_{N1} & \cdots & R_{NM} \end{bmatrix} \quad (5)$$

where * is the $(\circ, \wedge)$ operator. From Eq.(5), the m-th output of the system is expressed by the $(\circ, \wedge)$ operator as

$$Y_m = X_1 \circ R_{1m} \wedge X_2 \circ R_{2m} \wedge \cdots \wedge X_N \circ R_{Nm} \quad (6)$$

and

$$R_{nm} = \bigvee_{k=1}^{K} \{ X_{n(k)} \wedge Y_{m(k)} \}, \quad m = 1,2,\cdot\cdot,M. \quad (7)$$

The structure of the multivariable fuzzy system in the Gupta's method is shown in Fig. 2. The inference error of the method may increase in a certain system because it uses only simplified two-dimensional relation matrices. we now analyze the decomposition of control rules using the projection method which is described immediately.

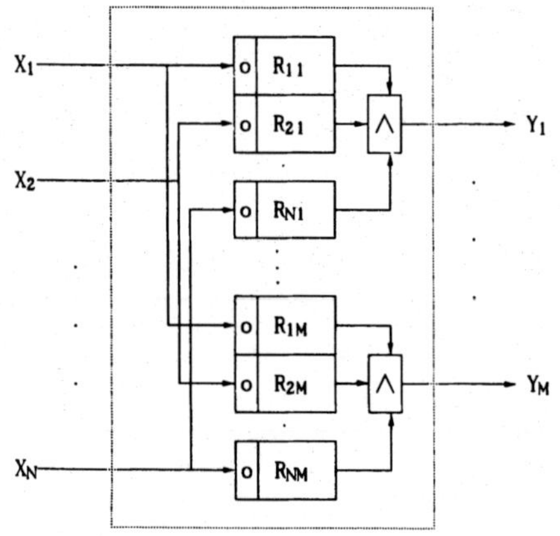

Fig. 2. Structure of a multivariable fuzzy system using two-dimensional relation matrices

In the multivariable fuzzy system with N inputs and M outputs, we define a projection which choose a maximum element of the relation matrix R_m along a fixed quantized point i_n of region X_n as follows:

$$P_{nm} = \text{Proj}_{\,x_n = i_n} R_m = \sup_{\substack{x_1 \in X_1 \\ x_2 \in X_2 \\ \vdots \\ x_n \in X_N}} R_m(x_1, \cdots, x_n = i_n, \cdots, x_N, Y_m = j_m), \quad (8)$$

$$m = 1,2,\cdots,M, \quad n = 1,2,\cdots,N$$

To explain the projection clearly, consider the following K linguistic control rules of a system with two inputs and one output as an example.

IF X_1 is $X_{1(1)}$ AND X_2 is $X_{2(1)}$ THEN Y_1 is $Y_{1(1)}$
·
ALSO ·
·
IF X_1 is $X_{1(k)}$ AND X_2 is $X_{2(k)}$ THEN Y_1 is $Y_{1(k)}$ (9)
·
ALSO ·
·
IF X_1 is $X_{1(K)}$ AND X_2 is $X_{2(K)}$ THEN Y_1 is $Y_{1(K)}$

Then, the relation matrix is constructed as follows:

$$R_1 = \bigvee_{k=1}^{K} \{ X_{1(k)} \wedge X_{2(k)} \wedge Y_{1(k)} \} \quad (10)$$

On the other hand, the two-dimensional relation matrices R_{11} and R_{21} of the system are obtained as follows:

$$R_{11} = \bigvee_{k=1}^{K} \{ X_{1(k)} \wedge Y_{(k)} \} \quad (11)$$

$$R_{21} = \bigvee_{k=1}^{K} \{ X_{2(k)} \wedge Y_{(k)} \} \quad (12)$$

To compare the results of inference obtained from the decomposed two-dimensional relation matrices R_{11} and R_{21} with the results calculated from original relation matrix R_1, we define the fuzzy sets A_1 and A_2 of don't care terms. These fuzzy sets mean that the membership values of all elements in the universe of discourse are unity.

Given the current inputs X_1 and X_2, the present output with the don't care terms is expressed as follows :

$$Y_1 = X_1 \circ R_{11} \wedge X_2 \circ R_{21}$$
$$= X_1 \circ A_2 \circ \{ \bigvee_{k=1}^{K} X_{1(k)} \wedge A_{2(k)} \wedge Y_{(k)} \}$$
$$\wedge A_1 \circ X_2 \circ \{ \bigvee_{k=1}^{K} A_{1(k)} \wedge X_{2(k)} \wedge Y_{(k)} \} \quad (13)$$

Each element of the relation matrices calculated from inputs, output, and the don't care terms is not less than that of the original relation matrix. Also, the elements of the don't care terms A_1 and A_2 are greater than or equal to those of the current inputs X_1 and X_2. Therefore, the following inequality holds

$$Y_1 \geq X_1 \circ X_2 \circ \{ \bigvee_{k=1}^{K} X_{1(k)} \wedge X_{2(k)} \wedge Y_{(k)} \} \quad (14)$$

where the right_hand side is to infer the output from the original relation matrix. The above equation implies that when the output is inferred from two-dimensional relation matrices, in some cases a considerable inference error can be originated.

To better understand the projection, let the normalized inputs X_1 and X_2 be fuzzy singletons instead of fuzzy sets. Since A_2 is the don't care term in the relation matrix R_{11}, we know that the result of inference is equal to the greatest value $\max_i$ obtained by taking projection of all elements of X_2 domain along the fixed quantized element i of the current input X_1 in the original relation matrix. In the same way, we can also obtain $\max_j$ by taking projection of all elements of X_1 domain. The inferred output calculated from the Gupta's decomposition method, Eq.(13), implies that it represents the smaller value of $\max_i$ and $\max_j$.

The projection method is depicted in Fig.3, where the direction of arrows indicates that of projection. It is easily understood that the following inferred output equations are obtained from the figure.

$$Y^p = \max_{i_p} \wedge \max_{j_p}, \quad p = 1,\cdots,P \tag{15}$$

$$Y_1 = [\, Y^1 \;\; Y^2 \;\; \cdots \;\; Y^P \,] \tag{16}$$

where Y^p is the p-th element of the fuzzy set of inference output and the dimension of the universe of discourse of the output Y_1 is P.

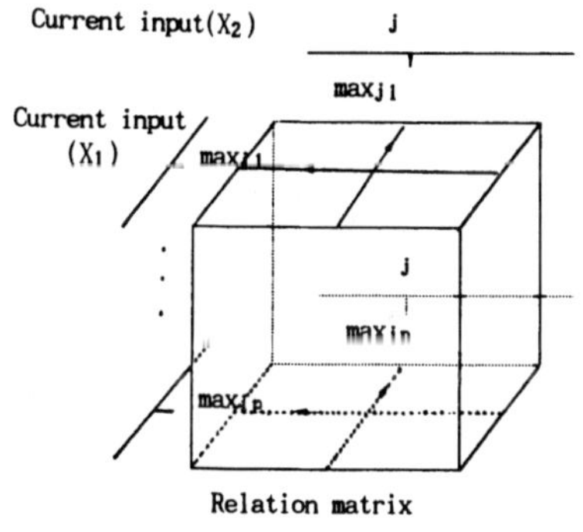

Fig. 3. Output inference by projection

In the above example, the result inferred from two-dimensional relation matrices which are composed of only single input and single output is equal to the result obtained by taking projection with fixing an element of the input variable. When this result is compared with the output inferred from the original relation matrix, it is noted that a good deal of inference error may occur depending on the system characteristics.

A NEW DECOMPOSITION METHOD OF RELATION MATRICES

Using the projection method introduced in the preceding chapter, we define an index C of applicability, which can classify whether the decomposition method of control rules can be applied to multivariable fuzzy systems with N inputs and M outputs or not:

$$C = \frac{1}{M}\sum_{m=1}^{M} C_m, \quad 0 \le C_m < 1 \tag{17}$$

and

$$C_m = \frac{\displaystyle\sum_{i_1,\cdots,i_N,j_m} \{ \bigwedge_{n=1}^{N} P_{nm} - R_m(i_1,i_2,\cdots,i_N,j_m) \}}{q_1 \times \cdots \times q_N \times p_m} \tag{18}$$

where

$$\bigwedge_{n=1}^{N} P_{nm} = P_{1m} \wedge P_{2m} \wedge \cdots \wedge P_{Nm}$$

$$\sum_{i_1,i_2,\cdots,i_N,j_m} = \sum_{j_m=1}^{p_m}\sum_{i_N=1}^{q_N}\cdots\sum_{i_2=1}^{q_2}\sum_{i_1=1}^{q_1}$$

From the definitions, Eqs.(17) and (18), it is easily shown that the indices C_m and C are less than unity. The index C_m represents the normalized area error between the result inferred from two-dimensional relation matrices and that obtained from the (N+1)-dimensional relation matrix without decomposition. Note that when the index C_m lies close to unity, the result of inference is incorrect.

Taking advantage of the index of applicability We propose a new decomposition method of the relation matrix in order to reduce the inference error possibily resulted from the over simplification of the control rules. In the method we adopt a coordinator which can determine the dimension of the relation matrix in the upper level. Fig. 4 shows the block diagram of the structure of the new decomposition method.

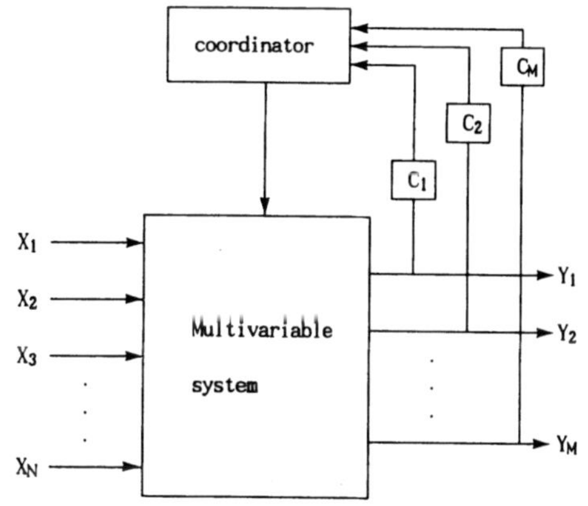

Fig. 4. Structure of new decomposion method with a coordinator.

The procedure to obtain the inference output by the new method can be summarized as follows.

Step 1: Assign a upper bound (ζ) of the indices of applicability C_m, m=1,2,$\cdots$,M in the upper level.

Step 2: Compute the indices $C_1, C_2, \cdots, C_M$ corresponding to each output and transfer them to the upper level.

Step 3: At the upper level, check whether the indices are within the pre-determined upper bound. If not, increase the dimension of the corresponding relation matrices to three-dimension. Otherwise, set the relation matrices to two-dimension and go to step 4.

Step 4: Given the current inputs, infer the output value from the relation matrices the dimensions of which are determined at the upper level.

As an explanatory example, we consider a multivariabe fuzzy system with three inputs and two outputs. Suppose that the index C_1 is less than the pre-assigned upper bound and C_2 is greater than that. In this case, the block diagram of a new decomposition method is shown in Fig. 5.

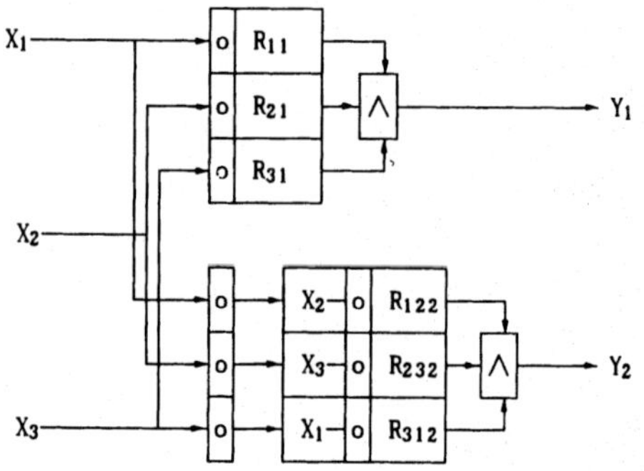

Fig. 5. Structure of multivariable fuzzy system using the new decomposition method.

With the (o, $\wedge$) operator, the proposed method is explained by following equations.

$$Y_1 = X_1 \circ R_{11} \wedge X_2 \circ R_{21} \wedge X_3 \circ R_{31} \quad (19)$$

$$Y_2 = X_1 \circ X_2 \circ R_{122} \wedge X_2 \circ X_3 \circ R_{232} \wedge X_3 \circ X_1 \circ R_{312} \quad (20)$$

where

$$R_{huv} = \bigvee_{k=1}^{K} \{ X_{h(k)} \wedge X_{u(k)} \wedge Y_{v(k)} \}$$

$$R_{uv} = \bigvee_{k=1}^{K} \{ X_{u(k)} \wedge Y_{v(k)} \}$$

NUMERICAL EXAMPLES AND RESULTS

The inference of output of an ENOR gate model with crisp set

To better understand the problem that the Gupta's decomposition method using only two-dimensional relation matrices may result in a consideable inference error, we apply this method to an ENOR gate model with crisp set. The logic symbol of the ENOR gate model is shown in Fig. 6.

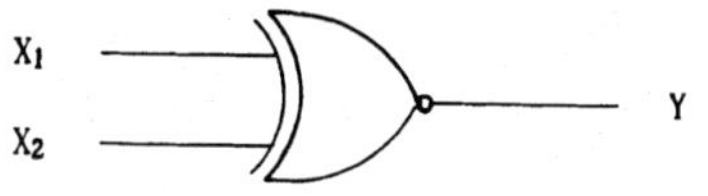

Fig. 6. Logic symbol of an ENOR gate

Using linguistic descriptions we will represent the case in which the gate input value is greater than 2.5v as high state(H). Otherwise, it will be expressed as low state(L).

TABLE 1 Truth Table for ENOR Gate

Input variables		Output variable
X_1	X_2	Y
1 0(L)	1 0(L)	0 1(H)
1 0(L)	0 1(H)	1 0(L)
0 1(H)	1 0(L)	1 0(L)
0 1(H)	0 1(H)	0 1(H)

Table 1 shows the relationship between the gate inputs and the output, where H and L are represented by [0 1] and [1 0] as crisp set, respectively. Given the current inputs X_1 and X_2 as high states H, the inference output from only two two-dimensional relation matrices is calculated as follows:

$$R_{11} = \begin{bmatrix} 1 & 1 \\ 1 & 1 \end{bmatrix}, \quad R_{21} = \begin{bmatrix} 1 & 1 \\ 1 & 1 \end{bmatrix}$$

$$Y = [\,1 \quad 1\,]$$

The inference output Y and the actual gate output is shown in Fig. 7. It is easily verified that the inferred output from the original relation matrix is exactly the same as the actual gate output.

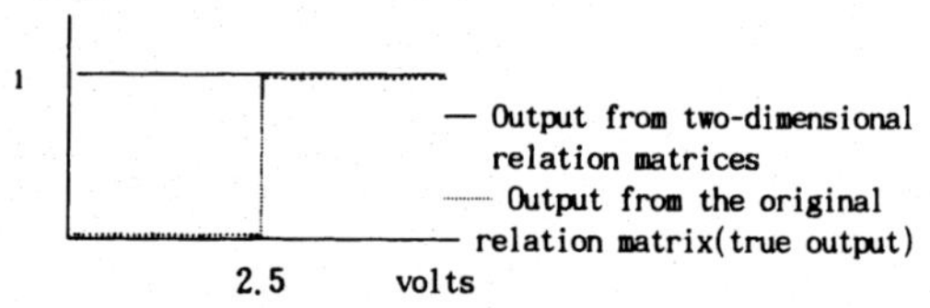

Fig. 7. Outputs of the ENOR gate

The index C_l of applicability for the ENOR gate model is computed as

$$C_l = \frac{\sum_{i_1, i_2, j_1} \{ \bigwedge_{n=1}^{2} P_{nl} - R_l(i_1, i_2, j_1) \}}{q_1 \times q_2 \times p_1} = 0.5$$

The index 0.5 means that the normalized area error between the inference value by the Gupta's method and true value is 50%, which is well illustrated in Fig. 7.

In implementing a practical system this error could be a main source of degradation of the performance of inference. And it is almost impossible to obtain the proper control input to the system since the inference output includes both the states H and L in any circumstances.

The inference of outputs of a multivariable fuzzy system

For another example, consider a multivariable fuzzy system with two inputs and two outputs. It can be extended to a system with N inputs and M outputs without loss of generality.

TABLE 2 Membership Functions of Fuzzy Sets for System Inputs and Outputs

Universe	Linguistic Fuzzy Sets[1]					
	Z	AZ	S	M	BMB	B
0	1	1	0.7	0.3	0	0
1	0.7	1	1	0.7	0.3	0
2	0.3	0.7	0.7	1	0.7	0.3
3	0	0.3	0.3	0.7	1	0.7
4	0	0	0	0.3	1	1
5	0	0	0	0	0.7	1

1. Z-Zero, AZ-Approximately Zero, S-Small, M-Medium, BMB-Between Medium and Big, B-Big.

The Membership functions of fuzzy sets of the system inputs X_1, X_2 and outputs Y_1, Y_2 are represented in Table 2. Assume that the upper bound of the index of applicability in the coordinator is 0.10. Let the system be expressed by the verbal description connected by the conjunction ALSO as follows:

IF X_1 is B AND X_2 is Z THEN Y_1 is Z AND Y_2 is M
ALSO
IF X_1 is AZ AND X_2 is Z THEN Y_1 is BMB AND Y_2 is S
ALSO
IF X_1 is B AND X_2 is BMB THEN Y_1 is BMB AND Y_2 is B
ALSO
IF X_1 is AZ AND X_2 is BMB THEN Y_1 is Z AND Y_2 is BMB

The indices C_1 and C_2 of applicability for this system are calculated in a similar manner as in the preceding example.

$$C_1 = 0.152, \quad C_2 = 0.079$$

Given the current inputs X_1 as B and X_2 as BMB, the output Y_1 from the Gupta's method is obtained as follows;

$$Y_1 = \min\{ Y_1^{*1}, Y_2^{*2} \}$$

$$= [\ 1.0\ 0.7\ 0.7\ 1.0\ 1.0\ 0.7\ ]$$

where

$$Y^{*1} = X_1 \circ R_{11} = [\ 1.0\ 0.7\ 0.7\ 1.0\ 1.0\ 0.7\ ]$$

$$Y^{*2} = X_2 \circ R_{21} = [\ 1.0\ 0.7\ 0.7\ 1.0\ 1.0\ 0.7\ ]$$

$$R_{11} = \begin{bmatrix} 1.00 & 0.70 & 0.70 & 1.00 & 1.00 & 0.70 \\ 1.00 & 0.70 & 0.70 & 1.00 & 1.00 & 0.70 \\ 0.70 & 0.70 & 0.70 & 0.70 & 0.70 & 0.70 \\ 0.70 & 0.70 & 0.70 & 0.70 & 0.70 & 0.70 \\ 1.00 & 0.70 & 0.70 & 1.00 & 1.00 & 0.70 \\ 1.00 & 0.70 & 0.70 & 1.00 & 1.00 & 0.70 \end{bmatrix}$$

$$R_{21} = \begin{bmatrix} 1.00 & 0.70 & 0.70 & 1.00 & 1.00 & 0.70 \\ 0.70 & 0.70 & 0.70 & 0.70 & 0.70 & 0.70 \\ 0.70 & 0.70 & 0.70 & 0.70 & 0.70 & 0.70 \\ 1.00 & 0.70 & 0.70 & 1.00 & 1.00 & 0.70 \\ 1.00 & 0.70 & 0.70 & 1.00 & 1.00 & 0.70 \\ 0.70 & 0.70 & 0.70 & 0.70 & 0.70 & 0.70 \end{bmatrix}$$

In the same way, output Y_2 is also computed as

$$Y_2 = [\ 0.3\ 0.3\ 0.7\ 0.7\ 1.0\ 1.0\ ].$$

The indices C_1 and C_2 in this example imply that the normalized area errors of Y_1 and Y_2 between the inference values by Gupta's method and true values are 16.2% and 2.9%, respectively.

Since the index C_1 is greater than the upper bound, the three-dimensional relation matrix R_{121} should be used to infer Y_1 in the proposed method. On the other hand, in order to infer Y_2 two-dimensional relation matrices R_{12} and R_{22} are sufficient because the index C_2 is within the bound. The inference equations and results of both outputs are following.

$$Y_1 = X_1 \circ X_2 \circ R_{121}$$

$$= [0.3\quad 0.3\quad 0.7\quad 1.0\quad 1.0\quad 0.7]$$

$$Y_2 = X_1 \circ R_{12} \wedge X_2 \circ R_{22}$$

$$= [0.3\quad 0.3\quad 0.7\quad 0.7\quad 1.0\quad 1.0]$$

For the comparision, outputs computed from three-dimensional relation matrices are obtained as

$$Y_1 = X_1 \circ X_2 \circ R_1$$

$$= [0.3\quad 0.3\quad 0.7\quad 1.0\quad 1.0\quad 0.7]$$

$$Y_2 = X_1 \circ X_2 \circ R_2$$

$$= [0.3\quad 0.3\quad 0.3\quad 0.7\quad 1.0\quad 1.0]$$

Figures 8 and 9 show the outputs inferred by three different methods. Note that the result of the proposed method is better than that of the Gupta's method in case that the index is greater than the bound.

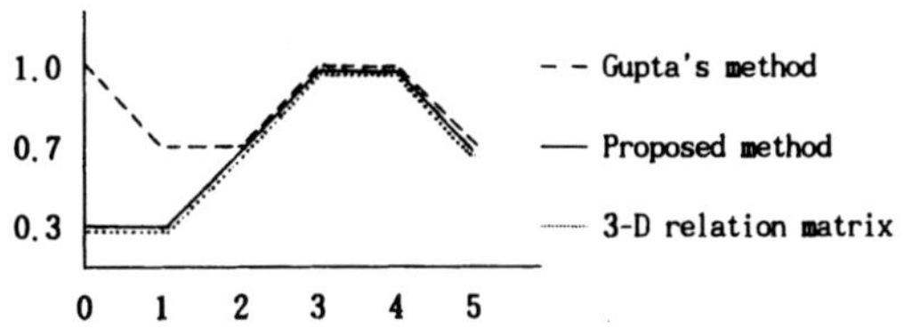

Fig. 8. Comparision of inference output, Y_1

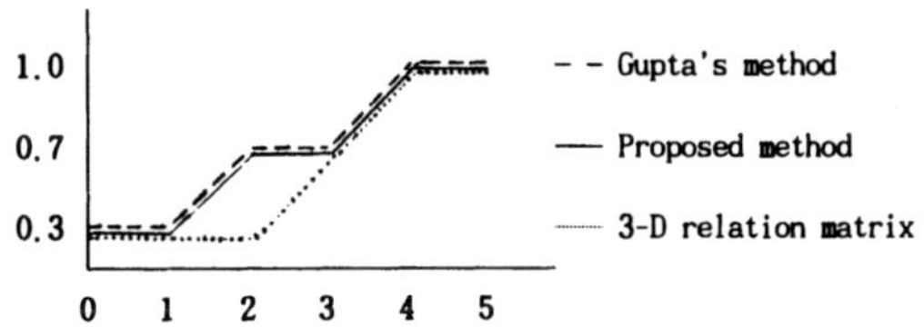

Fig. 9. Comparision of inference output, Y_2

CONCLUSION

In this paper we define an index of applicability which decides whether the Gupta's decomposition method of control rules can be applied to a multivariable fuzzy system or not. Using the index of applicability a new method of decomposition of relation matrices of multivariable fuzzy systems is proposed. Since the coordinator in the new decomposition method can judge the inference error of two-dimensional relation matrices, the method can be successfully applied to the design of multivariable autonomous fuzzy control systems. The utility of the index of applicability is verified by an illustrative example of an ENOR gate model.

Acknowledgement: This work was supported in part by Engineering Research Center for Advanced Control and Instrumentation of Seoul National University by Korea Science and Engineering Foundation.

REFERENCES

Gupta, M.M., J.B. Kiszka, and G.M. Trojan (1986). multivariable structure of fuzzy control systems. *IEEE Trans. Syst. Man Cybern.*, *16*, 638-655.

Kickert, W.J.M., and H.R. Van Nauta Lemke (1976). Application of a fuzzy controller in a warm water plant. *Automat.*, *12*, 301-306.

King, P.J., and E.H. Mamdani (1977). The application of fuzzy control system to industrial processes. *Automat.*, *13*, 235-242.

Zadeh, L.A. (1965). Fuzzy sets. *Inform. Contr.*, *8*, 338-353.

Identificaion of Fuzzy Control Rules Utilizing Genetic Algorithms and Its Application to Mobile Robot

Hee Soo Hwang*, Young Hoon Joo*, Hyun Ki Kim**,
Kwang Bang Woo*,

*Department of Electrical Engineering Yonsei University, Seoul 120-749, Korea
**Department of Electrical Engineering Suwon University, Suwon 445-743, Korea

Abstract. *In this paper, an approach to identify fuzzy control rules is presented. The decision of the optimal number of fuzzy control rules is made by means of soft c-means clustering which is a classification oriented approach based on quantitative infomation of a system and produces fuzzy clusters that are optimal in a generalized least squared error sense. The identification of the parameters of fuzzy implications is carried out by use of genetic algorithms which display an excellent robustness in complex optimization problems. For the efficient and fast parameter identification, the reduction technique of search areas of genetic algorithms is proposed. The feasibility of the proposed approach is evaluated through the identification of the fuzzy model to describe an input-output relation of Gas Furnace. Utilizing the navigation data of a mobile robot obtained from the expert's control actions to follow the center of a corridor, the fuzzy control rules are identified and applied to the navigation control.*

Keywords. *Fuzzy control; genetic algorithms; identification of fuzzy implication;*

INTRODUCTION

An approach to identify fuzzy control rules is presented in this paper. If fuzzy control rules are successfully implemented, an elaborate, iterative, error-prone, and time-consuming stage of model-building is eliminated. However, if fuzzy controller is to perform its job better, the scale factors of the input and output variables, the set of control rules, and the membership functions of the variables must be optimised. The scale factors have a strong influence on the dynamics of the closed loop systems. Nevertheless, they can not remove a few unsuitable rules, since they influence the entire control rules.

An effective way of identifying the fuzzy control rules can be implemented as a self-organizing controller. The structure of the controller compromises a performance table, a simplified process model, and a rule modification algorithm. Any required reinforcement given by performance table is transformed into a value of control by means of the process model, which contributes to the rule modification. The structure satisfying learning task is based on a simplified version of the process model. However, it is difficult to verify how far the model may preserve the property of self-organization. In this aspect, parameter identification by self-organization mechanism is restrictive.

Neural networks offer the advatage of deducing decision strategies directly from training data, as opposed to rule-based systems which derive their knowledge through expert consultation. Thus, approachs based on neural networks have been employed in order to identify fuzzy control rules. The identification is performed by means of back-propagation learning algorithm. However, these approaches require additional parameters, such as learning rates and slopes of sigmoid functions, and do not guarantee optimal solutions because of local minima problem.

In this paper, the identification of fuzzy control rules is classified into the decision of the opimal number of fuzzy implications and the identification of parameters defining membership functions in each fuzzy implication. The optimal number of the membership functions of each input variable is determined by soft c-means clustering which produces a fuzzy c-partition of input-output data, via iterative optimization of the squared distances weighted by the mth power of the membership. The optimal number of fuzzy rules is determined by the multiplication of the cluster numbers of each input variable. The identification of the parameters is carried out utilizing genetic algorithms which do not require additional parameters for running and which display an excellent robustness in complex optimization problems. For an efficient and fast identification of parameters, the reduction technique of search areas of genetic algorithms is proposed.

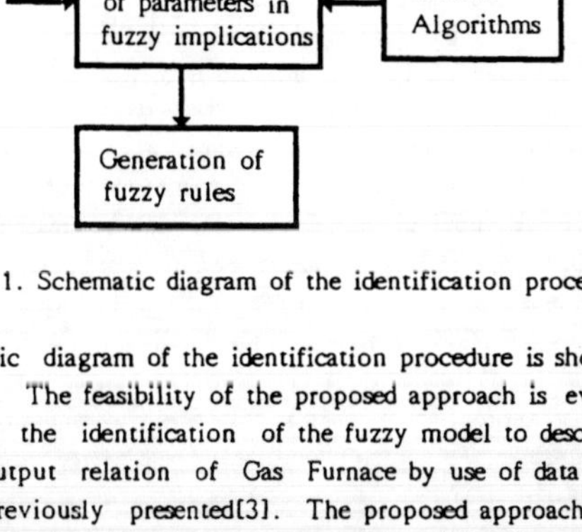

Fig. 1. Schematic diagram of the identification procedure

Schematic diagram of the identification procedure is shown in Fig. 1. The feasibility of the proposed approach is evaluated through the identification of the fuzzy model to describe an input-output relation of Gas Furnace by use of data which were previously presented[3]. The proposed approach is also employed using the navigation data of a mobile robot obtained from the expert's control actions in which fuzzy control rules to follow the center of the corridor in a confined area are identified and applied to the navigation control.

FUZZY IMPLICATION AND REASONING

The format of fuzzy implication and reasoning algorithm is described as follows:

Format of Implication

We consider an ith fuzzy implication, R^i.

R^i: If x_1 is Small and x_2 is Big, then $y_i = w_i \cdot a_i + b_i$

where Small and Big are fuzzy labels of x_1 and x_2, respectively, w_i the degree of fulfillment of the premise, and a_i and b_i consequent parameters.

Reasoning Algorithm

Suppose implications R^i(i=1,2) of the above format.

R^1: If x_1 is Small and x_2 is Big, then $y_1 = w_1 \cdot a_1 + b_1$

R^2: If x_1 is Big and x_2 is Medium, then $y_2 = w_2 \cdot a_2 + b_2$

Fig. 2 shows the procedure of reasoning, where w_1 and w_2 are calculated by eg.(1). Given input data x_1^0 and x_2^0, the final output y^* inferred from above two implications is obtained in terms of the average of y_1 and y_2 with the weights w_1 and w_2.

$$w_1 = \mu_{Small}(x_1^0) \cdot \mu_{Big}(x_1^0) \quad (1)$$

$$w_2 = \mu_{Big}(x_1^0) \cdot \mu_{Medium}(x_1^0)$$

$$y^* = \frac{w_1 \cdot y_1 + w_2 \cdot y_2}{w_1 + w_2} \quad (2)$$

$$= \frac{w_1 \cdot (w_1 \cdot a_1 + b_1) + w_2 \cdot (w_2 \cdot a_2 + b_2)}{w_1 + w_2}$$

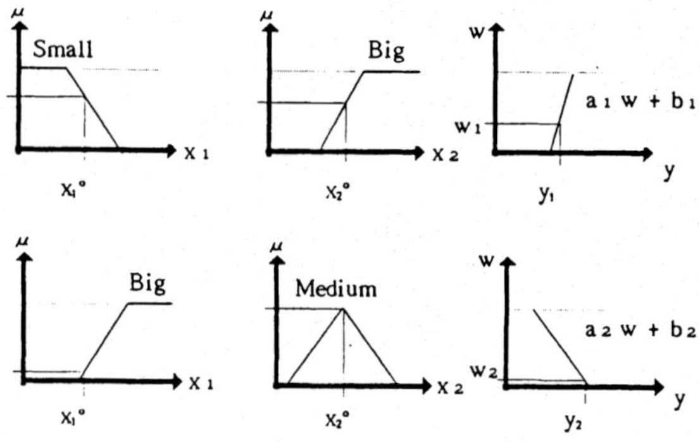

Fig. 2. Procedure of reasoning

STRUCTURE IDENTIFICATION

The identification of structure which corresponds to the decision of the optimal number of fuzzy implications to describe an input-output relation is carried out by means of soft c-means clustering which is a classification oriented approach based on quantitative information of a system and dynamic thretholding of the grades of membership, and produces a fuzzy c-partition of data set, via iterative optimization of the squared distances weighted by the mth power of the membership. Our purpose is to find the optimal or sub-optimal numbers of clusters to effectively describe the relation between each input-output value of a system. For the brevity, we consider a system composed of two inputs x_1 and x_2 and one output y. If the optimal or sub-optimal cluster numbers of x_1-y and x_2-y are c_1 and c_2, respectively, the optimal number of fuzzy implications is c_1c_2. The validity of partitioning is evaluated by the values of several indices, H, H1, F, H2, DME, F2, and NFI[1] according to the number of clusters. The appropriate values of clusters can be determined at higher values of F, F_2, NFI, and DME, and lower values of H, H_1, and H_2.

SCM Clustering Algorithm

The soft c-means(SCM) clustering algorithm produces a fuzzy c-partition of the data set $X=(X_1, X_2, \ldots, X_n)$. The basic steps of the algorithm used in this paper are given as follows[1]:

Step 1

① Fix the number of cluster c, $(2 \le c \le n)$, where n is number of data items.

Fix m, $(1 \le m \le \infty)$.

② Define D_{ik} as inner product norm.

③ Set p = 1.

④ Initialize the fuzzy c partition, $U^{(p-1)}$ where column j of c×n matrix U represents membership values of X_j in fuzzy c clusters.

Step 2

Calculate the c cluster centers $(v_i^{(p)})$ with $U^{(p-1)}$ and the formula(3) for the ith cluster center.

$$v_{iL}^{(p)} = \frac{\sum_{k=1}^{n} (\mu_{ik})^m x_{kL}}{\sum_{k=1}^{n} (\mu_{ik})^m}, \quad L = 1, \ldots, d \quad (3)$$

where d is dimension of data vector X, and $\mu_{ik}=\mu_i(X_k)$ is the membership grade of X_k in fuzzy set μ_i.

Step 3

Update $U^{(p)}$ for k = 1 to n.

① Calculate I_k and $\bar{I}_k$.

$I_k = \{i \mid 1 \le i \le c, D_{ik} = \| X_k - v \| = 0\}$

$\bar{I}_k = \{1, 2, \ldots, c\} - I_k$

② For data item k, comput new membership values.

i) If $I_k = 0$,

$\mu_{ik} = D_{ik}^{2/(1-m)}$

if $(\mu_{ik} < t(\mu))$ then $\mu_{ik} = 0$

$\mu_{ik} = \mu_{ik} / \sum_{i=1}^{c} \mu_{ik}$

ii) If $\bar{I}_k \ne 0$,

$\mu_{ik}=0$ for all $i \in \bar{I}_k$, and $\sum_{i \in I_k} \mu_{ik} = 1$.

③ Next k.

Step 4

$$J_m^{(p)} = \sum_{k=1}^{n} \sum_{i=1}^{c} (\mu_{ik})^m D_{ik}^2 \quad (4)$$

Compare $J_m^{(p)}$ and $J_m^{(p-1)}$. If $|J_m^{(p)} - J_m^{(p-1)}| \le \varepsilon$, stop; otherwise, set p = p + 1, and go to step 2.

In order to execute the SCM clustering algorithm, we initialize the fuzzy c partition, U^0 as 1/c and define ε as 10^{-6} and $t(\mu) = 0.5/c$. The reason we select $|J_m^{(p)} - J_m^{(p-1)}|$ as a criterion of convergence is that J_m is the objective function to be minimized.

PARAMETER IDENTIFICATION

The identification of parameters which define the membership functions of the premise and coefficients of the consequent is carried out using genetic algorithms which provide an excellent robustness in complex optimization problems. Genetic algorithms (GAs) are different from conventional optimization and search procedure in four ways[2]: (1) GAs work with a coding of the parameter set, not the parameter themselves; (2) GAs search from a population of points, not a single point; (3) GAs use objective function not derivatives or other auxiliary information; (4) GAs use probabilistic transition rules to guide their search. These four differences contribute to a genetic algorithm's robustness and resulting advantage over other more commonly used techiques.

Genetic algorithms: Genetic algorithms are iterative adaptive general purpose search strategies based on the principles of natural population genetics and natural selection. They maintain a population of knowlege structures, called a population, which is made to evolve. A structure, which is a string of symbols from a fixed set, is called chromosome. It represents a candidate solution to the problem. A simple genetic algorithm that yields good results in many practical problems is composed of three operators: reproduction, crossover, and mutation. Reproduction is a process in which individual strings are copied according to their objective funtion (fitness function) values which we want to maximize. Copying strings according to their fitness values means that strings with a higher value have a higher probability of contributing one more offspring in the next generation. This operator is an artificial version of natural selection, a Darwinian survival of the fitness among string creatures, and may be implemented in algorithmic form in a biased roulette wheel where each current string in the population has a roulette wheel slot sized in proportion to its fitness. After reproduction, simple crossover may proceed in two steps. First, members of the newly reproduced strings in the mating pool are mated at random. Second, each pair of strings is selected uniformly at random between 1 and string length less one, L-1. Two new strings are created by swapping all characters between positions k+1 and L inclusively. Mutation is a secondary operator whose use guarantees that the probability of searching a particular sub-region of the solution is never zero. These operators are simplicity itself, involving nothing more complex than random number generation, string copying, and partial string exchanging; yet, despite their simplicity, the resulting search performance is wide-ranging and impressive due to implicit parallelism of genetic algorithm.

Fitness Function: In many problems, the objective is stated as the minimization of some cost function rather than the maximization of some utility or profit. In the parameter identification, our purpose is to minimize the cost function, eq.(5) which is defined as the average of squared errors between target output data and inferred output data.

$$E = \frac{1}{N}\sum_{i=1}^{N}(y_i^0 - y_i^*)^2 \qquad (5)$$

where N is the total number of data, y_i^0 target output value, y_i^* output inferred from fuzzy implications.

With genetic algorithms, we use the following cost to fitness transformation:

Fitness function, f = 1.0/E (6)

Multiparameter Coding

One successful method employed in coding multiparameters of optimization problems is the concatenated, multiparameter, mapped, and fixed point coding. To construct a multi-parameter coding, we can simply concatenate many single parameter codings which define the membership functions of the premise and coefficients of the consequent. Each coding has its own sublength L, its own minimum and maximum values, P_{min} and P_{max}, respectively, and the precision of the decoded value is controlled by $(P_{max} - P_{min})/(2^L - 1)$.

Reduction of Search Areas

For the efficient and fast convergence of genetic algorithms, the reduction technique of search areas of the parameters is used. The reduction is carried out through the increment of P_{min} and the decrement of P_{max} of each parameter, where P_{max}'s and P_{min}'s are obtained from the decoded values from m chromosomes whose fitness values are greater than a value α. The sequence of genetic algorithms by reduction of search areas is shown in Fig. 3.

SIMULATION

In this section, the feasibility of the proposed approch is evaluated through initially, the identification of the fuzzy model to describe an input-output relation of Gas Furnace and subsequently, the application to the navigation control of the fuzzy control rules identified from the navigation data of a mobile robot based on expert's control actions.

```
Generate random populations, P(0);
Evaluate fitness of individuals in P(0); freq = 1;
While(t is less than max generation)
{
    Generate P(t+1) from P(t) as follows:
    {
        select the fittest individuals from P(t);
        recombine them by crossover and mutation;
    }
    Evaluate the fitness of individuals in P(t);
    While(fitness value is greater than reduction criterion)
    {
        decode individuals;
        decide the min and max ranges of parameters;
        if(freq is equat to m) reduce search areas; freq=1;
        else freq = freq + 1;
    }
    t = t + 1;
}
Retain the fittest population;
```

Fig. 3. Sequence of GAs by reduction of search areas.

Identification of Fuzzy Model

Our purpose is to identify the fuzzy model which describe the relation between a gas flow u(t) and the combusted CO_2 concentration y(t) of Gas Furnace using the 299 pairs of data presented by Box and Jenkins[3]. We consider u(t-4) and y(t-1) as input variables of fuzzy implications, and y(t) as output variable. The appropriate numbers of clusters which describe each input-output relation effectively are determined as 3 and 2 from the validity indices graphs shown in Fig. 4 and Fig. 5, respectively.

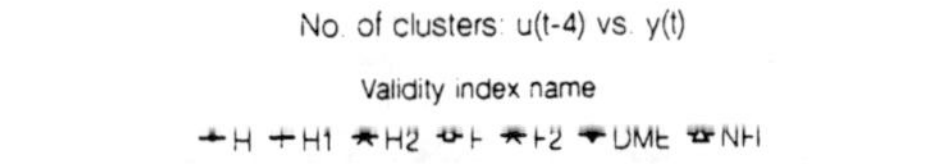

Fig. 4. Graph for the decision of cluster number in u(t-4) vs. y(t)

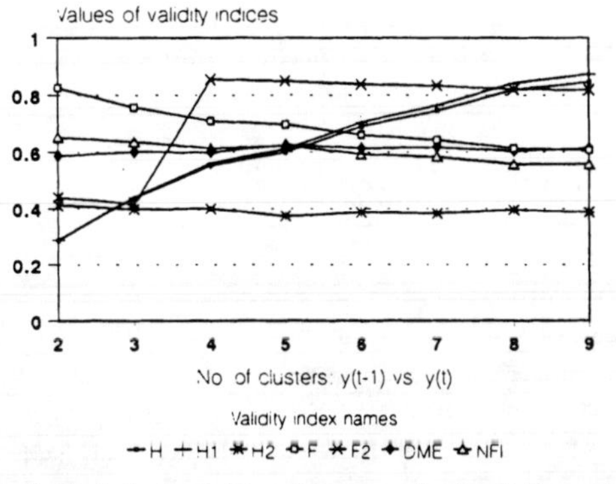

Fig. 5. Graph for the decision of cluster number in y(t-1) vs. y(t)

The graphs to reveal the relation of each input-output and the centroids of the decided clusters are shown in Fig. 6 and Fig. 7, respectively.

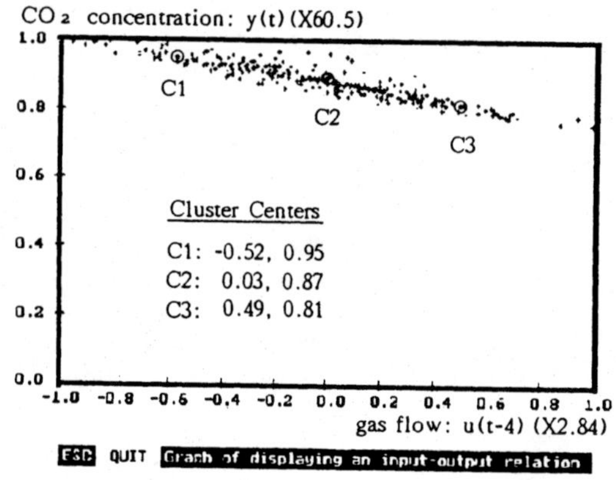

Fig. 6. Graph of input u(t-4) and output y(t) data having 3 clusters

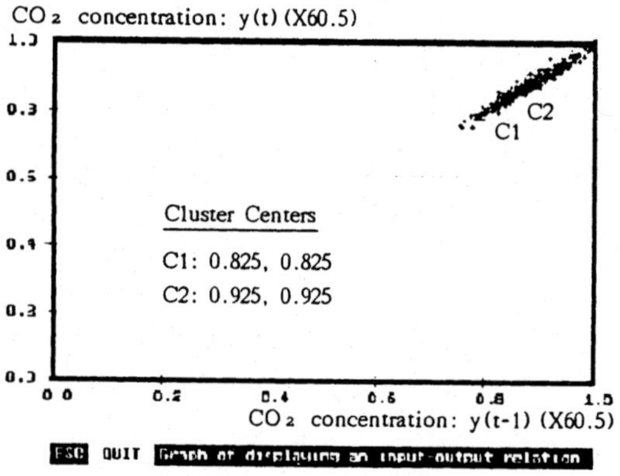

Fig. 7. Graph of input y(t-1) and output y(t) data having 2 clusters

From the clustering results, we define fuzzy labels of u(t-4) as Positive, Zero, Negative, and those of y(t-1) as Small and Big. The fuzzy implications for modeling are composed of as follows:

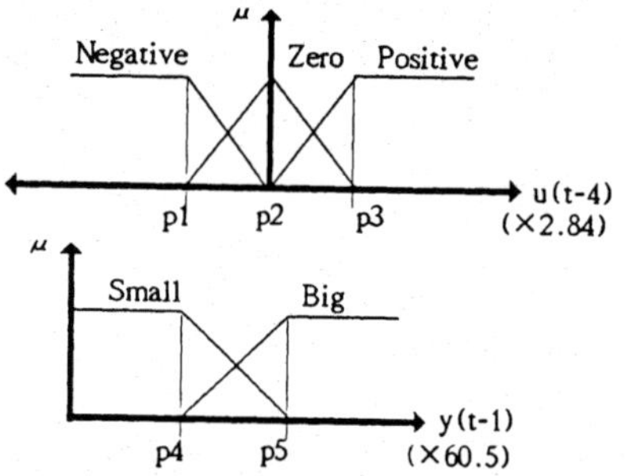

If u(t-4) is Negative & y(t-1) is Small, then $y = a_1 \cdot w_1 + b_1$
If u(t-4) is Negative & y(t-1) is Big, then $y = a_2 \cdot w_2 + b_2$
If u(t-4) is Zero & y(t-1) is Small, then $y = a_3 \cdot w_3 + b_3$
If u(t-4) is Zero & y(t-1) is Big, then $y = a_4 \cdot w_4 + b_4$
If u(t-4) is Positive & y(t-1) is Small, then $y = a_5 \cdot w_5 + b_5$
If u(t-4) is Positive & y(t-1) is Big, then $y = a_6 \cdot w_6 + b_6$

The parameter $p_1, p_2, \ldots, p_5$, and $a_1, b_2, \ldots, a_6, b_6$ are identified by use of GAs. Initial parameters for GAs are as follows:

Population size	50
Length of individuals	10
Crossover rate	0.6
Mutation rate	0.033

The reduction of search areas is carried out at generation number 63 and 82. Fitness values calculated from the best strings in evolving populations are shown in Fig. 8.

Identified parameters are shown in Table 1, and mean square errors for the comparison with other results in Table 2.

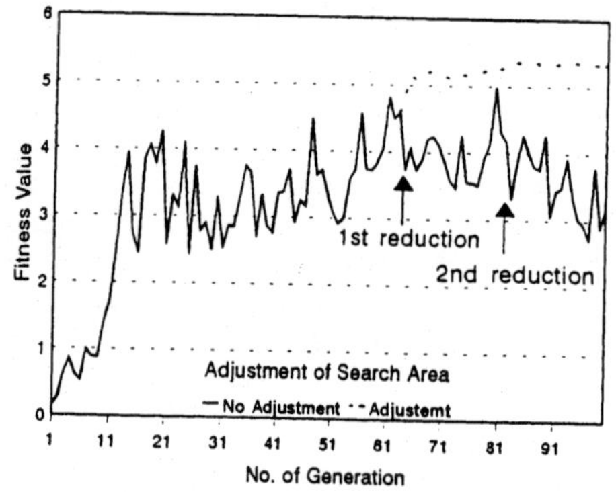

Fig. 8. Graph of fitness values

Table 1 Identified parameters of the membship functions

Premise		Consequent			
p1	-0.695	a1	6.654	b1	44.349
p2	-0.138	a2	1.951	b2	58.025
p3	0.906	a3	2.394	b3	39.94
p4	0.561	a4	0.15	b4	58.478
p5	0.989	a5	-4.86	b5	41.989
		a6	3.412	b6	51.274

Table 2 Comparison of mean square errors with other results

Model Name	Mean Square Error	Number of rules
Tong's model[4]	0.469	19
Pedrycz's model[5]	0.776	20
Xu's model[6]	0.328	25
Sugeno's model[7]	0.355	6
Our model	0.187	6

Application to Navigation Control

In the last few years a new approach in fuzzy logic has been used to control the navigation of a mobile robot. One method to derive the fuzzy sets and control rules is to model an experts' driving actions. Given a set of input-output data obtained from the expert's control actions, the membership functions of those fuzzy variables and the coefficients in the consequents of the rules are identified by the proposed method in this paper. The problem to solve is moving a mobile robot to follow freely the center of the corridor in a confined area. The structure of the overall system for fuzzy control is shown in Fig. 9.

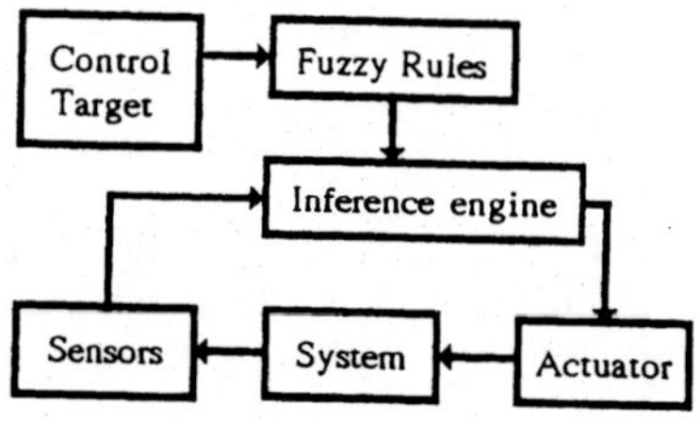

Fig. 9. Diagram of the fuzzy controller

At every sampling interval, we read the measurements of ultrasonic sensors to know the distances to the left and the right walls and calculate the orientation of the mobile robot from the measures distances. The fuzzy controller calculates the new steering wheel angle from the orientation of the mobile robot and the difference between the distances to the left and the right wall, and the mobile robot is moved with the new steering wheel angle.

System Configuration

The structure of the used mobile robot is shown in Fig. 10, and its main characteristics are as follows: The robot has a 0.28(m) height, a 0.75(m) length, a 0.7(m) width. The reference point is located in the center of 4 free wheels. Maximum velocity is 1(m/s) and maximum acceration 1(m/s). The steering wheel angle of front wheel is controlled by the difference between velocities of two driving wheels. Sampling time is 0.14(sec). All these characteristics are specified in the LABMATE made by TRC coporation. It receives the increments in the steering wheel angle, the constant velocity 0.3(m/s), and the steering wheel turning speed 0.2(m/s) as inputs in this experiment.

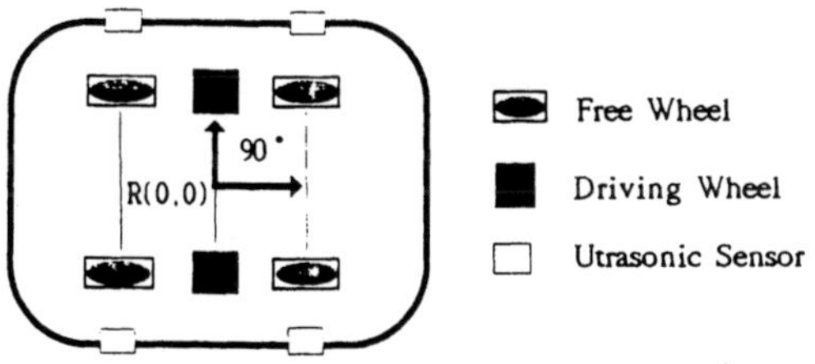

Fig. 10. Configuration of the mobile robot

The location of the four ultrasonic sensors is displayed with black boxes in Fig. 10, and the detection limit of ultrasonic sensors is 3.0(m). The fuzzy control rules identified from the expert's control actions is shown in Fig. 11.

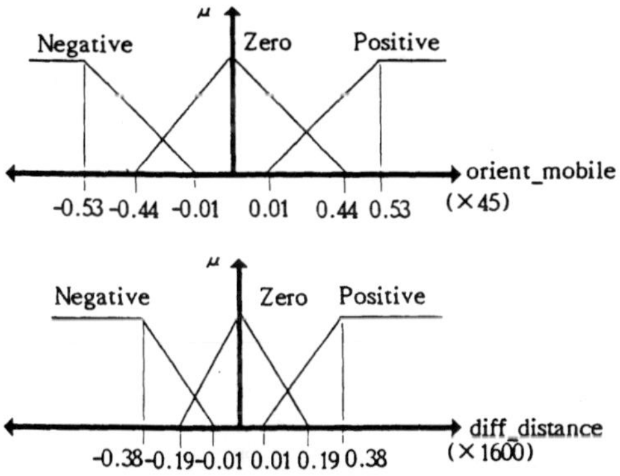

If diff_dist is Negative & orient_mobile is Negative,
then $y = -1.265 \cdot w_1 + 52.034$

If diff_dist is Negative & orient_mobile is Zero,
then $y = -6.050 \cdot w_2 + 26.440$

If diff_dist is Negative & orient_mobile is Positive,
then $y = 9.500 \cdot w_3 - 9.066$

If diff_dist is Zero & orient_mobile is Negative,
then $y = 0.114 \cdot w_4 + 14.819$

If diff_dist is Zero & orient_mobile is Zero,
then $y = -0.568 \cdot w_5 + 0.568$

If diff_dist is Zero & orient_mobile is Positive,
then $y = -2.335 \cdot w_6 - 13.785$

If diff_dist is Positive & orient_mobile is Negative,
then $y = -5.985 \cdot w_7 + 8.061$,

If diff_dist is Positive & orient_mobile is Zero,
then $y = 5.449 \cdot w_8 - 26.018$

If diff_dist is Positive & orient_mobile is Positive,
then $y = 3.473 \cdot w_9 - 53.904$

Fig. 11 The fuzzy control rules for the navigation

The navigation trajectory of the mobile robot controlled by the identified fuzzy control rules is displayed in Fig. 12.

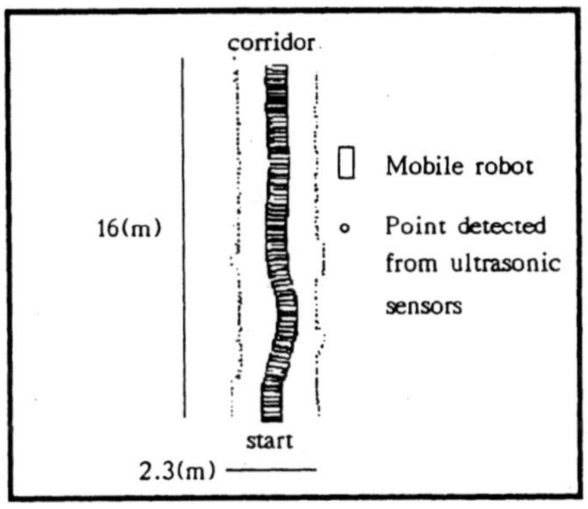

(a)

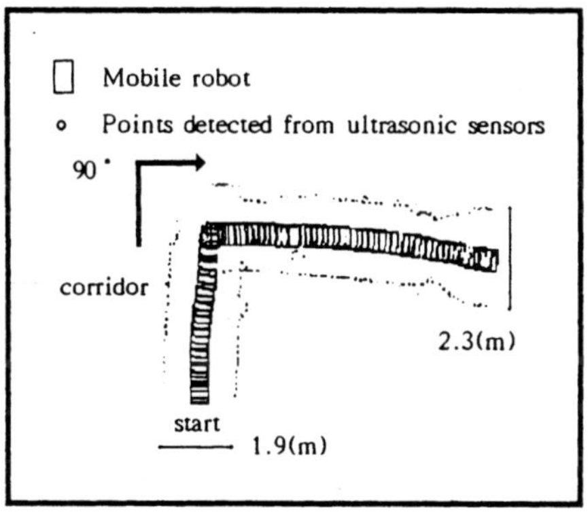

(b)

Fig. 12 The navigation trajectory of the mobile robot
(a) 'I' typed corridor (b) 'Γ' typed corridor

CONCLUSION

In spite of the simplicity of the proposed identification method, the accuarcy of the identified fuzzy model of gas furnace is superior as compared with that of other fuzzy models. A high degree of autonomy in the mobile robot behavior to follow the center of the corridor in a confined environment is reached by means of fuzzy control rules which model human drivers' control strategies. It can be improved by adding the routine to avoid obstacles in front of the mobile robot and complementary sensory system, such as vision. The robot shows flexible and robust behavior facing quite different situations, and recovering from unexpected inputs in spite of the scarce information intially available.

ACKNOWLEDGEMENTS

This work was supported in part by Engineering Research Center for Adv. Control and Instr.(of SNU) by KOSEF.

REFERENCE

[1] Ismail, M. A. (1988). Soft Clustering Algorithms and Validity of Solutions. In: Fuzzy Computing Theory, Hardware and Applications., Elsevier, North Holland, 445-471.

[2] Goldberg, D. E. (1989). Genetic Algorithms in Search, Optimization, and Machine Learning., Addison Wesley

[3] Box, G. E. P. and et al. (1970). Time Series Analysis, Forcasting and Control. Holden Day, Sanfrancisco

[4] Tong, R. M. (1980). The evaluation of fuzzy models derived from experimental data. Fuzzy Sets and Systems., Vol. 13, 1-12.
[5] Pedrycz, W. (1984). An identification algorithm in fuzzy relational systems. Fuzzy Sets and Systems. Vol. 13, 153-167.
[6] Chen-Wei Xu and Yong-Zailu. (1987). Fuzzy Model Identification and Self-Learning for Dynamic Systems. IEEE Trans. on Syst. Man and Cybern., SMC-17, No. 4, 684-689.
[7] Sugeno, M. and Yasukawa, T. (1991). Linguistic Modeling Based on Numerical Data. IFSA' 91 Brussels, Computer, Management & System Science., 264-267.

CONTROL OF FED-BATCH FERMENTATION PROCESS USING A SELF-ORGANIZING FUZZY CONTROLLER

Myeong Seok Park and Yong Keun Chang

BioProcess Eng. Res. Center, Dept. of Chem. Eng.,
Korea Advanced Institute of Science and Technology,
373-1, Kusung-dong, Yusung-gu, Taejon, 305-701, Korea

Abstract. An algorithm has been developed for controlling the substrate concentration at a predetermined set point or following a trajectory in a fed-batch fermentor. It is a model-independent self-organizing fuzzy controller with a backpropagation neural network. The neural network is used to classify process error patterns. The classified error patterns form a basis for rule formation and tuning of the fuzzy controller. Different tuning strategies are used for different patterns of the process error. The tuning operation is accomplished by varying scaling factors. Through simulation studies with a model system of lysine fermentation, the performance of the self-organizing fuzzy controller was evaluated and compared with that of a best tuned PID controller. The classified error patterns by neural nets are well corresponded to intuitive patterns. The overall performance of the self-organizing fuzzy controller was observed to be satisfactory for regulatory and servo problems despite measurement noise.

Keywords. Self-organizing fuzzy controller; fuzzy control; neural nets; pattern recognition; fed-batch fermentation;

INTRODUCTION

Because of its effectiveness in overcoming such problems as substrate inhibition, glucose effects, and catabolite repression, most industrial fermentation processes are carried out in a fed-batch mode to produce antibiotics, amino acids, microbial cells, enzymes, and organic acids. That is, whenever the specific rates of growth and/or production formation are non-monotonic functions of the substrate concentration, a fed-batch operation which is accomplished by manipulating substrate feed rate is in need.

The control of fermentation processes is quite system-specific. In addition, process dynamics vary with the fermentation time and the physiological state of the cells. In a consequence, an control algorithm for fermentation processes should have an adaptability. As a nonlinear adaptive controller, a self-organizing fuzzy controller in which fuzzy control rules are generated and tuned as system dynamics change, is best suited for this purpose.

Procyk and Mamdani(1979) suggested for the first time the concept of self-organizing fuzzy controller. The controller uses a meta rule base which has a human-like learning ability to create and modify fuzzy control rules based on the desired performance of the system. Recently, to add learning ability to the controller various learning theories have been adopted. A neural network performs parallel learning and operation. Nakanishi and colleagues(1990) proposed a self-organizing fuzzy controller with a neural network, and applied it to a simple power plant system model.

In this article, a neuro-fuzzy self-organizing control algorithm is proposed to control the substrate concentration at a predetermined set point or following a trajectory in a fed-batch lysine fermentation. The neural network classifies process error patterns, which is the basis of rule formation and tuning of the fuzzy controller. The capability of pattern classification of the neural network is demonstrated. The controller performance is compared with that of a best tuned PID controller.

SELF-ORGANIZING FUZZY CONTROLLER

The control system proposed in this study is shown in Fig. 1. It consists of three main components. The backpropagation neural network classifies error patterns and the rule modifier uses the result to generate or tune the fuzzy control rules.

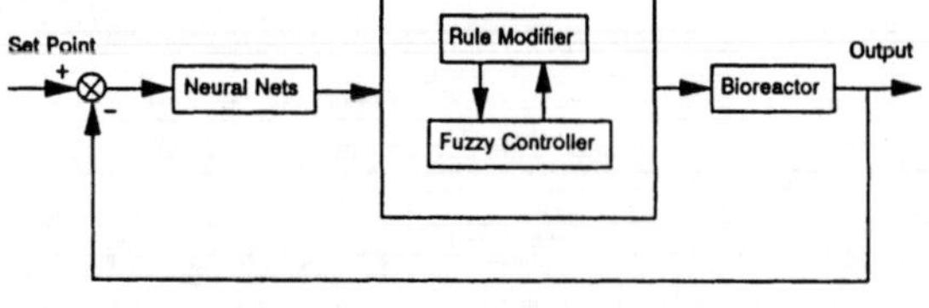

Fig. 1. Self-organizing control system.

The fuzzy controller acts on the fed-batch bioreactor with a set of updated fuzzy rules. Possible strategies include the on-line adjustment of: 1. scaling factors for the input and output variables, 2. the set of control rules, or 3. membership functions of fuzzy variables. We used the first method.

Fuzzy Controller

Generally a fuzzy controller uses a linguistic rule base of the form,

$$\text{If } E \text{ is } E_i, \text{ and } \Delta E \text{ is } \Delta E_i, \text{ then } U \text{ is } U_i \qquad (1)$$

where E is the error of the system output, ΔE is its rate of change , U is the control action, and E_i, ΔE_i, U_i are corresponding linguistic terms. Each membership function for E, ΔE, or U is partitioned into 7 linguistic terms and equally positioned in the domain of interest, and defined to be of triangular form. The linguistic terms(i.e. fuzzy variables) are NB(negative big), NM(negative medium), NS(negative small), ZO(zero), PS(positive small), PM(positive medium), and PB(positive big).

The schematic diagram of the fuzzy controller used in this study is shown in Fig. 2. Actual values of error and its rate of change(denoted by e and Δe in the figure) at the current sampling time are scaled with scaling factors, Ge and Gc, and fuzzified into E and ΔE, respectively. Fuzzy reasoning is accomplished with Zadeh's compositional rule of inference from the inputs and a set of fuzzy rules of the form in Eq. (1).

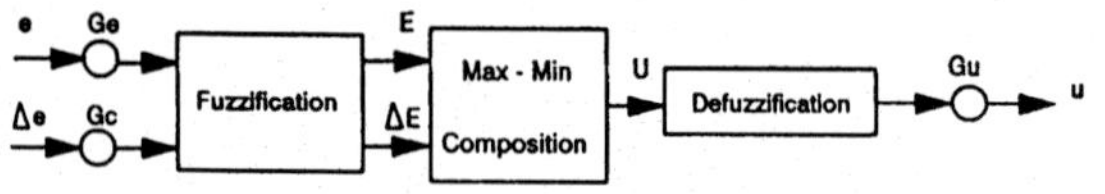

Fig. 2. Structure of fuzzy controller.

The fuzzy subset output, U is defuzzified by using the center-of-gravity procedure and again the defuzzified value is scaled with a scaling factor, Gu, to give a real controller output, u. The scaling factors are defined as follows,

$$E(nT) = F(Ge \cdot e(nT)) \qquad (2)$$

$$\Delta E(nT) = F(Gc \cdot \Delta e(nT)) \qquad (3)$$

$$U(nT) = F(Gu \cdot u(nT)) \qquad (4)$$

where n and T are the sample number and the sampling time, respectively, and F represents a fuzzification procedure.

Neural Network

A backpropagation neural network(Lippman, 1987) was used to classify error patterns which reflect changes in the process dynamics. Twenty learning patterns were used. Among them, six most typical ones are shown in Fig. 3. The error and its rate of change are negligible. Pattern B represents small deviations from the set point, Pattern C oscillations near the set point, Pattern D a converging trend, Pattern E a diverging trend, and Pattern F large deviations from the set point. When the neural network fails to identify the pattern, it is classified as Pattern N. Seven most recent data are used for pattern identification. The identification result is supplied to the rule modifier.

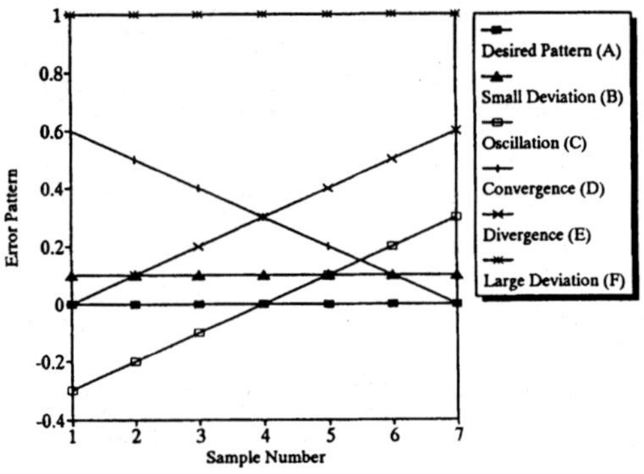

Fig. 3. Typical learning patterns for neural network.

Rule Modifier

The rule modifier performs three main tasks. These are the initialization of the fuzzy controller, rule generation, and rule tuning. The first task is necessary to provide the controller with an initial set of rules to start with. Minimum number of arbitrary but consistent fuzzy rules must be supplied. Intuitively speaking, it would be better if those initial rules reflect initial state of the process(Lee, 1990).

If the neural network classifies the current error pattern to A, the rule modifier generates new rule(s). Rule formation takes place when, at least, one of the fuzzified current error and its rate of change has a fuzzy value ZO. A new rule is formed as follows,

$$U = E + \Delta E \qquad (5)$$

where the arithmetic operation is concerned only on the fuzzy singleton. For example, if E is PB and ΔE is ZO, then U formed newly is PB.

For the other patterns except for Patterns A and N, the rule base is tuned by the rule modifier. No established methodology of rule tuning for fuzzy controller was not available. Therefore, we used rather empirical rules. Maximum values of the error and its rate of change out of 7 data points in the current time interval are used to tune the scaling factors, Ge and Gc. The Gu value is modified based on the maximum error. For Pattern B, to make the process output not too much sensitive to the control input, Ge, Gc are decreased and Gu increased. In the case of Pattern C, Gu must be increased further to surpress oscillations. Increasing Ge and Gc for Pattern D can result in a fast convergence. To prevent divergence from the set point, Ge and Gc values are significantly decreased for Pattern E. Finally, for Pattern F representing large deviations from the set point, the fuzzy controller output is made more sensitive by increasing Ge and Gc as well as decreasing Gu.

The proposed self-organizing fuzzy control algorithm is depicted in Fig. 4.

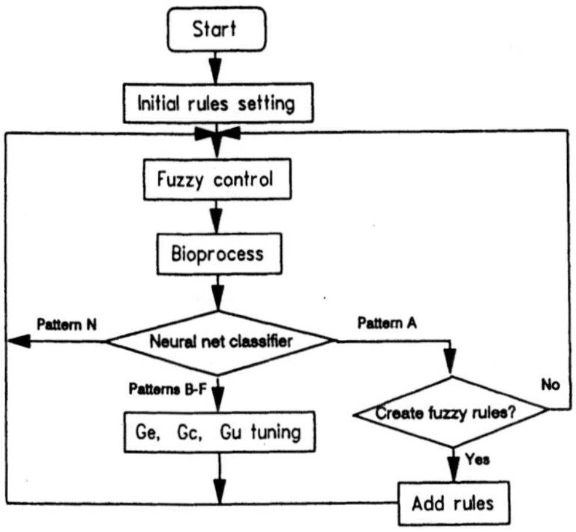

Fig. 4. Self-organizing fuzzy control algorithm.

APPLICATION TO FED-BATCH LYSINE FERMENTATION

Simulation studies were performed for a fed-batch lysine fermentation system. Our objective was to maintain substrate concentration at a predetermined set point or following an optimal trajectory by manipulating substrate feed flow rate. The fed-batch process model used is as follows,

$$\frac{dV(t)}{dt} = F(t) \quad (6)$$

$$\frac{dX(t)}{dt} = \mu(t)\cdot X(t) - \frac{F(t)}{V(t)}\cdot X(t) \quad (7)$$

$$\frac{dS(t)}{dt} = \frac{F(t)}{V(t)}\cdot [S_F - S(t)] - \sigma(t)\cdot X(t) \quad (8)$$

$$\frac{dP(t)}{dt} = -\frac{P(t)\cdot F(t)}{V(t)} + \pi(t)\cdot X(t)$$

$$V(0) = V_0,\ X(0) = X_0,\ S(0) = S_0,\ P(0) = P_0 \quad (9)$$

$$V(t) \le V_{max} \quad (10)$$

$$0 \le F(t) \le F_{max} \quad (11)$$

where, F, X, S, S_F, P and V represent feed flow rate, cell concentration, substrate concentration, feed substrate concentration, and bioreactor working volume, respectively. Subscript 0 denotes initial states and superscript max denotes maximum values. A kinetic model of Ohno, Nakanishi, and Takamatsu(1976) was used for the specific growth rate μ, the specific substrate consumption rate σ, and the specific lysine production rate π.

$$\mu(t) = 0.125\cdot S \quad (12)$$

$$\sigma(t) = \mu(t)/Y_{x/s} \quad (13)$$

$$\pi(t) = -384\cdot\mu(t)^2 + 134\cdot\mu(t) \quad (14)$$

$$Y_{x/s} = 0.135 \quad (15)$$

RESULSTS AND DISCUSSIONS

Pattern Recognition

For the given system, error patterns from an uncontrolled and a controlled processes were classified as in Fig. 5.

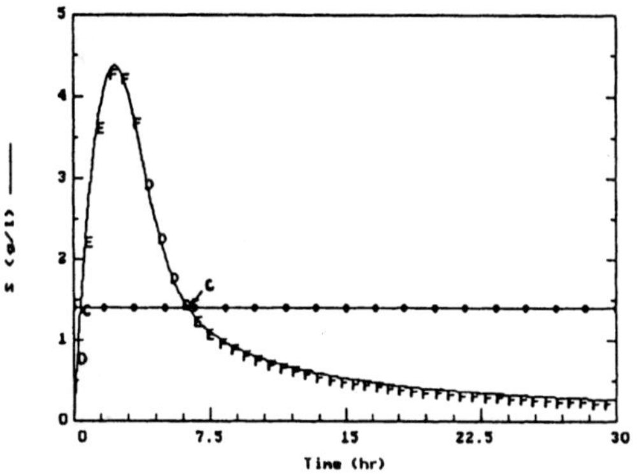

(a)

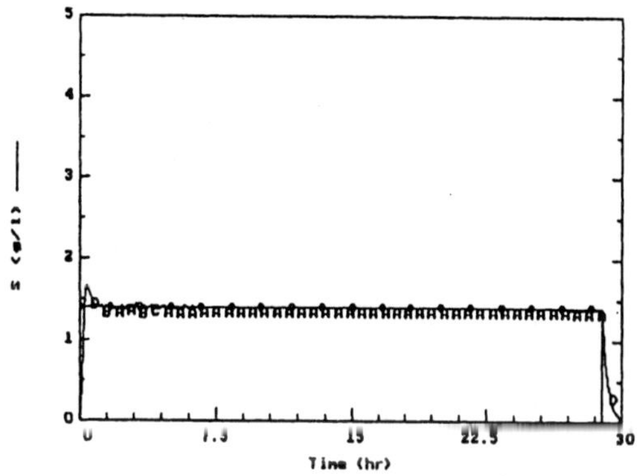

(b)

Fig. 5. Process error classification by neural network,
(a) A step response of uncontrolled process,
(b) A process response controlled by self-organizing fuzzy controller.
A-F: Patterns A-F, —•—: set point.

The figure shows that the neural network can effectively recognize the process state. Here, no measurement noise was assumed. From Fig 5b, we know that the classified patterns can be used as a local performance index for the controller.

Control Performance

The performance of the self-organizing fuzzy controller was compared that of a best tuned PID controller in fed-batch lysine fermentation for cases with 10% measurement noises. Fig. 6 shows controlled process responses to step changes in the set point. The PID controller showed a good control performance(Fig. 6a). The self-organizing fuzzy controller was initiated with arbitrary values of the scaling factors and an arbitrary set of fuzzy rules. At the beginning it showed a poor performance because of its ill-tuned state. However, its control performance rapidly improved as it became better tuned with time, and shortly became as good as that the PID controller(Fig. 6b). A case when the substrate concentration was controlled following an optimal trajectory which had been calculated by Modak(1985) is shown in Fig. 7. Initially, the self-organizing fuzzy controller showed a poor performance, but its performance shortly became satisfactory. This implies that the developed algorithm has a good auto-tuning capability and thus does not require *a priori* tuning of its parameters and necessary fuzzy rules. The tuning of controller parameters is a cumbersome and time-consuming task in using a conventional type of PID controller. In overall the self-organizing fuzzy controller was more noise-sensitive than the PID controller. Probably, it is because of difficulties in pattern recognition caused by noises.

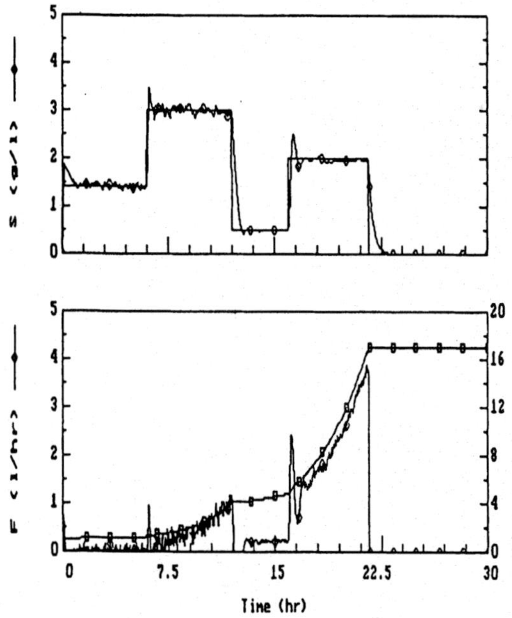

(a)

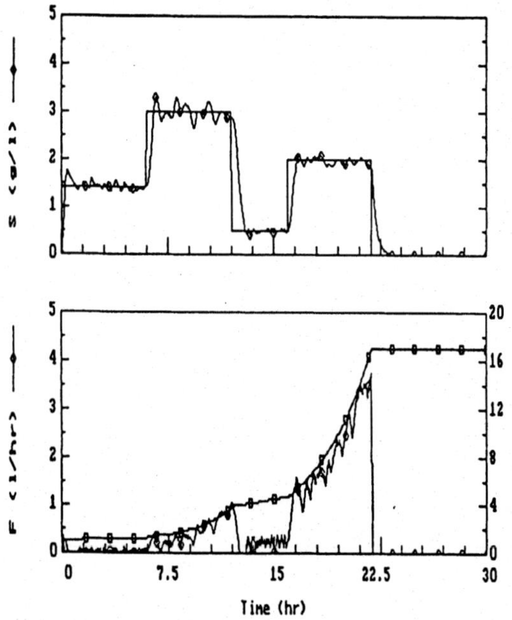

(b)

Fig. 6. Regulation of substrate concentration in lysine fermentation,
(a) using PID controller,
(b) using self-organizing fuzzy controller.

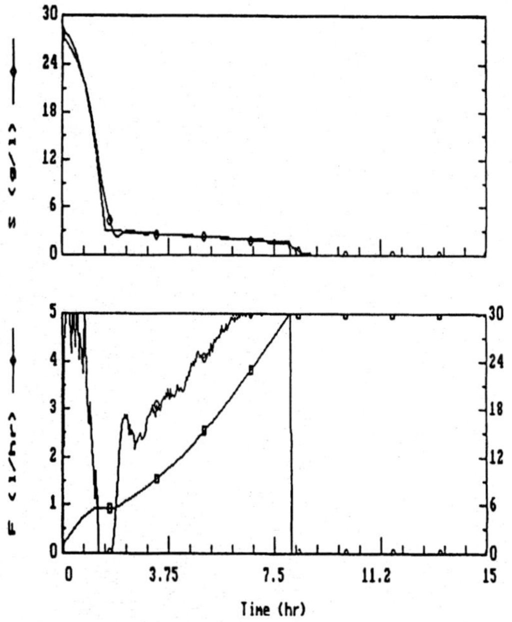

(a)

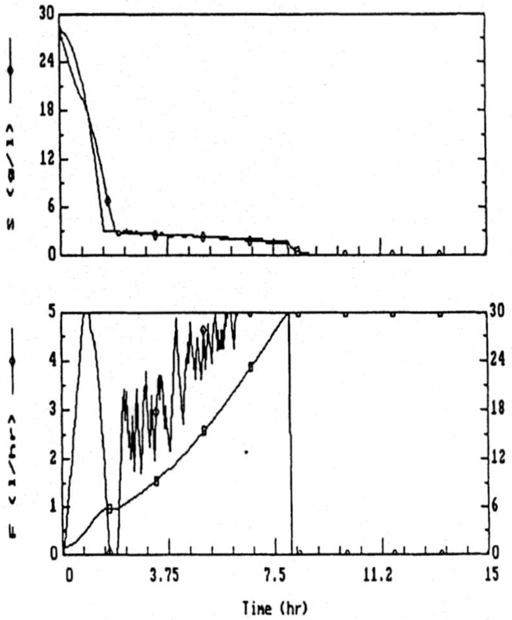

(b)

Fig. 7. Profile control of substrate concentration in lysine fermentation,
(a) using PID controller,
(b) using self-organizing fuzzy controller.

CONCLUSIONS

We have developed a model-independent self-organizing fuzzy controller with a backpropagation neural network. The neural network plays a role of guidance to the fuzzy controller by classifying the process error patterns. Different tuning strategies which are based on rather empirical rules are used for different patterns of the process error. The pattern recognition capability of the neural network and the performance of the proposed controller was evaluated for a model system of fed-batch lysine fermentation. There was a good agreement between the pattern recognition by the neural network and human observations. The self-organizing fuzzy controller as a whole showed a good adaptability for both regulatory and servo problems.

ACKNOWLEDGMENT

The authors are grateful to Korea Science and Engineering Foundation for the support of this research.

REFERENCES

Lee, C. C. (1990). Fuzzy logic in control systems: Fuzzy logic controller-Part I. *IEEE Trans. Syst. Man Cybern.*, *20*, 404-418.

Lippman, R. P. (1987). An introduction to computing with neural nets. *IEEE ASSP Magazine*, *4*, 4-22.

Modak, J. M. (1985). M.S. Thesis, Purdue University Indiana, USA.

Nakanishi, S, T. Takagi, K. Unehara, and Y. Gotoh (1990). Self-organizing fuzzy controllers by neural networks. *Proc. of Int. Conf. on Fuzzy Logic and N.N.*, *1*, 187-191.

Ohno, H., E. Nakanishi, and T. Takamatsu (1976). Optimal control of a semibatch fermentation. *Biotech. Bioeng.*, *18*, 847-864.

Procyk T. J., and E. H. Mamdani (1979). A Linguistic self-organizing process controller. *Automatica*, *15*, 15-30.

PARALLEL ADAPTIVE CONTROL FOR TURBOGENERATOR SYSTEMS

M. D. Brown, G. W. Irwin, E. Swidenbank, B. W. Hogg

Department of Electrical and Electronic Engineering, The Queen's University of Belfast, UK

Abstract. This paper describes the design and implementation of digital self-tuning controllers for turbogenerator AVR and governor systems. Practical aspects such as controller robustness, lack of sufficient processing power, and algorithm mapping issues are discussed. A parallel architecture using Inmos transputers combined with an advanced VME bus-based measurement system provides a solution to these problems and offers the control systems designer the opportunity to implement complex parallel algorithms. Simulation results, involving both SISO and multi-loop self-tuning regulators on a network of T800 transputers, are presented together with implementation results on a laboratory scale microalternator and comparisons with a fixed-gain AVR.

Keywords. Transputers ; power system control ; real-time systems ; parallel adaptive control

INTRODUCTION

In the power industry the need for improved control of generation equipment has been steadily increasing. Modern generating sets tend to be less stable than their predecessors, and therefore demand complex control systems for their successful operation. Control equipment manufacturers have recognised this increased need and are in the process of developing computer based systems. Digital governors have been in existence for some time now, and digital voltage regulators are becoming available (Ham, 1989).

Research has shown that self-tuning and adaptive control can offer advantages over conventional control implementations (Wu and Hogg, 1988). Simulation studies have proven the theoretical possibility of improving control performance, but implementation problems have hampered advancement on real systems. The primary implementation difficulty is the lack of processing power available from conventional microprocessors. Some success has been achieved by the use of specialised devices such as Digital Signal Processors (Bo, 1988), but the underlying architecture of these devices is suitable only for a limited number of applications. Alternative control architectures are therefore being developed, and one of the more hopeful schemes involves the use of Inmos transputers.

The transputer is a general purpose, single-chip high speed microprocessor, which is a member of the MIMD class of parallel architectures (Flynn, 1966). The T800 architecture combines a 32-bit, 10 MIPS RISC processor with an on-board 64-bit floating point unit, memory and communications. The key technical feature, which distinguishes the transputer from other high performance microprocessors, is the addition of four high-speed, duplex links. These allow a single transputer to be used as a node among any number of similar devices to and from a parallel processing system. Control engineers view the transputer as a powerful processing element capable of easy implementation in an embedded parallel processing system, and an increasing number of successful applications are being reported (Irwin and Fleming, 1992).

Increased computational speed is of course the primary benefit of parallel processing. This allows fast systems to be controlled and gives the control engineer the choice of added complexity in the control algorithm. Easy expansion, within a uniform hardware and software base, is another feature of concurrent control systems, since it is possible to add more processors as required. This implies reduced development and maintenance costs in real-time, computer control applications. Parallel processing also offers a closer relationship between the inherent parallelism expressed at the design stage, in the form of block diagram descriptions of the system, and the hardware implementation. Fault tolerance can be realised in a parallel processing system by organising computational operations in a distributed sense, so that an operational failure results in performance degradation rather than a complete breakdown of the controller. However, the full potential of parallel processing will not be realised in control engineering unless effective tools are developed to assist in the mapping of algorithms to architectures (Maguire, 1991).

This paper reports on transputer-based parallel implementation for the control of a laboratory turboalternator, and describes how the hardware and software schemes achieve the required real-time performance. Algorithm mapping issues are also discussed with respect to parallel implementation of SISO and MIMO controllers. The hardware system includes the industry standard VME bus architecture and employs Intel 8751, Motorola 68020 and Inmos transputer systems to perform various tasks in a pipelined structure. Advanced signal processing algorithms based on a Fourier technique are used to measure terminal quantities, such as voltage and current, and optical transducers are used to calculate machine speed and rotor angle.

The entire scheme was initially tested on a simulation of a three-phase turboalternator running in parallel with both the measurement and control algorithms on a network of T800 transputers (Brown and Swidenbank, 1991). Building on this work, further results involving both SISO and MIMO regulators are presented. These include real-time tests carried out on a scaled laboratory turbogenerator using a reduced-order SISO algorithm, and simulation tests using a multi-loop controller. The results show the improved performance gained when using the multi-loop controller.

MEASUREMENT SYSTEM

Micromchine System

The micromachine system consists of a small synchronous generator with an associated turbine simulator, tied to the laboratory busbar through a transformer and artificial transmission line. The synchronous generator is a 3 kVA, 220 V, 50 Hz, 1500 rpm 4-pole microalternator, whose parameters have been specifically designed to match those of a full size generator. The alternator is driven by a separately excited d.c. motor, whose field current is kept constant. The torque supplied by the motor is therefore directly proportional to its armature current, being controlled by an analogue turbine simulation.

Although a laboratory machine cannot be realistically expected to give results that entirely match those obtainable in a full scale power station, it does provide a means of verifying the behaviour of controllers which have proved to be successful in simulation. As well as allowing a particular control system to be tested under real-time constraints, the micromachine provides a more stringent testing environment, incorporating many features not possible with a computer simulation. Effects such as non-ideal transducer characteristics leading to limited resolution and noise, saturation and other non-linearities present on the real machine, computational delays and variations in busbar voltage and frequency, all contribute to the problems that any controller will face in a practical situation.

Fourier Measurement Algorithm

Conventional methods for determining machine terminal quantities like voltage and power rely on rectification and summation of the individual phase quantities to form an average RMS value. Since the voltage and current waveforms generated by a synchronous machine are rarely perfectly balanced, and are invariably contaminated by harmonics and noise, these traditional methods often produce poor representations of the true terminal quantities, especially during transient periods. Inaccurate measurement of feedback variables inevitably leads to a degradation in control performance and often means that the expected performance benefits from more advanced controllers do not accrue.

One method which can be used to extract the fundamental components from complex periodic waveforms, such as those produced by a synchronous generator, is based on a finite Fourier series. Fundamental waveform components derived using this method can be combined to produce accurate measurements of terminal quantities under both symmetrical and unsymmetrical transient and steady-state conditions. This Fourier method amounts to a time-series harmonic filter as follows (Brown and Swidenbank, 1991) :

If u_i (i = 0, ..., 11) are the collected samples of a complex waveform (sampled at 12 equally spaced intervals per cycle in this case), then the fundamental components, a_1 and b_1, of the sampled waveform can be expressed as an algebraic sum of past samples as :

$$a_1 = \frac{1}{6}\left[(u_0-u_6) + \frac{\sqrt{3}}{2}(u_1-u_5-u_7+u_{11}) + \frac{1}{2}(u_2-u_4-u_8+u_{10})\right] \quad (1)$$

$$b_1 = \frac{1}{6}\left[(u_3-u_9) + \frac{\sqrt{3}}{2}(u_2+u_4-u_8-u_{10}) + \frac{1}{2}(u_1+u_5-u_7-u_{11})\right] \quad (2)$$

The time series filter equations (1) and (2) are executed at every sample interval to give a moving average of the fundamental components of the waveform. If each of the three-phase voltage and current waveforms are subjected to this filtering, the fundamental components can be combined to give measurements of the generator terminal electrical quantities of RMS voltage, current, real and reactive power.

VME hardware System

In order to advance the theoretical work and to create a structurally open-ended environment, a VME bus-based system was chosen as the host hardware for the implementation of both the control and measurement algorithms. This standard bus system is already being used by power systems control manufacturers, and is well established in many industries.

Voltage and current transformers are attached directly to the outputs from the microalternator. Signals from these devices are passed through anti-aliasing low pass filters on a custom designed board and then on to the analogue i/o board. An 8751 microcontroller generates an interrupt signal at 12 times the system frequency (600 Hz). This interrupt triggers the master 68020 board to read in samples of the filtered electrical waveforms from the analog i/o board and to read in the values of speed and rotor angle calculated by the 8751. On completion of the read sequence, the raw values are then passed to shared DMA (Direct Memory Access) memory to be picked up by the IMS BO11 T800 transputer. The transputer then performs the Fourier measurement algorithm to produce the four electrical terminal quantities.

SELF-TUNING CONTROL

SISO Control Strategy

A minimum variance, self-tuning regulator (STR) with an auxiliary predictor is applied to the synchronous generator excitation loop, replacing the conventional automatic voltage regulator as shown below (Wu and Hogg, 1988) :

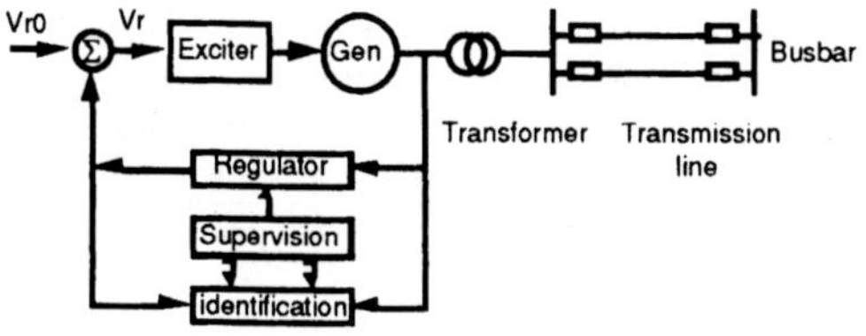

Fig. 1 Self-tuning AVR

The outputs and inputs of the controlled system are :

$y(t) = \Delta V_t(t) + \gamma\Delta\omega(t) \qquad -1 < \gamma < 0$
$u(t) = \Delta V_r(t)$

where $V_t(t)$ is the terminal voltage at time t, ω is the speed in rads^{-1}, $V_r(t)$ is the input to the exciter, and Δ is the deviation from the steady-state value.

In the SISO case, the generator can be described as :

$$A(q^{-1})y(k) = B(q^{-1})u(k-1) + C(q^{-1})\xi(k) \qquad (3)$$

where y(k) and u(k) are the output and input of the system at the sample instant k, A, B, and C are polynomials in the backward shift operator q^{-1} and $\xi(k)$ is an uncorrelated random sequence of zero mean disturbing the system.

Rewriting the equation in predictor form leads to the following general control law, where the parameters of the polynomials F, G and C are determined by the technique of Extended Recursive Least Squares :

$$u(k) = -\frac{g_0}{R}\left[\frac{Fy(k) + Gu(k)}{C}\right] \qquad (4)$$

If the order of C is chosen to be zero, the control law can be reduced to equation (5), where the parameters of the polynomials F and G are estimated using Ordinary Recursive Least Squares :

$$u(k) = -\frac{g_0}{R}(Fy(k) + Gu(k)) \qquad (5)$$

MIMO Control Strategies

Two approaches may be adopted when considering turbogenerator AVR and governing systems. One employs a separate self-tuner for both the governor and AVR and is referred to as multi-loop control. The other method integrates both the governor and AVR systems into a single multivariable controller. Figures 2 and 3 illustrate these two procedures.

In the multi-loop case, the control laws for both the governor and AVR are the same as the SISO system. The multivariable controller has the same form as the SISO algorithm, except that the associated vectors are replaced by matrices and multivariable least squares is used to identify the system. Although the multi-loop system is easier to design and requires much less computation time to implement, it is expected that the multivariable controller will better exploit the inherent coupling in the turbogenerator system (Ibrahim et al, 1989).

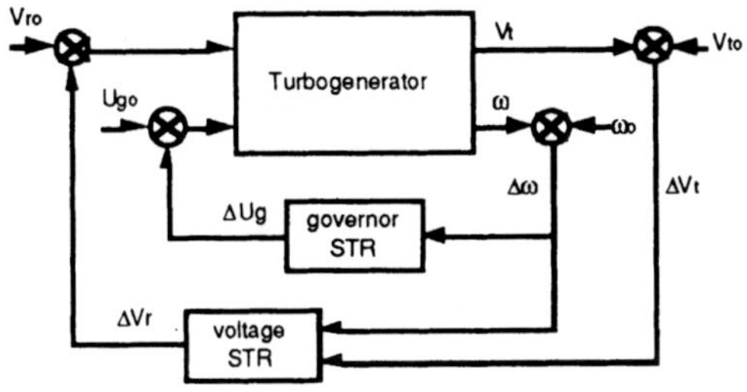

Fig. 2 Multi-loop self-tuning control

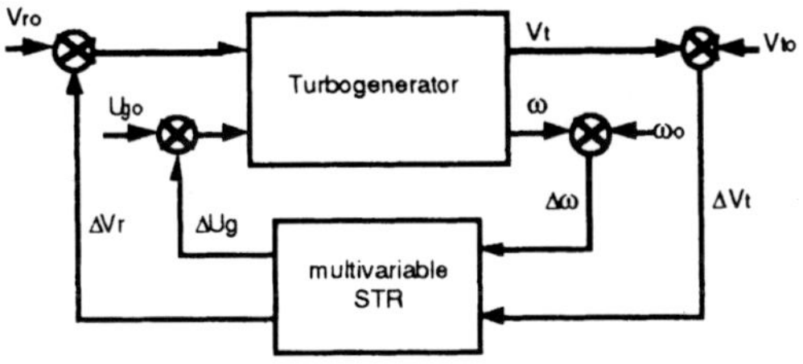

Fig. 3 Multivariable sekf-tuning control

Concurrent processing issues

Conventional control algorithms usually require a strict sequential data flow, which limits the scope for parallelism. This paper employs functional decomposition of the multi-loop controller shown in Figure 2. The parallelism of this system is obvious since the AVR and governor loops are considered distinct in this case. Two transputers can therefore be used to implement the controller, with no associated increase in the sequential computation time and hence no increase in the overall control sample interval.

Other work has shown that, although the control laws and associated supervision schemes are inherently sequential, the matrix-based parameter estimator is suitable for parallel implementation (Irwin and Maguire, 1991). Techniques such as Gantt chart partitioning onto multiple transputers and systolic array decomposition allow the estimator to be implemented in parallel. This will facilitate the real-time application of complex techniques, such as multivariable control, to fast-acting processes.

TEST RESULTS

Multi-task architecture

The configuration of the hardware and software systems needed to implement the various controllers follows closely that given in second section for the measurement system (Brown and Swidenbank, 1991). An extra task is required in this instance, however, to look after the control output to the generator excitation system. This takes the form of a simple program which converts the per unit output excitation signal from a particular controller into a suitable scaled value and then writes it to one of the digital-to-analogue converters on the analogue input-output board.

This program must run on the VME transputer since the D/A converters are accessed through DMA memory on the VME bus. The program therefore executes in parallel with the Fourier measurement program. The control algorithm runs on the PC based TRAM and communicates with the measurement program and the digital-to-analogue conversion program via an Inmos link wire. The entire process can be visualised in Fig 4.

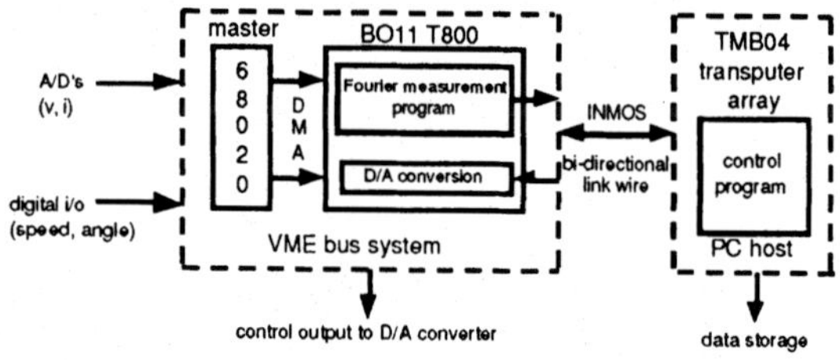

Fig. 4 Control system architecture

As can be seen from Figure 4, both the Fourier measurement algorithm and the digital-to-analogue conversion routine run on the same transputer. The processor must, therefore, schedule cpu time between the two programs. This does not cause any conflict, however, or affect the synchronism of the measurement system as the D/A conversion program takes only a negligible time to execute when compared to the computationally complex Fourier algorithm.

Since the Fourier measurement program samples the waveforms at 12 times per cycle, the fastest possible control sample rate will also be 12 times per cycle (1.67 ms). To enable proper operation and to avoid overrunning the sample time, the control algorithm must therefore be completed within 1.67 ms. In practice, however, control sample rates of 10-20 ms are more usual, leaving much more time for the control algorithm. If the control sample rate is 20 ms for example, the measurement program will continue to operate in parallel at the higher sample rate of 1.67 ms. This mode of operation would not be possible if the system was entirely sequential.

The multi-tasking parallel implementation of the measurement and control algorithms is facilitated by the use of transputers. The VME transputer can be linked, as in this application, to an external transputer system containing any number of processors. This creates a system with vastly increased computing power, and with the advent of the next generation of transputers will provide the control systems designer with the opportunity to implement virtually any advanced control strategy (the new H1 series T9000 transputer is capable of 10 Mflops compared to the 1.5 Mflops of the TX series T800 transputers used here (Dettmer, 1990)).

Run-time performance

Benchmark tests on both the Fourier measurement system and the control algorithms yielded a time of 6 ms for STR (eqn 4) and 9 ms for STR (eqn 5)

It should be mentioned that all measurement and control software is written in C using double-precision arithmetic. Using a medium-level language such as C not only reduces the development time of the software, but also means that the control programs can be moved from simulation to the physical system with little or no changes. The numerical robustness of the algorithms is also greatly improved by the use of double-precision arithmetic. This represents an improvement over more conventional implementations where a substantial part of the software may be sourced in some form of assembler due to hardware limitations.

Brown and Swidenbank (1991) showed that, with a single-loop STR on one transputer the control cycle time could be reduced to about 6 ms from the previous 40 ms value. If the control sample rate is kept at 20 ms (once per cycle, 50 Hz), the self-tuning controllers in equations (4) and (5) can therefore operate comfortably. The remaining time in each sample period is used either to perform further hardware and software supervisory functions, an essential feature if the self-tuning algorithms are to be used in a critical industrial application, or to add complexity to the controller in order to improve performance.

Test results with the SISO controller

Extensive simulation studies have proved the satisfactory performance of the self-tuning algorithms, and work has now progressed onto the physical generation system. This section presents results obtained when the hardware and software schemes outlined previously were used to control the behaviour of the synchronous machine during various transient disturbances such as short-circuits, line-switches, transformer tap-changes and changes in the turbine mechanical input power.

In this application, the SISO controller (equation 4) sampled the system at 20 ms intervals and ran on the PC-based T800 transputer in parallel with the VME system. A 7th order Pseudo Random Binary Sequence (PRBS) was superimposed on the exciter inputs for an initial period of 8 seconds to ensure proper convergence of the algorithm. This input was subsequently removed, leaving noise on the system to provide sufficiently exciting signals.

Extensive tests were conducted to ensure both the short-term transient and long-term dynamic stability of the control system. During both severe and minor disturbances the machine remained stable and power oscillations were well damped.

The graphs in Figure 5 (a) - (d) show the machine response to a 60 ms short circuit applied at the sending end of the transmission line at an operating point of 0.4 p.u. real power and 0.2 p.u. reactive power. Here, the machine remains well damped despite the major disturbance. Figure 5 (c) shows the control effort during the fault and Figure 5 (d) illustrates the behaviour of controller parameter g_1 during this short-term period. It can be seen that the parameter converges and that the estimator process is frozen during the transient period. This is due to the algorithm supervision scheme, one of whose duties is to monitor the condition of the machine and to switch off estimation if the machine is subjected to a transient disturbance.

Figure 6 shows the rotor angle response to a three-phase short circuit at the sending end of the transmission line, cleared after 60 ms with one transmission line switched out. Also included is the response obtained when the generator was controlled by the conventional fixed-gain AVR. Here, the STR provides a better transient response than the fixed-gain controller. It should be noted that the STR has little knowledge of the controlled system prior to the tune-in period, whereas the fixed-gain AVR was designed using a linearised model of the machine system with the alternator parameters known beforehand.

To test the long-term reliability of the control implementation, the generator was subjected to a series of events which might occur on a real power system (Brown and Swidenbank, 1991). Figure 7 compares the response of the STR to that obtained with the fixed-gain controller during a tap change of the transmission transformer in this period.

During this long-term period, the generator went from lagging phase to leading phase operation, from light load to heavy load, and was subjected to both minor and severe disturbances. The self-tuning controller remained stable and provided good transient response throughout, confirming the validity of both the self-tuning algorithm and the associated supervisory code.

Test results with the multi-loop controller

The multi-loop controller described in the third section was used to control a simulated power system. The algorithm was run in parallel on two T800 transputers and, since both the governor and AVR loops are essentially separate in the multi-loop case, the entire controller implementation was able to operate at the same sample rate as the SISO controller. Figure 8 illustrates the behaviour of the controlled system when subjected to a 300 ms short circuit at the sending end of the transmission line at an operating point of 0.8 p.u. real power and 0.2 p.u. reactive power. Figure 8 (a) compares the response of the multi-loop controller to that obtained using the SISO AVR. The multi-loop controller is seen to reduce the first rotor angle swing (an important criterion in power system stability) and to increase the damping in subsequent swings. Also shown in Figure 8 is the control effort.

DISCUSSION AND CONCLUSIONS

The self-tuning regulation strategy coupled with the advanced measurement system outlined in this paper provide effective control of both simulated power systems and a laboratory micromachine. Transputers are used to calculate the terminal electrical quantities and to perform the self-tuning control law. Extensive computer simulation studies, using a multi-task parallel model of the control architecture running on a network of transputers, have shown that improved control performance can be gained by using self-tuning regulators instead of a more conventional fixed gain regulator. An implicit 5-parameter self-tuning regulator, an extended 9-parameter self-tuning regulator and an associated algorithm supervision scheme have all proven to be robust in simulation.

All of the control programs were moved from the simulated system to the physical test bed with only minor modifications. Both the fixed gain and 5-parameter self-tuning controllers provided satisfactory performance during short term transient tests and over a more realistic longer term operating period on the micromachine. Work is progressing on the implementation of the 9-parameter self-tuning controller where improved performance should accrue from modelling of the noise process (Maguire, 1991).

The self-tuning regulators exhibited better damping, less power oscillations and quicker voltage stabilisation than the fixed gain AVR. This not only confirms the viability of the self-tuning algorithm but also validates the robustness of the supervision scheme. This should be viewed in the light of the fact that the self-tuning algorithms have no prior knowledge of the plant and are based on a model which could not be derived by any analytical means. Although the results achieved with the micromachine studies cannot be reasonably expected to match those obtainable on a full size generating set, they do serve to demonstrate the practical viability of both the measurement and control systems.

The parallel computer control system that has been developed in this paper constitutes a robust environment suitable for the implementation of both conventional and advanced control algorithms. The application of both SISO self-tuning regulators has proved to be successful and implementation of schemes such as expert control, self-tuning PID regulators, multivariable adaptive control and even artificial neural networks should be possible. Simulation results have already shown that improved performance can be obtained by using a coordinated multi-loop control scheme, and future research involving parallel processing techniques is continuing on implementing both multi-loop and multivariable controllers in real-time.

REFERENCES

Bo, Z. (1988). Development of a Multi-processor System for Control of a Laboratory Turbogenerator. PhD thesis, Dept of Elect. & Electron. Eng. The Queen's University of Belfast, U.K.

Brown, M.D. and Swidenbank, E. (1991). Multi-processor architecture for implementation of an automatic voltage regulator. Proc 1st IFAC symposium on 'Parallel Algorithms and Architectures for Real-time Control' Univ. of Bangor, Wales, pp 187 - 192.

Dettmer, R. (1990). Great expectations : the H1 transputer. IEE Review., December, pp 411 - 413.

Flynn, M.J. (1966). Very high speed computing systems. IEEE Proc., vol 54, pp 1901 - 1909.

Ham, P.A. (1989). Advances in steam turbine control, in 'Power Generation Technology'. Review of Primary Production . 1989, pp 239 - 243, Sterling Pub. Int. Ltd.

Ibrahim, A.S., Hogg, B.W. & Sharaf, M.M. (1989). Self-tuning controllers for turbogenerator excitation and governing systems. IEE Proc., vol 136, Pt D, no 5, pp 238 - 251.

Irwin, G.W. and Fleming, P.J. (eds) (1992). Transputers in Real-time Control. Research Studies Press (J. Wiley & Sons Ltd), 1992 (in press).

Irwin, G.W. and Maguire, L.P. (1991). Parallel adaptive control. Proc European Control Conference, Grenoble, France, July 2-5, pp 590 - 593.

Maguire, L.P. (1991). Parallel Architectures for Kalman Filtering and Self-tuning Control. PhD thesis, Dept of Elect. & Electron. Eng. The Queen's University of Belfast, U.K.

Wu, Q.H. and Hogg, B.W. (1988). Robust self-tuning regulator for a synchronous generator. IEE Proc., vol 135, Pt D, no 6, pp 463 - 473.

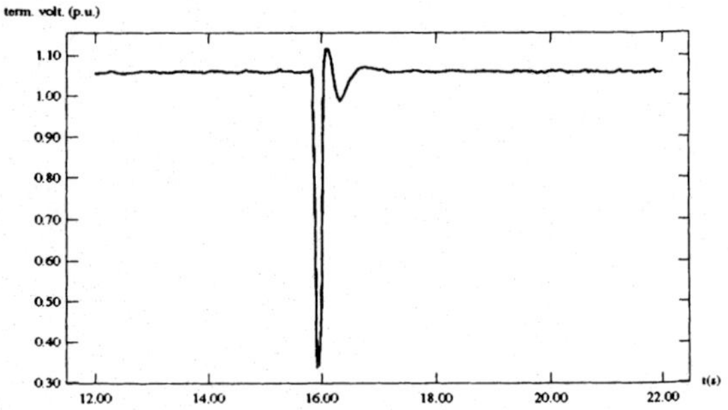

Fig. 5 (a) Voltage resposnse to 3-ph s/c with SISO control

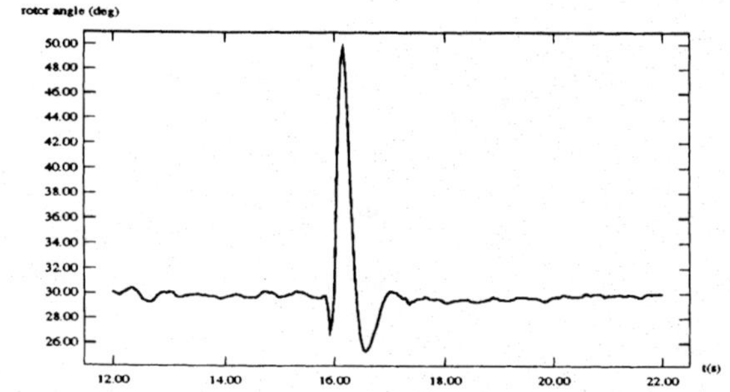

Fig. 5 (b) R. angle resposnse to 3-ph s/c with SISO control

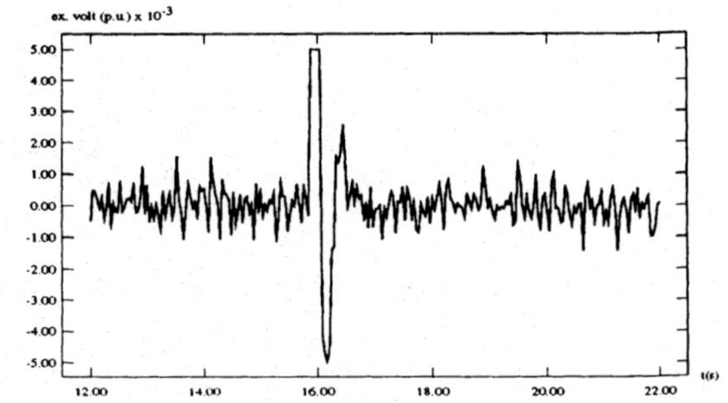

Fig. 5 (c) Exciter resposnse to 3-ph s/c with SISO control

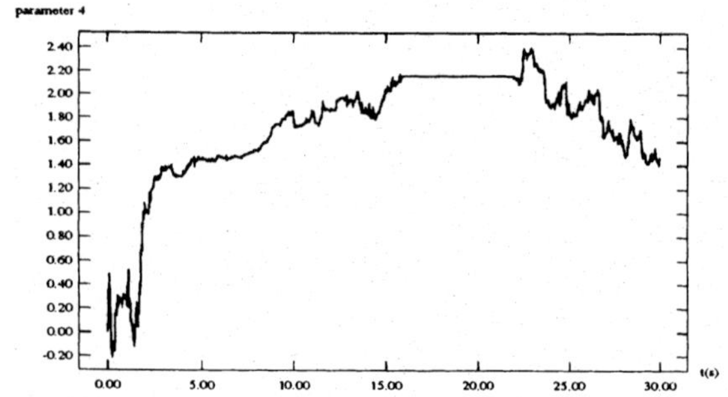

Fig. 5 (d) Parameter resposnse to 3-ph s/c with SISO control

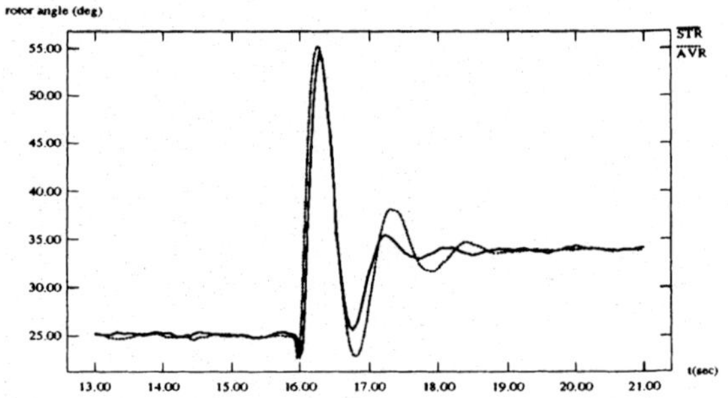

Fig. 6 R. angle resposnse to 3-ph s/c + line-out (SISO)

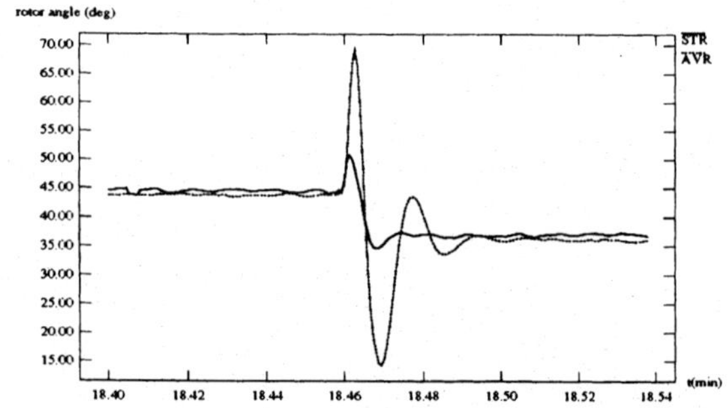

Fig.7 R. angle resposnse to tap change with SISO control

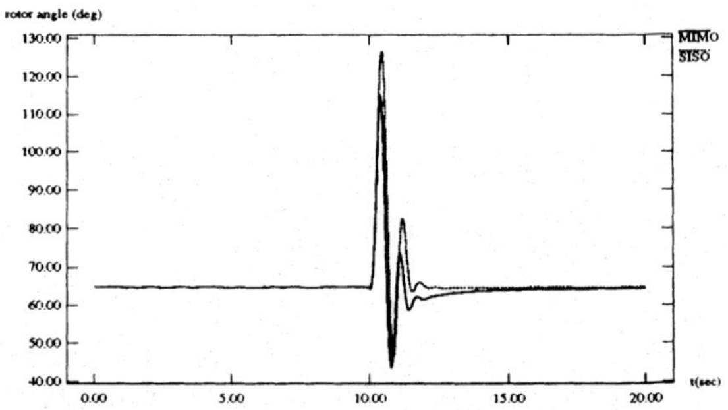

Fig.8 (a) R. angle resposnse to s/c (SISO+MIMO) control

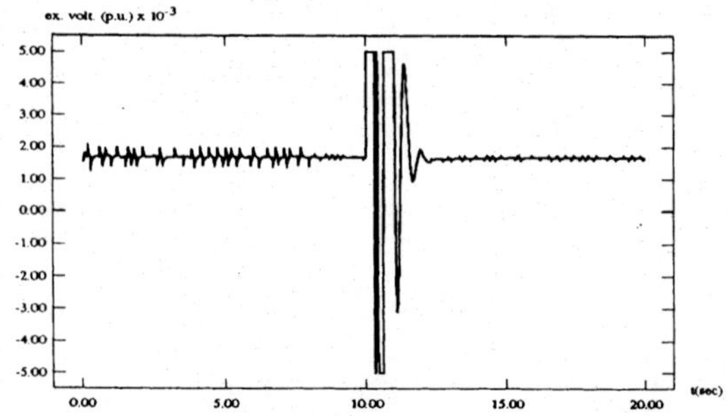

Fig.8 (b) Exciter input during s/c (multi-loop) control

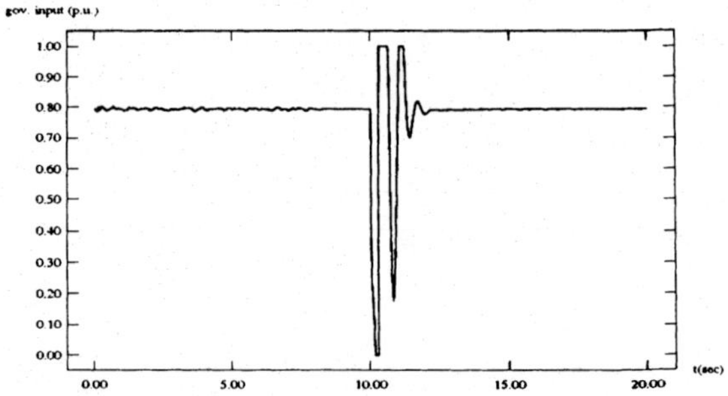

Fig.8 (c) Governor input during s/c (multi-loop) control

EXPERT AUTO-TUNER FOR MULTIVARIABLE CONTROL APPLICATIONS

Kar-Ann Toh and R. Devanathan

School of Electrical and Electronic Engineering
Nanyang Technological University
Nanyang Avenue, Singapore 2263

Abstract

The main idea of the proposed auto-tuning control methodology is to apply SISO (Single-Input-Single-Output) PID controllers with appropriate adaptation to the MIMO (Multi-Input-Multi-Output) environment. An expert tuning processor is to supervise the PID controller to provide good tuning parameters based on the observed MIMO response. A novel SISO process identification technique based on pattern recognition is described. Multivariable tuning is then activated using a detuning factor on the SISO settings. A new technique to identify the RDG (Relative Disturbance Gain) on-line is also included to assist the decision on tuning the controller upon natural load disturbances. Experimental results on the application of the auto-tuner are also included.

Keywords : Automatic control, expert systems, PID control, process control, threeterm control, multivariable systems.

1 INTRODUCTION

Most control problems involve uncertainty. Signals are corrupted by sensor noise as well as by distortion during quantization. Load disturbance which is usually non-measurable poses another obstacle towards good control. Moreover, many industrial processes are multivariable in nature. The existence of interactions between the control loops complicates the problem further. Typical examples include the flow-pressure monitoring of fluid flow systems, the well known distillation columns, the paper making process and many chemical processes. It is thus desired to find a multivariable PID controller for the system so that asymptotic regulation occurs for all constant reference signals and that the response is as 'good' as possible (Koivo and Tanttu, 1991). In this paper, an expert auto-tuning methodology for the PID controller for use in 2×2 multivariable processes is proposed. The PID controller is selected because it is robust enough and it has well balanced action to reduce offset and to improve response. More importantly, the PID controller has been widely accepted in the industry with its clear physical meaning of the controller parameters. Novel identification techniques for the required system information are also included.

2 TERMINOLOGY

For the 2×2 system of the p-canonical type as described in fig. 1, the well known concept of the Relative Gain (Bristol, 1966) is taken to be

$$\lambda = \frac{1}{1 - \frac{K_{p12}K_{p21}}{K_{p11}K_{p22}}} \tag{1}$$

where K_{pij}, $i = 1,2$ are the static gains of the process transfer functons denoted by G_{pij}, $i,j = 1,2$. The relative gain array (RGA) is given by

$$\Lambda = \begin{bmatrix} \lambda & 1-\lambda \\ 1-\lambda & \lambda \end{bmatrix} \tag{2}$$

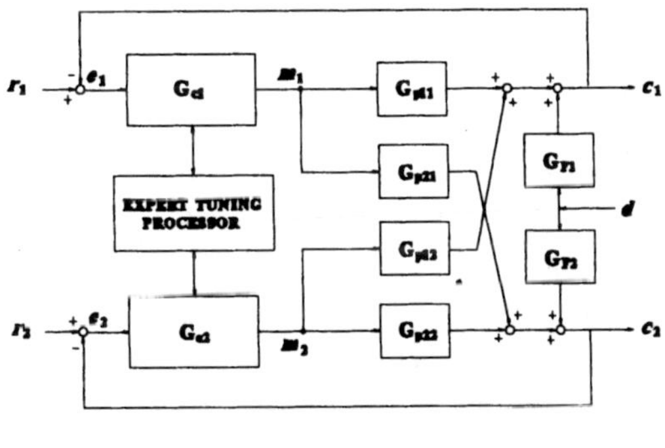

Figure 1: Block diagram of the proposed auto-tuner for a 2×2 system

To determine whether interaction resulting from a load disturbance is favorable or unfavorable, the Relative Disturbance Gain (RDG) (Stanley et al., 1985) is used. The RDG is defined as follows

$$\beta_i = \frac{\frac{\partial m_i}{\partial d}|_{x_i x_j}(perfect\ x_j\ control)}{\frac{\partial m_i}{\partial d}|_{x_i m_j}(no\ control\ over\ x_j)} \tag{3}$$

where $i, j = 1,2$ *and* $i \neq j$. Using the same control system modelled by that shown in fig. 1, the RDG for loop 1 can be written as in equation 4.

$$\beta_1 = \lambda\left(1 - \frac{K_{F2}K_{p12}}{K_{F1}K_{p22}}\right) \tag{4}$$

3 THE PROPOSED METHODOLOGY

Consider the system architecture of a 2×2 system as shown in Figure 1. G_{c1} and G_{c2} whose parameters are set by the expert tuning processor, are scalar PID controllers for loops 1 and 2 respectively. The auto-tuning control is performed in three stages : 1. single-loop process identification, 2. multivariable process identification using a SISO method under interaction control and 3. multivariable tuning using a detuning factor on the SISO case.

In the single-loop identification, a novel process identification technique based on the pattern recognition technique (Toh and Devanathan, 1991) is employed under closed loop control. The required steady state gains of the system are obtained through step tests. The Ziegler-Nichols tuning formula and the minimum ITAE criterion are applied to obtain the starting values of the PID parameters under single-loop control. The two controllers are then put to automatic under multi-loop control where the multivariable identification and tuning is followed. The multivariable identification uses a test pulse to observe the process response under stable interactive control. In the case of unstable control resulting from closing the multivariable loops, the unstable response patterns are recorded as part of the identification process for expert tuning. The multivariable tuning based on the observed patterns is subsequently triggered after the stage of process identification. The tuning mechansim can be divided into two phases : 1. pretuning phase, and 2. fine tuning phase. In the pretuning phase, the stability of the system under the SISO optimum settings is taken care of by detuning the proportional band alone. It is then followed by the fine tuning phase where the identified process is used for estimation of the detuning factors from a simple analytical tool. In the case where the analytical tool does not apply, the expert tuning procedure will take place in the fine tuning phase so that a complementary solution is obtained. The detuning procedure takes place in two separate steps for the two controllers.

Apart from using the estimated detuning factors, the RDG is used as part of the expert knowledge for detuning. A new on-line technique for estimating the RDG is proposed using the known steady state gains and the change of manipulated variables upon occurrence of the load disturbance. The decision on whether to detune the controllers from their SISO settings can thus be inferred based on the type of disturbance entering the system. The following sections present the details of the proposed expert auto-tuning methodology.

4 PROCESS IDENTIFICATION

The elements in the process transfer function matrix are assumed to be modelled by the FOPDT type (equation 5) which is known to be suitable for about 80% of industrial processes (Luyben, 1990).

$$G_{pij}(s) = \frac{K_{pij}\ e^{-T_{dij}s}}{1 + T_{1ij}s}, \quad i,j = 1,2 \tag{5}$$

Putting one loop in automatic and the other in manual, the loop in control is actually under SISO condition. Consider the closed-loop system shown in fig. 2. Under normal stable control action, a load step change in u gives a damped response as shown in fig. 3. The pattern recognition method used for process identification is similar to that employed by the Foxboro's EXACT controller (Kraus, 1984).

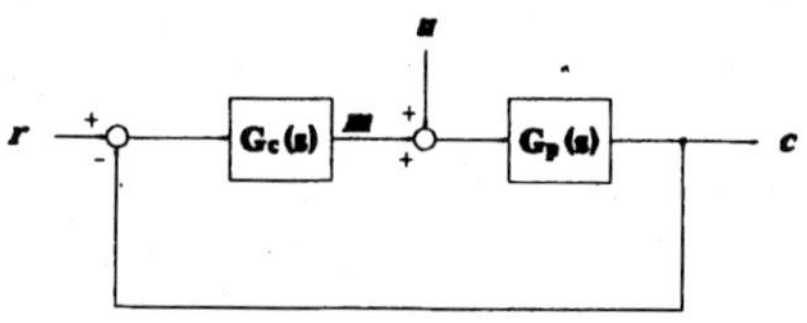

Figure 2: Block diagram of a closed-loop control system

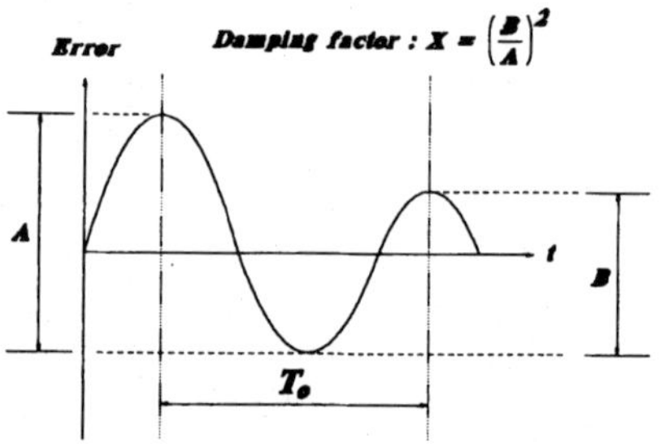

Figure 3: A closed-loop damped response

The damped response is recognized by its loop period, T_o and the damping factor, X as shown in fig. 3. Two independent load responses from two different settings of the PID parameters of the controller are required for the estimation of the process dead time, T_d and the time constant, T_1. The PID controller is assumed to be of the non-interacting type as that shown in equation 6.

$$G_c(s) = \frac{100}{PB}(1 + \frac{1}{Is} + Ds) \tag{6}$$

For an interacting PID controller, its effective values (Shinskey, 1988) relative to the non-interacting type are used instead. The effective values are obtained from the following equations.

$$I_{eff} = I + D \tag{7}$$

$$D_{eff} = \frac{1}{\frac{1}{I} + \frac{1}{D}} \tag{8}$$

$$\frac{100}{PB_{eff}} = \frac{100}{PB}(1 + \frac{D}{2}) \tag{9}$$

Taking T_{oi} to be the period measured from the ith damped response curve, the identification proceeds as follows

$$T_1 = \frac{tan\psi}{2\pi - tan\psi\ lnX_1} T_{o1} \tag{10}$$

$$T_d = (\pi + \theta_1 - \psi)\frac{T_{o1}}{2\pi} \quad (11)$$

where ψ, the process lag angle is given by

$$tan(\delta + \alpha\psi) = \frac{2\ \pi\ \alpha\ tan\psi}{2\pi + (\alpha\ lnX_2 - lnX_1)tan\psi} \quad (12)$$

$$\delta = (\pi + \theta_2) - \alpha(\pi + \theta_1) \quad (13)$$

$$\alpha = \frac{T_{o1}}{T_{o2}} \quad (14)$$

and the controller angles θ_1 and θ_2 are calculated from

$$\theta_i = tan^{-1}\left[\frac{\frac{2\pi D_i}{T_{oi}} - \frac{1}{\frac{\sigma_i{}^2\ I_i\ T_{oi}}{2\pi} + \frac{2\pi I_i}{T_{oi}}}}{1 + \sigma_i\ D_i + \frac{\sigma_i}{I_i[\sigma_i{}^2 + (\frac{2\pi}{T_{oi}})^2]}}\right] \quad (15)$$

and

$$\sigma_i = \frac{lnX_i}{T_{oi}}, \quad i = 1, 2 \quad (16)$$

□

Good estimation results could be achieved through monitoring the loop periods (Toh and Devanathan, 1991) of the two perturbed responses. However this identification technique applies only to the under-damped case. For a closed loop process with over-damped response, the controller gain has to be increased sufficiently so as to obtain an under-damped response.

5 ESTIMATION OF DETUNING FACTOR

If the reset time is assumed to be inversely proportional to the ultimate frequency, the change in the ultimate frequency can be viewed as the change in the reset time (Marino-Galarraga and McAvoy, 1987). The detuning factor, f for the loop concerned, under multi-loop control can then be obtained from the knowledge of change in the ultimate periods from the multi-loop interactive case (T_{uI}) to the SISO case (T_u).

$$f = \frac{T_{uI}}{T_u} \quad (17)$$

Using the dominant closed-loop complex conjugate poles approximation with an open-loop FOPDT model process transfer function and a PID controller as in section 4 above, the following equations can be written for the angle and the gain conditions at the closed-loop poles.

$$\psi + \frac{2\pi T_d}{T_o} = \pi + \theta \quad (18)$$

$$\frac{K_p\frac{100}{P}\sqrt{1 + tan^2\theta}}{\sqrt{1 + tan^2\psi}} = 1$$

$$ie.\quad K_p\frac{100}{P}sec\theta\ cos\psi = 1 \quad (19)$$

For the interacting case,

$$\psi_I + \frac{2\pi T_{dI}}{T_{oI}} = \pi + \theta_I \quad (20)$$

$$\frac{K_p}{\lambda}\frac{100}{P_I}sec\theta_I\ cos\psi_I = 1 \quad (21)$$

Dividing equation 19 by equation 21 and rearranging

$$cos\psi_I = \frac{P_I}{P}\frac{\lambda sec\theta\ cos\psi}{sec\theta_I} \quad (22)$$

Determine ψ_I using equation 22 and then determine $\Gamma_I = T_{dI}/T_{1I}$. from

$$\Gamma_I = \frac{(\pi + \theta_I - \psi_I)(2\pi - \sigma_I T_{oI}\ tan\psi_I)}{2\pi\ tan\psi_I} \quad (23)$$

Let

$$\frac{T_{uI}}{T_{dI}} = N_{uI} \quad (24)$$

the corresponding angle equation is thus

$$tan^{-1}\frac{2\pi}{N_{uI}\Gamma_I} + \frac{2\pi}{N_{uI}} = \pi \quad (25)$$

Knowing Γ_I, equation 25 can be solved for N_{uI}. Then,

$$T_{uI} = N_{uI}T_{dI} \quad (26)$$

From the response for the SISO identification, $\Gamma = T_d/T_1$ is known. Also, using equation 25 and equation 26 for the SISO case, T_u can be found.

The detuned reset time is thus obtained as

$$I_I = fI \quad (27)$$

If the derivative time is treated in the same manner, the following can be used

$$D_I = fD \quad (28)$$

For the case of the interacting controller, it is straight forward to show that the detuning factor exhibits a linear relationship (Toh, 1992).

6 ON-LINE IDENTIFICATION OF THE RDG

From the knowledge of the change in the manipulated variables with respect to a single load disturbance, the RDG can be determined by the equation below (Toh, 1992).

$$\kappa = \frac{K_{p12} + \eta K_{p11}}{K_{p22} + \eta K_{p21}}, \quad \Delta m_2 \neq 0 \quad (29)$$

where $\kappa = K_{F1}/K_{F2}$ and $\eta = \Delta m_1/\Delta m_2$. The κ value can be substituted into equation 4 to obtain the RDG value.

Recent research on the application of a SISO self-tuning controller (Foxboro EXACT) in the MIMO environment (Cao and McAvoy, 1990) shows that the controller has trouble with systems where the RDG is close to zero and the RGA is small. The controller for the loop with zero RDG will measure a damping coefficient that seems to be too low or equivalently an overshoot that seems too large. This controller will begin to lower its gain where it should not. The identified RDG serves as a guideline to determine whether interaction is favorable for a particular load disturbance in multi-loop control. The auto-tuning controller can then decide on whether to detune the proportional band of the controller.

7 IMPLEMENTATION AND RESULTS

The auto-tuning controller is implemented using an IBM PC/486 which acts as the expert tuning processor and two microprocessor-based SISO controllers regulating two interacting controlled variables. The two SISO controllers are realized using the Intel-8086 microprocessor with a pair of analog input/output ports and a serial asynchronous communication port. The 12-bit Analog-to-Digital and Digital-to-Analog converters are used to provide sufficient accuracy for the experiment. The communication port which is linked to the personal computer uses the RS232C communication standard. Display panels for the set-point, manipulated variables and controlled variables are also included. Additional functions such as adjustable sampling rate control, interrupt control, auto/manual switching and keyboard for set-point and controller output entry are also included. The block diagram for the hardware architecture of the SISO controller which is linked to the expert tuning processor is given in fig. 4.

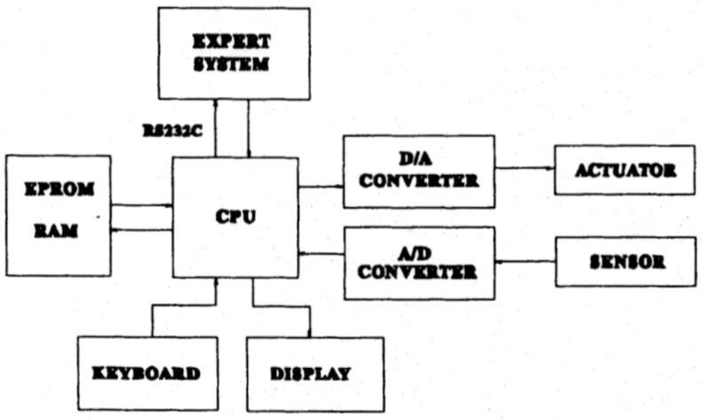

Figure 4: Block diagram for the SISO controller with interface to PC

The expert system is realized by applying meta-rules of the "If <SITUATION> then <ACTION>" form. The <SITUATION> part of the rule describes the current state of the process in which the rule applies. The <ACTION> part of the rule specifies the appropriate adjustments of the controller settings. The overall inference relation for the expert system is shown in fig. 5.

The rule base is classified under two major categories : rules for testing and rules for tuning. The testing rules ensure that the required conditions for both the SISO and the MIMO identifications are met before triggering the tuning rules. For instance, the accuracy of the SISO identification of the system dynamics depends on the ratio of the loop oscillation periods. The rules are set to check for fulfillment of the ratio condition as well as validity of the observed patterns. For the tuning rules, it can be further divided into expert tuning as well as analytical tuning. The expert tuning includes rules like detuning the proportional band to ensure closed loop stability of the multivariable system. The analytical tuning rules make use of the available information to estimate the required detuning factor.

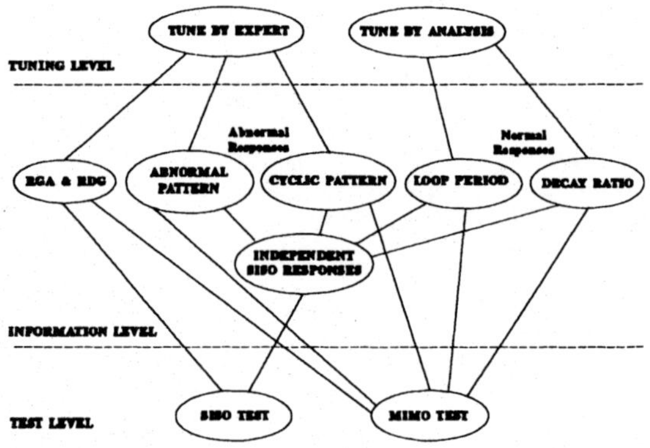

Figure 5: The inference relation of the system

The expert system methodology is studied over a range of processes characterized by their static relative gains and the relative speeds (given by R and difference in dead times) of the off-diagonal elements to the diagonal elements. R is defined as ratio of the relative time constants as in equation 30

$$R = \frac{T_{11}T_{22}}{T_{12}T_{21}} \tag{30}$$

The studies are limited to the class of processes that belongs to the group of small normalized dead time (Åström et al., 1992) values since the first order Padé approximation is used in the analog computer simulated process. The results which show fairly good tuning response for the experiments conducted are summarized in fig. 6- 9.

Fig. 6 and fig. 7 show the obtained detuning factor plotted against the RGA values for four types of processes characterized by the normalized dead time and the relative difference in dead time using the ZN tuning and the ITAE tuning formula. The continuous lines and the dotted lines indicate the detuning factor obtained for loop 1 and loop 2 respectively. The lower three sets of plots are obtained as the result of applying the analytical detuning estimation for relatively slow interactions. The upper set which corresponds to faster interactions is the result of expert tuning. In the former case, since linear control action is achieved, the detuning factor shows a higher value for RGA ≤ 1. The latter case indicates that the existence of the interacting components deteriorate controllability. The much higher detuning factor for larger RGA values is actually an effort to gain stable dynamic response under poor static controllability.

Fig. 8 and fig. 9 show the worstcase plots where there exist no dead time in the off-diagonal components. The results show satisfactory responses for RGA ≤ 1. The unity detuning factor seen in the plots for $\lambda \gg 1$ is the result of expert tuning which indicates that no improvements could be achieved through detuning the controllers. The controllability is poor in these cases. Decoupling control is hence recommended for these cases.

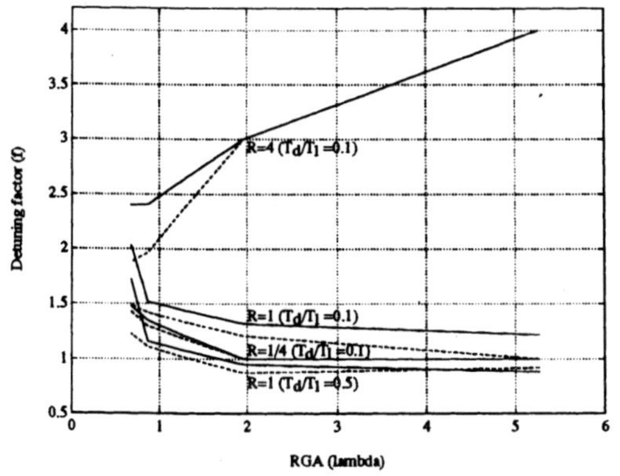

Figure 6: Detuning factor versus λ (ZN tuning)

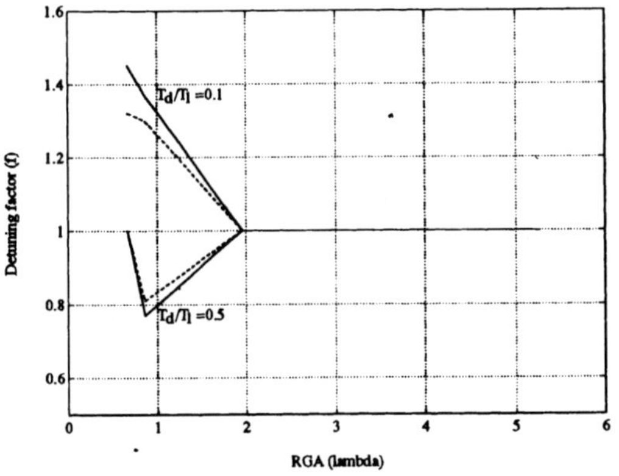

Figure 8: Detuning factor versus λ (ZN tuning) under worst case condition

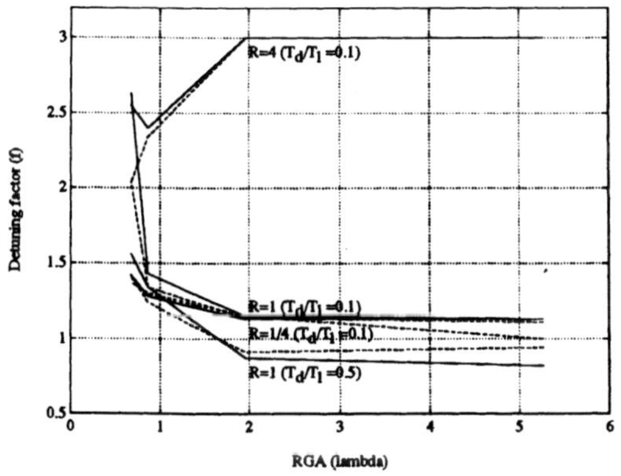

Figure 7: Detuning factor versus λ (ITAE tuning)

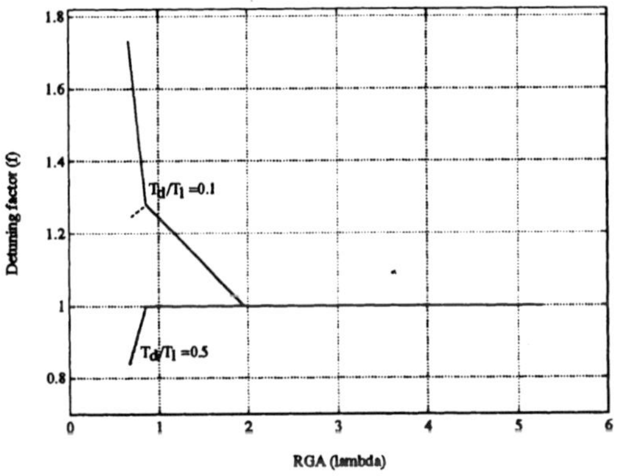

Figure 9: Detuning factor versus λ (ITAE tuning) under worst case condition

8 AN APPLICATION EXAMPLE

The expert auto-tuner is tested in a laboratory scale flow-pressure pilot plant. Figure 10 shows the block diagram for the setup of the flow-pressure process pilot plant. Two pneumatic valves as shown in the figure are used for the flow-pressure regulation where $c_i, i = 1,2$ and $m_i, i = 1,2$ are the controlled variables and manipulated variables respectively. PT and FT denotes the pressure transmitter and the flow transmitter used in the control system. Signal transmission uses the 4-20 mA current transmission such that minimum attenuation is achieved.

The results for the SISO identification including the step tests are tabulated in Table 1. Using the identified process dynamics, the SISO tuning parameters based on the Ziegler-Nichols PI formula are found to be 263/0.06/0 and 20/0.06/0 for loops 1 and 2 respectively. The results for the MIMO experiment are given in Table 2 and Figure 11. In this figure, loop 2 is disturbed by a load pulse of 2 volt for a duration of 5 seconds.

9 CONCLUSION

A simple control structure associated with a novel methodology which applies to a class of multivariable processes is proposed. The basic idea of the proposed methodology is to combine the expert system design with an analytical tool to bring about the merits of both components. The analytical tool includes the use of a new SISO identification technique which is further developed into the MIMO identification in terms of its interacting effect. The expert system design is to bring the process into the linear control region in the case where its stability is not acquired. A new estimation technique for the RDG is also proposed to aid the decision on the MIMO tuning upon load disturbances entering the system.

The expert auto-tuner is studied over a range of processes characterized by the RGA, the dead times and the time constants using analog computer simulation. It is also applied to a flow-pressure process pilot plant in which satisfactory result has been obtained. It is noted that much more research and development effort is to be built upon the basic idea to provide a well equipped practical controller for indus-

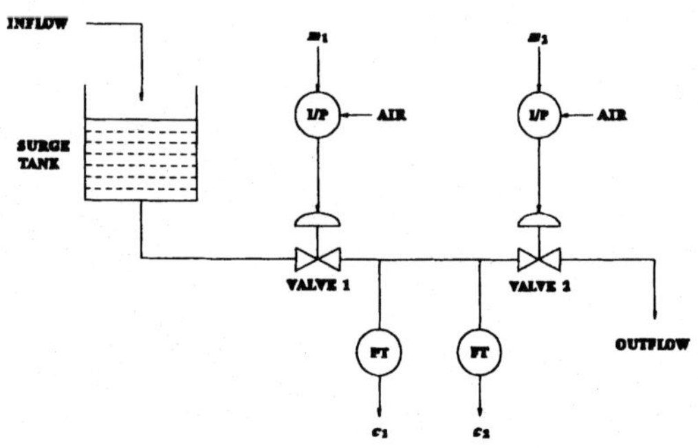

Figure 10: The flow-pressure pilot plant

Table 1: SISO identification results of loops 1 and 2 for the flow-pressure process

Loop 1		Loop 2	
PID_1	= 30/0.10/0	PID_1	= 40/0.01/0
PID_2	= 90/0.01/0	PID_2	= 80/0.01/0
T_{o1}	= 0.0583	T_{o1}	= 0.1450
T_{o2}	= 0.1000	T_{o2}	= 0.2200
X_1	= 0.4180	X_1	= 0.5900
X_2	= 0.3669	X_2	= 0.1800
K_{p11}	= 2.55	K_{p12}	= −1.05
K_{p21}	= 2.25	K_{p22}	= 0.50
θ_1	= −0.0920	θ_1	= −1.2320
θ_2	= −1.1194	θ_2	= −1.5369
δ	= 0.2432	δ	= 0.3461
ψ	= 1.2181	ψ	= 1.1802
Identified		Identified	
T_{d11}	= 0.0170 min	T_{d22}	= 0.0168 min
T_{11}	= 0.0183 min	T_{22}	= 0.0466 min

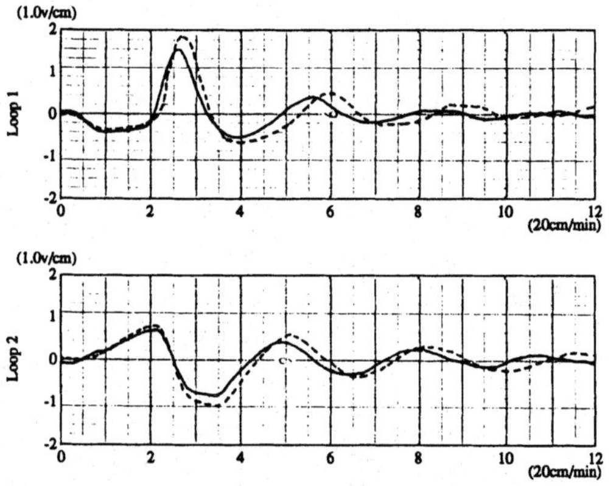

Figure 11: Error versus time plots for load pulse under pretuning (- - -) and after detuning (—)

Table 2: Estimation of detuning factor for the flow-pressure process

Loop 1		Loop 2	
PID	= 263/0.06/0	PID	= 30/0.06/0
T_{oI}	= 0.1700	T_{oI}	= 0.1650
X_I	= 0.1837	X_I	= 0.4050
θ_I	= −0.4427	θ_I	= −0.4287
ψ_I	= 0.2731	ψ_I	= 1.2934
Identified		Identified	
T_{dI}	= 0.0656 min	T_{dI}	= 0.0373 min
T_{1I}	= 0.0070 min	T_{1I}	= 0.0613 min
f_1	= 2.72	f_2	= 2.08

trial applications. However, the combination of the pattern recognition technique with the expert system approach is shown to be a potential candidate for good multivariable adaptive control.

References

Åström, K. J., Hang, C. C., Persson, P., and Ho, W. K. (1992). Towards intelligent PID control. *Automatica*, 28 no. 1:1–9.

Bristol, E. H. (1966). On a new measure of interaction for multivariable process control. *IEEE Trans. Automatic Control.*

Cao, R. and McAvoy, T. (1990). Evaluation of a pattern recognition adaptive PID controller. *Automatica*, 26, no. 4:797–801.

Koivo, H. N. and Tanttu, J. T. (1991). Tuning of PID controllers: Survey of SISO and MIMO techniques. In *Proceedings of IFAC International Symposium on Intelligent Tuning and Adaptive Control (ITAC 91)*, Singapore.

Kraus, T. (1984). Self-tuning control : An expert system approach. *Advances in Instrumentation.*

Luyben, W. L. (1990). *Process Modelling, Simulation and Control for Chemical Engineers.* McGraw-Hill, Chap. 14.

Marino-Galarraga, M. and McAvoy, T. J. (1987). Short-cut operability analysis. 2. estimation of f_i detuning parameter for classical control systems. *Ind. Eng. Chem. Res.*, 26:511–521.

Shinskey, F. G. (1988). *Process Control Systems.* McGraw-Hill.

Stanley, G., Marino-Galarraga, M., and McAvoy, T. J. (1985). Shortcut operability analysis. 1. the relative disturbance gain. *Ind. Eng. Chem. Process Des. Dev*, 24:1181–1188.

Toh, K. A. (1992). Expert auto-tuner for multivariable control applications. *M. Eng Thesis (School of EEE NTU).*

Toh, K. A. and Devanathan, R. (1991). PC-based simulation studies on a new closed loop process identification technique. In *Proceedings of the Asia-Pacific Conference on Measurement and Control (APMC)*, Guangzhou, PRC.

An Intelligent Self-Tuning Controller And Its Application To Load/Pressure Control System

Liu Jizhen, Liu Yanquan and D. Z. Li

Department of Thermal Power Engineering
North China Institute of Electric Power
Boading, Hebei, P. R. China 071003

Abstract: In this paper a novel self-tuning controller based on the concept of process supervision and gain adjustment are documented. The features of the design include leading a time varing factor $\lambda(k)$ to controller output and adopting estimater switch technique. The new design known as ISTC was extended to MIMO system under a particular process model assumption. Simulation study was done in a 2×2 load/pressure system of a turbine - boiler unit. Experiment results shown the efficiency of new design and possibility of industry application in practice.

Keywords: Self-tuning regulator; Intelligent control; Multivariable system Process control; Boiler-turbine unit.

1. Introduction

The advantage of self-tuning control is of its ability to online adjust controller parameters to adapt the characteristics of processes which are not well understood, are slowly time varying, or which have significant nonlinearities (Chien et al, 1985). As a minimum variance control algorithm, self-tuning control is easier for design and implimentation than other adaptive control algorithm. However, in practical industry process control area, self-tuning controllers can not achive good results as expected. Many researchers have been paid attention to improve the performance and robustness, including the controller structure robustness and process parameter estimation robustness. There is much effort leading to significant advancements in theory and practical application (Wittenmark and Astrom, 1984, 1989, Seborg et al 1986).

In this paper, we develop a self - tuning controller version based on the traditional self-tuning regulator (STR) (Astrom and Wittenmark , 1973) and self-tuning controller (STC) (Clarke and Gawthrop , 1975) . A time varying right factor was developed and can be online adjustment by given rules. In this case, the controller has a structure of mult- layer control strategy. The term intelligent self-tuning controllers (ISTC) was adopted to hightlight the features of the algorithm design.

The ISTC algorithm was extended to MIMO system by using a CARMA process model. A load/pressure control system of a 200MW oil fired boiler-turbine power unit was documented, preliminary simulation results show the possibility and the efficiency of ISTC in practical application.

2. ISTC algorithm

For simplification, a ISTC agorithm design of SISO system was documented, then it is extended to a MIMO system.

2.1 SISO algorithm design

The process to be controlled is described in a discrete time model :

$$A(q^{-1})y(k) = q^{-d}B(q^{-1})u(k) + C(q^{-1})\zeta(k) \quad (1)$$

$$A(q^{-1}) = 1 + a_1q^{-1} + \ldots + a_nq^{-n} \quad (2)$$

$$B(q^{-1}) = b_1q^{-1} + \ldots + b_nq^{-n} \quad (3)$$

$$C(q^{-1}) = 1 + C_1q^{-1} + \ldots + C_nq^{-n} \quad (4)$$

where y(k) and u(k) are the system output and input, and $\zeta(k)$ is disturbance with a white gaussian distribution. If it is assumed that only uncoloured noise enters the process ($C(q^{-1}) = 1$), then, Clarke and Gawthrop (1975) have developed the self-tuning controller (STC) which, when servo action is included, gives the control law:

$$u(k) = \frac{y_r(k+d+1) - \phi^T(k)\,\theta(k)}{\beta_0 + \rho/\beta_0} \quad (5)$$

where d is the estimated system dead time and $y_r(k)$ is reference input or setpoint. The other components of the control law given by equation (5) are found from:

$$\phi(k) = [-y(k)\ -y(k-1)\ldots -y(k-n-1)$$

$$u(k-1)\ u(k-2)\ldots u(k-p)]^T \qquad (6)$$

$$\theta(k) = [\alpha_0 \alpha_1 \ldots \alpha_{n-1} \beta_1 \beta_2 \ldots \beta_p]^T \qquad (7)$$

where $p = n + d - 1$. The STC design will minimize the following cost function:

$$J(r) = \sum_{k=0}^{\infty}([y(k) - y_r(k)]^2 + \rho u^2(k)) \qquad (8)$$

If $\rho = 0$ then the STC control objective of equation (5) equates to that for the self-tuning regulator (STR) design of Astrom and Wittenmark (1973). The process parameter $\theta(k)$ in equation (5) can be recursive estimated using RLS method.

As wellknown that STR can not be adapted to non-minimum phase system and may results in control active u(k) overmuch or ocsillation.STC can give out suitable u(k) by selection of a resonable right factor ρ.However, a steady state error of system output may be resulted. In other hand, the convergence of estimated parameter $\theta(k)$ will affect the amplitude and stability of control action u(k).In order to make an academic or theoretical algorithm become an industrially practical algorithm, it is important to consider some practical limitations in implimentation.For many industrial process, such as thermal process discussed in this paper, there are long periods of time and slow time variation in process parameters. Only when the variation of process parameters and operating condition reach a "significant" level, it is necessary to modify estimated controller parameters and to adapt the variation of process characteristics. A steady control variable is desireable to avoid frequent active or ocsilation in actuator. In most cases, zero steady state offset is needed.

Based on the consideration mentioned above, the ISTC design will be developed under following procedures:

1. Leading a time varying right factor $\lambda(k)$ to the STC control law of equation (5).

$$u(k) = \frac{y_r(k+d+1) - \phi^T(k)\theta(k)}{\beta_0 \cdot \lambda(k)} \qquad (9)$$

where $\lambda(t) \geqslant 1$. $\phi^T(k)$ and $\theta(k)$ are given in equation (6) and (7). As a particular, taking the $\lambda(k)$ as constant:

$$\lambda(k) = (1 + \rho/\beta_0^2) \qquad (10)$$

Then the equation (9) equals to the equation (5).

2. An "estimator switch" can be implimented by monitoring the input and output of process directly. When the input and output of process lies within a range dicated by the noise inherent in the process, the estimator should be turned off (ES = 0). In this case, the controller becomes a constant linear feedback controller. Otherwise, the estimator should be turned on (ES = 1), and y(k) and u(k) are used to update $\theta(k)$.

3. When the "estimator switch" is turn on, $\lambda(k)$ is a time varying right factor, it is converged to a ultimate value from a given larger initial value. If ultimate value of $\lambda(k)$ is a unique, the ISTC algorithm converges to STR algorithm, which can guarantee that process output steady state offset is zero. Depend upon the idea motioned above, some rules for ISTC are summarzed as follow:

Rule 1:

IF $|y(k) - y(k - h1)| > \delta_1$

OR $|u(k) - u(k - h1)| > \delta_2$

THEN ES = 1 (estimator is turned on)

AND $\lambda(k) = \lambda_0$

Rule 2:

IF ES = 1

THEN N samples are used for parameters updating.

Rule 3:

IF $\lambda(K) > \lambda_{min}$

THEN $\lambda(K) = c * \lambda(K-1)$ $(0 < c < 1)$

The constants δ_1, δ_2, h_1, h_2 and N can be specified by user. Critical values for δ_1 and δ_2 indicate a "significant" time varying for process characteristics and are related to noise level. h_1 and h_2 are used for detecting a slow variation in process input and output such as low frequency draft. A reasonable value of N is in the range 10 to 20 for a convergent parameter update.

The variable λ_0, λ_{min} is initial and ultimate value of time varying factor $\lambda(k)$, and c is a convergent coefficent .

Based on the basic STC and STR algorithm, ultimate value λ_{min} can be designed according to traditional method given in equation (10) . Simulation experience has shown good result, where the λ_0 is 5 to 10 times of λ_{min} and c is given by

$$c^N = \frac{\lambda_{min}}{\lambda_0} \qquad (11)$$

The ISTC control system contructure is show in Fig. 1.

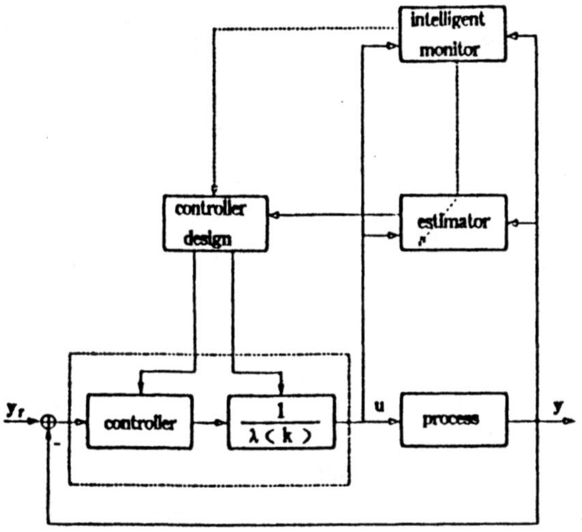

Fig. 1. ISTC control system structure

2.2 MIMO algorithm design

Some MIMO adaptive algorithm have been developed as a natural extension of SISO adaptive control (K. Ichikawa, 1987) . For practical purpose, it is hoped to reduce the process model parameters and a prior information needed on the system. Refering to the modelling assumption, by Goodwin et al (1980), the process model is represented in following CARMA forms as :

$$A(q^{-1})Y(k) = q^{-d}B(q^{-1})U(k) + C(q^{-1})E(k) \qquad (12)$$

with

$$A(q^{-1}) = \mathrm{diag}[A_1(q^{-1})\dots A_m(q^{-1})]$$

$$A_i(q^{-1}) = 1 + a_{i1}q^{-1} + \dots + a_{in}q^{-n}$$

$$B(q^{-1}) = [\dots B_{ij}(q^{-1})]$$

$$B_{ij}(q^{-1}) = b_{ij}^{o} + b_{ij}^{1}q^{-1} + \dots + b_{ij}^{n}q^{-n}$$

$$b_{ij}^{o} \neq 0 \quad i = 1,2,\dots m; \; j = 1,2,\dots m$$

$$C(q^{-1}) = I$$

The process can be rewritten as following m equations with single output and m input:

$$A_i(q^{-1})y_i(k) = q^{-1}B_i(q^{-1})U(k) + e_i(k) \qquad (13)$$

$$B_i(q^{-1}) = [B_{i1}(q^{-1}), B_{i2}(q^{-1})\dots B_{im}(q^{-1})]$$

The process is reparametrized in terms of an explicit self- tuning method using the m identities.

$$I = A_i(q^{-1})F_i(q^{-1}) + q^{-d}G_i(q^{-1}) \qquad (14)$$

So the process can be rewritten as the following predict equations:

$$y_i(k+d) = G_i(q^{-1})y_i(k) + F_i(q^{-1})U(k) + $$
$$+ F_i(q^{-1}) \cdot e_i(k+d) \qquad (15)$$

By minimizing the cost function:

$$J_i = E\{[y_i(k+d) - y_{Mi}(k+d)]^2\} \qquad (16)$$

It is now easy to compute the control law such that $y_i(k)$ tracks the ith reference input or setpoint $y_{Mi}(k)$ by choosing:

$$G_i(q^{-1})y_i(k) + F_i(q^{-1})B_i(q^{-1})U(k) -$$
$$- y_{Mi}(k) = 0 \qquad i=1,2\dots m \qquad (17)$$

Equation (17) can be rewritten as a matrix form:

$$\mathrm{diag}[G_1(q^{-1})\dots G_m(q^{-1})]Y(k) +$$
$$+ [F_1(q^{-1})\dots F_m(q^{-1})][B_{ij}(q^{-1})]U(K) -$$
$$- Y_M(k+d) = 0 \qquad (18)$$

Multipling $F_i(q^{-1})$ with $B_i(q^{-1})$ together and notice that:

$$F_i(q^{-1}) = 1 + f_{i1}q^{-1} + \dots + f_{i,d-1}q^{-d+1} \qquad (19)$$

The equation (18) can be reformed as :

$$\mathrm{diag}[G_1(q^{-1})\dots G_m(q^{-1})Y(k) + B_o U(k) +$$
$$B(q^{-1})U(k-1) - Y_m(k+d) = 0 \qquad (20)$$

with $B_o = [\dots b_{ij}^{o}]$

$$B(q^{-1}) = [\dots B_{ij}(q^{-1})]$$

$$B_{ij}(q^{-1}) = \beta_{ij}^{1}q^{-1} + \dots \beta_{ij}^{n+k-1}q^{-n-k+1}$$

$$i=1,2\dots m, \qquad j=1,2,3\dots m$$

It is required that the matrix B_o is non-singular, and the control law can be given:

$$U(k) = -B_o^{-1}\{\mathrm{diag}[G_1(q^{-1})\dots G_m(q^{-1})]Y(k) +$$
$$+ B(q^{-1})U(k-1) - Y_M(k+d)\} \qquad (21)$$

Controller parameters can be direct estimated based on the process output and input at time k.

$$\Phi_i^T(k) = [y_i(k)\dots y_i(k-n+1), U^T(k), \dots$$
$$\dots, U^T(k-n-d+1)]$$

$$\Theta_i^T = [g_i^{o}\dots g_i^{n-1}, \beta_{i1}^{1}\dots\beta_{im}^{1}, \beta_{i1}^{2}\dots$$
$$\beta_{im}^{2}, \dots, \beta_{i1}^{n+k-1}\dots\beta_{im}^{n+k-1}]$$

For the closed loop identifiablty, B_o can be given by desiger.

In the case of non-minimum phase sytems, the design procedure has to be slightly modified by adding a weight matrix Q on the control cost function. In so doing, the resulting control law (STC) become:

$$U(k) = [B_o.Q]^{-1}\{Y_M(k+d) - \mathrm{diag}[G_1(q^{-1})\dots$$
$$G_m(q^{-1})]Y(k) - B(q^{-1})U(k-1)\} \qquad (22)$$

The ISTC design discussed in previor section can be extended to MIMO system by leading a time varying diagonal coffecient matrix to the controller output of equation (22).

$$U_{ISTC}(k) = U(k) * \Lambda(k) \quad (23)$$

$$\text{where} \quad \Lambda(k) = [\, 1 / \lambda_1(k) \ldots 1 / \lambda_m(k)] \quad (24)$$

The rules 1 to 3 can be used to goverer the varying of $\lambda_i(k)$. In some cases, partial of $\lambda_i(k)$ can be constant 1.

3. System Description

The process to be controlled is a 200MW oil fired boiler-turbine unit. The subsequent increase in the price of fuel oil dicated that the boiler should no longer be operated at base load, but should be flexible load and, if necessary, shut down and restarted in sliding pressure operating mode. The new operating regime would lead to time varying and nonlinear characteristics of controlled process.

The main controlled variables of a boiler turbine unit are the generated load W and steam pressure P at "before governor value". Load and pressure are affected by several inputs such as oil flow B, air-to-fuel ratio, and the position of governor value u. Consider the air flow is controlled by local loop to keep the air-to-fuel ratio, so that load W and pressure P can be controlled by two master signals, namely V and B. Thus, the problem has been reduced to a two-input / two-output problem.

Model of plant has been obtained at based load from off-line identification (H.T. Zhou, 1989).

$$\begin{bmatrix} W \\ P \end{bmatrix} = \begin{bmatrix} G_{11}(s) & G_{12}(s) \\ G_{21}(s) & G_{22}(s) \end{bmatrix} \begin{bmatrix} V \\ B \end{bmatrix} \quad (25)$$

$$G_{11}(s) = \frac{118s}{(200s+1)(6.94s+1)}$$

$$G_{12}(s) = \frac{1}{(100s+1)(20s+1)}$$

$$G_{21}(s) = -\,0.25 - \frac{0.5}{(170s+1)(10s+1)}$$

$$G_{22}(s) = \frac{0.714}{(160s+1)(9s+1)}$$

where the variables scales are processed, such as $P = \Delta P/P_S$, $P_S = 9.807 \times 10^6$ Pa, $W_S = 100$ MW, $V_S = 100\%$, $B_S = 14.286$ t/h.

Teaditional boiler-turbine coordinated control strategy relates to the inability to adapt the timing and nonlinear characteristics. The ISTC load/pressure control strategy is applied to improve the robustness of self-tuning controller and to gain steady and non-vibrating control actions.

4. Simulation Results

Process model (25) is implemented in computer program for simulating controlled plant. Refering to the equation (12), controller model structure is determented by n = 2, m = 2 and d=1. From off-line identification, parameter matrix B_0 is given:

$$B_0 = \begin{bmatrix} 0.3 & 0.0057 \\ -0.125 & 0.0049 \end{bmatrix}$$

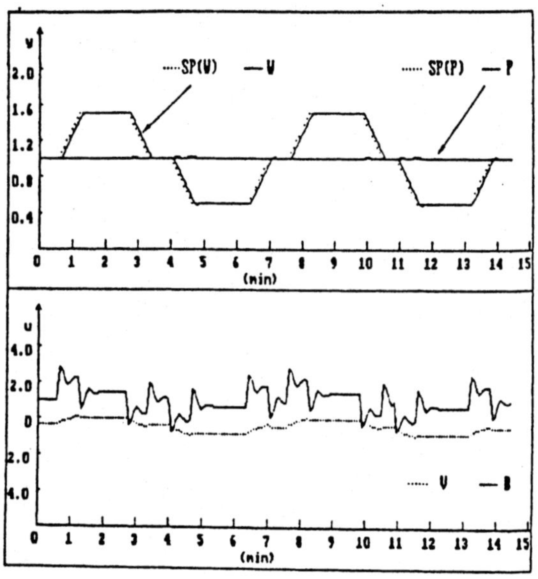

Fig.2 System response to load setpoint ($\lambda_1 = \lambda_2 = 1$)

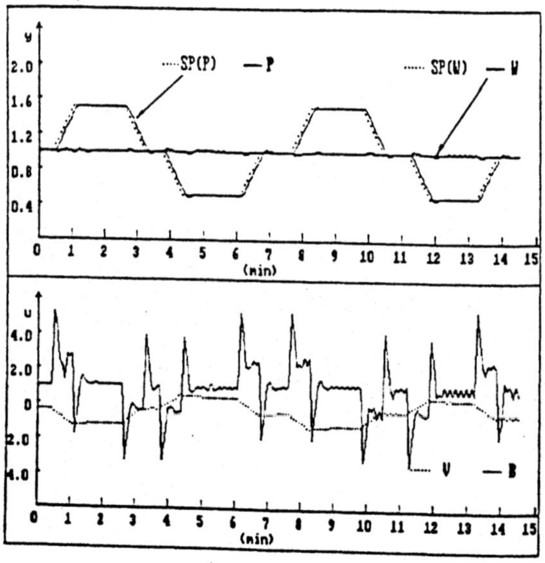

Fig.3 System response to pressure setpoint ($\lambda_1 = \lambda_2 = 1$)

The weighting matrix Q is selected as:

$$Q = \begin{bmatrix} 1 & 0 \\ 0 & 2 \end{bmatrix}$$

Some practical measures proposed by previous research (Liu et al, 1991), such as controller start-up, bumpless transfer etc, were adopted. Rule 1 and Rule 2 are used at first for leading estimator switch to controller, but let $\lambda_1(k)=\lambda_2(k)=1$. In this case, process inputs and outputs response curves at setpoints step change are shown in Fig. 2 and Fig. 3. It is obvious that satisfactory control preformance can be obtained, e.g. the decoupling ability, the dead - beat dynamic response feature, and keeping the load and pressure as close as to the desired setpoints. However, the change of fuel oil B is too vigorous at the pressure setpoint step change. It is undesirable in practical application.

By leading Rule 3, and let

$$\lambda(k) = [\ 1\ ,\ 1/\lambda_2(k)\ ]$$

With $\lambda_{2min} = 1$, $\lambda_{20} = 5$

$$c = 0.9$$

To repeat the experiments discussed above, the response curves are shown in Fig. 4 and Fig. 5. Comparing Fig. 4, Fig. 5 with Fig.3 and Fig. 4, it is apparent that control action B become acceptable in practical by leading $\lambda_2(k)$ to the controller. System output can track the setpoint with good dynamic and steady state performance. However, the dynamic limitation to the control action lead to a slight worse in decoupling ability. Simulation experiments under controller output disturbance and model parameters changing are also done. It is appropriate to design $\Lambda(k)$ by a compromise between steady controller output and system decoupling performance.

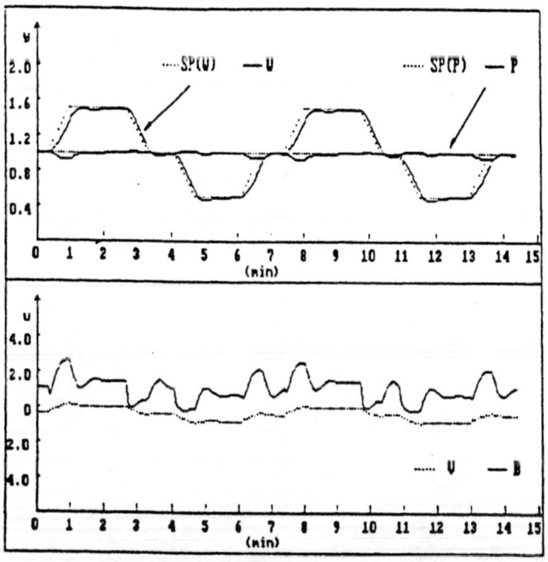

Fig.4 System response to load setpoint ($\lambda_1 = 1$)

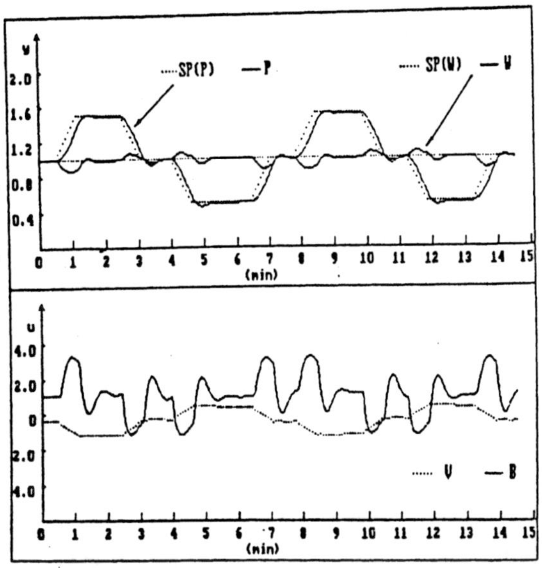

Fig.5. System response to pressure setpoint ($\lambda_1 = 1$)

5: Summary and conclutions

In this paper, a novel self -tuning control design known as ISTC was documented. Against to the practical problems of normal self-tuning control design, a time varing factor $\lambda(k)$ and estimater switch technique are applied. A set of experiential rules governed by a process suppervisor are proposed. The purpose of the new design is to improve the robusness of controller, and make the control variables more suitable in practice.

ISTC design was extended to multivariable system based on a particular process model assumption. It was applied in a 2 × 2 load/pressure control system of a 200 MW oil fired boiler-turbine unit. Simulation results expressed obvious improvement compared with traditional coodinated power unit control system in the capabilities of decoupling, setpoint following and overcoming unknown disturbance. Control variables are smooth and steady which is special required in practice. It shown the efficiency of ISTC design and the possibility of industry application.

Implementation guidelines have been developed to ensure that the ISTC design is not too complex. The rules introduced into the design are simple in nature and introduce negligible computational expense. Tuning parameters are, for the most part, intuituve and easy to select. However, a understanding of process operating condition and noise environment is necessary for control system design and on-line tuning. The stability of the overall system need further study.

Refercnces

Astrom, K.J. AND WITTENMARK, B. (1973): "On Self -Tuning Regulaters," Automatica, Vol.9, pp. 185-199.

CHIEN, I.L., SEBORG, D.E. AND MELLICHAMP, D. A. (1985): "A Self-Tuning Controller for Systems with Unknown or Varying Time Delays," Int. Jrnl. Control, Vol. 42, No. 4, pp. 949-964.

CLARK, D.W. AND GAWTHROP, P.J.(1975): "A Self -Tuning Controller," Proceedings IEEE, Vol. 122, No. 9, pp. 929-934.

Goodwin, G.C. end Long, R.S. , 1980, IEEE Trans, Autom. Control, Vol. 25, pp. 1241.

H. T. Zhou (1988): "A General Method for Coordinated Control System Designing of Tehermal Generating Unit." Proceeding of the CSEE Vol.8. No. 1.

J.Z. Liu, B.W.Surgenor and J.K.Pieper (1991) : "Practical Adaptive Level Control of A Thermal Hydranlic Process," Proceeding of IFAC ITAC'91, Singapore

SEBORG, D.E. , EDGAR, T.F. AND SHAH, S.L.(1986): "Adaptive Control Strategies for Process Control: A Survey," AICHE Journal,Vol. 32, No.6 · pp. 881 -912.

WITTENMARK, B. AND ASTROM, K.J.(1984):"Practical Issues in the Implementation of Self- Tuning Control," Automatica, Vol. 20, No. 5, pp.595-605.

SYNTHESIS OF ADAPTIVE CONTROL SYSTEMS FOR ROBOTS

V. F. Filaretov

Department of Robotics, Far Eastern Polytechnic Institute, Vladivostok, USSR

Abstract. Some new original methods of synthesis of adaptive control system and formulated regulations for choosing these methods in dependance on the features and requirements to the manipulator to be designed are considered in this paper. The peculiarity is that with the help of CAD system it is possible to fast choose one method or several methods of designing and to synthesize regulators, ensuring the given dynamic accuracy of control at a given velocity and trajectory of manipulator movement by using the simplest ones with the aim to put into practice devices or algorithms of control. The problems of identification of driver loading parameters and realization of adaptive control systems real-time are considered. The results of the usage of the system developed confirmed its high effectiveness in designing arbitrary manipulation systems.

Keywords. Adaptive control; control systems; synthesis; CAD; controllers; difference equations; invariance; drives; robots.

INTRODUCTION

To raise manipulated robot productivity or to use them in performing operations on speedy production lines it is necessary to develop such control systems that could provide the dynamic precision of contour control at significant influences between manipulator degrees of freedom (DOF). At present there are different ways of synthesis of adaptive control systems, permitting to solve the problem mentioned above (Vucobratovic, 1982), however, all of them possess different properties, different labour-consuming nature, but regulators of different practical realization complexity can be synthesized with their help. As a result, designer has a problem of choosing a method of synthesis, that, depending on manipulator configuration, its velocity movement, the type of the actuator used, requirements to the transient process quality and others, permits to synthesize such types of regulators that could ensure the definite (high) accuracy of dynamic control and could have the least complexity of practical realization.

At first, in this paper a generalized approach to the synthesis and new original methods of synthesis of adaptive control systems are considered. After that the problems of identification of loading parameters and realization of adaptive regulators at real-time are discussed.

THE CHOICE OF WAYS TO SOLVE SYNTHESIS PROBLEM

The researches carried out showed, that at present time there is not any universal approach, that could give a possibility to synthesize effective, but rather easy realized control systems for any manipulator structures. Probably, in designing manipulation systems (MS) it is expedient to have a number of methods, each of them ensuring the synthesis of effective regulators for definite class or for even DOF of MS.

The analysis done showed that for control of MS is better to use algorithms for constructing adaptive systems (AS) with the open loop of self-adapting. They are effective for synthesis of control systems with the fast change of parameters and realized with the help of rather simple technical means, however, it requires information about all current parameters of the objects controlled.

A generalized scheme of suggested ways for solving the problem of control manipulator sythesis is represented in Fig. 1. Here P_i are the generalized forces (driving torques) acting on the i-th DOF; H_{ii}, h_i, Q_i are inertial, speedy and gravitational constituents, accordingly; M_i is the action of external perturbation, being not a function of the coordinate q_i and its derivatives; $F_i(q_i)$ is nonlinear constituent P_i; $\dot{q}_i$, q_i are the velocity and acceleration of the generalized coordinate q_i, accordingly; i is tho number of the generalized coordinate.

On the basis of that scheme one of the approaches to regulator synthesis, meeting all the requirements of dynamic accuracy and quick- response with the aid of easy realized AS, is chosen.

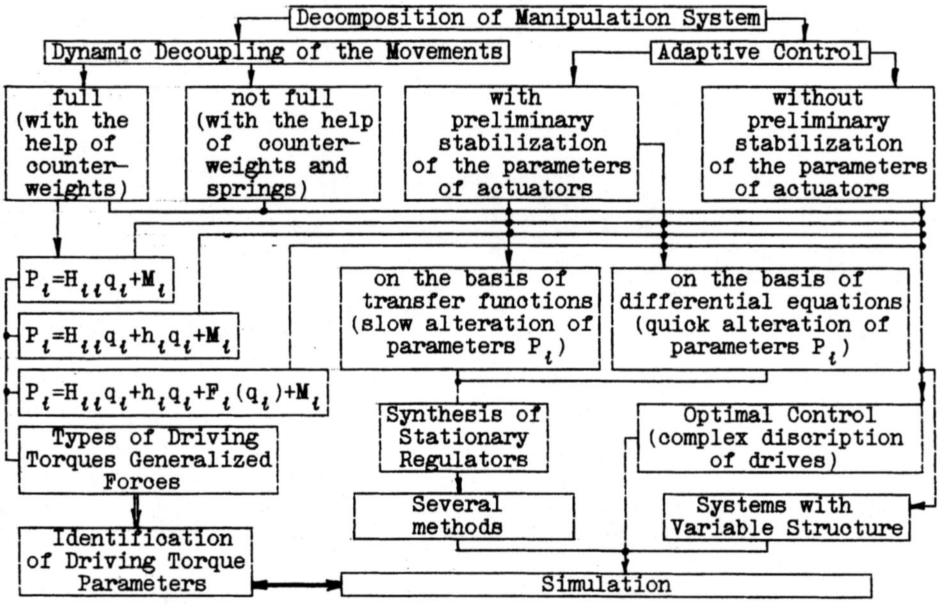

Fig. 1. A Generalized Scheme to Solve the Automated Design Problem of Effective Control Systems for any Manipulators

However, this choice depends on requirements to control dynamic accuracy and peculiarities of actuators used, interaction moment types between the DOF and velocities of their parameter changes, as well as transient process desired.

In developing AS synthesis methods the decomposition principle known was used. Due to this principle the complex MS was decomposed into several local subsystems (separate actuators), for which the regulator synthesis was realized.

However, a complete or at least dominating interaction between all the DOF was kept and taken into account. That decentralization allows to significantly simplify the regulator structure and to keep the high control performance effect.

To simplify the form P_i, and, therefore, to synthesize AS, where it is possible, it is necessary to choose either kinematic schemes of MS or with the help of automatic mechanical devices to realize a dynamic decoupling of its DOF. Several methods of automatic mechanical devices constructions are considered in this paper. They secure dynamic decoupling actuators of MS. As a result, in robot actuators the change of inertia moments and external moments in the function of load gripped mass m_g takes place only. In that case P_i takes the simplest form

$$P_i = H_{ii}\ddot{q}_i + M_i.$$

In dependance on peculiarities of MS and the type P_i in synthesizing AS two approaches to be applied are offered.

The first approach is, that, first of all, with the use of adaptive correction all the parameters of MS actuators are stabilized at the nominal level. Then, using the known methods, stationary regulators, securing the given accuracy for the actuators with the parameters having been stabilized before, are synthesized.

The second approach does not mean the preliminary stabilization of robot actuator parameters at the nominal level, and allows to elaborate regulators taking into account all current changes of those parameters. In Fig. 1 two different versions of that approach are considered.

The first version is based on the security of invariant property to the changeable parameters of load. This property is ensured by means of construction AS with variable structure, in which the sliding regimes with maximum possible velocity for the current parameters of the load appears. Thus, the monotonous character of the transient processes is ensured. The second variant is based on the assumption, that when MS moves, the control object parameters are quasi-stationary. In this case for all possible combination of changeable parameters of actuator load at the design stage the synthesis and computer storage of AS parameters are realized, that change in further changing of load parameters.

The mentioned approaches will be considered below.

METHODS OF SYNTHESIS OF ADAPTIVE CONTROL SYSTEMS

Synthesis of Adaptive Control Systems, Stabilizing Actuator Parameters at Nominal Level

The paper presents two methods of adaptive control synthesis, stabilizing actuator parameters. The first method is based on stabilization of transfer function parameters, while another one - on stabilization of differential equation parameters describing the actuators of corresponding DOF of MS (see Fig. 1).

In using the first method the hypothesis about quasi-stationary parameters of actuators is assumed. That enable to use the traditional transfer functions. In this case at the process of synthesis of AS the conception of compensation of pole parameter being changed by means of adjustment of "mobile" null parameter is used.

By introducting an adaptive corrective element with the transfer function $W_k(s)$ into the direct servodrive branch with essentially variable parameters and transfer function $W_p(s)$, the transfer function of the adaptive servodrive takes the form

$$W_{as}(s) = W_k(s)W_p(s) = \frac{k}{(T_1s+1)(T_2s+1)},$$

where k=const, T_1=const, T_2=const.

The paper presents the transfer functions obtained and on their basis structural schemes of adaptive correcting elements (ACE) are created for various d.c. electro-drives. The transfer function image and corresponding ACE are also shown here, as well as their practical realization. All these significantly depend on the value of electric time constant.

For example, when $(H^*+J)R >> (h^*+k_v)L$, $T_m >> L/R$, the transfer function of ACE for robot turn drive, kinematic scheme of which is in Fig. 2, can be written in the form

$$W_k(s) = \frac{1}{k_pk_w} \cdot \frac{(T_ms+1)}{(T_1s+1)}, \qquad (1)$$

where $T_m = (H^* + J)R \,/\, [(h^*+k_v)R + k_mk_w]$; $k_p=k_m/[(h^*+k_v)R+k_mk_w]$; $h^*=h/i^2$; $H^*=H_{tt}/i_r^2$; $H^*= H_{tt}/i_r^2$; J is inertia moment of engine rotor and rotating parts of reductor; R, L are the active resistance and inductivity, accordingly; k_w, k_m are coefficients of counter-electromotor force end torque, accordingly; k_v is the viscous friction coefficient; i_r is the reducing ratio of reductor.

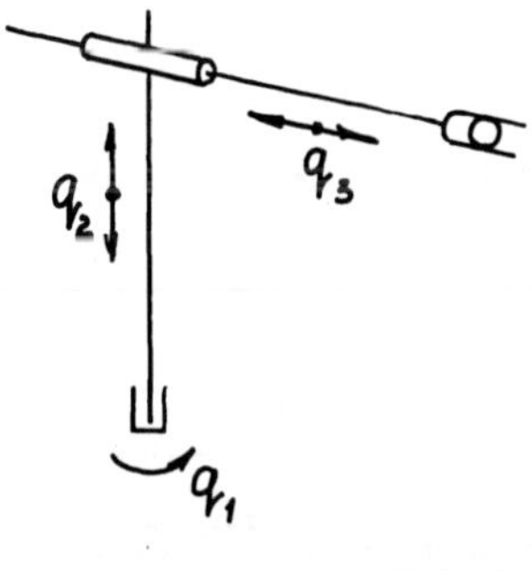

Fig. 2. Kinematic Scheme of Robot

The numbers 1 and 2 in Fig. 3 stand for the errors $\mathcal{E}$ of the servoactuator of the turning (see Fig. 2) respectively without adaptation and with the developed ACE (1), when q_1 and q_3 coordinates have been changed according to the laws

$$q_1=0.5(1-\cos(4t)), \qquad (2)$$
$$q_3=0.25(1-\cos(16t)).$$

From Fig. 3 you can see, that the developed ACE enables to considerably increase dynamic accuracy of actuators by the significant interactions between DOF.

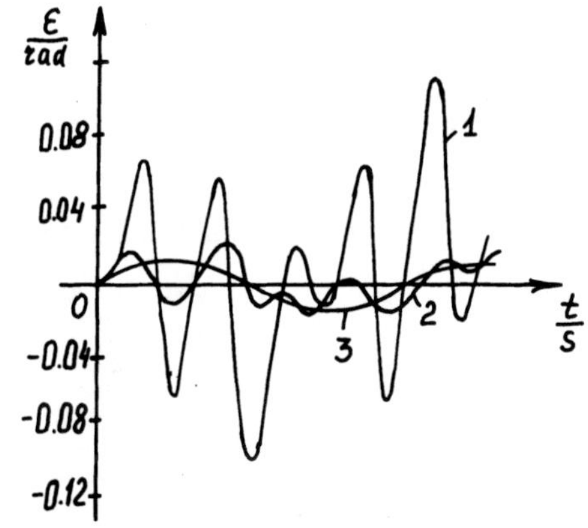

Fig. 3 Errors of Rotation Servoactuators

If the velocity of the gripper movement increases (becomes more than 1.5-2 m/s), then the transfer function will not ensure the required dynamic accuracy. In this case, to keep the required control quality is better to use the second method, that allows to stabilize the differential equation coefficients.

That is, the task is to choose such control law U_k, for which the differential equation with variable parameters, for example, for rotational servoactuator of robot (see Fig. 2)

$$L(H^*+J)i_rq_1+[R(H^*+J)+L(h^*+\dot{H}^*+k_v)]i_rq_1+ +[k_mk_w+R(h^*+k_v)+L(\dot{h}^*+\dot{k}_v)]i_r\dot{q}_1+RM_f+L\dot{M}_f= = k_uk_mU_k,$$

transforms into a differential equation with constant desired parameters in the form

$$\frac{LJ_n}{k_m}\cdot i_r\cdot \dddot{q}_1 + \frac{RJ_n}{k_m}\cdot i_r\cdot \ddot{q}_1 + k_w\cdot i_r\cdot \dot{q}_1 = k_uU,$$

where k_u is the coefficient of the power amplifier; M_f is a dry friction moment; U_k, U are signals at exit and input ACE, accordingly.

Such control law has the form

$$U_k = \frac{L(2h^*+k_v)}{k_mk_u} i_r\ddot{q}_1 + \left[\frac{k_w}{k_u}\left(1 - \frac{H^*+J}{J_n}\right) + \frac{R(h^*+k_v)+L\dot{h}^*}{k_mk_u}\right] i_r\dot{q}_1 + \frac{R}{k_mk_u} M_f + \frac{H^*+J}{J_n} U. \qquad (3)$$

Control law (3) differs from ACE (1) because it stabilizes precisely dynamic properties of drivers at any velocity of changing h^* and H^* (see curve 3 in Fig. 3).

After the stabilization of actuator manipulator parameters at nominal level one may use various stationary correcting elements with the purpose to further increase the dynamic control accuracy. The peculiarities of SCE introduction are considered with the use of ACE.

Synthesis of Adaptive System with Variable Structure

In this chapter the problem of ACE synthesis is solved. Those ACE don't allow to have overshoot at MS. In contrast to traditional approach, this paper considers AS with variable structure (ASVS), providing robot drives maximum fast action (in sliding regime) taking into account current values of load parameters. The idea of self-adaptation is that, when changing the parameters of actuators one need to change the position of switching hypersurface with the purpose of securing both the sliding regime and maximum fast action. There were attempts of creating ASVS before, however, the current information about system parameters have not been used. As a result, synthesized regulators proved to be either rather complicated or less effective.

This paper presents a new method of constructing ASVS of different types. They are designed for various operation regimes of MS. The structure complexity of ASVS being created depends on the complexity of equations, which are used to describe electrodrives, on type of moment influence, as well as, permissible value of linear zone of power amplifier.

For robot turn drive (see Fig. 2) adaptive control law has the form ($L \cong 0$)

$$U_k=(|\mathcal{E}|+k_s|M_f|+k_g|\ddot{q}_1^*+a\dot{q}_1^*|)\mathrm{sign}(S), \qquad (4)$$

where $\mathcal{E}=q_1^*-q_1$; $\ddot{q}_1^*$, $\dot{q}_1^*$ are the desired values of coordinates $\ddot{q}_1$ and $\dot{q}_1$, accordingly; $S=(\dot{q}_1^*-\dot{q}_1)+c(q_1^*-q_1)$; c=var.; $a=[k_m k_w+ R(k_v+h^*)][R(J+H^*)]^{-1}$.

For meeting requirements of existence of the sliding regime in the ASVS it is necessary to fulfil the correlations

$$\dot{c}-c^2+ac = k_m k_u[Ri_r(J+H^*)]^{-1}\cdot\mathrm{sign}(S\cdot\mathcal{E}),$$
$$k_s \geqslant R(k_u k_m)^{-1};\; k_g = Ri_r(J+H^*)(k_m k_u)^{-1}. \qquad (5)$$

In Fig. 4 are presented the modelling results of Robot (see Fig. 2) with AVSV according to control laws (4) and (5), when q_3 has been changed according to the laws 0.25(1-cos10t), 0.25(1+cos10t), 0.5(1-cos10t), 0.5(1+cos10t) (numbers 1, 2, 3 and 4, accordinqly).

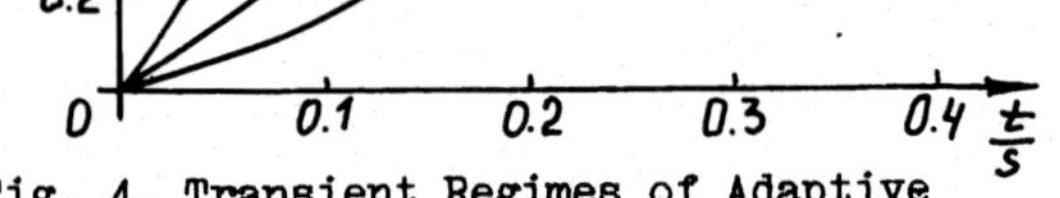

Fig. 4. Transient Regimes of Adaptive Variable Structure Systems

Mathematical simulation and experimental research showed that by introducing adaptation it is possible to increase fast action of drive ≈2 times (in sliding regime). In this case it is not required to ensure preliminary stabilization of their dynamic properties, that enable maximum usage of energetic capacity of engines with the purpose to increase fast action of MS.

Synthesis of Adaptive Control Systems According to Quadratic Criterion of Quality without Preliminary Parameter Stabilization

This approach is an approximate one, however, it is well used for general (complex) form P_i, as well as for robot drives, which are described with the help of non-linear differential equations above of the third order. In this case, the use of the methods, considered above, resulted in considerable structure complication and realization of ACE. The method suggested allows to take into account all basic non-linearities P_i.

The researches carried out showed, that d.c. electro-drives of any DOF of any manipulators taking into account or not gaps and tensions can be described in the form of matrix differential equation

$$\dot{\mathcal{E}}=A\mathcal{E}+BU+\tilde{F}(\mathcal{E})+W, \quad \mathcal{E}(t_0)=\mathcal{E}_0, \qquad (6)$$

where $A \in R^{n\times n}$ is matrix of the system dynamic properties, $B \in R^n$ is vector of amplification coefficient at control action, $U \in R^m$ is vector of control actions, $\mathcal{E} \in R^n$ is vector of errors of system phase coordinates, $\tilde{F}(\mathcal{E}) \in R^n$ is vector of any nonlinear constituents, $W \in R^n$ is vector of external perturbations.

Moreover, matrix A elements depending on the choice of drive phase coordinates are functions of separate elements of moments of interactions and, in some cases, their derivatives. It becomes the main criterion in choosing the phase coordinates, since it defines the complexity of practical realization of AS.

For the system (6) the problem of control law synthesis minimizing typical quadratic performance criterion is solved.

Synthesized drive control law ot the manipulator has the form

$$U_i = D_i^1(H_{ii},h_i)\mathcal{E}_i + D_i^n(H_{ii},h_i)\tilde{F}(\mathcal{E}_i) + D_i^v(H_{ii},h_i)W_i, \qquad (7)$$

where $D_i^1 \in R^n$, $D_i^n=D_i^v \in R^n$ are vectors of corresponding amplification coefficients.

Thus, due to expession (7) the coefficients of amplification AS for any robots depend on at maximum of 3 elements P_i. These elements are calculated in advance at the stage of regulator synthesis in the function of values H_{ii}, h_i, and are stored in computer memory in the form of tables or approximated by functional dependencies. Further, in the process of control depending on the current values of H_{ii}, h_i it is necessary to retrieve only the required amplification coefficients from computer memory.

Mathematical simulation showed that for the servoactuator of the turning (see Fig. 2), when q_1 and q_3 have been changed according to the laws (2), due to self-adaptation it is possible to increase 6-8 times the operation accuracy for MS with complex kinematic scheme.

SOLUTION OF INVERSE DYNAMIC TASK

The solution this task to form adaptive systems mentooned above are used.

This paper considers manipulators, a kind of open kinematic chains, consisting of absolutely hard members jointed by fifth class hinges. Mass of each link m_i and its tensor of inertia T_i concerning mass centre are known. Main vectors used are shown in Fig. 5, where $O_0X_0Y_0Z_0$ are absolute coordinate system connected with the manipulator platform.

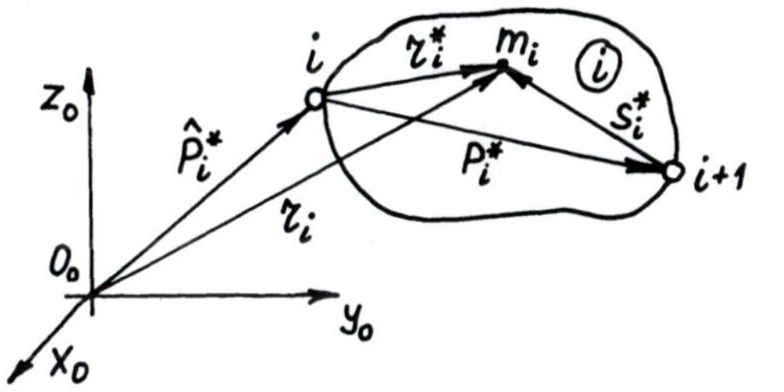

Fig. 5. Presentation of Basic Vector

In solving inverse dynamic task (IDT) all calculations are done within the coordinate system connected with the corresponding manipulator links. As a result of usage of properties of vector product and modificated systems of coordinates named Denavit-Hartenberg, as well, because of introduction of intermediate vector $\hat{P}_i^*$ on the basis of well known approaches the following calculation algorithm was finally formed

$$\omega_i = A_i^{i-1}\omega_{i-1} + \dot{q}_i l_i \bar{\delta}_i, \ \omega_0=0, \quad (i=\overline{1,n}),$$

$$\dot{\omega}_i = A_i^{i-1}\dot{\omega}_{i-1} + [(A_i^{i-1}\omega_{i-1})\mathbf{x}(\dot{q}_i l_i) + \ddot{q}_i l_i]\bar{\delta}_i, \ \dot{\omega}_0=0, \quad (i=\overline{1,n}),$$

$$\ddot{\hat{P}}_i^* = A_i^{i-1}(\ddot{\hat{P}}_{i-1}^* + \Omega_{i-1}P_{i-1}^*) + (2\dot{q}_i\omega_i \mathbf{x} l_i + q_i l_i)\delta_i, \ \ddot{\hat{P}}_i^*=-g, \ (i=\overline{1,n}), \tag{8}$$

$$\ddot{r}_i = \ddot{\hat{P}}_i^* + \Omega_i r_i^*, \ (i=\overline{1,n}),$$

$$\hat{f}_i = A_i^{i+1}\hat{f}_{i+1} + m_i\ddot{r}_i - \tilde{F}_i, \ \hat{f}_{n+1}=0, \quad (i=\overline{n,1}),$$

$$\hat{n}_i = A_i^{i+1}\hat{n}_{i+1} + P_i^* \mathbf{x}(A_i^{i+1}\hat{f}_{i+1}) + r_i^* \mathbf{x}(m_i\ddot{r}_i) + \tau_i\dot{\omega}_i + \omega_i \mathbf{x}(\tau_i\omega_i) - c_i^* \mathbf{x}\tilde{F}_i - R_i, \ (i=\overline{n,1}),$$

$$P_i = l_i^T(\hat{f}_i\delta_i + \hat{n}_i\bar{\delta}_i).$$

At present this algorithm possesses the least computational complexity among all known algorithms (n_m=95n-101, n_a=84n-97), where n_m, n_a are the number of multiplication and addition operations, correspondently; n is the number of DOF of manipulator; $A_i^j \in R^{3x3}$ is the direction cosines matrix; $\omega_i \in R^3$, $\dot{\omega}_i \in R^3$ are angular velocity and angular acceleration of the link i, respectively; δ_i=1, if the joint i is prismatic and δ_i=0, if it is revolving; $l_i=[0\ 0\ 1]^T$ is vector directed along the joint i axis; $g \in R^3$ is gravitational acceleration (given in absolute system of coordinates); $\hat{f}_i \in R^3$, $\hat{n}_i \in R^3$ are reaction force and reaction moment, acting at joint i, respectively; $\tilde{F}_i \in R^3$, $R_i \in R^3$ are the main vector and the main moment of external forces, respectively (without gravitation), acting on the link i; $c_i^* \in R^3$ is the vector of the point of acting vector $\tilde{F}_i$ according to joint i,

$$\Omega = \begin{bmatrix} -(\omega_2^2+\omega_3^2) & \omega_1\omega_2-\dot{\omega}_3 & \omega_1\omega_3+\dot{\omega}_2 \\ \omega_1\omega_2+\dot{\omega}_3 & -(\omega_1^2+\omega_3^2) & \omega_2\omega_3-\dot{\omega}_1 \\ \omega_1\omega_3-\dot{\omega}_2 & \omega_2\omega_3+\dot{\omega}_1 & -(\omega_1^2+\omega_2^2) \end{bmatrix}$$

(in expression for Ω index i is omitted and lower indices show at the number of elements of corresponding vectors).

The problem of identifying unknown m_r is solved on the basis of developed algorithm IDT.

The idea of approach is that to separate terms with unknown m_r in common mathematical relations. Then, using the information on moments in corresponding drives and moments which would take place if there were no any load, calculation m_r is done due to information either from one drive only or (more exactly) from the drives k of the DOF. If at the beginning of manipulator operation short duration stops are possible, the simpler method of identification m_r is suggested.

On the basis this algorithm for solving IDT yhe problem of determining the individual components of moment action at real-time in adaptive systems can be solved.

Belov the solution this problem with the help of simple technical means are discussed.

DETERMINING THE INDIVIDUAL COMPONENTS OF MOMENT ACTIONS

It is not difficult to show that, if

1. all linkes and axes of all prismatic DOF of the manipulator lie in the same plane in which lies the first joint axis,

2. the axis of all revolute DOF (except the first one), are ortogonal to the mentioned plane,

3. links are axially symmetrical,

then generalized moment P_1 has the form

$$P_1 = H_{11}\ddot{q}_1 + h_1\dot{q}_1 + G_1(q), \ h_1 = \frac{dH_{11}}{dt} \tag{9}$$

or

$$P_1 = \frac{d}{dt}(H_{11}\,\dot{q}_1) + G_1(q), \qquad (10)$$

where $G_1(q)$ is the gravitational force. From (10) it follows that

$$H_{11} = \dot{q}_1^{-1} \int_{t_0}^{t} (P_1 - G_1)dt. \qquad (11)$$

By knowing H_{11} from (9) it is easy to determine h_1, provided $G_1(q)$ is known and P_1, $\dot{q}_1$ and q_1 are measured. If the gravitational component $G_1(q)$ is also annuled, this is the case with manipulation mechanism in which the rotational DOF have vertical axes, end the robots, working in weightless state, eq. (11) is even more simplified

$$H_{11} = \dot{q}_1^{-1} \int_{t_0}^{t} P_1 dt \qquad (12)$$

and the value h_1, due to (9) is determined from the relation

$$h_1 = \dot{q}_1^{-1}(P_1 - H_{11}\ddot{q}_1). \qquad (13)$$

If term $G_1(q) \neq 0$, then, for its determination effective algorithm for solving the IDT shown before should be used, setting $\ddot{q}_1 = \dot{q}_1 = 0$.

In the given case, the following relations hold:

$$\ddot{r}_i = A_i^{i-1}\,\ddot{r}_{i-1}, \qquad \ddot{r}_0 = -g, \quad (i=\overline{1,n}),$$

$$\hat{f}_i = A_i^{i+1}\,\hat{f}_{i+1} + m_i\ddot{r}_i, \quad (i=\overline{n,1}), \qquad (14)$$

$$\hat{n}_i = A_i^{i+1}\hat{n}_{i+1} + P_i^{*}x(A_i^{i+1}\,\hat{f}_{i+1}) + r_i^{*}x(m_i\ddot{r}_i),$$

$$n_{i+1} = 0, \quad (i=\overline{n,1}),$$

$$G_i = l_i^{T}(\hat{f}_i\delta_i + \hat{n}_i\bar{\delta}_i), \quad (i=\overline{n,1}).$$

Calculation of G_1 according to algorithm (14) for a case, when the axes of the neigboring joints are parallel or orthogonal (most frequent case), requires (29n-20) multiplications and (19n-19) additions/sub tractions.

In this way, for a continuous determination of the parameters H_{11} and h_1 for use in self-adjusting control systems of manipulators, it is necessary to fulfill the following:

1. Determine G_1 according to algorithm (14),

2. Determine H_{11} from relation (11) or (12),

3. Determine h_1 from relation (9) or (13).

The general structural scheme for determining the values H_{11}, h_1 at real time is given in Fig. 6.

It should be noted that in determining the parameters H_{11} and h_1 it is necessary to exclude the possiblity of division by $\dot{q}_1 = 0$ (Fig. 6). The mentioned situation can be excluded, if the following known

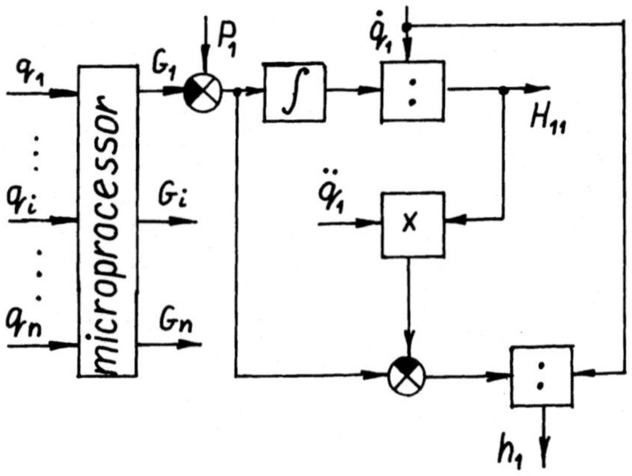

Fig. 6. Determining the Values H_{11}, h_1

procedure is used. For $|\dot{q}_1| < \mathcal{E}$, where $\mathcal{E}$ is a small value chosen in advance, the calculation process of H_{11} and h_1 is stopped and the last values of H_{11} and h_1 are memorized, and remain unchanged till the time instant when the condition $|\dot{q}_1| > \mathcal{E}$ is fulfilled. It is clear that for $|\dot{q}_1| \leqslant \mathcal{E}$ the current values of H_{11} and h_1 can be changed and can differ from the memirized ones, i.e. from those used for forming the control signals in the considered situation. However, the working regimes, for which $|\dot{q}_1| \leqslant \mathcal{E}$ are, as a rule, of short duration, if, of course, the actuators are not in the blocked state, when there is no need for their control. For small time intervals in the case of manipulation mechanism motion, the values H_{11} and h_1 do not change significantly and, the investigations have shown that the calculation scheme proposed for the working regimes ensures the obtaining of fully acceptable identification results for H_{11} and h_1.

CONCLUSION

The results of practical usage of system developed confirmed its effectiveness in designing manipulators of various purposes. Along with it, new methods of synthesis of adaptive control systems allowed to considerably increase dynamic control accuracy of any manipulators at high velocities of their operations. Moreover, the synthesized adaptive control systems ensure not only high control quality, but have the least complexity of practical realization.

The developed algorithm to solve inverse dynamic task and technical means allow to ensure identification of parameter variables of drives loaded at real time. It is necessary for practical use of the synthesized adaptive regulators.

REFERENCES

Vucobratovic M., D.Stokic (1982). Scientific Foundations of Robotics, Vol 2. Control of Manipulation Robots. Springer-Verlag.

A SIMPLIFIED ROBUST PREDICTIVE CONTROLLER IN THE PRESENCE OF EXOGENOUS BOUNDED DISTURBANCES

Guo-Hua Wu and Xing-Yuan Gu

Dept. of Automatic Control
Northeast Univ. of Technology
Shenyang. 110006, P.R. China

ABSTRACT. It is shown that a simple adaptive predictive controller can be designed based on a strongly simplified model in the presence of exogenous disturbances. The adaptive predictive controller requires to know some parameters which are easy to choose from the impulse response of the open-loop system by simple experiment and has only one tuning parameter. Moreover the asymptotic stability and convergence of resulting closed-loop system are established and the upper bound of tracking error is also given.

KEY WORDS. self-tuning; predictive control; impulse response; identification.

INTRODUCTION

In the process control field, since complex processes can be controlled surprisingly well by simple regulators. It is convenient and useful to approximate a real but unknown plant by a strongly simplified model, In this regard, Astrom (1980) has proposed a new design method for any unknown single input single output stable linear time-invariant system with a monotone step response. Lu and Kumar (1984) have extended Astrom's approach and given corresponding results for the multivariable system.

However, For Astrom's approach, the sampling period might be quite long for some plants resulting in poor controlled response. For Lu's method, the number of tuning parameters in the controller might be too more to self-tuning control for real-time control.

The basic idea of the paper is that the controller is designed for a strongly simplified model of the process, which is approximated triangle impluse response model. The controller has only one tuning parameter so that there exist plant modelling error. At the same time, there exist exogenous disturbances. It is useful and necessary to distinguish the one from another. So the parameter estimation schme used here updates conditionally the parameter estimates and prevents effectively the "blowing up" of the covariance matrix.

TRIANGLE MODEL AND DESIGN OF EXTENDED HORIZON ADAPTIVE PREDICTIVE CONTROLLER

Consider a stable system with positive impulse responses, See Fig.1, The system is approximated with the following triangle crude model.

$$y(t) = \sum_{i=1}^{\infty} h(i)u(t-i) + d(t)$$

$$= \sum_{i=1}^{N} h(i)u(t-i) + d(t) \qquad (1)$$

where d(t) is a zero-mean bounded disturbance. Suppose M to be a known upper bound of $|d(t)|$, i.e.

$$\sup |d(t)| < M$$

and

$$h(i) = \begin{cases} H\ i/m & ,\ 0 < i < m \\ H\ (N-i)/(N-m) & ,\ m < i < N \\ 0 & ,\ i > N \end{cases} \quad (2)$$

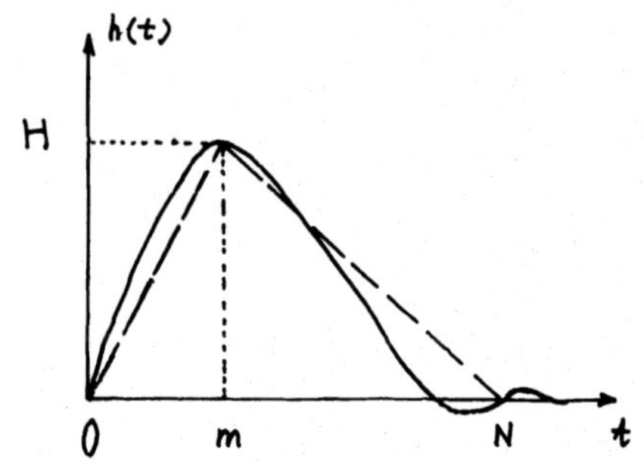

Fig.1 Impulse response and its approximation model

Note that m is selected with respect to the rule which h(m) should near to maximum of impulse responses by experiment.
By (1) and (2) one has

$$y(t) = H\left[\frac{1}{m}\sum_{i=1}^{m} i\ u(t-i) + \frac{1}{N-m}\sum_{i=m+1}^{N}(N-i)u(t-i)\right] + d(t) \quad (3)$$

The extended input $w(t)$ is defined as

$$w(t) = \left[\frac{1}{m}\sum_{i=1}^{m} i\ u(t-i) + \frac{1}{N-m}\sum_{i=m+1}^{N}(N-i)\ u(t-i)\right] + d(t) \quad (4)$$

So the equation (1) is expressed as following

$$y(t) = H\ w(t) + d(t) \quad (5)$$

It is shown that only one parameter requires to be estimated.We can design the controller based on this strongly simplified model of the process.
In the extended horizon predictive control, the objective to keep the k-step ahead prediction of the plant output as close as possible to a given setpoint $y^*$is pursued by designing $u(t)$ under the assumption $u(t)=u(t+1)=....=u(t+k-1)$, Then from (3), it follows that (Scattolini and Bittanti, 1991).

$$y(t+k) = H\left[\frac{1}{m}\sum_{i=1}^{m} i\ u(t+k-i) + \frac{1}{N-m}\sum_{i=m+1}^{N}(N-i)u(t+k-i)\right] + d(t+k) \quad (6)$$

Generally speaking, $k>m$, so one has

$$\hat{y}(t+k) = H\left[\frac{m+1}{2} + \frac{(k-m)(2N-k-m-1)}{2(N-m)}\right]u(t) + \frac{H}{N-m}\sum_{i=k+1}^{N}(N-i)u(t+k-i) \quad (7)$$

By setting $\hat{y}(t+k)=y^*(t+k)$, the receding horizon control law is

$$u(t) = \frac{2(N-m)/H}{(m+1)(N-m)+(k-m)(2N-k-m-1)}\left[y^*(t+k) - \frac{H}{N-m}\sum_{i=1}^{N-k}(N-k-i)u(t-i)\right] \quad (8)$$

By using the RLS algorithm of Gu and Shao(1989) as following:

$$H(t) = H(t-1) + \frac{\lambda(t)P(t-2)w(t-1)}{\mu + P(t-2)w^2 t-1}e(t) \quad (9)$$

$$e(t) = y(t) - w(t-1)H(t-1) \quad (10)$$

$$P(0) = k_0$$

$$P(t-1) = P(t-2) - \frac{\lambda(t)P^2(t-2)w^2(t-1)}{\mu + P(t-2)w^2(t-1)} \quad (11)$$

$$\lambda(t) = \begin{cases} 0 & \text{if } |e(t)| < 2M \\ v & \text{otherwise } v \in [\varepsilon, 3(1-\varepsilon)/4] \\ & 9/25 < \varepsilon < 3/7 \end{cases} \quad (12)$$

Remark 1: The choice of v guarantees the existence of $u(t)$ and BIBO stability of the closed-loop system.
Remark 2: The recursive least-squares estimation scheme suggested here prevents effectively

the "blowing-up" of the covariance matrix.

Remark 3: The μ is independent of the stability of the closed-loop system and can be chosen according to the convergence rate the parameter estimations.

The above considerations motivate the following step of adaptive predictive control algorithm:

Step 1 : Compute extended input $w(t)$ using (4)

Step 2 : Estimate on-line the impulse response H using (9)-(12).

Step 3 : Compute the approximation $h(\cdot)$ using (2).

Step 4 : Compute the control law $u(t)$ from (8)

Step 5 : Let k=k+1 and repeat.

STABILITY AND CONVERGENCE ANALYSES

It will now be investigated what happens when the extended horizon adaptive controller which is designed using the simplified model, is applied to a stable system, The following result holds:

Theorem 1 : Consider a stable linear system with monotone step response, provided that

$$k > N - \sqrt{N(N-m)/2} \tag{13}$$

Is satisfied, the extended horizon adaptive predictive controller when applied to the system (1) leads to

S1: $\{u(t)\}$ and $\{y(t)\}$ are bounded for all t.

S2: exists $T < \infty$, for $t > T$, $|y(t)-y^*(t)| < 2M$

Proof:

1) The closed-loop system is thus characterized by equations (8) and (1). Equation (8) gives

$$H\left[\frac{N+2Nk-k^2-k-Nm}{2(N-m)} + \frac{1}{N-m}\sum_{i=k+1}^{N}(N-i)z^{-i}u(t)\right] = y^*(t+k) \tag{14}$$

It follows from a theorem of Wiener that the closed-loop system is asymptotically stable if the function

$$A(z) = \sum_{n=0}^{\infty} a_n z^{-n}$$

has the property

$$\inf |A(z)| > 0 \quad \text{for } |z| > 1$$

Note that

$$\frac{N+2Nk-k^2-k-Nm}{2(N-m)} > \frac{(N-k)^2-(N-k)}{2(N-m)}$$

so $\quad 2(N-k)+N(N-m) > N(N-m) > 2(N-k)^2$

i.e. $\quad k > N - \sqrt{N(N-m)/2}$

It is shown that if the condition (13) is satisfied , the resulting closed-loop system is aymptotically stable . we prove $\{u(t)\}$ is bounded. Certainly $\{y(t)\}$ is too bounded.

2) The key equations are as following and define

$$\hat{H} = H(t-1) \qquad \bar{H} = H(t-k-1) \tag{15}$$

$$e(t) = y(t) - \hat{H}w(t-1)$$

$$= y(t) - \hat{H}\left[\frac{1}{m}\sum_{i=1}^{m} i\,u(t-i) + \frac{1}{N-m}\sum_{i=m+1}^{N}(N-i)u(t-i)\right] \tag{16}$$

$$0 = y^*(t) - \bar{H}\left[\frac{m+1}{2} + \frac{(k-m)(2N-m-k-1)}{2(N-m)}\right]u(t-k) + \frac{\bar{H}}{N-m}\sum_{i=K+1}^{N}(N-i)u(t-i) \tag{17}$$

By subtracting (17) from (16) one obtains

$$e(t) = y(t) - y^*(t)$$

$$+ (\bar{H}-\hat{H})\frac{1}{N-m}\sum_{i=k+1}^{N}(N-i)u(t-i)$$

$$+ (\bar{H}-\hat{H})\left[\frac{m+1}{2} + \frac{(k-m)(2N-k-m-1)}{2(N-m)}\right]u(t-k)$$

$$+\hat{H}\left[\frac{1}{m}\sum_{i=1}^{m} i z^{-k} - \frac{1}{m}\sum_{i=1}^{m} i z^{-1} u(t)\right.$$

$$\left. + \hat{H}\left[\frac{1}{N-m}\sum_{i=m+1}^{k} (N-i) z^{-k} - \frac{1}{N-m}\sum_{i=m+1}^{k} (N-i) z^{-1} u(t)\right]\right. \qquad (18)$$

The identification parameter satisfy the following Lemma 1 (Goodwin and Sin, 1984)

Lemma 1 The estimation scheme (9)-(12) has the following properties

(1) $$\lim_{t\to\infty} [\, h(t) - h(t-k) \,] = 0 \qquad (19)$$

(2) $$\lim_{t\to\infty} \frac{\lambda(t) e^2(t)}{\mu + P(t-2) w^2(t-1)} = 0 \qquad (20)$$

(3) $V(t)-V(t-1) < 0$
where $V(t)=P(t-1)h^2(t)$ (21)

From (20) and according to S1, it follows

$$\lim_{t\to\infty} \lambda(t) e^2(t) = 0 \qquad (22)$$

Define $H= \{t \mid \lambda(t) \neq 0 \; t \in N\}$. We can prove that H must be a finite set. Otherwise a series of $\{t_n\}$ can be obtained and $t_n (n\to\infty)$. From (22) it follows readily that

$$\lim_{n\to\infty} e(t_n) = 0 \qquad (23)$$

Then when n is large enough, it must be

$$| e(t) | < 2M$$

From (12) it follows that $\lambda(t_n) = 0$. This contradicts the definition of H. Since H is a finite set, there exists $t1 > 0$ and when $t > t1$, $\lambda(t)=0$, Then when $t > t1+k$, it yields

$$h(t-k)=h(t-1)= h(0) \quad \text{(const vector)}$$

Let $T=t1+k$ and from (18)

$$| y(t)-y^*(t) | < 2M \quad t>T \qquad (24)$$

Thus we complete the proof.

CONCLUSION

This paper gives a very simple way of designing a robust controller for stable systems in the presence of exogenous disturbance and it overcomes the drawbacks of Astrom and Lu's methods. It is shown that this adaptive controller can distinguish between the exogenous disturbance and plant modelling error.

REFERENCE

Astrom, K.J. (1980). A robust sampled regulator for stable systems with monotone step responses, Automatica, Vol.16, 313-315.

Goodwin, G.C. and K. S. Sin (1984). Adaptive filtering, prediction and control, Prentice-hall, Inc.

Gu, X. Y. and C. Shao (1989). A self-tuning controller using conditional updating techniques, ACTA Automatica SINICA. Vol.15, 232-241.

Lu, W.S. and K.S.Kumar (1984). A staircase model for unknown multivariable systems and design of regulators, Automatica, Vol.20, 109-112.

Scattolini, R and S.Bittanti (1990). On the choice of the horizon in long-range predictive control--- some simple criteria, Automatica, Vol.26, 915-917.

A Hybrid Method for Real-time Simulation of Continuous-time Bilinear Systems

X. Yin and H. Schwarz

Department of Measurement and Control
(Prof. Dr. -Ing. H. Schwarz)
Faculty of Mechanical Engineering
University of Duisburg, P.O. Box 101503
W-4100 Duisburg, Germany

Abstract. In this paper we present a method for the simulation of continuous-time bilinear systems $\dot{\boldsymbol{x}}(t) = \boldsymbol{A}\boldsymbol{x}(t) + \boldsymbol{N}\boldsymbol{x}(t)u(t) + \boldsymbol{b}u(t)$, that can be used to realize bilinear observers for real-time control systems. Our method attempts to approximate the state variable $\boldsymbol{x}(t)$ in the bilinear term $\boldsymbol{x}(t)u(t)$ with an interpolation polynomial and treats this term as an independent input signal, then solves the bilinear state space equation like a linear system equation. The method yields, especially for large integrating step sizes, a more exact solution than a 4th order Runge-Kutta method. Even if the Runge-Kutta method becomes unstable the method works well.

Keywords. Bilinear control, Numerical method, Observers, Interpolation.

INTRODUCTION

Bilinear systems are a special class of nonlinear systems which have numerous applications in engineering, biology, and economics (Mohler 1970, Espana and Landau 1978, Guo and Schwarz 1990). Control of bilinear systems with state feedback often involves the problem that the state vector of a process has to be determined from input/output measurement. This requires an observer for bilinear systems. Analogous to the Luenberger-Observer for linear systems the observers for bilinear systems often have a bilinear form (Grasselli and Isidori 1981).

The implementation of a bilinear observer on a real-time computer system deals with the numerical solution of a bilinear state space differential equation. Classical numerical integrating methods such as Euler and Runge-Kutta algorithms may yield relative large errors or even become unstable for large integrating step sizes. Reducing the integrating step sizes will improve the results, but the computational burden increases and may exceed the restriction of the maximal admissible sampling interval. When controlling a continuous system with a digital computer the input remains constant during one sampling period and we can solve the bilinear system equation exactly for each time point. This method avoids the disadvantages of Euler and Runge-Kutta methods with large integrating step-sizes, but computational burden to calculate the transition matrix for each step is too large.

In this paper we present a new method for real-time simulation of bilinear systems without the disadvantages mentioned above. The proposed method requirse the determination of some constant matrices only once at the beginning of the simulation.

The remainder of this paper is organized as follows. In Section 2 we discuss the problem formulation and solution strategy. Subsequently, in Section 3 we present the simulation algorithms. We then provide in Section 4 a preliminary error analysis of the method. In Section 5 we present numerical examples, and finally in Section 6 we draw some conclusions.

SOLUTION STRATEGY

We consider the following continuous-time bilinear system:

$$\begin{aligned} \dot{\boldsymbol{x}}(t) &= \boldsymbol{A}\boldsymbol{x}(t) + \boldsymbol{N}\boldsymbol{x}(t)u(t) + \boldsymbol{b}u(t) \quad (1) \\ y(t) &= \boldsymbol{c}^T\boldsymbol{x}(t) \\ \boldsymbol{x}(0) &= \boldsymbol{x}_0 \end{aligned}$$

where $\boldsymbol{x}(t) \in \boldsymbol{R}^n$ is the state variable, $u(t) \in \boldsymbol{R}^1$ is the input variable, $\boldsymbol{A}, \boldsymbol{N}, \boldsymbol{b}$ and $\boldsymbol{c}^T$ are system matrices with appropriate dimensions and $\boldsymbol{x}_0$ is the initial state.
It is assumed that the input $u(t)$ remains constant during the sampling period:

$$\begin{aligned} u(t) &= u(t_k), \quad (2) \\ t_k &\leq t < t_{k+1}, \quad k = 0, 1, 2, \cdots, \end{aligned}$$

where $t_{k+1} = t_k + h$ and h is the sampling period, which can be regarded as the integrating step size. In the following $x(t_k)$ and $u(t_k)$ are abbreviated to x_k and u_k respectively.

The task now is to determine the state variable $x(t)$ at the time point t_{k+1} from the previously determined state vector x_k and the known input u_k.

Our main idea is that the state variable $x(t)$ in the bilinear term $x(t)u(t)$ in the system Eq.(1) is treated as an independent input. Then the bilinear system can be seen as a time invariant linear system in the interval $t_k \le t < t_{k+1}$ and has a solution

$$x_{k+1} = e^{Ah}x_k + \int_{t_k}^{t_{k+1}} e^{A(t_{k+1}-t)}bu_k dt + \int_{t_k}^{t_{k+1}} e^{A(t_{k+1}-t)}Nx(t)u_k dt. \quad (3)$$

where e^{At} is the transiton matrix. With the substitute $\tau = t - t_k$, we have

$$x_{k+1} = e^{Ah}x_k + \int_0^h e^{A(h-\tau)}bu_k d\tau + \int_0^h e^{A(h-\tau)}Nx(\tau + t_k)u_k d\tau. \quad (4)$$

The first integral in (4)

$$I_1 = \int_0^h e^{A(h-\tau)}bu_k d\tau \quad (5)$$

has an explicit solution. The second integral in (4)

$$I_2 = \int_0^h e^{A(h-\tau)}Nx(\tau + t_k)u_k d\tau \quad (6)$$

involves an unknown variable $x(\tau + t_k)$. We assume that the state variable at the time point t_{k+1} has a value of x_{k+1}, then using the system Eq.(1) we can calculate $\dot{x}_k$ and $\dot{x}_{k+1}$. With x_k, $\dot{x}_k$, x_{k+1} and $\dot{x}_{k+1}$ we are able to construct a 3rd order polynomial of τ in the interval $0 \le \tau < h$ to approximate the unknown state variable $x(\tau + t_k)$. Thus, we solve the second integral I_2 explicitly and transform the integral Eq.(4) into a linear algebraic equation in x_{k+1}. Solving this linear algebraic equation we can determine an approximated value of the state variable at the time point t_{k+1}.

ALGORITHM DESCRIPTION

Consider a polynomial of the form

$$p(\tau) = \alpha\tau^3 + \beta\tau^2 + \gamma\tau + \delta \quad (7)$$

Our objective here is to approximate the state variable $x(\tau + t_k)$ with this polynomial in the interval $0 \le \tau < h$

$$x(\tau + t_k) \approx \alpha\tau^3 + \beta\tau^2 + \gamma\tau + \delta \quad (8)$$
$$0 \le \tau < h,$$

where $\alpha, \beta, \gamma, \delta \in R^n$ are coefficient vectors and $\tau = t - t_k$. Clearly, the polynomial (7) must satisfy

$$\begin{aligned} p(0) &= x_k \quad \dot{p}(0) = \dot{x}_k \\ p(h) &= x_{k+1} \quad \dot{p}(h) = \dot{x}_{k+1}. \end{aligned} \quad (9)$$

From (7) , (9) and the system equation (1), after some straightforward manipulations, we get

$$\begin{aligned} \alpha &= (\frac{A + Nu_k}{h^2} - \frac{2E}{h^3})x_{k+1} + \\ &+ (\frac{2E}{h^3} + \frac{A + Nu_k}{h^2})x_k + \frac{2b}{h^2}u_k \\ \beta &= (\frac{3E}{h^2} - \frac{A + Nu_k}{h})x_{k+1} + \\ &- [\frac{2(A + Nu_k)}{h} + \frac{3E}{h^2}]x_k - \frac{3b}{h}u_k \\ \gamma &= \dot{x}_k \\ \delta &= x_k \end{aligned} \quad (10)$$

where $E \in R^{n \times n}$ is the identity matrix.
Now the second integral in (4) can be approximated as

$$\begin{aligned} I_2 &\approx \int_0^h e^{A(h-\tau)}Np(\tau)u_k d\tau + \\ &= \int_0^h e^{A(h-\tau)}\tau^3 d\tau N\alpha u_k + \\ &+ \int_0^h e^{A(h-\tau)}\tau^2 d\tau N\beta u_k + \\ &+ \int_0^h e^{A(h-\tau)}\tau d\tau N\gamma u_k + \\ &+ \int_0^h e^{A(h-\tau)}d\tau N\delta u_k. \end{aligned} \quad (11)$$

We define

$$\begin{aligned} \Phi &= e^{Ah} = \sum_{i=0}^{\infty} \frac{(Ah)^i}{i!} \\ \Psi_l &= \int_0^h e^{A(h-\tau)}\tau^l d\tau, \quad l = 0, 1, 2, 3. \\ \psi &= \Psi_0 b. \end{aligned} \quad (12)$$

Applying partial integration yields

$$\Psi_l = l! \sum_{i=0}^{\infty} \frac{A^i h^{i+l+1}}{(i + l + 1)!}. \quad (13)$$

Using (4),(11) and (13), we obtain

$$x_{k+1} = e^{Ah}x_k + \psi u_k + (\Psi_0 N\delta + \Psi_1 N\gamma + \Psi_2 N\beta + \Psi_3 N\alpha)u_k \quad (14)$$

Collecting the terms which involve x_{k+1} in α and β together, we have

$$\begin{aligned} [E - (Q_1 &+ Q_2 u_k)u_k]x_{k+1} = \Phi x_k + \\ +[\psi &+ (Q_3 + Q_4 u_k)x_k + q_5 u_k]u_k, \end{aligned} \quad (15)$$

where

$$\begin{aligned} Q_1 &= \frac{\Psi_2 N(3E - Ah)}{h^2} + \frac{\Psi_3 N(Ah - 2E)}{h^3} \\ Q_2 &= (\frac{\Psi_3}{h^2} - \frac{\Psi_2}{h})N^2 \end{aligned}$$

$$Q_3 = \Psi_0 N + \Psi_1 NA - \frac{\Psi_2 N(3E+2Ah)}{h^2} + \frac{\Psi_3 N(Ah+2E)}{h^3}$$

$$Q_4 = (\Psi_1 - \frac{2\Psi_2}{h} + \frac{\Psi_3}{h^2})N^2$$

$$q_5 = (\Psi_1 - \frac{3\Psi_2}{h} + \frac{2\Psi_3}{h^2})Nb. \qquad (16)$$

Defining

$$R_k = E - (Q_1 + Q_2 u_k)u_k \qquad (17)$$

$$r_k = \Phi x_k + [\psi + (Q_3 + Q_4 u_k)x_k + q_5 u_k]u_k$$

we rewrite (15) as

$$R_k x_{k+1} = r_k. \qquad (18)$$

Solving x_{k+1} from this equation, we obtain an approximated solution of (1) for the time point t_{k+1}. Thus, the original problem (4) reduces to solving a linear algebraic equation. Notice that the matrices Φ and Q_i, $i = 1, 2, \cdots, 4$ and the vectors ψ, q_5 are independent from t, so they need to be determined only once at the beginnig of a simulation. The coefficient vectors α, β, γ and δ vanish in Eq.(15), so we only need to compute R_k and r_k directly with (17) to determine x_{k+1}.

ERROR ANALYSIS

In fact, the approximation of the state vector $x(t)$ with the polynomial (7) is a Hermite interpolation problem (Stoer and Bulirsch, 1980). For the interpolation error we have following theorem :

Theorem 1: If the state vector $x(t)$ is at least 4 times differentiable on the interval $\tau \in [0, h]$, then to every $\bar{\tau} \in [0, h]$ there exists a vector ξ : $\xi_i \in [0, h]$, such that

$$x(\bar{\tau} + t_k) - p(\tau) = \frac{x^{(4)}\{\xi\}\omega(\bar{\tau})}{24}, \qquad (19)$$

where

$$\omega(\tau) = \tau^2(\tau - h)^2$$

$$\xi \equiv [\xi_1 \ \xi_2 \cdots \xi_n]^T \qquad (20)$$

$$x^{(4)}\{\xi\} \equiv [x_1^{(4)}(t_k+\xi_1) \ x_2^{(4)}(t_k+\xi_2) \cdots x_n^{(4)}(t_k+\xi_n)]^T. \qquad (21)$$

Proof: Let $p(\tau)$ be the polynomial which interpolates the state vector $x(\tau + t_k)$ in the interval $[0, h]$ and suppose $\bar{\tau} \in (0, h)$ (for $\bar{\tau} = 0$ and $\bar{\tau} = h$, there is nothing to show). Denoting i–th element of $x(\tau + t_k)$ as $x_i(\tau + t_k)$ and i–th element of $p(\tau)$ as $p_i(\tau)$, we can find a constant k_i such that the function

$$f_i(\tau) = x_i(\tau + t_k) - p_i(\tau) - k_i\omega(\tau) \qquad (22)$$

$$i = 1, 2, \cdots, n$$

vanishes for $\tau = \bar{\tau}$

$$f_i(\bar{\tau}) = 0. \qquad (23)$$

Consequently, $f_i(\tau)$ has at least 3 zeros: 0, h, $\bar{\tau}$ in the interval$[0, h]$. By Rolle's theorem, $\dot{f}_i(\tau)$ has in each interval $(0, \bar{\tau})$ and $(\bar{\tau}, h)$ at least one zero. Considering condition (9), $\dot{f}_i(\tau)$ has at least 4 zeros in the interval $[0, h]$. Applying Rolle's theorem repeatedly, $\ddot{f}_i(\tau)$ has at least 3 zeros, $f_i^{(3)}(\tau)$ has at least 2 zeros, and finally $f_i^{(4)}(\tau)$ has at least 1 zero $\xi_i \in [0, h]$. Since $p_i^{(4)}(\tau) \equiv 0$,

$$f_i^{(4)}(\xi_i) = x_i^{(4)}(\xi_i + t_k) - 24\, k_i = 0, \qquad (24)$$

or

$$k_i = \frac{x_i^{(4)}(\xi_i + t_k)}{24}, \qquad (25)$$

thus,

$$x_i(\bar{\tau} + t_k) - p_i(\tau) = \frac{x_i^{(4)}(\xi_i + t_k)\omega(\bar{\tau})}{24}$$

$$i = 1, 2, \cdots, n. \qquad (26)$$

Using definition (21) we can rewrite Eq.(26) as a vector form

$$x(\bar{\tau} + t_k) - p(\tau) = \frac{x^{(4)}\{\xi\}\omega(\bar{\tau})}{24}. \qquad (27)$$

This proves the proposition. □

Now we use this theorem to analyze the local descretization error of the hybrid method. We denote the approximated solution of x_{k+1} as $\bar{x}_{k+1}$, then for $\bar{x}_k = x_k$ using (4), (11) and theorem 1

$$x_{k+1} - \bar{x}_{k+1} \qquad (28)$$

$$= \int_0^h e^{A(h-\tau)}N[x(\tau + t_k) - p(\tau)]u_k d\tau$$

$$= \int_0^h e^{A(h-\tau)}Nu_k \frac{x^{(4)}\{\xi\}\tau^2(\tau - h)^2}{24}d\tau.$$

Denoting $e^{A(h-\tau)}Nx^{(4)}\{\xi\}$ as $s(\tau)$

$$x_{k+1} - \bar{x}_{k+1} = \int_0^h \frac{s(\tau)}{24}\tau^2(\tau - h)^2 d\tau u_k. \qquad (29)$$

Applying the mean-value theorem for each element of $s(\tau)$, we obtain

$$x_{k+1} - \bar{x}_{k+1} = \frac{s\{\zeta\}u_k}{24}(\frac{1}{5} - \frac{1}{2} - \frac{1}{3})h^5$$

$$= \frac{s\{\zeta\}u_k}{720}h^5, \qquad (30)$$

where $\zeta = [\zeta_1 \ \zeta_2 \cdots \zeta_n]^T$, $\zeta_i \in [0, h]$ and

$$s\{\zeta\} = [s_1(\zeta_1) \ s_2(\zeta_2) \cdots s_n(\zeta_n)]^T. \qquad (31)$$

As a conclusion of above discussion, we have following theorem

Theorem 2: Let $\bar{x}_{k+1}$ be an approximation of the exact solution x_{k+1} by the hybrid method and $x^{(4)}(t)$ be bound in the interval $[t_k, t_{k+1}]$, then the local descretization error of the hybrid method is

$$x_{k+1} - \bar{x}_{k+1} = \frac{s\{\zeta\}u_k}{720}h^5 \qquad (32)$$

Remark 1: From Eq.(32) one sees that the error of the hybrid method converges to zero with a rate h^4 for $h \to 0$ and the error of the Runge-Kutta method with a rate $\leq (h^4)$ (Engeln-Mülles and Reuter,1990). Therefore both methods has equivalent convergence properties for small step sizes.

Remark 2: For the large step sizes h the error of the hybrid method increases with a rate h^4, but the error term of the Runge-Kutta method involves $h^5,\ h^6,\ h^7,\ \cdots$, so the error inceases more rapidly than that of the hybrid method. Therefore the hybrid method is less sensitive than the Runge-Kutta method to increasing step sizes.

Remark 3: With Eq.(28) we can give an upper bound of the descretization error of the hybrid method:

$$\begin{aligned} & \|\boldsymbol{x}_{k+1} - \bar{\boldsymbol{x}}_{k+1}\| \\ & \leq \frac{\|e^{A\zeta}\|_{max}\|\boldsymbol{N}\|\ \|\boldsymbol{x}^{(4)}\{\xi\}\|_{max}|u_k|}{720}h^5. \end{aligned} \quad (33)$$

Evidently, we have

$$\lim_{\|N\|\to 0} \|\boldsymbol{x}_{k+1} - \bar{\boldsymbol{x}}_{k+1}\| = 0. \quad (34)$$

This means, for weak nonlinearity of a bilinear system we can achieve more precise solutions with the method.

We complete this section with a brief comparison of computational demands of the hybrid method and the Runge-Kutta method.
The Runge-Kutta method requires four evaluations of the right hand function of (1) per step, inclusive of other necessary operations, that is about $8n^2 + 10n$ operations. Here n is the order of a bilinear system. The hybrid method requires $2n^2$ operations to evaluate $\boldsymbol{R}_k$, $3n^2 + 2n$ operations to evaluate $\boldsymbol{r}_k$ and $\frac{n^3}{3} - \frac{n}{3} + n^2$ operations to solve the Eq.(18) (Engeln-Mülles and Reuter,1990). Alltogether it requires about $6n^2 + \frac{n^3}{3} + 2n - \frac{n}{3}$ operations per step. Therefore for a bilinear system with $n < 9$, the hybrid method requires less operations than the Runge-Kutta method per step.

NUMERICAL EXAMPLES

In this Section, we will apply our hybrid method to two examples to demonstrate its performance.

Example 1

In this example, we simulate a bilinear system (1)

$$\boldsymbol{A} = \begin{bmatrix} 0 & 1 & 0 \\ 0 & 0 & 1 \\ -1 & -2 & -4 \end{bmatrix}$$

$$\boldsymbol{N} = \begin{bmatrix} -0.5 & 0 & 0 \\ 1.2 & 0 & 0 \\ -3.5 & 0 & 0 \end{bmatrix}$$

$$\begin{aligned} \boldsymbol{b} &= \begin{bmatrix} 0 & 0 & 3 \end{bmatrix}^T \\ \boldsymbol{c}^T &= \begin{bmatrix} 1 & 0 & 0 \end{bmatrix} \end{aligned} \quad (35)$$

with an initial state $\boldsymbol{x}(0) = [0\ \ 0\ \ 0]^T$ for different integrating step sizes. The input signal is

$$\begin{aligned} u(t) &= \sin(2\pi f t_k) \\ t_k &\leq t < t_{k+1}, \quad k = 0,1,2,\cdots, \end{aligned} \quad (36)$$

where $f = 0.01$. Obviouslly, for each sampling point the bilinear system can be exactly solved by

$$\begin{aligned} \boldsymbol{x}_{k+1} &= \boldsymbol{\Phi}_k \boldsymbol{x}_k + \boldsymbol{\psi}_k u_k \\ k &= 0,1,2,\cdots \end{aligned} \quad (37)$$

where

$$\begin{aligned} \boldsymbol{\Phi}_k &= \sum_{i=0}^{\infty} \frac{[(\boldsymbol{A} + \boldsymbol{N}u_k)h]^i}{i!} \\ \boldsymbol{\psi} &= \sum_{i=0}^{\infty} \frac{(\boldsymbol{A} + \boldsymbol{N}u_k)^i h^{i+1}}{i!}\boldsymbol{b}. \end{aligned} \quad (38)$$

We simulate the system both with the Runge-Kutta and the hybrid methods and compare the results with the exact solution from (37). By $h = 0.2$ both methods works well (see Fig.1). But the Runge-Kutta method already becomes unstable for $h = 0.8$ (Fig.2) while the hybrid method provides acceptable results even for $h = 8.0$ (Fig.3). The maximal errors are shown in the table 1.

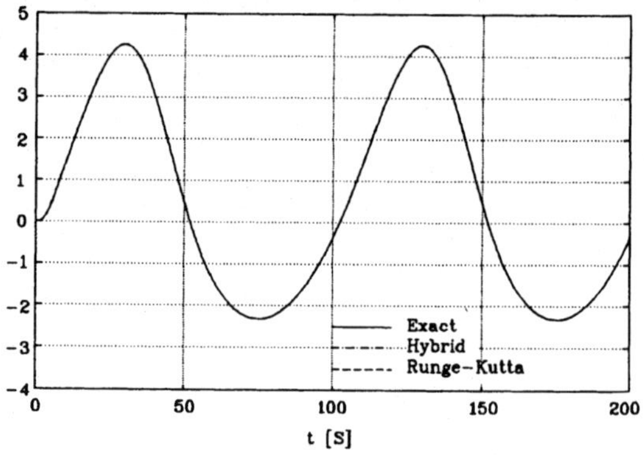

Figure 1: Simulation results: h=0.2

Table 1

Maximal errors of the simulation results		
h	Hybrid	Runge-Kutta
0.2	4.2903×10^{-8}	1.3118×10^{-6}
0.8	9.6163×10^{-6}	7.2550×10^{6}
8.0	0.2049	1.2055×10^{102}

We see that the maximal error of the hybrid method increases with a rate about h^4, but the maximal error of the Runge-Kutta method increases with growing rates about 10^{12} (from $h = 0.2$ to $h = 0.8$) and 10^{96} (from $h = 0.8$ to $h = 8.0$). So the hybrid method is more robust than the Runge-Kutta

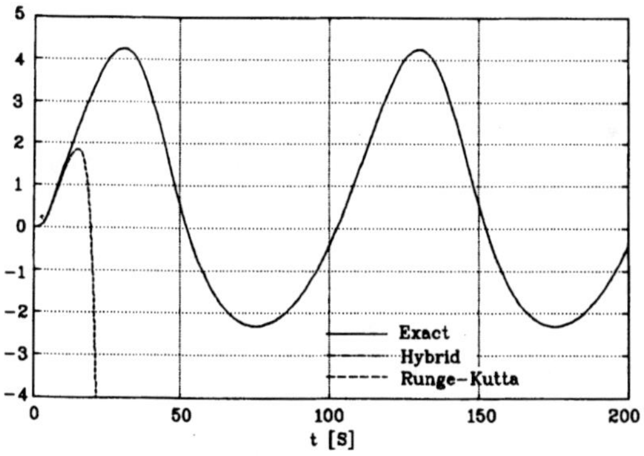

Figure 2: Simulation results: h=0.8

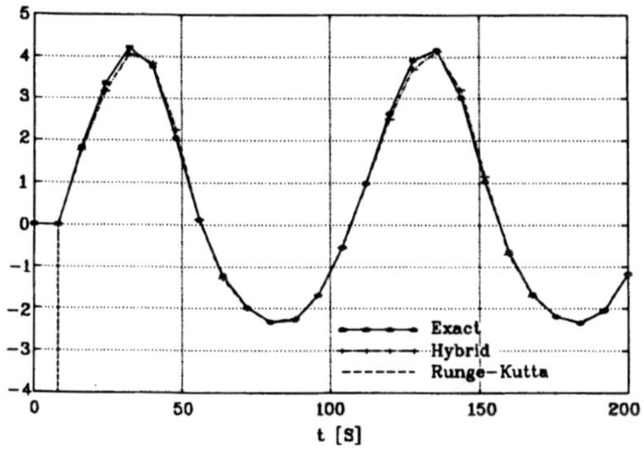

Figure 3: Simulation results: h=8.0

method against increasing integrating step sizes.

Example 2

In this example, we apply our method to an observer realization problem. A bilinear system identified from input/output measurements of a translatory electro-hydraulic drive (Yin, 1992) has a form

$$\begin{aligned} \dot{\boldsymbol{x}}(t) &= \boldsymbol{A}\boldsymbol{x}(t) + \boldsymbol{N}\boldsymbol{x}(t)|u(t)| + \boldsymbol{b}u(t) \\ y(t) &= \boldsymbol{c}^T\boldsymbol{x}(t) \end{aligned} \tag{39}$$

where

$$\boldsymbol{A} = \begin{bmatrix} 0 & 0 & 0 & 0 & 0 \\ 1 & 0 & 0 & 0 & -2.500\times10^{-4} \\ 0 & 1 & 0 & 0 & -1.567\times10^{-3} \\ 0 & 0 & 1 & 0 & -8.553\times10^{-2} \\ 0 & 0 & 0 & 1 & -4.457\times10^{-1} \end{bmatrix}^T$$

$$\boldsymbol{N} = \begin{bmatrix} 0 & 0 & 0 & 0 & 0 \\ 0 & 0 & 0 & 0 & -5.760\times10^{-5} \\ 0 & 0 & 0 & 0 & -2.421\times10^{-3} \\ 0 & 0 & 0 & 0 & -1.836\times10^{-2} \\ 0 & 0 & 0 & 0 & -2.910\times10^{-2} \end{bmatrix}^T$$

$$\begin{aligned} \boldsymbol{b} &= \begin{bmatrix} 0 & 0 & 0 & 0 & 2.218\times10^{-4} \end{bmatrix}^T \\ \boldsymbol{c}^T &= \begin{bmatrix} 1 & 0 & 0 & 0 & 0 \end{bmatrix}. \end{aligned} \tag{40}$$

And $y(t)$ is the position of the piston. A bilinear observer designed for this system is

$$\begin{aligned} \dot{\hat{\boldsymbol{x}}}(t) &= \boldsymbol{A}\hat{\boldsymbol{x}}(t) + \boldsymbol{N}\hat{\boldsymbol{x}}(t)|u(t)| + \boldsymbol{b}u(t) \\ &+ \boldsymbol{k}_o(y(t) - \boldsymbol{c}\hat{\boldsymbol{x}}(t)) \end{aligned} \tag{41}$$

where

$$\boldsymbol{k}_o = \begin{bmatrix} 1.4 & 0.6 & 0.08 & -3.6\times10^{-3} & 7.0\times10^{-6} \end{bmatrix}^T \tag{42}$$

is the observer matrix. The control law is

$$\begin{aligned} u &= -\boldsymbol{k}_c(\hat{\boldsymbol{x}} - \boldsymbol{x}_r) \qquad (43) \\ \boldsymbol{k}_c &= \begin{bmatrix} 0.1533 & 5.109 & 90.28 & 370.14 & 310.28 \end{bmatrix}. \end{aligned}$$

where $\hat{\boldsymbol{x}}(t)$ is the observed state vector, $\boldsymbol{x}_r$ is the reference input and $\boldsymbol{k}_r$ is the feedback matrix. The structure of the state feedback control system is shown in Fig.4.

Now we simulate the closed loop system. To illustrate the influences of the different methods on the control, we simulate the system equation (39) with (37) exactly. The observer is simulated with the exact, the hybrid and the Runge-Kutta method. The initial states of the system and the observer are

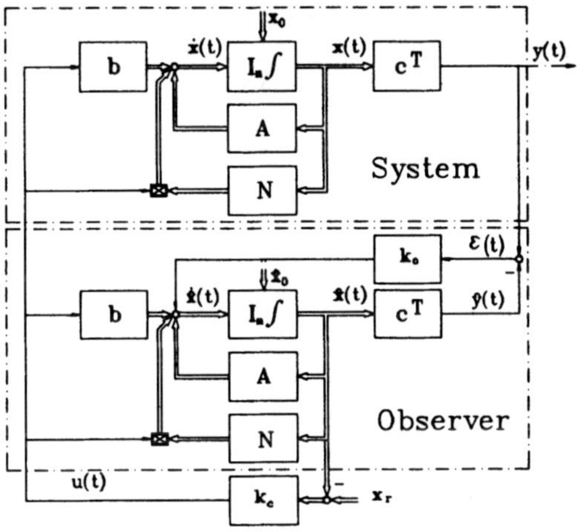

Figure 4: Bilinear system and bilinear observer.

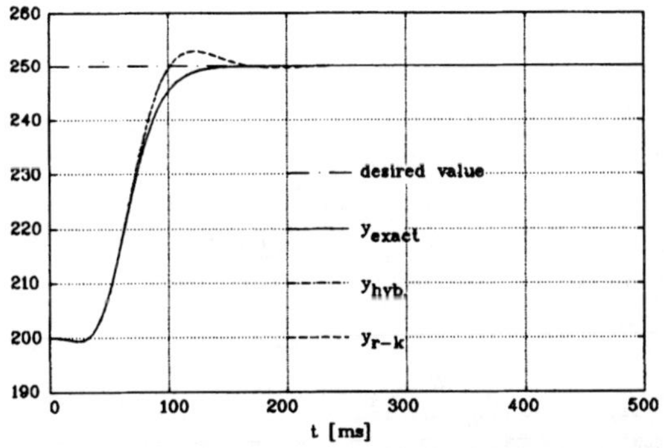

Figure 5: Simulation results: h=1 ms

$$\begin{aligned} \boldsymbol{x}_0 &= \begin{bmatrix} 200 & 0 & 0 & 0 & 0 \end{bmatrix}^T \qquad (44) \\ \hat{\boldsymbol{x}}_0 &= \begin{bmatrix} 150 & 3.0 & 1.0 & -0.03 & 0.002 \end{bmatrix}^T. \end{aligned}$$

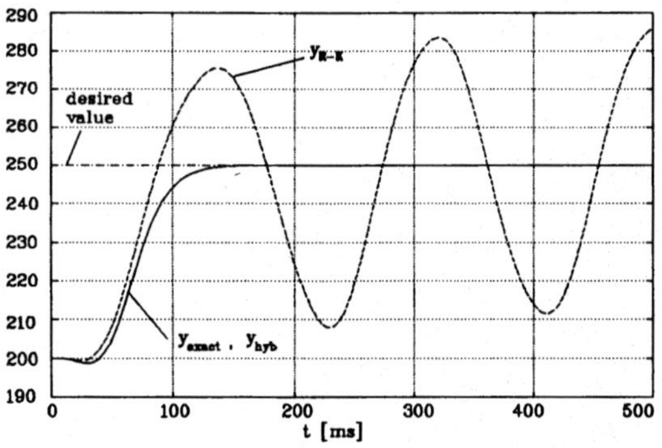

Figure 6: Simulation results: h=2 ms

respectively. For $h = 1ms$ the closed loop systems, where the observers were realized both with the exact and the hybrid methods, yield about similar results (Fig.5). But the result with the Runge-Kutta method is already different. For $h = 2ms$,

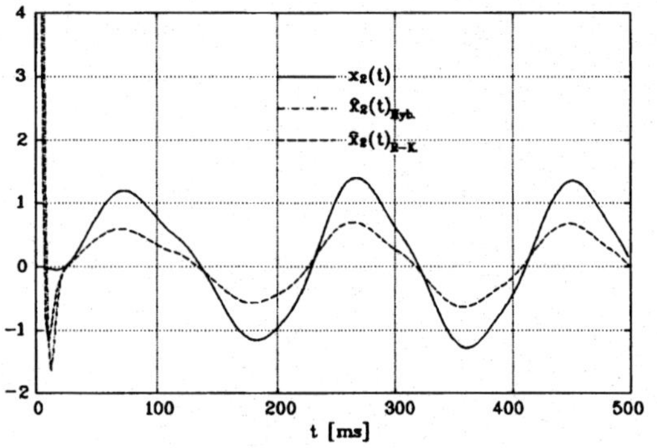

Figure 7: Real and observed state variable x_2

the system with the observer implemented with the Runge-Kutta method becomes unstable (Fig.6). The reason is that the observer with the Runge-Kutta method does not work adequatly. For example, the observed state variable $\hat{x}_2(t)$ does not converge to the real state variable $x_2(t)$ because of the numerical errors of the Runge-Kutta method (Fig.7). But it converges with the hybrid method. Fig.8 shows the simulation results for $h = 10ms$.

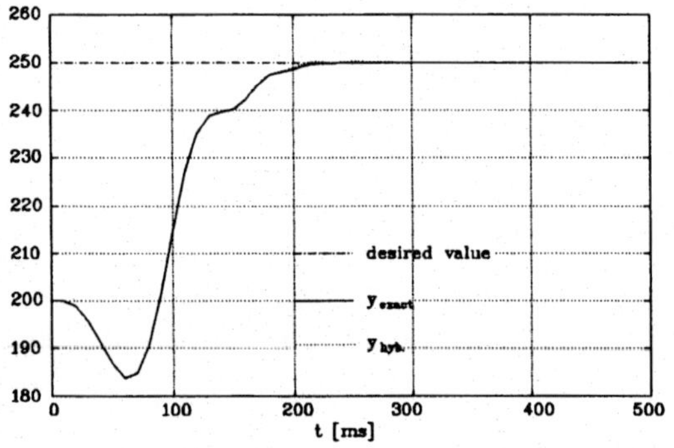

Figure 8: Simulation results: h=10 ms

We see that the closed loop system with the hybrid method provides almost same results as the system with the exact method.

CONCLUSIONS

In this paper we have presented a new method for real-time simulation of continuous-time bilinear systems that can be applied to on line implementation of bilinear observers. We have shown that the method provides more exact and more stable results than the 4th order Runge-Kutta method for large integrating step sizes. We have also given two numerical examples to examine the performance of the method. For real-time control of nonlinear systems with state feedback based on identified bilinear models, the bilinear observer must be implemented on line. The implementation with the Runge-Kutta method often fails because of its relative large computational burden and its high sensitivity to large integrating step sizes, unless we use high performance computers. But if we use the hybrid method to realize the observer, we can choose relative large sampling periods (of cause the closed loop system must be stable without regarding the numerical errors of the observer implementation for the choosen sampling period), then we have enough time to solve the observer equation with the hybrid method and need not a high performance computer.

REFERENCES

Engeln-Müllges G. and Reutter F. 1990. *Formelsammlung zur Numerischen Mathematik mit C-Programmen.* Mannheim a. o.: Wissenschaftsverlag.

Espana M. and Landau I. D. 1978. Reduced Order Bilinear Models for Distillation Columns. *Automatica, 14,* 345-355.

Guo L. and Schwarz H. 1990. Beobachterauslegung für bilineare Systeme und Anwendung an einem sekundärgeregelten hydraulischen Antrieb. *Automatisierungstechnik at 38 (1990) 3.*

Grasselli O. M. and Isidori A. 1989. An Existence Theorem for Observers of Bilinear Systems. *IEEE Trans. Aut. Control 26,* 1299-1300.

Mohler R. R. 1973. *Bilinear Control Processes.* New York: Academic Press.

Stoer J. and Bulirsch R. 1980. *Introduction to Numerical Analysis.* New York a. o.: Springer-Verlag.

Yin X. 1992. Bilinear Modelling and State Feedback Control of an Electro-Hydraulic Drive.*IFAC/IMACS/IEEE Workshop on Motion Control For Intelligent Automation.* 27-29 Octb. 1992, Perugia (To appear).

A STUDY ON THE MINIMAL-TIME CONTROL PROBLEM OF A QUANTIZED LINEAR DISCRETE SYSTEM WITH RATIONAL COEFFICIENTS

Tsunehiro MUNAKATA

Department of Mechanical Engineering, Hiroshima-Denki Institute of Technology
20-1, Nakano 6-chome, Aki-ku, Hiroshima-city, 739-03 Japan

Abstract : In quantized control systems, the control values can take only given discrete (e.g. integer) values. Optimization of such systems has conventional been performed by mathematical programming methods including dynamic programming or integer programming. To linear quantized, time-invariant discrete systems with rational coefficients, a minimal-time control problem is formulated, relating to the transition between two points of the state space. Considering the properties attainable with similar systems and the correspondence between quantized control and mathematical programming, the method on solutions with optimality (RSM) of this control problem is presented, along with two numerical examples.

Keywords: minimal-time optimal quantized discrete control problem, linear time-invariant discrete system, system with rational coefficients, mathematical programming technique, algebraic integer, Gauss-Jordan method

1. INTRODUCTION

Control systems whose controls can take only given discrete-level values (i.e. integer values) are called quantized control systems. Such systems have been optimized by mathematical programming methods such as dynamic programming (DP)[1),2)] or integer programming (IP)[3)] .

In comparison of quantized control systems with nomal continuous control systems, it can be stated that the accuracy is smaller, but reverse control is simplified and fixed anticipating acquisition control can be realized. These points are considered essential for the problem of switching systems, ecological systems handling a number of individuals, investment, traffic control, etc. which can be described by state equation (1) given in 2.1 . However, to the best of the author's knowledge, there are not so many researches on this quantized system from control engineering aspcts except studies about a structure of this system: reachability,controllability, realization, state estimation and dicrete-time control etc. by Anzai[4),5)] ,Miyamichi[6)] and Munakata[7)] .

In this paper, refered to results of Anzai and Miyamichi, the author will investigate the minimal-time control problem of a quantized linear discrete-time system with rational coefficients related to the reachability of that system, and will give a method of control with mathematical programming technique (R.S.M.), along with numerical examples.

2. MINIMAL-TIME CONTROL OF DISCRETE QUANTIZED LINEAR CONTROL SYSTEMS WITH RATIONAL COEFFICIENTS

2.1 Minimal-time control problem

The state equation for this single input systems is:

$$x(i+1) = Ax(i) + bu(i) \quad i = 0,1,2,\cdots \qquad (1)$$

where,

- x is the state vector, element of the n-dimensional real space R^n;
- u is the control vector, element of the subspace Z^1 (set of the admissible discrete values*) of control vector) of the 1-dimensional real space R^1;
- A is an n-dimensional square matrix with ratioal elements; and
- b is an n-dimensional vector with rational elements.

Let x_1 and x_2 denote 2 elements of the state space R^n where $x_1 = 0$ (origin), and x_2 is given as a vector that differs from x_1.

Let G_n denote the following matrix:

$$G_n = [A^{n-1}b \quad A^{n-2}b \quad \cdots \quad Ab \quad b] . \qquad (2)$$

*) Herein the admissible control vectors are integer value vectors, therefore Z^1 is a set of vectors with scalar integer values.

It is assumed that

$\text{rank } G_n = n$.

Then, the minimal-time control problem for discrete quantized case is given as follows.

Determine the minimum time (the minimum number of steps) i and the integer value control u(i) until the time (step) when point x_2 is attained under the conditions that the starting state is $x(0) = x_1 = 0$, state equation (1) is satisfied, and $u(i) \in Z^1$.

See Fig. 1.

The value i satisfying the foregoing control problem is called optimal minimum time (optimal minimum number of steps), and control u(i) until that time (step) is called optimal integer control vector; they will be denoted by $\hat{i}$ and $\hat{u}(i)$, $i = 1,2,\cdots,\hat{i}$, respectively.

2.2 Reachability in linear discrete time-invariant quantized control systems with rational coefficients

Let the system equation be given in the same form as (1):

$$x(i+1) = Ax(i) + bu(i) \quad i = 0,1,2,\cdots$$

Similarly, matrix G_i is given as

$$G_i = [A^{i-1}b \quad A^{i-2}b \quad \cdots \quad Ab \quad b] ,$$

and it is assumed that

$\text{rank } G_n = n$.

In this case, with x(0) = 0 as the starting point, the foregoing state equation yields the set of reachable points in i time (step) as

$$M_i = \{x \mid x \in R^n,\ ^{\exists}u = (u(0),u(1),\cdots,u(i))^T,\ u(j) \in Z^1,\ ^{\forall}j,\ x = G_i \cdot u\} . \qquad (3)$$

Here, T denotes the transpose of vector.

Definition on the reachability.

If $x \in M_\infty$, then x is defined as reachable from the origin ($0 \in R^n$). The set M_∞ is obtained from M_i by extending the set M_i in index (time) i to infinity. That is,

$$M_\infty = \bigcup_{i \geq 0} M_i . \qquad (4)$$

The case $\overline{M_\infty} = R^n$ is defined as complete reachability**). $\overline{M_\infty}$ denotes the enclosure of set M_∞.

Lemmas, Corollary and Theorems on the reachability.

Lemma 1.

If matrix A in quantized system holds eigenvalues which are not algebraic integers, arbitrary $x \in M_\infty$ is an accumulation point in M_∞.

Now, algebraic integer is the solution of polynomial that coefficient of highest degree is 1 and another coefficients are integers.

Proof)

See Reference 6).

**) The reachable set M_∞ of a discrete quantized system contains not only the countable state points; so it is not the complete dense reachability in the usual sense, that is,in the sense of "almost reachable[5),6)]" .

The next Corollary will hold as the result of Lemma 1.

Corollary

If algebraic integers are not contained in eigenvalues of matrix A in quantized system, this quantized system is complete reachability.

Lemma 2.

Let A be matrix with rational elements. Then, next two conditions are equivalent.

① All eigenvalues of A are algebraic integers.

② Matrix A can be resulted in matrix with integer elements.

Proof)

See Reference 6).

Lemma 3.

In quantized system given in (1), let both A and b be matrix and vector with rational elements, respectively. Then, if all eigenvalues of A are algebraic integers, M_∞ is constructed by only isolated points.

Proof)

By using transformation of coordinates on some nonsingular matrix T with rational elements based on Lemma 2., system matrix A can be resulted in matrix with integer elements. And, the relation $M_\infty = M_n$ holds in quantized system whose matrix A is with integer elements. Therefore, M_∞ is constructed by only isolated points (state vectors).

Theorem 1.

The necessary and sufficient condition that quantized system with rational elements will be complete reachability is that algebraic integers do not contain in the eigenvalues of matrix A.

Proof)

Sufficiency: if algebraic integers are contained in the eigenvalues of matrix A, matrix A can be decomposed A_1 and A_2 with direct-sum forms by nonsingular matrix, that is, A_1 is the matrix whose eigenvalues are only algebraic integers and A_2 is whose eigenvalues are not algebraic integers. And, by Lemma 3., a subspace to matrix A_1 is discrete subspace. So, reachable sets of states in this system are not dense partially, that is, this system is not complete reachability.

Necessity: it is evident by Corollary of Lemma 1.

According to these results, an accumulation on M_∞ in quantized system with rational elements depends on the characteristics on eigenvalues of matrix A only.

2.3 Relations between minimal-time control problem[8),9)] and reachability and solution to this problem

Based on Theorem 1., the next Theorem 2. will hold.[10)]

Theorem 2.

In quantized system given in (1),

[1] if algebraic integers are not contained in eigenvalues of system matrix A, there exists $\hat{i}$ in $0 \leq i < \infty$ for arbitrary $x_2 \in R^n$.

[2] if algebraic integers are contained in eigenvalues of system matrix A,

1) in case of $x_2 \notin M_2$, $\hat{i}$ does not exists.

2) in case of $x_2 \in M_2$, there exists $\hat{i}$ in $0 \leq i \leq t-1$.

Now, t-1 is i= t-1 such that $M_t = M_{t+1}$ holds and $M_t = M_\infty$.

Proof)

[1]: regarded as Theorem 1., if algebraic integers are not contained in eigenvalues in matrix A, arbitary x_2 ($\in R^n$) is representd as infinitely generated integer submodule. Therefore, there exists $\hat{i}$ in $0 \leq i < \infty$.

[2]: by Theorem 1., if algebraic integers are contained in eigenvalues in matrix A, the relation $M_t = M_\infty$ holds and an increase of M_t on t stops. And, M_t is the set of isolated points. Therefore, if x_2 is not contained in $M_\infty (x_2 \notin M_\infty)$, then x_2 can not be reached from x_1, so $\hat{i}$ does not exist. If x_2 is contained in $M_\infty (x_2 \in M_\infty)$, then $M_\infty = M_t$ holds, so there exists $\hat{i}$ in $0 \leq i \leq t-1$.

Based on Theorem 2., if there exists $\hat{i}$ (optimal minimal-time) in this control problem, from system equation:

$$x(i+1) = Ax(i) + bu(i) \quad i = 0,1,2,\cdots \quad ,$$

$$x(i+1) = A^{i+1}x(0) + \sum_{j=0}^{i} A^{i-j}bu(j) \ , \tag{5}$$

and substituting $x_1 = x(0) = 0$, $x_2 = x(i+1)$ for (5), we will obtain

$$x_2 = \sum_{j=0}^{i} A^{i-j}bu(j) \ . \tag{6}$$

Let

$\gamma_j = A^{i-j}b$, j = 0,1,···,i

$z_j \equiv u(j)$, j = 0,1,···,i ,

(6) will be the next:

$$x_2 = \sum_{j=0}^{i} \gamma_j \cdot z_j \ . \tag{7}$$

Mathematical programming (M.P.) problem on integer minimization:

min i

subject to;

$$x_2 = \sum_{j=0}^{i} \gamma_j \cdot z_j$$

$z_j \in Z^1$.

By solving the above M.P. problem, we can obtain $\hat{i}$.

And, in this paper, we will solve this problem by using the recursive solving method (R.S.M.).[8]

2.4 Numerical examples

Two numerical examples are given in this section.

Example 1.

A system is given by the state equation:

$$x(i+1) = \begin{bmatrix} 2 & 0 \\ -1/2 & 1/3 \end{bmatrix} x(i) + \begin{bmatrix} 3/2 \\ 1 \end{bmatrix} u(i), \quad i = 0,1,2,\cdots$$

n= 2, and

state variable: $x(i) = \begin{bmatrix} x^1(i) \\ x^2(i) \end{bmatrix} \in R^2$,

control variable: $u(i) \in Z^1$.

Let

$$x_1 = x(0) = \begin{bmatrix} 0 \\ 0 \end{bmatrix}, \quad \text{and } x_2 = x(i+1) = \begin{bmatrix} 3 \\ 181/18 \end{bmatrix}.$$

Determine minimal-time i and integer-valued controls u(i) so that a state x_1 can reach x_2 in this system.

Result:

rank G_2 = 2, so the assumption is satisfied.

The equation of eigenvalues on system matrix is,

$f(\lambda) = \lambda^2 - (7/3)\lambda + (2/3) = 0$.

So, eigenvalues are not algebraic integers.

M. P. problem:

min i

subject to;

$$\begin{bmatrix} 3 \\ 181/18 \end{bmatrix} = \sum_{j=0}^{i} \begin{bmatrix} 2 & 0 \\ -1/2 & 1/3 \end{bmatrix}^{i-j} \cdot \begin{bmatrix} 3/2 \\ 1 \end{bmatrix} z_j$$

$z_j \in Z^1$.

Thus, solving the above M.P. problem, we can obtain the solutions of this control problem. By using R.S.M., $\hat{i}$ = 3, and then $\hat{u}(0) = -3$, $\hat{u}(1) = -12$, $\hat{u}(2) = 39$ and $\hat{u}(3) = -4$.

Example 2.

A system is given by the state equation:

$$x(i+1) = \begin{bmatrix} 1 & -2/3 \\ 0 & 2 \end{bmatrix} x(i) + \begin{bmatrix} 1 \\ 1/2 \end{bmatrix} u(i) \quad i = 0,1,2,\cdots \ ,$$

n= 2, and

state variable: $x(i) \in R^2$, control variable: $u(i) \in Z^1$.

Let

$$x_1 = x(0) = \begin{bmatrix} 0 \\ 0 \end{bmatrix}, \quad \text{and } x_2 = x(i+1) = \begin{bmatrix} -13/3 \\ 6 \end{bmatrix} \text{ or } \begin{bmatrix} 1/2 \\ 7/4 \end{bmatrix}.$$

Determine i and u(i) so that x_1 can reach x_2 in the minimal-time.

Result:

rank G_2 = 2, so the assumption is satisfied.

The equation on system matrix is,

$f(\lambda) = \lambda^2 - 3\lambda + 2 = 0$.

So, eigenvalues are algebraic integers.

To $x_2 = \begin{bmatrix} 13/3 \\ 6 \end{bmatrix}$,

$x_2 \in M_2$ $(=M_\infty)$, and $\hat{i}$ = 1. And we will obtain $\hat{u}(0) = 7$, and $\hat{u}(1) = -2$.

To
$$x_2 = \begin{bmatrix} 1/2 \\ \\ 7/4 \end{bmatrix},$$

$x_2 \notin M_2$ $(=M_\infty)$, and $\hat{i}$ does not exist. See Fig. 2.

3. CONCLUSION

The minimal-time control problem is investigated in the case of discrete quantized linear time-invariant systems with rational coefficients. In relation to the reachability properties of such systems, one method is shown which is based on integer minimization by mathematical programming, along with two numerical examples.

It is an investigation for the future that R.S.M. in this study is not necessarily an efficient method and so the author has to give more efficient one.[11),12)]

Acknowledgement

The author would like to appreciate Professors: Y.Hayashi, K.Shimizu and Y.Anzai in Keio University and H.Yamada at the same Institute of Technology for their valuable and helpful comments on this study.

References

1) J.B.Lewis and J.T.Tou: Optimum sampled-data systems with quantized control signals, AIEE, 82, pp.229-233, 1963.

2) R.M.Havira and J.B.Lewis: Computation of quantized controls using differential dynamic programming, IEEE, Trans. AC-17, 2, pp.191-196, 1972.

3) M.Kim and K.Djadjuri: Optimum sampled-data control systems with quantized control function, ISA Trans. 6, 1, pp.65-73, 1967.

4) Y.Anzai: Concepts of controllability of quantized control systems, Keio engineering reports, Vol.25, No.12, 1972.

5) Y.Anzai: A note on reachability of discrete-time quantized control systems, IEEE, Trans. AC-19, 5, pp.575-577, 1974.

6) J.Miyamichi and T.Fukao: Controllability of quantized control systems (in Japanese), Trans. SICE in Japan, Vol.12, No.2, pp.145-150, 1976.

7) T.Munakata: A study on the discrete quantized optimal control systems and its applications (in Japanese), Ph.D. Dissertation, Keio University, 1977.

8) E.Polak et al: Theory of optimal control and mathematical programming, Chap. 8, McGraw-Hill. 1970.

9) E.Shimemura: Optimal control theory and its computation methods (in Japanese), Lectures on optimization methods, SICE in Japan, pp.15-34, 1973.

10) T.Munakata: Minimal-time control for discrete linear quantized systems —— in case of systems on field of rational number (in Japanese), Proc. of Autumn conference on Japan OR Gakkai, 1978.

11) T.Munakata and Y.Nishida: Suboptimal minimal-time control to discrete quantized linear control systems with integer coefficients, Electronics and communications in Japan, Part 1, Vol.70, No.1, pp.42-49, Scripta Technica, Inc. 1987.

12) T.Munakata: Optimal minimal-time control to discrete quantized linear control systems with rational coefficients (in Japanese), Memories of Hiroshima-Denki institute of technology and Hiroshima junior college of automotive engineering, Vol.21, pp.101-106, 1988.

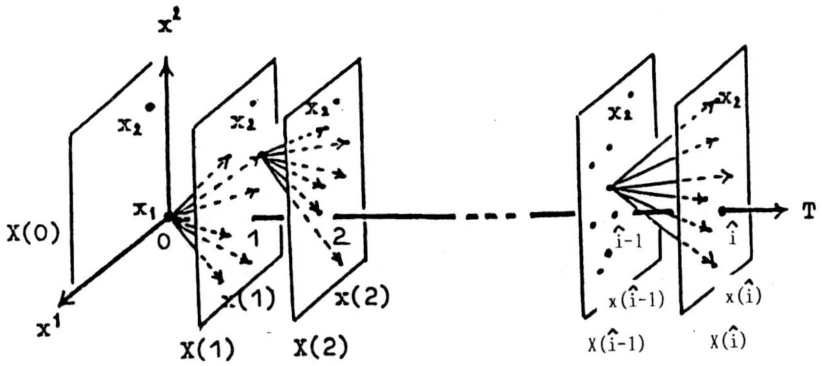

Fig. 1. A transition of state vectors in each time (step) based on state equation and a check of the terminal point in case of two-dimensional systems.

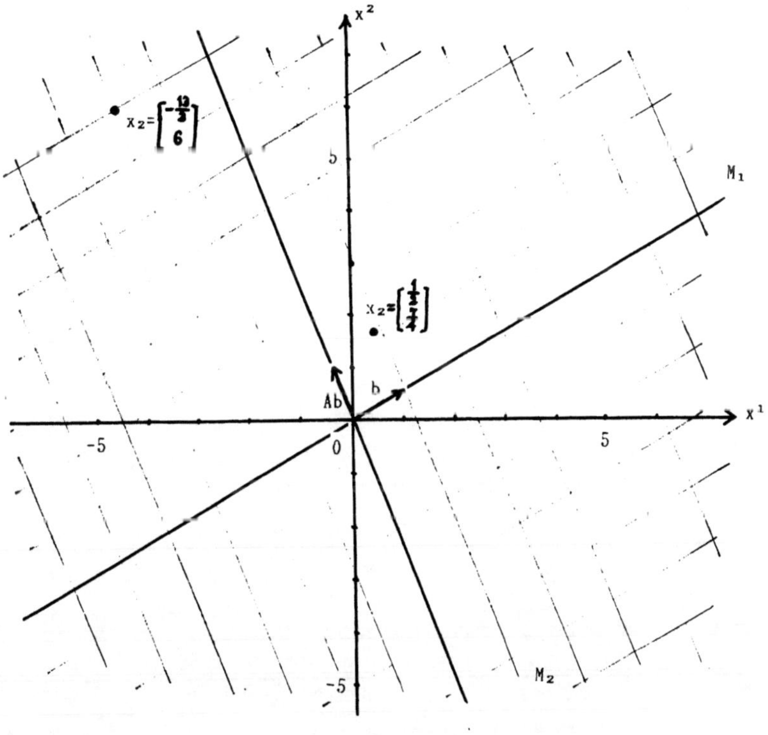

Fig. 2. A reachable set M_2 and the relation between two terminal states and minimal-time optimal controls in Example 2.

PREDICTIVE CONTROL OF SYSTEMS WITH MULTIPLEXED MEASUREMENTS

Gyu-In Jee* and Rob Roy**

**Department of Electronics Engineering, Konkuk University*
Seoul 133-701, Korea
***Biomedical Engineering Department, RPI*
Troy, NY 12180, U.S.A.

Abstract

This paper considers the design of predictive control systems characterized by a multiplexed measurements and multirate sampling mechanism. Several control systems share one common instrument to measure their output signals. For each controller the output measurements are only available at infrequent rate less than the rate adopted for inputs updating. To resolve the problem of infrequent output measurements, intermediate control inputs between actual output samplings is calculate based on output prediction. A dynamic polling method for multiplexing is proposed and shows an improved control performance.

1 Introduction

In designing sampled-data control systems, it is usually assumed that both the system inputs updating and the system outputs measurements are performed simultaneously at a constant rate. However, it is not often realistic or necessary for technical and/or economical reasons[2],[3],[4]. For example, this situation can be seen in a closed circuit anesthesia controller based on a multiplexed mass spectrometer system commonly used in hospitals[1]. The object is to control alveolar oxygen and halothane concentrations. To build cost effective system the measurement of gas concentration is accomplished by *a* centrally located multiplexed mass spectrometer which is shared between several operating rooms for their own controllers. For each controller the output measurements are only available at infrequent rate less than the inputs updating rate.

In this paper we consider certain sampled-data control systems where *an* instrument is shared between several controllers. The block diagram of this control systems would appear as in Figure 1. At each sampling moment only *one* controller is allowed to use the instrument to measure it's system output information. A multiplexing mechanism is used to connect several systems/controllers to an instrument.

In this framework we consider the following two things.

- How to resolve the problem of infrequent output measurements: The control inputs are allowed to change at any sampling period h, while the system outputs are measured at $T_s = mh \quad (m \geq 1)$ interval because of multiplexed output measurement. Output prediction is used to obtain pseudo measurements between actual output sampling.
- How to construct a multiplexing mechanism: A straightforward method is cyclic polling. Systems/controllers are connected to an instrument sequentially, that is, system 1, system 2, ..., system m, are connected sequentially and this procedure is repeated. The other method can be considered is dynamic polling. At a certain instant each control system can have different status. Some systems are in steady state but the others are in transient state. Priority to access to the instrument is determined by taking each system's current status into account at each sampling time.

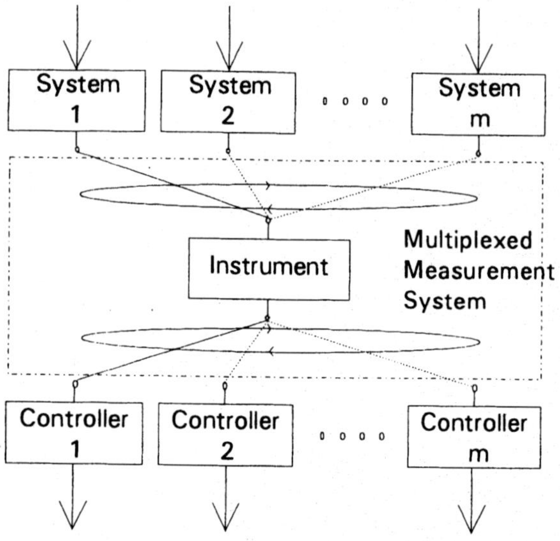

Figure 1: Control Systems with Multiplexed Measurement

2 System Model

Consider the following SISO continuous linear system:

$$y(t) = G_c(\rho)u(t) = \frac{b(\rho)}{a(\rho)}u(t) \tag{1}$$

where ρ represents the differential operator, $\rho \triangleq d/dt$, $u(t)$ and $y(t)$ are the input and output, respectively. Define T_s as the interval between two successive output measurements, and h as the interval between allowable input updates. We assume that T_s and h have the following relation:

$$T_s = mh, \quad m = 1, 2, \ldots \tag{2}$$

If we convert this input-output model to state space form and then discretize assuming a zero-order hold input and a sampling period of h, we obtain the following discrete input-output model in backward shift operator form:

$$A(h, q^{-1})y_t = B(h, q^{-1})u_{t-1} \tag{3}$$

where $y_t \triangleq y(th)$, and $u_t \triangleq u(th)$, $t = 0, 1, \ldots$. Note that $A(h, q^{-1})$ and $B(h, q^{-1})$ are the functions of sampling period h. The derivation of this form from equation (1) is straightforward and well known [5].

3 Predictive Control

We consider a predictive control problem with infrequent output measurements and propose a control method based on Generalized Predictive Control (GPC) [7] and intermediate output prediction between output samplings.

A model of the system is assumed to be given in the following CARIMA (Controlled AutoRegressive Integrated Moving Average) model:

$$A(q^{-1})y_t = B(q^{-1})u_{t-1} + \frac{T(q^{-1})}{\Delta}\xi_t \tag{4}$$

where Δ is the differencing operator $1 - q^{-1}$, and ξ_t represents the general disturbance. The disturbance term can be used to model the model-plant mismatch [6]. In this case $T(q^{-1})$ can be thought of as a design polynomial which can result in improved robustness of the controller in the presence of unmodelled dynamics. To compute the j-step ahead output prediction the following two Diophantine equations need to be solved.

$$T(q^{-1}) = E_j(q^{-1})A\Delta + q^{-j}F_j(q^{-1}) \tag{5}$$

$$E_j(q^{-1})B(q^{-1}) = G_j(q^{-1})T(q^{-1}) + q^{-j}\Gamma(q^{-1}) \tag{6}$$

After some algebraic manipulations the prediction of y_{t+j} can have the following expression:

$$\hat{y}_{t+j} = G_j(q^{-1})\Delta u_{t+j-1} + \hat{y}_{t+j|t} \tag{7}$$

where $\hat{y}_{t+j|t}$ is the prediction of y_{t+j} assuming that control increments after time $t-1$ will be zero,

$$\hat{y}_{t+j|t} = \Gamma_j(q^{-1})\frac{\Delta}{T}u_{t-1} + F_j(q^{-1})\frac{\Delta}{T}y_t \tag{8}$$

From the above prediction equation the predicted input-output relationship of the plant can be written as the vector equation,

$$\hat{y} = G\tilde{u} + f \tag{9}$$

where

$$\hat{y} = [\hat{y}_{t+1}, \hat{y}_{t+2}, \ldots, \hat{y}_{t+N_2}]^T, \tag{10}$$

$$\tilde{u} = [\Delta u_t, \Delta u_{t-1}, \ldots, \Delta u_{t+N_u-1}]^T, \tag{11}$$

$$f = [\hat{y}_{t+1|t}, \hat{y}_{t+2|t}, \ldots, \hat{y}_{t+N_2|t}]^T, \tag{12}$$

$$G = \begin{bmatrix} g_0 & 0 & \cdots & 0 \\ g_1 & g_0 & \cdots & 0 \\ \vdots & \vdots & \ddots & \vdots \\ g_{N_u-1} & g_{N_u-2} & \cdots & g_0 \\ \vdots & \vdots & & \vdots \\ g_{N_2-1} & g_{N_2-2} & \cdots & g_{N_2-N_u} \end{bmatrix}, \tag{13}$$

N_2, N_u are the costing horizon, and the control horizon respectively. The cost function to be minimized is:

$$J_{GPC} = (\hat{y} - r)^T(\hat{y} - r) + \lambda\tilde{u}^T\tilde{u} \tag{14}$$

where r is defined from the reference signal as

$$r = [r_{t+1} r_{t+2} \dots r_{t+N_2}]^T \quad (15)$$

The solution giving the optimal control increment is

$$\tilde{u} = (G^T G + \lambda I)^{-1} G^T (r - f) \quad (16)$$

Even though future control inputs are calculated, often only the current input increment Δu_t, the first element of $\tilde{u}$, is actually implemented, and then a new set of controls are calculated again at the next step, using newly measured output information. This is called a receding horizon control policy.

Now consider a predictive control problem with multiplexed output measurements. Assume that m control systems are arranged to share one common instrument to measure their outputs as shown in Figure 1. If each control system is permitted to access to the instrument sequentially with even priority, and control inputs are allowed to change at sampling period h, then for each control system the system outputs can be measured at most once per every $T_s = mh$ interval.

A straightforward implementation of GPC on this sampled-data system is, to discretize the continuous system $G_c(s)$ using a sampling period of T_s and calculate $\tilde{u}$ in equation (16) based on model parameters $A(T_s, q^{-1})$ and $B(T_s, q^{-1})$ at T_s period. Although control inputs can be updated more frequently at h period, $u(t)$ is held constant over T_s sampling interval,

$$u(\tau) = u_t, \quad th < \tau < (t+m)h. \quad (17)$$

Due to it's long sampling time the resulting controller would suffer performance degradation as m become larger.

One remedy for this problem is to update control input u_t at every h period rather than long T_s period. Then $\tilde{u}$ in (16) needs to be calculated based on model parameters $A(h, q^{-1})$ and $B(h, q^{-1})$ at h interval. But, since the actual output measurement y_t is available only when $t = im, i = 1, 2, \dots$, a sort of pseudo output measurement is required for replacing y_t in (8) when $t \neq im$. One-step ahead output prediction $\hat{y}_t$ based on informations up to $t-1$, is used here for pseudo output measurement at t. From (7), when $t = t - 1$ and $j = 1$

$$\hat{y}_t = g_0 \Delta u_{t-1} + \hat{y}_{t|t-1} \quad (18)$$

Note that $\hat{y}_t$ is equal to y_t when system is noise free and there is no model-plant mismatch. Among the GPC equations (5)-(16), only y_t term in (8) is modified as

$$y_t = \begin{cases} y_t, & \text{real measurement}, \quad t = im \\ \hat{y}_t, & \text{pseudo measurement}, \quad t \neq im \end{cases} \quad (19)$$

where $i = 1, 2, \dots$. Here $u(t)$ is held constant over h sampling interval,

$$u(\tau) = u_t, \quad th < \tau < (t+1)h. \quad (20)$$

4 Multiplexing Algorithm

So far, a cyclic polling for output measurement multiplexing is assumed. That is, m control systems are connected to an instrument sequentially in predetermined order. Now we consider dynamic polling method. Priority to access to an instrument is determined based on the current systems status.

Although m control systems are arranged to share one common instrument, they are running independently in different operating conditions. For example, they have different values for GPC tuning knobs (N_1, N_2, N_u, and λ), design polynomial $T(q^{-1})$, and reference signal $r(t)$, *etc.* Therefore at certain instant each control system can have different status. If a control system show a large tracking error at some moment, it is urgent to feed it a updated control input calculated from the newly measured output value. It needs true output measurement more desperately than all the others.

Define J_i as the GPC cost function of i-th control system and i^* as the index of control system which has the maximum cost function value over all m control systems:

$$i^* \triangleq \arg\max_i (J_1, J_2, \dots, J_m) \quad (21)$$

A dynamic polling method based on the control system cost function is proposed as follows:

1. At each sampling time at period of h, evaluate cost functions $J_1, J_2, \dots, J_m$ and determine i^*.

2. Select control system i^* and allow it to access to a common instrument to obtain output measurement.

3. Repeat this procedure on next sampling time.

Combining intermediate input updating between output samplings and dynamic polling gives the following GPC algorithm for multiplexed measurement system:

1. Discretize the system models using a sampling period of h to obtain model parameters $A_i(h, q^{-1})$, $B_i(h, q^{-1})$, $i = 1, 2, \dots, m$.

2. Select control system i^* which will take output measurement by dynamic polling.

3. For each control system i, assign y_t in (8) as follows:

$$y_t^i = \begin{cases} y_t^i, & i = i^* \\ \hat{y}_t^i, & i \neq i^* \end{cases} \tag{22}$$

4. Calculate $\tilde{u}_i$ and implement u_t^i.

5 Simulations

In order to exhibit the performance of the proposed control algorithms the following examples are considered.

Example 1: Controllers based on long output sampling time T_s (equation (17)) and intermediate input update at h interval (equation (20)) are compared. The real plant to be controlled is [6]

$$G_c(s) = \frac{458}{s^3 + 31s^2 + 259s + 229} \tag{23}$$

It is assumed that there exists plant-model mismatch. The plant is modelled as

$$G(s) = \frac{440}{s^3 + 40^2 + 220s + 280} \tag{24}$$

Input update interval, h, is 0.25. Four control systems share multiplexed measurement system by cyclic polling, so output sampling interval for each control system, T_s, is 1 ($m = 4$). As shown in Figure 2, intermediate control input updates result in improved output response.

Example 2: Control systems with cyclic polling and dynamic polling are compared. In both case intermediate input update is used. The same real plant model as example 1 is used. Three control systems share multiplexed measurement system by dynamic polling. To give them different operating conditions, the following mismatched models are assumed for each controller:

$$G_1(s) = \frac{500}{s^3 + 40^2 + 220s + 280} \tag{25}$$

$$G_2(s) = \frac{440}{s^3 + 30^2 + 200s + 350} \tag{26}$$

$$G_3(s) = \frac{400}{s^3 + 20^2 + 220s + 300} \tag{27}$$

Figure 3 shows the system (1, 2, or 3) which is allowed to access to the common instrument. Possible improvement of output responses is observed in Figure 4.

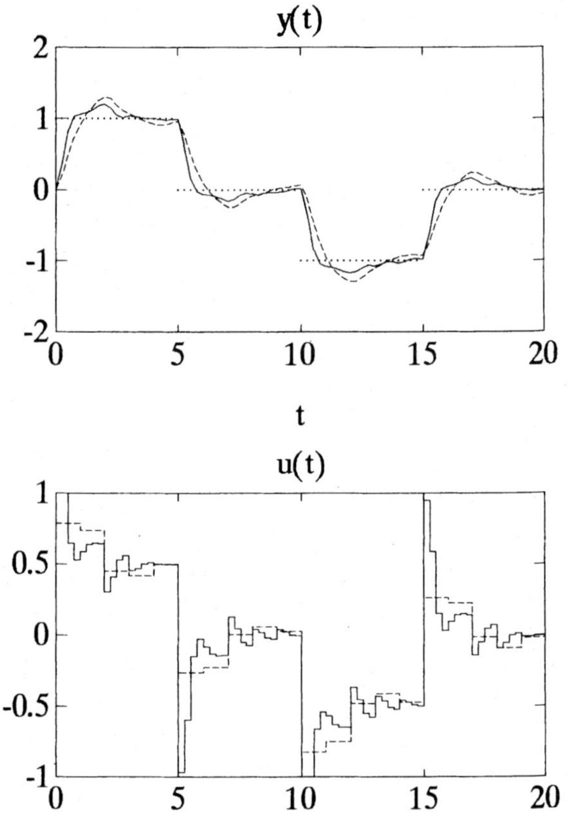

Figure 2: Input and output responses of example 1. solid line: controller (20), dashed line: controller (17).

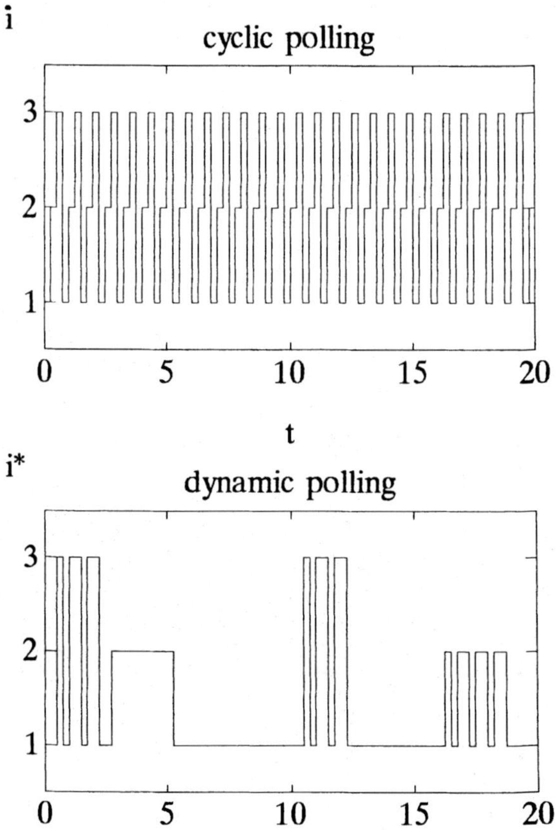

Figure 3: System selected by multiplexer in example 2

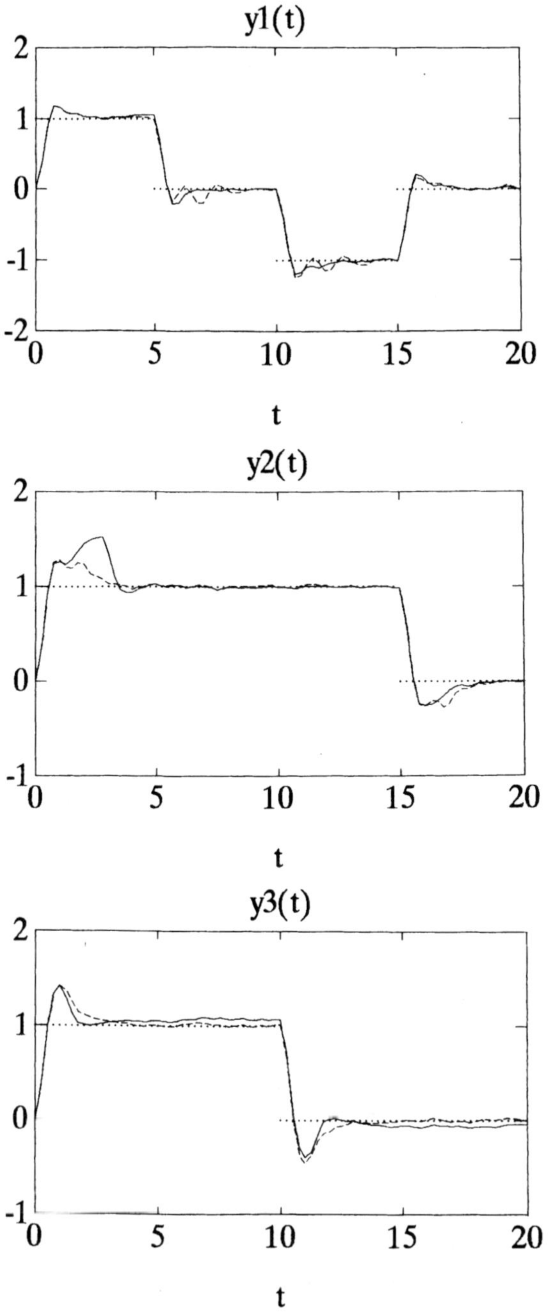

Figure 4: Output responses of example 2. solid line: dynamic polling , dashed line: cyclic polling.

6 Conclusions

In this paper we have considered the design of predictive control system characterized by a multiplexed measurements and multirate sampling mechanism. To resolve the problem caused by infrequent output measurements, output prediction is used to obtain pseudo measurements between actual output samplings. To construct a multiplexing mechanism cyclic polling and dynamic polling methods are considered. In dynamic polling the priority to access to the common instrument is determined by cost function evaluation of each control systems. Simulations show improved output responses of proposed control algorithm. We can expect to build an cost effective control system using multiplexed measurement system which will use only one expensive instrument.

References

[1] Jee, G.-I., and R. Roy (1992). "Adaptive Control of Multiplexed Closed Circuit Anesthesia", *IEEE Transactions on Biomedical Engineering*, to appear in October, 1992.

[2] R. Scattolini (1988). "Self-tuning Control of Systems with Infrequent and Delayed Output Sampling", *IEE PROCEEDINGS*, Vol. 135, Pt. D., No. 4, pp. 213-221.

[3] R. Scattolini (1989). "Multirate Self-Tuning Control of Multivariable Systems", *3rd IFAC Symposium on Adaptive Systems in Control and Signal Processing*, Glasgow, UK, pp. 61-66.

[4] Colaneri, P., R. Scattolini, and N. Schiavoni (1990). "Stabilization of Multirate Sampled-data Linear Systems", *Automatica*, Vol. 26, No. 2, pp. 377-380.

[5] Middleton, R.H., and G.C. Goodwin (1990). *Digital Control and Estimation*, Prentice-Hall.

[6] Shook, D.S., C. Mohtadi, and S.L. Shah (1991). "Identification for Long-Range Predictive Control", *IEE PROC-D*, Vol. 138, No. 1, pp. 75-84.

[7] Clarke, D.W., and C. Mohtadi (1989). "Properties of Generalized Predictive Control", *Automatica*, Vol. 25, No. 6, pp. 859-875.

Real-Time Expert Intelligent Control System REICS

Yao-Nang Wang*, Tiao-Sheng Tong*, Zi-Xing Cai**,
Jing Zhang*, Xian-Ming Zhang*

*Department of Electrical Engineering, Hunan University,
Changsha, Hunan 410082, P.R. China
**Department of Automatic Control Engineering, Central South Univ.
of Tech. , Changsha, Hunan 410083, P.R. China

Abstract. A new developed and designed Real-time Expert Intelligent Control System,REICS,for industrial process control has been introduced in this paper.The global architecture and the diagram of the hardward and softward of REICS have been proposed.In the global design of REICS, a multi-level method of the knowledge representation based on the combination of the frame,production rules and real-time procedure has been discussed,and used to represent the control knowledge and algorithms of the control system.Meanwhile,the reasoning mechanisms of the three control levels—organization level,coordination level and real-time control level—are described in detail;the control strategies,the flow charts of the reasoning program and the related rules and algorithms are expounded.The proposed development tool and control method have been simulated and executed,and applied to the speed control of a D.C. motor which control performances are superior to the ones of the variable structure control.REICS is specially available to intelligent control of the middle and large industrial process control systems with imprecisely and uncertainly mathematical model.

Keywords. real-time system,expert system,knowledge engineering, intelligent control,fuzzy inference.

INTRODUCTION

Real-time expert intellegent control is a typical combination of artificial intelligence,expert system,fuzzy set and control theory,and a new developing direction with promising application for intelligent control(Astrom 1986).This control method is based on the following techniques: utilization of expertise,knowledge model, knowledge base,knowledge inference,control decision and control strategy;combination of knowledge model and tranditional mathematical model,and combination of knowledge information processing technique and control technique;simulation of human intellectual behaviors etc.This method can solve the control problems of large-scale and complex systems with time-varying, nonlinearity,multi-disturbance in real-time control process.Meanwhile,this control system possesses strong adaptive capability and learning function.Therefore,this control method has been used in industry control more and more widely.Some expert controllers have been used to industry process control recent years(Arezen 1989, Peel 1987,and Swami 1989).However,most of this expert controllers have only applied simple production rules of knowledge model and simple reasoning strategies,so that the control knowledge with uncertainty is hard to handle;the knowledge base and the reasoning machine are firmed into a unit that makes the softward and hardward of the control system to be revised once the control process changes and knowledge refreshes.In order to reduce the period of development and application,extend the application field,enlarge the flexibility for the control system,and overcome the difficulties of the expert system of control in environment knowledge acquisition and real-time response,it is necessary to develop an expert control system that possesses certain versatility and capability for processing the control knowledge with uncertainty.When developing and using this expert control system,the system designer can quickly build an expert intelligent control system for certain control domains after one simply gives out some rules and facts of the related domains.A new and more general development tool for real-time expert intelligent control system,REICS-SHELL, has been designed,developed and introduced in this paper.

ARCHITECTURE AND KNOWLEDGE REPRESENTATION OF REICS

Structure of REICS

The global structure of the hardward and softward of the REICS are shown in Fig.1 and Fig.2 respectively.

For the system hardward,the microprocessor IBM-PC/AT is used as decision and computation in the high-level expert system, and a single chip microprocessor MCS-51, A/D and D/A converters as well as the interfaces are used in the low-level circuits.The softward of REICS consists of the modules of dynamic knowledge acquisition,knowledge base and its

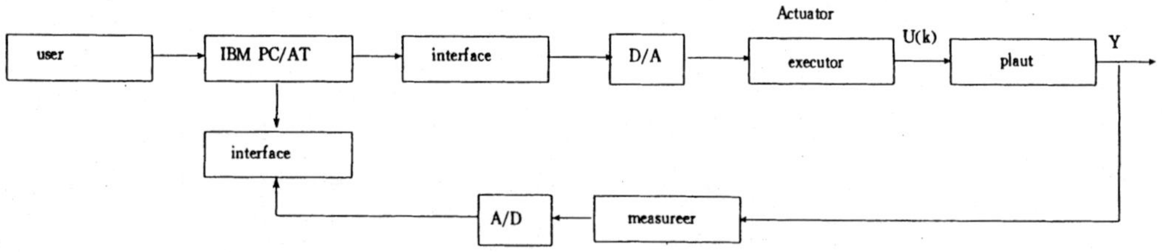

Fig.1. Structure of Hardward for REICS

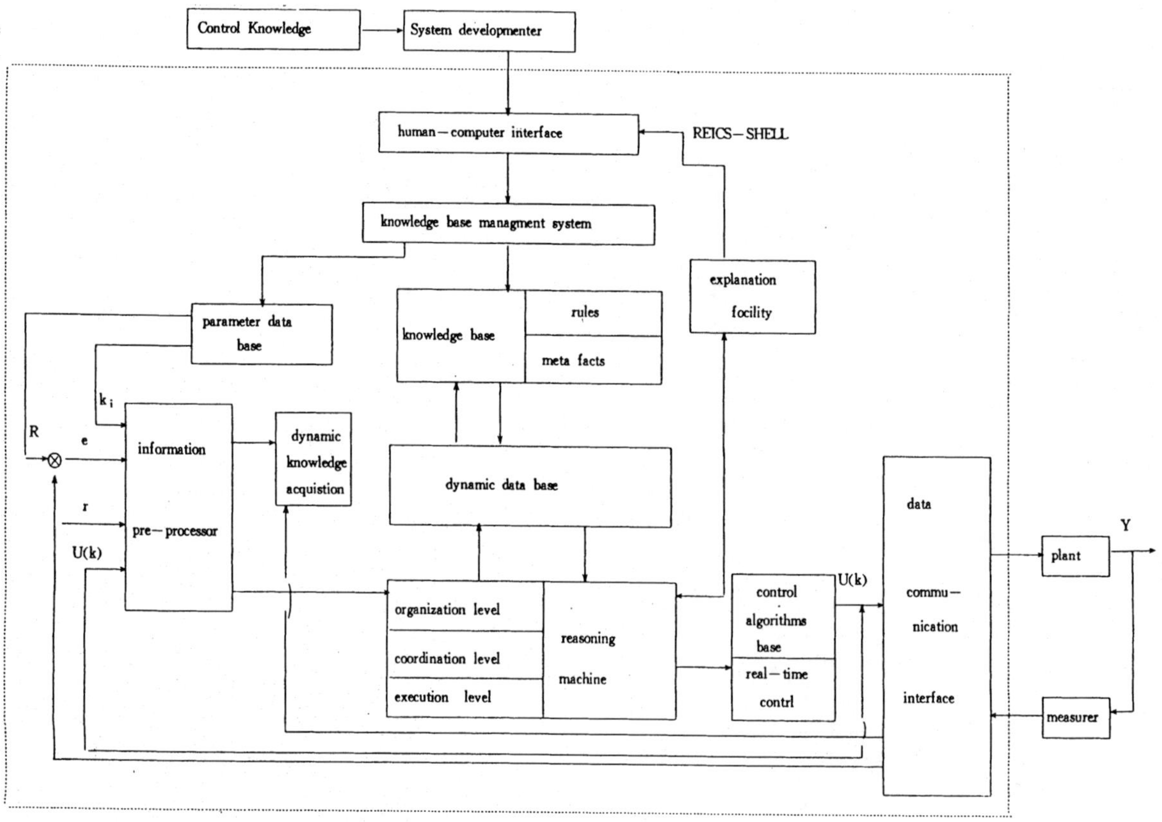

Fig.2. Structure of Softward for REICS

management system,parameter data base, dynamic database,real-time inference engine (reasoning machine),information pre-processor,interpreter,set of control algorithms,data communication interface and human-machine interface etc.The inference engine and knowledge representation are implemented by programming language Turbo-prolog,and the programming language C is used to computate the set of control algorithms and data communication.

In the global design of the REICS,we have proposed a "generalized production system" based on the frame,production rules and real-time procedure,and used to represent control knowledge and algorithms for the control system.In the highest control level,the organization level,the reasoning knowledge is represented by the frame,and the imprecisely reasoning and the choice of the control strategy are made by the fuzzy approximate reasoning.In the coordination control level,the coordination control knowledge is represented by the generalized production rules;the choice of control algorithms and the parametr self-tuning rules,the parameter evaluation are performed by backward reasoning,the refined uncertainty reasoning algorithm and the breadth-first reasoning principle.In the real-time control level,the forward real-time reasoning is done according to the reasoning and decision-making results of the above two higher control levels and the sampled real-time data (facts);as a result,the real-time control input of the controlled object at any time is given.

Knowledge Representation for REICS

It is the key of expert intelligent control that to make intelligent decision through problem-solving.Therefore,it is important for implementing intelligent control how to build a fine model of knowledge representation for problem-solving.Owing to the knowledge of object characteristics and parameter rules in control domain are very complex,the different control object may use different control knowledge.As metioned above,a generalized production system tool,REICS-SHELL,has been used to represent the multi-level knowledge.

Definition 1

A generalized production rule has the form:

$$\text{IF}\{\langle \text{frame}(X_1 \text{ is } a_1 \wedge X_2 \text{ is } a_2 \wedge \cdots \wedge X_n \text{ is } a_n \longrightarrow Y_i \text{ is } B_i)\rangle\}$$

$$\text{THEN}\{\text{IF}\langle \text{rule}(K_1 \wedge K_2 \wedge \cdots \wedge K_n \longrightarrow P_i \ CF(P_i, K_i)\rangle$$

$$\text{THEN}\langle \text{process}((S_1(t_1) \wedge S_2(t_2) \wedge \cdots \wedge S_n(t_n)) \longrightarrow U_i(t_i))\rangle\}$$

where, $Y_i \in$ control strategies, and includes ES-PID, ES-Fuzzy, ES-Predict, and ES-Adaptive that stand for the expert intelligent PID control, expert fuzzy control, expert intelligent predictive control and expert intelligent adaptive control respectively; $X_i \in$ characteristic knowledge of control object; $B_i, a_i \in$ variable value of fuzzy set; $K_i \in$ precodition of rule including facts, parameters and real-time sample data etc; $P_i \in$ combination of control algorithms, evaluated algorithms, control rules and parameter regulation rules; $CF(P_i, K_i) \in$ imprecise reasoning function; $S_i(t_i) \in$ control process at a given time; $U_i(t_i) \in$ output of the control system REICS at a given time.

Definition 2

The frame rule of the knowledge representation in organization level has the form:

(<plant>
(<mathematical model X_1>(<character>))
(<nonlinearity X_2>(<character>))
(<time-delaying X_3>(<character>))
(<time-varying X_4>(<character>))
(<parameter variation X_5>(<character>))
(<inertia X_6>(<character>))
(<disturbance X_7>(<character>))
(<dynamic response X_8>(<character>))
⋮ ⋮
(<control strategies>(<ES-PID>))
(<ES-Fuzzy>))
(<ES-Predict>))
(<ES-Adaptive>)))

where, plant stands for controlled object.

The structure of knowledge base in the organization level that corresponds to the following form

$$\text{frame}([X_1(A_1), X_2(A_2), \cdots, X_n(A_n)], Y(B_i)); \text{predicate}$$

where, $A_i, B_i \in \{BO, MO, SO, ZO, DO\}$ — {big, middle, small, zero, undefined}, and are variable value of fuzzy statement.

For example, the frame rule corresponded to characteristics of an object is:

IF(unknown mathematical model(ZO),
middle nonlinearity (MO),
big time-delaying (BO),
big variation of process parameter (BO))
THEN(big probability for using control strategy ES-PID (BO))

This frame rule can be rewritten as

$$\text{frame}([X_1(ZO), X_2(MO), X_3(BO), X_4(\_), X_5(BO), \cdots], Y(\text{ES-PID}), B_i(BO)); \text{predicate}$$

Knowledge representation for coordination level has the form

$$\text{rule IF}(A, C(A)) \text{ THEN}(B, C(B)) \ CF(B, A)$$

$$A \in \{K_1 \wedge K_2 \wedge \cdots \wedge K_n\},$$

$$B \in \{P_1, P_2, \cdots, P_m\},$$

$$CF(B, A): A \text{--} B \in [-1, 1],$$

$$C(\cdot): U \longrightarrow [0, 1]$$

where, $C(\cdot)$ represents the uncertain level of evidence; $CF(B,A)$ describes the knowledge uncertainty as confidence (certainty factor) value of rule, i.e., the confidence with which the facts in antecedent A supports the conclusion in consequent B. $CF(B,A) \in [-1,1]$, if $CF(B,A)=-1$, then A is false, and the operation in B is negative absolutely; if $CF(B,A)=1$, then A is true, and the operation in B is positive absolutely; if $CF(B,A)=0$, then B is independent to A.

In REICS, the knowledge about facts, rules and procedures is stored in knowledge base with the form of predicate, and is independent of inference engine.

The inner form of fact:

data(N,Exp,Value)

where, N stands for the rule number, Exp—parameter expression, Value—real-time measured data of parameter.

The inner form of rule:

rule(N,Ex,Action,P,
[premise 1, ··· ,premise N],CF)

where, Ex—rule type, Action—rule conclusion, premise i—premise condition, P—conclusion type.

The procedure knowledge:

$$\text{process:- } S(t), S(t+1), \cdots, S(t+n)$$

The knowledge base has the hierarchical and modular structure that can decrease the reasoning and searching space, and make knowledge base much more flexible.

INFERENCE MECHANISM OF REICS

Infrence engine is the core of REICS. Inference engine can select related knowledge according to reasoning strategy, and make inference for control algorithms, facts, evidences provided by control experts and data acquired from real-time sampling. As a result, the corresponding decision for intelligent control can be given and used to guide the control action. Inference mechanism includes inference methode and control strategy; it possesses the features: (1) can process imprecise knowledge, (2) can make real-time reasoning rapidly, (3) highly roliability of control in the on line operation, (4) widely versatility.

The reasoning consists of three levels which global flow diagram of control strategy is shown in Fig.3. We will discuss the reasoning mechanism for the three levels in detail as follow.

Reasoning in organization Control Level

In the first control level, organization level, the approximate reasoning of fuzzy set is used to handle the fuzzy knowledge in frame, and the searching technique of reduction heuristic matching in AI is used to search for the goal. The algorithm for handling the fuzzy knowledge is as follow:

Frame rules for searching match in reasoning to be:

$$\text{frame}((X_1 \text{ is } a_{11} \wedge X_2 \text{ is } a_{12} \wedge \cdots \wedge X_n \text{ is } a_{1n}) \longrightarrow Y_1 \text{ is } B_1)$$

$$\text{frame}((X_1 \text{ is } a_{21} \wedge X_2 \text{ is } a_{22} \wedge \cdots \wedge X_n \text{ is } a_{2n}) \longrightarrow Y_2 \text{ is } B_2)$$

$$\vdots$$

$$\text{frame}((X_1 \text{ is } a_{n1} \wedge X_2 \text{ is } a_{n2} \wedge \cdots \wedge X_n \text{ is } a_{nn}) \longrightarrow Y_n \text{ is } B_n)$$

where, $a_{ij} \in$ [ZO, SO, MO, BO, DO], and the computation of subordinative function is listed at TABLE1

TABLE 1

Subordinative Function

μ_i	0	1	2	3	4	5
ZO	1.0	0.6	0.1	0	0	0
SO	0.6	1.0	0.6	0.1	0	0
MO	0.1	0.6	1.0	0.8	1.0	0
BO	0.0	0.1	0.4	0.8	1.0	0
DO	1.0	1.0	1.0	1.0	1.0	1.0

frame relation:

$$R = \bigvee_{i=1}^{n} (X_1 * X_2 * \cdots * X_n * Y_i)$$

approximate reasoning synthesis:

$$Y_i = (X_1 * X_2 * \cdots * X_n) \circ R$$

criterion for applying control strategy:

$$Y_i = \sum_{i=1}^{n} K_i * i \Big/ \sum_{i=1}^{n} K_i$$

$$K_i = \begin{cases} \mu_i & \mu_i \geq 0.6 \\ 0 & \mu_i < 0.6 \end{cases}$$

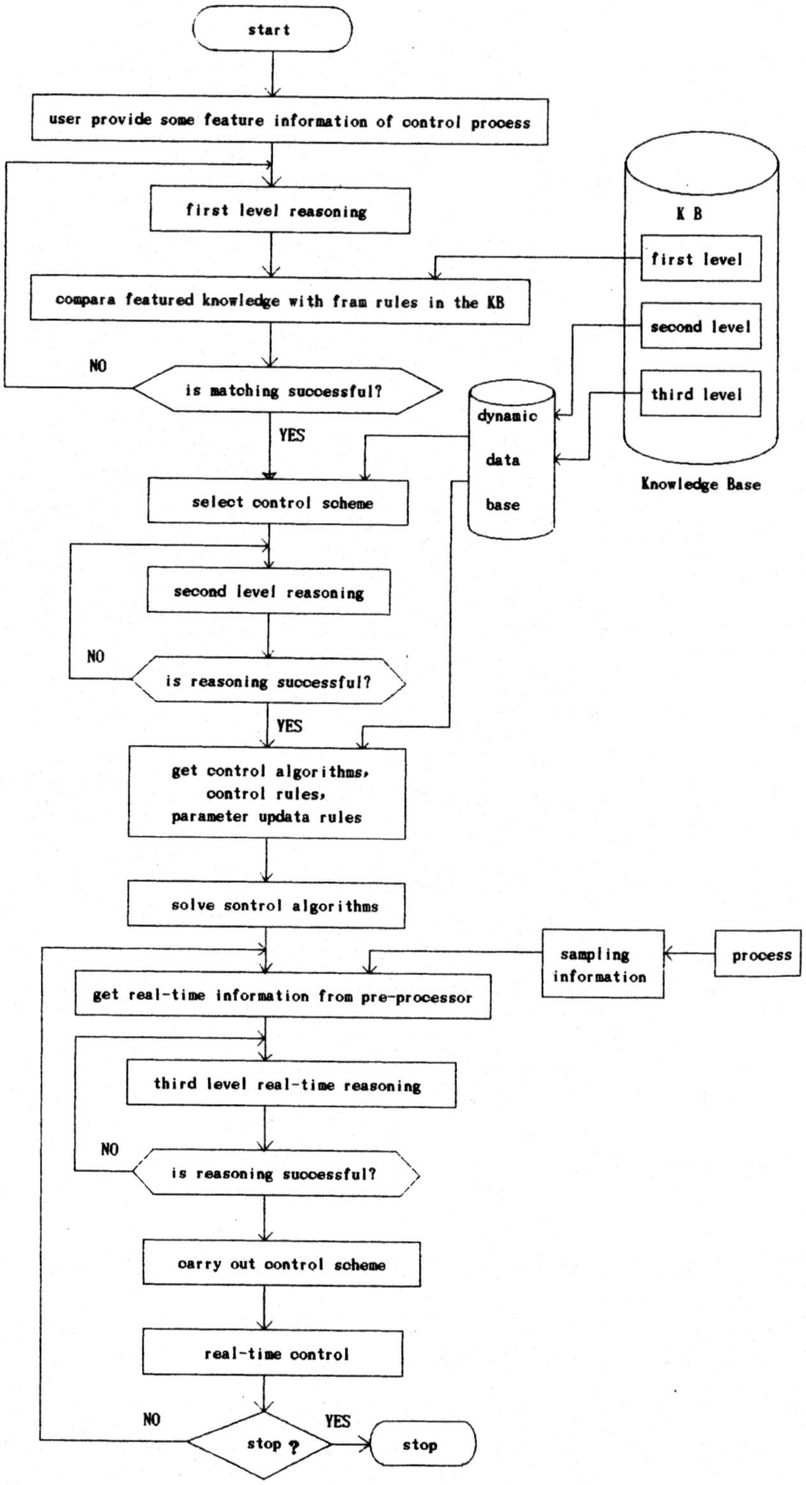

Fig.3. Flow Diagram of Reasoning Program in REICS

Reasoning in Coordination Control Level

In the second control level, coordination control level, the reasoning process is as follow: initially to access conclusion from the first level in dynamic data base, and to use it as supposed goal of the second level; to search for rules that can reach the goal, then to draw a set of conclusions; to find a premise of optimal rule by use of "refined algorithm" with imprecise reasoning, and to use it as a new supposed goal to make further reasoning until a solution of the problem is found. During operation, user can exhibit the record of every step in reasoning process by using predicate "why". After a conclusion is drawn, the user can inquire how the conclusion is drawn by means of predicate "how".

The"refined algorithm" with imprecise reasoning is that the value of the uncertainty about conclusion,i.e.,the uncertainty of the dynamic strength and knowledge,is derived by using the uncertainty about evidance.

Refined Algorithm

1. If the promise evidance of knowledge rule is a combination of multiple AND,i.e.

$$\text{IF } E_1 \wedge E_2 \wedge \cdots \wedge E_n \quad \text{THEN C } CF(C,E)$$

then

$$CF(E)=CF\{E_1 \wedge E_2 \wedge \cdots \wedge E_n\}$$
$$=\min\{CF(E_1),CF(E_2),\cdots,CF(E_n)\}$$
$$CF(C)=CF(C,E)*\max\{0,CF(E)\}$$

2. If evidance is a combination of multiple OR,i.e.

$$\text{IF}(E_1 \vee E_2 \vee \cdots \vee E_n) \quad \text{THEN C} \quad CF(C,E)$$

then

$$CF(E)=CF\{E_1 \vee E_2 \vee \cdots \vee E_n\}$$
$$=\max\{CF(E_1),CF(E_2),\cdots,CF(E_n)\}$$
$$CF(C)=CF(C,E)*\max\{0,CF(E)\}$$

3. If multiple knowledge rules have same conclusion,then two rules with same conclusion are computed,and a recurrent computation is made afterward,i.e.

$$\text{IF } E_1 \quad \text{THEN C} \quad CF(C,E_1)$$
$$\text{IF } E_2 \quad \text{THEN C} \quad CF(C,E_2)$$
$$\text{THEN } CF_1(C)=CF(C,E_1)*\max\{0,CF(E_1)\},$$
$$CF_2(C)=CF(C,E_2)*\max\{0,CF(E_2)\},$$

$$CF_{12}(C)=\begin{cases} CF_1(C)+CF_2(C)-CF_1(C)*CF_2(C), & \text{if } CF_1(C) \text{ and } CF_2(C) \geqslant 0 \\ CF_1(C)+CF_2(C)+CF_1(C)*CF_2(C), & \text{if } CF_1(C) \text{ and } CF_2(C) \leqslant 0 \\ CF_1(C)+CF_2(C)/\{1-\min(|CF_1(C)|,|CF_2(C)|)\}, & \text{otherwise} \end{cases}$$

where,CF(E) stands for certainty factor about evidance;CF(C)—certainty factor about conclusion.

Reasoning in Real-time Control Level

During real-time operation in REICS,the forward reasoning ,reasoning from initial data to control goal,is used.First at all, based on the reasoning in the above two higher control levels,the current information $E=\{e,\dot{e},Y,U,R,K_p,\cdots\}$ provided by information pre-processor and dynamic data base is used as promise condition. Then,control rule that matches with this promise condition in dynamic data base is searched.If the matching is successful, then the state goal is found,and a series of control actions about the conclusion of rule ,such as computation of current control parameters and control value,D/A and A/D conversion,data sampling,control action transfer,and information reception etc.,would be carried out;if the matching is no successful,then a seraching for matched rule should be continued.

Design of Evaluation Interpreter

In order to overcome the difficulty in reasoning for the real-time expert control and to speed up the control response,an evaluation interpreter is designed and used to evaluate the expression of the computation-typed control rules,to judge the logic function and to make matching for the real-time reasoning during the forward and backward reasoning process.As the promise of rule is tenable,the value of conclusion,control volume,can be derived directly.For example,there are rules:

rule 100 $\text{IF}((-R<e) \wedge (e<R))$
$\quad\quad\text{THEN } U_n=K_p*e+K_p\sum e_i+K_d*\dot{e}$

rule 200 $\text{IF}(a+b*c/d \geqslant 0.5*e)$
$\quad\quad\text{THEN } K_p(n+1)=0.98*K_p(n-1)$

call in major predicate expr-eval

("(a+b*c/d ⩾ 0.5*e)",TF)

return with TF=1 representing that the promise of rule is tenable;with TF=-1— the promise of the rule is not tenable.If tenable,then the value can be derived by calling

infix-eval("$0.98*K_p$",A_n)

return with A_n=(expression value).

SYSTEM SIMULATION & APPLICATION

REICS-SHELL is a new development tool with multi-level reasoning and generalized knowledge base for intelligent control system.In order to examine the range and effect of application after design and implementation,we have made experimental study to the system in the IBM-PC/AT microprocessor by inserting a control interface with multiple functions.

The main menu and submenu of the REICS are shown at TABLE 2 and TABLE 3.

TABLE2 Main Menu of REICS

1.Instruction for system usage
2.Subsystem of knowledge acquisition
3.Management system of knowledge base
4.Knowledge storage
5.Install knowledge rule
6.Off-line expert system
7.On-line real-time expert control
8.End

TABLE3 Submenu of REICS

1.Insert rule
2.Add rule
3.Delete rule and/or fact
4.Edit rule
5.Display and print rule
6.Search rule
7.Return

For a special application,the item 2 in main menu has to be input to build its knowledge base,which includes input control rule,set system parameter and initial value.Then items 5,6 and 7 should be keyed in respectively for installing knowledge base,self optimizing parameters, defining control algorithm,computing control volume in real-time and outputing on-line intelligent control.

Four kinds of intelligent control algorithms have been equipped into the REICS, they are ES-PID, ES-Fuzzy, ES-Predict, and ES-Adaptive that are installed into control algorithm base. Meanwhile, some control rules should be input such as:

rule 1 IF($e > \alpha R$) THEN($U_n = U_m$) CF=0.9

rule 2 IF($e < -\alpha R$) THEN($U_n = -U_m$) CF=0.9

rule 3 IF($|e| < a_1 \wedge |\dot{e}| < \delta_1$)
THEN($U_n = U_{n-1} + K_p \dot{e} + K_i e$) CF=0.8

rule 4 IF($|e| < \delta_1 \wedge |\dot{e}| < \delta_2$)
THEN($U_n = U_{n-1}$) CF=0.95

rule 5 IF($e * \dot{e} \leq 0$)
THEN(IF($|\dot{e}| \leq a_2 \wedge |e| \leq b_2$)
THEN PID(K_p, K_i, K_d) CF=0.94

⋮

rule 50 IF($e * \dot{e} \geq 0$)
THEN(IF($|e| > a_2 \wedge |\dot{e}| > b_2$)
THEN FC(K_e, K_c, K_i) CF=0.95

Parameter rules:

rule 51 IF(big time-delay)
THEN(K_e is small, K_c is small)
CF=0.89

rule 52 IF(divergent oscillation response)
THEN(K_e, K_c, K_p, K_i decrease,
a_i, b_i increase) CF=0.78

⋮

rule 110 IF($c_1 = 0 \wedge c_2 = 0 \wedge c_3 = 0$)

$$\text{THEN}\begin{cases} K_p(n) = 0.98 * K_p(n-1) \\ T_i(n) = 0.618 * T_i(n-1) \\ T_d(n) = 0.25 * T_d(n-1) \end{cases}$$

where, $e, \dot{e}$ — error and its rate for a control object; $\delta_1, \delta_2, a_i, b_i$ — parameters given by experts; c_1, c_2, c_3 — the first, second and third peak of the dynamic response curve measured in real-time.

For example, the step response curves of control object $G(S) = 20/S(0.5S+1)$ under three different control schemes are shown in Fig.4, and for object $G(S) = e^{-0.85}/(S+1)(S+2)$ — in Fig.5.

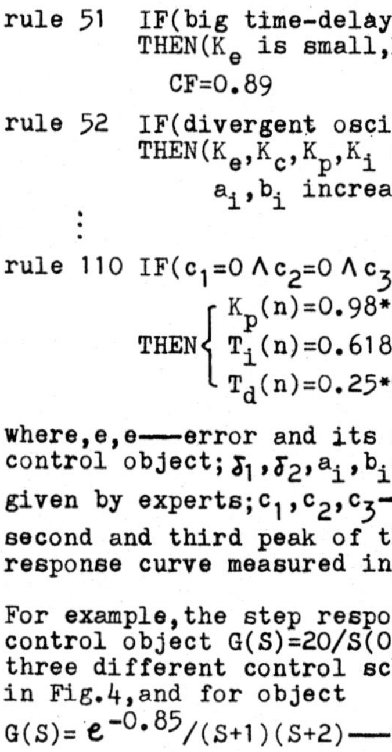

Fig.4. Step Response Curves for $G(S)=20/S(0.5S+1)$
①-PID, ②-Fuzzy, ③-REICS

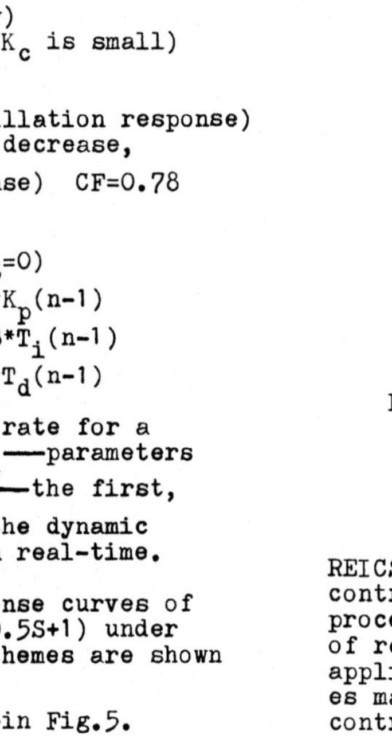

Fig.5. Step Response Curves for $G(S)=e^{-0.85}/(S+1)(S+2)$
①-PID, ②-Fuzzy, ③-REICS

According to (Yuan-Yih Hsu, 1984), a variable structure speed (VSS) control of D.C. motor which state equation is as follow:

$$\dot{X} = AX + BU + rZ$$

$$X(0) = 0$$

where,

$$X = [x_1, x_2, x_3, x_4]^T$$

$x_1 = \int(\Delta\omega_{ref} - \Delta\omega)dt$, $x_2 = \Delta\omega$, $x_3 = \Delta i_a$, $x_4 = \Delta i_f$, $Z = \Delta\omega_{ref}$, A, B — system matrix

$$A = \begin{bmatrix} 0 & -1 & 0 & 0 \\ 0 & -0.5336 & 5.769 & 24.46 \\ 0 & -120 & -120 & -5497.8 \\ 0 & 0 & 0 & -1 \end{bmatrix}$$

$$B = \begin{cases} [0,0,100,0]^T & u_f = 240V \\ [0,0,0,0.016]^T & u_a = 240V \end{cases}$$

This control object has been used to test and compare REICS with VSS. Given refence speed $\Delta\omega = 9.16 \sim 10$ rad/s, then a simulation result can be gotten in Fig.6. The Fig.6 shows that the control by REICS is superior to the control by VSS.

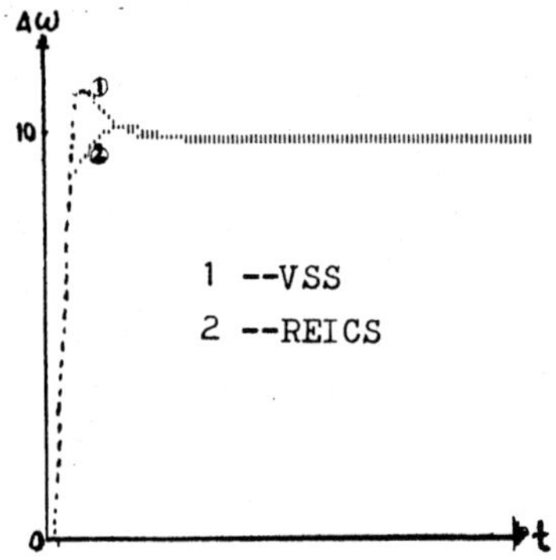

Fig.6. Comparison of Simulations for REICS and VSS

CONCLUDING REMARK

REICS is a development tool for real-time control expert system operated in microprocessor system IBM-PC/AT 386. The results of research both in experiment and application have shown that REICS possesses many advantages such as flexible control strategy, easily knowledge acquisition, able to process control knowledge with uncertainty, good real-time control effect, adaptivity and versatility etc. When using this tool one simply adds control rules, revises rules, facts and control algorithm parameters in knowledge base. Therefore, this tool can be used to some industrial process control systems with uncertainty and imprecise mathematical models.

REFERENCES

Arezen, K-E. (1989). An architecture for expert system based feedback control. Automatica, 25.

Astrom, K.J. (1986). Expert control. Automatica, 22.

Cai, Z-X. (1990). Intelligent Control. Electronic Industry Press, Beijing.

Hsu, Y-Y. (1984). Optimal variable-structure controller for D.C. motor speed control. IEE Proc131.

Peel, D. (1987). Adaptive control and AI. Journal A, 28.

Swami, R.D. (1989). Performance monitoring in ECS. Automatica, 25.

Control Environment Implementation Issues - with Case Studies

Robert E. Betz*, Peter J. Moylan*, Len J. Sciacca**

*Department of Electrical and Computer Engineering, University of Newcastle, Callaghan, NSW,2308, Australia

*Department of Electrical Engineering, University of Melbourne, Parkville, Vic, 3052, Australia

Abstract: This paper discusses a number of practical issues related to the software environment required for the successful implementation of medium to large scale control systems. It is argued that for all but trival systems it is desirable to implement the control system within a software environment which formally supports real-time multi-tasking.

Keywords: Real-time computer systems, Operating Systems

Introduction

The development of a computer control system involves a number of aspects: the design of a control algorithm, the selection of suitable hardware, the software design and implementation, the development of actuator and sensor mechanisms, and so on. Looking at the control theory literature, one can get the impression that the algorithm design is the biggest part of the job, or at least the most intellectually demanding. In practice, one tends to find that the software development phase is where most of the development time goes; and that the complexity of the work is such that it cannot be trusted to "average" computer programmers.

Typically, the control algorithm itself can, even if it is a complex algorithm, be expressed in just a few pages of straightforward code in a high level language. The total software package, in contrast, is likely to be hundreds of pages long; and its design involves some complex and often subtle problems.

The majority of the code in most medium to large control systems is related to functions such as:–

(a) Operator interface - usually a large section of code. Important to the user as it is the means of controlling and monitoring the system.

(b) Logical decisions. Usually surrounds the controller code. Governs control transitions such as the change from linear to slewing mode.

(c) Sequencing of events. Usually a large and complex section of code. Contains logical decisions which also contain timing constraints. For example, timed sequences are required to start and stop most control systems.

(d) Error handling and reporting. This code interacts with most of the other code in the system. Context sensitive error handling important. In some system consequential suppression is also supported.

(e) Task support. This code provides the underlying multi-tasking support for the system.

(f) Communications support. This code provides the software interface to serial channels, networks etc. that the system must communicate with.

(g) Plant interface modules. This code provides the software interfaces to the hardware devices which directly interact with the environment. Examples are A/D converters, relay control logic and other more specialised I/O devices.

The form of the underlying concurrent task support is often an area of confusion. A Computer Scientist for example assumes that a real-time system is any system that has to react to external asynchronous events. However there needs to be a differentiation between, say an airline booking system, and a aircraft control system. An airline booking ystem is an example of a *soft* real-time system, because the consequences of slower response are not significant. An aircraft control system is an example of a *hard* real-time system. If timing constraints are not met all the time then the control system will not achieve acceptable performance, or catastrophic failure may result. Soft real-time applications can usually be supported with conventional operating systems, whereas hard real-time applications require special operating system kernels which can offer deterministic task response times.

The remainder of this paper will concentrate on the concurrent task support requirements for hard real-time operating system kernels (RTOSK). Section 2 will discuss the facilities that are required in a RTOSK. Section 3 will describe the structure of two RTOS's developed at the University of Newcastle.

Finally, section 4 will discuss two control system case studies using one of these kernels.

Real-time Kernel Requirements

A real-time kernel should provide the following features:–

(i) Multi-tasking support - i.e. the ability to run a number of separate asynchronous tasks.
(ii) A scheduling algorithm which attempts to ensure that the timing constraints required for correct control system operation can be satisfied at all times.
(iii) It should not adversely affect the interrupt latency times for the system.
(iv) A mechanism to precisely time and trigger software execution.
(v) If timing constraints are not being satisfied then the kernel should provide basic monitoring support to allow the reason for the timing failure to be identified.
(vi) The kernel should be backed up by a good development environment which amongst other things should allow debugging in a multitasking environment.

Most commercial real-time kernels do not offer all of these facilities, or only offer some in a limited way.

Example Real-time Kernels

Over the last 4 years two real-time operating systems kernels have been developed at the University of Newcastle.

The UNOS kernel

The development the University of Newcastle Operating System (UNOS) was primarily motivated by the requirement to have a real-time kernel for a large project. One of the main design objectives of the UNOS kernel was that it be portable across different hardware platforms. The first version was written in PL/M, but was subsequently translated into 'C'.

UNOS is essentially a conventional pre-emptive priority based real-time kernel [1]. However it does contain a few unusual features which are not found in most commercial real-time kernels.

One problem that commonly arises in any moderate to large size real-time programming task is the sharing of resources between low priority and high priority tasks. Generally it is a good idea to try and avoid these situations, but in some cases this cannot be achieved. The result of such an interaction is that the high priority task's priority can transiently become that of the low priority task. This can occur when the low priority task claims a resource and at some time prior to its release a high priority task requests the same resource. The high priority task then becomes blocked on the resource and must wait until the low priority task releases the resource. The length of time for which the effective priority alteration occurs is very much application dependent.

The solution adopted in UNOS to overcome this problem is to implement dynamic priorities for tasks. When a task is created it is assigned two priorities – a dynamic priority and a static priority. Initially these two are the same. However in the situation when a low priority task has claimed a resource which a high priority task requires the dynamic priority of the low priority task is raised to that of the high priority requesting task. The low priority task will then continue to execute at this new priority until it releases the resource, at which time its dynamic priority will drop back to its static priority. All task scheduling is carried out based on the task dynamic priority.

The inclusion of dynamic priorities adds considerable complexity to the kernel. Another undesirable property of the dynamic priority mechanism is that there is a possibility of multiple recursion inside the kernel. Consider the following example. A low priority task claims a resource and then is blocked itself on another resource, prior to releasing the first resource. A high priority task then requests the first resource. This means that the priority of the claiming task must be raised to that of the high priority task. Because the first task is blocked on a second resource this means that its position in a priority ordered semaphore queue has to change to reflect the new dynamic priority. This scenario can then be repeated again for the task that is claiming the second resource, and so on. Obviously such behaviour leads to uncertainty about the time that a semaphore operation takes. The solution taken in UNOS to solve this problem is to only allow a limited number of recursions to occur.

Another feature which is currently being added to the UNOS kernel is support for timing violation monitoring. If a task has timing violation monitoring turned on then a time window is defined (around an absolute time) which the task has to execute within, else it has not satisfied the desired timing. If the timing is violated then at the time of violation the state of the system is saved into a timing violation circular buffer, which can be interrogated to see what tasks were running prior to the violation. This should be an invaluable aid in debugging timing violations that result from unanticipated task sequences.

The timing of tasks in UNOS can be achieved in two ways. Since the operating system supports time slicing then some tasks can be run every time slice. This situation often occurs in control systems applications since the control algorithm task almost always fall into the category of tasks that operate this way. More sophisticated task timing requires another mechanism. In UNOS an object called a timer is provided by the operating system. There are two types of timers - single shot timers, and repetitive timers. When a timer is created by a task the time to time out and the task to be executed at time out are initialised. In the case of a single shot timer when time out occurs the time out function is executed and the timer is placed back into a queue of inactive timers. Repetitive timers on the other hand place the timer immediately back onto the active time queue with the time out period re-initialised and then execute the time out function. This means that repetitive timers can support very accurate repetitive timing. The clock tick used for the timers is the basic clock tick for the system. However UNOS allows the tick frequency to be separated from the time slice frequency, therefore subject to tick routine overheads arbitrary timing accuracy can be achieved.

Timers are user transparently by the operating for such operations as "timed waits" on semaphores and time constraints on the receipt of messages from mail boxes (see below).

UNOS uses mail boxes as the inter-task communication mechanism. Although mail boxes are a common technique of inter-task communication in many operating systems, the implementation in UNOS has some novelty.

When a task is created a mail box is created and bound to that task. This means that any task can always communicate with any other task because the mail box always exists. In order to locate the mail box for another task it has associated with it an address, which is the task name for the task bound to the mail box. A mail exchange agent within the operating system kernel routes sent messages to the correct task using this address.

The mail boxes themselves are conventional in structure. They contain a user definable of message slots of user definable size. Messages are read from the mail boxes in a FIFO fashion, except in the case of an express message (for which there is a special message location). The mail boxes are designed to support concurrent operation of tasks, and do not force a "rendezvous" synchronization between the sending and receiving tasks. A return address is always attached to any sent message.

The PMOS kernel

The second kernel, known as PMOS (Peter Moylan's Operating System), was motivated primarily from an educational requirement. This kernel is being used in a number of undergraduate courses within the department, and is also used as a vehicle to test new scheduling ideas. PMOS is much larger than UNOS in that it supports file systems and has extensive display text and graphics support. The implementation language is Modula-2 [2, 3].

The key feature of PMOS is its modularity. It is presumed that the applications programmers will at times need to make modifications to the system software - adding new device drivers, for example, or changing the scheduling algorithm for a special application - and the system is designed to make this easy to do. Users can add, remove, or modify modules with confidence that there will be no unexpected side-effects in the rest of the system.

The modularity ensures, too, that applications programs get only the features that they need. Some applications require features such as a file system, graphics, screen menus, and the like; for others, such features would represent an unacceptable overhead. To allow for this, PMOS is implemented as a large number of small modules, which can be linked in with the applications programs as needed.

The file system, for example, does not import anything from the device driver modules. Instead, each device driver has a procedure call (in its initialization code) to "install" itself into the file system. This means that only those devices whose driver modules are imported at a higher level will become known to the file system. The programmer also has the option of omitting the file system and calling device drivers directly. This is usually a better option in embedded system applications, where special-purpose devices are more common than file-oriented devices.

The modularity extends to the point where the notion of a "task" or "process" is largely hidden from the programmer. There are, for example, no kernel mailboxes. Tasks are synchronized using semaphores, and they communicate via shared data structures. The shared data structures, implemented through "abstract data type" modules, are typically not visible at the application level. To the application programmer, intertask communication occurs through procedure calls.

Scheduling methods in PMOS are currently under review. The present arrangement is a pure priority-based preemptive scheduler. We are now working on methods for implementing priority inheritance and - if possible - deadline inheritance. There are still some unsolved theoretical problems in this area.

Case Studies

The Pipeline Crack Investigation Vehicle

This project, which began in 1988, involved the design and development of a computer controlled vehicle to travel down underground gas pipelines and search for areas of stress corrosion cracking. The mechanics of the system were designed by an industrial partner, and the University of Newcastle built the computing and electronics hardware and wrote the software.

Due to the vague design specification during the initial stages of the development it was decided to base the system on a real-time operating system. The main reason for this was the flexibility offered by the operating system approach. The ease with which new tasks can be added to an operating system based software strucutre meant that unforeseen design changes would not result in significant changes to the existing software. It was decided to develop our own operating system, and hence the UNOS operating system was born.

The hardware for PCIV was based on an Intel 80C186/187 microprocessor and a Texas Instruments TMS320C25 DSP. The software for the system could be divided into three main sections:–

(i) The user interface
(ii) The signal processing and crack detection.
(iii) The vehicle control.

A simple menu based user interface was used to monitor and program the system. Ancillary modules added to aid in debugging and validation of the software were:–

a) A context sensitive error logging and display system.
b) An operating system monitor which allowed one to view the status of the tasks in the system.

The completed software system contained approximately 30 tasks and was 60,000 lines of documented code. The main task was the sequencer task whose function was to control the other tasks in the system by sending mail box messages to them. The communications was carried out on a send - acknowledge basis. If an acknowledgment did not arrive in a specified time then an error action would occur. Additional sequencer complexity resulted from the fact that it could accept asynchronous messages from other tasks. Figure 1 shows the relationship between the sequencer and the other tasks in the system.

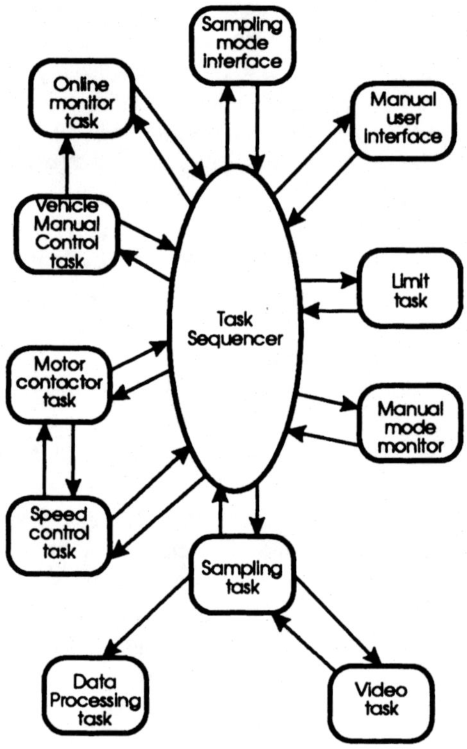

Figure 1 : PCIV task relationships.

The sampling task was the major periodic task. It fired the ultrasonic heads used to detect the cracking in the pipes, and then collected the return echo data, pre-processed and stored it for later retrieval and post processing. It was executed every 5 msec. The pre-processing and storage was carried out concurrently with sampling by the data processing task. The vehicle speed was controlled using a PI controller and a 2 quadrant reversing DC drive system. The limit task was also running concurrently when the vehicle was operating in sampling mode, constantly checking to see if the sampling start and stop distances had occurred, if the batteries still have sufficient capacity to retrieve the vehicle, or if there was enough water couplant to lubricate the ultrasonic heads.

Of the tasks shown in Figure 1, six of them operate concurrently when sampling is in progress. In addition to the tasks shown here, there are two tasks handling the serial communications via a radio modem link. This degree of concurrency would be difficult to handle in a traditional looping executive environment. The use of the multi-tasking environment also aided in code developement, since it was easy to write tasks that were able to simulate such things as the vehicle's motion. This allowed code development and testing to be carried out on a bench.

The version of UNOS used for the PCIV project was primitive in relation to the current UNOS versions. The experience gained during this project influenced the development of the subsequent 'C' UNOS version.

Antenna Tracking Control System

This project involved the design and implementation of a real-time control and supervisory system for four medium sized (approximately 26m diameter) satellite tracking antennae for low-inclined orbit satellites. Capability for star tracking was also required to carry out gain/temperature calibration of the receiver.

The design process involved a quality assurance plan, rigid project management, and software development procedures to military standards. Also tight budget and timing constraints meant a development path of least risk was needed. All of these factors highlighted how closely coupled a control system design is to other aspects of a project. As well as these project management issues, we had to take a systems view of the project ensuring the proper integration of structural and mechanical components.

Although many commercially available real-time operating systems were evaluated, it was decided early in the project to use UNOS. The reason for this was that other operating systems available at the time did not satisfy our requirement for real-time control. Also, although they offered apparently convenient graphic user interfaces, they failed to address such issues as high resolution timer support, deadlock in intertask communication, preemptive kernel task scheduling, and a suitable development environment.

To further reduce risk and maintain a policy of a multi-sourced product, industrial grade 80386/80387 PCs were used with commercially available bus interfaces. A Programmable Logic Controller (PLC) was requested by the client to provide monitoring of other mechanical and electrical components around the structure. The Antenna Control and Tracking Unit (ATCU) was required to communicate with other computers via serial lines. One of the supervisory systems was located hundreds of meters away, communicating via a fibre-optic cable.

An advantage of using an in-house kernel was that it could be tailored to the precise needs of the project. Having the source code for the RTOSK also meant bugs could be quickly found and rectified without having to wait on a third party supplier to solve problems.

The major components in this system are:

(a) Supervisory Control and Data Acquisition (SCADA) systems for operator use.
(b) Antenna Control and Tracking Unit (ATCU) This monitored critical antenna parameters and carried out closed loop position control. Control output was via a digital-to-analogue converter to two sets four-quadrant regenerative DC motor drives, one for each axis.
(c) PLC to monitor less important antenna structural and electrical parameters.

Altogether there were some 23 tasks in the final system comprising communication serial drivers, PLC protocol converters, parsers, the sequencer, Kalman filter updates for the satellite tracking algorithms, and operating system monitoring tasks.

Only one month after starting the project a working system was implemented with a sequencer shell, dummy tasks, and antenna models for simulation. This proved to be an extremely good software development model in which to approach a medium size project. From the very beginning we could see the system evolve and, as new software modules were written, tasks could be added, and bugs detected and removed at each stage of software writing. In taking this approach, the momentum of the project development was sustained, keeping morale high in the development team, and highlighting possible problems that even critical design principles could not foresee.

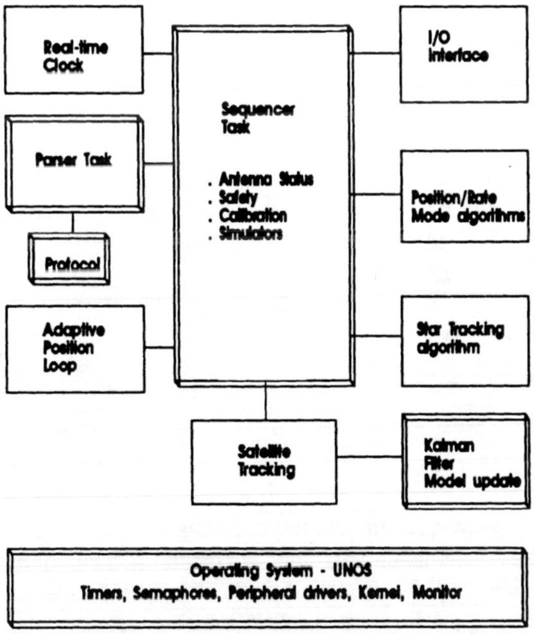

Figure 2 : Antenna controller task relationships.

As with the previous case study, a sequencer was employed to supervise antenna control. Figure 2 provides an overview of sequencer operation. The sequencer in this case, however, was a single task modelled as a sequential state machine driven by various internal states on each control sampling period. To reduce jitter in control sampling and digital-to-analogue output, the sequencer was signalled every control interval using the UNOS repetitive timer facility and was designated as the highest priority task in the system. This resulted in the sequencer being guaranteed to run with only the timing jitter due to background interrupt routines and kernel task switch times. Being a sequential state machine, state changes were synchronised with the sampling period making controller changeovers and integrator and adaptive filter resetting simpler. This was particularly important considering that there were over 100 states and modes that the sequencer could be in and smooth controller changeover was essential.

The sequencer was primarily concerned with the state of the control system, whereas other house-keeping duties were performed by other tasks. The sequencer contained no resource waits. This ensured that, regardless of the status of other tasks - and provided that the kernel itself did not crash - the sequencer could maintain antenna control, and take appropriate action on system failure. As a further precaution a hardware watchdog had to be refreshed by the sequencer at regular intervals. On sequencer task failure the system would reset, and the antenna motor drives could be automatically shut down.

The sequencer was also responsible for calling the position control loops. These consisted of adaptive feedforward PI controllers. The adaptive algorithm is described elsewhere [4].

An extensive simulation facility was built into the system from the start. This facility was built into the sequencer and other tasks enabling the entire system to be run on any PC. NVRAM, digital inputs and outputs, antenna structural resonance, wind disturbances, serial input, and satellite beacon locations were all simulated. This proved advantageous as no changes in software were required, even after 12 months development without a real antenna, during the commissioning stage.

What this case study has told us is that a controller can not be thought of in isolation of other system components, even the more intangible aspects of project management and software management become factors in its design. The boundaries of where the linear controller code ends and the operating system begins is a vague one. The ATCU alone consisted of over 100,000 lines of documented code, of which only 0.1% was for the adaptive controller. However the behaviour of the controller in this case could only be properly studied with the entire real-time system running - something present day simulation packages can not achieve easily and modern controller theory has not addressed.

We found the use of a simulation facility built into the code beneficial from a development viewpoint. With a well thought out simulation capability, commissioning becomes merely a fine tuning exercise. This raises another important point: in practice, the plant is usually well known beforehand. All *a priori* knowledge of the plant, even before it is built, should be used in the design. Controller analysis does form part of the design process, especially where adaptive loops are involved; but most of the control loop coding is related to engineering a solution based on hundreds or even thousands of non-linear conditions, boolean states, as well as computer and operating system issues.

Conclusions

Experience with the above-mentioned as well as other projects leads to the following conclusions:–

(i) All but trival real-time programming projects should be implemented using a operating system kernel specifically designed for real-time applications. The use of a RTOSK enhances the modularity of the software, and allows structured development even if there is an ill-defined specification.

(ii) Access to the source of the kernel is very important since is allows the kernel to be customized and bugs to be fixed.

(iii) Provision of debugging aids in the form of error display, timing violation and operating system monitor facilities is important in large projects. The latter two imply special RTOSK support for their implementation.

References

[1] Betz, R., and Sciacca L. (1992). UNOS User Manual (version 1.5a). *Tech. Rept.*, University of Newcastle, Australia.

[2] Moylan, P. (1991). The PMOS definition modules. *Tech. Rept. EE9107*, University of Newcastle, Australia.

[3] Moylan, P. (1992). The PMOS real-time kernel. Intl. Workshop on Real-Time Programming, Bruges, Belgium.

[4] Sciacca L. J. , Evans R. J., Betz R. E.,and Cooper D. N. (1990). Adaptive Control for the Australia Telescope, 11th IFAC World Congress, Tallinn, Estonia.

REAL-TIME CELL CONTROL FOR FLEXIBLE MANUFACTURING

Atef Fahim, Kyunghyun Choi

Department of Mechanical Engineering
University of Ottawa, Ottawa, Canada

Abstract. Flexible Manufacturing Cells (FMCs) may be considered the most significant development in small batch manufacturing. Setting up and operating costs prove to be the most major hinderance to large scale implementation and use of the FMCs, particularly by small and medium size industries. Incompatibility between the different components constituting the cells, and the lack of a unified language/approach to programming and coordinating it, is cited as the cause of the complexity of setting up and subsequent operating the cells. An object oriented software environment, UniSet, is being developed at the University of Ottawa to address the shortcomings of setting-up and operating FMCs comprising incompatible components from different suppliers. The environment provides the users with a consistent platform to configure, program, simulate, and control an FMC irrespective of the nature of its constituents. This paper presents the object oriented cell representation approach, and the interaction between the objects in UniSet. This approach allows the development of task/high level cell programs, and the simulation and verification of these programs. Real-time cell control and coordination in UniSet benefit a great deal from the object based method of representation.

Keywords. Flexible manufacturing; Industrial production systems; Programable controllers; Object oriented programming; Real time flexible manufacturing cell control.

INTRODUCTION

A manufacturing cell consists of a number of programmable devices with suitable sensors interacting harmoniously to do useful work and produce a product. The cell is considered flexible if it can be reconfigured to produce differing products through software changes. Justification of the FMC approach is provided by the speeded flow of parts, inventory control, reduction of labour costs, and reduction of change-over time from one part style to another (Bjorke, 1979). A machining cell would typically include one or more machine tools serviced by a suitable robot. An assembly cell may include one or more robots interacting with automatic clamps and fixtures, and a suitable in-feeder and output buffer. Device coordination in the cell can be carried out by any of the machine controllers, or by a separate controller. In general, the constituents of the cell are produced by different equipment manufacturers. Due to the lack of standard communication protocols and programming environments and languages, the problems associated with programming and coordinating the cell with all its incompatible components is left to the user.

Several approaches have been developed over the years to deal with the problems of programming, scheduling and controlling flexible manufacturing cells and systems. The vast majority of these approaches are based on Petri nets, extended Petri nets, or timed Petri nets (Besombers, Marcon, and Alla, 1992; Egilmez, and Kim, 1989, 1990; Harhalakis, Nagi, Proth, and Vernadat, 1992; Jeng, and DiCesare, 1990; Ravichandran, and Chakravarty, 1990). These allow the behaviour of the system to be represented as event graphs for which rules governing operation of the system can be derived and verified. Other approaches utilizing knowledge bases and concepts of artificial intelligence have also been proposed (O'Grady, and Lee, 1990). E-nets, which are essentially variations of Petri nets have also been used as means of representing and controlling FMCs and FMSs (Nof, Whinston, and Bullers, 1980). E-nets prove to be a powerful tool in the operation stage where the structure of manufacturing environment is invariant, however, would require physical and logical mapping when a change in the structure of the manufacturing environment occurs. Lately object oriented approaches have emerged. The reported approaches were very closely geared to a specific application, e.g. tool loading in an FMS (Bu-Halaiga, and Chakravarty, 1988), or mould filling (Joannis, and Krieger, 1992), and would require a great deal of knowledge about OOP to set up an FMS or FMC.

The approach reported in this paper uses object oriented concepts. Rather than tackling different aspects of construction and operation of an FMC separately, a comprehensive cell environment, UniSet (Unified Instruction Set), is developed.

This paper deals with object oriented cell representation approach adopted for UniSet. The approach provides the backbone for carrying out program integrity check, simulation, as well as the entire operation of the cell control mode.

THE UniSet ENVIRONMENT

UniSet has been developed as a consequence of a comprehensive study of automated machine programming languages, and to the problems encountered when setting up and reconfiguring FMCs. The study showed that similar types of machines, (e.g. lathes) from different machine-tool manufacturers, as well as different types of machines (i.e. machining centers, lathes, robots, etc.) have a great deal of commonalities in their abstract functionalities. The differences between the command languages for the same abstract function are mainly lexical. This commonality between the abstract functions forms the first basis of UniSet. UniSet uses a single nonexclusive command and command modifiers for each function irrespective of the machine. A machine can be programmed in UniSet or in its native language. The UniSet environment carries out the translation from UniSet to the native machine language. The cell controller, however, is programmed in UniSet. UniSet lexical is nonstatic. Commands can be added to the set locally at any installation in order to take advantage of unique features of machines in the cell. Furthermore, most of the problems encountered when setting up a FMC arise from non-homogeneous external communication methods, and communication protocols. To operate under the UniSet environment the cell constituents have to be connected in a star network with the hub being a dedicated computer as shown in Fig. 1. This computer acts as cell programming and coordinating station, as well as carrying out the task of harmonization and translation between the different communication protocols of the devices.

The general structure of an FMC program is that of interleaved device program fragments synchronized by appropriate sensor signals. Some of these program fragments may be concurrent. In UniSet a cell program is developed in a serial fashion with the program fragments threaded together by the synchronizing signals. Parts of code obtained from other sources like CAD software tool path programs are also imbedded in the cell program as code fragments. This approach of serial code-fragment development is both natural and intuitive, and greatly simplifies FMC programming. Once an FMC program is developed, the UniSet environment checks its integrity, and decomposes it into its code-fragments. The code-fragments for each machine are collected in a single file. Depending on the machine controller, the code-fragments are sent one at a time as required, or the entire file is downloaded and the code-fragments executed upon commands from the cell controller. A simple example of a UniSet program and the resulting code-fragments are shown in Fig. 1.

The UniSet environment provides four modes of operation. A cell setup mode, a cell programming mode, a simulation mode, and a cell control mode. Description of these modes follow:

Setup mode: This is a protected mode accessible only to supervisory engineers. A cell designer uses this mode to define the number and type of devices forming the cell, as well as their characteristics and placement within the cell space. The characteristics include such salient features as number of axes, programming language, etc. The physical placement is used in a subsequent stage of the cell programming mode to carry out an integrity check on the program.

Programming mode: A cell operator uses this mode of operation to program the cell. As mentioned earlier the cell program is developed entirely in UniSet or as a combination of UniSet and native languages of the cell constituents. The environment carries out the translation between UniSet and the different native languages of the cell constituents. Two levels of program integrity checks are carried out in this mode. The first occurs during program entry, and is intended to detect syntactic errors and the validity of the command for the intended machine. The command interpreter imbedded in the program editor notifies the cell programmer immediately of the error. The second integrity check is carried out at the end of the program and is intended to detect such subtle errors like collision between machines, excessive metal removal, etc.

The ability to add commands locally to UniSet, and to present the programmer with only the commands pertinent to the machine for which the code-fragment is being developed provide UniSet with a unique feature. Task level commands can be developed and imbedded in the instruction set of different machines. Depending on how a task-level instruction was coded, in the translation phase UniSet expands these commands to a UniSet low-level instruction set or to native machine command set. Fig. 2 shows the task level UniSet program and its equivalent low level program.

Simulation Mode: In this mode, the environment uses the cell program to simulate the cell operation. This allows programs to be developed off-line and evaluated before implementation.

Control mode: Portions of the cell program for a machine are collected, and if coded in UniSet are translated to the native language of that machine. The cell controller dispatches the code to the machine controller, but retain supervisory tasks like start, stop, and receipt and transmission of external signals in order to coordinate the cell.

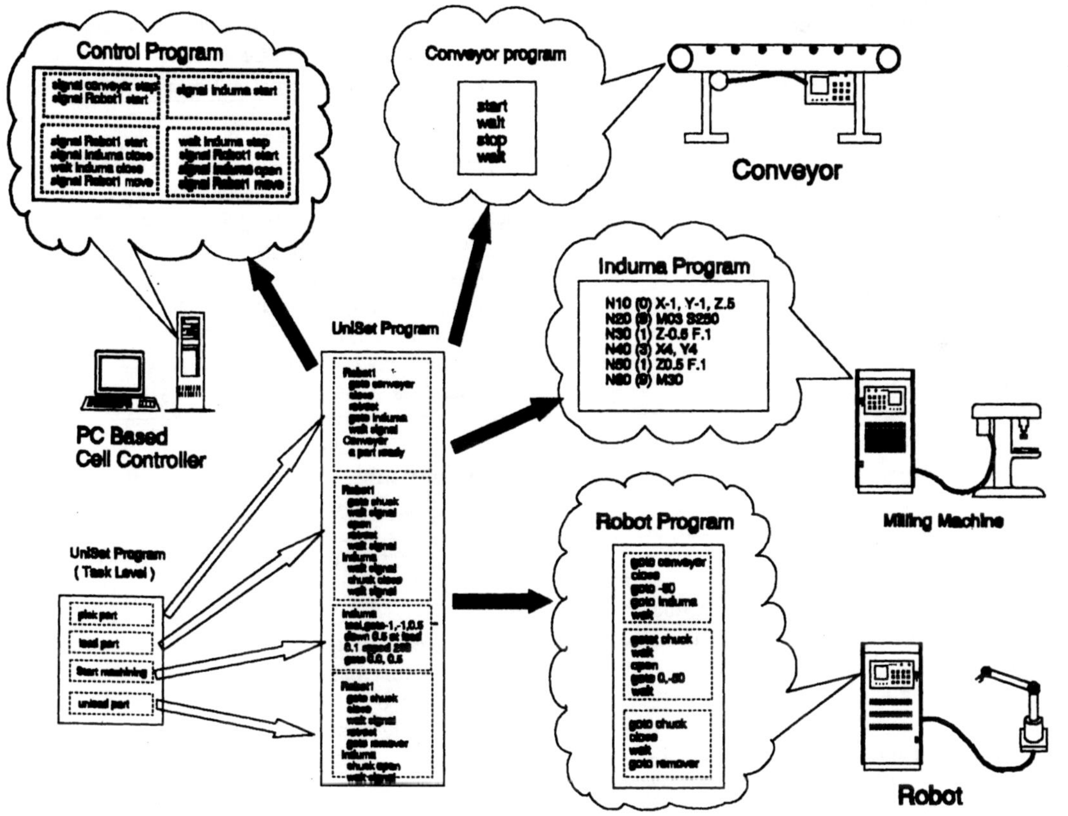

Fig. 1. Cell network and UniSet program translation.

OBJECT ORIENTED CELL REPRESENTATION APPROACH

The cell representation scheme was developed in such a way that it would be able to reflect the physical structure of FMCs, and to portray their functional and behavioral characteristics using a coherent and closely coupled set of tools. Object Oriented Philosophies (OOPs) lend themselves to the development of such a scheme. The adoption of OOPs also proves beneficial since UniSet is developed using Smalltalk, an object oriented programming language/environment. The object oriented cell representation approach adopted for UniSet is fundamental for the operation of the cell program integrity check and the real-time control mode.

Using the notation adopted by Booch (1991), a description of the object oriented approach developed for cell representation in UniSet follows.

Machines: Each machine type constituting the cell is represented by a "real" machine class, and a set of virtual device classes. The "real" machine class is in fact the communication link between the controlling computer at the hub of the cell network and the machine controller. The set of virtual device classes is constructed as a shadow machine that has a structure mimicking in abstraction the physical machine. To that end, a machine in the cell is decomposed to a main component and a number of other components which interact or interface with other constituents of the cell. One class is created for the main component, and as many classes as required are created for the individual components. The virtual classes are used as holders of information, and states concerning the respective machine components. With reference to Fig. 2, two examples of the machine representation follow:

- For robots in the cell;
  A "real" robot class *Robot is created.* This class is used for communication and inherits its characteristics from a device driver super class *Dev_Drv*.
  A set of virtual classes is created. One is the virtual robot class *Robot_V*. This class holds all the information concerning the robot arm itself as reported to it by the class *Robot*. The other two classes for the tool gripper, *Tl_Grp_V*, and the part gripper, *Prt_Grp_V*, hold information concerning these two devices. Both *Tl_Grp_V* and *Prt_Grp_V* inherit their characteristics from the super-class *Clamp*.
- For the milling machine;
  A "real" milling machine class *Mill* is created. *Mill* is also a subclass of Dev_Drv.
  A virtual machine class *Mill_V* is created to hold information concerning the main component of the milling machine. Three other classes are created for the virtual door *Door_V*, for the

virtual part holder *Prt_Hldr_V*, and for the tool holder *Tl_Hldr_V*. Both *Prt_Hldr_V* and *Tl_Hldr_V* inherit their characteristics from the super class *Clamp*.

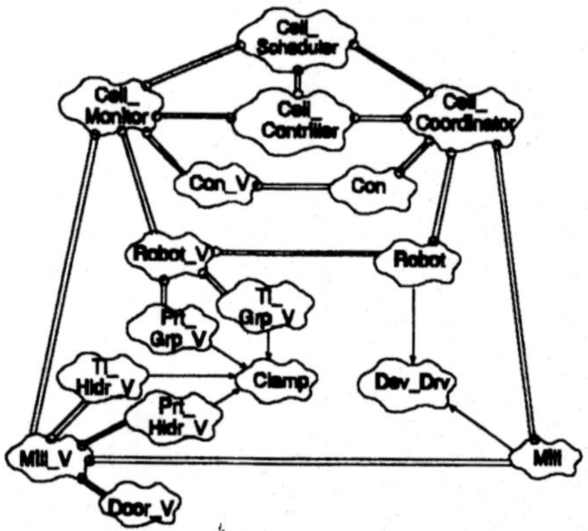

Fig. 2. Class diagram for the cell.

Cell Monitoring: The cell status and functionality are monitored using a class *Cell_Monitor*. This class uses all the virtual component classes to gather the required information for its functioning.

Cell Scheduling: Scheduling the different tasks of the cell is carried out by the class *Cell_Scheduler*. Whenever the cell state changes, *Cell_Monitor* informs *Cell_Scheduler* which uses a set of rules and the task level program of the cell to determine the next task to be executed.

Cell Coordination: *Cell_Coordinator* is responsible for coordinating the activities of the cell constituents. When it receives requests to carry out a task from the *Cell_Scheduler*, it fetches the appropriate cell control code-fragment and executes it. It uses event schedules and direct sensory information from the "real" machine classes to sequence and pace the machines. Depending on the characteristics of the machine controller, *Cell_Coordinator* may down-load machine code-fragments to the machines. *Cell_Coordinator* class deals only with the routine operation of the cell.

Cell Control: Overall control of the cell is carried out by a cell controller class *Cell_controller*. This class uses information from *Cell_Monitor*, *Cell_Scheduler*, and *Cell_Coordinator*, as well as a set of rules to continually assess the integrity of the cell operation, detect malfunctions, and deal with exceptional conditions. All communication between the outside world and the cell or its components takes place via *Cell_Controller*.

TASK INITIATION DIAGRAMS

Most of the approaches that have been developed so far to represent the cell activities are adaptations of ones used in digital logic or electric circuitries. In these approaches the cell or machine state is considered primal. The state of the cell or one of its constituents change as a result of firing of an event. This approach is not suitable for control or verification of flexible manufacturing cells or systems. For such systems the task to be performed rather than the state must be regarded as primal. A task is executed when a particular set of conditions are met.

The methodology developed for UniSet regards the tasks to be performed by the cell or any of its constituent machines as being primal. Sensory signals indicating the change of state of machines are used to trigger or initiate tasks. Tasks may be simple and require a relatively short time to execute, or may be complex and lengthy. The methodology may be depicted by a set of diagrams we dubbed "task initiation diagrams", and their accompanying rules.

Conventions and terminology used for the task initiation diagrams:

The task initiation diagrams are multi-layered. The topmost layer shows the task level program of the cell. For example, the task level UniSet program of Fig. 1 is shown diagrammatically in Fig. 3. For conciseness, in this paper signals/states and tasks will be designated by symbols rather than their verbose description. The verbose description of the symbols are shown in Tables 1 to 4. Description of the components and rules for these diagrams follow:

Tasks: Tasks in the top level diagram would generally consist of a concerted group of subtasks or operations involving more than one constituent of the cell, and in such a case are termed composite tasks. These are shown by the framed boxes in Fig. 3. As an example TE1 (load part) involves concerted operations involving the milling machine and the robot. Tasks involving a single machine are called simple tasks and are shown by a box, e.g. TE2 (machine part). For complex cell programs, more than one top level diagram may be required. Furthermore, the second level diagrams may include composite tasks that would necessitate a third level of diagrams.

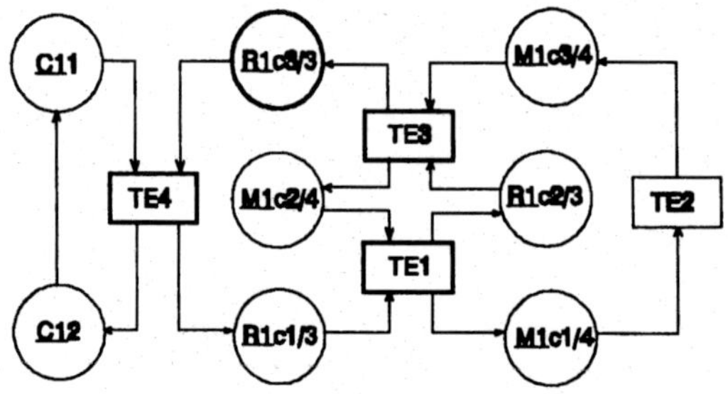

Fig. 3. Task initiation diagram.

Composite states: The cell state is given at any instant by the collection of states of its constituents. For initiation of a given task, a composite state consisting of a subset of the cell state needs to exist. These composite states are shown in Fig. 3 and Fig. 4 by thick line circles, e.g. R1c3/3 or M1c3/4. The last number of the symbols show how many individual states are required to determine this composite state. A task level UniSet program does not include explicit information about the states necessary to initiate the tasks. This is because such information is imbedded in the implementation of each task. In second level diagrams, composite states like R1c1/3 and M1c2/4 (Fig. 4) get decomposed to their individual states at one end, and new ones like M1c1/4 and R1c2/3 are composed at the other end. States that require time to get established are symbolized by a loop back to itself and a bubble indicating the upper bound on time. As an example, C12 (part not available) in Fig. 3 require some time for the conveyor to move a part to the robot pickup location and change the state to C11 (part available).

Operations: Second level diagrams show details of the composite tasks. Each task is shown on a separate diagram. As an example, the details of task TE1 (load part) of Fig. 3 are shown in the diagram of Fig. 4. Tasks in these second level diagrams generally involve single machines, and as such will be termed operations so as to distinguish them from the tasks of the top level diagrams.

Minimal state set: The initiation of an operation requires a specific set of states to exist. This minimal state set is necessary to ensure the successful and safe completion of the operation. The states that constitute the minimal state set are indicated by thick lines leading into the operation box as shown in Fig. 4. At least one state of that set will change as the result of the operation. With reference to the figure, OP1 (robot: approach to load part) for example, requires; (a) that the robot be at location "load" (R15) near the machine, this state would change to R19 (location: vise) after the operation; (b) that the machine is stopped (M17); (c) that the machine door is open (M13).

Complimentary state sets: The states constituting these sets are indicated by solid lines leading into an operation box. Complementary sets exist in pairs, the set and its complement. If more than one pair exists, numbers are used to indicate the complementary pair. For loosely controlled cells, or during dry runs the absence of a complementary pair would not prevent an operation from being initiated. As an example, OP1 (robot approach to load part) has a set comprising R13 (gripper closed) and R11 (part in gripper), and its complement M16 (part not in vise) and M11 (vice open). The presence of one of those state sets requires the existence of the other before OP1 can be initiated. This pair of complementary state sets ensure that the robot would not jam one part against another in the vise.

Optional state set: These states are used for confirmation only. They are indicated on the diagrams by dashed lines leading into operation boxes. Failure to detect the presence of an optional set of states would not prevent an operation from being initiated, but would trigger a warning or information message. As an example M15 (part in vise) is not essential for safe closure of the vise (OP2), but may be used to warn about a potentially useless operation.

State transition: State that undergo transitions as a result of an operation are indicated by solid lines leading out of the operation box. All other states that do not undergo transitions as a result of an operation are indicated by dashed lines leading out of the operation box.

Auxiliary states: Auxiliary states are set by the *Cell_Coordinator*. They are external signals used to start operations that would otherwise automatically start due to the presence of all other states necessary for the initiation of the task. Auxiliary states or signals are indicated on the operation transition diagrams by a circle with a token leading into an operation box. With reference to Fig. 4, all the states necessary for the initiation of OP2 originate from the milling machine. To prevent an automatic initiation of the operation when the state conditions are satisfied, an auxiliary signal is introduced.

Conservation of the number of states: The number of states across an operation box must be constant. This rule is necessary for ease of verification of cell program integrity.

Continuity of state transitions: A continuous path must exist for each state between composite states across an operation initiation diagram. This ensures that operations will exist in the cell program to cycle a state through its different phases.

CONCLUSION

This paper presents the object oriented approach developed for an FMCs environment, UniSet, developed at the University of Ottawa. Using this environment all aspects of cell setup, programming, simulation, and control are carried out. Contrary to the commonly used representation approaches for cell activity which concentrate on the states of the cell constituents and their change of states, the approach presented here focuses on the tasks and operations that need to be performed and regards the states of the cell constituents as mere triggers to the cell activities.

User defined task level instructions can be added to the UniSet instruction set. Using these instructions, a programmer would be able to program the cell efficiently, view graphically the cell program, and simulate the cell operation.

Based on a set of rules derived from the task initiation diagrams and imbedded in the implementation of *Cell_Controller*, *Cell_Scheduler*, and *Cell_Coordinator*, tasks and operations are coordinated and executed.

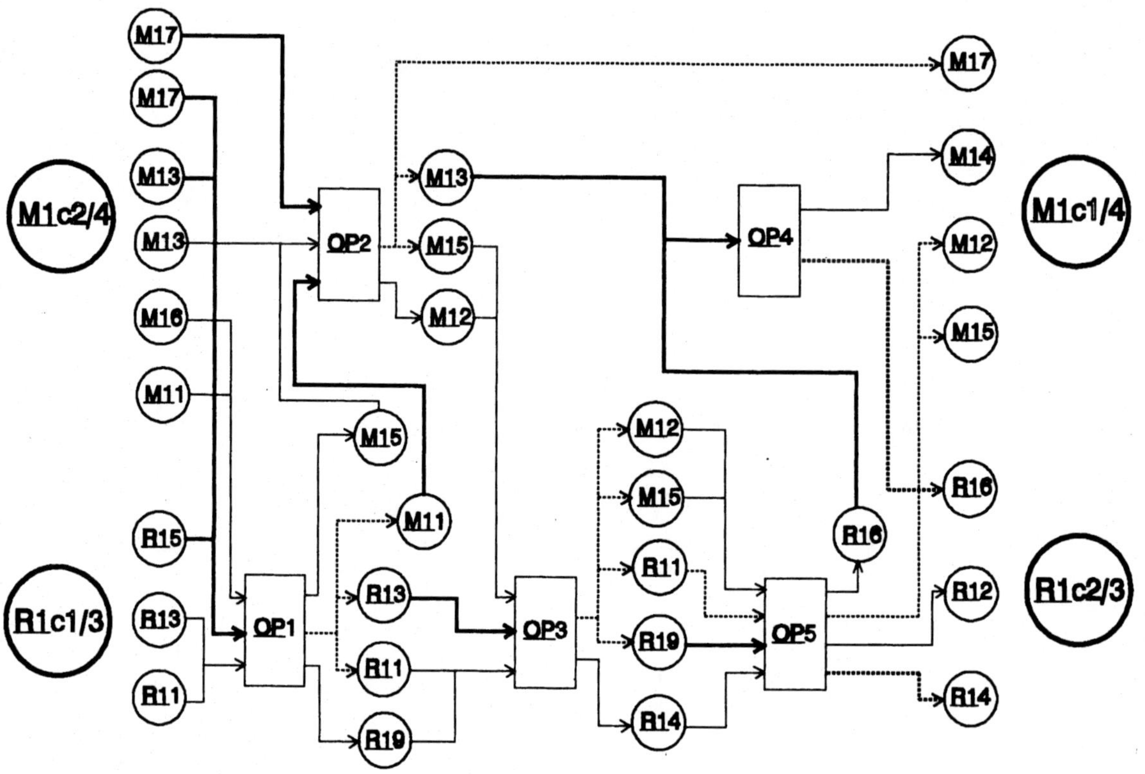

Fig. 4. Operation initiation diagram for task TE1 of Fig. 3.

Table 1 Tasks and Operations

Cell Tasks		Machine Operations	
Name	**Description**	**Name**	**Description**
TE1	Load part	OP1	Robot approach to load part
TE2	Start machining	OP2	Machine close vise
TE3	Unload part	OP3	Robot open gripper
TE4	Pick part	OP4	Machine close door
		OP5	Robot retract

Table 2 Robot Composite States

STATE	Gripper	Part	Location	Name
R1c1/3	Closed (R13)	Present (R11)	Load (R15)	Ready to laod part
R1c2/3	Open (R14)	Absent (R12)	Unload (R16)	Ready to Unload part
R1c3/3	Open (R14)	Absent (R12)	Pick	Ready to pick part from convayor
R1c4/3	Closed (R13)	Present (R11)	Deliver	Ready to deliver part to out bin

Table 3 Milling Machine States

Robot States		Mill States		Convayor States	
Name	**Description**	**Name**	**Description**	**Name**	**Description**
R11	Part in gripper	M11	Vise open	C11	Part present
R12	Part not in gripper	M12	Vise closed	C12	Part not available
R13	Gripper closed	M13	Door open		
R14	Gripper open	M14	Door closed		
R15	Location: load	M15	Part in vise		
R16	Location: unload	M16	Part not in vise		
R19	Location: vise	M17	Machine stopped		

Table 4 Mill Composite States

STATE	Vise	Part	Door	Activity	Name
M1c1/4	Closed (M12)	Present (M16)	Closed (M14)	Stopped (M17)	Ready to laod part
M1c2/4	Open (M11)	Absent (M15)	Opened (M13)	Stopped (M17)	Ready to Unload part
R1c3/4	Open (M12)	Present (R16)	Open (M13)	Stopped (M17)	Ready to pick part from convayor

Reference

Besombers, B., E. Marcon, and A. Alla (1992). Application of Generalised Bond-Graphs and Continuous Petri Nets to Modelling Industrial Processes and Manufacturing Systems. IFAC-INCOM'92, 617-622.

Bjorke, O. (1979). Computer aided part Manufacturing. Computers in industry, 1, 3-9.

Booch, G. (1991). Object oriented design with applications. Benjamin/Cummings Pub. Co., Calif.

Bu-Hulaiga, M. I., and A. K. Chakravarty (1988). An object-oriented knowledge representation for hierarchical real-time control of flexible manufacturing. International Journal of the Production Research, Vol.26, No.5, 777-793.

Egilmez, K., and S. H. Kim (1990). A logical approach to knowledge based control. Journal of Intelligent Manufacturing, 1, 59-76.

Elilmez, K., and S. H. Kim (1989). Design and Control of Manufacturing Systems: An Integrated Approach based on Modal Logic and Zonecharts. Robotics & Computer-Integrated Manufacturing, Vol.6, No.3, 209-228.

Harhalakis, G., R. Nagi, J. M. Proth, and F. Vernadat (1992). Design of Manufacturing Systems: A Bottom-Up Approach Based on Petri Nets", IFAC-INCOM'92, 678-682.

Jeng, M. D., and F. DiCesare (1990). A review of synthesis techniques for Petri Nets. Proceedings of the second International Conference on CIM, 348-355.

Joannis Robert, and Krieger Moshe (1992). Object-Oriented approach to the specification of manufacturing systems. Computer-Integrated Manufacturing Systems, Vol.5, No.2, 113-145.

Nof, S., A. Whinston, and W. Bullers (1980). Control and decision support in automatic manufacturing systems. AIIE Transactions, 12, 156-167.

O'Grady, P. J., and K. H. Lee (1990). An Intelligent Cell Control System for automated manufacturing. Knowledge-based Systems in manufacturing(Edited by Kusiak.A.,), 151-172.

Ravichandran, R., and A. K. Chakravarty (1987). Decision Support in Flexible Manufacturing Systems using Timed Petri Nets. Journal of Manufacturing System, Vol.5, No.2, 89-101.

THE DIGITAL CONTROL SYSTEM REALIZATION OF A SHORT RANGE SURFACE TO AIR MISSILE

Yangsik Baek, Jeongwhan Kim, Byungchul Kwak, Kyupil Cho

Agency for Defence Development, Taejeon, Korea

Abstract. This paper presents the digital control system realization of a Short Range Surface to Air Missile(SRSAM). The digital control system consists of on-board computer, control sensors and input/output data interface, whose functions are processing of autopilot algorithm, control gain scheduling, fin command mixing, and data interfacing. Design requirements of the on-board computer are to execute the autopilot algorithm within 5 msec, to meet the interface protocol with other subsystems, and to minimize the weight and size. In order to verify the performance of the realized digital control system, real time Hardware-In-the-Loop Simulation(HILS) using ADI100/10 computer and Carco Flight Motion Simulator(FMS) is performed.

Keyword. Digital control system; Autopilot; On-board computer; Real time HILS

INTRODUCTION

This paper describes a hardware realization of digital control system for missile. We choose a cruciform SRSAM which is aerodynamically controlled for this study. A cruciform SRSAM system generally requires fast response and high stability for wide environmental conditions.

The control loop schematic diagram of this missile is shown in Fig. 1. For this missile model, we design the digital control system and implement hardware including on board computer. Decision procedures of microprocessor, memory size, input/output data interface according to system requirements is discussed in detail.

A standard test and evaluation method to verify a developed missile control system performance before real flight test is Hardware-In-the-Loop Simulation(HILS), whose usefulness is already known.

In order to verify the realized digital control system in hardware and software, we performed real time HILS and compared the results with those of mathematical model simulation.

CONFIGURATION OF CONTROL SYSTEM

Digital control system of the SRSAM consists of pitch, yaw, and roll autopilots. In case of cruciform missile, the pitch and yaw controller are identical.

The role of pitch and yaw control system are to make the missile follow the acceleration guidance commands. The role of roll control system is to maintain the roll positon angle of missile body to zero. Control gain scheduling method is used for a fast and stable missile control system to adapt to a wide variation of flight environment.

The role of G-Command resolver is to convert the guidance command in guidance coordinate into that of body coordinate. The role of δ-Command limitter and mixer is to limit actuator control command and to convert pitch, yaw, and roll actuation commands to 4 control surface commands.

The block diagram of digital control system is shown in Fig. 2.

Pitch/Yaw Autopilot

Pitch/yaw autopilot is designed to get the time response as fast as possible while satisfying the specification of stability margin. The designed pitch/yaw autopilot block diagram is shown in Fig. 3. The pitch/yaw autopilot consists of control sensors and lead/lag compensator. The sensors are the rate gyro and accelerometer for measuring the body motion.

Where,

$$COM1 = L_1 * W_i + \frac{L_2 * W_i}{1 - Z^{-1}}$$

(L_1, L_2 are constants)

$$COM2 = K_1 + \frac{K_3 * Z^{-1}}{1 - K_2 * Z^{-1}} + \frac{K_5 * Z^{-1}}{1 - K_4 * Z^{-1}}$$

($K_1 \sim K_5$ are constants)

In pitch/yaw autopilot, the control gain Ka, Kv, Kr, and W_i are the functions of mach(M) and height(H) of the missile. The control gains are obtained from Eq. (1).

$$K = M * (A * H + B) + C * H + D \tag{1}$$

In Eq.(1), A, B, C, and D are gain table constants and decided by M and H.

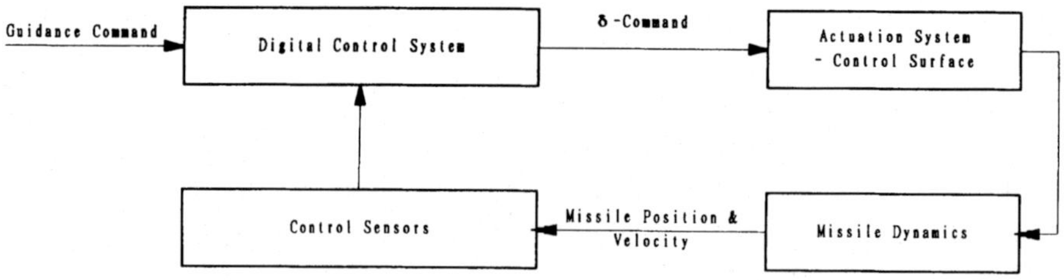

Fig. 1. Control Loop Schematic Diagram of a SRSAM

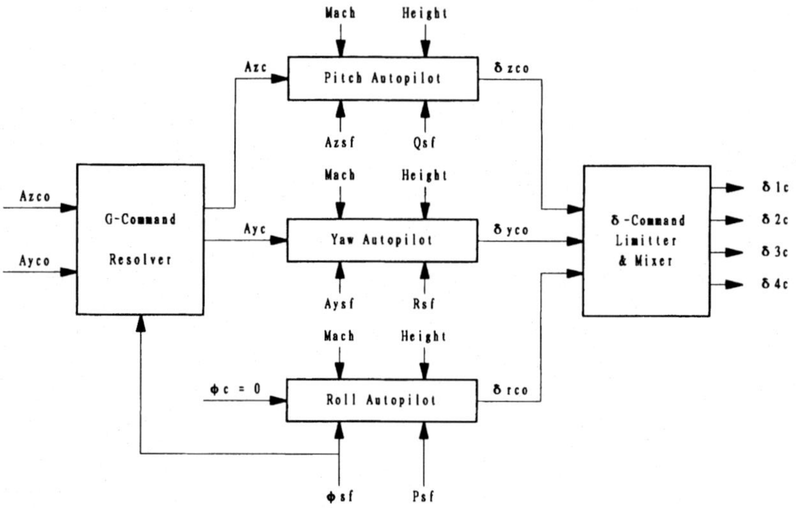

Azco : Pitch Guidance Command(G)
Ayco : Yaw Guidance Command(G)
Azc : Pitch Guidance Command in Body Axis(G)
Ayc : Yaw Guidance Command in Body Axis(G)
Azsf : Pitch Accelerometer Output(G)

Aysf : Yaw Accelerometer Output(G)
Qsf : Pitch Rate Gyro Output(Deg/sec)
Rsf : Yaw Rate Gyro Output(Deg/sec)
φ sf : Roll Position Gyro Output(Deg)
Psf : Roll Rate Gyro Output(Deg/sec)

Fig. 2. Digital Control System Block Diagram

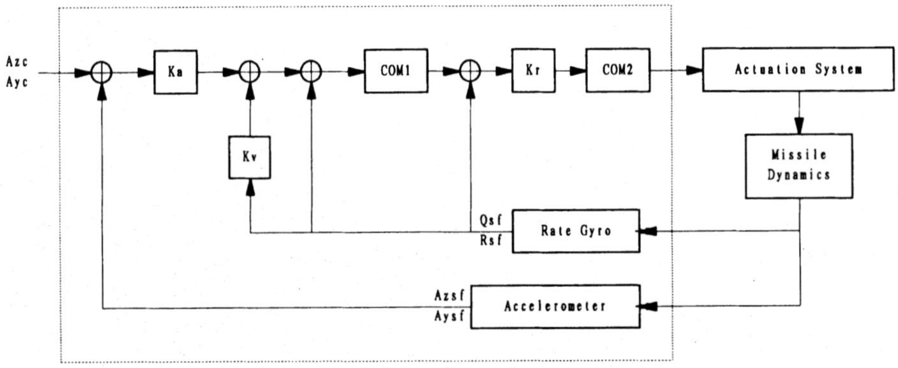

Fig. 3. Pitch/Yaw Autopilot Block Diagram

<u>Roll Autopilot</u>

Roll autopilot is designed to minimize the effect of external disturbance and induced roll moment on the missile. The designed roll autopilot block diagram is shown in Fig. 4. The roll autopilot consists of control sensors and compensator. The sensors are free gyro and rate gyro. The adaptive control gain K_{phi} and K_p are obtained from Eq. (1).

Where,

$$COM3 = K_1 + \frac{K_4 * Z^{-1} + K_5 * Z^{-1}}{1 + K_2 * Z^{-1} + K_3 * Z^{-1}} + \frac{K_7 * Z^{-1}}{1 - K_6 * Z^{-1}}$$

($K_1 \sim K_7$ are constants)

<u>G-Command Resolver</u>

G-Command resolver in Fig.2 is used to convert the guidance commands in guidance coordinate into those of body coordinate, and each command is expressed in Eq. (2), (3).

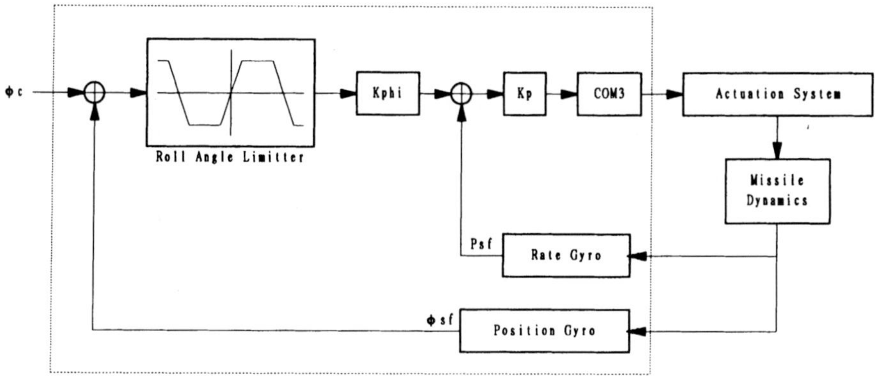

Fig. 4. Roll Autopilot Block Diagram

$$A_{zc} = A_{zco} * \mathrm{Cos}(\phi_{sf}) - A_{yco} * \mathrm{Sin}(\phi_{sf}) \quad (2)$$

$$A_{yc} = A_{zco} * \mathrm{Sin}(\phi_{sf}) + A_{yco} * \mathrm{Cos}(\phi_{sf}) \quad (3)$$

<u>δ-Command Limitter and Mixer</u>

δ-Command limitter and mixer is to limit actuator command and to convert δ_{zco}, δ_{yco}, and δ_{rco} commands to 4 control surface commands, and performed like Eq. (4)~(10).

$$\delta_{rc} = \mathrm{Min}(\ \delta_{rco},\ \pm 10_{deg}) \quad (4)$$

$$\delta_{zc} = \mathrm{Min}(\ \delta_{zco},\ \delta_{max}) \quad (5)$$

$$\delta_{yc} = \mathrm{Min}(\ \delta_{yco},\ \delta_{max}) \quad (6)$$

$$(\delta_{max} = 20_{deg} - \mathrm{Abs}(\ \delta_{rc}\ ))$$

$$\delta_{1c} = \delta_{rc} - \delta_{zc} \quad (7)$$

$$\delta_{2c} = \delta_{rc} + \delta_{yc} \quad (8)$$

$$\delta_{3c} = \delta_{rc} + \delta_{zc} \quad (9)$$

$$\delta_{4c} = \delta_{rc} - \delta_{yc} \quad (10)$$

HARDWARE DESIGN AND IMPLEMENTATION

For the hardware implementation of the digital control system, we design the on-board computer, I/O data interface circuits, and sensor package. Also programs are developed to execute the autopilot algorithm, firing sequence, and hardware tests.

In the design of on-board computer, arithmetic calculation speed, accuracy, total program size, input/output data interface, and automatic testability are considered.

In hardware implementation, manufacturing process, operational environmental condition, size, and weight limitations are also considered.

Components are selected to meet the required military specifications.

<u>On-Board Computer Construction</u>

The schematic diagram of digital on-board computer which realizes digital control system is shown in Fig. 5.

The microprocessor controls the firing sequence during launch phase and caculates δ-Command during flight phase. The coprocessor is used for the floating point arithmetic calculation.

On-board computer contains telemetry interface circuit for ground test and flight test results analysis.

<u>Microprocessor & coprocessor selection.</u> In the selection of microprocessor and coprocessor, military specification, available instruction set, calculation speed, and accuracy are considered.

To program autopilot algorithm, floating point arithmetic(multiplication, division, addition, and subtraction), control transfer(jump and branch), bit manipulation, and data transfer instructions are needed.

The word length requirement for floating point arithmetic is more than 32 bits(short real).

The loop calculation time requirement for the autopilot program is less than 5 msec.

With these constraints, we select Intel i80c186/187 processor. Intel i80c186/187 processor has 80 bit floating point arithmetic capability and 12Mhz clock speed.

<u>Memory selection.</u> In the selection of memory, total program length, access time, and slim-size are considered.

Static RAM(SRAM) is selected for simplicity of circuit. SRAM is used for data buffer. The required word size is about 1K.

Erasable programmable ROM(EPROM) is selected for easy change of program. EPROM is used for program and data table. The required word size is about 23K (initialization program 2K, digital control program 3K, test/communication program 8K, and gain/data table 10K).

We select 2K word SRAM and 32K word EPROM to meet the increasing program requirements in future.

<u>Input interface.</u> The input data types of on-board computer are analog, serial digital, and discrete signal.

Initial condition of on-board computer is transmitted

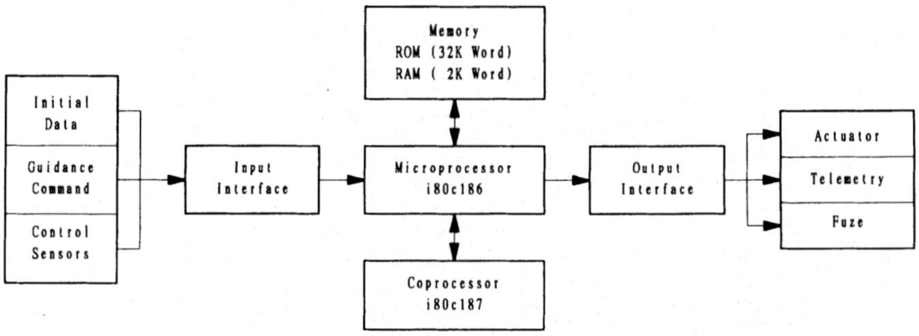

Fig. 5. On-board Computer Schematic Diagram

from Missile Firing Unit(MFU) using Synchronous Data Link Control (SDLC) protocol. For SDLC, Intel 8274 Multi-Protocol Serial Controller(MPSC) chip is used. EIA RS-422A interface circuit is used between on-board computer and MFU.

Guidance commands, mach, and height are received by SRSAM's RF receiver. RF receiver transmit these information to on-board computer using RS-422A method.

Sensor outputs are analog signal. Analog signal is converted to digital data by 12 bit Analog to Digital(A/D) converter. Converted digital data is stored in the dual port RAM. The converting period is 800μsec for each sensor output.

Output interface. The output data types of on-board computer are analog, parallel digital, and discrete signal.

Actuator command(δ) is analog signal. Digital data is converted to analog signal by 12 bit Digital to Analog(D/A) converter.

Parallel digital data for the telemetry is stored in the dual port RAM. Telemetry can read the dual port RAM data at any time.

Discrete signals for the fuze are transmitted through the open collector interface circuit.

On-board Computer Programming

On-board computer program consists of three parts, as shown in Fig. 6.

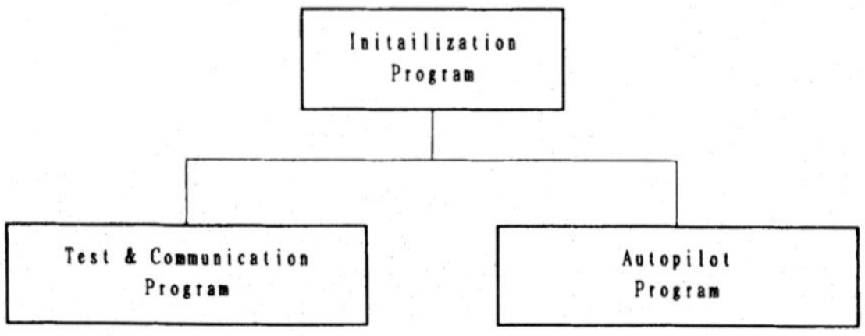

Fig. 6. On-Board Computer Program

Initialization program intializes microprocessor, MPSC, and I/O interface circuits, and clears SRAM.

Test/communication program is used by Automatic Test Set(ATS)/MFU to execute the hardware testing on the ground, and the firing sequence during launch phase.

Autopilot program executes digital control algorithms.

These programs are written in assembly language. Assembly language reduces program execution time and minimizes program size.

Modular programming method is used for easy debugging and changing.

Autopilot program execution time and memory size. Each module's function and execution time of autopilot program are shown in TABLE 1.

TABLE 1 Function, Size & Execution Time of On-Board Program Module

Module	Function	Size (byte)	Execution Time (μsec)
Roll Sensor	Roll Rate & Position Reading	68	67
Pitch/Yaw Sensor	Pitch/Yaw Acc. & Rate Reading	160	125
Pitch Autopilot	Pitch Autopilot Calculation	82	140
Yaw Autopilot	Yaw Autopilot Calculation	78	140
Roll Autopilot	Roll Autopilot Calculation	245	302
Command Resolver	Body Axis Command Calculation	89	209
δ Limit & Mixing	$\delta_{1c,2c,3c,4c}$ Calculation	394	447
δ Output	Actuator Command Output	106	129
Gain	Adaptive Gain Calculation	593	840
Filter	Pitch/Yaw/Roll Filter Calculation	216	368
Telemetry	Telemetry Digital Data Output	289	85

Execution time is measured by ICE186 emulator. Floating point arithmetic is programmed in short real data format(32 bits). If long real data format(64 bits) is used, the execution time is increased about 20 percents.

Program emulation. For the program emulation, Intel ICE186 emulator is used. Emulator checks program operation and measures execution time.

Total autopilot program execution time is about 3.5 msec and meets digital control system requirement (less than 5 msec).

Hardware Test

Electrical performance of on-board computer is checked by ATS. Test item is shown in TABLE 2.

TABLE 2 ATS Test Item

Test Item	Test Action
Memory	ROM, RAM Read/Write & Checksum
Power Supply	+5, ±15 V DC & 22/9 V AC
A/D , D/A	Data Conversion
Interface	Telemetry & Fuze Interface
Sensor Package	Sensors Output
Autopit Program	Autopilot Program Running

REAL TIME HILS

Static performance and operational function of the on-board computer can be checked by program emulation and ATS testing. It's real time computing capability, dynamic performance, and especially sensor performance can be proved only by real time HILS.

Real time HILS block diagram is shown in Fig. 7.

AD100/10 computer calculates aerodynamic, actuator, accelerometer, and roll position gyro model equation, generates angular motion command for FMS. Also AD100/10 transmits acceleration and roll position to on-board computer.

Rate gyros mounted on the FMS measure angular rate(p, q, and r). On-board computer caculates actuator command(δ) using this rate signal, acceleration, and roll position received from AD100/10. AD100/10 computer calculates model equation and then transmits acceleration and roll position to on-board computer.

This simulation process is repeated in real time until predefined mission time for HILS.

Typical flight profile is shown in Fig. 8. Mathematic model simulation results are shown in Fig. 9. Real time HILS results are shown in Fig. 10. Acceleration, actuator command, and pitch rate are shown in Fig. 9 and Fig. 10. For comparison purpose, we only show these results.

Real time HILS results and mathematic model simulation results are almost same. The small difference comes from sensor noise, FMS dynamic characteristics, quantization error of on-board computer, and word length difference (mathematic model simulation uses 64 bits word length).

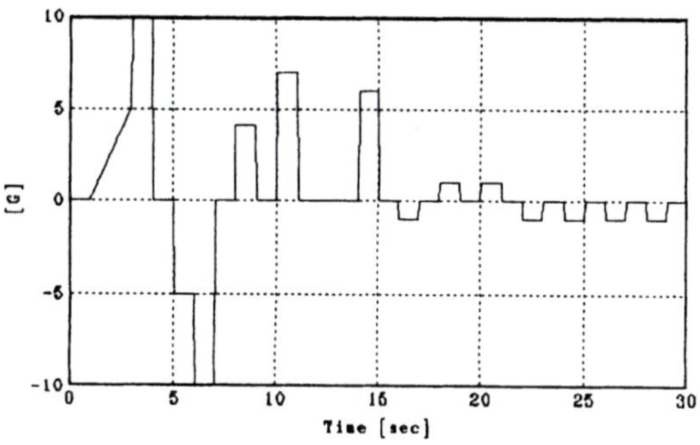

Fig. 8. Flight Profile (Azco)

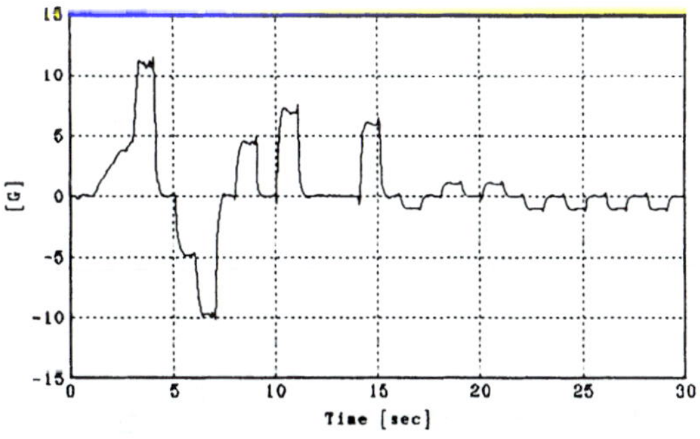

Fig. 9.1. Acceleration (Azsf - S/W)

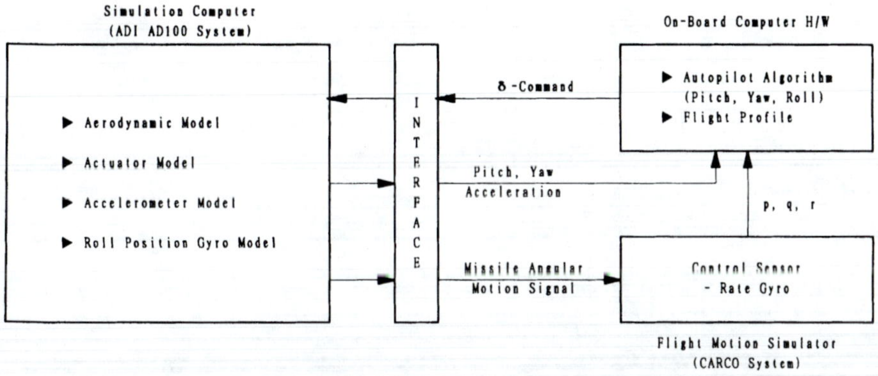

Fig. 7. Real Time HILS Block Diagram

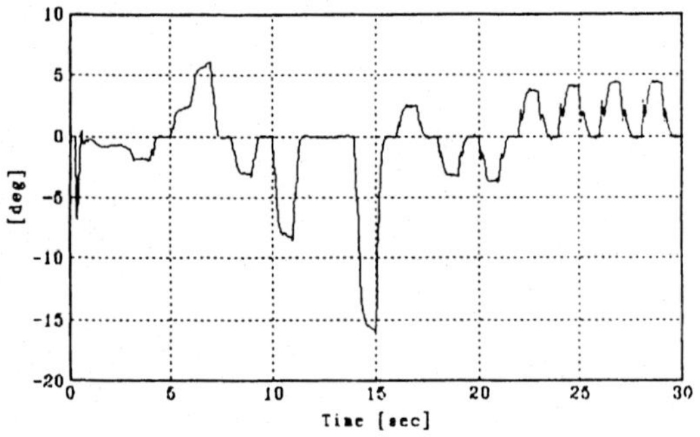

Fig. 9.2. Actuator Command ($\delta 1c$ - S/W)

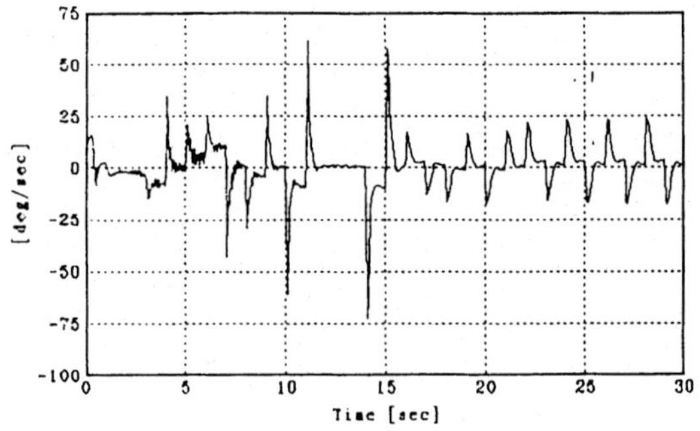

Fig. 10.3. Pitch Rate (Qsf - H/W)

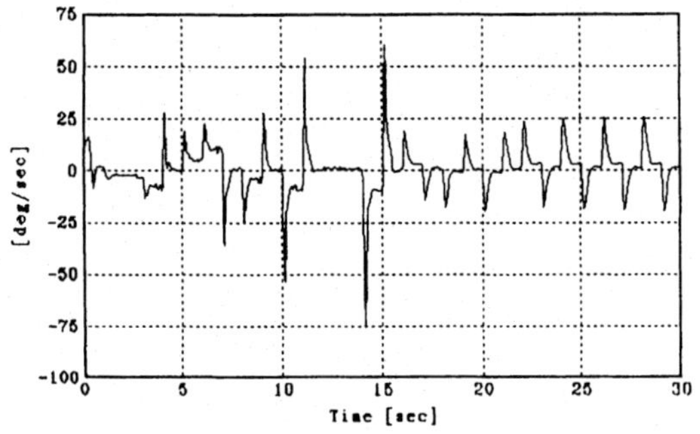

Fig. 9.3. Pitch Rate (Qsf - S/W)

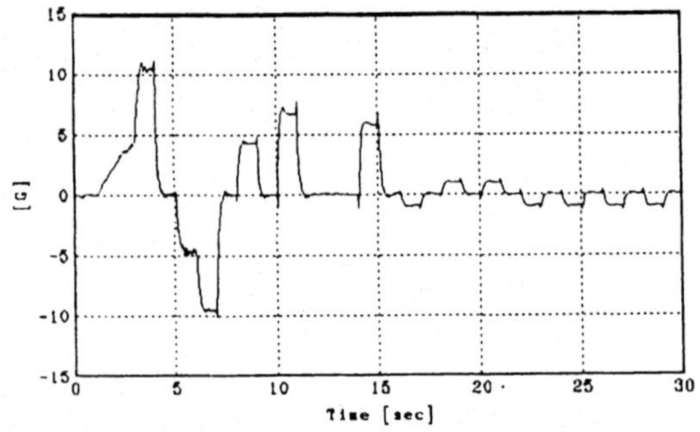

Fig. 10.1. Acceleration (Azsf - H/W)

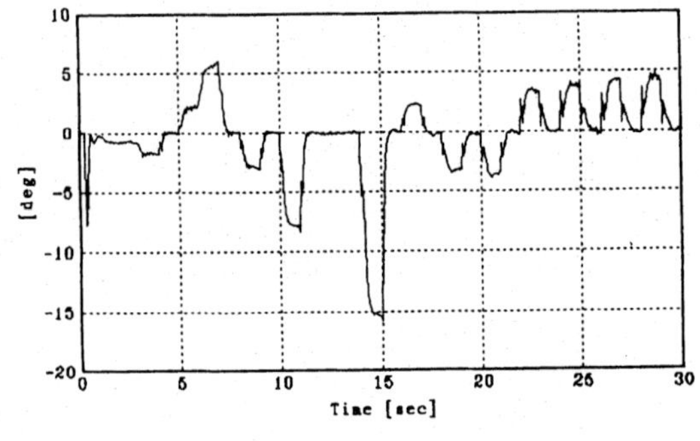

Fig. 10.2. Actuator Command ($\delta 1c$ - H/W)

CONCLUSION

In this paper, a digital realization of a SRSAM control system has been presented.

The on-board computer is designed based on 16 bits microprocessor.

The autopilot algorithm is programmed in assembly language. The calculation time for the autopilot is about 3.5 msec, which is fast enough to control a cruciform SRSAM.

The dynamic performance of the control system has been confirmed by a real time HILS. The HILS results are compared with mathematic model simulation results.

It is found that the designed control system is fast and stable.

REFERENCE

Eichblatt, E. J. (1989). Test and Evaluation of the Tactical Missile. AIAA, Washington DC.

Garnell, P. (1977). Guided Weapon Control Systems. Pergamon Press, London.

Lin, C. F. (1991). Modern Navigation, Guidance, and Control Processing. Pretice Hall, New Jersey.

Middleton, R. H. and Goodwin, G. C. (1990). Digital Control & Estimation. Prentice Hall, New Jersey.

Intel Data Book, (1991). 16/32 Bit Embedded Processors. ISBN 1-55512-122-5.

Military Standard MIL-STD-1562. List of Standard Microcircuits. Department of Deffence, U.S.A.

Military Specification MIL-M-38510. Microcircuits, General Specification for. Department of Deffence, U.S.A.

Military Standard MIL-STD-810C. Environmental Test Methods. Department of Deffence, U.S.A.

Continuation Method for Dynamic Economic Load Dispatch with Emission Constraints

Taiji Satoh, Hiroshi Fujii and Tsuyoshi Okita

Department of Electrical and Electronic Engineering, Faculty of Engineering, Yamaguchi University, 2557, Tokiwadai, Ube, 755, Japan

Abstract. This paper presents a method for solving a dynamic economic load dispatch problem of electric power systems. Since the expected system load varies over a given time horizon, the sum of generator outputs must follows the system load subject to power-rate limits on generators. In order to solve the dynamic ELD problem, the continuation method is employed in this paper. It is shown that each generator output is expressed as a piecewise-linear function of a continuation parameter, and this fact enables us to obtain the optimal solution efficiently. Numerical tests for a model system indicate the validity and effectiveness of the proposed method.

Keywords. Power system control; dynamic economic load dispatch; emission constraints; optimization; mathematical programming; continuation method.

INTRODUCTION

The economic load dispatch (abbreviated as ELD, hereafter) is an optimization problem to find the most economical schedule of thermal generators in electric power systems. In the references (Happ, 1977; Chowdhury, 1989), the literature of the ELD problem and its solution methods are surveyed. The most fundamental problem among them is the classic ELD problem. Since the classic ELD is an optimization problem subject to the system demand at a single time period only, Aoki and Satoh (1984) presented efficient algorithms for finding the optimal dispatching schedule to the classic ELD problem.

Since the expected system load varies over a given time horizon, the dynamic properties of power systems must be taken into consideration. In the dynamic ELD problem, therefore, the optimization must be made subject to such constraints as upper and lower bounds on generator outputs, system power balance equations and power-rate limits on thermal generators. Moreover, the environment constraints such as NOx and SOx emission should also be taken into consideration. Many methods have been proposed for solving the dynamic ELD problem; for example, Ross and Kim (1980) reported the successive approximation dynamic programming algorithm, Isoda (1982) proposed the equal incremental fuel cost principle combined with backward and forward dispatching procedures, van den Bosch (1985) employed the gradient-projection method using conjugate search direction.

As described in the survey paper by Richter and DeCarlo (1983), the continuation method is a powerful approach for solving the system of nonlinear equations. Huneault and co-workers (1985) applied the continuation method to the economic dispatch problem of single period, and it is used for the parametric analysis when some system constants are changed. In this paper, the continuation method is employed for solving the dynamic ELD problem of multiple periods. It is shown that the generator output is expressed as a piecewise-linear function of a continuation parameter. Thus the optimal solution can be found very efficiently.

In the followings, the mathematical formulation of the dynamic ELD problem is given firstly. Next, the solution algorithm based on the continuation method is presented. Then the validity of the proposed algorithm is shown through numerical tests. The final chapter states the conclusions and open questions.

FORMULATION

The dynamic ELD problem (it is called "P" for simplicity, hereafter) is formulated as follows:

$$P : Min. \quad \sum_{t\in T}\sum_{i\in I^t} f_i(g_i^t) \tag{1}$$

$$s.t. \quad g_{LBi} \leq g_i^t \leq g_{UBi},\ i \subset I^t,\ t \subset T \tag{2}$$

$$\Delta_{DWi} \leq g_i^t - g_i^{t-1} \leq \Delta_{UPi},\ i \in I^t,\ t \in T \tag{3}$$

$$\sum_{t\in T} d_i g_i^t \leq H_i,\ i \in I \tag{4}$$

$$\sum_{i\in I^t} g_i^t = D^t,\ t \in T \tag{5}$$

where

T : index set of time periods,
I^t : index set of on-line generators,
g_i^t : output of generator-i at period-t,
g_{UBi}, g_{LBi} : upper and lower bounds of g_i^t,
$\Delta_{UPi}, \Delta_{DWi}$: ramping up and down rate of g_i^t ($\Delta_{DWi} < 0 < \Delta_{UPi}$),
D^t : electric power demand at period-t,
d_i : NOx or SOx emission coefficient,
H_i : upper bound on NOx or SOx emission of generator-i.

The fuel cost $f_i(g_i^t)$ of thermal generators can be approximated by the following quadratic function:

$$f_i(g_i^t) = a_i + b_i g_i^t + \frac{1}{2}c_i\{g_i^t\}^2 \tag{6}$$

where a_i is a constant term, and b_i and c_i are linear and quadratic coefficients, respectively.

The problem P defined above is a quadratic programming (QP) problem; *i.e.*, the objective function is quadratic and all the constraints are linear. However, it is not a wise way to solve P directly by using a general purpose QP algorithm (van de Panne,

1975). This is because the problem P has the following special problem structure:

1. the objective function is separable, and
2. the constraints are sparse and they have a block structure.

In this paper, therefore, exploiting the fact that the ELD problem of single period can be solved efficiently, the continuation method is employed for solving the dynamic ELD problem by continuously transforming the problem of single period to the desired dynamic one.

CONTINUATION METHOD

Parameterization

In the continuation method (Richter and DeCarlo, 1983), the continuation parameter θ is introduced into the original problem P and the parameterized problem $P(\theta)$ is constructed. Let us assume that the optimal solution to $P(\theta)$, for $\theta = 0$, can be trivially or easily found, and the optimal solution to $P(\theta)$, for $\theta = 1$, will be the desired solution to the original problem P. If the trajectory of the solution exists and moves in a continuous manner, then the optimal solution to the original problem can be found by increasing the continuation parameter θ from 0 to 1. In this paper, it is shown that the trajectory is expressed as a piecewise-linear function of θ. Therefore the optimal solution can be found very efficiently by following the piecewise-linear trajectory.

In this paper, the continuation parameter θ is introduced into the power balance equation (5), and the parameterized problem $P(\theta)$ is defined as follows:

$$P(\theta) : Min. \sum_{t\in T}\sum_{i\in I^t} f_i(g_i^t) \quad (7)$$

$$s.t. \ g_{LBi} \le g_i^t \le g_{UBi}, \ i \in I^t, \ t \in T \quad (8)$$

$$\Delta_{DWi} \le g_i^t - g_i^{t-1} \le \Delta_{UPi}, \ i \in I^t, \ t \in T \quad (9)$$

$$\sum_{t\in T} d_i g_i^t \le H_i, \ i \in I \quad (10)$$

$$\sum_{i\in I^t} g_i^t = (1-\theta)D_0^t + \theta D^t, \ t \in T \quad (11)$$

where D_0^t is the basic power demand defined by

$$D_0^t = \sum_{i\in I^t} g_{LBi}, \ t \in T \quad (12)$$

It is evident that the parameterized problem $P(\theta)$ coincides with the original problem P when $\theta = 1$, and that the optimal solution to $P(\theta)$, for $\theta = 0$, is given by

$$g_i^t(\theta)\Big|_{\theta=0} = g_{LBi}, \ i \in I_t, \ t \in T \quad (13)$$

Therefore the original problem P can be solved by following the trajectory of $g_i^t(\theta)$ when θ is increased from 0 to 1.

In the followings, the active set strategy (Gill and Murray, 1977) is applied to the problem $P(\theta)$, and only the active ramping up and down rate constraints and the active emission constraints for the current value of θ are taken into consideration. Let us assume that the following constraints are active:

$$g_i^t - g_i^{t-1} = \Delta_{UPi}, \ i \in J_{UP}^t, \ t \in T \quad (14)$$

$$g_i^t - g_i^{t-1} = \Delta_{DWi}, \ i \in J_{DW}^t, \ t \in T \quad (15)$$

$$\sum_{t\in T} d_i g_i^t = H_i, \ i \in K_{ENV} \quad (16)$$

where J_{UP}^t and J_{DW}^t are the set of active ramping up and down rate constraints from time period $t-1$ to t, and K_{ENV} is the index set of generators of which emission constraint is active. Then the parameterized problem becomes

$$PA(\theta) : Min. \sum_{t\in T}\sum_{i\in I^t} f_i(g_i^t) \quad (17)$$

$$s.t. \ g_{LBi} \le g_i^t \le g_{UBi}, \ i \in I^t, \ t \in T \quad (18)$$

$$g_i^t - g_i^{t-1} = \Delta_{UPi}, \ i \in J_{UP}^t, \ t \in T \quad (19)$$

$$g_i^t - g_i^{t-1} = \Delta_{DWi}, \ i \in J_{DW}^t, \ t \in T \quad (20)$$

$$\sum_{t\in T} d_i g_i^t = H_i, \ i \in K_{ENV} \quad (21)$$

$$\sum_{i\in I^t} g_i^t = (1-\theta)D_0^t + \theta D^t, \ t \in T \quad (22)$$

Optimal Condition

In this section, the generator output g_i^t is expressed as a function of the continuation parameter θ. Firstly, the Lagrangian function of $PA(\theta)$ is formulated, next the optimal conditions are derived by differentiating the Lagrangian function. Then the desired function $g_i^t(\theta)$ is obtained by solving the linear system of optimal conditions.

The Lagrangian function of the problem $PA(\theta)$ is defined by

$$\begin{aligned} L(\theta) = & \sum_{t\in T}\sum_{i\in I^t} f_i(g_i^t) \\ & + \sum_{t\in T} \lambda^t\{(1-\theta)D_0^t + \theta D^t - \sum_{i\in I^t} g_i^t\} \\ & + \sum_{t\in T}\sum_{i\in J_{UP}^t} \mu_i^t(g_i^t - g_i^{t-1} - \Delta_{UPi}) \\ & + \sum_{t\in T}\sum_{i\in J_{DW}^t} \xi_i^t(\Delta_{DWi} - g_i^t + g_i^{t-1}) \\ & + \sum_{i\in K_{ENV}} \alpha_i \sum_{t\in T} d_i(g_i^t - H_i) \\ & + \sum_{t\in T}\sum_{i\in I^t} \gamma_i^t(g_{LBi} - g_i^t) \\ & + \sum_{t\in T}\sum_{i\in I^t} \rho_i^t(g_i^t - g_{UBi}) \end{aligned} \quad (23)$$

By differentiating $L(\theta)$ with respect to g_i^t, the following stationary point condition is obtained.

$$\begin{aligned} & b_i + c_i g_i^t - \lambda^t + \mu_i^t - \mu_i^{t+1} - \xi_i^t + \xi_i^{t+1} \\ & \quad + d_i\alpha_i - \gamma_i^t + \rho_i^t = 0 \end{aligned} \quad (24)$$

where λ^t is the Lagrange multiplier associated with the power balance equation (22), γ_i^t and ρ_i^t associated with the upper and lower bounds on generator outputs (18), μ_i^t and ξ_i^t associated with the ramping up and down rate constraints (19) and (20), and α_i associated with the emission constraints (21).

The above equation (24) can be solved for g_i^t, and g_i^t is expressed as a linear function of the Lagrange multipliers.

$$\begin{aligned} g_i^t = & (\lambda^t - \mu_i^t + \mu_i^{t+1} + \xi_i^t - \xi_i^{t+1} \\ & - d_i\alpha_i + \gamma_i^t - \rho_i^t - b_i)/c_i \end{aligned} \quad (25)$$

For simplicity, as in the case with the algorithm for the classic ELD problem, the Lagrange multipliers γ_i^t and ρ_i^t associated with the simple bounds on variables are implicitly treated in the algorithm and they are eliminated from eq.(25). By substituting eq.(25) into the power balance equation (22), the following linear equation is obtained:

$$\begin{aligned} & \sum_{i\in I_{MID}^t} \{(1/c_i)\lambda^t - (1/c_i)\mu_i^t + (1/c_i)\mu_i^{t+1} \\ & \quad + (1/c_i)\xi_i^t - (1/c_i)\xi_i^{t+1} - (d_i/c_i)\alpha_i - b_i/c_i\} \\ & = (1-\theta)D_0^t + \theta D^t - \sum_{i\in I_{LB}^t} g_{LBi} - \sum_{i\in I_{UB}^t} g_{UBi} \end{aligned} \quad (26)$$

where I_{MID}^t is the set of free generators which take an intermediate value between lower and upper bounds; I_{LB}^t and I_{UB}^t are the sets of generators fixed at lower and upper bounds, respectively.

By substituting eq.(25) into the ramping up and down rate constraints (19) and (20), the following linear equations are obtained:

$$\begin{aligned} g_i^t - g_i^{t-1} &= -(1/c_i)\lambda^{t-1} + (1/c_i)\lambda^t \\ &\quad +(1/c_i)\mu_i^{t-1} - (2/c_i)\mu_i^t + (1/c_i)\mu_i^{t+1} \\ &= \Delta_{UPi} \qquad (27) \\ g_i^t - g_i^{t-1} &= -(1/c_i)\lambda^{t-1} + (1/c_i)\lambda^t \\ &\quad -(1/c_i)\xi_i^{t-1} + (2/c_i)\xi_i^t - (1/c_i)\xi_i^{t+1} \\ &= \Delta_{DWi} \qquad (28) \end{aligned}$$

In the same way, the following equation is derived from eq.(21):

$$\begin{aligned} &\sum_{t\in T} d_i\{(1/c_i)\lambda^t - (1/c_i)\mu_i^t + (1/c_i)\mu_i^{t+1} \\ &\quad +(1/c_i)\xi_i^t - (1/c_i)\xi_i^{t+1} - (d_i/c_i)\alpha_i - b_i/c_i\} \\ &= H_i \qquad (29) \end{aligned}$$

In order to demonstrate the structure of the equations (26)–(29), let us consider the following small-scale example; the system composed of three generators over three time periods ($I = \{1,2,3\}$, $T = \{1,2,3\}$), five active ramping up constraints ($J_{UP}^1 = \{1,2,3\}$, $J_{UP}^2 = \{3\}$, $J_{UP}^3 = \{2\}$), and two active emission constraints ($K_{ENV} = \{1,2\}$). In this case, the number of Lagrange multipliers are ten, *i.e.*, three for power balance equations, five for active ramping constraints and two for active emission constraints. Since all the equations (26)–(29) are linear, the linear system of equations to be solved can be written as follows:

$$Ax = b \qquad (30)$$

where A is the 10×10 coefficient matrix shown in Table 1, and x is a column vector defined by

$$x = \begin{bmatrix} \lambda^1 & \lambda^2 & \lambda^3 & \mu_1^1 & \mu_1^2 & \mu_1^3 & \mu_2^3 & \mu_3^2 & \alpha_1 & \alpha_2 \end{bmatrix}^T \qquad (31)$$

The right-hand-side vector b is given by

$$b = \begin{bmatrix} (1-\theta)D_0^1 + \theta D^1 - \sum_{i\in I_{LB}^1} g_{LBi} - \sum_{i\in I_{UB}^1} g_{UBi} \\ (1-\theta)D_0^2 + \theta D^t - \sum_{i\in I_{LB}^2} g_{LBi} - \sum_{i\in I_{UB}^2} g_{UBi} \\ (1-\theta)D_0^3 + \theta D^t - \sum_{i\in I_{LB}^3} g_{LBi} - \sum_{i\in I_{UB}^3} g_{UBi} \\ \Delta_{UP1} + g_1^0 \\ \Delta_{UP1} \\ \Delta_{UP1} \\ \Delta_{UP2} \\ \Delta_{UP3} \\ H_1 \\ H_2 \end{bmatrix}$$

$$= \begin{bmatrix} D_0^1 - \sum_{i\in I_{LB}^1} g_{LBi} - \sum_{i\in I_{UB}^1} g_{UBi} \\ D_0^2 - \sum_{i\in I_{LB}^2} g_{LBi} - \sum_{i\in I_{UB}^2} g_{UBi} \\ D_0^3 - \sum_{i\in I_{LB}^3} g_{LBi} - \sum_{i\in I_{UB}^3} g_{UBi} \\ \Delta_{UP1} + g_1^0 \\ \Delta_{UP1} \\ \Delta_{UP1} \\ \Delta_{UP2} \\ \Delta_{UP3} \\ H_1 \\ H_2 \end{bmatrix} + \begin{bmatrix} D^1 - D_0^1 \\ D^2 - D_0^2 \\ D^3 - D_0^3 \\ 0 \\ 0 \\ 0 \\ 0 \\ 0 \\ 0 \\ 0 \end{bmatrix} \cdot \theta \qquad (32)$$

As is evident from the definition of A and b, the continuation parameter θ appears only in the right-hand-side of the equation, eq.(30) can be solved *numerically* for Lagrange multipliers λ^t, μ_i^t, ξ_i^t and α_i by using an appropriate subroutine package for the system of linear equations. Thus the Lagrange multipliers are expressed as a linear function of θ and they can be written as follows:

$$\begin{aligned} \lambda^t &= \lambda_C^t + \lambda_L^t\theta & (33) \\ \mu_i^t &= \mu_{Ci}^t + \mu_{Li}^t\theta & (34) \\ \xi_i^t &= \xi_{Ci}^t + \xi_{Li}^t\theta & (35) \\ \alpha_i &= \alpha_{Ci} + \alpha_{Li}\theta & (36) \end{aligned}$$

where subscripts "C" and "L" respectively denote constant term and linear coefficients.

By substituting eqs.(33)–(36) into eq.(25), the generator output g_i^t is expressed as a linear function of the continuation parameter θ.

$$\begin{aligned} g_i^t &= (\lambda^t - \mu_i^t + \mu_i^{t+1} + \xi_i^t - \xi_i^{t+1} - d_i\alpha_i - b_i)/c_i \\ &= \{\lambda_C^t + \lambda_L^t\theta - (\mu_{Ci}^t + \mu_{Li}^t\theta) + (\mu_{Ci}^{t+1} + \mu_{Li}^{t+1}\theta) \\ &\quad +(\xi_{Ci}^t + \xi_{Li}^t\theta) - (\xi_{Ci}^{t+1} + \xi_{Li}^{t+1}\theta) \\ &\quad -d_i(\alpha_{Ci} + \alpha_{Li}\theta) - b_i\}/c_i \\ &= g_{Ci}^t + g_{Li}^t\theta, \ i \in I_{MID}^t \qquad (37) \end{aligned}$$

where

$$\begin{aligned} g_{Ci}^t &= (\lambda_C^t - \mu_{Ci}^t + \mu_{Ci}^{t+1} + \xi_{Ci}^t - \xi_{Ci}^{t+1} \\ &\qquad -\alpha_{Ci} - b_i)/c_i & (38) \\ g_{Li}^t &= (\lambda_L^t - \mu_{Li}^t + \mu_{Li}^{t+1} + \xi_{Li}^t - \xi_{Li}^{t+1} - \alpha_{Li})/c_i & (39) \end{aligned}$$

Update of Active Set

Equation (37) indicates that the generator output g_i^t is an increasing (decreasing) function of θ, if the linear coefficient $g_{Li}^t > 0$ ($g_{Li}^t < 0$). Thus the value of θ at which the generator output g_i^t becomes active on its upper or lower bound can be directly calculated from eq.(37). By substituting eq.(37) into the ramping up constraint which is not active at the current value of θ, the value of θ when the ramping up constraints becomes active is calculated by the following equation:

$$\theta_{UP} = \{\Delta_{UPi} - (g_{Ci}^t - g_{Ci}^{t-1})\}/(g_{Li}^t - g_{Li}^{t-1}) \qquad (40)$$

In the same way, the value of θ when the inactive emission constraints becomes active can also be calculated.

TABLE 1 Coefficient Matrix

	λ^1	λ^2	λ^3	μ_1^1	μ_1^2	μ_1^3	μ_2^3	μ_3^2	α_1	α_2
D^1	$1/c_1 + 1/c_3$			$-1/c_1$	$1/c_1$			$1/c_3$	$-d_1/c_1$	
D^2		$1/c_1 + 1/c_2 + 1/c_3$			$-1/c_1$	$1/c_1$	$1/c_2$	$-1/c_3$	$-d_1/c_1$	$-d_2/c_2$
D^3			$1/c_1 + 1/c_2 + 1/c_3$			$-1/c_1$	$-1/c_2$		$-d_1/c_1$	$-d_2/c_2$
$g_1^1 - g_1^0$	$1/c_1$			$-1/c_1$	$1/c_1$				$-d_1/c_1$	
$g_1^2 - g_1^1$	$-1/c_1$	$1/c_1$		$1/c_1$	$-2/c_1$	$1/c_1$				
$g_1^3 - g_1^2$		$-1/c_1$	$1/c_1$		$1/c_1$	$-2/c_1$				
$g_2^3 - g_2^2$		$-1/c_2$	$1/c_2$				$-2/c_2$			
$g_3^2 - g_3^1$	$-1/c_3$	$1/c_3$						$-2/c_3$		
H_1	d_1/c_1	d_1/c_1	d_1/c_1	$-d_1/c_1$					$-3d_1^2/c_1$	
H_2		d_2/c_2	d_2/c_2							$-2d_2^2/c_2$

Next, let us discuss the case that the active ramping up and down constraints become inactive. The value of θ when the generator at lower or upper bound becomes a free generator which takes an intermediate value can be determined from the equal incremental fuel cost principle (Aoki and Satoh, 1984). For example, the generator at lower bound becomes a free generator when

$$\lambda^t \geq \lambda_{LBi} \tag{41}$$

where λ_{LBi} is the value of the incremental fuel cost at lower bound defined by

$$\lambda_{LBi} = b_i + c_i \cdot g_{LBi} \tag{42}$$

From the comparison between the Kuhn-Tucker conditions for the problems $P(\theta)$ and $PA(\theta)$, only the optimal conditions which may be violated by the change of θ are the non-negativity of the Lagrange multipliers associated with the active ramping up and down and emission constraints; *i.e.*, $\mu_i^t \geq 0$, $\xi_i^t \geq 0$ and $\alpha_i \geq 0$.

Since the Lagrange multipliers μ_i^t, ξ_i^t and α_i defined by eqs.(34)-(36) are linear, the Lagrange multipliers are decreasing (increasing) functions of θ, if the linear coefficients are negative (positive). From the viewpoint of the active set strategy, the negative value of the Lagrange multiplier indicates that the objective function can be improved if the constraint is removed from the current active set. Therefore the active ramping up and down and emission constraint must be removed from the active set when the value of $\mu_i^t \geq 0$, $\xi_i^t \geq 0$ and $\alpha_i \geq 0$ decrease and become to zero. Then the new problem $PA(\theta)$ is constructed for the updated active set of J_{UP}^t and J_{DW}^t.

Algorithm

From the above discussion, the algorithm of the continuation method for the dynamic ELD problem with emission constraints can be summarized as follows:

Step-0: Solve the parameterized problem $P(\theta)$ for $\theta = 0$, and express the generator output g_i^t as a function of the continuation parameter θ.

Step-1: Calculate the value θ_{min} by

- **i)** Find the value of θ when the intermediate value of g_i^t becomes lower or upper bounds.
- **ii)** Find the value of θ when the inactive ramping up or ramping down constraint become active.
- **iii)** Find the value of θ when the inactive emission constraint become active.
- **iv)** Find the value of θ when the generators at lower or upper bounds take intermediate values.
- **v)** Find the value of θ when the Lagrange multipliers μ_i^t and ξ_i^t become to zero.
- **vi)** Find the value of θ when the Lagrange multiplier α_i becomes to zero.
- **vii)** Set the minimum value of θ calculated in **i)**–**vi)** above as θ_{min}.

Step-2: If $\theta_{min} > 1$, the optimal solution is found for $\theta = 1$; terminate the procedure.

Step-3: Increase the value of θ to θ_{min} and update the active set. Next, solve the system of linear equations and return to Step-1.

NUMERICAL EXAMPLE

In order to check the validity of the proposed method, a small-scale model system (3-generators, 3-time periods) and a real-scale model system (15-generators, 6-time periods) are solved in this paper.

The generator characteristics of the small-scale model system are shown in Table 2. In this example, the unit of the generator output and system demand are expressed in [MW], and the length of one period is 10 minutes. The system demand at each period is shown in Table 3. The program of the proposed algorithm is written in FORTRAN and it is executed on APOLLO DN-3500 (4-MIPS) computer. In the Step-0 of the algorithm, the system is solved for $\theta = 0$ by using the equal incremental fuel cost principle. The initial solution and the final solution obtained by the proposed method are shown in Table 4. In this table, the first line shows the initial solution, and the subsequent lines are the optimal generator outputs at each time period. Fig. 1-(a), (b) and (c) show the output of each generator as a function of the continuation parameter θ. In these figures, the value of θ at each breakpoint designates the value at which the shape of the function changes. As shown in these figures, the generator output is expressed as a piecewise-linear function of θ. In the small scale example, only the ramping up and down constraints are active at the optimal solution.

For the real-scale model. the following three cases are solved:

- **i)** none of the emission constraints is active at the optimal solution
- **ii)** one emission constraint for the generator #7 is active at the solution
- **iii)** two emission constraints for the generators #7 and #10 are active at the solution

Fig. 2-(a) and (b) respectively show the trajectory of the generator output at time periods #3 and #4 for the case iii) just above. In this case, the emission constraints on the generators #10 and #7 become active when $\theta = 0.917$ and $\theta = 0.940$, respectively. Since the total generator output is restricted by these active emission constraints, the corresponding trajectory becomes flat for the value of $\theta \geq 0.917$ and 0.940.

Table 5 shows the computation time for the real-scale models. The same problem is solved by using the general purpose QP program (van de Panne, 1975), and the QP program finds the same solution to the continuation method. This fact indicates that the optimal solution is obtained by using the continuation method proposed in this paper. As shown in this table, the com-

TABLE 2 Model System

i	a_i	b_i	$c_i * 10^3$	g_{LBi}	g_{UBi}	Δ_{UPi}	Δ_{DWi}
1	181.2	6.430	3.317	80.0	156.0	20.0	-20.0
2	143.6	6.750	4.272	95.0	156.0	20.0	-20.0
3	288.7	5.471	7.842	150.0	250.0	40.0	-40.0

TABLE 3 System Demands

t	D^t
1	350.0
2	420.0
3	480.0

TABLE 4 Initial and Final Solution

g_1^0	g_2^0	g_3^0
80.0	95.0	150.0
g_1^1	g_2^1	g_3^1
100.0	95.0	155.0
g_1^2	g_2^2	g_3^2
120.0	105.0	195.0
g_1^3	g_2^3	g_3^3
140.0	114.5	225.5

TABLE 5 Computation Time

	Quadratic Programming	Continuation Method
Case i)	143.699 [s]	1.379 [s]
Case ii)	264.967 [s]	1.675 [s]
Case iii)	262.601 [s]	1.779 [s]

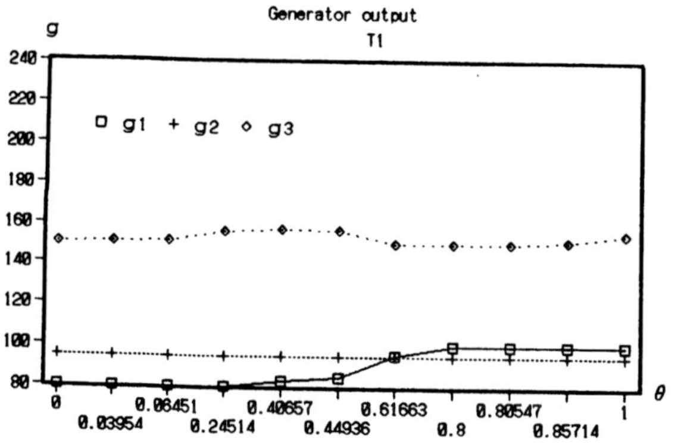

Fig. 1-(a). Generator output of small-scale system (t=1).

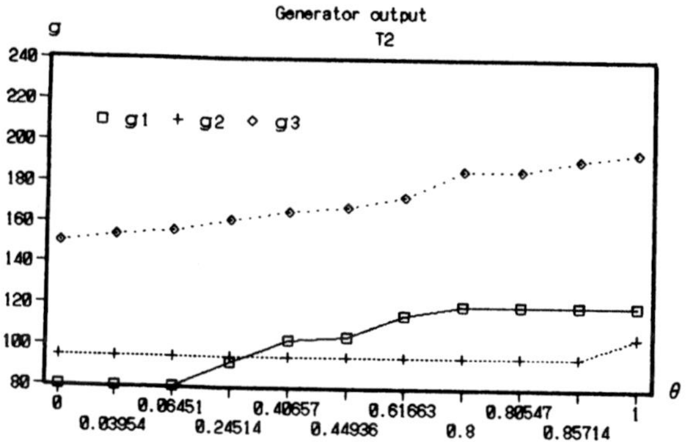

Fig. 1-(b). Generator output of small-scale system (t=2).

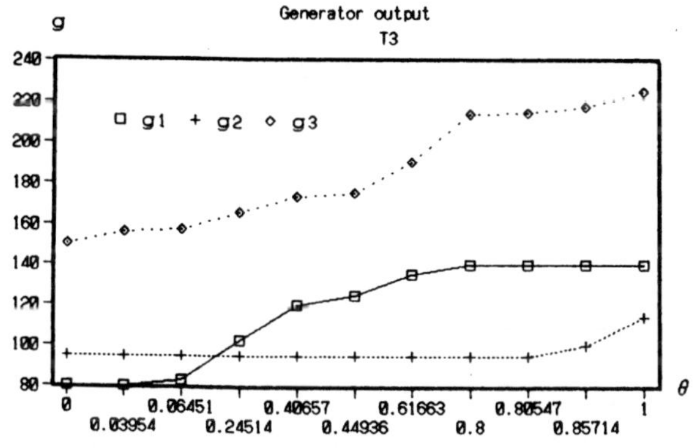

Fig. 1-(c). Generator output of small-scale system (t=3).

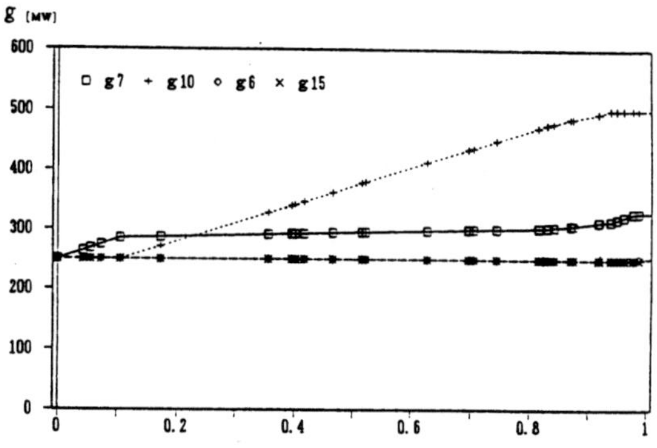

Fig. 2-(a). Generator output of real-scale system (t=3).

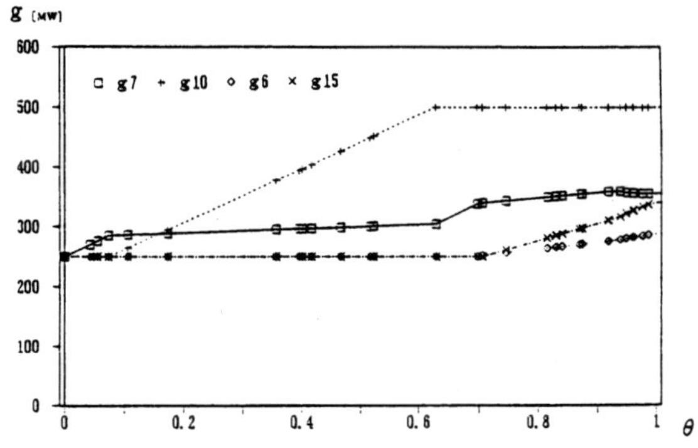

Fig. 2-(b). Generator output of real-scale system (t=4).

putation times of the proposed method are only a few seconds if it is run on an engineering work station. This fact demonstrates that the proposed method can be applied to the real-time control of electric power systems.

CONCLUSIONS

In this paper, the continuation method for solving the dynamic ELD problem is proposed. It is shown that the solution to the parameterized problem is expressed as a piecewise-linear function of the continuation parameter. In order to demonstrate the validity and effectiveness of the proposed method, the computer program is written and it is executed for the real-scale model systems. The results show the proposed method can be applied to the real-time operation of electric power systems. The extension of the proposed continuation method for the security constrained ELD problem may be the subject of future research.

REFERENCES

Aoki, K. and T. Satoh (1984). New algorithms for classic economic load dispatch. *IEEE Trans. Power App. Syst.*, Vol.PAS-103, No.6, pp.1423-1431.

Chowdhury, B. H. and S. Rahman (1989). A review of recent advances in economic dispatch. *IEEE/PES 1989 Summer Meeting*, Paper No. 89 SM 696-6 PWRS.

Gill, P. E. and W. Murray (1977). Linearly constrained problem including linear and quadratic programming. In *The State-of-the-art in Numerical Analysis*, Academic Press, pp.313-363.

Happ, H. H. (1977). Optimal power dispatch - A comprehensive survey. *IEEE Trans. Power App. Syst.*, Vol.PAS-96, No.3, pp.841-854.

Huneault, M., A. Fahmideh-Vojdani, M. Juman, R. Calderon and F. D. Galiana (1985). The continuation method in power system optimization: Applications to economy - security functions. *IEEE Trans. Power App. Syst.*, Vol.PAS-104, No.1, pp.114-124.

Isoda, H. (1982). On-line dispatching method considering load variation characteristics and response capabilities of thermal units. *IEEE Trans. Power App. Syst.*, Vol.PAS-101, No.8, pp.2925-2930.

Richter, S. L. and R. A. DeCarlo (1983). Continuation methods: Theory and applications. *IEEE Trans. Automat. Contr.*, Vol.AC-28, No.6, pp.660-665.

Ross, D. W. and S. Kim (1980). Dynamic economic dispatch of generation. *IEEE Trans. Power App. Syst.*, Vol.PAS-99, No.6, pp.2060-2068.

van den Bosch, P. P. J. (1985). Optimal dynamic dispatch owing to spinning-reserve and power-rate limits. *IEEE Trans. Power App. Syst.*, Vol.PAS-104, No.12, pp.3395-3401.

van de Panne, C. (1975). Methods for linear and quadratic programming. In *Studies in Mathematical and Managerial Economics*, Vol.17, North-Holland, pp.261-296.

REAL-TIME ARCHITECTURE OF AN ACTIVE NOISE ATTENUATION SYSTEM USING MICROPROCESSOR DSP96002

W. K. Yung and K. F. Man

Department of Electronic Engineering,
City Polytechnic of Hong Kong, Hong Kong

Abstract. The Key element contributes to the success of an active noise attenuation ANA scheme is its capability of identifying the noise dynamics in real-time, and the use of adaptive optimal control laws to generate the corresponding anti-phase acoustic signal (from a loudspeaker) for the ultimate noise cancellation. In the context of robustness, convergence and tracking capability of the parameter estimating process, where the ANA is very much dependent on, a new robust recursive least squares (RLS) algorithm of which an optimally varied forgetting factor is derived. To further enhance its real-time performance, the Motorola's DSP96002 floating-point arithmetic digital processor (DSP) chip has been used. Due to its design in software and hardware architecture, the accuracy of arithmetic computation and the speed of processing of the algorithm has been greatly improved and has been found that it is well-suited for the application of ANA system.

Keywords. Acoustic Control; Adaptive Control; Parameter Estimation; Robustness; Supervisory Control.

1. INTRODUCTION

In the past, the environmental noise (undesired acoustic signal) problem has always been tackled passively. The traditional method of noise reduction is by installing noise absorbing materials or erecting barriers to block-off the undesirable noise. The noise reduction is about 5 - 12 dB(A). However, the cost of overall construction for this type of noise reduction scheme is phenomenal in order to achieve an acceptable standard. Since the passive means of noise reduction is ineffective in the low frequency region of 6Hz 200Hz, then the cost of construction for this range of noise spectrum is either causing the cost to accelerate an unrealistic figure or its ultimate achievable effectiveness can never be met.

Considering the advancement of today's VLSI technology where cheaper and fast computation microprocessors are easily available on the market, to tackle the noise reduction problem by means of active control device has become a viable proposition. It is not only less expensive in terms of cost but its effectiveness in the low frequency region can never be in doubt. The key element contributes to the success of this type of active noise attenuation ANA scheme is the capability of identifying the noise dynamics in real time and the use of adaptive optimal control laws to generate the corresponded anti-phase signal (from a loudspeaker) for the ultimate noise cancellation.

However, to be able to identify the noise dynamics where the frequency of interest that significantly affects the human ears is in range of 6Hz to 3KHz. A real-time processing of such range of signal, it is not only generating computational and numerical problems in hardware, but the critical choices of the software algorithm also seriously affect the system response performance. Therefore, in order to made an electronic ANA system to become a reality, the current well provened technique in signal processing, the advanced adaptive control laws design together with the state-of-the-art microprocessor product have to be formed as an integrated unit for the development of this ANA system. In this paper, the real-time issues for both software and hardware development of the ANA system will be addressed and results obtained experimentally are presented.

2. ACTIVE NOISE ATTENUATION CONTROLLER

Active noise control system is based on the principle of the real time system identification. The unknown acoustical plant is modeled on-line by an adaptive IIR filter, so that the adaptive optimal control laws can be used to generate the corresponded anti-phase signal to minimized the undesired noise. The basic arrangement of active noise control system expressed in the complex frequency s domain is shown in Figure 1.

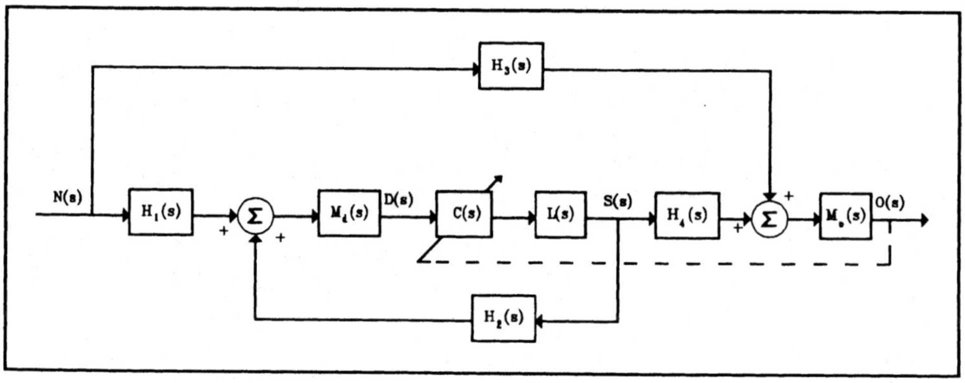

Figure 1. Block diagram of active noise control system.

where

N(s) = primary source signal
S(s) = secondary source signal
D(s) = signal measured by the detector
O(s) = signal measured by the observer
C(s) = transfer function of the controller
L(s) = transfer function of loudspeaker
$M_d(s)$ = transfer function of the detector microphone
$M_o(S)$ = transfer function of the observer microphone
$H_1(S)$ = transfer function of the acoustic path through primary source to detector
$H_2(S)$ = transfer function of the acoustic path through secondary source to detector (acoustic feedback path)
$H_3(S)$ = transfer function of the acoustic path through primary source to observer
$H_4(S)$ = transfer function of the acoustic path through secondary source to observer

Then the detector D(S) and observer O(s) signal can be stated as the following :

$$D(s) = M_d(s)\ [\ H_1(s)N(s) + H_2(s)L(s)C(s)D(s)\ ] \qquad (1)$$

$$O(s) = M_o(s)\ [\ H_3(s)N(s) + H_4(s)L(s)C(s)D(s)\ ] \qquad (2)$$

To obtain an optimum performance O(s)=0, the required transfer function of the controller is

$$C(s) = \frac{H_3(s)}{M_d(s)L(s)\ [H_2(s)H_3(s) - H_1(s)H_4(s)]} \qquad (3)$$

Hence the transfer function of a controller is dependent on the geometric configuration to give optimal cancellation. The ANA system can be divided into two equivalent subsystem under two different conditions such as $H_{31}(s)$ and $H_{od}(s)$ (Tokhi and Leitch, 1991) where

$$H_{31}(s) = \frac{O(s)}{D(s)}\bigg|_{S(s)=0} \qquad (4)$$

$$H_{od}(s) = \frac{O(s)}{D(s)}\bigg|_{S(s)\neq 0} \qquad (5)$$

Hence, the controller transfer function is given in equation (6).

$$C(s) = \frac{1}{(\ 1 - \dfrac{H_{od}(s)}{H_{31}(s)}\ )} \qquad (6)$$

or,

$$C(z) = \frac{1}{(\ 1 - \dfrac{H_{od}(z)}{H_{31}(z)}\ )} \qquad (7)$$

The adaptive controller will be based on the models of these subsystem. Real time parameter identification can be carried out based on the detector signal 'D' and observer signal 'O' which are sampled at detector microphone and observer microphone respectively. The two distinct transfer functions are identified on real time under two different conditions of S(t)=0 and S(t)≠0. The transfer function $H_{31}(z)$ can be identified by keeping the output signal (from loudspeaker) to zero such as S(k)=0, whereas the transfer function of $H_{od}(z)$ can be obtained by activating the output signal same as received input signal, S(k)≠0.

Traditionally, recursive least square (RLS) algorithm with the use of a fixed forgetting factor $\lambda<1$ (Ljung and Söderström, 1983) is commonly used for modeling and tracking the time-varying acoustical dynamics. However, the convergence, stability and noise sensitivity of the algorithm is poor (Åström and Wittermark,1980) when the measurement errors follow some non-normal distribution, particularly one that has longer or heavier tails than the normal, such as outliers. These outliers may have a strong influence on the least squares estimates due to an invalid Gaussian noise assumption. The conventional estimation method with the unknown information is potentially unstable and produces suboptimal parameter estimates in both the convergence (parameter fluctuation) and steady state (bias) conditions. In addition, the desired noise to be attenuated is usually corrupted by other noise such as speech which affects the convergence and stability of the algorithm during the on-line identification. So the use of the ordinary RLS algorithm is not the best choice for the real time active noise reduction system. The other aspects of algorithm such as robustness and tracking capability must also be addressed if ANA system is successfully applied in practice.

In order to identify the acoustical plant effectively, a new robust recursive least squares (RLS) algorithm incorporated with variable optimal forgetting factor have been used for real time parameter identification so that the robustness, convergence and tracking capability can be enhanced. This algorithm is based on the derivation of an optimally varied forgetting factor that its variation can be adjusted according to the pre-set lower and upper bound of the noise variance information. The criteria for the selection and the update of the forgetting factor are governed by a three sigma rule together with a Huber's model (Huber,1981). It has been demonstrated experimentally that this algorithm is very robust and its speed of tracking performance is found to be much faster than the existing algorithms (Man et al,1992). Hence it is well suited for this application.

3. ROBUST RLS ALGORITHM

To identify the two subsystems $H_{31}(z)$ and $H_{od}(z)$ with respect to two different conditions $S(k)=0$ and $S(k) \neq 0$ respectively, a new robust RLS algorithm is used to estimate the dynamics which can be represented by

$$O(t) = \theta^T(t)\phi(t) + e(t) \tag{8}$$

where

$\theta^T(t) = [a_1 \ldots a_n b_1 \ldots b_m]$ *is the parameter vector,*

$\phi^T(t) = [\, O(t-1) \ldots O(t-n)\ D(t-1) \ldots D(t-m)\,]$

is the observation vector.

and e(t) is prediction error and/or external disturbance. It is assumed that the noise e(t) due to external disturbance and measurement error defined as

$$\Gamma_\epsilon = \{\, F/F = (1-\epsilon)G + \epsilon H,\ H \in \Omega\,\} \tag{9}$$

and $F \in \Gamma_\epsilon$

where Ω is the space of all probability measures and ϵ reflects the believed accuracy of the prior assessment. Equation (9) is also called gross error model (Huber,1981).

The parameter vector can be recursively calculated by the following three equations

$$\theta(t) = \theta(t-1) + \frac{P(t-1)\phi(t)[O(t) - \theta^T(t-1)\phi(t)]}{\lambda(t) + \phi^T(t)P(t-1)\phi(t)} \tag{10}$$

$$P(t) = \frac{1}{\lambda(t)}[\, P(t-1) - \frac{P(t-1)\phi(t)\phi^T(t)P(t-1)}{\lambda(t) + \phi^T(t)P(t-1)\phi(t)}\,] \tag{11}$$

and

$$K(t) = P(t)\phi(t) \tag{12}$$

And the forgetting factor (Man and Yung,to be published) becomes the form

$$\lambda(t) = \frac{(1-e(t))\sigma_1^2(t) + e(t)\sigma_2^2(t)}{(1-e(t-1))\sigma_1^2(t-1) + e(t-1)\sigma_2^2(t-1)} \tag{13}$$

where

$$e(t) \sim \Gamma_\epsilon(\bullet/0,\sigma_1^2,\sigma_2^2) = (1-\epsilon)N(\bullet/0,\sigma_1^2) + \epsilon N(\bullet/0,\sigma_2^2), \tag{14}$$

and

$$\sigma_2^2 > \sigma_2^1$$

implies that

$$E[e(t)] = 0$$

and

$$E[e(t)e(t)] = \sigma^2 = (1-\epsilon)\sigma_1^2 + \epsilon\sigma_2^2 \tag{15}$$

Then σ_1^2 and σ_2^2 can be updated based on pre-assigned bound $(3\sigma_1)$ based on three sigma rule. The updating processes for these variance (Puthenpura and Sinha,1990) are shown as the followings :

We can set

For $|e(t)| \leq 3\sigma_1(t-1)$,

$$\sigma_1^2(t) = \sigma_1^2(t-1) + \frac{1}{t-k}[\, e^2(t) - \sigma_1^2(t-1)\,]$$

$$\sigma_2^2(t) = \sigma_2^2(t-1) \tag{16}$$

otherwise

$$\sigma_1^2(t) = \sigma_1^2(t-1)$$

$$\sigma_2^2(t) = \sigma_2^2(t-1) + \frac{1}{k}[e^2(t) - \sigma_2^2(t-1)] \tag{17}$$

3.1 Huber Model for Time-varying Systems

Lemma 1.

Using a suitable value of ϵ, the Huber's gross error model can be applied for time-varying system. The derivation is based on the assumption of the prior prediction error residues v(t) and the posterior prediction error e(t) together with the measurement noise. In this way, the algorithm can be proved to be suitable for time-varying system application.

Proof.

The least squares criterion can be interpreted as minimization of the error between the predicted and the measured values of the output. Considering the prediction error form

$$O(t) = \theta^T(t-1)\phi(t) + v(t) \tag{18}$$

where $\nu(t)$ is the prior prediction error based on estimates up to time, t-1.

And,

$$O(t) = \theta^T(t)\phi(t) + e(t) \tag{19}$$

where e(t) is the residue or posterior prediction error based on estimates up to time, t.

Assuming that $\{\nu(t)\}$ and $\{e(t)\}$ are sequences of independent random variables with zero mean values.

And

$$E[v^2(t)] = r_2$$

$$E[e^2(t)] = \sigma^2$$

Then we also assume that the true value of the system parameters varies according to

$$\theta(t) = \theta(t-1) + w(t-1) \quad (20)$$

where w(t) can be assumed to be independent Gaussian random vectors such that w(t) has zero mean and covariance matrix $R_1(t)$.

Under these circumstances, we have

$$E[w(t)w^T(s)] = R_1(t)\ \delta_{ts} ,$$

$$E[w(t)v(s)] = 0 ,$$

$$E[w(t)e(s)] = 0 ,$$

$$E[v(t)v(s)] = R_2(t)\ \delta_{ts} ,$$

and

$$E[e(t)e(s)] = \sigma^2\ \delta_{ts} .$$

where δ_{ts} is Kronecker's delta.

We can re-write the equation (18) using equation (20) to give

$$O(t) = (\theta^T(t) - w^T(t-1))\phi(t) + v(t)$$

$$O(t) = \theta^T(t)\phi(t) + (v(t) - w^T(t-1)\phi(t))$$

Compare with (19), we have

$$e(t) = v(t) - w^T(t-1)\phi(t) \quad (41)$$

Since, $E[e(t)]=0$ and $E[\nu(t)]=0$, we have

$$E[w^T(t-1)\phi(t)] = 0$$

It also implies that w(t) must be independent with $\phi(t)$ without contradiction.

Since $\{\nu(t)\}$, $\{w(t)\}$ and $\{\phi(t)\}$ are assumed to be a sequence of independent Gaussian random vectors with zero mean values. So we have

$$Var[e(t)] = E[e^2(t)] = E[(v(t) - w^T(t-1)\phi(t))^2]$$

$$\sigma^2 = r_2 + r_1 \quad (22)$$

where

$$E[(w(t-1)^T\phi(t))^2] = r_1 ,$$

$$E[(v(t)w^T(t-1)\phi(t))] = 0.$$

The variance of gross error is the function of the time-varying parameters' variance and the measurement noise variance. It can be extensively considered that

$$e(t) \sim \Gamma_\varepsilon(\bullet/0,\sigma_1^2,\sigma_2^2) \text{ as shown in } (14).$$

It completes the proof.

The resulting algorithm has an ability to forget ($\lambda < 1$) for time-varying system, and yet it also has an ability to emphasize ($\lambda > 1$) old data as compared with the new data which is considered to be an outlier. The analysis of robustness, convergence and tracking ability of this proposed algorithm is to be submitted for publication.

4. ALGORITHM OF AN ACTIVE NOISE CONTROL SYSTEM

After the two subsystem (4)-(5) stated as previous section have been identified, the controller (7) can be calculated and implemented into digital signal processor. The controller can be updated adaptively based on the prespecified performance criterion. When the characteristics of acoustic system is changed, the change in performance of the control system is detected based on the testing of observer signal O(s). If level of the noise $O^2(s)$ is greater than the prespecified value q, the acoustical modelling is carried on again in order to adjust the parameters of the controller to give optimum performance. This scheme is implemented into the supervisory control level in order to monitor the overall attenuation performance. The algorithm of the active noise control system is shown in figure 2.

5. ARCHITECTURE OF AN ACTIVE NOISE ATTENUATION SYSTEM

In order to identify the broadband noise dynamics (6Hz to 3kHz) in real time, not only the choice of the parameter estimation algorithm is important, but the computational and numerical problems in hardware must be carefully considered.

During the real time parameter estimation, the three equations (10)-(12) are recursively processed, it involves a large dynamic range of computation, say from integer to fractional values with a large order. The common method to handle the arithmetic calculation for a fixed point processor is either implementing floating point arithmetic (mantissa and exponent) or splitting the variables into integral and decimal parts for manipulation.

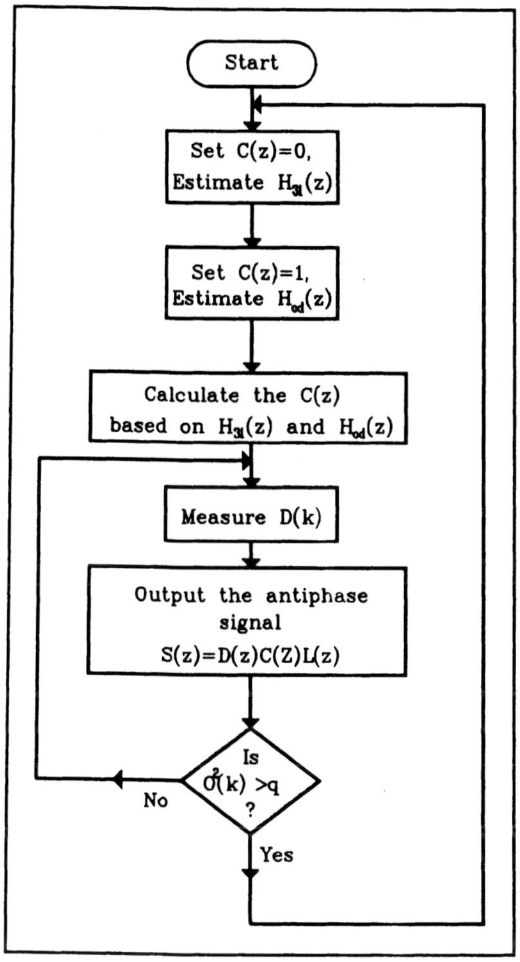

Figure 2. Algorithm of self-tuning controller

However, both these methods increase the total computational load and cause truncation error. Moreover, these suffer the computational overflow problem due to the limit of memory word length. The previous solution for minimizing the chance of overflow is to scale down the variables during the processing. However, the arithmetic accuracy is still poor and hence decrease the convergence of RLS algorithm. In addition, both these methods involve more data and program memory for storing data and additional arithmetic routines respectively. As the result, the corresponding hardware architecture become complex with less flexibility of software algorithm.

In order to enhance the flexibility and efficiency of the software algorithm, the up-to-date Motorola's DSP96002 floating-point arithmetic digital signal processor (DSP) chip have been used to simplify the hardware architecture. The main characteristics of the DSP96002 digital signal processing chip are the support of IEEE single precision unit, the single extended precision floating point and the 32 bit signed and unsigned fixed point arithmetic which coupled with two identical external memory expansion ports. It can be manipulated to execute 49.5 Million Floating Point Instructions per Second (MFLOPS) and 16.5 Million Instructions per Second (Mips) with a 33 MHz clock. Due to this powerful DSP processor, the software algorithms for proposed acoustical parameter estimation and the use of adaptive feedback control theory to generate the anti-phase signal can thus be easily implemented into the processor. Finally, the accuracy of arithmetic computation and the processing speed of the algorithms have been greatly significantly. Hence the bandwidth of the being attenuated signal can be increased.

The whole architecture of the active noise attenuation system consists of a high level general purpose computer as the host and the DSP96002 processor as the controller. The Host computer serves the purpose of supervisory control and the main numerical computational process will be carried out by the DSP controller. As the Host computer, an AT386 has been used. It provides a mass memory (RAM) for storing data which can be analyzed off-line. The Host computer also provides a programming environment for the DSP96002 including an editor, assembler and the C cross-compiler to generate the executable codes.

A schematic diagram of the hardware architecture of the overall ANA system is shown in figure 3.

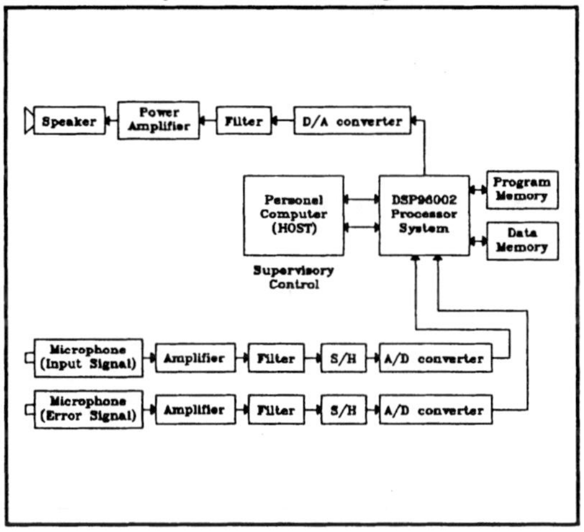

Figure 3. Architecture of overall active noise attenuation system

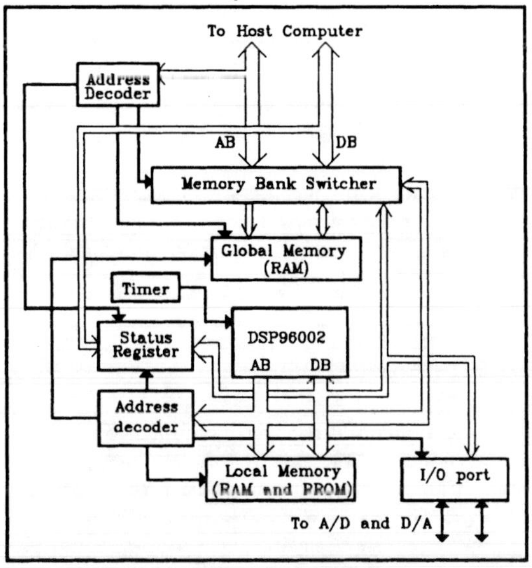

Figure 4. Block diagram of a DSP controller

Figure 4 shows the structure of a DSP96002 system. The data and program memory of the system are organized into global and local memory. The host computer and the digital signal processor both can access the global memory for data communication. The synchronization between the host computer and DSP processor depend on the status of 'semaphore flag' in a status register. When the host computer is accessing the memory, say semaphore = '1', the digital signal processor cannot read from or write into the same memory or vice versa (Hayes,1988). The local program memory, which holds an initialization and synchronization program, is accessible only from the DSP96002. During parameter estimation and real time control, all parameters of the algorithms are stored in local data memory. Besides, the digitized data of the observer signal are stored in global memory to be accessed by Host computer for supervisory control scheme. The graphic user interfaces program is implemented into the host in order to download the executable program codes (parameter estimation and optimum controller) into global memory of DSP. All the functions are programmed in C. The D/A and A/D converters are directly connected into I/O port so that the DSP96002 controller can address the input and output the signals easily and quickly.

Based on the design of the hierarchical structure for the active noise controller, the use of real-time operating system was avoided. And the DSP controller can work under interrupt or polling environment.

6. EXPERIMENTAL RESULTS.

The algorithm of the active noise control system has been implemented in real-time with the proposed hardware system. The performance of the system was investigated using a loudspeaker as the primary source. The noise source generates an acoustic sinusoidal wave with a frequency range from 100 Hz to 1kHz. The actuator (secondary loudspeaker) is identical to the noise source and a detector microphone was placed very close to it. The attenuation performance of the ANA system was measured by observer microphone at a distance of about 2m from the source. This is shown in table 1. A maximum attenuation is achieved at 400 Hz. The overall average amount of attenuation is about 23 dB.

Further investigation on the performance of the ANA system under a real broadband noise which is generated from an AC generator and a hand tamping machine will be conducted. The results will be published later.

7. CONCLUSIONS

The main contribution of the success of the active noise attenuation system is the use of the new robust RLS algorithm. The robustness, convergence and tracking capability of the ordinary RLS algorithm are improved significantly. So, it can track the changes in the acoustic characteristics properly even there exists external disturbance. The proposed self-tuning control system is upgraded by the use of the floating point DSP96002 processor which solves the computational and numerical problems. This system is designed to have a minimum human-machine interaction and therefore, it is a fully adaptive and automatic system which can be used in different working noisy environment.

Frequency / Hz	ANA Off / dB	ANA On / dB	Attenuation / dB
100	80	70	10
200	80	58	22
300	80	54	26
400	80	46	34
500	80	48	32
600	80	49	31
700	80	53	27
800	80	55	25
900	80	65	15
1000	80	73	7

Table 1. Attenuation performance of ANA system

REFERENCE

Åström, K.J. and B. Wittermark (1980). Self-Tuning Controller Based on Pole-Zero Placement,.IEE Proc.D, Control Theory and Application, 127, 120-130.

Hayes, J.P. (1988). Computer Architecture and Organization, McGraw Hill, London.

Huber, P. J. (1981). Robust Statistics, John Wiley and Sons, N.Y..

Ljung, L. and T. Söderström (1983). Theory and Practice of Recursive Identification , MIT Press, MA..

Man, K. F. and W. K. Yung (to be published). A Robust Parameter Identification Algorithm with an Optimal forgetting Factor.

Man, K.F., W.K. Yung, Benedict Y.W. Chan and C.Y.Chan (1992). A Stand-Alone Microprocessor Based Architecture for Intelligent Control, CAI Symposium 92 on Advanced Control and Automation Application, Hong Kong.

Motorola Inc. (1989). DSP96002 IEEE Floating-point Dual-Port Processor User's Manual.

Puthenpura, S. and N.K. Sinha (1990). A Robust Recursive Identification Method, Control Theory and Advanced Technology, Vol.6, No.4, December, 683-695.

Tokhi, M.O. and R.R.Leitch (1991). Design and implementation of self-tuning active noise control system, IEE Proc.-D, Vol.138, No.5, 421-430.

A STUDY ON NONLINEAR ROBUST ADAPTIVE CONTROL SYSTEM DESIGN OF INDUSTRIAL ROBOTIC MANIPULATOR

Sung Hyun Han* and Man Hyoung Lee**

*Department of Mechanical Design and Mechanical Engineering,
Kyung Nam University, Korea
**Department of Mechanical and Control Engineering,
Pusan National University, Korea

Abstract This paper presents new and simple robust adaptive control system for the implementation of real time conrol of the multi-joint robot manipulator. Based-on the linearized dynamic equation, the adaptive control scheme used a least square identification(LSAC) is designed to perform the on-line parameter estimation of the feedback controller gains. Next, the nonlinear robust adaptive control system(NRAC) is designed to perform the robust and real-time control of desired trajectory tracking under the existence of the unknown payload variation and inertia parameter uncertainties. In LSAC scheme, a recursive least-square identification scheme is used to estimate the feedback gain matrix directly, not to estimate the inertia parameters of manipulator. The feedback gains of the controller are updated and adjointed in each sampling time to achieve the desired control performance. In NRAC, the proposed adaptive control algorithm is developed within the framework of adaptive model following control theory using the Lyapunov Direct Method. The performance of the proposed adaptive control algorithm is compared with a similar adaptive control scheme. To demonstrate the effectiveness of proposed control system, simulation is performanced under the existence of error of inertia parameter and unknown payload.
KeyWords direct adaptive control, real-time control, robotic manipulator, least square identification, parameter uncertainties, robustness, stability, payload variation, controller estimation, Lyapunov direct method

1. INTRODUCTION

From now on, the control system design of the robot manipulator is depend upon the use of an independent servomechanism for each joint, ie, the feedback controllers dedicated to each joint are designed to move the manipulator joints independently. These controllers (Middleton and Goodwin,1987) do not attempt to model the detailed dynamics of the interacting manipulator joint in determining actuator signals. Such independent SISO servo design inevitably results in sacrifices of performance since a robot manipulator is intriniscally a multivariable coupled the nonlinear system. To improve the performance of robot manipulator control, a variety of approaches(Lee, 1982; Luh and Lin,1982) have been proposed and promising preliminary results have been obtained. However, all the previously discussed design method of manipulator control are based on the assumption that an accuracy dynamic control is available. The modelling errors such as the unmeasurable friction and the uncertainty in the system parameters may render the dynamic model in accurate. Therefore, the adaptive control methods(Dubowksy and DesForges,1979; Horowitz and Tomizuka,1980; Koivo and Guo,1983; Lee and Choi,1989)may be applied in which the dynamic model parameter assumed unknown and a suitable adaptation law is used to adjust the controller parameters on-line based on the input/output informaton of the controlled dynamic model. In this paper, new adaptive control schemes applicable to robotic manipulators is presented for the real-time control. In the adaptive control with estimation scheme, a recursive least square identification is used to estimate the feedback gain matrix. The feedback gains of the controller are updated and adjusted in each sampling step to obtain the desired control effort. Next, the direct adaptive control schemes based on the adaptive model following control theory has also been proposed for the real-time control implementaton of robotic manipulator. For the consideration of the overall system stability, the proposed direct adaptive control system has been designed using Lyapunov Direct method. The purpose of this paper is to achieve the real-time control of the multi-joint robot manipulator. The proposed control law is adaptive time-varying and does not require knowledge of the complex dynamic model or parameter values of the manipulator or unknown payloads. The control schemes are computationally very fast, making it specially suitable for the real-time control of redundant robot manipulator. This paper is organized as follows. In section 2, the dynamics analysis of robot manipulator is introduced. In section 3, the adaptive control approaches are proposed for the real-time control of robot mainpulator. The performance of designed adaptive control scheme is demonstrated in Section 4 by simulation of the PUMA robot under the inertia parameter uncertainties and unknown payload variations. Section 5 discusses the simulation results and draws some conclusion of this paper.

2. DYNAMICS OF ROBOT MANIPULATOR

It is noted the dynamic equations for robot manipulators are highly nonlinear due to the presence of Coriolis and centrifugal acceleration terms. The essential importance of the dynamics of the robot manipulators is clearly pronounced in the control of the robot motion, forces, and couples. Obviously, the control of robot manipulator is one of the most important subjects in robotics and bears heavily on the dynamics of the robot manipulators. There are essentially three different approaches available for the derivation of the governing equations pertaining to robot arm dynamics:

- The Newton-Euler modelling method (N-E)
- The Lagrange-Euler modelling method (L-E)
- The bond graph modelling method

In the L-E formulation, the real-time control is nearly impossible to achieve due to the large computational time. If the nonlinear terms are neglected, then the problem is manageable for implementation of real-time control. However, the resulting control is suboptimal and only valid for very slow motion. Deriving the dynamic model of a robot manipulator by means of the N-E formulation is advantageous for the real-time control because of its speed and accuracy. Thus the dynamic equation of n-joint robot manipulator by the N-E formulation is represented by the form as follows:

$$A(q)\ddot{q} + B(q,\dot{q}) + C(q) = T(t) \tag{1}$$

Where $A(q)$ is the nxn symmetric positive-definite inertia matrix, $B(q,\dot{q})$ is the nxl coriolis and centrifugal torque vector, and $C(q)$ is the nxl gravitational loading vector. The first and second derivative of n-dimensional vector of joint variable q are, respectively, the velocity and acceleration of each joint, and $T(t)$ is the n-dimensional vector of joint torques and forces.

The end-effector cartesian position, velocity, and acceleration vectors related to the joint variables is defined as follows:

$$X(t) = f(q)$$
$$\dot{X}(t) = J(q)\,\dot{q}(t)$$
$$\ddot{X}(t) = \dot{J}(q(t),\,q(t))\,\dot{q}(t) + J(q(t))\,\ddot{q}(t)$$

Where, $f(q)$ is the nxl vector representing the forward kinematics of manipulator, and $J(q) = [\partial f(q)/\partial q]$ is the nxn Jacobian matrix of the manipulator.

3. ROUBST ADAPTIVE CONTROL OF THE MANIPULATOR

3.1 LEAST SQUARE ADAPTIVE CONTROL SCHEME

A common feature of many adaptive control schemes is that they aim at identifying the inertia parameters of the manipulator and from estimates of these, either the constant or the time-varying feedback gains are computed.

The proposed manipulator control problem is to find a real time robust control law such that the closed loop control system is asymptotically stable and tracks a desired trajectory as closely as possible over a wide range of payload for all the times. Therefor, it is essential to adaptive the gains of feedback controller and the operating point term so as to ensure stability and trajectory by the total control law.

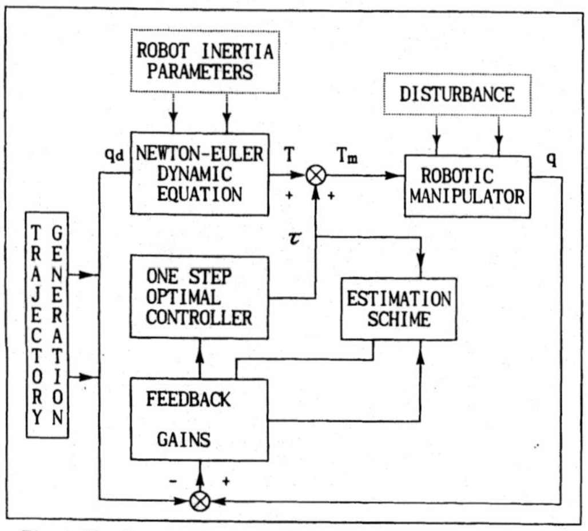

Fig.1 The least square adaptive control system with estimation schime.

The highly coupled nonlinear dynamic equations of the manipulator are linearized about a prescribed nominal trajectory. Under the assumption of small position, velocity, and acceleration derivations from nominal values, a linear time varying controller is designed so as to render the linearized system a closed-loop, second order, stationary system of a prescribed form.

In this section, a recursive least square identification scheme is used to perform the on-line parameter estimation for the feedback gains of the controllers directly, instead of estimating the inertia parameters of the robotic manipulator themselves.

The feedback gains of the controller are updated and adjusted in each sampling step to obtain the desired control purpose. A recursive least square identification scheme is used to estimate the feedback gain matrix. And the time-varying controller gain is estimated instead of estimating the inertia parameters of the robotic manipulator. The dynamic equation of manipulator with payload be expressed as follows:

$$T_m(t) = A_m(q)\,\ddot{q}(t) + B_m(q,\dot{q})\dot{q}(t) + C_m(q) \tag{2}$$

where

$$A_m(q) = A(q) + mJ^T(q)J(q)$$
$$B_m(q,\dot{q}) = B(q,\dot{q}) + mJ^T(q)\dot{J}(q,\dot{q})\dot{q}$$
$$C_m(q) = C(q) + mJ^T(q)g$$

It is noted that the elements of A_m, B_m, and C_m are highly complex nonlinear functions of the link configuration q, the speed of motion $\dot{q}$, and payload mass m, and modified inertia matrix A_m is symmetric and positive definite .

It is desirable for the manipulator control system to achieve trajectory tracking irrespective of the payload mass and parameter uncertainies so that the performance of the manipulator is insensitive to the payload and parameter uncertainty. In order to derive the feedback control for this purpose, the aforementioned the nonlinear dynamic equation is first the linearized about nominal trajectory.

The pertubation equation is obtained as follows:

$$A_p(\varphi)\delta\ddot{q} + B_p(\varphi,\phi)\delta\dot{q} + C_p(\varphi,\phi,\nu)\delta q = \delta T(t) \tag{3}$$

Where

$$A_p(\varphi) = \partial[A(q)\dot{q}]/\partial\ddot{q}\ ,\quad B_p(\varphi,\phi) = [\partial A(q)/\partial\dot{q}]$$
$$C_p(\nu) = \partial[A(q)\ddot{q}]/\partial q + [\partial A(q)/\partial q]$$

$\varphi(t)$, $\phi(t)$ and $\nu(t)$ is denote the time histories of vectors $q(t)$, $\dot{q}(t)$, and $\ddot{q}(t)$ along the prescibed trajectory, which are either given directly or obtained from inverse kinematics. Since $\varphi(t)$, $\phi(t)$, and $\nu(t)$ are all the functions of time, equation(3) can be expressed as follows:

$$A_p(t)\ddot{\theta}(t) + B_p(t)\dot{\theta}(t) + C_p(t)\theta(t) = \tau(t) \tag{4}$$

Where, $\theta(t)$ and $\tau(t)$ are the deviations of the joint angle and joint torque vectors from their nominal values $\dot{q}(t_i)$ and $T(t_i)$ at operating point $p=\{T(t_i),\ q(t_i),\ q(t_i)\}$:

$$\theta(t) = q(t) - q(t_i),\quad \tau(t) = T(t) - T(t_i).$$

The three nxn matrice A_p, B_p, C_p is the theoretical parameter matrices of the robotic manipulator model at the nomimal operating point.

The proposed control problem is reduced to determining a suitable $\tau(t)$ which derives the deviation $\theta(t)$ to zero. To achieve this purpose, the design and estimation of the manipulator controller have to be performed in this section. Now it is required to find a robust control scheme to achieve real time control that renders the linearized equation a predefined second-order system of the form as follows:

$$\ddot{\theta} + 2\,Q^T R\,Q\,\dot{\theta} + R^T R\,\theta = 0 \tag{5}$$

where, R is the canstant nonsingular matrix of undamped natural fequency and Q is the nondimensional constant positive-definite damping matrix of the system.

The above two matrices can be chosen as diagonal, which will allow one to meet the said condition more easily.

In adaptive control system, let

$$\tau(t) + K_d(t)\theta(t) + K_v(t)\dot{\theta}(t) = 0 \quad (6)$$

where K_d and K_p are the time varying feedback matrices(nxn) that are to be first determined in terms of the manipulator parameters and next, estimated in terms of input and output measurements. From the equation (4) and (6), following second-order equation is derived :

$$A_p(t)\ddot{\theta}(t) + [K_v(t)+B_p(t)]\dot{\theta}(t)+[K_d(t)+C_p(t)]\theta=0 \quad (7)$$

Therefore, the time-varying feedback gain matrices derived:

$$K_v(t) = 2A_p(t)Q^TRQ- B_p(t) \quad (8\text{-a})$$

$$K_d(t) = A_p(t)R^TR - C_p(t) \quad (8\text{-a})$$

Both $K_d(t)$ and $Kv(t)$ depend on the parameters of the system. If the precise parameters in the pertubation equation were known perfectly, It could be done to achieve real time control by using the feedback matrices.

But the design of the foregoing feedback control scheme is still unrealistic because the parameters of the manipulator can not be known perfectly. Thus parameter identification techniques and adaptive control are used to account for the parameter uncertainties. It may be possible to identify the unknown parameter of the controller with these techniques. The feature of proposed estimation method is aimed at the estimation of controller directly. Therefore, it may be possible to produce a very simple control scheme with a relatively low computational complexity.

The pertubation equation can be represented in the state space vector equation as follows:

$$\dot{\delta}_1(t) = \delta_2(t) \quad (9)$$

$$\dot{\delta}_2(t) = A_p(t)^{-1}[I(t) - B_p(t)\delta_2(t) - C_p(t)\delta_1(t)] \quad (10)$$

Where, $\delta_1(t)$, $\delta_2(t)$, and $I(t)$ are all n-dimension vectors, defined as follows:

$$\delta_1(t) = \theta(t) \quad (11)$$

$$\delta_2(t) = \dot{\theta}(t) \quad (12)$$

$$I(t) = \tau(t) \quad (13)$$

From the forementioned (8-a),(8-b), and(10), the next equation is derived as follows:

$$A_p(t)\delta_2(t) + [I(t)R^TR - K_d(t)]\delta_1(t) + [2A_p(t)Q^TRQ - K_v(t)]\delta_2(t) = I(t) \quad (14)$$

In (14), the unknown matrices $K_d(t)$ and $Kv(t)$ will be estimated via a least square identification scheme.

Thus we obtain the following time varying discrete-time system for the robotic manipulator.

$$A_p(k)\delta_2(k)=\lambda[I(k)-2A_p(k)Q^TRQ\ \delta_2(k) - A_p(k)R^TR\ \delta_1(k) + K_d(k)]\delta_1(k) + K_v(k)]\delta_2(k)] + A_p(k)\delta_2(k) + V(k) \quad (15)$$

Hence, (15) can be represented in the following form :

$$X_i(k+1) = Y(k)^T J_i(k), \quad i = 1, 2, \cdots, n \quad (16)$$

Where

$$J_i(k) = [K_d(k)_{i1}, \cdots, K_d(k)_{in}, K_v(k)_{i1}, \cdots, K_v(k)_{in}, \rho_i(k)]^T \quad (17\text{-a})$$

$$K(k) = [K_d(k), K_v(k), \varepsilon(k)] \quad (17\text{-b})$$

$$Y(k) = [Y_1(k), \cdots, Y_i(k), \cdots, Y_{2n}(k), \bar{\eta}]^T \quad (17\text{-c})$$

$$\varepsilon(k)\eta = \mu(k) + V(k) \quad (17\text{-d})$$

$$\delta_1(k)\lambda = [Y_1(k), \cdots, Y_n(k)]^T \quad (17\text{-e})$$

$$\delta_2(k)\lambda = [Y_{n+1}(k), \cdots, Y_{2n}(k)]^T \quad (17\text{-f})$$

$$X(k+1) = A_p(k)\ \delta_2(k+1) \quad (17\text{-g})$$

$$\mu(k) = \lambda[I(k) - A_p(k)R^TR\delta_1(k) - 2A_p(k)Q^TRQ\ \delta_2(k)] + A_p(k)\delta_2(k) \quad (17\text{-h})$$

Where $\varepsilon(k)$ is an unknown n-dimension vector and η is an estimating weighting factor which can be adjusted during the recursive identification procedure. And $V(k)$ is a arbritary error vector added in order to compensate for inertia parameter uncertainties. In the next, a least square approach is introduced that allow us to minimize the estimation error in Euclidean norm. To this end, a recursive scheme is derived. The procedure is reffered to as recursive identification. In the basic framwork of the above formulation, the model output of manipulator is identified from the recursive equations, which are readily derived from (16). The robotic manipulator is time varying system, and it is required to eliminate the infuence old data. This can be done by defining a loss function with exponential weighting as follows:

$$L(J) = \sum_{r=1}^{n} e^{n-r}[X(k+1) - \delta(k+1)^TJ(k)]^2, \quad 0<e<1 \quad (18)$$

Thus we obtain the next recursive least square estimation scheme for time varying robot system. The recursive least square formulation can be represented as the following form:

$$\hat{J}_i(k+1) = \hat{J}_i(k) + P(k+1)H(k)S(k+1)\cdot[X_i(k+1) - S^T(k+1)J(k)] \quad (19)$$

$$H(k+1)=[H(k)-P(k+1)H(k)S(k+1)\cdot S(k+1)^TH(k)][e]^{-1} \quad (20)$$

$$P(k+1) = [S(k+1)^TH(k+1)S(k+1) - e]^{-1} \quad (21)$$

In the above equation, J is estimated from the minimization of loss function to measure of how fast old datas are forgotten. And the ith component is estimated from the minimization of the loss function L, under no contraint. And the estimates of J is indicated as $\hat{J}$, and the ith component of n-dimensional vector $X(k)$ is defined as $S(k)TJ_i(k)$ in the forementioned least square algorithm.

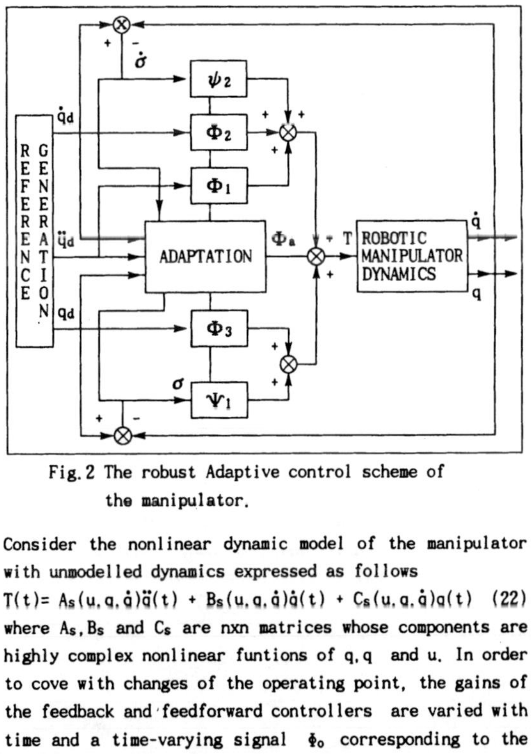

Fig.2 The robust Adaptive control scheme of the manipulator.

Consider the nonlinear dynamic model of the manipulator with unmodelled dynamics expressed as follows

$$T(t)= A_s(u,q,\dot{q})\ddot{q}(t) + B_s(u,q,\dot{q})\dot{q}(t) + C_s(u,q,\dot{q})q(t) \quad (22)$$

where A_s, B_s and C_s are nxn matrices whose components are highly complex nonlinear funtions of q,q and u. In order to cove with changes of the operating point, the gains of the feedback and feedforward controllers are varied with time and a time-varying signal Φ_o corresponding to the operating point term is included in the control law, where Φ_o is considered as an auxiliary signal to be synthesized by the adaptation scheme. In the adaptive control scheme time-varying adaptive control law is expressed as follows:

$$T(t)=[\Psi_1(t)\sigma(t) + \Psi_2(t)\dot{\sigma}(t)] + \Phi_3(t)q_d(t) + \Phi_2(t)\dot{q}_d(t) + \Phi_1(t)\ddot{q}_d(t)] + \Phi_o(t) \quad (23)$$

On applying the proposed adaptation law(23) to the nonlinear dynanmic model(22), We obtain the error differential equation reresented as follows:

$$A_e\ddot{\sigma}(t) + Be\dot{\sigma}(t) + Ce\sigma(t) + \phi_o(t)+ \Phi_3q_d + \Phi_2\dot{q}_d + \Phi_1\ddot{q}_d = \Psi_1\dot{\sigma}(t)+ \Psi_2\sigma(t) + A_s\ddot{q}_d(t) + B_s\dot{q}_d + C_sq_d \quad (24)$$

where $\sigma(t)=(q_d-q)$ is the tracking error of position and is the tracking error of velocity. Now, the position and velocity error vector $E(t)=[\sigma(t),\dot{\sigma}(t)]^T$ is defined. Thus above equation(24) is represented in the state-space equation:

$$E(t)=D_e(t)\dot{E}(t) + A_e(t)\ddot{q}_d(t) + B_e\dot{q}_d(t) + C_e(t)q_d(t) + D_o \quad (25)$$

where,

$$D_e(t)=\begin{bmatrix} 0 & I_n \\ -[A_s]^{-1}[C_s + \Psi_1] & -[A_s]^{-1}[B_s + \Psi_2] \end{bmatrix}$$

$$A_e(t)=\begin{bmatrix} 0 \\ [A_s]^{-1}[C_s - \Phi_1] \end{bmatrix},\quad B_e(t)=\begin{bmatrix} 0 \\ [A_s]^{-1}[B_s-\Phi_2] \end{bmatrix}$$

$$C_e(t)=\begin{bmatrix} 0 \\ [A_s]^{-1}[A_s-\Phi_1] \end{bmatrix},\quad D_o=\begin{bmatrix} 0 \\ [A_s]^{-1}[-\phi_o] \end{bmatrix}$$

And $E=[\sigma , \dot{\sigma}]^{-1}$ is $2n\times1$ position-velocity error vector. In the desigened adaptive control scheme, state-space equation(25) constitutes the "adjustable system" in the adaptive model following control framework. We shall now select the reference model which embodies the desired performance of the manipulator in terms of the tracking-error $\sigma(t)$. The desired control performance is that each tracking error $\sigma_i(t)$ is decoupled from the others and a second-order homogeneous dfferential equation of the form as follows:

$$\ddot{\sigma}_i(t) + 2\zeta_i W_i\dot{\sigma}_i(t) + W_i^2\sigma_i(t) = 0,\ i=1,2,3,\cdots,n \quad (26)$$

where W_i and ζ_i are the undamped netural frequency.and the damping ratio. Now we define the $2n\times1$ vector of desired position and velocity error, $E_m=[\sigma_m, \sigma_m]^T$, then equation(26) can be represented in the vector equation as follows:

$$\dot{E}_M = \begin{bmatrix} 0 & I_n \\ -M_2 & -M_1 \end{bmatrix} E_M = ME_M(t) \quad (27)$$

where $M_1 = \mathrm{diag}(2\zeta_i W_i)$ and $M_2=\mathrm{diag}(W_i^2)$ are constant nxn diagonal matrices. Since the reference model(27) is stable, there exists a symmetric positive-definite 2nx2n constant matrix as follows:

$$N = \begin{bmatrix} N_1 & N_2 \\ N_3 & N_4 \end{bmatrix} \quad (28)$$

which satisfies the Lyapunov equation as follows:

$$F + MN + N^TM = 0 \quad (29)$$

where F is a symmetric positive-definite 2nx2n constant matrix. Now we shall derive the adaptation laws which ensure that for any reference trajectory $q_d(t)$, the state of the adjustable system $E = [\sigma(t),\dot{\sigma}(t)]$ tends to that of the reference model $E_m(t)=0$ asymptotically. In this section, It is required to design the adaptive control scheme based-on the framework of adaptive model following control theory using the lyapunov Direct Method.

$$V=\Delta M\Delta^T + [\Lambda_0-\Lambda_0]\ W_0[\Lambda_0-\Lambda]^T + T_r\{[\Lambda_1-N_1-\Lambda_1]^T W_1[\Lambda_1-N_1-\Lambda_1]\} + T_r\{[\Lambda_2-N_2-\Lambda_2]^T W_3[\Lambda_2-N_2-\Lambda_2]\} + T_r\{[\Lambda_3-N_3-\Lambda_3]^T W_3[\Lambda_3-N_3-\Lambda_3]\} + T_r\{[\Lambda_4-N_4-\Lambda_4]^T W_4[\Lambda_4-N_4-\Lambda_4]\} + T_r\{[\Lambda_5-\Lambda_5]^T W_5[\Lambda_5-\Lambda_5]\} \quad (30)$$

where $\{\Lambda_0,\cdots\cdots,\Lambda_5\}$ are function of time which will be selected suitable value later. M is the solution of the Lyapunov equation(29) for the reference model, and $\{W_0,\cdots,W_5\}$ are arbitrary symmetric positive-definite constant nxn matrices. From derivative above defined Lyapunov function (30),

$$\dot{V} = -\Delta^T M\Delta-2\Pi^T\omega_0\Pi-2(\Pi^T\Pi)\sigma^T\omega_1\sigma-2(\Pi^T\Pi)\sigma^T\omega_2\sigma-2(\Pi^T\Pi)q_d\omega_3 q_d - 2(\Pi^T\Pi)q_d{}^T\omega_4 q_d- 2(\Pi^T\Pi)q_d\omega_5 q_d \quad (31)$$

where $\Pi = [N_3,N_4]\Delta = [N_3,N_4] = E = N_3\sigma + N_4\sigma$.

The derivative of Lyapunov function V is a negative function of Δ, in view of the positive semi-definiteness of $W_0,\cdots,W_5$. Consequently the error differential equation (31) is asymptotically stable, implying that $\Delta(t) \to 0$ as $t \to\infty$. A forementioned adjustable system (25) can be represented as following state space equation.

$$\dot{E}(t) = \begin{bmatrix} 0 \\ \Lambda_0 \end{bmatrix} + \begin{bmatrix} 0 & I_n \\ -\Lambda_2 & -\Lambda_1 \end{bmatrix} E(t) + \begin{bmatrix} 0 \\ \Lambda_3 \end{bmatrix} \ddot{q}_d + \begin{bmatrix} 0 \\ \Lambda_4 \end{bmatrix} \dot{q}_d + \begin{bmatrix} 0 \\ \Lambda_5 \end{bmatrix} q_d \quad (32)$$

where

$\Lambda_1=-[A_s]^{-1}[B_s+K_2]$, $\Lambda_2=-[A_s]^{-1}[C_s+K_1]$, $\Lambda_0=[A_s]^{-1}[- S_a]$

$\Lambda_3=[A_s]^{-1}[C_s-C]$, $\Lambda_4=[A_s]^{-1}[B_s-B]$, $\Lambda_5=[A_s]^{-1}[A_s-A]$

The adaptive model following control techique is to adjust the controller terms so that for any $q_d(t)$, the robot system state $E(t)$ approaches the reference model state $E_m(t)$ asymptotically; ie. $E(t)$ --> as t --> ∞

Let us define the adaptation error $\Delta(t)=E_m(t)-E(t)$ and, from equation (25) and (32), define the error differntial equation as follows

$$\dot{\Delta}(t) = Z_o + Z_1\Delta(t) + Z_2E(t) + Z_3\ddot{q}_d + Z_4\dot{q}_d + Z_5 q_d \quad (33)$$

where

$$Z_1 = \begin{bmatrix} 0 & I_n \\ -M_2 & -M_1 \end{bmatrix},\quad Z_2 = \begin{bmatrix} 0 & 0 \\ \Lambda_2-M_2 & \Lambda_1-M_1 \end{bmatrix}$$

$$Z_o= \begin{bmatrix} 0 \\ -\Lambda_0 \end{bmatrix},\ Z_3= \begin{bmatrix} 0 \\ -\Lambda_3 \end{bmatrix},\ Z_4= \begin{bmatrix} 0 \\ -\Lambda_4 \end{bmatrix},\ Z_5= \begin{bmatrix} 0 \\ -\Lambda_5 \end{bmatrix}$$

Now the adaptation laws are derived by ensuring the stability of the error dynamics (32). As shown in the adaptive control scheme, proposed adaptive control scheme is developed within the basic strcture of adaptive model following control theory, and the adaptive control law is composed of three components, the feedback controller, feedforward, and auxiliary signal. The proposed adaptive control law is expresened as follows:

$$T(t) = \Phi_o(t) + \Psi_1(t)\dot{\sigma}(t) + \Psi_2(t)\sigma(t) + \Phi_1(t)\ddot{q}_d(t) + \Phi_3(t)\dot{q}_d(t) + \Phi_3(t)q_d(t) \quad (34)$$

where, Ψ_o is the auxiliary signal synthesised by the adaptation scheme, the term $[\Psi_1(t)\sigma(t) + \Psi_2(t)\sigma(t)]$ is due to propotional-plus-derivative feedback controller, and the term $[\Phi_1(t)\ddot{q}_d(t) + \Phi_2(t)\dot{q}_d(t) + \Phi_3(t)q_d(t)]$ is the contribution of the proportional-plus-double derivative feedforward controller. In the proposed adaptive control scheme, the feedback controller acts on the position tracking error $\sigma(t)$ and velocity error $\dot{\sigma}(t)$. The feedforward controller operates on the desired position, the desired velocity, and the desired acceleration. The auxiliary signal $\Phi_o(t)$ improves transient performance while resulting in better tracking and providing more flexibility. Using the defined based-on the improved Lyapunov, We derive the adaptation gain. The required feedback controller gains, feedforward controller gains, and the auxiliary adaptation gains are updated based on the weighted error vector is defined as follows:

$$\exists(t) = D_p\sigma(t) + D_v\dot{\sigma}(t) \quad (35)$$

From the following simple adaptation laws

$$\Phi_0(t) = \alpha_1\int \exists(t)dt + \alpha_2\ \sigma(t)+ \Phi_0(0) \quad (36\text{-a})$$

$$\Psi_1(t)= \beta_1\int \exists(t)\sigma(t)^T dt + \beta_2\ V(t)\sigma(t)^T + \Psi_1(0) \quad (36\text{-b})$$

$$\Psi_2(t)= \gamma_1\int \exists(t)\dot{\sigma}(t)^T dt + \gamma_2\ V(t)\sigma(t)^T + \Psi_2(0) \quad (36\text{-c})$$

$$\Phi_1(t)= \Lambda_1\int \exists(t)\ddot{q}_d(t)^T dt + \Lambda_2\ V(t)\sigma(t)^T + \Phi_1(0) \quad (36\text{-d})$$

$$\Phi_2(t)= \nu_1\int \exists(t)\dot{q}_d(t)^T dt + \nu_2\ V(t)\sigma(t)^T + \Phi_2(0) \quad (36\text{-e})$$

$$\Phi_3(t)= \xi_1\int \exists(t)q_d(t)^T dt + \xi_2\ V(t)\sigma(t)^T + \Phi_3(0) \quad (36\text{-f})$$

where, the values of the adaptation gains and weighting matrices determine the rate at which tracking errors conterge to Zero. The matrix D_p and D_v are nxn diagonal weighting matrices chosen by the designer to reflect the relative significance of the position and velocity errors $\sigma(t)$ and $\dot{\sigma}(t)$ in forming the vector q. And, $\{\alpha_1,\beta_1,\gamma_1,\Lambda_1,\nu_1,\xi_1\}$ and $\{\alpha_2,\beta_2,\gamma_2,\Lambda 2,\nu_2,\xi_2\}$ are, respectively, any positive scalar integral adaptation gains and any positive scalar propotional adaptation gains. The auxiliary signal $\Phi_o(t)$ can be generated by a proportional-integral-derivative feedback contoller driven by position tracking error $\sigma(t)$ as follows:

$$\Phi_o(t) = \Phi_o(0)+ \alpha_2\int^t [N_2\sigma(t) + N_3\dot{\sigma}(t)]dt = \Phi_o(0) + [\alpha_1 N_2]\int^t\sigma(t)dt + [\alpha_2 N_2 + \alpha_1 N_3]\sigma(t)$$

Thus the adaptive control law is obtained as follows :

$$T(t)=[\Psi_1(t)\sigma(t) + \Psi_2(t)\dot{\sigma}(t)] + \Phi_3(t)q_d(t) + \Phi_2(t)\dot{q}_d(t) + \Phi_1(t)\ddot{q}_d(t)] + \Phi_o(t) \quad (37)$$

4. SIMULATION

In order to demonstrate the effectiveness of the proposed new robot manipulator controller, the PUMA arm is performenced in the dynamic simulation . The task is to illustrate the capability of trajectry tracking with payload and parameter uncertainties.

The joint trajectories are prescribed as smooth function times of time with zero velocity and acceleraction at the both ends. suppose that the basic task to be performed requireds to track the desired trajectory $q_d(t)$ given $q_i=[t-0.15\ \sin(2\pi/t_s)]q_{if}$. Where t_s, time period ellapsed when tracing the whole trajectory is defined as 1 sec and q_{if} is the final value of each joint variable.

The parameter of robotic manipulator inertia and the denavit-harteuberg are shown in the Tables 1 and Table 2.

Table.1 Denavit-Hartenberg Parameters of PUMA 560.

관절(i)	q_i	α_i[deg]	α_i[m]	a_i[m]
1	q_1	-90	0	0
2	q_2	0	0	0.4318
3	q_3	90	0.1505	-0.0191
4	q_5*	-90	0.4331	0
5	q_6*	90	0	0
6	q_7*	0	0	0

Table 2. Inertia parameters of PUMA 560.

	mass	center of mass			radius of gyration		
Link	m_i[kg]	x_i[m]	y_i[m]	z_i[m]	k_{ixx}	k_{iyy}	k_{izz}
1	12.96	0.0000	0.3088	0.3089	0.1816	0.0152	.1811
2	22.37	-0.3289	0.0050	0.2038	0.0596	0.1930	.1514
3	5.01	0.0204	0.0137	0.0037	0.0151	0.0155	.0021
4	1.18	0.0000	0.0863	-0.0029	0.0119	0.0029	.0118
5	0.62	0.0000	-0.0102	0.0013	0.0009	0.0009	.0009
6	0.16	0.0000	0.0000	0.0029	0.0008	0.0008	.0004
3·*	6.97	0.0136	0.0092	0.1522	0.0783	0.0786	.0021

The unit of radius of gyration is kg-m² in the Table2. The forgotten factor e and weighting factor η were chosen to be 0.96 and 0.00015 repectively. The simulation sampling time was chosen as 0.001 sec and the Q and R matrix were chosen as the identity matrix, which represents that the linearized feedback plant is desired to have a set of second-order systems. The performance of control algorithm is tested by introducing the payload variation(5kg, 10kg) and inertia parameter error(5%, 10%) in the dynamic model. The final values each joint are slected as $q_{1f} = 45°$, $q_{2f} = 45°$, and $q_{3f} = -30°$. and $\{\alpha_1, \beta_1, \gamma_1, \Lambda_1, \nu_1, \xi_1\}$ and $\{\alpha_2, \beta_2, \gamma_2, \Lambda_2, \nu_2, \xi_2\}$ are chosen as {0.6, 1.8, 1.8, 0.6, 0.5, 0.5} and {0, 0, 0, 0, 0, 0} respectively. The D_p and D_v are 50000 and 6000, and the innitial values of the controller terms are all chosen as zero. The results of simulation are shown in Fig.3-9, and represent that the joint angles track their responding desired reference trajectries very closely under parameter uncertainties and payload variation. Thus the proposed adaption controller represents good control performance.

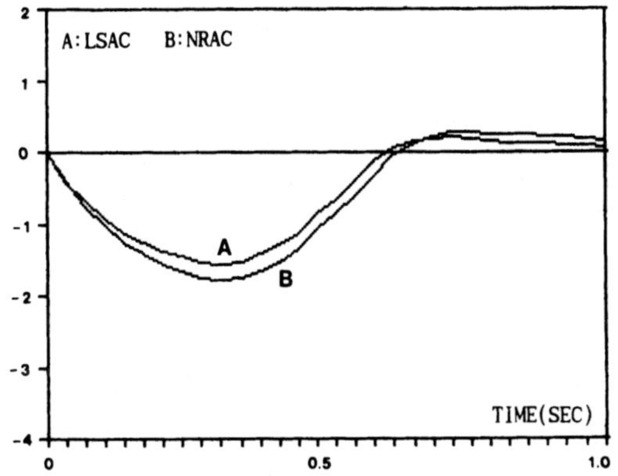

Fig.3 Comparision of control performance of NRAC and LSAC for joint 2 with payload(5Kg)

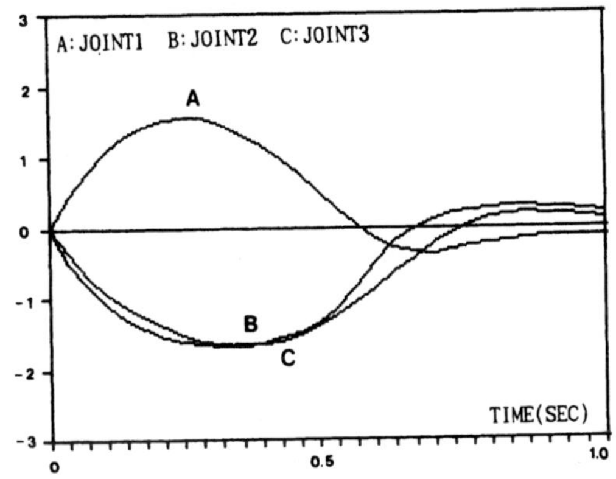

Fig.4 Control performance of LSAC scheme for each joint with parameter uncertainties(5%)

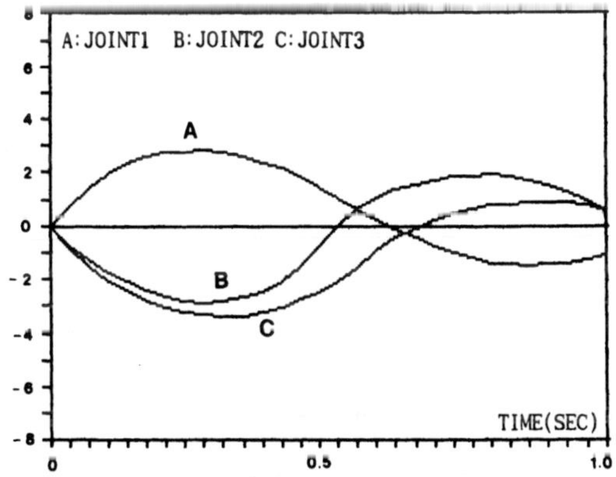

Fig.5 Control performance of LSAC for each joint with inertia parameter uncertainties(10%) and payload(5Kg)

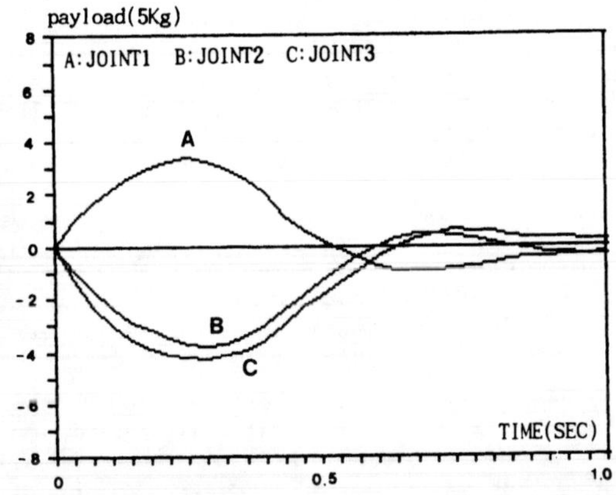

Fig.6 Control performance of NRAC with parameter uncertainties(10%)

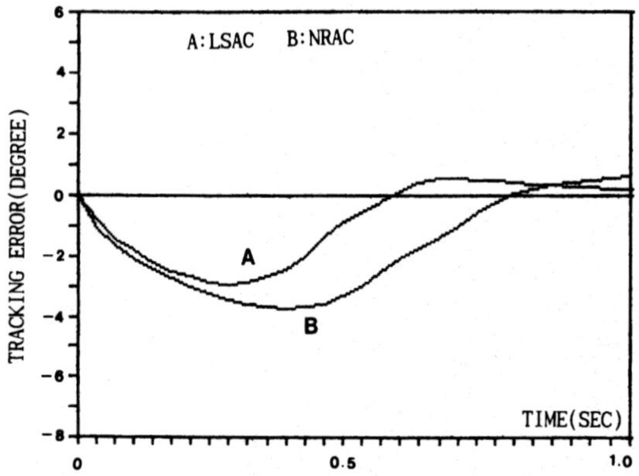

Fig.7 Comparision of control performance for LSAC and NRAC with payload

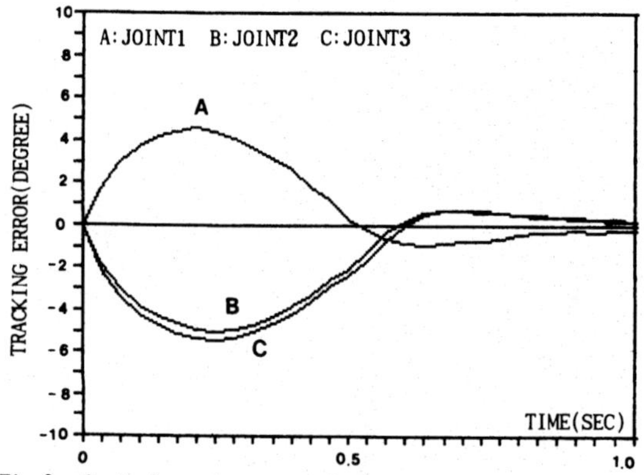

Fig.8 Control performance of NRAC for each joint with parameter uncertainties(10%)

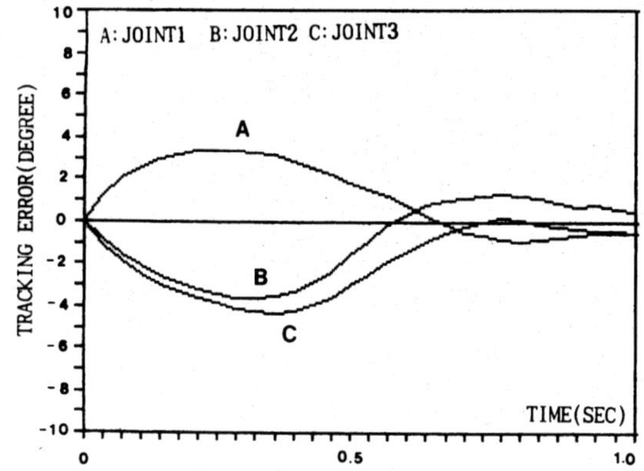

Fig.9 Control performance of NRAC for each joint with parameter uncertainties(5%) and payload(10Kg)

5. CONCLUSION

In this paper, the new adaptive control technique is represented for the real-time control of the robotic manipulator. The adaptive control scheme developed in this paper has the simpler structure, and the computation time required to calculate the control algorithm has reduced much than any other similar scheme. The attractive feature of the proposed adaptation laws is that they do not need the knowledge of complex dynamic model or the unknown payload and inertia parameter value of robotic manpulator, are based entirely on the actuarl and reference trajectory which are directly available.

These attributes make the proposed adaptive control scheme suitable for implementation of the real-time control of robotic manipulator. The special feature of proposed NRAC scheme is the utilization of the time-varying adaptive feedforward controller, the feedback controller and the auxiliary adaptation signal $\Phi_0(t)$ synthesized by the adaptation scheme. In the unpredictable gross payload variations and inertia parameter uncertainties, it has been demonstrated that the NRAC scheme has the more excellent adaptation and robustness than the LASC from the simulation result.

REFERENCE

Valavanis, k.P., Leahy, N.B., and Saridis, G.N. (1985), " Real Time Evalutaion Control Method," in Proc. IEEE Int. Conf. Robotics and Automation, St. Louis, Missouri, pp. 644-649.

Astrom, K.J. and B. Whittenmark (1980), " Self-Tunning Control Based on Pole - Zero Placement," IEE. Proc., Vol. 127, No.3, pp.120-130.

Li, W. and Slotine, J.J.E.(1988), "Globally Convergent Indirect Adaptive Robot Control," in Proc. IEEE Int. Conf. on Robotics and Automation, Philadelphia, Pennsylvania.

Dubowsky, S.,and DesForges, D.T.(1979), " The Application of Model Reference Adaptation Control to Robot Manipulators," ASME J.Dyn. Syst., Meas., Contr., Vol, 101, pp. 193-200. A. Balestrino, G.D. Maria, and Sciavco, L.(1983), " An Adaptive Model Following Control for Robotic Manipulators," ASME J. Dyn. Syst., Meas., Contr., Vol.105, pp.143-151.

Parks, P.C. (1976), "Lyapunov Redesign of Model Reference Adaptive Control Systems," IEEE Trans. Auto. Contr.,Vol. AC-11. No.3, pp.362- 367.

Seraji, H.(1987), " An Approach to Multivariable Control of Manipulator," ASME J. Dyn. Syst., Meas., Contr., Vol. 109, pp.485-493.

Landau, I.D. and Courtiol, B. (1976)," Design of Multivariable Adaptive Model Following Control System," Automatica, Vol.10, pp.483-494.

Koivo, A.J. and Guo, T.H. (1983), " Adaptive Linear Control for Robotic Manipulators," IEEE Trans. Auto. Contr., Vol. AC-29, pp.162-171.

AN OBJECT-ORIENTED CONCEPTION OF A REAL-TIME CONTROL OF FMS

Jaroslav Fogel, Jolana Sebestyénová

Institute of Control Theory and Robotics, Slovak Academy of Sciences, Bratislava, Czechoslovakia

Abstract The conception of the object-oriented database and the knowledge system represented by production rules is described. The design of the database is problem oriented to FMS and contains the predefined objects' classes. The production rules are divided into three groups: process knowledge rules (they create a dynamic state model of the resources), production plan rules (they define precondition of the operations), and expert knowledge rules (they create a reasoning model for solving a faulty situation). The real-time temporal logic as an assertion language for the specification of the behaviour of the manufacturing processes is described, and the verification of these behaviour in the stage of simulation is described.
Keywords control system, manufacturing processes, knowledge engineering, object-oriented database, temporal logic

INTRODUCTION

In FMS, three to five hierarchical levels are usually defined. The top level is the strategic planning level. The tactical factory planning level is below this. The shop floor is below the factory level. It consists of a processing stations'group or of the workcells connected together by a material handling system. This level is sometimes called the coordination level of the shop floor. The coordination level of the resources in the workcell is below that. The bottom level is the local control level of the machines.

A detailed analysis of three cell controllers is given in (Boulet,1991). It concerns the current research of AMRF (the Automated Manufacturing Research Facility) at the National Institute of Standarts and Technology, ESPRIT Project 932, and MADEMA (MAnufacturing DEcision MAking) at the Computer Aided Manufacturing International.

We would like to contribute to the study of the real time cell controller by our paper. The paper describes a component of our work directed to the development of a software tools which can be used to generate control actions for shop floor level of manufacturing system, in normal and unexpected situations.

The structure of the workcell controller modul is depicted in Fig. 1.

First, in Section 2 the object database with the problem orientation to FMS is presented. It consists of the predefined classes of objects which contain the static (attributes, functions) and the dynamic (production rules) parts. In Sections 3, 4 the dynamic part of the database is explained in detail. In Section 3, the process knowledge rules are described which define the possibility of the resources and create the dynamic model of the resources. In Section 4, the production plan represented by the rules is described. The description of the production controller by a set of tasks is also given. In Section 5, the role of the expert knowledge rules for a solution of unexpected situations is explained. In Section 6, the real-time temporal logic as an assertion language for the specification of the system behaviour together with their verification mechanism is described. A visual formalism statecharts which are extension of finite-state machine and their visual counterpart, state-transition diagrams about depth, ortogonality and broadcast communication is used as description machanism for the dynamic model of the resources, the production controller, and for verification of the system behaviour. For more detailed information about syntax and semantics of statecharts see (Harel,1987).

THE DESIGN OF AN OBJECT DATA-BASE WITH A PROBLEM ORIENTATION TO FMS

Object-oriented data-base systems are base software technology for supporting the development of data-intensive applications (CAD, CAE, CASE, CAM, knowledge-based systems, scientific model-

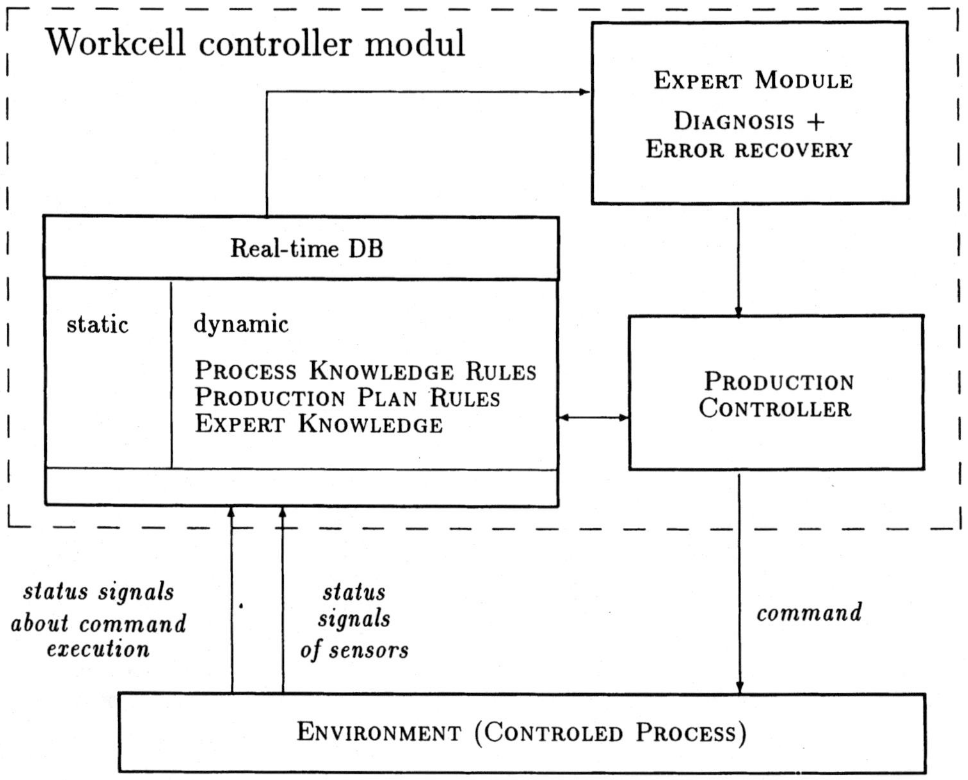

Figure 1: Workcell controller modul

ing and analysis programs).
The scope of modeling capabilities for object-oriented data models includes:

- knowledge (constraints, rules)
- data (attributes, functions)
- model management

The object-oriented approach to developing a production control system of FMS requires to invest considerable means into an object database design. It must be sufficiently complex for its use in general application. During an analysis and in a low-level design, objects are choosen as basic run-time entities in the object-oriented system. A set of procedures and functions working on them is joined to each object. It means, the object comprises data and behaviour. The objects are organized into classes and their relations are expressed by inheritance. The characteristics of object-oriented programming such as polymorphism and dynamic binding are utilized in the high-level application design and during the implementation.
The components of FMS, that have to be described in ODB, are of two basic groups: the resources and the activities (see Fig.2). The resources consist of the active and passive equipment (e.g., machines, their tools, ...), further of all material entered the production process and the products (material flow and output).
Objects of a *class machinery* are characterized by:

- attributes (name, position, replaceability, set of tools, set of states, e.g., working, waiting, chainging a tool, free,...)
- functions (e.g., virtual tool_change(),...) The functions that can be carried out on the machines are given here. They may be specialized in subclasses.
- **process knowledge** (rules describing all facilities of all machine)

Subclasses derived from the class machinery: machining centers, positioner, pallet buffer, transporter means, robot, welding equipment, working space. Objects of a *class tools* are characterized by attributes: name, parameters, position, states (blunt, suspicious,...).
Objects of a *class product* are characterized by: attributes (name, position, mass, set of states, e.g., machined, being machined, free,...) and functions (e.g., virtual take_hold(),... which function describe how to take some semiproduct to cary it over; it may be modified by specialization to such functions as holding by pneumatic plant or take in robot gripper, ...)
A class product has subclasses: input material (raw, preprepared goods, parts) and final product.
Objects of a class final product contain **production plan** rules describing the production sequence of that product.
The activities consist of the operations that can be carried out on the machine tools, further events

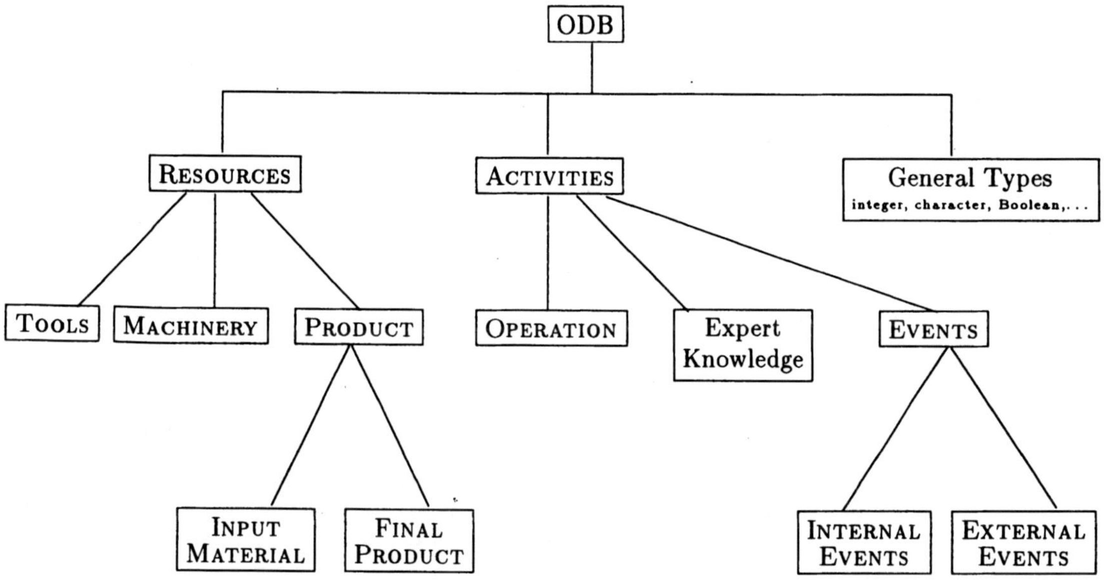

Figure 2: Object data-base for FMS

(using of them is explained later) and expert knowledge, which don't start any operation but they cause the change of a system's states.
Objects of a *class operation* are characterized by:

- attributes (name of the operation, name of a machine, interrupting ability, reversibility)
- functions (start of a proper program of some machine, i.e., local control of the machine, which itself is the lowest level of control hierarchy)
- **timing constraints** They express limitation on the operation's duration and timing conditions of the coherence and collaboration among activities; using of them is helpfull for error detection.

A class operation has subclasses: machining, transport, welding, cutting, assembling.
In addition to these basic classes, ODB contains some general types (e.g., integer) to enable expression of those attributes that are not contained in ODB, in spite of its extension.

PROCESS KNOWLEDGE RULES

The process rules are defined in the class of resources for both the active and the passive objects, and they create a dynamic model of the state's changes for the production resources. The active objects are the objects of the class machinery on which the technological operations are actively carried out. The passive objects are: working spaces, machine tools, pallet buffers, etc.
The set of states can be defined for the active objects as follows:
S = { IDLE, WAIT, ACTIVE }

The object executes an operation in its active state. The operations are of the following kind:

- the operations for performing the normal technological process to produce the final product (these operations are defined in the class *final product* as a production plan).
- the operations that express all facilities of the machine; they need not to be described in the production plan. For example, to this kind belong here: change of tool (if stage of the tool's wear is unsuitable), change of buffer (if buffer is empty).
- the operations useful for solution of error-recovery situations, e.g., reversible operations of the machine, replace operations for a spoiled machine, etc.

Concisely speaking, those operations belong to process knowledge, which will be used by expert system for generation of the recovery actions as well as the operations secure the fulfilment of the conditions for the normal production plan. For example: the robot supplies NC-machine by preprepared goods from the buffer, in the case of empty buffer, the robot secures the restoration of the one, although the operation change of buffer do not belong among the operations of the production plan.
The process rules are of the form:

$$ON < event > WHEN < IN(state) > \quad DO < action > \qquad (1)$$

where
The set E of events consists of: *start*(O_N),

finish(O_N), where O_N is an operation of the production plan, *change*(data) if data is a database information, *true*(c), *false*(c) if c is a condition, *true*(p), *false*(p) if p is a premise of the expert rule, and *timeout*(e,n) if e is an event and n is a time unit. The set A of actions consists of: *null* action, *start*(O_P) where O_P is an operation from the machine's posibilities, setting the variable to the value from the expression, etc.
in(state) is the condition of the resource being in the given state.
The set of the passive object's states depends on the type of the object and for the typical object's type it is as follows:
working space= { FREE,OCCUPY }
machine tool= $\{ wearing\_stage_i \}_{i=1}^{n}$
buffer= $\{ number\_of\_elements \}$
The models of the active and passive resources in the statecharts syntax are given in (Fogel,1992).

PRODUCTION PLAN

The production plan consists of the sequence of operations performed on the basic production components as a product, a semiproduct, etc. These sequential operations enable to describe the state of the production process by the help of a finite discrete set of the abstract states s_0^i, s_1^i,...,s_n^i, where s_0^i is an initial state of the product i, and the transition from the state s_{j-1}^i to the state s_j^i is caused by application of the operation O_j^i. We mark a set of operations $\{O_1^i, O_2^i, ..., O_n^i\}$ as a task on the product i. We will further assume about the task:

1. $O_2, O_3, ..., O_{n-1}$ is a subsequence of operations independent of the operations from other tasks and the operations are coupled in a strictly sequential way.
2. Synchronization with other tasks is confined to just state s_0 and s_n so before the execution of O_1, and after the execution of O_n.
3. In the inner states the task may use the mutual exclusion mechanism to solve a conflict associated with resource sharing.
4. The task may be repeated in the cycle.

In this moment it is necessary to emphasize that by synchronization we mean the synchronization of products' flow in the assemblage process.
The operation O_i is executed in dependance upon the state s_{i-1} and on the precondition of the operation. The condition is evalueted on the base of the dynamic state model of the resources (process rules).
The preconditions of the operations are given in the rule's form:

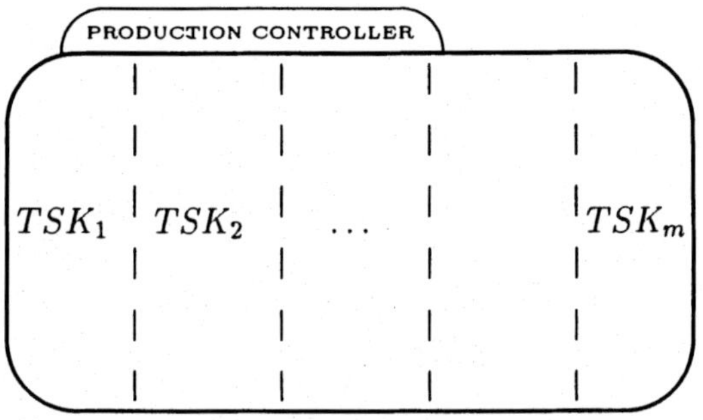

Figure 3: Production controller

$$IF < condition > THEN < operation > \quad (2)$$

where conditions are defined on the set of resources' states, and for every operation they concern:

- the freedom of the active resource
- the state of the passive resource, e.g., if the buffer is empty or not
- the technological parameters of the operation, e.g., if the tool is correct, if the robot's head is suited for assembly operation, etc.
- the state of the sensor signal

It is obvious that the tasks are running concurrently, and in the case of cyclic process some of them arise, some of them disappear, others are waiting for the certain event, others are running, etc. For this reason, we will define the set of tasks' states as follows:
ACTIVE - task is in some of the states $s_0, s_1, ..., s_n$
WAIT - task is waiting for an event or the condition being true
IDLE - task is idle
READY - task is ready to run
RECOVERY - task executes the recovery sequence of the actions
The control logic of production controller can be specified by a graphic means in the statecharts formalism. The state of production controller is represented by the states of all concurrent tasks. This fact is expressed in statecharts' syntax on Fig.3, where state of production controller is AND-decomposed into substates $TSK_1, ..., TSK_m$. The single task is XOR-decomposed into the states defined before. The statecharts' description of the task depends on the type of the task. Three type of tasks are given in (Fogel,1992).

EXPERT SYSTEM

The solution of a fault situation in the expert system can be characterized as a reasoning process which is based on the implicite data information and declarative knowledge, in contrast with the procedural solution of such situation in a non expert system. The knowledge in the expert system can be represented by the knowledge rules. The rules will be used in three types of moduls:

- error detection
- error diagnosis
- error recovery

In the error detection modul the fault is detected on the base of the timing constraints of the operations (they are given in the database), and of the status signal about a command execution from a real environment.

The main purpose of the diagnosis modul is to deduce causes of a failure from the context of the detected error (machines, tools, etc.). In (Chang,1989) the author shows that, in manufacturing systems, the causes of the failure may be very often found in operations previously executed. So by build up of the failure propagation tree, the fault may be found by analysis of the precondition of past operations performed in the production control modul.

The aim of recovery modul is to remove the cause of a fault and its consequences by the recovery sequence of actions that is generated in this modul. From the control point of view, a problem of the generation of the recovery sequence can be investigated as the problem of the trajectory determination in the state space from the initial state to the goal state. The purpose of the diagnosis modul is to determine both the actual (initial) and goal states.

A form of the knowledge rules is as follows:

$$IF \qquad p_1 \wedge p_2 \wedge \ldots p_n \qquad THEN \qquad q \qquad (3)$$

where p_i, $1 \leq i \leq n$ are premises an q is a conclusion

These rules are used for the deduction of new facts. They work separately from the process knowledge and the production plan rules, that only execute the recovery and production actions, respectively, and so they cause the change of the states (facts' values), but do not create new facts. The rules are evalueted on the set of facts from the database.

In the phase of the error recovery, the actions are generated usually by the algorithm based on the backward chaining mechanism of the rules. The backward chaining method contains the following steps:

1. to find the rule whose conclusion is a desired goal
2. if the premise of the rule from the previous step is true or can be fulfilled by process rules: go to 4.step, if it is not: to find the rules whose conclusions (subgoals) are parts of the premise of the rule from the previous step

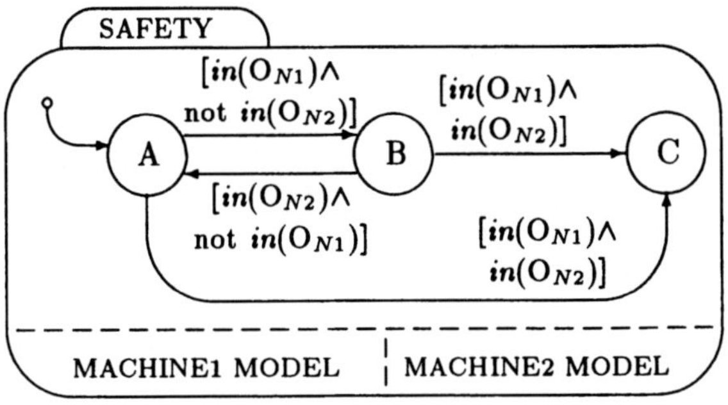

Figure 4: Temporal constraint in Statecharts syntax of property 1

3. the second step is repeated as soon as the rule is found whose premise is true or can be made true by process rules
4. if all rules were found that secure the successive fulfilling of the next rule's conditions, starting from the rules whose premises are true, up to the rule whose conclusion is a goal, the rules are executed by a forward chaining method to obtain the conditions for reaching the goal proposition, if not, the information in the set of facts is unsufficient

TEMPORAL LOGIC AS AN ASSERTION LANGUAGE FOR SPECIFICATION OF SYSTEM'S BEHAVIOUR

In real-time systems, it is necessary to specify the controled process characteristics and hold them during a control process. Several various classifications of these characteristics exist. One of them is as follows:

1. Safety- the characteristic which must be hold during the hole process control; it will guarantee collision free process.
2. Liveness- the characteristic guarantees the realization of certain action if its conditions are fulfilled.
3. Real- time condition- the characteristic is stating some timing constraints on the operation's duration, the freguency of the event's occurrence, the system response time, etc.

In the paper, we will introduce the characteristics'specification of the manufacturing system by the help of RTTL.

In this part we briefly describe the syntax and semantics of the temporal operators that are suitable for specification of the properties in the models of FMS. Some significant temporal operators referring to the future are:

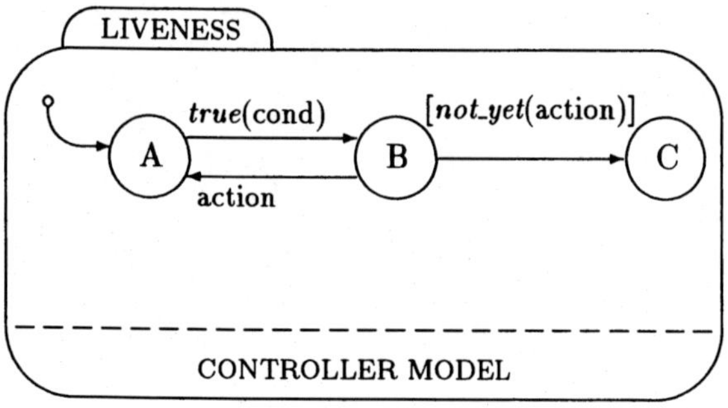

Figure 5: Temporal constraint in Statecharts syntax of property 2

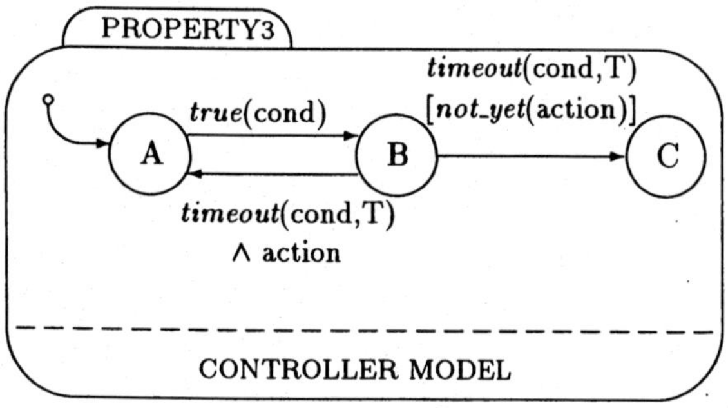

Figure 6: Temporal constraint in Statecharts syntax of property 3

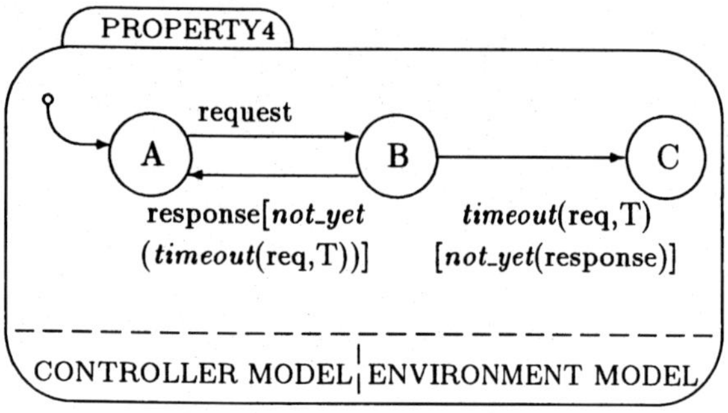

Figure 7: Temporal constraint in Statecharts syntax of property 4

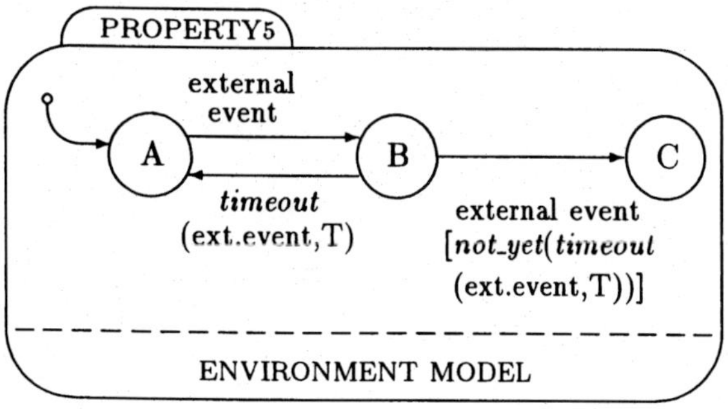

Figure 8: Temporal constraint in Statecharts syntax of property 5

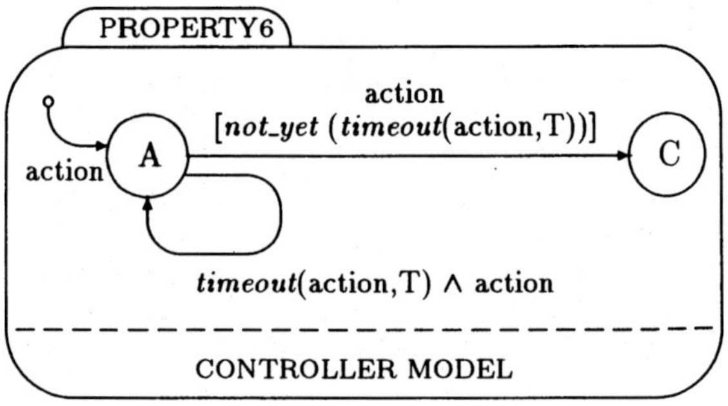

Figure 9: Temporal constraint in Statecharts syntax of property 6

ALWAYS / $\Box$ / p - p holds true in all state
EVENTUALLY / $\Diamond$ / p - p will hold true in some state
NEXT / $\bigcirc$ / p - p will be true in the next state
p UNTIL / U / q - eventually q will hold and until then p holds continuously
In / / there are abbreviated versions of these oprators.

The restricted versions of the future operators are defined recursively in (Descart,1988):
$\bigcirc < t > p$ ::= if t=0 then p else
$\bigcirc (\bigcirc < t\text{-}1 > p)$
$\Box < t > p$ ::= if t=0 then p else
$(p \wedge \bigcirc (\Box < t\text{-}1 > p))$
$\Diamond < t > p$ ::= if t=0 then p else
$(p \vee \bigcirc (\Diamond < t\text{-}1 > p))$
$p\ U < t > q$ ::= if t=0 then q else
$(q \vee (p \wedge \bigcirc (p\ U < t\text{-}1 > q)))$

The meaning of restricted versions of these operators is
ALWAYS $< t >$ p - p holds true in the next t time units
EVENTUALLY $< t >$ p - p will hold true in some state within t time units
NEXT $< t >$ p - p will be true at the state which follows t time units after current one
p UNTIL $< t >$ q - eventually q will hold and until then p holds t time units

For the manufacturing systems some real-time characteristics can be expressed by temporal formulas.

1. Mutual exclusion property of collision operations
   $\Box (\neg O_{N1} \vee \neg O_{N2})$

2. Liveness property
   $\Diamond$ (condition $\Rightarrow$ action)

3. Every condition is followed by action in exactly t time units
   $\Box$ (condition $\Rightarrow \bigcirc < t >$ action)

4. After each request, response will come in

maximum t time units
$\Box$ (request $\Rightarrow \Diamond < t >$ response)

5. The maximum frequency of the event's occurence must be t time units
$\Box$ (event $\Rightarrow \neg \Diamond < t >$ event)

6. Periodicity of the action is t time units
$\Diamond$ act $\wedge \Box$ (act $\Rightarrow \bigcirc$ ($\neg$ act U $< t >$ act))

The idea of using statecharts as an alternative way of expressing temporal properties is from the reference (Descart,1988). The idea consists of the fact that the temporal constraints expressed in the statecharts syntax serve as a kind of "watch-dog" aside the modeled system (e.g., modeled by statecharts too). In such statecharts, there are no generation of action at all. It only accept events and values of conditions coming from the modeled environment and from the model itself, and reacts by performing transitions whose goal is to record the computation history. Of course, only events and conditions participating in the considered temporal formulas are recorded. Such "watch-dog" is added to the modeled system and behaves as a top-level orthogonal component. In this manner it is possible to verify the temporal properties of the modeled system in the stage of simulation.
The verification of the properties 1-6 is shown in the fig.4-9. The boxes represent the top level orthogonal decomposition of the state into its substates marked by dashed lines. These are a "watch dog" substate expressing temporal properties and substates represented by models on the top hierarchical level to which the temporal constraints are concerned. In all cases the temporal formula holds iff ALWAYS(not *in*(C)).
The meaning of the special statecharts function expressed in the figures are the following:
timeout(event,t) represents the event that occurs precisely when the specified number of time units have elapsed from the occurrence of the specified event.
in (state) is the condition of the model being in the given state.
not_yet (event) is the condition which stands for event has not yet occurred.
The transitions in diagrams are labeled by e[c] where e is an event and c is a condition.
e and c are optional, and if $e_1, e_2 \in E$ then $e_1 \wedge e_2$, $e_1 \vee e_2$, $\neg e_1 \in E$
and if $c_1, c_2 \in C$ then $c_1 \wedge c_2$, $c_1 \vee c_2$, $\neg c_1 \in C$
E and C are sets of events and conditions, respectively.
The transition is triggered if the source state of the transition is active, event occurs and condition is fulfilled.

CONCLUSIONS

In this paper the workcell controller for shop floor level of discrete manufacturing was presented. The programming realization of the system is under development in this time. Besides described parts, it will contain a user friendly interface for the objects definition in the classes of database and the graphical tools for control logic specification in the statecharts syntax.

REFERENCES

Boulet, B., B. Chhabra, G. Harhalakis,I. Minis,J. Proth (1991). Cell controllers: Analysis and comparison of three major projects, *Computers in Industry 16*, 239-253

Chang, S.J., F. Dicesare, G. Goldbogen (1989). An algorithm for constructing a failure propagation tree in manufacturing systems, *IEEE International Symposium on Intelligent Control*, Albany, New York.

Descartes project (1988). Expressing global temporal constraints in Statemate, DESCARTES Project Ref. 937, Rehovot, Israel, 1- 13

Fogel, J., J. Sebestyénová (1992). Knowledge system a tool for specification of control logic in FMS, *Systems Science XI*, Int. Conf. Wroclaw, in press, (21 pages).

Harel, D., A. Pnueli, J.P. Schmidt, R. Sherman (1987). On the formal semantics of statecharts, *Proc. 2nd IEEE Symp. Logic in Computer Science*, New York, 1987, 54-64

AUTHOR INDEX

KEYWORD INDEX

Made in United States
North Haven, CT
05 November 2026

21778846R00204